SECOND EDITION

# CRIMINAL LAW

## DAVID C. BRODY, JD, PHD
Associate Professor
Criminal Justice Program
Washington State University
Spokane, Washington

## JAMES R. ACKER, JD, PHD
Distinguished Teaching Professor
School of Criminal Justice
University at Albany

JONES AND BARTLETT PUBLISHERS
*Sudbury, Massachusetts*
BOSTON    TORONTO    LONDON    SINGAPORE

*World Headquarters*
Jones and Bartlett Publishers
40 Tall Pine Drive
Sudbury, MA 01776
978-443-5000
info@jbpub.com
www.jbpub.com

Jones and Bartlett Publishers Canada
6339 Ormindale Way
Mississauga, Ontario L5V 1J2
Canada

Jones and Bartlett Publishers International
Barb House, Barb Mews
London W6 7PA
United Kingdom

Jones and Bartlett's books and products are available through most bookstores and online booksellers. To contact Jones and Bartlett Publishers directly, call 800-832-0034, fax 978-443-8000, or visit our website, www.jbpub.com.

Substantial discounts on bulk quantities of Jones and Bartlett's publications are available to corporations, professional associations, and other qualified organizations. For details and specific discount information, contact the special sales department at Jones and Bartlett via the above contact information or send an email to specialsales@jbpub.com.

**Production Credits**
Publisher, Higher Education: Cathleen Sether
Associate Editor: Megan R. Turner
Production Director: Amy Rose
Production Assistant: Julia Waugaman
Associate Marketing Manager: Jessica Cormier
Manufacturing and Inventory Control Supervisor: Amy Bacus
Composition: CAE Solutions Corporation
Cover Design: Kristin E. Parker
Cover and Title Page Image: © Olaru Radian-Alexandru/ShutterStock, Inc.
Printing and Binding: Courier Westford
Cover Printing: Courier Westford

**Library of Congress Cataloging-in-Publication Data**
Brody, David C.
 Criminal law / By David C. Brody and James R. Acker. -- 2nd ed.
   p. ; cm.
 Includes bibliographical references and index.
 ISBN 978-0-7637-5913-1 (casebound)
 1. Criminal law--United States. I. Acker, James R., 1951- II. Title.
 KF9219.B73 2009
 345.73--dc22
                              2009011052
6048
Printed in the United States of America
13 12 11 10     10 9 8 7 6 5 4 3 2

To my family, Wendy, Seth, Joey, Kelly, and Garth for their patience, assistance, encouragement, and support behind this endeavor.

—D.B.

To Jenny, Elizabeth, and Anna. The substantive criminal law—involving as it does real people, replete with their idiosyncrasies and immersed in life circumstances with the most uncanny twists and turns—continuously reaffirms my belief that the truth is far more exciting and spectacular than even the most inventive fiction. I know a few good people about whom the same may be said. You are at the top of my list.

—J.A.

# CONTENTS

The substantive criminal law embodies and reflects the most fundamental rules of organized society. Some of those rules are so basic they have endured essentially unchanged over time—for example, the familiar prohibition against killing other people ("thou shalt not kill"). Others have changed subtly over time in response to changing social conditions and values. The criminal law is unique, in part, because of the consequences that follow violations of its rules: fines, imprisonment, and sometimes even death. This book focuses on the fundamental principles of the substantive criminal law, as applied to a fascinating array of conduct as richly diverse as flag-burning, stalking, and corporate fraud, as well as more traditional offenses involving interpersonal violence and harm to property. The subjects are intriguing, intellectually challenging, and of great practical significance.

Although the criminal law was initially prescribed and refined over time by judges, it now is largely defined by statutes. Criminal legislation has been enacted in each of the 50 states and by the federal government. Students thus should anticipate encountering considerable variability in the criminal laws, owing to the diversity of statutes and the different approaches adopted by courts in interpreting those statutes.

This book covers a multitude of crimes and rules of criminal law. It is organized to provide a unifying framework for the study of those rules.

Chapters 1–3 present a general introduction to the criminal law and the operation of the criminal justice system. Chapter 1 introduces the general principles of the criminal law, discusses the theories and goals of criminal punishment, and addresses essential procedural matters such as how a criminal case is initiated and progresses through the court system. It also touches upon burden of proof issues, trial by jury, and other related matters. Chapter 2 explores the many crucial constitutional limitations on criminal laws and punishment that are designed to help guard against abuses of the criminal sanction. Chapter 3 provides a more focused examination of the general principles of the criminal law, including the core necessities of voluntary conduct (in the form of an act or omission), *mens rea*, the concurrence principle, causation, and harm. Mastery of these fundamental principles is essential preparation for study of their application in different contexts in ensuing chapters.

Chapters 4 and 5 cover defenses to criminal liability. Chapter 4 presents justification defenses, and Chapter 5 involves excuse defenses. Students are introduced not only to the formal rules governing these defenses, but also to the underlying principles that explain their availability, and their application to particular fact patterns.

The remainder of the text covers specific crimes, including different types of criminal homicide, rape and sexual assault, other crimes against the person, crimes against property and habitation, and white collar and inchoate crimes. Issues of vicarious liability are also explored.

The book relies on statutory materials and carefully selected and edited judicial opinions to present the fundamental principles, doctrines, and rules of the criminal law. By studying case materials, students will develop the analytical skills essential to understanding how legal principles have developed over time and how they are best applied to ever-changing factual situations. It has been our collective experience that students also enjoy reading about true events involving real people and incidents, and that the "case method" of study results in lively and rewarding class discussions. In deciding which among the thousands of available cases at our disposal to include in this book, we strove to present those containing clear discussions of key principles of the criminal law, and also those that will prove interesting to students and make their studies more enjoyable. We have carefully edited the cases to omit confusing and unessential aspects of decisions, and to reduce the opinions to manageable length. We have omitted most citations and footnotes. Concurring and dissenting opinions are included when the views expressed in them are helpful to a fuller understanding of the cases.

In preparing this *Second Edition*, we set out to strengthen the core feature of the book, the direct presentation of case law, while making pedagogical-based changes stemming from the feedback we received from our students and colleagues over the past eight years.

As an initial step in developing this *Second Edition*, we reviewed each case contained in the *First Edition* with an eye toward continued validity, mode of presentation, and over inclusion of unnecessary portions of opinions. Based on this review, a number of changes were made to how specific topics covered in the *First Edition* were presented in this *Second Edition*. In some instances, where more recent cases did a better job of explaining the point of law, or when the point of law in a case was directly or indirectly overruled, the case was replaced. In other instances, the judicial opinion was reedited to read better and remove unnecessary dicta or verbiage.

In addition to the changes involving presentation of the substantive law, several new areas of law not covered before were included in this new edition. The most notable of these additions involves the inclusion of case law and text related to the "war on terrorism." The *Second Edition* presents materials that illustrate important developments in the substantive criminal law as well as constitutional limitations stemming from the war on terrorism embarked on subsequent to the events of September 11, 2001.

Beyond materials related to homeland security issues, a number of areas were strengthened and updated. These items include updates due to recent precedent in the areas of assisted suicide prohibitions, anti-sodomy laws, and the constitutionality of implementing the death penalty for crimes not involving homicide. Additional materials were also added related to anti-stalking legislation as well as the prosecution of cybercrimes.

Perhaps more than revising the substantive materials discussed above, preparing a *Second Edition* gave us the opportunity to add a number of pedagogical features not included in the *First Edition*. New features that have been added in response to suggestions from students and faculty include the following items.

- To help students and instructors focus on the main themes and points presented in each chapter, bulleted chapter objectives are provided at the beginning of individual chapters.
- To help students retain the definition of key terms, a glossary is included that explains key terms contained in judicial opinions and supplemental text.
- To assist students in developing their critical thinking skills and grasping the material, a number of review questions are presented at the end of each chapter.

This new edition of the text was prepared to facilitate and encourage students and instructors to make use of the wealth of information available on the Internet. With this in mind we have developed a Web site containing additional case materials, updated case law, links to unedited Supreme Court opinions and oral arguments, and links to current events related to the text.

We are grateful to the many professionals at Jones and Bartlett who have been so very helpful in making this book a reality, especially Jeremy Spiegel, who has consistently provided us with motivation, support, and encouragement. We also are indebted to Wayne Logan, our co-editor on the *First Edition* of this volume. Wayne's duties at the Florida State University College of Law, where he is the Gary & Sallyn Pajcic Professor of Law and Associate Dean for Academic Affairs, prevented him from being able to participate in the preparation of this *Second Edition*. We could not have begun this venture without Wayne, our good friend and colleague. His many contributions remain evident throughout this volume. We would also like to acknowledge the effort and input provided by Professor Craig Hemmens of Boise State University.

In addition, David Brody is thankful for the assistance provided by Charles Johnson, Melinda York, Shane Griffith, Tuyen Truong, and Nichole Lovrich for their assistance in preparing this *Second Edition*. Most of all, he is grateful to his wife, Wendy, for her eye for detail and editorial prowess in helping mold ideas into this text.

James Acker expresses his appreciation to Carmine Pesca who, while a graduate student at the University at Albany, searched cheerfully, diligently, and successfully for judicial decisions suitable to illustrate principles of the substantive criminal law. Carmine has since assumed a faculty position at Hudson Valley Community College, where he teaches substantive criminal law, among other subjects, and where students now are able to take advantage of his considerable talents as an educator. Finally, although it may be self-evident, the edited cases reprinted in this volume do not on their own initiative purge themselves of nonessentials, pare down to sufferable proportions, and present themselves in the form in which they are reprinted. Professor Acker had the best of assistance with these editorial essentials. He owes an immeasurable debt to his daughter Elizabeth, the chief cutter-and-paster, and to his daughter Anna, the chief-assistant cutter-and-paster. They provided support unlike any he could possibly have derived from other sources. Thank you, ladies.

# Introduction to the Study of Criminal Law

## CHAPTER 1

## Chapter Objectives

- Understand the general principles of criminal law
- Learn the definition of *crime*
- Learn the justifications for punishment
- Understand the criminal court process
- Learn the sources of the criminal law
- Distinguish between a *misdemeanor* and a *felony*
- Understand the roles of various types and levels of courts
- Learn the importance of the *burden of proof*

## On Crimes and Punishment: The Domain of Substantive Criminal Law

### General Principles of Criminal Law

The substantive criminal law defines specific conduct as crimes and assigns corresponding punishment. Although this description is generally accurate, it falls short as a definition of our subject matter. We advance our understanding of the criminal law quite modestly by acknowledging that it pertains to crimes and punishment. We nevertheless should be reluctant to rush too quickly to define the scope of the criminal law and its dependent construct, "crime"; those tasks command a great deal of our attention in the pages that lie ahead.

Our preliminary description of the substantive criminal law focuses on what conduct is prohibited and made subject to punishment. This emphasis helps distinguish a related field of study, *criminal procedure,* which primarily is concerned with *how* questions. Criminal procedure law addresses a variety of issues relating to how far the police may go in investigating criminal activity, how suspected offenders are processed and charged before trial, and how their cases are tried and reviewed in the courts. To a much greater extent than substantive criminal law, criminal procedure is governed by constitutional rules—rules such as the Fourth Amendment's prohibition against unreasonable searches and seizures, the Fifth Amendment's privilege against compelled self-incrimination, the Sixth Amendment's right to counsel, and so on. This is not to say that the substantive criminal law is oblivious to constitutional constraints, for it is not. Indeed, in Chapter 2 we consider several constitutional principles that place limits on the government's power to enact crimes and punishments. Yet the states and the federal government are given considerable latitude within those constitutional restraints to make and enforce rules of substantive criminal law.

Because of this leeway and the multiplicity of values and policies that different jurisdictions choose to promote through their criminal laws, rules of substantive criminal law differ markedly throughout the country. This diversity of rules also helps distinguish criminal law and criminal procedure, and sometimes is a source of frustration to students. Because of the overarching significance of U.S. Supreme Court rulings in criminal procedure matters, the fundamentals of that area of the law are far more similar than different among the states. Much less consistency exists among rules of substantive criminal law adopted by the states and the federal government. ✱

For example, there is no guarantee that the crime of "burglary" means the same thing in Iowa as it does in Mississippi, or that the "felony murder" rule is applied in Massachusetts as it is in Maryland, or that the "insanity" defense has the same requirements in New York as in Arizona—or that it even is recognized at all. The substantive criminal law reflects many different approaches to how crimes are defined, when defenses will be accepted, and what punishments follow on conviction. Some rules may seem more sensible, or at least more defensible, to you than others. But you should not expect to find—and you should not be perplexed when you do not find—a universal rule that applies to the specific crimes, defenses, ✱ and punishments within the substantive criminal law.

In part because of the futility of trying to catalogue and master the nuances of the specific rules of criminal law that proliferate throughout the United States, but for substantive reasons as well, the focus of our studies lies elsewhere. It is essential that students of the criminal law come to grips with the basic principles that underlie particular rules of law and with the history, values, and policy choices that breathe life into those fundamental principles. Having such a foundation will equip you to analyze and reflect critically about the particular form that different rules take and to assess their relative strengths and weaknesses.

✱ know the amendments!

Jerome Hall identified five principles of criminal law that constitute the "essential elements of crime."[1] Using Hall's terminology, we generally should expect crimes to be organized around the following general principles: "(1) *mens rea,* (2) act (effort), (3) the 'concurrence' (fusion) of *mens rea* and act, (4) harm, [and] (5) causation."[2]

*Mens rea* is a Latin term that normally is translated as "guilty mind." The legal construct "crime" is based on conduct (an act, accompanied by a guilty mind) that causes harm. As you might imagine, each of these defining principles is laden with conceptual and operational intricacies, which we shall consider in due course.

Hall appended two additional principles to the five that correspond to the essential elements of crimes. At the front end of Hall's scheme is the *legality* principle, sometimes stated as *nullum crimin sin lege, nulla poena sin lege* (no crime without law, no punishment without law). The legality principle grounds the definition and punishment of crime within the tradition of the law. It serves multiple purposes, not the least of which are to help ensure that people have a fair opportunity to know what is expected of them, and thus to avoid running into problems with the law, and to guard against official arbitrariness and abuses of power in enforcing the criminal law.

Hall's seventh and last principle, *punishment,* becomes operative at the rear end of the scheme. In one sense state-sanctioned punishment is perhaps the criminal law's single most distinguishing characteristic. After all, many forms of harm-causing conduct, even when inspired by a guilty mind, do not result in punishment at the hands of the government and do not qualify as crimes. For example, contracts regularly are breached, often to the considerable detriment of one of the parties as well as to the general public, yet such conduct is not of concern to the criminal law. Instead, we make the injured party seek a civil remedy (generally money damages) as compensation (not "punishment") for the suffered loss. Torts—wrongs such as medical malpractice, injuries inflicted through automobile accidents, uncontrolled vicious dogs, defective products, libel, and countless others—involve harms that generally are redressed through civil and not criminal means.

In civil actions "punitive" damages sometimes are awarded, which closely resemble fines imposed in the name of criminal punishment. Similarly, the involuntary confinement of a mentally ill person who has been civilly committed to a secure hospital, or of a person held in jail pending trial because of an inability to post bail, or of a juvenile who has been adjudicated delinquent may be hard to distinguish from a jail or prison sentence resulting from a criminal conviction. Most criminal laws and punishments come clearly labeled. When we are confronted with a trial for murder, rape, kidnapping, or theft and when that trial is conducted in the general manner of a criminal trial—for example, beginning with an indictment and entailing the right of confrontation and cross-examination, the privilege against compelled self-incrimination, court-appointed counsel for indigent defendants, and the possibility of a criminal sentence—we

would have little or no trouble identifying the area of law involved. Characterization issues more often arise from liberty-depriving, stigma-imposing legal interventions that resist the label "criminal" in the hope of resisting, among other consequences, the more demanding procedural and doctrinal requirements of the criminal law.

## Meaning of "Punishment"

If **punishment** is an essential attribute of the criminal process, then it is important to be able to define that term with some precision. If we cannot, we have accomplished little more than converting questions about the content of the criminal law to questions about the meaning of punishment. We will not attempt an exhaustive definition at this point but rather first offer a few general observations and then inquire into the essential purposes of criminal punishment.

Legal philosopher H. L. A. Hart identified five central elements of punishment:

**(i)** It must involve pain or other consequences normally considered unpleasant.

**(ii)** It must be for an offense against legal rules.

**(iii)** It must be of an actual or supposed offender for his offense.

**(iv)** It must be intentionally administered by human beings other than the offender.

**(v)** It must be imposed and administered by an authority constituted by a legal system against which the offense is committed.[3]

Hart's criteria eliminate several commonly used references to punishment. For example, when we refer to the devoted husband who administers a lethal poison to his long-suffering, terminally ill wife at her request and say, "Don't prosecute this man because he has been 'punished' enough," we fail to conform to Hart's criteria. Human sacrifice would not constitute "punishment" (it is not imposed on an "offender" for his or her "offense"); nor would a retaliatory underworld slaying (it is not imposed and administered by authority of a legal system for a violation of legal rules); and nor, for similar reasons, would a host of other painful and unpleasant experiences to which we might be tempted to apply the term. Hart's definition thus is helpful in excluding many outcomes, although more ambiguity is introduced when it is applied to others such as "civil forfeiture" proceedings, preventive detention, the confinement of juvenile delinquents, the coercive "treatment" of drug addicts and alcoholics, and even quarantining those with contagious diseases.[4]

Several U.S. Supreme Court decisions have struggled to define criminal punishment. For example, in *Kennedy v. Mendoza-Martinez* the justices ruled that a federal law that stripped Americans of their citizenship for fleeing the country to evade military service was a punitive measure requiring procedural safeguards equivalent to those recognized in criminal trials. The Court found that Congress clearly intended to punish draft evaders by enacting the statute and relied on that finding to conclude that the loss of citizenship mandated by the law was properly classified as punishment.

## Kennedy v. Mendoza-Martinez, 372 U.S. 144, 83 S. Ct. 554, 9 L. Ed. 2d 644 (1963)

The punitive nature of the sanction here is evident under the tests traditionally applied to determine whether an Act of Congress is penal or regulatory in character, even though in other cases this problem has been extremely difficult and elusive of solution. Whether the sanction involves an affirmative disability or restraint, whether it has historically been regarded as punishment, whether it comes into play only on a finding of scienter, whether its operation will promote the traditional aims of punishment—retribution and deterrence, whether the behavior to which it applies is already a crime, whether an alternative purpose to which it may rationally be connected is assignable for it, and whether it appears excessive in relation to the alternative purpose assigned are all relevant to the inquiry, and may often point in differing directions. Absent conclusive evidence of congressional intent as to the penal nature of a statute, these factors must be considered in relation to the statute on its face. Here, although we are convinced that application of these criteria to the face of the statutes supports the conclusion that they are punitive, a detailed examination along such lines is unnecessary, because the objective manifestations of congressional purpose indicate conclusively that the provisions in question can only be interpreted as punitive.

Using the factors considered by the Court in *Mendoza-Martinez,* you should be able to distinguish between these two cases: (1) a court orders Martin Shepard, a convicted offender, to wear an electronic bracelet and remain confined in his home for 3 months, and (2) a court orders Martina Robbins, who is afflicted with a highly contagious and dangerous disease, to remain quarantined in her home for a minimum of 90 days while the illness runs its course. We should have no hesitation in calling Martin's court-ordered confinement punishment. We probably will agree that Martina's confinement is a health-related measure serving legitimate regulatory ends and is not a form of punishment. Martin and Martina may find their circumstances equally restrictive and objectionable. On the other hand, Martin may even report being happy to be at home instead of in jail. Although Martin and Martina objectively may appear to be experiencing similar hardships and deprivations, not everything that is unpleasant is punishment.

In *Bell v. Wolfish,* 441 U.S. 520 (1979), the Supreme Court ruled that detainees who had been arrested and were being held in jail pending trial—who were subject to

the identical regimen and same restrictive conditions as offenders serving jail sentences as punishment for crimes—were not being "punished." Using the principles set forth in *Mendoza-Martinez,* can you justify this decision? Consider the following circumstances.

Leroy Hendricks served 10 years in prison for sexually molesting two 13-year-old boys. He had a long record of similar sex offenses committed against children. In 1994, immediately before he was scheduled for release from prison, the State of Kansas took steps to have Hendricks indefinitely civilly committed under its Sexually Violent Predator Act on the grounds that he suffered from a "mental abnormality" that made him likely to engage in future "predatory acts of sexual violence." Hendricks objected that he had paid his debt to society by serving a prison sentence for his crimes and that Kansas' attempt to prolong his confinement subjected him to additional "punishment," in violation of his constitutional rights. The Supreme Court narrowly rejected Hendricks' argument in a 5–4 decision that further elaborated on the definition of criminal punishment.

## Kansas v. Hendricks, 521 U.S. 346, 117 S. Ct. 2072, 138 L. Ed. 2d 501 (1997)

The thrust of Hendricks' argument is that the Act establishes criminal proceedings; hence confinement under it necessarily constitutes punishment. He contends that where, as here, newly enacted "punishment" is predicated upon past conduct for which he has already been convicted and forced to serve a prison sentence, the Constitution's Double Jeopardy and Ex Post Facto Clauses are violated. We are unpersuaded by Hendricks' argument that Kansas has established criminal proceedings. The categorization of a particular proceeding as civil or criminal "is first of all a question of statutory construction." We must initially ascertain whether the legislature meant the statute to establish "civil" proceedings. If so, we ordinarily defer to the legislature's stated intent. Here, Kansas' objective to create a civil proceeding is evidenced by its placement of the Sexually Violent Predator Act within the Kansas probate code, instead of the criminal code, as well as its

*(Continues)*

description of the Act as creating a *"civil commitment procedure."* Kan. Stat. Ann., Article 29 (1994) ("Care and Treatment for Mentally Ill Persons"), § 59-29a01 (emphasis added). Nothing on the face of the statute suggests that the legislature sought to create anything other than a civil commitment scheme designed to protect the public from harm. Although we recognize that a "civil label is not always dispositive," we will reject the legislature's manifest intent only where a party challenging the statute provides "the clearest proof" that "the statutory scheme [is] so punitive either in purpose or effect as to negate [the State's] intention" to deem it "civil." United States v. Ward, 448 U.S. 242, 248–249 (1980). In those limited circumstances, we will consider the statute to have established criminal proceedings for constitutional purposes. Hendricks, however, has failed to satisfy this heavy burden.

As a threshold matter, commitment under the Act does not implicate either of the two primary objectives of criminal punishment: retribution or deterrence. The Act's purpose is not retributive because it does not affix culpability for prior criminal conduct. Instead, such conduct is used solely for evidentiary purposes, either to demonstrate that a "mental abnormality" exists or to support a finding of future dangerousness. . . .

In addition, the Kansas Act does not make a criminal conviction a prerequisite for commitment—persons absolved of criminal responsibility may nonetheless be subject to confinement under the Act. . . . An absence of the necessary criminal responsibility suggests that the State is not seeking retribution for a past misdeed. . . . Moreover, unlike a criminal statute, no finding of scienter is required to commit an individual who is found to be a sexually violent predator; instead, the commitment determination is made based on a "mental abnormality" or "personality disorder" rather than on one's criminal intent. The existence of a scienter requirement is customarily an important element in distinguishing criminal from civil statutes. The absence of such a requirement here is evidence that confinement under the statute is not intended to be retributive.

Nor can it be said that the legislature intended the Act to function as a deterrent. Those persons committed under the Act are, by definition, suffering from a "mental abnormality" or a "personality disorder" that prevents them from exercising adequate control over their behavior. Such persons are therefore unlikely to be deterred by the threat of confinement. And the conditions surrounding that confinement do not suggest a punitive purpose on the State's part. The State has represented that an individual confined under the Act is not subject to the more restrictive conditions placed on state prisoners, but instead experiences essentially the same conditions as any involuntarily committed patient in the state mental institution. . . . Although the civil commitment scheme at issue here does involve an affirmative restraint, "the mere fact that a person is detained does not inexorably lead to the conclusion that the government has imposed punishment." United States v. Salerno, 481 U.S. 739, 746 (1987). The State may take measures to restrict the freedom of the dangerously mentally ill. This is a legitimate non-punitive governmental objective and has been historically so regarded. . . . Hendricks focuses on his confinement's potentially indefinite duration as evidence of the State's punitive intent. That focus, however, is misplaced. Far from any punitive objective, the confinement's duration is instead linked to the stated purposes of the commitment, namely, to hold the person until his mental abnormality no longer causes him to be a threat to others. . . .

If, at any time, the confined person is adjudged "safe to be at large," he is statutorily entitled to immediate release. . . .

Hendricks next contends that the State's use of procedural safeguards traditionally found in criminal trials makes the proceedings here criminal rather than civil. . . .

The numerous procedural and evidentiary protections afforded here demonstrate that the Kansas Legislature has taken great care to confine only a narrow class of particularly dangerous individuals, and then only after meeting the strictest procedural standards. That Kansas chose to afford such procedural protections does not transform a civil commitment proceeding into a criminal prosecution. Finally, Hendricks argues that the Act is necessarily punitive because it fails to offer any legitimate "treatment" Without such treatment, Hendricks asserts, confinement under the Act amounts to little more than disguised punishment. . . .

We have already observed that, under the appropriate circumstances and when accompanied by proper procedures, incapacitation may be a legitimate end of the civil law. . . . Accordingly, the Kansas court's determination that the Act's "overriding concern" was the continued "segregation of sexually violent offenders" is consistent with our conclusion that the Act establishes civil proceedings, especially when that concern is coupled with the State's ancillary goal of providing treatment to those offenders, if such is possible. While we have upheld state civil commitment statutes that aim both to incapacitate and to treat, we have never held that the Constitution prevents a State from civilly detaining those for whom no treatment is available, but who nevertheless pose a danger to others. A State could hardly be seen as furthering a "punitive" purpose by involuntarily confining persons afflicted with an untreatable, highly contagious disease. Similarly, it would be of little value to require treatment as a precondition for civil confinement of the dangerously insane when no acceptable treatment existed. To conclude otherwise would obligate a State to release certain confined individuals who were both mentally ill and dangerous simply because they could not be successfully treated for their afflictions. . . .

Where the State has "disavowed any punitive intent"; limited confinement to a small segment of particularly dangerous individuals; provided strict procedural safeguards; directed that confined persons be segregated from the general prison population and afforded the same status as others who have been civilly committed; recommended treatment if such is possible; and permitted immediate release upon a showing that the individual is no longer dangerous or mentally impaired, we cannot say that it acted with punitive intent. We therefore hold that the Act does not establish criminal proceedings and that involuntary confinement pursuant to the Act is not punitive. Our conclusion that the Act is non punitive thus removes an essential prerequisite for both Hendricks' double jeopardy and ex post facto claims.

As *Hendricks* suggests, the legal consequences of labeling a particular deprivation "criminal punishment" as opposed to a "civil" or "regulatory" remedy or sanction can be profoundly significant. For example, the Fifth Amendment's double jeopardy clause prohibits multiple "punishments" for a single offense; the constitutional prohibition against *ex post facto* laws forbids "punishment" for conduct not clearly defined as illegal when committed; and numerous procedural rights apply to "criminal" proceedings that do not necessarily apply in other contexts. Using a test similar to the one used in *Hendricks,* the Supreme Court ruled in *Smith v. Doe,* 538 U.S. 84 (2003) that requiring convicted sex offenders to register and be subjected to community notification regarding information about their offenses, their residence, and other descriptive matter was not a form of "punishment"; accordingly, retroactive application of those "Megan's Law" provisions did not violate *ex post facto* principles. After considering rulings such as *Mendoza-Martinez, Hendricks,* and *Smith v. Doe,* how much closer are we to identifying what is distinctive about the criminal law?

## Henry M. Hart, Jr. "The Aims of the Criminal Law."

We can get our broadest view of the aims of the criminal law if we look at them from the point of view of the makers of a constitution—of those who are seeking to establish sound foundations for a tolerable and durable social order. From this point of view, these aims can be most readily seen, as they need to be seen, in their relation to the aims of the good society generally.

In this setting, the basic question emerges: Why should the good society make use of the method of the criminal law at all?

### A. What the Method of the Criminal Law Is

The question posed raises preliminarily an even more fundamental inquiry: What do we mean by **"crime"** and **"criminal"**? Or, put more accurately, what should we understand to be "the method of the criminal law," the use of which is in question? This latter way of formulating the preliminary inquiry is more accurate, because it pictures the criminal law as a process, a way of doing something, which is what it is. A great deal of intellectual energy has been misspent in an effort to develop a concept of crime as "a natural and social phenomenon" abstracted from the functioning system of institutions which make use of the concept and give it impact and meaning. But the criminal law, like all law, is concerned with the pursuit of human purposes through the forms and modes of social organization, and it needs always to be thought about in that context as a method or process of doing something.

What then are the characteristics of this method?

1. The method operates by means of a series of directions, or commands, formulated in general terms, telling people what they must or must not do. Mostly, the commands of the criminal law are "must-nots," or prohibitions, which can be satisfied by inaction. "Do not murder, rape, or rob." But some of them are "musts," or affirmative requirements, which can be satisfied only by taking a specifically, or relatively specifically, described kind of action. "Support your wife and children," and "File your income tax return."

2. The commands are taken as valid and binding upon all those who fall within their terms when the time comes for complying with them, whether or not they have been formulated in advance in a single authoritative set of words. They speak to members of the community, in other words, in the community's

*Criminal law vs. private law/torts, etc.

behalf, with all the power and prestige of the community behind them.

3. The commands are subject to one or more sanctions for disobedience which the community is prepared to enforce.

Thus far, it will be noticed, nothing has been said about the criminal law which is not true also of a large part of the noncriminal, or civil, law. The law of torts, the law of contracts, and almost every other branch of private law that can be mentioned operate, too, with general directions prohibiting or requiring described types of conduct, and the community's tribunals enforce these commands. What, then, is distinctive, about the method of the criminal law?

Can crimes be distinguished from civil wrongs on the ground that they constitute injuries to society generally which society is interested in preventing? The difficulty is that society is interested also in the due fulfillment of contracts and the avoidance of traffic accidents and most of the other stuff of civil litigation. The civil law is framed and interpreted and enforced with a constant eye to these social interests. Does the distinction lie in the fact that proceedings to enforce the criminal law are instituted by public officials rather than private complainants? The difficulty is that public officers may also bring many kinds of "civil" enforcement actions—for an injunction, for the recovery of a "civil" penalty, or even for the detention of the defendant by public authority. Is the distinction, then, in the peculiar character of what is done to people who are adjudged to be criminals? The difficulty is that, with the possible exception of death, exactly the same kinds of unpleasant consequences, objectively considered, can be and are visited upon unsuccessful defendants in civil proceedings.

If one were to judge from the notions apparently underlying many judicial opinions, and the overt language even of some of them, the solution of the puzzle is simply that a crime is anything which is called a crime, and a criminal penalty is simply the penalty provided for doing anything which has been given that name. So vacant a concept is a betrayal of intellectual bankruptcy. Certainly, it poses no intelligible issue for a constitution-maker concerned to decide whether to make use of "the method of the criminal law." Moreover, it is false to popular understanding, and false also to the understanding embodied in existing constitutions. By implicit assumptions that are more impressive than any explicit assertions, these constitutions proclaim that a conviction for crime is a distinctive and serious matter—a something, and not a nothing. What is that something? . . .

5. The method of the criminal law, of course, involves something more than the threat (and, on due occasion, the expression) of community condemnation of antisocial conduct. It involves, in addition, the threat (and, on due occasion, the imposition) of

unpleasant physical consequences, commonly called punishment. But if Professor Gardner is right, these added consequences take their character as punishment from the condemnation which precedes them and serves as the warrant for their infliction. Indeed, the condemnation plus the added consequences may well be considered, compendiously, as constituting the punishment. Otherwise, it would be necessary to think of a convicted criminal as going unpunished if the imposition or execution of his sentence is suspended.

In traditional thought and speech, the ideas of crime and punishment have been inseparable; the consequences of conviction for crime have been described as a matter of course as "punishment." . . .

At least under existing law, there is a vital difference between the situation of a patient who has been committed to a mental hospital and the situation of an inmate of a state penitentiary. The core of the difference is precisely that the patient has not incurred the moral condemnation of his community, whereas the convict has.

*Source:* "The Aims of the Criminal Law," by Henry M. Hart, Jr., (1958) Law and Contemporary Problems, 23, pp. 401, 402–406. Used by permission of Duke University Law School.

Professor Hart argues that "a formal and solemn pronouncement of the moral judgment of the community"—a "judgment of community condemnation"—is indispensable to a proper understanding of the criminal law. There is no disputing that notions of *blame* and *blameworthiness* are central to the substantive criminal law and punishment. Before ascribing blame we generally make implicit assumptions about the moral responsibility of the person whose conduct is at issue. Thus, although we might identify lightning as the source of a fire that destroyed a building, we would be hard pressed to pass judgment "blaming" anyone or anything (in the moral sense) for causing the harm. For related reasons we undoubtedly would be reluctant to attempt to invoke the criminal law or to inflict punishment in response to such a disaster.

Similarly, under most circumstances we would not resort to the criminal law if a fire broke out from an unforeseeable accident traceable to an individual, even if significant property damage or personal injury ensued. We normally do not "blame" people for accidents that truly are unpredictable. Although perhaps more controversial, our reluctance to "blame" a 3-year-old child who plays with matches and thus ignites a building, or a seriously mentally ill person who has little conception of what fire is or that it burns, would generally restrain us from having those actors subjected to criminal punishment. Nor would we blame, or desire to punish, a firefighter who intentionally causes a farmer's field to burn when necessary to create a firebreak so that occupied dwellings will be spared the ravages of a fast-approaching forest fire. We would be inclined to excuse the child and the mentally ill person from criminal liability and consider the firefighter's conduct to be justified. In none of these cases would we believe the "judgment of community condemnation" to be appropriate.

A very different case is presented by the arsonist who deliberately sets fire to a building to collect insurance proceeds or to carry out a vendetta against the building's owner or its occupants. Such conduct is appropriately defined as blameworthy. Because we have no inhibitions against blaming the arsonist, neither are we inhibited about concluding that the behavior should be punished. Blame and punishment are closely intertwined. The same reasons for assigning blame help account for our feeling justified in inflicting punishment.

This is not to say that moral blameworthiness invariably is a prerequisite for punishment. We will encounter "strict liability" crimes and other instances where it may be difficult to conclude that an offender acted with a "guilty mind" or engaged in behavior that in most people's judgment reasonably can be condemned as "wrongful." Nevertheless, if criminal punishment embodies the "formal and solemn pronouncement of the moral condemnation of the community," as Professor Hart has suggested, we generally should expect that the reasons or justifications for that punishment will be linked to a judgment that a wrongdoer fairly can be blamed for his or her conduct. *strict liability,

## Purposes of Punishment

At this juncture we pause to consider *why* we punish people who violate the criminal laws. Are we simply trying to pay them back for the harm they have caused? Or, are we trying to accomplish something in addition to giving offenders their just deserts? For example, is punishing lawbreakers necessary so others will not be tempted to commit crimes? Is it required as a signal to individual offenders that their unlawful conduct will not be tolerated, and that they had better not commit future crimes? Do we use punishments like imprisonment and the death penalty to help keep society safe from the people locked up or executed? When we punish, do we simultaneously hope to treat or reform criminal offenders and thus help them turn their lives around? If there are multiple goals of punishment, do they always point to the same conclusion regarding the amount and type of punishment that should be imposed in an individual case? Should there be any limits—either minimum or maximum—on the punishment imposed against specific offenders or for particular offenses?

We begin analyzing questions of this nature, first with the assistance of scholarly commentary and then by considering a classic 19th century legal decision that requires close examination of the application of the criminal law and punishment.

---

Kent Greenawalt, "Punishment."

**Moral justifications and legal punishment.** Since punishment involves pain or deprivation that people wish to avoid, its intentional imposition by the state requires

justification. The difficulties of justification cannot be avoided by the view that punishment is an inevitable adjunct of a system of criminal law. If *criminal law* is defined to include punishment, the central question remains whether society should have a system of mandatory rules enforced by penalties. . . .

If actual punishment never or very rarely followed threatened punishment, the threat would lose significance. Thus, punishment in some cases is a practical necessity for any system in which threats of punishment are to be taken seriously; and to that extent, the justification of punishment is inseparable from the justification of threats of punishment. Punishment ≈ justification

The dominant approaches to justification are retributive and utilitarian. Briefly stated, a retributivist claims that punishment is justified because people deserve it; a *utilitarian believes that justification lies in the useful purposes that punishment serves. . . .

**Retributive justification.** Why should wrongdoers be punished? Most people might respond simply that they deserve it or that they should suffer in return for the harm they have done. Such feelings are deeply ingrained, at least in many cultures, and are often supported by notions of divine punishment for those who disobey God's laws. A simple retributivist justification provides a philosophical account corresponding to these feelings: someone who has violated the rights of others should be penalized, and punishment restores the moral order that has been breached by the original wrongful act. The idea is strikingly captured by Immanuel Kant's claim that an island society about to disband should still execute its last murderer. Society not only has a right to punish a person who deserves punishment, but it has a duty to do so. In Kant's view, a failure to punish those who deserve it leaves guilt upon the society; according to G.W.F. Hegel, punishment honors the criminal as a rational being and gives him what it is his right to have. In simple retributivist theory, practices of punishment are justified because society should render harm to wrongdoers; only those who are guilty of wrongdoing should be punished; and the severity of punishment should be proportional to the degree of wrongdoing, an approach crudely reflected in the idea of "an eye for an eye, a tooth for a tooth.". . .

*Moral guilt and social judgment.* One fundamental question is whether people are ever morally guilty in the way that basic retributive theory seems to suppose. If all our acts are consequences of preceding causes over which we ultimately have had no control, causes that were set in motion before we were born—if, in other words, philosophical determinism is true—then the thief or murderer is, in the last analysis, more a victim of misfortune than a villain on the cosmic stage. Although he may be evil in some sense and able to control his actions, his character has been formed by forces outside himself, and that ultimately determines the choices he makes. . . . It requires some notion of free will that attributes to humans responsibility for doing wrong in a way that is not attributed to other animals. . . .

*Violations of social norms and fairness.* A rather different retributive approach is that criminals deserve punishment because they violate norms established by society, the magnitude of the violation being measured by the seriousness with which society treats the offense. In this form, the theory sidesteps the objection that correcting moral wrongs is not the business of the criminal law, and it does not impose upon officials the impossible burden of ascertaining subtle degrees of moral guilt. This version of the theory fits better with existing (and conceivable) practices of criminal punishment, but in doing so, it no longer connects moral guilt so strongly to justifiable punishment and does not resolve the question of why morality demands that society punish those who violate its norms simply for the sake of punishing them.

One answer to this question is that fairness to citizens who make sacrifices by obeying the law requires that violators be punished rather than reap benefits for disregarding legal standards. What is crucial and debatable about this view is the claim that law-abiding members of the community will suffer an actual injustice if the guilty go unpunished. . . . *innocent suffer if guilty isn't punished.*

**Utilitarian justification.** Utilitarian theories of punishment have dominated American jurisprudence during most of the twentieth century. According to Jeremy Bentham's classical utilitarianism, whether an act or social practice is morally desirable depends upon whether it promotes human happiness better than possible alternatives. Since punishment involves pain, it can be justified only if it accomplishes enough good consequences to outweigh this harm. . . .

The catalogs of beneficial consequences that utilitarians have thought can be realized by punishment have varied, but the following have generally been regarded as most important.

1. *General deterrence.* Knowledge that punishment will follow crime deters people from committing crimes, thus reducing future violations of right and the unhappiness and insecurity they would cause. The person who has already committed a crime cannot, of course, be deterred from committing that crime, but his punishment may help to deter others. In Bentham's view, general deterrence was very much a matter of affording rational self-interested persons good reasons not to commit crimes. With a properly developed penal code, the benefits to be gained from criminal activity would be outweighed by the harms of punishment, even when those harms were discounted by the probability of avoiding detection. Accordingly, the greater the temptation to commit a particular crime and the smaller the chance of detection, the more severe the penalty should be.

   Punishment can also deter in ways more subtle than adding a relevant negative factor for cool calculation. Seeing others punished for certain behavior can create in people a sense of association between punishment and act that may constrain them even when they are sure they will not get caught. Adults, as well as children, may subconsciously fear punishment even though rationally they are confident it will not occur.

2. *Norm reinforcement.* For young children, the line may be very thin between believing that behavior is wrong and fearing punishment. Adults draw the distinction more plainly, but seeing others punished can still contribute to their sense that actions are wrong, helping them to internalize the norms society has set. Practices of punishment can thus reinforce community norms by affecting the dictates of individual consciences. Serious criminal punishment represents society's strong condemnation of what the offender has done, and performs a significant role in moral education. . . .

3. *Individual deterrence.* The actual imposition of punishment creates fear in the offender that if he repeats his act, he will be punished again. Adults are more able than small children to draw conclusions from the punishment of others, but having a harm befall oneself is almost always a sharper lesson than seeing the same harm occur to others. To deter an offender from repeating his actions, a penalty should be severe enough to outweigh in his mind the benefits of the crime. For the utilitarian, more severe punishment of repeat offenders is warranted partly because the first penalty has shown itself ineffective from the standpoint of individual deterrence.

4. *Incapacitation.* Imprisonment puts convicted criminals out of general circulation temporarily, and the death penalty does so permanently. These punishments physically prevent persons of dangerous disposition from acting upon their destructive tendencies.

5. *Reform.* Punishment may help to reform the criminal so that his wish to commit crimes will be lessened, and perhaps so that he can be a happier, more useful person. Conviction and simple imposition of a penalty might themselves be thought to contribute to reform if they help an offender become aware that he has acted wrongly. However, reform is usually conceived as involving more positive steps to alter basic character or improve skills, in order to make offenders less antisocial. . . . *Not practiced -drug courts . . .*

6. *Vengeance.* The utilitarian, in contrast to the retributivist, does not suppose that wrongful acts intrinsically deserve a harsh response, but he recognizes that victims, their families and friends, and some members of the public will feel frustrated if no such response is forthcoming. Satisfying these desires that punishment be imposed is seen as one legitimate aim in punishing the offender. In part, the point is straightforwardly to increase the happiness, or reduce the unhappiness, of those who want the offender punished, but formal punishment can also help increase their sense of respect for the law and deflect unchanneled acts of private vengeance.

Unlike a basic retributive theory, the utilitarian approach to punishment is compatible with philosophical determinism. . . .

From the utilitarian perspective, the acts for which criminal punishment should be authorized are those with respect to which the good consequences of punishment can outweigh the bad; the persons who should be punished are those whom it is useful to punish; and the severity of punishment should be determined not by some abstract notion of deserts but by marginal usefulness. Each extra ingredient of punishment is warranted only if its added benefits outweigh its added harms. . . .

✱**Philosophical objections to utilitarianism.** Utilitarian programs for systems of punishment are subject to two kinds of objections: those which challenge basic philosophical premises, and those which claim that different systems would better accomplish social aims. . . .

The most fundamental objection is to treating the criminal as a means to satisfy social purposes rather than as an end in himself. This objection bears on why, and how, guilty offenders may be punished; but the most damaging aspect of the attack is that utilitarianism admits the possibility of justified punishment of the innocent. The retributivist asserts that such punishment is morally wrong even when it would produce a balance of favorable consequences. . . . *utalitarian vs. retributivists*

The utilitarian may answer that his theory will certainly not support any announced practice of punishing the innocent. The purposes of punishment would not be served if people knew a person was innocent, and even to establish a general policy that officials would at their discretion occasionally seek punishment of those they know are innocent would cause serious insecurity. . . .

**Mixed theory.** Given these problems with unalloyed utilitarian theory, some mixture of utilitarian and retributive elements provides the most cogent approach to punishment. The basic reasons for having compulsory legal rules backed by sanctions are utilitarian; these reasons should dominate decisions about the sorts of behavior to be made criminal. Moral wrongs should not be subject to legal punishment unless that is socially useful, and behavior that is initially morally indifferent may be covered by the criminal law if doing so serves social goals. Notions of deserts, however, should impose more stringent constraints on the imposition of punishment than pure utilitarianism acknowledges. . . . *the fuck?*

Source: "Punishment," by Kent Greenawalt from Encyclopedia of Crime and Justice, 2E. © 2001 Gale, a part of Cengage Learning, Inc. Reproduced by permission. www.cengage.com/permissions.

---

One excellent treatise, or "hornbook," analyzing the substantive criminal law is *Criminal Law* by Wayne R. LaFave.[5] This book can be a useful reference if you have questions or seek additional information about the workings of the criminal law. A useful discussion of the purposes of criminal punishment is presented in this hornbook on pages 22 through 30. Other handy references are *Understanding Criminal Law*,[6] by Joshua Dressler, and *Criminal Law*,[7] by Perkins and Boyce.

## Principles of Punishment in Application

Although the justifications offered in support of criminal punishment can be described in relatively straightforward terms, applying those general principles to specific

*cogent: logical/convincing*

cases can be exceedingly difficult. No convenient scale exists to inform us what type of and how much punishment is appropriate for particular offenses. In addition, although the different justifications for punishment may seem rational when considered separately, they may be almost impossible to reconcile when punishment is considered in individual cases. For instance, the risk that some offenders will commit future crimes may be practically nonexistent. In fact, we might confidently predict that an individual committing an offense will be an upstanding and productive citizen in the future and have no further interaction with the criminal law. There would be little need to incarcerate such a person in the name of **specific deterrence**, incapacitation, or to reform the offender. On the other hand, if a serious crime has been committed, not imposing a prison sentence may disserve the retributive and general deterrence functions of punishment.

Consider the following offenses and offenders. Try to rank-order them according to their relative seriousness. Then decide the punishment you would consider appropriate in each case, selecting between the death penalty, a specific term of imprisonment, a particular fine, or probation. Be prepared to defend your choices in light of the commonly accepted justifications for criminal punishment.

a. Three college students, intent on finding a novel wall hanging for their dormitory room, remove a stop sign from an intersection on a residential street. Two hours later a motorist unfamiliar with the road proceeds without slowing down into the intersection from which the stop sign was removed. A pickup truck with the lawful right of way hits her car broadside; she is killed. *involuntary manslaughter.*

b. Same facts as above except that a police officer spots the students and arrests them immediately after they have taken down the stop sign. The officer replaces the sign before any cars drive through the intersection. No accidents occur and no one is injured.

c. A college student celebrating the conclusion of final exams has six mixed drinks at a bar and then, quite intoxicated, drives her car home. Proceeding the wrong way down a one-way street and running through a red light, she smashes into another car lawfully being driven on the same road. The driver of the other vehicle, who was not wearing his seat belt, in violation of state law, is thrown from his car and killed. The intoxicated college student has no prior traffic infractions and no criminal record. *reckless driving, DUI,*

d. A man with four prior convictions for driving while intoxicated, and whose driver's license has been suspended, is driving the wrong direction on a one-way street over 20 mph in excess of the speed limit. An alert police officer orders the driver to pull his car over. A breathalyzer test reveals the man is highly intoxicated. The officer arrests him. *Jail!*

e. A man with no prior criminal record is caught in his car with 672 grams of cocaine (over 1 ½ pounds) having a street sale value of roughly half a million dollars.

f. A man with two prior felony convictions—one for fraudulently using a credit card to obtain goods worth $80 and another for passing a forged check in the amount of $28.36—is convicted of yet another felony, obtaining $120.75 by false pretenses. Total property value associated with his three crimes is just over $229. *Damn*

g. A teenager steals the one and only car owned by the Phelps family, which Mr. and Mrs. Phelps use to drive to work and transport their two small children to and from day care. The car is a 1997 Plymouth with a market value of $900. *GTA*

h. A teenager steals one of four cars owned by Reginald G. Porter III, a wealthy socialite. The car is a new Mercedes valued at more than $50,000.

i. A woman takes a loaded gun, aims it at point-blank range at her sleeping archenemy, and, with the intent to kill him, pulls the trigger. The gun jams and fails to discharge. The intended victim is unaware after awakening that the episode occurred. *Premeditated mrdr.*

j. A woman takes a loaded gun, points it at her arch-enemy, and, with the intent to kill him, pulls the trigger. Her aim is bad and the bullet misses the man, but the incident scares him badly.

k. A woman takes a loaded gun, points it at her arch-enemy, and, with the intent to kill him, pulls the trigger. Her aim is bad and the bullet strikes the man in the shoulder. A bystander summons an ambulance, and the man is treated at a hospital and later released. *Felony 3, prem. murder.*

l. A woman takes a loaded gun, points it at her arch-enemy, and, with the intent to kill him, pulls the trigger. Her aim is bad and the bullet strikes the man in the shoulder. No bystanders witness the shooting or are in the area. Unable to summon an ambulance, the man loses a large quantity of blood. He lapses into unconsciousness and subsequently dies. Cause of death is loss of blood resulting from the shooting. *MURDER!*

m. A woman takes a loaded gun, points it at her arch-enemy, and, with the intent to kill him, pulls the trigger. The bullet strikes the man in the heart, killing him instantly. *DEATH SENTENCE!*

n. The man described immediately above is the woman's sleeping husband. He has beaten her mercilessly for the past several years. She has decided that there is no escape from him and that she cannot take any more beatings.

o. The man described in paragraph m, above, is not the woman's husband and he has never laid a hand on her. He is the president of the United States.

p. A man steals a watch worth $800 from a department store.

q. A bank teller embezzles $800 from the bank at which he is employed.

r. A man threatens to reveal damaging secrets about another unless he is paid $800.

## A Case Example

*Briefing a Case. The Queen v. Dudley and Stephens,* which follows, is the first full judicial decision we have encountered. It presents particularly vexing questions regarding blame, criminal responsibility, and punishment. We will regularly read case decisions in the pages that lie ahead. To assist your understanding of these decisions, we strongly suggest that you develop the habit of "briefing" cases. *Dudley and Stephens* is different from most cases presented in this book in that it was decided by the highest court in England, the Queen's Bench division of the high court of justice (called the "King's Bench" during the reign of a king). British court decisions are reported in somewhat different form than American court decisions, but Lord Coleridge's opinion for the unanimous five-judge panel in *Dudley and Stephens* is sufficiently similar to the American state and federal court opinions you will encounter throughout this book that those differences need not detain us for now.

When appellate courts decide the issues presented to them, they explain and attempt to justify their decisions through the written opinions of the judges who heard the case. Of course, all judges sitting on a court do not always agree with one another. Thus, in addition to the "majority" opinion—the prevailing opinion in the case joined by a majority of the judges on the court—"dissenting" opinions frequently will be written or joined in by one or more members of the court. Judges sometimes write "concurring" opinions as well, in which they agree with the result reached in the case but nevertheless want to explain their own reasons for reaching that result. Sometimes a majority of the judges on a court cannot agree on the reasoning used in support of a decision. In such cases a "plurality" opinion is written, along with the concurring and dissenting (if any) opinions of the other judges. Because plurality opinions do not represent the views of a majority of the court, they carry significantly less weight as precedent for subsequent decisions. Between three and nine judges, but occasionally more, typically comprise the courts that decide criminal appeals.

We study judicial opinions to gain an understanding of the reasons underlying a court's decision. Those reasons typically involve a combination of the idiosyncratic case facts and the general rules of law that apply to a case. In turn, prior judicial decisions (precedent), history, social policy choices, constitutional principles, statutes, and a host of other factors may shape the rules.

It is imperative to understand the reasons offered in support of judicial decisions. Very few cases involve identical facts. If we do not understand the rationale behind a rule of law, we may be hopelessly confused about the application of that rule in different factual settings. Through studying judicial opinions we also gain an appreciation for how the law changes over time in response to novel issues and changing social circumstances. Furthermore, as you analyze the reasons offered in support of a decision you may find those reasons to be strained or unpersuasive. You should not be reluctant to criticize a judge's opinion if you can point to problems in logic or other flaws in the judge's reasoning.

*always quest.*

A technique commonly used to dissect and analyze judicial opinions is *briefing* a case. Although you should not be reluctant to fine-tune the briefing format we suggest, your case briefs should include the following essential parts:

1. *Case name, citation, and page where the case begins in the book.* This much should be easy; simply copy down the name of the case and its citation. The citation will enable you to locate the decision in the library or electronically if you ever want to do so. It also identifies the court making the decision and the decision date, which may be of interest. Writing down the page at which the opinion begins in the book is just a convenience for future reference purposes.

2. *Facts.* The first section of the brief should be a statement of the case facts. What happened in the case? Who did what to whom, and why? Provide a sufficiently complete statement of the facts so that a reader who has not been exposed to the case decision will understand what has happened. However, present only the *relevant* facts. Remember, you are "briefing" the case. Be brief, without losing facts that are important to the resolution of the case. Your statement of the facts should include the *judicial history* of the case. That is, you should identify the court in which the case originated, any other courts that previously issued rulings in the case, and what action those courts took as the case worked its way to the court writing the decision that you are briefing.

3. *The issue.* State the precise question (or questions) the court has been called on to answer. Being able to pinpoint the specific issue decided by the court is crucial to your understanding of a case. Courts sometimes make this task straightforward by clearly stating the question presented by a case; at other times courts seem intent on obscuring the precise issue before them. There is probably no more important part of your brief than identifying the question presented for decision. State the question in a form that would enable a naive listener—one who has not read the court opinion—to grasp its significance upon hearing it. To convey the issue in a meaningful way you must incorporate enough background information into the question to give our hypothetical listener a fair chance of grasping its legal significance. For example, stating the issue in *Dudley and Stephens* as, "Are Dudley and Stephens guilty of murder?" is *not* adequate for purposes of a case brief. Phrase the issue so that important case facts are built into the question and so that the answer represents a general rule of law that can be applied in other circumstances.

4. *Holding and rationale.* What did the court decide? How did it answer the question you have just identified? Who won, and who lost? The holding in a case should be relatively clear. Of far more importance is your explanation of the court's rationale, or reasons in support of its holding. The investment you make of time and intellectual energy in setting forth the judges' reasoning will pay off handsomely in promoting your understanding of the decision. Also make note of interesting points or arguments raised in concurring or dissenting opinions. Jot down questions or criticisms you have about the court's opinion, and consider raising those matters during class discussions.

A case brief is a tool used to gain a better understanding of a judicial decision. You should regularly brief cases and use the briefs during class discussions and as study aids in preparation for examinations. The briefing process will help hone your critical reasoning skills, which are the essential building blocks for legal analysis. Brief the decision in *Dudley and Stephens* after you finish reading Lord Coleridge's opinion.

*criminal responsibility & punishment*

### The Queen v. Dudley and Stephens, 14 Q.B.D. 273 (1884)

. . . Indictment for the murder of Richard Parker on the high seas within the jurisdiction of the Admiralty.

At the trial before Huddleston, B., at the Devon and Cornwall Winter Assizes, November 7, 1884, the jury, at the suggestion of the learned judge, found the facts of the case in a special verdict which stated "that on July 5, 1884, the prisoners, Thomas Dudley and Edward Stephens, with one Brooks, all able-bodied English seamen, and the deceased also an English boy, between seventeen and eighteen years of age, the crew of an English yacht, a registered English vessel, were cast away in a storm on the high seas 1600 miles from the Cape of Good Hope, and were compelled to put into an open boat belonging to the said yacht. That in this boat they had no supply of water and no supply of food, except two 1 lb. tins of turnips, and for three days they had nothing else to subsist upon. That on the fourth day they caught a small turtle, upon which they subsisted for a few days, and this was the only food they had up to the twentieth day when the act now in question was committed. That on the twelfth day the remains of the turtle were entirely consumed, and for the next eight days they had nothing to eat. That they had no fresh water, except such rain as they from time to time caught in their oilskin capes. That the boat was drifting on the ocean, and was probably more than

*(Continues)*

1000 miles away from land. That on the eighteenth day, when they had been seven days without food and five without water, the prisoners spoke to Brooks as to what should be done if no succour came, and suggested that some one should be sacrificed to save the rest, but Brooks dissented, and the boy, to whom they were understood to refer, was not consulted. That on the 24th of July, the day before the act now in question, the prisoner Dudley proposed to Stephens and Brooks that lots should be cast who should be put to death to save the rest, but Brooks refused to consent, and it was not put to the boy, and in point of fact there was no drawing of lots. That on that day the prisoners spoke of their having families, and suggested it would be better to kill the boy that their lives should be saved, and Dudley proposed that if there was no vessel in sight by the morrow morning the boy should be killed. That next day, the 25th of July, no vessel appearing, Dudley told Brooks that he had better go and have a sleep, and made signs to Stephens and Brooks that the boy had better be killed. The prisoner Stephens agreed to the act, but Brooks dissented from it. That the boy was then lying at the bottom of the boat quite helpless, and extremely weakened by famine and by drinking sea water, and unable to make any resistance, nor did he ever assent to his being killed. The prisoner Dudley offered a prayer asking forgiveness for them all if either of them should be tempted to commit a rash act, and that their souls might be saved. That Dudley, with the assent of Stephens, went to the boy, and telling him that his time was come, put a knife into his throat and killed him then and there; that the three men fed upon the body and blood of the boy for four days; that on the fourth day after the act had been committed the boat was picked up by a passing vessel, and the prisoners were rescued, still alive, but in the lowest state of prostration. That they were carried to the port of Falmouth, and committed for trial at Exeter. That if the men had not fed upon the body of the boy they would probably not have survived to be so picked up and rescued, but would within the four days have died of famine. That the boy, being in a much weaker condition, was likely to have died before them. That at the time of the act in question there was no sail in sight, nor any reasonable prospect of relief. That under these circumstances there appeared to the prisoners every probability that unless they then fed or very soon fed upon the boy or one of themselves they would die of starvation. That there was no appreciable chance of saving life except by killing someone for the others to eat. That assuming any necessity to kill anybody, there was no greater necessity for killing the boy than any of the other three men. But whether upon the whole matter by the jurors found the killing of Richard Parker by Dudley and Stephens be felony and murder the jurors are ignorant, and pray the advice of the Court thereupon, and if upon the whole matter the Court shall be of opinion that the killing of Richard Parker be felony and murder, then the jurors say that Dudley and Stephens were each guilty of felony and murder as alleged in the indictment."

The learned judge then adjourned the assizes until the 25th of November at the Royal Courts of Justice. On the application of the Crown they were again adjourned to the 4th of December, and the case ordered to be argued before a Court consisting of five judges. . . .

Lord Coleridge, C.J. The two prisoners, Thomas Dudley and Edwin Stephens, were indicted for the murder of Richard Parker on the high seas on the 25th of July in the present year. They were tried before my Brother Huddleston at Exeter on the 6th of November, and, under the direction of my learned Brother, the jury returned a special verdict, the legal effect of which has been argued before us, and on which we are now to pronounce judgment.

The special verdict as, after certain objections by Mr. Collins to which the Attorney General yielded, it is finally settled before us is as follows. [His Lordship read the special verdict as above set out.] From these facts, stated with the cold precision of a special verdict, it appears sufficiently that the prisoners were subject to terrible temptation, to sufferings which might break down the bodily power of the strongest man, and try the conscience of the best. Other details, yet more harrowing, facts still more loathsome and appalling, were presented to the jury, and are to be found recorded in my learned Brother's notes. But nevertheless this is clear, that the prisoners put to death a weak and unoffending boy upon the chance of preserving their own lives by feeding upon his flesh and blood after he was killed, and with the certainty of depriving *him* of any possible chance of survival. The verdict finds in terms that "if the men had not fed upon the body of the boy they would *probably* not have survived," and that "the boy being in a much weaker condition was *likely* to have died before them." They might possibly have been picked up next day by a passing ship; they might possibly not have been picked up at all; in either case it is obvious that the killing of the boy would have been an unnecessary and profitless act. It is found by the verdict that the boy was incapable of resistance, and, in fact, made none; and it is not even suggested that his death was due to any violence on his part attempted against, or even so much as feared by, those who killed him. Under these circumstances the jury say that they are ignorant whether those who killed him were guilty of murder, and have referred it to this Court to determine what is the legal consequence which follows from the facts which they have found. . . .

There remains to be considered the real question in the case—whether killing under the circumstances set forth in the verdict be or be not murder. The contention that it could be anything else was, to the minds of us all, both new and strange, and we stopped the Attorney General in his negative argument in order that we might hear what could be said in support of a proposition which appeared to us to be at once dangerous, immoral, and opposed to all legal principle and analogy. . . .

First it is said that it follows from various definitions of murder in books of authority, which definitions imply, if they do not state, the doctrine, that in order to save your own life you may lawfully take away the life of another, when that other is neither attempting nor threatening yours, nor is guilty of any illegal act whatever towards you or anyone else. But if these definitions be looked at they will not be found to sustain this contention. The earliest in point of date is the passage cited to us from Bracton, who lived in the reign of Henry III. . . . But in the very passage as to necessity, on which reliance has been placed, it is clear that Bracton is speaking of necessity in the ordinary sense—the repelling by violence, violence justified so far as it was necessary for the object, any illegal violence used towards oneself. . . .

It is, if possible, yet clearer that the doctrine contended for receives no support from the great authority of Lord Hale. . . . It is not possible to use words more clear to shew that Lord Hale regarded the private necessity which justified, and alone justified, the taking the life of another for the safeguard of one's own to be what is commonly called "self-defence." . . .

But if this could be even doubtful upon Lord Hale's words, Lord Hale himself has made it clear. For in the chapter in which he deals with the exemption created by compulsion or necessity he thus expresses himself:—"If a man be desperately assaulted and in peril of death, and cannot otherwise escape unless, to satisfy his assailant's fury, he will kill an innocent person then present, the fear and actual force will not acquit him of the crime and punishment of murder, if he commit the fact, for he ought rather to die himself than kill an innocent; but if he cannot otherwise save his own life the law permits him in his own defence to kill the assailant, for by the violence of the assault, and the offence committed upon him by the assailant himself, the law of nature, and necessity, hath made him his own protector cum debito moderamine inculpatæ tutelæ." (Hale, Pleas of the Crown, vol. i. 51.)

But, further still, Lord Hale in the following chapter deals with the position asserted by the casuists, and sanctioned, as he says, by Grotius and Puffendorf, that in a case of extreme necessity, either of hunger or clothing; "theft is no theft, or at least not punishable as theft, as some even of our own lawyers have asserted the same." "But," says Lord Hale, "I take it that here in England, that rule, at least by the laws of England, is false; and therefore, if a person, being under necessity for want of victuals or clothes, shall upon that account clandestinely and animo furandi steal another man's goods, it is felony, and a crime by the laws of England punishable with death." (Hale, Pleas of the Crown, i. 54.) If, therefore, Lord Hale is clear—as he is—that extreme necessity of hunger does not justify larceny, what would he have said to the doctrine that it justified murder? . . .

Is there, then, any authority for the proposition which has been presented to us? Decided cases there are none. . . .

The American case cited by my Brother Stephen in his Digest, from Wharton on Homicide, in which, it was decided, correctly indeed, that sailors had no right to throw passengers overboard to save themselves, but on the somewhat strange ground that the proper mode of determining who was to be sacrificed was to vote upon the subject by ballot, can hardly, as my Brother Stephen says, be an authority satisfactory to a court in this country. . . .

The one real authority of former time is Lord Bacon, who, in his commentary on the maxim, "necessitas inducit privilegium quoad jura privata," lays down the law as follows—"Necessity carrieth a privilege in itself. Necessity is of three sorts—necessity of conservation of life, necessity of obedience, and necessity of the act of God or of a stranger. First of conservation of life; if a man steal viands to satisfy his present hunger, this is no felony nor larceny. So if divers be in danger of drowning by the casting away of some boat or barge, and one of them get to some plank, or on the boat's side to keep himself above water, and another to save his life thrust him from it, whereby he is drowned, this is neither se defendendo nor by misadventure, but justifiable." On this it is to be observed that Lord

*(Continues)*

*(Continued)*

Bacon's proposition that stealing to satisfy hunger is no larceny is hardly supported by Staundforde, whom he cites for it, and is expressly contradicted by Lord Hale in the passage already cited. And for the proposition as to the plank or boat, it is said to be derived from the canonists. At any rate he cites no authority for it, and it must stand upon his own. Lord Bacon was great even as a lawyer; but it is permissible to much smaller men, relying upon principle and on the authority of others, the equals and even the superiors of Lord Bacon as lawyers, to question the soundness of his dictum. There are many conceivable states of things in which it might possibly be true, but if Lord Bacon meant to lay down the broad proposition that a man may save his life by killing, if necessary, an innocent and unoffending neighbour, it certainly is not law at the present day. . . .

Now, except for the purpose of testing how far the conservation of a man's own life is in all cases and under all circumstances, an absolute, unqualified, and paramount duty, we exclude from our consideration all the incidents of war. We are dealing with a case of private homicide, not one imposed upon men in the service of their Sovereign and in the defence of their country. Now it is admitted that the deliberate killing of this unoffending and unresisting boy was clearly murder, unless the killing can be justified by some well-recognised excuse admitted by the law. It is further admitted that there was in this case no such excuse, unless the killing was justified by what has been called "necessity." But the temptation to the act which existed here was not what the law has ever called necessity. Nor is this to be regretted. Though law and morality are not the same, and many things may be immoral which are not necessarily illegal, yet the absolute divorce of law from morality would be of fatal consequence; and such divorce would follow if the temptation to murder in this case were to be held by law an absolute defence of it. It is not so. To preserve one's life is generally speaking a duty, but it may be the plainest and the highest duty to sacrifice it. War is full of instances in which it is a man's duty not to live, but to die. The duty, in case of shipwreck, of a captain to his crew, of the crew to the passengers, of soldiers to women and children, as in the noble case of the *Birkenhead*; these duties impose on men the moral necessity, not of the preservation, but of the sacrifice of their lives for others, from which in no country, least of all, it is to be hoped, in England, will men ever shrink, as indeed, they have not shrunk. It is not correct, therefore, to say that there is any absolute or unqualified necessity to preserve one's life. "Necesse est ut eam, non ut vivam," is a saying of a Roman officer quoted by Lord Bacon himself with high eulogy in the very chapter on necessity to which so much reference has been made. It would be a very easy and cheap display of commonplace learning to quote from Greek and Latin authors, from Horace, from Juvenal, from Cicero, from Euripides, passage after passage, in which the duty of dying for others has been laid down in glowing and emphatic language as resulting from the principles of heathen ethics; it is enough in a Christian country to remind ourselves of the Great Example whom we profess to follow. It is not needful to point out the awful danger of admitting the principle which has been contended for. Who is to be the judge of this sort of necessity? By what measure is the comparative value of lives to be measured? Is it to be strength, or intellect, or what? It is plain that the principle leaves to him who is to profit by it to determine the necessity which will justify him in deliberately taking another's life to save his own. In this case the weakest, the youngest, the most unresisting, was chosen. Was it more necessary to kill him than one of the grown men? The answer must be "No"—

> "So spake the Fiend, and with necessity,
> The tyrant's plea, excused his devilish deeds."

It is not suggested that in this particular case the deeds were "devilish," but it is quite plain that such a principle once admitted might be made the legal cloak for unbridled passion and atrocious crime. There is no safe path for judges to tread but to ascertain the law to the best of their ability and to declare it according to their judgment; and if in any case the law appears to be too severe on individuals, to leave it to the Sovereign to exercise that prerogative of mercy which the Constitution has intrusted to the hands fittest to dispense it.

It must not be supposed that in refusing to admit temptation to be an excuse for crime it is forgotten how terrible the temptation was; how awful the suffering; how hard in such trials to keep the judgment straight and the conduct pure. We are often compelled to set up standards we cannot reach ourselves, and to lay down rules which we could not ourselves satisfy. But a man has no right to declare temptation to be an excuse, though he might himself have yielded to it, nor allow compassion for the criminal to change or weaken in any manner the legal definition of the crime. It is therefore our duty to declare that the prisoners' act in this case was wilful murder, that the facts as stated in the verdict are no legal justification of the homicide; and to say that in our unanimous opinion the prisoners are upon this special verdict guilty of murder.

THE COURT then proceeded to pass sentence of death upon the prisoners.

[This sentence was afterwards commuted by the Crown to six months' imprisonment.]

## Notes and Questions

1. Were the actions taken by Dudley and Stephens "blameworthy"? In the words of Professor Hart, do they merit "the moral condemnation of the community"? If so, where did Dudley and Stephens go wrong? Was it in their decision that "some one should be sacrificed to save the rest" (i.e., *whether* one of the four on the lifeboat should be killed)? Or rather was it in the selection process they used to decide *who* among the four should be killed? Does the court's opinion intimate what *should* have been done under the circumstances? What action, if any, do you believe should have been taken?

2. What purpose(s) is (are) served by punishing Dudley and Stephens? Do they seem to be dangerous people from whom society needs protection? Under the extreme conditions confronting them in the lifeboat, including the perceived likelihood that their death was imminent, is the law likely to have significant deterrent value? Are they evil people, or at least people whose conduct was so clearly wrong that their punishment is justified under the name of retribution? Is the normative or educative force of the law promoted by punishing them? Is it fair to consider utilitarian justification for their punishment without the prior conclusion that their punishment is deserved?

3. If Dudley and Stephens were properly adjudged guilty of murder, and thus are suitable candidates for punishment, what measure of punishment should be imposed? Under English law then in effect, the mandatory punishment for murder was death, and Dudley and Stephens originally were ordered to hang. As you read, however, the Queen commuted their death sentences to 6 months' imprisonment. Does this adjustment of their sentence seem fair in light of the circumstances? What if the Queen, shocked by the conduct of two men in killing and cannibalizing a helpless boy of 17, had refused to intervene and Dudley and Stephens died at the gallows? Would that result seem fair in light of the circumstances?

4. What do you understand to be the meaning of the following passage from Lord Coleridge's opinion?

> Though law and morality are not the same, and many things may be immoral which are not necessarily illegal, yet the absolute divorce of law from morality would be of fatal consequence; and such divorce would follow if the temptation to murder in this case were to be held by law an absolute defence of it.

Do you agree with the passage?

## *Dudley and Stephens* and the General Principles of Criminal Law

We earlier introduced seven general principles of the criminal law, which we encounter repeatedly in the cases and materials presented in this book. We presage our later consideration of these principles by applying them to the facts of *Dudley and Stephens*, with a few variations.

a. *The legality principle.* Dudley and Stephens are presumed to know that murder is a crime and that it was punished automatically by death. However, is it fair to hold them to the knowledge that a killing committed under the extreme circumstances confronting them in the lifeboat constituted murder? Recall that through their special verdict the 12 members of the jury, after the opportunity for calm reflection and deliberation in the safe haven of a courthouse, professed to be "ignorant" about whether "the killing of Richard Parker by Dudley and Stephens be felony and murder." That question was not definitively resolved until the judges of the Queen's Bench division issued their decision, months after the killing occurred. Can arguments reasonably be advanced to support the position that Dudley and Stephens should not even be expected to know that their conduct was a crime?

b. *Mens rea.* In *Dudley and Stephens* there is no doubt that Dudley intended to cause Richard Parker's death by slitting his throat with a knife, and that Stephens had agreed to the deed and possessed full knowledge that Dudley was to embark on it. Still, not all intentional killings are murder. Lord Coleridge's opinion specifically "exclude[s] from . . . consideration all the incidents of war" and rejects the analogy to homicides committed in self-defense—another type of intentional killing—argued on the defendants' behalf. As we will see, some conduct that normally is prohibited by the criminal law (*e.g.*, killing another person) can be "justified" or "excused"— and thus exempted from criminality—even if committed intentionally.

What if Dudley, delirious from 3 weeks in the lifeboat and several days without food or fresh water, began hallucinating and honestly believed that Richard Parker was a giant turtle and not a boy at all? Or, what if Dudley actually did see a large turtle in the sea and, clumsily lunging at it with his knife, mistakenly struck Parker in the throat, killing him? Under either set of facts, would Dudley have the _mens rea_ needed to support a conviction for murder? For any crime at all?

Should the criminal law distinguish, based on their respective states of mind, between (i) Dudley, who entered an agreement that Parker should be killed and _intended_ to kill Parker; (ii) Stephens, who entered an agreement that Parker should be killed and had _knowledge_ that Dudley would kill Parker; and (iii) Brooks, who _refused to agree_ that Parker should be killed yet had _knowledge_ that Dudley would kill Parker?

c. _Act._ The criminal law generally reserves punishment for individuals' voluntary acts. Dudley's act of slitting Richard Parker's throat with a knife unambiguously brings his conduct within this general principle. We would have a different case if Dudley had placed his knife on the side of the lifeboat, only to have a gale force wind sweep up the knife and hurl it into Parker's throat. We similarly would be hesitant to attribute the stabbing to an act voluntarily committed by Dudley if the wind picked up Dudley when he had knife in hand, flung him against the unfortunate Parker, and the knife penetrated Parker's throat. However, external forces of this nature were not at work in this case.

But, what of Stephens, who also was convicted of murder? Stephens "agreed to" the planned sacrifice and gave his "assent" before Dudley's stabbing Parker but appears never to have laid a hand on the cabin boy. Where is his "act" of murder? Even if his agreement and assent to the killing distinguish his conduct from that of Brooks, are not his actions also distinguishable from Dudley's? Is there not a difference between talking about and even agreeing with plans for a killing and actually picking up a lethal weapon and committing the killing? Does talk inevitably escalate into action? In terms of the specific acts committed by the two men, what might justify treating Dudley and Stephens differently under the criminal law? What might justify treating them the same?

If the criminal law is predicated on acts voluntarily committed by individuals, consider the following possibilities. Let us say that instead of the inhabitants of the lifeboat facing starvation and dehydration, the lifeboat was precariously overloaded and in danger of sinking under the weight of the four men. To lighten the load Dudley, with the assent of Stephens, throws Richard Parker overboard and Parker drowns. Would you have any trouble concluding that Dudley's act of throwing Parker out of the boat is analogous to his act of slitting Parker's throat? [Facts of this general nature were presented in another lifeboat episode, _United States v. Holmes_, 26 F. Cas. 360 (No. 15,383) (C.C.E.D. Pa. 1842), which is "[t]he American case" referred to somewhat incredulously in Lord Coleridge's opinion in _Dudley and Stephens_.]

If the rule announced in _Dudley and Stephens_ likely would apply regardless of whether a knifing or a forcible ejection from the boat was the act in question, what if Parker accidentally fell from the lifeboat into the sea? Assume that Dudley, at no risk of danger to himself, easily could have tossed a life preserver to Parker and returned him to the lifeboat. Instead, however, Dudley does nothing but watch the helpless lad bob in the ocean and then drown. If Dudley could have saved Parker but chose not to, should he be guilty of murder? How does this scenario differ, if at all, from the one in which Dudley pushes Parker overboard? Can (should) the failure to act ever be construed as an "act" for purposes of the criminal law? Would it matter if Dudley were the captain of the ship that had gone down and Parker was a passenger on that ship? If Dudley had been Parker's father? His best friend?

d. _Concurrence._ Brooks "dissented" from the plan to kill Richard Parker but after Parker was slain joined Dudley and Stephens in feeding on Parker's body and blood. Let us imagine that after the men were rescued, Brooks thanked Dudley and Stephens for going forward with their plan, admitted that it was the only thing to do under the circumstances, and acknowledged that they all faced almost certain death if Dudley and Stephens had refrained from action. Through such belated agreement with the plan, should Brooks be held accountable for Parker's murder, much as Stephens was held accountable based on his prior agreement? The "concurrence" principle requires "the fusion of the legally material thought and effort in conduct."[8] How different are Brooks and Stephens, in terms of their culpability, under the hypothesized facts? Can you think of other examples where the concurrence principle might come into play?

e. _Harm._ Imagine that Richard Parker had succumbed to the elements just moments before Dudley slit his throat but that Dudley was oblivious to the fact that Parker had died and was under the impression the boy was still alive. Is Dudley's conduct any less "blameworthy" or "culpable" if it is later discovered that Parker was dead before Dudley's knife pierced his throat? Is Dudley guilty of murder under these changed facts?

Assume that on the 24th of July Dudley and Stephens agreed to kill Parker if they were not rescued by the morning of the 25th, and that Dudley began sharpening his knife in preparation. Further assume that a ship miraculously appeared near sundown on the 24th, and all aboard the lifeboat were saved. Have Dudley and Stephens committed a crime? Does conspiracy to commit murder loom as a possibility? If Parker in fact was not killed, what would be the "harm" associated with this or any other possible "conspiracy"?

What if Parker remained conscious and coherent to the end and, realizing his condition and the desperate plight of the others on the lifeboat, beseeched Dudley to end his suffering and heroically suggested that the others should extend their lives by consuming his remains after his death. After confirming that Parker was of sound mind and after unsuccessfully attempting to get him to reconsider, Dudley reluctantly complies by swiftly slitting Parker's throat. If Parker consents to—and even requests—being killed, has Dudley committed murder?

**f.** *Causation.* What "caused" Richard Parker's death? The knife wound? The effects of thirst, hunger, and the elements that had weakened Parker so irretrievably? The shipwreck? If the *Mignonette*—the ship on which the four crew members originally set sail—had not gone down, the tragedy ending in Parker's death never would have unfolded. In light of this fact, can Dudley fairly be identified as "the cause" of the cabin boy's death?

Assume for a moment that Richard Parker suffered from hemophilia, the propensity to bleed uncontrollably from even minor wounds due to deficiencies in the blood's clotting agents. Further assume that Dudley was unaware of Parker's condition and that the instant before he was to slit Parker's throat he realized he could not follow through with the plan. He withdrew his knife but not before nicking Parker's neck with a cut no more significant than the kind a man might encounter while shaving. Despite attempts to stop the bleeding, the wound bled profusely and the loss of blood led to Parker's death. Under these circumstances, did Dudley cause Parker to bleed to death? Did the hemophilia? Did some combination of the two factors? If death resulted from the combined effect of the knife wound and Parker's hemophilia, should Dudley's act be isolated as the legally significant cause of death?

What if Dudley delivered a stab wound to Parker's (no longer a hemophiliac) shoulder that ordinarily would not have been fatal. Stephens, a physician, binds the wound but uses filthy dressing despite the ready availability of sterile bandages in the lifeboat's medical kit. Owing to this grossly negligent treatment, the wound becomes infected, complications develop, and Parker dies. Who (or what) caused Parker to die: Dudley's act of stabbing him with a knife? Stephens' negligent treatment of the wound? Some combination of these factors? If the combination, which act(s) should be identified as the legally significant cause(s) of death?

**g.** *Punishment.* We already have raised questions about the propriety of punishing Dudley and Stephens for their conduct and have pointed out the official ambivalence evidenced about the punishment when the original death sentences were commuted to 6 months of imprisonment. If Dudley and Stephens are guilty of murder, as the court's opinion holds, does a sentence of just 6 months of incarceration send a message that what they did really was not so condemnable? If they are not to be (fully) condemned for their conduct, is it appropriate to brand their conduct murder, or even a crime? Does the compromise punishment imposed in the case reflect Solomon-like legal wisdom; is it inexcusably lenient; or is it—as well as the men's conviction—intolerably severe in light of the circumstances confronting the men on the lifeboat?

For a provocative essay exploring the issues presented by *Dudley and Stephens* see Lon L. Fuller, "The Case of the Speluncean Explorers."[9]

## Criminal Process

We now consider how criminal cases originate and progress through the court system. We also discuss different sources of law, focusing on legislation, constitutions, and judicial decisions. Finally, we introduce the Model Penal Code, which has been influential in helping to shape the criminal law in many jurisdictions.

## Criminal Courts

Criminal charges are officially filed through written accusatory instruments, which typically take the form of a **complaint**, an **indictment**, or an **information**. For example, let us say you own a lap-top computer you use to take class notes, prepare case briefs, write papers, and so on. One day after a class break during which you left your computer unattended, you return to find the computer missing. The student who had been sitting next to you, Phoebe Hitchcock, also is gone. Three of your classmates volunteer that they saw Phoebe place your computer in her book bag and leave the classroom. You report these facts to a police officer, who also takes statements from the three witnesses.

Depending on the value of your computer, the apparent theft may be a **misdemeanor** or a **felony**. Misdemeanors are crimes that are considered to be relatively minor. They generally are punishable by a fine or a jail term of not more than 6 months to 2 years, depending on the jurisdiction. Felonies are more serious crimes that normally can be punished by fine, a prison term, and sometimes even death. Prison sentences generally are based on the seriousness of the crime and/or the offender's prior record and may range from several months, to a few years, to life.

Assume that in your state thefts involving property worth less than $500 are misdemeanors and are felonies if the property value is $500 or more. If your lap-top computer is old or was relatively inexpensive to begin with and its present value is $450, the crime is a misdemeanor. In that case the police officer may ask you to go to the local courthouse and appear before a magistrate or a judge to swear out a complaint charging Phoebe Hitchcock with theft. You would appear before the magistrate or judge and then, while under oath, explain what happened. If the judicial officer believed that the facts you recited supplied "probable cause" that Phoebe took your computer valued at $450 without your permission, intending to keep it for her own, or made out a "prima facie case" to that effect, he or she would issue the complaint formally accusing Phoebe of committing the crime.

A warrant or order authorizing Phoebe's arrest would be signed, and Phoebe thereafter would be arrested and served with a copy of the complaint accusing her of stealing your computer. In due course she would be brought to court and asked how she pleaded to the charge. An attorney would be appointed to represent her, if necessary. The complaint would serve as the document formally charging Phoebe with misdemeanor theft through the time of her trial or the entry of a guilty plea.

If the computer is worth $500 or more, the alleged theft would be a felony under the hypothetical law in effect. In felony cases the police frequently appear before the judge or magistrate themselves to swear out the complaint, relying on information they have learned from their investigation. Although a complaint usually is the initial charging instrument in a felony case, it is not always. Some charges originate by indictment or information. When a complaint is used in a felony case, it later is superseded by an indictment or an information before the suspect stands trial.

In the federal courts, and in roughly half of the states, felony charges must be reviewed and approved by a grand jury before a suspect can be brought to trial. When a minimum number of grand jurors (at least a majority) determines that sufficient evidence backs up the charge to justify a trial—"probable cause" is the standard normally used—the grand jury returns a true bill of indictment. Like a complaint, the indictment alleges that the named defendant committed the crime that is described. In the states that do not rely on grand juries, prosecutors are authorized to charge suspects through an "information" prepared by the prosecutor's office. Defendants charged by an information generally have a right to a preliminary hearing before a judge, who determines whether sufficient evidence supports the charge to require the suspect to stand trial.

Misdemeanor charges are heard in lower level courts, which may be called "**police courts**," "**district courts**," or another name. Felony cases are resolved in higher level trial courts commonly referred to as "**superior courts**," "county courts," or a similar name. The federal Constitution guarantees criminal defendants the right to trial by jury in all cases where the maximum punishment for the crime charged exceeds 6 months. *Baldwin v. New York,* 399 U.S. 117 (1970); *Duncan v. Louisiana,* 391 U.S. 145 (1968). Although most defendants have a right to a jury trial, this right is infrequently exercised. Large numbers of accused offenders—typically, upward of 90 percent of those whose cases are decided—plead guilty, frequently after entering into a plea bargain. Among those who contest their guilt, defendants in most states can and often do waive the right to trial by jury and have their cases heard by a judge presiding over a "**bench trial**."

We will assume that the value of your lap-top computer exceeded $500, making the alleged theft a felony, and that Phoebe Hitchcock has pleaded not guilty to an indictment charging that crime and is exercising her right to a jury trial. The jury's job is to listen to the evidence offered during the trial and to the arguments made by the prosecutor and defense attorney (which do not constitute "evidence"). The jury also hears the judge's instructions defining the crime of felonious larceny (theft) and advising them about other important matters of law, such as defenses and burden-of-proof requirements. During deliberations the jurors must decide whether the charge against Phoebe has been proven. They do so by resolving factual issues and applying the law they have been charged to those facts.

Jury deliberations are secret. In criminal trials juries rarely do anything more than announce their ultimate verdict of "guilty" or "not guilty." Because of the procedures followed in trial courts and the function those courts are expected to serve, trial judges almost never write published opinions explaining the decisions reached in their courts. Occasionally, they do write such opinions, and we encounter a few of them in later chapters. But for the most part we focus on the decisions reached and the opinions written by *appellate courts.*

Defendants found guilty of a crime in a trial court have a right to appeal their conviction (and sometimes their sentence). Because of double jeopardy principles, a prosecutor typically cannot appeal a judge's or a jury's acquittal of a defendant. Court systems are structured similarly in different jurisdictions, although important differences may exist as well. We begin our description of the appeals process by following Phoebe Hitchcock's hypothetical case through successive stages of judicial review.

We will assume, as the facts of the case suggest, that the case was brought to trial in a *state* court and not a *federal* court. Federal trial courts (called **U.S. district courts**) only have jurisdiction over cases involving federal crimes or crimes alleged to have been committed against the U.S. government. The overwhelming majority of crimes and criminal trials in this country involve state offenses. The larceny of a lap-top computer is a crime we should expect to see prosecuted in the state courts; we would not expect such a crime to be a federal offense.

Let us assume that Phoebe vehemently denied her guilt at the trial, claiming that she mistakenly believed that the lap-top computer she tucked into her book bag was her own, thus negating the *mens rea* required for larceny. She also contested the admissibility of a statement taken from her by the police in which she made a damaging admission about taking the computer. She claimed that the statement had been taken in violation of her rights under *Miranda v. Arizona,* 384 U.S. 436 (1966), which help safeguard the constitutional right against compelled self-incrimination.

On appeal, Phoebe's argument that she mistook your lap-top computer for her own is likely to go nowhere. Appellate courts do not rehash issues of fact that have been decided by a judge or jury. The jury at Phoebe's trial had the benefit of listening to the testimony of witnesses, observing witnesses' demeanor, and being able to assess their credibility first-hand. An appellate court has only a written record to review, consisting of the trial transcript and related items of evidence. It only makes sense that appellate courts do not second-guess the facts found in trial courts, although they sometimes are asked to evaluate whether sufficient evidence was presented at the trial to support the fact-finder's verdict.

Rather than reviewing findings of fact that explicitly or implicitly support a guilty verdict in a criminal trial, the job of appellate courts is to review the legal errors allegedly committed during the trial. In Phoebe's case an error of law has been alleged: that the statement taken from her by the police was admitted into evidence in violation of her constitutional right against compelled self-incrimination.

Most state court systems have two levels of appellate courts: an intermediate appellate court, typically (but not always) called the **court of appeals**, and the high appellate court, traditionally (although not invariably) called the state **supreme court**. Smaller, less populous states may have only a supreme court and no intermediate-level court of appeals. We will assume that Phoebe's conviction has occurred in a state with a two-tier appellate court system.

Phoebe's one and only appeal as a matter of right is likely to be to the state court of appeals. The intermediate-level appellate court has no choice; it must hear her appeal. Its principal function is *error correction;* that is, to determine if Phoebe has identified errors of law committed in the trial court that entitle her to a new trial. In our example it must listen to—though it need not agree with—Phoebe's argument that her *Miranda* rights were violated. Intermediate-level appeals courts occasionally are bypassed, with appeals being taken directly from the trial court to the state supreme court in cases of special consequence, such as those in which the death penalty has been imposed. Ordinarily, however, appeals from trial court convictions are made to the state court of appeals.

In most cases a defendant's *right* to have an appellate court decide the issues raised in his or her case ends after the court of appeals has spoken. Subsequent review by the state supreme court is granted only at the supreme court's *discretion*. The state's high court ordinarily assumes that the court of appeals has faithfully fulfilled its obligation to detect and correct errors committed at the trial. The supreme court thus does not sit to duplicate the court of appeals' error-correction duties. Rather, its principal function ordinarily is loosely described as "rule making." The state's highest court typically hears cases to announce and clarify rules of law that are likely to be of interest or significance throughout the state.

The progression of a case from the intermediate to the high appeals courts usually does not involve an appeal but rather takes place on a *petition* for a **writ of *certiorari*** (*certiorari* is a Latin term meaning "to be made certain") or more simply by way of a petition for discretionary review. When a supreme court declines to review a case under its discretionary review authority (*i.e.,* when it

denies *certiorari* or denies *cert.*), such a decision is *not* a signal that the high court agrees with or is affirming the lower court's decision. It simply means that the supreme court has declined the invitation to review the court of appeals' decision in the case. Either the prosecution, on behalf of the state, or the defendant can request a state supreme court to hear a case that was decided adversely by the court of appeals.

Unlike trial court judges state court of appeals and supreme court judges commonly do write opinions explaining their case decisions. These appellate court opinions are the essential materials for our study throughout the remainder of this book.

## Sources of the Criminal Law

We already have introduced the *legality* principle, a basic premise of the criminal law signifying that there shall be "no crime without law, no punishment without law." We explore the implications of this principle in greater detail in Chapter 2. At this point we briefly consider the types or sources of law that are most relevant to our subject matter: *legislation, constitutions,* and *judicial decisions.* We also discuss the **Model Penal Code** and its significance to the substantive criminal law.

### Legislation

Look in the statute books of any state or the federal government and you will find comprehensive legislation defining crimes and their corresponding punishment. Criminal codes cover a wide range of offenses against persons, property, public morality and decency, the public peace, and the administration of government. Legislation is the primary source of the criminal law today in all 50 states and in the federal jurisdiction, and for good reason.

The minimal conditions required for social living would be impossible without enforcement of the rules embodied by the criminal law. It is only appropriate that a democratically elected, representative body such as a state legislature or the U.S. Congress articulate those rules, as well as the consequences for their violation. In addition, the definition of crimes and their punishment through legislation provides fair notice to people governed by the rules about what conduct is prohibited, and it helps ensure that those rules will not be created or applied arbitrarily.

### Constitutions representatives; seen as one its weaknesses.

Legislation, of course, is subject to constitutional constraints. One of the strengths of statutory law—its origination in legislatures comprised of representatives elected through the political process—is also one of its potential weaknesses. The framers of the state constitutions and the U.S. Constitution knew that legislative power could be abused. Unchecked majoritarian rule makes vulnerable individuals and groups who do not share the majority group's interests or characteristics. The criminal law in particular, with its awesome sanctions of imprisonment

intermediate → high ⊕ no appeal
writ of certiorari

and death, requires checks and balances in the form of constitutional safeguards. Several constitutional provisions apply specifically to crimes and punishment.

For example, the U.S. Constitution, as adopted in its original form in 1788, provides for the "Privilege of the Writ of Habeas Corpus" (Art. I, § 9 [2]), prohibits *ex post facto* laws (Art. I, § 10 [1]), requires that "[t]he trial of all Crimes. . . shall be by Jury" (Art. III, § 2 [3]), regulates the trial and punishment of treason (Art. III, § 3 [1], [2]), and governs extradition of fugitives between the states (Art. IV, § 2 [2]). In 1791, 10 amendments to the U.S. Constitution were ratified. Collectively known as the Bill of Rights, these amendments were designed to place additional limits on the power of the newly created federal government. The Fourth, Fifth, Sixth, and Eighth Amendments, in particular, focus on individuals' rights in the context of crimes and their investigation, prosecution, and punishment. Other Bill of Rights protections, including those protecting free speech and religious freedoms in the First Amendment, the Second Amendment's "right to bear arms," the Ninth Amendment's provision regarding "enumerated" rights, and the Tenth Amendment's reservation of undelegated powers to the states, also have potential relevance to the substantive criminal law.

The Fourteenth Amendment was adopted in 1868, following the conclusion of the Civil War. Its provisions apply to *state* governments and *state* actors, not just to the *federal* government and its representatives. In significant part, for our purposes, the Fourteenth Amendment provides

> nor shall any State deprive any person of life, liberty, or property, without due process of law; nor deny to any person within its jurisdiction the equal protection of the laws.

State constitutions also represent an important limitation on the criminal law within individual states. State constitutions sometimes impose greater restrictions on the reach of the criminal law than does the federal Constitution, thus affording individuals in a state more extensive rights than they are guaranteed under federal law. We will encounter examples of this principle in later chapters.

## Judicial Decisions (Case Law)

Throughout early English history crimes were not defined by statute but instead were recognized through judicial decisions made in individual cases. The principles announced in those decisions acquired precedential value when similar cases arose, and controlling rules of law gradually were refined as necessary to apply to unique case facts and changing social conditions. The result of this process of declaring and modifying the law through case adjudication, in the absence of governing statutes, became known as the "common law."

Thus at a time before legislation effectively could be enacted and publicized across the land, the common law, as "discovered" and announced through judicial decisions,

defined the criminal laws of England. English legal principles significantly influenced developing law in this country, and several of the states expressly carried forward the common law when they entered the union. Although most U.S. jurisdictions now have abolished common law crimes and they are not recognized under federal law, a minority of states still retain common law offenses. Even in the overwhelming majority of cases in which crimes are defined statutorily, common law principles can be quite important. For example, legislation might prohibit the commission of "manslaughter" or specify that "malice" is an element of murder yet provide no definition of those crucial terms. Reference must be made to the common law—the early judge-made law—to interpret and apply many statutes.

Even when statutes or constitutional provisions are written in relatively straightforward language, their application to specific case facts may be far from clear. Judicial decisions are necessary to clarify and explain the meaning of legislation and constitutions in the almost infinite variety of factual circumstances to which those other sources of law apply. Thus the bulk of materials presented in this book consists of judicial decisions that explain the rules and general principles of the substantive criminal law.

## Model Penal Code

Although not itself a form of binding law, few initiatives have had as profound an influence on the shape of modern American criminal law as the Model Penal Code (MPC). As its name implies, the MPC was designed as "model" criminal legislation. The American Law Institute, a private organization consisting of legal academicians, judges, and practitioners, began work on drafting the MPC in 1952. Professor Herbert Wechsler of Columbia Law School played a leading role in the development of the MPC, which went through 13 tentative drafts before the American Law Institute approved a proposed official draft in 1962.

Before the publication of the MPC, the criminal statutes of many states were little more than a hodgepodge of laws that displayed little coherence. The American Law Institute undertook its work with the objective of drafting model criminal legislation that could be adopted, in whole or in part, by states willing to consider reforming their criminal codes. The project was a tremendous success. Between 1962 and 1984 no less than 34 states enacted new penal codes, and several additional states took similar action or substantially revised their criminal laws in the mid-1980s and thereafter. Although the MPC was nowhere adopted *en toto,* it proved to be highly influential. Many of the MPC's underlying principles and specific provisions were incorporated in the new state legislation.[10]

Among the major innovations and accomplishments of the MPC was its conceptualization of four basic types of *mens rea: purposely, knowingly, recklessly,* and *negligently.* These concepts substituted for the confusing array of terms describing mental states that existed previously. The MPC also used the classification of theft to consolidate and simplify a number of offenses dealing with the wrongful taking of property. It implemented numerous other important reforms.[11] We have frequent occasion to highlight specific provisions of the MPC throughout this book.

## Burden of Proof

Defining a crime in conceptual terms and proving that a crime actually was committed are quite different undertakings. In this section we ask how confident we must be that a person accused of a crime in fact committed the offense before we authorize conviction and punishment.

Assume that a serial killer has claimed the lives of several people in a medium-size city over a 6-month period. The people of the community live in constant fear that the killer, who seemingly selects victims at random, will soon strike again. One night, the 11:00 news announces that yet another slaying has occurred and that the police have arrested a drifter they caught rifling the pockets of the victim. The *modus operandi* of the homicide is consistent with the previous rash of killings. The arrested man admits to looking for the deceased's wallet but claims the victim was already dead when he stumbled upon him. He can neither produce any alibi witnesses nor account for his whereabouts at the time of the other killings. He has a long criminal record including several assault convictions. He is charged with multiple counts of murder and brought to trial on his plea of not guilty. He faces the death penalty or life imprisonment if convicted.

How much proof of the man's guilt should we demand before concluding that he is guilty as charged? What would be the consequences of convicting the man only to learn, perhaps years later, that he had nothing to do with the killings? On the other hand, what would be the consequences of acquitting him and releasing him, only to learn later that he was responsible for the killings? Which type of error is the more serious one? Should we generally be more anxious to avoid the risk of erroneously convicting a person who in fact is innocent of wrongdoing or to guard against the risk of erroneously exonerating someone who is guilty of a charged crime? If we must admit that both types of error are possible in a criminal trial, how much proof should we demand—and of whom—before deciding whether a suspected offender is guilty as charged? Human institutions, including the courts, are fallible. Between 1973 and 2008 no fewer than 129 people found guilty of capital murder and sentenced to death in this country subsequently were determined to have been wrongfully convicted and ordered released from death row. Some of those people came perilously close to execution.[12] Many people incarcerated for noncapital crimes also have been wrongly convicted and released after their innocence was established.[13] It is impossible to know how many innocent people have been wrongly convicted without the error ever being uncovered.

By the same token there is no question that guilty people can and sometimes do avoid conviction in the criminal courts. Erroneous acquittals undermine community confidence in the criminal process, disserve justice, and potentially leave society at risk to predatory criminals.

In an effort to strike an appropriate balance between too lax a standard, under which there would be a high risk of erroneous convictions, and too high a standard, under which guilty offenders would escape punishment, in *In re Winship*, 397 U.S. 358 (1970), the Supreme Court held that before a person may be convicted of a crime, the accused's guilt must be established "beyond a **reasonable doubt**." Although you undoubtedly have heard that prosecutions require proof beyond a reasonable doubt to obtain a conviction, precisely what is meant by these words is far from simple and has been considered by courts on a number of occasions.

The U.S. Supreme Court addressed this general question when ruling on due process challenges to state court criminal convictions obtained in California and Nebraska. The Court joined the California and Nebraska cases for decision in *Victor v. Nebraska*, 511 U.S. 1 (1994). In the California case, *Sandoval v. California*, the trial judge instructed the jury on the government's **burden of proof** as follows:

> A defendant in a criminal action is presumed to be innocent until the contrary is proved, and in case of a reasonable doubt whether his guilt is satisfactorily shown, he is entitled to a verdict of not guilty. This presumption places upon the State the burden of proving him guilty beyond a reasonable doubt.
>
> Reasonable doubt is defined as follows: It is *not a mere possible doubt;* because everything relating to human affairs, and depending on moral evidence, is open to some possible or imaginary doubt. It is that state of the case which, after the entire comparison and consideration of all the evidence, leaves the minds of the jurors in that condition that they cannot say they feel an abiding conviction, *to a moral certainty,* of the truth of the charge." . . . [emphasis added]

The Court, through an opinion authored by Justice O'Connor, found no constitutional error in the trial court's definition of proof beyond a reasonable doubt. It defined the "constitutional question" as "whether there is a reasonable likelihood that the jury understood the instructions to allow conviction based on proof insufficient to meet the *Winship* standard." The Court found no such likelihood in *Sandoval*.

In the companion case from Nebraska, the trial judge instructed the jury as follows:

> [t]he burden is always on the State to prove beyond a reasonable doubt all of the material elements of the crime charged, and this burden never shifts.

The charge continued:

> Reasonable doubt is such a doubt as would cause a reasonable and prudent person, in one of the graver and more important transactions of life, to pause and hesitate before taking the represented facts as true and relying and acting thereon. It is such a doubt as will not permit you, after full, fair, and impartial consideration of all the

evidence, to have an abiding conviction, *to a moral certainty,* of the guilt of the accused. At the same time, absolute or mathematical certainty is not required. You may be convinced of the truth of a fact beyond a reasonable doubt and yet be fully aware that possibly you may be mistaken. You may find an accused guilty upon the *strong probabilities of the case,* provided such probabilities are strong enough to exclude any doubt of his guilt that is reasonable. A reasonable doubt is an *actual and substantial doubt* reasonably arising from the evidence, from the facts or circumstances shown by the evidence, or from the lack of evidence on the part of the State, as distinguished from a doubt arising from mere possibility, from bare imagination, or from fanciful conjecture. [emphasis added]

As in *Sandoval,* the Court rejected Victor's challenge to the instruction. It also rebuffed Victor's other arguments.

The Court summarized its holding on the constitutional issues surrounding the definition of proof beyond a reasonable doubt as follows: race of defendants?

The beyond a reasonable doubt standard is a requirement of due process, but . . . so long as the court instructs the jury on the necessity that the defendant's guilt be proved beyond a reasonable doubt . . . the Constitution does not require that any particular form of words be used in advising the jury of the government's burden of proof. Rather, "taken as a whole, the instructions [must] correctly conve[y] the concept of reasonable doubt to the jury." *Holland v United States,* 348 U.S. 121, 140 (1954). . . . The Due Process Clause requires the government to prove a criminal defendant's guilt beyond a reasonable doubt, and trial courts must avoid defining reasonable doubt so as to lead the jury to convict on a lesser showing than due process requires. In these cases, however, we conclude that "taken as a whole, the instructions correctly conveyed the concept of reasonable doubt to the jury." *Holland v United States,* supra, at 140. There is no reasonable likelihood that the jurors who determined petitioners' guilt applied the instructions in a way that violated the Constitution.

## Conclusion

This chapter introduced several fundamental issues concerning the study of substantive criminal law. We described the scope of our subject matter, distinguishing substantive criminal law from the related topic of criminal procedure. Then, relying on Jerome Hall's classic book, *General Principles of Criminal Law,* we identified seven principles or characteristics of the criminal law that help organize our studies: *the legality principle, mens rea, the act, concurrence, harm, causation,* and *punishment.* We dwelled a bit on the meaning of "punishment" and discussed its close (but not inevitable) link to individual blameworthiness. The initial judicial decision presented, *The Queen v. Dudley and Stephens,* provided a fittingly challenging opportunity to speculate about the propriety of invoking the criminal law to punish two of the three men rescued from the lifeboat.

We build on this foundation in Chapter 2, in which we examine important constitutional limitations on the reach of the criminal law.

## Key Terms

bench trial
burden of proof
complaint
court of appeals
crime
criminal
district courts
felony
general deterrence
indictment
information
misdemeanor

Model Penal Code
police courts
punishment
reasonable doubt
retributive justification
specific deterrence
superior courts
supreme court
U.S. district courts
utilitarian justification
writ of certiorari

## Review Questions

1. List the five elements of punishment, as outlined by H. L. A. Hart. Do you agree with them?
2. What is a crime? Describe in detail using material from this chapter.
3. What justifications for punishment support holding Dudley and Stephens responsible for their actions?
4. What is the difference between the *retributive* and *utilitarian* justifications for punishment?
5. What are the beneficial consequences of utilitarian justification?
6. What is the difference between a *misdemeanor* and a *felony*? Include a brief discussion of the court in which each might be tried.
7. What is the Model Penal Code, and what purpose does it serve?
8. What is meant by *burden of proof*?
9. Describe what is meant by *reasonable doubt* in the context of a criminal case. Is it a necessary component of the criminal justice system?

## Notes

1. Hall J. *General Principles of Criminal Law,* 2nd ed. Indianapolis, IN: Bobbs-Merrill; 1960:18
2. Hall, p. 18.
3. Hart HLA. Prolegomenon to the principles of punishment. In: *Punishment and Responsibility: Essays in the Philosophy of Law*, Vol. 1. New York: Oxford University Press; 1968:4–5.
4. *See* Cohen F. *The Law of Deprivation of Liberty,* 2nd ed. Durham, NC: Carolina Academic Press; 1991.
5. LaFave WR. *Criminal Law,* 3rd ed. St. Paul, MN: West Group; 2000.
6. Dressler J. *Understanding Criminal Law,* 4th ed. New York: Matthew Bender & Co.; 2006. *See* pp. 11–24 for an insightful discussion of the justifications for criminal punishment.
7. Perkins RM, Boyce RN. *Criminal Law.* Mineola, NY: The Foundation Press; 1982.
8. Hall, p. 185.
9. Fuller LL. The case of the speluncean explorers. *Harvard Law Rev.* 1949;62:616.
10. *See* Dressler, pp. 32–33.
11. *See* McClain C. Criminal law reform: Historical development in the United States. In: Kadish SH, ed. *Encyclopedia of Crime and Justice.* New York: The Free Press; 1983:510–512.
12. Death Penalty Information Center. *Innocence and the Death Penalty.* Available at: http://www.deathpenaltyinfo.org/article.php?did=412&scid=6. Accessed June 1, 2008. *See also* Gross SR. The risks of death: Why erroneous convictions are common in capital cases. *Buffalo Law Rev.* 1996;44:469; Radelet ML, Lofquist WS, Bedau HA. Prisoners released from death rows since 1970 because of doubts about their guilt. *Thomas M. Cooley L. Rev.* 1996;13:907.
13. *See* Gross SR, *et al.* Exonerations in the United States 1989 through 2003. *J. Criminal Law Criminol.* 2005;95:523; Huff CR, Rattner A, Sagarin E. *Convicted But Innocent: Wrongful Conviction and Public Policy.* Thousand Oaks, CA: Sage Publications; 1996.

# Constitutional Limits on the Definition and Punishment of Crimes

## Chapter Objectives

- Understand the jurisdictional limits of federal and state courts
- Understand the concepts of vagueness and overbreadth with regard to the constitutionality of criminal statutes
- Understand the concept of *ex post facto* laws
- Understand the role the Constitution plays in limiting the reach of the criminal law
- Understand the source and limits of the constitutional right to privacy
- Distinguish between constitutionally protected speech and speech that is not constitutionally protected
- Learn how the guarantee of equal protection of the laws applies to criminal law
- Consider constitutional issues regarding anti-terrorism laws
- Understand the concept of proportionality of punishment

In this chapter we examine the constitutional limits surrounding the substantive criminal law. We noted in Chapter 1 that the framers of the U.S. Constitution and the Bill of Rights were well aware of the potential for legislative and judicial abuses in matters involving crimes and punishments. The federal Constitution, as well as state constitutions, includes both specific provisions and broad principles (*e.g.*, due process of law, the prohibition against cruel and unusual punishments) that constrain governmental power to define crimes and impose penal sanctions.

The day-to-day administration of the criminal law infrequently triggers constitutional issues. As we explained in Chapter 1, the criminal law primarily revolves around statutes and common law principles that are of unquestionable constitutionality as applied in the great majority of cases. Nevertheless, the fundamental constitutional checks on the government's power to enact and enforce criminal laws remain vitally significant.

We begin by addressing jurisdictional issues that determine the respective spheres of authority of the federal government and the states to adopt criminal legislation. We next

investigate different dimensions of the legality principle, including vagueness, fair notice, and *ex post facto* issues. Then we examine the outer limits of the government's power to define conduct as crime, reviewing cases that present substantive due process, First Amendment, and equal protection issues. The chapter concludes by addressing the "proportionality" principle implicit in the Eighth Amendment's prohibition against cruel and unusual punishments.

## Federal and State Jurisdiction and the Criminal Law

Although offenses such as murder, assault, larceny, arson, and many others cause injury to specific victims and infringe on individual property rights, crimes formally involve wrongs committed against the government or the general body politic. Most, if not all, crimes against persons and personal property also are **torts**, meaning that the injured individual (at least in theory) can sue the perpetrator in a civil action and recover money damages. In the eyes of the law, however, crimes involve harms committed against the public at large. Criminal offenses thus are prosecuted and punished to vindicate the public interest rather than directly to redress the injuries suffered by private parties.

Centuries ago, before an effective criminal law had developed in England, private parties could not look to government for protection or redress when they suffered harm at the hands of another. Retaliatory action was common, often inspiring long-term feuding between the involved parties and their families and creating ever-expanding circles of victims and offenders. With the gradual evolution of the criminal law, the Crown stepped in to punish breaches of the peace, thus helping to quell the destructive cycle of blood feuds. James Fitzjames Stephen described this move as "the transition from the view that homicide was a wrong to the survivors, to the view that it was an offence against the state."[1]

The people in this country have long been accustomed to criminal cases being prosecuted on behalf of the government. Thus a criminal prosecution is not brought in the name of the person directly injured. Instead, cases typically are called *State v. _____* or *People v. _____* in the state courts and *United States v. _____* in the federal

*How diff. cases/courts are titled.*

courts. In a federalist country such as the United States, it becomes important to identify which government—state, federal, or both—is entitled to prosecute an alleged offender for a crime.

Sometimes the subject matter jurisdiction of the state and federal governments overlaps. For example, if Lee Harvey Oswald had lived to be prosecuted for shooting President Kennedy in Dallas, he could have been tried in the state courts of Texas for murder and in federal court for assassinating the President of the United States. Similarly, Timothy McVeigh's bombing of the federal building in Oklahoma City involved both federal and state crimes. The U.S. Supreme Court has ruled that there is no double jeopardy barrier against an individual being prosecuted in both the state and federal courts, or in the courts of two different states, for conduct that is an offense against the separate political sovereigns. *See Heath v. Alabama,* 474 U.S. 82, 106 S. Ct. 433, 88 L. Ed. 2d 387 (1985); *United States v. Lanza,* 260 U.S. 377, 43 S. Ct. 141, 67 L. Ed. 314 (1922). Multiple courts x no D.J.

Complications occasionally arise, however, concerning whether a legislative body has the power, or jurisdiction, to define conduct as a crime against the government for which it speaks. These complications can have constitutional dimensions. Jurisdictional issues have become increasingly important in light of the willingness of Congress to classify an expansive range of conduct that traditionally has been within the province of state criminal laws as federal crimes.

For example, in *United States v. Lopez,* 514 U.S. 549, 115 S. Ct. 1624, 131 L. Ed. 2d 626 (1995), the authority of Congress to enact the Gun-Free School Zone Act of 1990 was challenged. This Act made it a federal crime "for any individual knowingly to possess a firearm" in "a school zone" (*i.e.,* within 1,000 feet of any public, private, or parochial school). 18 U.S.C.A. §§ 922(q)(1)(A), 921(a)(25). Lopez, a 12th grade student, was charged with violating the federal law for carrying a .38-caliber handgun onto the grounds of Edison High School in San Antonio, Texas. The U.S. district court rejected Lopez's contention that the federal government had no jurisdiction to make his conduct a federal crime. The court concluded that Congress had the power to enact the Gun-Free School Zone Act under its constitutional grant of authority "[t]o regulate Commerce . . . among the several States." U.S. Const., Art. I, § 8[3]. However, the U.S. court of appeals disagreed and vacated Lopez's conviction.

The U.S. Supreme Court ruled (5–4) that Congress had exceeded its power under the Commerce Clause to make carrying a gun onto school grounds a federal crime. Chief Justice Rehnquist's opinion for the Court explained that such conduct could only be penalized by state law: "[T]he scope of the interstate commerce power 'must be considered in the light of our dual system of government and may not be extended so as to embrace effects upon interstate commerce so indirect and remote that to embrace them, in view of our complex society, would effectually obliterate the distinction between what is national and what is local and create a completely centralized government,'" quoting *NLRB v. Jones & Laughlin Steel,* 301 U.S.

1, 37 (1937). The Court majority ruled that "possession of a firearm in a local school zone" does not "substantially affect interstate commerce," and hence Congress had no constitutional authority to make that conduct a federal crime.

> The possession of a gun in a local school zone is in no sense an economic activity that might, through repetition elsewhere, substantially affect any sort of interstate commerce. Respondent was a local student at a local school; there is no indication that he had recently moved in interstate commerce, and there is no requirement that his possession of the firearm have any concrete tie to interstate commerce.
>
> To uphold the Government's contentions here, we would have to pile inference upon inference in a manner that would bid fair to convert congressional authority under the Commerce Clause to a general police power of the sort retained by the States. Admittedly, some of our prior cases have taken long steps down that road, giving great deference to congressional action. The broad language in these opinions has suggested the possibility of additional expansion, but we decline here to proceed any further. To do so would require us to conclude that the Constitution's enumeration of powers does not presuppose something not enumerated and that there never will be a distinction between what is truly national and what is truly local. This we are unwilling to do.

*Lopez* marked the first time since the 1930s that the U.S. Supreme Court had ruled that Congress exceeded its jurisdiction under the Commerce Clause to enact legislation affecting the states. Frequently described as a potential landmark ruling, the decision called into question Congress' authority to legislate in many additional areas and raised fundamental questions about the respective scope of federal and state governmental authority in matters involving the criminal law. It also has important implications regarding how deferential the courts must be when reviewing Congress' power to legislate. Five years later, signaling that *Lopez* was not an aberration, the Court again ruled that Congress lacked authority under the Commerce Clause to enact legislation (a provision of the Violence Against Women Act, 42 U.S.C. § 13981) that authorized victims of gender-motivated criminal violence to bring civil damage actions in federal court. *United States v. Morrison,* 529 U.S. 598, 120 S. Ct. 1740, 146 L. Ed. 2d 658 (2000). More recently, however, the justices upheld Congress' power against a Commerce Clause challenge to criminalize the cultivation and consumption of marijuana, even though California had made such practices lawful for medicinal purposes under appropriate circumstances. *Gonzales v. Raich,* 545 U.S. 1, 125 S. Ct. 2195, 162 L. Ed. 2d 1 (2005).

In sheer numbers, offenses processed in the state criminal courts dwarf federal criminal prosecutions. In 2004 more than 94 percent of felony convictions involved state offenses (including 98.7 percent of violent crimes and 96.3 percent of property offenses), with the remainder occurring in the federal courts (see Table 2-1, which presents a breakdown of state and federal felony convictions). This is not to suggest that federal crimes are relatively

**Table 2-1**

**Felony Convictions in State and Federal Courts by Offense, United States, 2004**

| Most Serious Conviction Offense | Felony Convictions | | | Federal Felony Convictions as Percent of Total |
| --- | --- | --- | --- | --- |
| | Total | State | Federal | |
| All offenses | 1,145,438 | 1,078,920 | 66,518 | 5.8 |
| Violent offenses | 197,138 | 194,570 | 2,568 | 1.3 |
| Murder[a] | 8,590 | 8,400 | 190 | 2.2 |
| Sexual assault[b] | 33,605 | 33,190 | 415 | 1.2 |
| Rape | 12,409 | 12,310 | 99 | 0.8 |
| Other sexual assault | 21,196 | 20,880 | 316 | 1.5 |
| Robbery | 40,230 | 38,850 | 1,380 | 3.4 |
| Aggravated assault | 94,845 | 94,380 | 465 | 0.5 |
| Other violent[c] | 19,868 | 19,750 | 118 | 0.6 |
| Property offenses | 322,501 | 310,680 | 11,821 | 3.7 |
| Burglary | 93,923 | 93,870 | 53 | 0.1 |
| Larceny[d] | 120,705 | 119,340 | 1,365 | 1.1 |
| Motor vehicle theft | 16,968 | 16,910 | 58 | 0.3 |
| Other theft | 103,737 | 102,430 | 1,307 | 1.3 |
| Fraud | 107,873 | 97,470 | 10,403 | 9.6 |
| Fraud[e] | 57,883 | 48,560 | 9,323 | 16.1 |
| Forgery | 49,990 | 48,910 | 1,080 | 2.2 |
| Drug offenses | 387,322 | 362,850 | 24,472 | 6.3 |
| Possession | 163,112 | 161,090 | 2,022 | 1.2 |
| Trafficking | 224,210 | 201,760 | 22,450 | 10 |
| Weapon offenses | 41,092 | 33,010 | 8,082 | 19.7 |
| Other offenses[f] | 197,385 | 177,810 | 19,575 | 9.9 |

[a]Includes non-negligent manslaughter.
[b]Includes rape.
[c]Includes offenses such as negligent manslaughter and kidnapping.
[d]Includes motor vehicle theft.
[e]Includes embezzlement.
[f]Composed of nonviolent offenses such as receiving stolen property and vandalism.

Source: U.S. Dept. of Justice, Bureau of Justice Statistics. *State Court Sentencing of Convicted Felons, 2004—Statistical Tables.* Available at: http://www.ojp.usdoj.gov/bjs/pub/html/scscf04/tables/scs04110tab.htm. Accessed June 12, 2008.

*commerce clause:*

unimportant; to the contrary, federal prosecutions may involve offenses that are especially important to national policies. Rulings like *Lopez* may limit federal law enforcement, but they of course do not affect the states' traditional prerogative of enacting and enforcing their own criminal laws.

## The Legality Principle

As we discussed in Chapter 1, the maxim, *Nullum crimin sine lege, nulla poena sine lege* (No crime without law, no punishment without law), is connected with **the legality principle**, one of the foundational principles of the crimi-

nal law. In this section we consider different dimensions of the legality principle, each of which implicates a constitutional limitation on state or federal power to enforce criminal laws.

We first take up the issue of **vagueness**. Then we examine whether an ordinance or statute that suffers no vagueness problems, and clearly announces what conduct is prohibited or required, automatically fulfills the "**fair notice**" requirement. We conclude this section by considering the prohibition against *ex post facto* laws.

## Vagueness

---

### *Papachristou v. City of Jacksonville*, 405 U.S. 156, 92 S. Ct. 839, 31 L. Ed. 2d 110 (1972)

Mr. Justice Douglas delivered the opinion of the Court.

This case involves eight defendants who were convicted in a Florida municipal court of violating a Jacksonville, Florida, vagrancy ordinance. (*See* Footnote 1.) Their convictions were affirmed by the

*(Continues)*

(*Continued*)

Florida Circuit Court in a consolidated appeal, and their petition for certiorari was denied by the District Court of Appeal on the authority of Johnson v State, 202 So 2d 852. The case is here on a petition for certiorari. . . . For reasons which will appear, we reverse. . . .

Jacksonville's ordinance and Florida's statute were "derived from early English law," Johnson v State, 202 So 2d, at 854, and employ "archaic language" in their definitions of vagrants. Id., at 855. The history is an often told tale. The breakup of feudal estates in England led to labor shortages which in turn resulted in the Statutes of Laborers, designed to stabilize the labor force by prohibiting increases in wages and prohibiting the movement of workers from their home areas in search of improved conditions. Later vagrancy laws became criminal aspects of the poor laws. The series of laws passed in England on the subject became increasingly severe. The conditions which spawned these laws may be gone, but the archaic classifications remain.

This ordinance is void for vagueness, both in the sense that it "fails to give a person of ordinary intelligence fair notice that his contemplated conduct is forbidden by the statute," United States v Harris, 347 US 612, 617, and because it encourages arbitrary and erratic arrests and convictions. Thornhill v Alabama, 310 US 88.

Living under a rule of law entails various suppositions, one of which is that "[all persons] are entitled to be informed as to what the State commands or forbids." Lanzetta v New Jersey, 306 US 451, 453.

Lanzetta is one of a well-recognized group of cases insisting that the law give fair notice of the offending conduct. In the field of regulatory statutes governing business activities, where the acts limited are in a narrow category, greater leeway is allowed. The poor among us, the minorities, the average householder are not in business and not alerted to the regulatory schemes of vagrancy laws; and we assume they would have no understanding of their meaning and impact if they read them. Nor are they protected from being caught in the vagrancy net by the necessity of having a specific intent to commit an unlawful act.

The Jacksonville ordinance makes criminal activities which by modern standards are normally innocent. "Nightwalking" is one. Florida construes the ordinance not to make criminal one night's wandering, Johnson v State, 202 So 2d, at 855, only the "habitual" wanderer or, as the ordinance describes it, "common night walkers." We know, however, from experience that sleepless people often walk at night, perhaps hopeful that sleep-inducing relaxation will result. . . .

"[P]ersons able to work but habitually living upon the earnings of their wives or minor children" may also embrace unemployed people out of the labor market, by reason of a recession or disemployed by reason of technological or so-called structural displacements.

Persons "wandering or strolling" from place to place have been extolled by Walt Whitman and Vachel Lindsay. The qualification "without any lawful purpose or object" may be a trap for innocent acts. Persons "neglecting all lawful business and habitually spending their time by frequenting . . . places where alcoholic beverages are sold or served" would literally embrace many members of golf clubs and city clubs.

Walkers and strollers and wanderers may be going to or coming from a burglary. Loafers or loiterers may be "casing" a place for a holdup. . . .

The difficulty is that these activities are historically part of the amenities of life as we have known them. They are not mentioned in the Constitution or in the Bill of Rights. These unwritten amenities have been in part responsible for giving our people the feeling of independence and self-confidence, the feeling of creativity. These amenities have dignified the right of dissent and have honored the right to be non-conformists and the right to defy submissiveness. They have encouraged lives of high spirits rather than hushed, suffocating silence. . . .

This aspect of the vagrancy ordinance before us is suggested by what this Court said in 1876 about a broad criminal statute enacted by Congress: "It would certainly be dangerous if the legislature could set a net large enough to catch all possible offenders, and leave it to the courts to step inside and say who could be rightfully detained, and who should be set at large." United States v Reese, 92 US 214, 221.

While that was a federal case, the due process implications are equally applicable to the States and to this vagrancy ordinance. Here the net cast is large, not to give the courts the power to pick and choose but to increase the arsenal of the police. . . .

Where the list of crimes is so all-inclusive and generalized as the one in this ordinance, those convicted may be punished for no more than vindicating affronts to police authority. . . .

We allow our police to make arrests only on "probable cause," a Fourth and Fourteenth Amendment standard applicable to the States as well as to the Federal Government. Arresting a person on suspicion, like arresting a person for investigation, is foreign to our system, even when the arrest is for past criminality. Future criminality, however, is the common justification for the presence of vagrancy statutes.

A direction by a legislature to the police to arrest all "suspicious" persons would not pass constitutional muster. A vagrancy prosecution may be merely the cloak for a conviction which could not be obtained on the real but undisclosed grounds for the arrest.

Those generally implicated by the imprecise terms of the ordinance—poor people, nonconformists, dissenters, idlers—may be required to comport themselves according to the lifestyle deemed appropriate by the Jacksonville police and the courts. Where, as here, there are no standards governing the exercise of the discretion granted by the ordinance, the scheme permits and encourages an arbitrary and discriminatory enforcement of the law. It furnishes a convenient tool for "harsh and discriminatory enforcement by local prosecuting officials, against particular groups deemed to merit their displeasure." Thornhill v Alabama, 310 US 88,97-98. It results in a regime in which the poor and the unpopular are permitted to "stand on a public sidewalk . . . only at the whim of any police officer." Shuttlesworth v Birmingham, 382 US 87, 90. . . .

The implicit presumption in these generalized vagrancy standards—that crime is being nipped in the bud—is too extravagant to deserve extended treatment. Of course, vagrancy statutes are useful to the police. Of course, they are nets making easy the roundup of so-called undesirables. But the rule of law implies equality and justice in its application. Vagrancy laws of the Jacksonville type teach that the scales of justice are so tipped even-handed administration of the law is not possible. The rule of law, evenly applied to minorities as well as majorities, to the poor as well as the rich, is the great mucilage that holds society together.

The Jacksonville ordinance cannot be squared with our constitutional standards and is plainly unconstitutional. . . .

## Notes and Questions

1. Four of the eight defendants convicted under the Jacksonville vagrancy ordinance were charged with "prowling by auto," a judicially created subcategory of "wandering or strolling around from place to place without any lawful purpose or object." Margaret Papachristou, Betty Calloway, Eugene Melton, and Leonard Johnson were on their way to a nightclub in Calloway's car shortly after midnight. Papachristou and Calloway were white, and Melton and Johnson were black. The arresting officer claimed that their car had stopped near a used car lot that had been broken into on several occasions. Of the two principal dangers of vague criminal legislation identified in Justice Douglas's opinion, which might be inferred from the above facts?

2. In *Kolender v. Lawson*, 461 U.S. 352 (1983), the U.S. Supreme Court considered a due process challenge to a California statute that raised vagueness issues. As interpreted by the state courts, the statute required persons who "loiter or wander on the streets to provide a 'credible and reliable' identification and to account for their presence when requested by a peace officer," when "the surrounding circumstances are such as to indicate to a reasonable man that the public safety demands such identification." Edward Lawson, a tall black man who frequently walked in predominantly white neighborhoods, was arrested for violating the statute roughly 15 times in less than 2 years. He was prosecuted on two occasions and convicted once. Lawson brought a civil action seeking damages and a declaration that the statute was unconstitutional. Focusing on the interpretation of the statute that required the production of "credible and reliable" identification, the Court ruled (7–2) that the legislation was unconstitutionally vague.

3. The history and application of vagrancy and loitering statutes reveal that one of the central purposes of such laws is to allow the police to arrest suspicious persons before they have a chance to engage in conduct that results in injury, property damage, the use or sale of drugs, prostitution, or another consummated crime. This feature of vagrancy and loitering statutes creates a potential benefit by allowing the criminal law to intervene before conduct matures into what traditionally would be recognized as a "completed" crime involving social harm. However, it simultaneously risks allowing the criminal law to be invoked too soon. Drawing inferences about future conduct from observed "suspicious" behavior can be hazardous. The observed conduct might be quite innocent. In addition, people contemplating crimes can always change their minds. The danger of arbitrary enforcement, as recognized in *Papachristou*, also is a concern. We confront similar issues when we discuss inchoate crimes, including solicitation, conspiracy, and attempts, in Chapter 12.

## Fair Notice

The maxim, "ignorance of the law is no excuse," is a familiar one. At first blush this doctrine appears to be in tension with the *mens rea* principle. If a person honestly did not know that his or her conduct was against the law and had no conscious awareness of wrongdoing, we may be hard-pressed to find evidence of the "guilty mind" that normally is a prerequisite for ascribing blame and imposing criminal punishment.

For example, a motorist may have deliberately exceeded the speed limit by 10 or 15 mph to try to avoid being late to work. Receiving a ticket under such circumstances may not make the driver happy, but at least it does not seem unfair. On the other hand, the driver may have been making every effort to abide by the law, religiously not allowing his or her car to exceed 55 mph, only later to find out that the speed limit on that particular stretch of road was 40 mph. Despite the motorist's lack of an intention to violate the law—indeed, despite earnest efforts to be in compliance with it—a protest to the effect that "I thought the speed limit was 55" is likely to fall on deaf ears. Ignorance of the law is no excuse.

The policy reasons behind failing to excuse the law-breaker who honestly does not know his or her conduct violates the law are relatively straightforward. To recognize such an excuse in some respects would be "rewarding" people who are not diligent enough to take necessary steps to find out what the law is. We expect people to make that effort. Their failure to do so supplies the element of a guilty mind, at least in an objective sense. Thus we ordinarily will not listen to claims (even honest ones) that a person did not know that his or her conduct was in violation of a criminal law.

The vagueness doctrine we just considered requires that all people have a fair opportunity to know what the law is. It does not protect people who actually do not know about a law that is fairly "knowable." We now consider whether there may be criminal laws that, although duly enacted, published, and not void for vagueness, nevertheless are constitutionally infirm because they violate fair notice principles.

---

### Lambert v. California, 355 U.S. 225, 78 S. Ct. 240, 2 L. Ed. 2d 228 (1958)

Mr. Justice Douglas delivered the opinion of the Court.

Section 52.38(a) of the Los Angeles Municipal Code defines "convicted person" as follows:

"Any person who, subsequent to January 1, 1921, has been or hereafter is convicted of an offense punishable as a felony in the State of California, or who has been or who is hereafter convicted of any offense in any place other than the State of California, which offense, if committed in the State of California, would have been punishable as a felony."

Section 52.39 provides that it shall be unlawful for "any convicted person" to be or remain in Los Angeles for a period of more than five days without registering; it requires any person having a place of abode outside the city to register if he comes into the city on five occasions or more during a 30-day period; and it prescribes the information to be furnished the Chief of Police on registering.

Section 52.43(b) makes the failure to register a continuing offense, each day's failure constituting a separate offense.

Appellant, arrested on suspicion of another offense, was charged with a violation of this registration law. The evidence showed that she had been at the time of her arrest a resident of Los Angeles for over seven years. Within that period she had been convicted in Los Angeles of the crime of forgery, an offense which California punishes as a felony. Though convicted of a crime punishable as a felony, she had not at the time of her arrest registered under the Municipal Code. At the trial, appellant asserted that § 52.39 of the Code denies her due process of law and other rights under the Federal Constitution, unnecessary to enumerate. The trial court denied this objection. The case was tried to a jury which found appellant guilty. The court fined her $250 and placed her on probation for three years. . . .

The Appellate Department of the Superior Court affirmed the judgment, holding there was no merit to the claim that the ordinance was unconstitutional. The case is here on appeal. . . . [W]e now hold that the registration provisions of the Code as sought to be applied here violate the Due Process requirement of the Fourteenth Amendment.

The registration provision, carrying criminal penalties, applies if a person has been convicted "of an offense punishable as a felony in the State of California" or, in case he has been convicted in another State, if the offense "would have been punishable as a felony" had it been committed in California. No element of willfulness is by terms included in the ordinance nor read into it by the California court as a condition necessary for a conviction.

We must assume that appellant had no actual knowledge of the requirement that she register under this ordinance, as she offered proof of this defense which was refused. The question is whether a registration act of this character violates due process where it is applied to a person who has no actual knowledge of his duty to register, and where no showing is made of the probability of such knowledge.

We do not go with Blackstone in saying that "a vicious will" is necessary to constitute a crime, 4 Bl. Comm. *21, for conduct alone without regard to the intent of the doer is often sufficient. There is wide latitude in the lawmakers to declare an offense and to exclude elements of knowledge and diligence from its definition. But we deal here with conduct that is wholly passive—mere failure to register. It is unlike the commission of acts, or the failure to act under circumstances that should alert the doer to the consequences of his deed. The rule that "ignorance of the law will not excuse" is deep in our law, as is the principle that of all the powers of local government, the police power is "one of the least limitable." On the other hand, due process places some limits on its exercise. Engrained in our concept of due process is the requirement of notice. Notice is sometimes essential so that the citizen has the chance to defend charges. Notice is required before property interests are disturbed, before assessments are made, before penalties are assessed. Notice is required in a myriad of situations where a penalty or forfeiture might be suffered for mere failure to act. But the principle is equally appropriate where a person, wholly passive and unaware of any wrongdoing, is brought to the bar of justice for condemnation in a criminal case.

Registration laws are common and their range is wide. Many such laws are akin to licensing statutes in that they pertain to the regulation of business activities. But the present ordinance is entirely different. Violation of its provisions is unaccompanied by any activity whatever, mere presence in the city being the test. Moreover, circumstances which might move one to inquire as to the necessity of registration are completely lacking. At most the ordinance is but a law enforcement technique designed for the convenience of law enforcement agencies through which a list of the names and addresses of felons then residing in a given community is compiled. The disclosure is merely a compilation of former convictions already publicly recorded in the jurisdiction where obtained. Nevertheless, this appellant on first becoming aware of her duty to register was given no opportunity to comply with the law and avoid its penalty, even though her default was entirely innocent. She could but suffer the consequences of the ordinance, namely, conviction with the imposition of heavy criminal penalties thereunder. We believe that actual knowledge of the duty to register or proof of the probability of such knowledge and subsequent failure to comply are necessary before a conviction under the ordinance can stand. As Holmes wrote in The Common Law, "A law which punished conduct which would not be blameworthy in the average member of the community would be too severe for that community to bear." Id., at 50. Its severity lies in the absence of an opportunity either to avoid the consequences of the law or to defend any prosecution brought under it. Where a person did not know of the duty to register and where there was no proof of the probability of such knowledge, he may not be convicted consistently with due process. Were it otherwise, the evil would be as great as it is when the law is written in print too fine to read or in a language foreign to the community.

Reversed.

Mr. Justice Frankfurter, whom Mr. Justice Harlan and Mr. Justice Whittaker join, dissenting.

The present laws of the United States and of the forty-eight States are thick with provisions that command that some things not be done and others be done, although persons convicted under such provisions may have had no awareness of what the law required or that what they did was wrongdoing. The body of decisions sustaining such legislation, including innumerable registration laws, is almost as voluminous as the legislation itself. The matter is summarized in United States v. Balint, 258 U.S. 250, 252: "Many instances of this are to be found in regulatory measures in the exercise of what is called the police power where the emphasis of the statute is evidently upon achievement of some social betterment rather than the punishment of the crimes as in cases of *mala in se*."

Surely there can hardly be a difference as a matter of fairness, of hardship, or of justice, if one may invoke it, between the case of a person wholly innocent of wrongdoing, in the sense that he was not remotely conscious of violating any law, who is imprisoned for five years for conduct relating to narcotics, and the case of another person who is placed on probation for three years on condition that she pay $250, for failure, as a local resident, convicted under local law of a felony, to register under a law passed as an exercise of the State's "police power." Considerations of hardship often lead courts, naturally enough, to attribute to a statute the requirement of a certain mental element—some consciousness of wrongdoing and knowledge of the law's command—as a matter of statutory construction. . . .

(Continues)

But what the Court here does is to draw a constitutional line between a State's requirement of doing and not doing. . . .

If the generalization that underlies, and alone can justify, this decision were to be given its relevant scope, a whole volume of the United States Reports would be required to document in detail the legislation in this country that would fall or be impaired. I abstain from entering upon a consideration of such legislation, and adjudications upon it, because I feel confident that the present decision will turn out to be an isolated deviation from the strong current of precedents—a derelict on the waters of the law. Accordingly, I content myself with dissenting.

## Notes and Questions

1. How, precisely, did the Los Angeles registration ordinance fail to provide Lambert with fair notice of what was expected of her? Despite its affirmance of the deep-rooted nature of "[t]he rule that 'ignorance of the law will not excuse,'" how faithful is the majority opinion to that principle?
2. Is Justice Douglas correct when he asserts that "circumstances which might move one to inquire as to the necessity of registration are completely lacking"? Under the ordinance, what triggers the registration requirement?
3. Does the Court invalidate the Los Angeles ordinance, or rather does it rule that it cannot constitutionally be applied under the facts in Lambert's case? Could a jurisdiction take any steps to rescue an ordinance of this nature from constitutional challenge?
4. Justice Frankfurter's dissent predicted that *Lambert* "will turn out to be an isolated deviation from the strong current of precedents—a derelict on the waters of the law." And, indeed, *Lambert* has been narrowly confined to its facts and has had rather limited impact.[2] *Lambert* remains good law. However, you should be hesitant to extend the holding beyond the relevant facts.

## Ex Post Facto Laws

The U.S. Constitution forbids both Congress and the states from passing any "ex post facto Law." U.S. Constitution, Art. I, § 9 [3]; Art. I, § 10 [1]. The conflict between *ex post facto* laws and the legality principle is clear: a law passed "after the fact" of an individual's committing an act, or raising the punishment for it, could not possibly provide fair notice that the person's conduct would be in violation of that same law or that it would be subject to the greater punishment. The prohibition against *ex post facto* laws promotes other policies in addition to the fair notice principle. At this point we consider the primary reasons behind the constitutional disapproval of *ex post facto* laws and identify the general circumstances that trigger *ex post facto* clause violations.

*ex post facto laws:*

### *Miller v. Florida*, 482 U.S. 423, 107 S. Ct. 2446, 96 L. Ed. 2d 351 (1987)

Justice O'Connor delivered the opinion of the Court.

At the time petitioner committed the crime for which he was convicted, Florida's sentencing guidelines would have resulted in a presumptive sentence of 3½ to 4½ years' imprisonment. At the time petitioner was sentenced, the revised guidelines called for a presumptive sentence of 5½ to 7 years in prison. The trial court applied the guidelines in effect at the time of sentencing and imposed a 7-year sentence. The question presented is whether application of these amended guidelines in petitioner's case is unconstitutional by virtue of the Ex Post Facto Clause. . . .

Article I of the United States Constitution provides that neither Congress nor any State shall pass any "ex post facto Law." See Art I, § 9, cl 3; Art I, § 10, cl 1. Our understanding of what is meant by ex post facto largely derives from the case of Calder v Bull, 3 Dall 386, 1 L Ed 648 (1798), in which this Court first considered the scope of the ex post facto prohibition. In Calder, Justice Chase, noting that the expression "ex post facto" "had been in use long before the revolution," id., at 391, summarized his understanding of what fell "within the words and the intent of the prohibition":

"1st. Every law that makes an action done before the passing of the law, and which was innocent when done, criminal; and punishes such action. 2d. Every law that aggravates a crime, or makes it greater than it was, when committed. 3d. Every law that changes the punishment, and inflicts a greater punishment, than the law annexed to the crime, when committed. 4th. Every law that alters the legal rules of evidence, and receives less, or different testimony, than the law required at the time of the commission of the offense, in order to convict the offender." Id., at 390, [emphasis omitted].

Justice Chase explained that the reason the Ex Post Facto Clauses were included in the Constitution was to assure that federal and state legislatures were restrained from enacting arbitrary or vindictive legislation. Justices Paterson and Iredell, in their separate opinions in Calder, likewise emphasized that the Clauses were aimed at preventing legislative abuses. In addition, the Justices' opinions in Calder, as well as other early authorities, indicate that the Clauses were aimed at a second concern, namely, that legislative enactments "give fair warning of their effect and permit individuals to rely on their meaning until explicitly changed." Weaver v Graham, 450 U.S. 24, 28–29 (1981). Thus, almost from the outset, we have recognized that central to the ex post facto prohibition is a concern for "the lack of fair notice and governmental restraint when the legislature increases punishment beyond what was prescribed when the crime was consummated." Weaver, 450 U.S., at 30.

Our test for determining whether a criminal law is ex post facto derives from these principles. As was stated in Weaver, to fall within the ex post facto prohibition, two critical elements must be present: first, the law "must be retrospective, that is, it must apply to events occurring before its enactment"; and second, "it must disadvantage the offender affected by it." Id., at 29. We have also held in Dobbert v Florida, 432 U.S. 282 (1977), that no ex post facto violation occurs if a change does not alter "substantial personal rights," but merely changes "modes of procedure which do not affect matters of substance." Id., at 293. Respondent contends that the revised sentencing law is neither impermissibly retrospective, nor to petitioner's disadvantage; respondent also contends that the revised sentencing law is merely a procedural change. We consider these claims in turn.

A law is retrospective if it "changes the legal consequences of acts completed before its effective date." Weaver, supra, at 31. Application of the revised guidelines law in petitioner's case clearly satisfies this standard. . . .

Here . . . the statute in effect at the time petitioner acted did not warn him that Florida prescribed a 5½- to 7-year presumptive sentence for that crime. Petitioner simply was warned of the obvious fact that the sentencing guidelines law—like any other law—was subject to revision. The constitutional prohibition against ex post facto laws cannot be avoided merely by adding to a law notice that it might be changed.

It is "axiomatic that for a law to be ex post facto it must be more onerous than the prior law." Dobbert, supra, at 294. . . .

Respondent maintains that the change in guidelines laws is not disadvantageous because petitioner "cannot show definitively that he would have gotten a lesser sentence." This argument, however, is foreclosed by our decision in Lindsey v Washington, 301 U.S. 397 (1937). [I]n Lindsey, the law in effect at the time the crime was committed provided for a maximum sentence of 15 years, and a minimum sentence of not less than six months. At the time Lindsey was sentenced, the law had been changed to provide for a mandatory 15-year sentence. Finding that retrospective application of this change was ex post facto, the Court determined that "we need not inquire whether this is technically an increase in the punishment annexed to the crime," because "[i]t is plainly to the substantial disadvantage of petitioners to be deprived of all opportunity to receive a sentence which would give them freedom from custody and control prior to the expiration of the 15-year term." Id., at 401–402. Thus, Lindsey establishes "that one is not barred from challenging a change in the penal code on ex post facto grounds simply because the sentence he received under the new law was not more onerous than that which he might have received under the old." Dobbert, supra, at 300.

Petitioner plainly has been "substantially disadvantaged" by the change in sentencing laws. To impose a 7-year sentence under the old guidelines, the sentencing judge would have to depart from the presumptive sentence range of 3½ to 4½ years. As a result, the sentencing judge would have to provide clear and convincing reasons in writing for the departure, on facts proved beyond a reasonable doubt, and his determination would be reviewable on appeal. By contrast, because a 7-year sentence is within the presumptive range under the revised law, the trial judge did not have to provide any reasons, convincing or otherwise, for imposing the sentence, and his decision was unreviewable. . . .

Finally, even if a law operates to the defendant's detriment, the ex post facto prohibition does not restrict "legislative control of remedies and modes of procedure which do not affect matters of

(Continues)

substance." Dobbert, 432 U.S., at 293. Hence, no ex post facto violation occurs if the change in the law is merely procedural and does "not increase the punishment, nor change the ingredients of the offence or the ultimate facts necessary to establish guilt." Hopt v Utah, 10 U.S. 574, 590 (1884). See Dobbert, supra, at 293–294 ("The new statute simply altered the methods employed in determining whether the death penalty was to be imposed; there was no change in the quantum of punishment attached to the crime"). On the other hand, a change in the law that alters a substantial right can be ex post facto "even if the statute takes a seemingly procedural form." Weaver, 450 U.S., at 29, n 12.

Although the distinction between substance and procedure might sometimes prove elusive, here the change at issue appears to have little about it that could be deemed procedural. The 20% increase in points for sexual offenses in no wise alters the method to be followed in determining the appropriate sentence; it simply inserts a larger number into the same equation. The comments of the Florida Supreme Court acknowledge that the sole reason for the increase was to punish sex offenders more heavily: the amendment was intended to, and did, increase the "quantum of punishment" for category 2 crimes. . . .

The law at issue in this case, like the law in W[e]aver, "makes more onerous the punishment for crimes committed before its enactment." Weaver, 450 U.S., at 36. Accordingly, we find that Florida's revised guidelines law, 1984 Fla Laws, ch 84–328, is void as applied to petitioner, whose crime occurred before the law's effective date. We reverse the judgment of the Supreme Court of Florida, and remand the case for further proceedings not inconsistent with this opinion.

## Notes and Questions

1. As the Court explains in Miller, in addition to serving the fair notice principle, the prohibition against ex post facto laws "was to assure that federal and state legislatures were restrained from enacting arbitrary or vindictive legislation." In this same regard Alexander Hamilton argued that, "The creation of crimes after the commission of the fact, or, in other words, the subjecting of men to punishment for things which, when they were done, were breaches of no law, and the practice of arbitrary imprisonments, have been, in all ages, the favorite and most formidable instruments of tyranny."[3]

2. In Weaver v. Graham, 450 U.S. 24 (1981), which is cited and relied on in Miller, the U.S. Supreme Court ruled that a Florida statute reducing the availability of prisoners' "good time credit" (i.e., a reduction in sentence in recognition of good conduct and obedience to prison rules) violated ex post facto principles when applied to prisoners whose crimes were committed before the statute's enactment. The Court again invalidated a Florida statute on ex post facto grounds in Lynce v. Mathis, 519 U.S. 433 (1997). In that case the Florida legislature attempted to cancel "early release credits" that state prisoners already had accumulated.

3. In Collins v. Youngblood, 497 U.S. 37 (1990), the Court rejected an ex post facto challenge to a change in Texas law, occurring after Youngblood committed his crime, that allowed a state appellate court to modify an improper jury verdict assessing a punishment not authorized by law. Chief Justice Rehnquist's opinion distinguished cases in which retrospective legislation altered the definition of a crime or increased punishment for a crime. The opinion characterized the change in Texas law as "procedural" and as not threatening ex post facto principles because it did not "deprive one charged with crime of any defense available according to law at the time when the act was committed."

4. Before 1993 California observed a 3-year statute of limitations for the prosecution of crimes including sexual abuse of a child. That year the legislature enacted a new statute of limitations that allowed the prosecution of sex-related child abuse within 1 year of the victim's reporting of an offense, even if it had allegedly been committed at a time when prosecution would have been barred by the previously existing 3-year limitation period. In 1998 Martin Stogner was indicted for sex-related child abuse that allegedly had taken place between 1955 and 1973 and thus clearly would not have been eligible for prosecution under the prior statute. Stogner moved to dismiss the indictment, claiming that to allow his prosecution would violate ex post facto and due process principles. By a 5–4 vote, the Supreme Court agreed that "a law enacted after expiration of a previously applicable limitations period violates the Ex Post Facto Clause when it is applied to revive a previously time-barred prosecution." Justice Breyer's majority opinion explained that "the new statute threatens the kind of harm that . . . the Ex Post Facto Clause seeks to avoid. . . . [T]he government has refused 'to play by its own rules.' It has deprived the defendant of the 'fair warning' that might have led him to preserve exculpatory evidence. . . . And a Constitution that permits such an extension, by allowing legislatures to pick and choose when to act retroactively, risks both 'arbitrary and potentially vindictive legislation,' and erosion of the separation of powers . . . ." Stogner v. California, 539 U.S. 607 (2003) [internal citations omitted].

5. The *ex post facto* clause prohibits retroactive application of a law that increases an offender's "punishment." In Chapter 1 we considered the definition of "punishment." We noted, for example, the U.S. Supreme Court's ruling in *Kansas v. Hendricks*, 521 U.S. 346 (1997), which found no *ex post facto* clause violation when Kansas applied an involuntary civil commitment statute to a convicted sex offender after the man's release from prison, even though the statute was enacted after he had committed his crime. The majority opinion reasoned that although Hendricks was confined against his will pursuant to the civil commitment statute, such confinement was not a form of "punishment," thus rendering *ex post facto* prohibitions inapplicable. The Supreme Court similarly has ruled that "Megan's Law" statutes, which require sex offenders to register with the authorities and frequently require or permit community notification regarding a previously convicted sex offender's presence in the neighborhood, do not present *ex post facto* problems because neither registration nor community notification constitutes "punishment." *See Connecticut Dept. Public Safety v. Doe,* 538 U.S. 1 (2003); *Smith v. Doe,* 538 U.S. 84 (2003).

## Constitutional Limits on the State's Power to Define Conduct as a Crime

Legislative bodies have broad authority to define behaviors that are detrimental to society and bring them within the sweep of the criminal laws. Although that power is vast, it is not unlimited. In this section we consider constitutional limitations on the legislative power to make specified conduct a crime. We first explore "**substantive due process**" issues. We then turn to First Amendment protections involving freedom of speech. We conclude by examining equal protection issues involving the definition of crimes.

## Right to Privacy and the Criminal Law

In *Griswold v. Connecticut,* 381 U.S. 479 (1965), the U.S. Supreme Court was asked to declare unconstitutional Connecticut statutes that made it a crime, punishable by 60 days to 1 year in jail and a fine of not less than $50, either to use contraceptives or to assist or counsel another to use contraceptives. The director of the state Planned Parenthood League and a physician were found guilty of violating the law because "[t]hey gave information, instruction, and medical advice to *married persons* as to the means of preventing conception" [emphasis in original]. Each was fined $100. The state appellate courts affirmed their convictions.

Nowhere in the body of or in the amendments to the U.S. Constitution is there an express right to use or counsel others in the use of contraceptives. The Connecticut law under challenge dated back to 1879. In the assessment of Justice Stewart, who dissented in *Griswold,* "this is an uncommonly silly law." He continued, "But we are not asked in this case to say whether we think this law is unwise, or even asinine. We are asked to hold that it violates the United States Constitution."

There is a profound difference between disagreeing with a law—concluding that it is "silly," unwise, or even offensive—and ruling that a law is unconstitutional. When a statute runs afoul of a specific provision of the Constitution—for example, a law forbidding political rallies measured against the First Amendment's prohibition that "Congress shall make no law . . . abridging the freedom of speech," or a law making it a crime for blacks but not whites to ride a bus reviewed under the Fourteenth Amendment's guarantee of "**the equal protection of the laws**"—the judiciary's authority (or duty) to invalidate the enactment is clear. But, what of the converse situation? If no explicit constitutional provision outlaws a statute, is the legislation thereby immune from constitutional challenge? Should the legislative process be relied on to change "uncommonly silly" laws, with judges keeping their hands off legislation that is not inconsistent with the specific commands of the Constitution?

The Court in *Griswold* did strike the Connecticut statute as unconstitutional. The Constitution speaks through specific commands and prohibitions as well as through general clauses that acquire meaning through interpretation. One of those general principles is enshrined in the Fifth and Fourteenth Amendments, which provide, respectively, that the federal and state governments shall not deprive persons of "life, liberty, or property, without due process of law." The justices voting to invalidate the anti-contraceptives law in *Griswold* offered different sources for their authority to do so.

Justice Goldberg's concurring opinion relied heavily on the Ninth Amendment, which reads: "The enumeration in the Constitution, of certain rights, shall not be construed to deny or disparage others retained by the people." He argued that the Ninth Amendment was meant to recognize that the Constitution's "specifically enumerated rights could not be sufficiently broad to cover all essential rights and that the specific mention of certain rights would [not] be interpreted as a denial that others were protected." The right of marital privacy, including the right of married couples to use and receive counseling about contraceptives, in his view was one of the unenumerated fundamental rights recognized by the Ninth Amendment.

Justice Douglas's majority opinion offered a theory about constitutional "penumbras." He said, "[S]pecific guarantees in the Bill of Rights have penumbras, formed by emanations from those guarantees that help give them life and substance. Various guarantees create zones of privacy." He cited the specific protections of the First, Third, Fourth, Fifth, and Ninth Amendments as creating an implicit **right of privacy** that applies to the marriage relationship, with sufficient breadth to doom Connecticut's anti-contraceptives law.

While defending their conclusions that the Constitution denied Connecticut the power to prohibit married couples from gaining access to contraceptives, Justices Goldberg and Douglas both invoked the broad protections of the Fourteenth Amendment's due process clause. *Griswold* now generally is interpreted as an important precedent evidencing the Court's willingness to rely on "substantive due process" rights to invalidate laws that threaten fundamental individual liberties although they do not run afoul of more specific constitutional constraints. *Griswold* was not the first in this line of cases. Justice Goldberg's opinion explained that "[t]he Court stated many years ago that the Due Process Clause protects those liberties that are 'so rooted in the traditions and conscience of our people as to be ranked as fundamental,'" quoting *Snyder* v. *Massachusetts*, 291 U.S. 97, 105 (1934). The Court's later decision in *Roe v. Wade*, 410 U.S. 113 (1973), recognizing a woman's constitutional right to terminate a pregnancy through abortion before the fetus' viability, and thereafter if the woman's life or health is in serious jeopardy, also relies on "substantive due process" grounds.

Justice Black's dissenting opinion in *Griswold* expressed one of the chief criticisms of the Court's relying on unarticulated or nonspecific constitutional "rights" to invalidate legislation:

> I do not believe that we are granted power by the Due Process Clause or any other constitutional provision or provisions to measure constitutionality by our belief that legislation is arbitrary, capricious or unreasonable, or accomplishes no justifiable purpose, or is offensive to our own notions of "civilized standards of conduct." Such an appraisal of the wisdom of legislation is an attribute of the power to make laws, not of the power to interpret them. The use by federal courts of such a formula or doctrine or whatnot to veto federal or state laws simply takes away from Congress and states the power to make laws based on their own judgment of fairness and wisdom and transfers that power to this Court for ultimate determination—a power which was specifically denied to federal courts by the convention that framed the Constitution.

The same themes evidenced in *Griswold*, involving the clash between the legislative power to enact criminal statutes designed to promote the public good—as defined by democratically elected lawmakers—and the judiciary's role in safeguarding individual liberties by application of "substantive due process" principles, are apparent in the cases that follow.

---

## Lawrence v. Texas, 539 U.S. 558, 123 S. Ct. 2472, 156 L. Ed. 2d 508 (2003)

Justice Kennedy delivered the opinion of the Court. . . .

### I.

. . . In Houston, Texas, officers of the Harris County Police Department were dispatched to a private residence in response to a reported weapons disturbance. They entered an apartment where one of the petitioners, John Geddes Lawrence, resided. . . . The officers observed Lawrence and another man, Tyron Garner, engaging in a sexual act. The two petitioners were arrested, held in custody overnight, and charged and convicted before a Justice of the Peace.

The complaints described their crime as "deviate sexual intercourse, namely anal sex, with a member of the same sex (man)." The applicable state law is Tex. Penal Code Ann. § 21.06(a) (2003). It provides: "A person commits an offense if he engages in deviate sexual intercourse with another individual of the same sex." The statute defines "[d]eviate sexual intercourse" as follows:

"(A) any contact between any part of the genitals of one person and the mouth or anus of another person; or

"(B) the penetration of the genitals or the anus of another person with an object." § 21.01(1).

[Following rejection of their constitutional claims on a trial *de novo* in Harris County Criminal Court, the petitioners were convicted on their pleas of *nolo contendere*. Each was fined $200 and assessed court costs. Their convictions were affirmed on appeal and the Supreme Court granted certiorari.] . . .

### II.

We conclude the case should be resolved by determining whether the petitioners were free as adults to engage in the private conduct in the exercise of their liberty under the Due Process Clause of the Fourteenth Amendment to the Constitution. For this inquiry we deem it necessary to reconsider the Court's holding in *Bowers* [v. *Hardwick*, 478 U.S. 186, 106 S.Ct. 2841, 92 L.Ed.2d 140 (1986)]. . . .

The Court began its substantive discussion in *Bowers* as follows: "The issue presented is whether the Federal Constitution confers a fundamental right upon homosexuals to engage in sodomy and hence invalidates the laws of the many States that still make such conduct illegal and have done so for a very long time." *Id.*, at 190. That statement, we now conclude, discloses the Court's own failure to appreciate the extent of the liberty at stake. To say that the issue in *Bowers* was simply the right to engage in certain sexual conduct demeans the claim the individual put forward, just as it would demean a married couple were it to be said marriage is simply about the right to have sexual intercourse. The laws involved in *Bowers* and here are, to be sure, statutes that purport to do no more than prohibit a particular sexual act. Their penalties and purposes, though, have more far-reaching consequences, touching upon the most private human conduct, sexual behavior, and in the most private of places, the home. The statutes do seek to control a personal relationship that, whether or not entitled to formal recognition in the law, is within the liberty of persons to choose without being punished as criminals.

This, as a general rule, should counsel against attempts by the State, or a court, to define the meaning of the relationship or to set its boundaries absent injury to a person or abuse of an institution the law protects. It suffices for us to acknowledge that adults may choose to enter upon this relationship in the confines of their homes and their own private lives and still retain their dignity as free persons. When sexuality finds overt expression in intimate conduct with another person, the conduct can be but one element in a personal bond that is more enduring. The liberty protected by the Constitution allows homosexual persons the right to make this choice.

Having misapprehended the claim of liberty there presented to it, and thus stating the claim to be whether there is a fundamental right to engage in consensual sodomy, the *Bowers* Court said: "Proscriptions against that conduct have ancient roots." *Id.*, at 192 . . .

[T]he historical grounds relied upon in *Bowers* are more complex than the majority opinion and the concurring opinion by Chief Justice Burger indicate. Their historical premises are not without doubt and, at the very least, are overstated.

It must be acknowledged, of course, that the Court in *Bowers* was making the broader point that for centuries there have been powerful voices to condemn homosexual conduct as immoral. The condemnation has been shaped by religious beliefs, conceptions of right and acceptable behavior, and respect for the traditional family. For many persons these are not trivial concerns but profound and deep convictions accepted as ethical and moral principles to which they aspire and which thus determine the course of their lives. These considerations do not answer the question before us, however. The issue is whether the majority may use the power of the State to enforce these views on the whole society through operation of the criminal law. "Our obligation is to define the liberty of all, not to mandate our own moral code." *Planned Parenthood of Southeastern Pa. v. Casey,* 505 U.S. 833, 850 (1992). . . .

Two principal cases decided after *Bowers* cast its holding into even more doubt. In *Planned Parenthood of Southeastern Pa. v. Casey, [supra],* the Court reaffirmed the substantive force of the liberty protected by the Due Process Clause. The *Casey* decision again confirmed that our laws and tradition afford constitutional protection to personal decisions relating to marriage, procreation, contraception, family relationships, child rearing, and education. In explaining the respect the Constitution demands for the autonomy of the person in making these choices, we stated as follows:

> "These matters, involving the most intimate and personal choices a person may make in a lifetime, choices central to personal dignity and autonomy, are central to the liberty protected by the Fourteenth Amendment. At the heart of liberty is the right to define one's own concept of existence, of meaning, of the universe, and of the mystery of human life. Beliefs about these matters could not define the attributes of personhood were they formed under compulsion of the State." *Id.* [at 851].

Persons in a homosexual relationship may seek autonomy for these purposes, just as heterosexual persons do. The decision in *Bowers* would deny them this right.

The second post-*Bowers* case of principal relevance is *Romer v. Evans,* 517 U.S. 620 (1996). There the Court struck down class-based legislation directed at homosexuals as a violation of the Equal Protection Clause. *Romer* invalidated an amendment to Colorado's Constitution which named as a solitary class persons who were homosexuals, lesbians, or bisexual either by "orientation, conduct, practices or relationships," *id.,* at 624, and deprived them of protection under state antidiscrimination laws. We concluded that the provision was "born of animosity toward the class of persons affected" and further that it had no rational relation to a legitimate governmental purpose.

As an alternative argument in this case, counsel for the petitioners and some *amici* contend that *Romer* provides the basis for declaring the Texas statute invalid under the Equal Protection Clause. That is a tenable argument, but we conclude the instant case requires us to address whether *Bowers* itself has continuing validity. Were we to hold the statute invalid under the Equal Protection Clause some might question whether a prohibition would be valid if drawn differently, say, to prohibit the conduct both between same-sex and different-sex participants. . . .

*(Continues)*

To the extent *Bowers* relied on values we share with a wider civilization, it should be noted that the reasoning and holding in *Bowers* have been rejected elsewhere. . . . The right the petitioners seek in this case has been accepted as an integral part of human freedom in many other countries. There has been no showing that in this country the governmental interest in circumscribing personal choice is somehow more legitimate or urgent.

The doctrine of **stare decisis** is essential to the respect accorded to the judgments of the Court and to the stability of the law. It is not, however, an inexorable command. . . .

*Bowers* was not correct when it was decided, and it is not correct today. It ought not to remain binding precedent. *Bowers v. Hardwick* should be and now is overruled.

The present case does not involve minors. It does not involve persons who might be injured or coerced or who are situated in relationships where consent might not easily be refused. It does not involve public conduct or prostitution. It does not involve whether the government must give formal recognition to any relationship that homosexual persons seek to enter. The case does involve two adults who, with full and mutual consent from each other, engaged in sexual practices common to a homosexual lifestyle. The petitioners are entitled to respect for their private lives. The State cannot demean their existence or control their destiny by making their private sexual conduct a crime. Their right to liberty under the Due Process Clause gives them the full right to engage in their conduct without intervention of the government. "It is a promise of the Constitution that there is a realm of personal liberty which the government may not enter." *Casey, supra,* at 847. The Texas statute furthers no legitimate state interest which can justify its intrusion into the personal and private life of the individual.

Had those who drew and ratified the Due Process Clauses of the Fifth Amendment or the Fourteenth Amendment known the components of liberty in its manifold possibilities, they might have been more specific. They did not presume to have this insight. They knew times can blind us to certain truths and later generations can see that laws once thought necessary and proper in fact serve only to oppress. As the Constitution endures, persons in every generation can invoke its principles in their own search for greater freedom.

The judgment of the Court of Appeals for the Texas Fourteenth District is reversed, and the case is remanded for further proceedings not inconsistent with this opinion.

*It is so ordered.*

Justice O'Connor, concurring in the judgment.

The Court today overrules *Bowers v. Hardwick, [supra]*. I joined *Bowers,* and do not join the Court in overruling it. Nevertheless, I agree with the Court that Texas' statute banning same-sex sodomy is unconstitutional. Rather than relying on the substantive component of the Fourteenth Amendment's Due Process Clause, as the Court does, I base my conclusion on the Fourteenth Amendment's Equal Protection Clause. . . .

The statute at issue here makes sodomy a crime only if a person "engages in deviate sexual intercourse with another individual of the same sex." Tex. Penal Code Ann. § 21.06(a) (2003). Sodomy between opposite-sex partners, however, is not a crime in Texas. . . .

Texas attempts to justify its law, and the effects of the law, by arguing that the statute satisfies rational basis review because it furthers the legitimate governmental interest of the promotion of morality. . . .

Moral disapproval of a group cannot be a legitimate governmental interest under the Equal Protection Clause because legal classifications must not be "drawn for the purpose of disadvantaging the group burdened by the law." [*Romer v. Evans,* 517 U.S., at 633.] Texas' invocation of moral disapproval as a legitimate state interest proves nothing more than Texas' desire to criminalize homosexual sodomy. . . . The Texas sodomy law "raise[s] the inevitable inference that the disadvantage imposed is born of animosity toward the class of persons affected." *Id.,* at 634. . . .

I therefore concur in the Court's judgment that Texas' sodomy law banning "deviate sexual intercourse" between consenting adults of the same sex, but not between consenting adults of different sexes, is unconstitutional.

Justice Scalia, with whom the Chief Justice and Justice Thomas join, dissenting. . . .

## III.

Having decided that it need not adhere to *stare decisis,* the Court still must establish that *Bowers* was wrongly decided and that the Texas statute, as applied to petitioners, is unconstitutional.

Texas Penal Code Ann. § 21.06(a) (2003) undoubtedly imposes constraints on liberty. So do laws prohibiting prostitution, recreational use of heroin, and, for that matter, working more than 60 hours per week in a bakery. But there is no right to "liberty" under the Due Process Clause, though today's opinion repeatedly makes that claim. . . . The Fourteenth Amendment *expressly allows* States to deprive their citizens of "liberty," *so long as "due process of law" is provided* . . . .

Our opinions applying the doctrine known as "substantive due process" hold that the Due Process Clause prohibits States from infringing *fundamental* liberty interests, unless the infringement is narrowly tailored to serve a compelling state interest. We have held repeatedly . . . that *only* fundamental rights qualify for this so-called "heightened scrutiny" protection—that is, rights which are "'deeply rooted in this Nation's history and tradition,' *Washington v. Glucksberg*, 521 U.S [702, 721 (1997)] . . . .

*Bowers* held, first, that criminal prohibitions of homosexual sodomy are not subject to heightened scrutiny because they do not implicate a "fundamental right" under the Due Process Clause . . . . *Bowers* concluded that a right to engage in homosexual sodomy was not "'deeply rooted in this Nation's history and tradition,'" *id.,* at 192.

The Court today does not overrule this holding. Not once does it describe homosexual sodomy as a "fundamental right" or a "fundamental liberty interest," nor does it subject the Texas statute to strict scrutiny. Instead, having failed to establish that the right to homosexual sodomy is "'deeply rooted in this Nation's history and tradition,'" the Court concludes that the application of Texas's statute to petitioners' conduct fails the rational-basis test, and overrules *Bowers'* holding to the contrary. "The Texas statute furthers no legitimate state interest which can justify its intrusion into the personal and private life of the individual." *Ante,* at 2484. . . .

## IV.

I turn now to the ground on which the Court squarely rests its holding: the contention that there is no rational basis for the law here under attack. This proposition is so out of accord with our jurisprudence—indeed, with the jurisprudence of *any* society we know—that it requires little discussion.

The Texas statute undeniably seeks to further the belief of its citizens that certain forms of sexual behavior are "immoral and unacceptable," *Bowers, supra,* at 196—the same interest furthered by criminal laws against fornication, bigamy, adultery, adult incest, bestiality, and obscenity. *Bowers* held that this *was* a legitimate state interest. The Court today reaches the opposite conclusion. . . . This effectively decrees the end of all morals legislation. If, as the Court asserts, the promotion of majoritarian sexual morality is not even a *legitimate* state interest, none of the above-mentioned laws can survive rational-basis review. . . .

* * *

Today's opinion is the product of a Court, which is the product of a law-profession culture, that has largely signed on to the so-called homosexual agenda, by which I mean the agenda promoted by some homosexual activists directed at eliminating the moral opprobrium that has traditionally attached to homosexual conduct. . . .

One of the most revealing statements in today's opinion is the Court's grim warning that the criminalization of homosexual conduct is "an invitation to subject homosexual persons to discrimination both in the public and in the private spheres." *Ante,* at 2482. It is clear from this that the Court has taken sides in the culture war, departing from its role of assuring, as neutral observer, that the democratic rules of engagement are observed. Many Americans do not want persons who openly engage in homosexual conduct as partners in their business, as scoutmasters for their children, as teachers in their children's schools, or as boarders in their home. They view this as protecting themselves and their families from a lifestyle that they believe to be immoral and destructive. The Court views it as "discrimination" which it is the function of our judgments to deter. So imbued is the Court with the law profession's anti-anti-homosexual culture, that it is seemingly unaware that the attitudes of that culture are not obviously "mainstream"; that in most States what the Court calls "discrimination" against those who engage in homosexual acts is perfectly legal; that proposals to ban such "discrimination" under Title VII have repeatedly been rejected by Congress; that in some cases such "discrimination" is *mandated* by federal statute, see 10 U.S.C. § 654(b)(1) (mandating discharge from the Armed Forces of any service member who engages in or intends to engage in homosexual acts); and that in some cases such "discrimination" is a constitutional right, see *Boy Scouts of America v. Dale,* 530 U.S. 640 (2000).

Let me be clear that I have nothing against homosexuals, or any other group, promoting their agenda through normal democratic means. Social perceptions of sexual and other morality change over time, and every group has the right to persuade its fellow citizens that its view of such matters is the best. That homosexuals have achieved some success in that enterprise is attested to by the fact

*(Continues)*

that Texas is one of the few remaining States that criminalize private, consensual homosexual acts. But persuading one's fellow citizens is one thing, and imposing one's views in absence of democratic majority will is something else. I would no more *require* a State to criminalize homosexual acts—or, for that matter, display *any* moral disapproval of them—than I would *forbid* it to do so. What Texas has chosen to do is well within the range of traditional democratic action, and its hand should not be stayed through the invention of a brand-new "constitutional right" by a Court that is impatient of democratic change. It is indeed true that "later generations can see that laws once thought necessary and proper in fact serve only to oppress," *ante,* at 2484; and when that happens, later generations can repeal those laws. But it is the premise of our system that those judgments are to be made by the people, and not imposed by a governing caste that knows best. . . .

At the end of its opinion—after having laid waste the foundations of our rational-basis jurisprudence—the Court says that the present case "does not involve whether the government must give formal recognition to any relationship that homosexual persons seek to enter." *Ante,* at 2484. Do not believe it. More illuminating than this bald, unreasoned disclaimer is the progression of thought displayed by an earlier passage in the Court's opinion, which notes the constitutional protections afforded to "personal decisions relating to *marriage,* procreation, contraception, family relationships, child rearing, and education," and then declares that "[p]ersons in a homosexual relationship may seek autonomy for these purposes, just as heterosexual persons do." *Ante,* at 2482 (emphasis added). Today's opinion dismantles the structure of constitutional law that has permitted a distinction to be made between heterosexual and homosexual unions, insofar as formal recognition in marriage is concerned. If moral disapprobation of homosexual conduct is "no legitimate state interest" for purposes of proscribing that conduct, *ante,* at 2484; and if, as the Court coos (casting aside all pretense of neutrality), "[w]hen sexuality finds overt expression in intimate conduct with another person, the conduct can be but one element in a personal bond that is more enduring," *ante,* at 2478; what justification could there possibly be for denying the benefits of marriage to homosexual couples exercising "[t]he liberty protected by the Constitution"? Surely not the encouragement of procreation, since the sterile and the elderly are allowed to marry. This case "does not involve" the issue of homosexual marriage only if one entertains the belief that principle and logic have nothing to do with the decisions of this Court. Many will hope that, as the Court comfortingly assures us, this is so. . . .

## Notes and Questions

1. What is the fairest statement of the issue before the Court? Is it, as Justice White and Chief Justice Burger suggest in *Bowers v. Hardwick,* whether there is "a fundamental right to engage in homosexual sodomy"? Or, as Justice Blackmun suggests in dissent in that case, is it whether the Georgia statute prohibiting sodomy violates "constitutionally protected interests in privacy and freedom of intimate association"? Is Justice Kennedy's approach in *Lawrence v. Texas* preferable, involving "whether the petitioners were free as adults to engage in the private conduct in the exercise of their liberty under the Due Process Clause of the Fourteenth Amendment to the Constitution"? Sometimes, how a question is asked largely dictates how it will be answered.

2. Does the majority opinion in *Lawrence* concern itself with whether the Texas law infringes on a "fundamental" right? Why is that classification important? What does Justice Scalia have to say about this matter in his dissenting opinion in *Lawrence*?

3. What reasons can be offered in support of the statutes challenged in *Hardwick* and *Lawrence*? Are they sufficient to outweigh the individual interests affected?

4. Justice O'Connor's concurring opinion in *Lawrence* suggests an alternative theory for invalidating the Texas statute. Do you find merit in her analysis? What is the likelihood that a legislature would enact and prosecuting authorities would enforce a law prohibiting sodomy between opposite sex partners?

5. Justice White's majority opinion in *Hardwick* is explicit about the Court's reluctance "to take a more expansive view of our authority to discover new fundamental rights imbedded in the Due Process Clause. The Court is most vulnerable and comes nearest to illegitimacy when it deals with judge-made constitutional law having little or no cognizable roots in the language or design of the Constitution." Justice Scalia echoes these concerns in his dissent in *Lawrence.* Are the issues confronted in these cases properly within the realm of the judiciary in its role as guardian of constitutional rights? Are they better defined as policy issues that should be entrusted to democratically elected legislatures? What other types of legislation might be vulnerable to similar challenges? Is it possible to distinguish laws prohibiting the possession of drugs or firearms in the home, laws forbidding adultery or incest, or laws denying same-sex couples the right to marry?

# First Amendment and Free Speech

## Overbreadth Doctrine

We already examined the constitutional problems associated with vague criminal laws. Statutes challenged on grounds of vagueness frequently are simultaneously alleged to be unconstitutionally "overbroad." Although they are related in certain respects, the vices of vagueness and overbreadth are conceptually distinct. It is possible for a law to be quite precise, thus avoiding vagueness problems, yet still be overly broad and thus unconstitutional. The **overbreadth doctrine** usually is invoked in connection with the First Amendment's prohibition against laws "abridging the freedom of speech."

*[handwritten: ...extremely detailed (no vagueness) & still be overbroad.... HOW?]*

---

## State v. Lilburn, 265 Mont. 258, 875 P.2d 1036 (1994), cert. denied, 513 U.S. 1078, 115 S. Ct. 726, 130 L. Ed. 2d 630 (1995)

### Trieweiler, Justice.

Defendant John Lilburn was charged in the Gallatin County Justice Court with the offense of hunter harassment in violation of § 87-3-142(3), MCA. He was convicted of that charge following a jury trial and appealed his conviction to the District Court for the Eighteenth Judicial District in Gallatin County. The District Court held that § 87-3-142, MCA, in its entirety, is facially unconstitutional in that it is both overbroad and vague, impermissibly infringing on the First Amendment right to free speech and the Fourteenth Amendment right to due process guaranteed by the United States Constitution. . . .

In March 1990, the Department of Fish, Wildlife, and Parks (DFWP) allowed three persons whose names had been drawn from a permit pool to hunt bison which had migrated from Yellowstone National Park. One of the persons who received a permit was Hal Slemmer.

On the morning of the hunt, when the DFWP personnel located the bison, a group of 11 persons on snowmobiles and cross-country skis were seen attempting to herd the bison back into the park. The demonstrators were warned that this was a legal hunt, and were told not to interfere with the hunters. The hunters were also warned about the presence of the demonstrators and were cautioned to conduct the hunt safely.

Warden David Etzwiler of the DFWP accompanied Slemmer to a clearing where the bison were crossing. When one of the animals was in sight, Slemmer sighted his rifle and prepared to pull the trigger. At that time, John Lilburn, one of the protesters, moved in front of Slemmer, placing himself between Slemmer and the targeted bison at a distance of 10 to 12 feet from the muzzle of Slemmer's rifle. Slemmer lifted his rifle when he saw Lilburn's head and shoulders come into the scope of the gun. Warden Etzwiler approached Lilburn and told him that this was a lawful hunt and not to interfere. Slemmer moved about six feet to his left and selected another bison from the group. He raised his rifle and took aim through the scope. Lilburn again moved in front of Slemmer. Slemmer testified that when he saw Lilburn's face in his scope, he "jerked the gun up quickly because I had been squeezing on the trigger." . . .

No arrests were made at that time. However, after DFWP officials conferred with the Gallatin County Attorney, Lilburn was charged with the offense of harassment, a misdemeanor, in violation of § 87-3-142(3), MCA. The complaint filed against Lilburn in the Gallatin County Justice Court alleged that he disturbed a hunter with the intent to dissuade or prevent the taking of a bison when he placed himself between the bison and the hunter who was aiming a loaded rifle at the animal. . . .

Lilburn was convicted following a jury trial in Gallatin County Justice Court. He appealed his conviction to the District Court and alleged that the harassment statute was unconstitutionally overbroad and vague. . . . [T]he District Court reversed the conviction and dismissed the complaint brought against Lilburn based on its determination that § 87-3-142, MCA, is unconstitutional on its face, and therefore, is invalid. The State appeals.

### ISSUE 1

Is Montana's Hunter Harassment Law, found at § 87-3-142, MCA, void because it is overbroad in violation of the First Amendment to the United States Constitution?

The statute at issue in this appeal, commonly known as Montana's Hunter Harassment Law, provides as follows:

87-3-142. Harassment prohibited.
(1) No person may intentionally interfere with the lawful taking of a wild animal by another. . . .
(3) No person may disturb an individual engaged in the lawful taking of a wild animal with intent to dissuade the individual or otherwise prevent the taking of the animal. . . .

*(Continues)*

---

Lilburn was convicted of violating subsection (3) of this statute because he twice disturbed Slemmer's attempt to lawfully shoot a bison when he placed his body between Slemmer and the animal. The District Court, in its analysis of subsection (3) for overbreadth, concluded that § 87-3-142(3), MCA is "obviously content-based" because it "prohibits communication which is intended to dissuade a person from hunting, while allowing communication which encourages hunting." The court further concluded that the statute's prohibition would encompass "all verbal and expressive conduct which has the intention to dissuade from hunting," and therefore, such things as "prayer vigils at trailheads, the singing of protest songs or the burning of hunting maps, if done with the intent to dissuade a hunter, would be violations of the statute." Therefore, the court held that to the extent the statute "implicates constitutionally protected speech and expressive conduct, it is overbroad."

On appeal, the State contends that the court erred when it invalidated § 87-3-142(3), MCA, on the basis of overbreadth because the statute primarily proscribes conduct rather than speech, and to the extent that protected expression is reached, it regulates on a content-neutral basis only the time, place, and manner of expression. The State asserts that the statute is not overbroad because any potential unconstitutional applications are speculative and insubstantial when judged against the plainly legitimate scope of this statute which is to promote safety in sport hunting and protect those engaged in lawful activities from interference. We agree. . . .

In his overbreadth challenge, Lilburn disputes the State's assertion that the statute primarily regulates conduct but contends that it criminalizes a broad category of speech and expressive conduct based on its content. He claims that the law reaches primarily conduct which conveys an anti-hunting sentiment, while allowing, under exactly the same circumstances, conduct and speech which conveys any other message. Lilburn cites *R.A.V. v. St. Paul*, 505 U.S. 377 (1992) for the proposition that the statute is "facially unconstitutional in that it prohibits otherwise permitted speech solely on the basis of the subjects the speech addresses."

*Broadrick*, 413 U.S. 601 (1973), is the leading case addressing the First Amendment overbreadth doctrine. In *Broadrick*, the Supreme Court clarified that a statute or government regulation should be invalidated on the basis of facial overbreadth in only limited situations:

> In such cases, it has been the judgment of this Court that the possible harm to society in permitting some unprotected speech to go unpunished is outweighed by the possibility that protected speech of others may be muted and perceived grievances left to fester because of the possible inhibitory effects of overly broad statutes. . . .
>
> Application of the overbreadth doctrine in this manner is, manifestly, strong medicine. It has been employed by the Court sparingly and only as a last resort. Facial overbreadth has not been invoked when a limiting construction has been or could be placed on the challenged statute. [Citations omitted.]

*Broadrick*, 413 U.S. at 612-13. The Court in *Broadrick* adopted limitations on the overbreadth doctrine "particularly where conduct and not merely speech is involved," and held that a statute which has constitutional applications may be facially invalidated for overbreadth only if its overbreadth is "real, but substantial as well, judged in relation to the statute's plainly legitimate sweep." *Broadrick*, 413 U.S. at 615. The Court made clear that the existence of imaginary potential unlawful applications does not by itself render a statute facially overbroad.

In *Members of the City Council v. Taxpayers for Vincent* (1984), 466 U.S. 789, 800-01, the Supreme Court further explained the parameters of the overbreadth doctrine:

> It is clear . . . that the mere fact that one can conceive of some impermissible applications of a statute is not sufficient to render it susceptible to an overbreadth challenge. On the contrary, the requirement of substantial overbreadth stems from the underlying justification for the overbreadth exception itself—the interest in preventing an invalid statute from inhibiting the speech of third parties who are not before the Court. . . . In short, there must be a realistic danger that the statute itself will significantly compromise recognized First Amendment protections of parties not before the Court for it to be facially challenged on overbreadth grounds.

After reviewing the statute in question and the arguments set forth by Lilburn, we conclude that this is precisely the type of situation where the limitations imposed by the Supreme Court on the overbreadth doctrine must be carefully applied. Before the hunter harassment statute can be invalidated on its face, there must be a clear showing that the potential invalid applications of the statute be both "real and substantial." For the reasons stated below, we conclude that no such showing has been made in this case.

Under the tests articulated in *Broadrick* and *Taxpayers for Vincent*, we must determine whether there is a real and substantial probability that § 87-3-142(3), MCA, will compromise recognized First Amendment protections when judged in relation to any "plainly legitimate sweep" of the statute. . . .

Section 87-3-142(3), MCA, prohibits a person from disturbing another individual engaged in the lawful taking of a wild animal with intent to dissuade the individual or otherwise prevent the taking of the animal. The term "wild animal" is defined to mean "any game animal, fur-bearing animal, or predatory animal," and a "taking" is defined to include "pursuit, hunting, trapping, shooting, or killing of a wild animal on land upon which the affected person has the right or privilege to pursue, hunt, trap, shoot, or kill the wild animal." Section 87-3-141, MCA.

The plain language of the statute, considered in light of these limiting definitions, makes clear that the statute's proscriptions reach only activities which occur in the field during an otherwise lawful hunt. While the disturbance which is prohibited may, under other circumstances, result from a verbal utterance, it makes no difference what the content of the verbal utterance is. The language of the statute does not support the assertion that the statute is aimed primarily at pure speech and expressive conduct conveying only an anti-hunting sentiment. The disturbance could just as well be caused by shouting "fire!"

Lilburn disagrees that the statute regulates primarily conduct and claims that the Legislature's inclusion of the word "dissuade" demonstrates that the intent of the statute is to proscribe only a very small class of expression which is uttered or carried out with the intent to dissuade a hunter from taking an animal. He contends it is the Legislature's use of the term "dissuade" that renders this a content-based regulation.

The Supreme Court has provided clear guidelines for distinguishing a content-neutral regulation from one which is impermissibly content-based:

The principal inquiry in determining content neutrality . . . is whether the government has adopted a regulation of speech because of disagreement with the message it conveys. The government's purpose is the controlling consideration. A regulation that serves purposes unrelated to the content of expression is deemed neutral, even if it has an incidental effect on some speakers or messages but not others. [Citation omitted.]

*Ward v. Rock Against Racism* (1989), 491 U.S. 781, 791. The determination of whether a regulation is content-based turns not on whether its incidental effects fall more heavily on expression of a certain viewpoint, but rather on whether the governmental purpose to be served by the regulation is not motivated by a desire to suppress the content of the communication. *City of Renton v. Playtime Theatres, Inc.* (1986), 475 U.S. 41. Therefore, even if a statute has a discriminatory impact, it can be determined to be content-neutral if its objective neither advances nor inhibits a particular viewpoint.

Neither of the parties dispute the fact that safety and an orderly regulation of hunting are legitimate state goals. This Court has made clear that "[i]n the area of wildlife regulation, it is sufficient to state the Legislature may impose such terms and conditions as it sees fit, as long as constitutional limitations are not infringed." *State v. Jack* (1975) 539 P.2d 726, 728.

Here, the legislative history demonstrates a motivation for adoption of this statute which is unrelated to the suppression of speech based on content. The Legislature was aware that confrontations between hunters and opponents of sport hunting, particularly with respect to the controversial bison hunts, could occur in the field when hunters were armed and actively pursuing their prey. It was recognized that this posed a serious danger to both the hunters and those interfering with their activities.

Contrary to the court's conclusion that the legislation was obviously content-based because it was prompted by past activities opposing the bison hunts, the legislative history supports a conclusion that the motivation was to prevent violent confrontations and to prevent interference with lawful activities. Moreover, we do not find any support in the legislative history for the contention that this was an attempt to silence the views of those opposed to the bison hunt or other types of sport hunting. It was recognized that persons opposed to sport hunting had the right to express their views, but that there were other forums more suited to political discourse. . . .

Lilburn contends that there are a significant number of situations where the law could be applied in an unconstitutional manner and urges the Court to "use our imagination to think of the various ways the statute might be applied against speech or expressive conduct." However, the test is not whether hypothetical remote situations exist, but whether there is a significant possibility that the law will be unconstitutionally applied.

*(Continues)*

*(Continued)*

Based on our conclusion that the plain language of the statute is directed primarily at conduct and if at speech, then without regard to its content, we conclude, in the absence of evidence otherwise, that Lilburn has not shown that any overbreadth of the statute is "substantial . . . judged in relation to the statute's plainly legitimate sweep." *Broadrick,* 413 U.S. at 615. . . .

We hold that § 87-3-142(3), MCA, is not unconstitutionally overbroad. To the extent that the statute may reach constitutionally protected expression, we conclude, as did the Supreme Court in *Broadrick,* that whatever overbreadth may exist should be cured through case-by-case analysis of the fact situations where the statute is assertedly being applied unconstitutionally. . . .

## Notes and Questions

1. The portion of the opinion in *Lilburn* discussing *Broadrick v. Oklahoma,* 413 U.S. 610 (1973), helps explain one of the significant features of the overbreadth doctrine. Under appropriate circumstances a defendant charged with violating an overly broad statute may succeed in having the law declared unconstitutional even if his or her own conduct does not benefit by the protection of the First Amendment. The policy reasons in support of this exception to the normal rule of "standing" are explained in the quoted portions of *Broadrick,* as are the circumstances under which the exception is recognized. In light of these principles why did Lilburn not prevail on his claim that Montana's Hunter Harassment law is overly broad?

## "Offensive" Speech

How robust are the First Amendment protections relating to free expression? When the First Amendment says that "Congress shall make no law . . . abridging the freedom of speech, or of the press," how literally are those words to be interpreted? Through case decisions and after the adoption of the Fourteenth Amendment, the reach of the First Amendment has been extended to include the states. The Amendment also applies to individuals such as police officers, town managers, high school principals, and other official representatives of government.

How absolute are the terms of the First Amendment, in particular the admonition that "no law" shall be made abridging the freedom of speech? Are all forms of speech, no matter how offensive, immunized by the Constitution from criminal punishment? We consider these issues in the follow case involving **offensive speech**.

## *Cohen v. California,* 403 U.S. 15, 91 S. Ct. 1780, 29 L. Ed. 2d 284 (1971)

Mr. Justice Harlan delivered the opinion of the Court.

This case may seem at first blush too inconsequential to find its way into our books, but the issue it presents is of no small constitutional significance.

Appellant Paul Robert Cohen was convicted in the Los Angeles Municipal Court of violating that part of California Penal Code § 415 which prohibits "maliciously and willfully disturb[ing] the peace or quiet of any neighborhood or person . . . by . . . offensive conduct. . . ." He was given 30 days' imprisonment. The facts upon which his conviction rests are detailed in the opinion of the Court of Appeal of California, Second Appellate District, as follows:

"On April 26, 1968, the defendant was observed in the Los Angeles County Courthouse in the corridor outside of division 20 of the municipal court wearing a jacket bearing the words 'Fuck the Draft' which were plainly visible. There were women and children present in the corridor. The defendant was arrested. The defendant testified that he wore the jacket knowing that the words were on the jacket as a means of informing the public of the depth of his feelings against the Vietnam War and the draft.

"The defendant did not engage in, nor threaten to engage in, nor did anyone as the result of his conduct in fact commit or threaten to commit any act of violence. The defendant did not make any loud or unusual noise, nor was there any evidence that he uttered any sound prior to his arrest." 1 Cal App 3d 94, 97–98, 81 Cal Rptr 503, 505 (1969).

In affirming the conviction the Court of Appeal held that **offensive conduct** means "behavior which has a tendency to provoke *others* to acts of violence or to in turn disturb the peace," and that the State had proved this element because, on the facts of this case, "[i]t was certainly reasonably foreseeable that such conduct might cause others to rise up to commit a violent act against the person of the defendant or attempt to forcibly remove his jacket." 1 Cal App 3d, at 99–100, 81 Cal Rptr, at 506. The California Supreme Court declined review by a divided vote. . . .

We now reverse. . . .

## I.

In order to lay hands on the precise issue which this case involves, it is useful first to canvass various matters which this record does *not* present.

The conviction quite clearly rests upon the asserted offensiveness of the *words* Cohen used to convey his message to the public. The only "conduct" which the State sought to punish is the fact of communication. Thus, we deal here with a conviction resting solely upon "speech," not upon any separately identifiable conduct which allegedly was intended by Cohen to be perceived by others as expressive of particular views but which, on its face, does not necessarily convey any message and hence arguably could be regulated without effectively repressing Cohen's ability to express himself. Further, the State certainly lacks power to punish Cohen for the underlying content of the message the inscription conveyed. At least so long as there is no showing of an intent to incite disobedience to or disruption of the draft, Cohen could not, consistently with the First and Fourteenth Amendments, be punished for asserting the evident position on the inutility or immorality of the draft his jacket reflected. Yates v United States, 354 US 298 (1957).

Appellant's conviction, then, rests squarely upon his exercise of the "freedom of speech" protected from arbitrary governmental interference by the Constitution and can be justified, if at all, only as a valid regulation of the manner in which he exercised that freedom, not as a permissible prohibition on the substantive message it conveys. This does not end the inquiry, of course, for the First and Fourteenth Amendments have never been thought to give absolute protection to every individual to speak whenever or wherever he pleases, or to use any form of address in any circumstances that he chooses. In this vein, too, however, we think it important to note that several issues typically associated with such problems are not presented here.

In the first place, Cohen was tried under a statute applicable throughout the entire State. Any attempt to support this conviction on the ground that the statute seeks to preserve an appropriately decorous atmosphere in the courthouse where Cohen was arrested must fail in the absence of any language in the statute that would have put appellant on notice that certain kinds of otherwise permissible speech or conduct would nevertheless, under California law, not be tolerated in certain places. See Edwards v South Carolina, 372 US 229, 236–237, and n. 11, (1963); Cf. Adderley v Florida, 385 US 39 (1966). No fair reading of the phrase "offensive conduct" can be said sufficiently to inform the ordinary person that distinctions between certain locations are thereby created. (*See* Footnote 3.)

In the second place, as it comes to us, this case cannot be said to fall within those relatively few categories of instances where prior decisions have established the power of government to deal more comprehensively with certain forms of individual expression simply upon a showing that such a form was employed. This is not, for example, an obscenity case. Whatever else may be necessary to give rise to the States' broader power to prohibit obscene expression, such expression must be, in some significant way, erotic. It cannot plausibly be maintained that this vulgar allusion to the Selective Service System would conjure up such psychic stimulation in anyone likely to be confronted with Cohen's crudely defaced jacket.

This Court has also held that the States are free to ban the simple use, without a demonstration of additional justifying circumstances, of so-called "fighting words," those personally abusive epithets which, when addressed to the ordinary citizen, are, as a matter of common knowledge, inherently likely to provoke violent reaction. Chaplinsky v New Hampshire, 315 US 568 (1942). While the four-letter word displayed by Cohen in relation to the draft is not uncommonly employed in a personally provocative fashion, in this instance it was clearly not "directed to the person of the hearer." Cantwell v Connecticut, 310 US 296, 309 (1940). No individual actually or likely to be present could reasonably have regarded the words on appellant's jacket as a direct personal insult. Nor do we have here an instance of the exercise of the State's police power to prevent a speaker from intentionally provoking a given group to hostile reaction. Cf. Feiner v New York, 340 US 315 (1951); Terminiello v Chicago, 337 US 1 (1949). There is, as noted above, no showing that anyone who saw Cohen was in fact violently aroused or that appellant intended such a result.

*(Continues)*

Finally, in arguments before this Court much has been made of the claim that Cohen's distasteful mode of expression was thrust upon unwilling or unsuspecting viewers, and, that the State might therefore legitimately act as it did in order to protect the sensitive from otherwise unavoidable exposure to appellant's crude form of protest. Of course, the mere presumed presence of unwitting listeners or viewers does not serve automatically to justify curtailing all speech capable of giving offense. While this Court has recognized that government may properly act in many situations to prohibit intrusion into the privacy of the home of unwelcome views and ideas which cannot be totally banned from the public dialogue, we have at the same time consistently stressed that "we are often 'captives' outside the sanctuary of the home and subject to objectionable speech." The ability of government, consonant with the Constitution, to shut off discourse solely to protect others from hearing it is, in other words, dependent upon a showing that substantial privacy interests are being invaded in an essentially intolerable manner. Any broader view of this authority would effectively empower a majority to silence dissidents simply as a matter of personal predilections.

In this regard, persons confronted with Cohen's jacket were in a quite different posture than, say, those subjected to the raucous emissions of sound trucks blaring outside their residences. Those in the Los Angeles courthouse could effectively avoid further bombardment of their sensibilities simply by averting their eyes. And, while it may be that one has a more substantial claim to a recognizable privacy interest when walking through a courthouse corridor than, for example, strolling through Central Park, surely it is nothing like the interest in being free from unwanted expression in the confines of one's own home. Given the subtlety and complexity of the factors involved, if Cohen's "speech" was otherwise entitled to constitutional protection, we do not think the fact that some unwilling "listeners" in a public building may have been briefly exposed to it can serve to justify this breach of the peace conviction where, as here, there was no evidence that persons powerless to avoid appellant's conduct did in fact object to it, and where that portion of the statute upon which Cohen's conviction rests evinces no concern, either on its face or as construed by the California courts, with the special plight of the captive auditor, but, instead, indiscriminately sweeps within its prohibitions all "offensive conduct" that disturbs "any neighborhood or person."

## II.

Against this background, the issue flushed by this case stands out in bold relief. It is whether California can excise, as "offensive conduct," one particular scurrilous epithet from the public discourse, either upon the theory of the court below that its use is inherently likely to cause violent reaction or upon a more general assertion that the States, acting as guardians of public morality, may properly remove this offensive word from the public vocabulary.

The rationale of the California court is plainly untenable. At most it reflects an "undifferentiated fear or apprehension of disturbance [which] is not enough to overcome the right to freedom of expression." Tinker v Des Moines Indep. Community School Dist., 393 US 503, 508 (1969). We have been shown no evidence that substantial numbers of citizens are standing ready to strike out physically at whoever may assault their sensibilities with execrations like that uttered by Cohen. There may be some persons about with such lawless and violent proclivities, but that is an insufficient base upon which to erect, consistently with constitutional values, a governmental power to force persons who wish to ventilate their dissident views into avoiding particular forms of expression. The argument amounts to little more than the self-defeating proposition that to avoid physical censorship of one who has not sought to provoke such a response by a hypothetical coterie of the violent and lawless, the States may more appropriately effectuate that censorship themselves.

Admittedly, it is not so obvious that the First and Fourteenth Amendments must be taken to disable the States from punishing public utterance of this unseemly expletive in order to maintain what they regard as a suitable level of discourse within the body politic. We think, however, that examination and reflection will reveal the shortcomings of a contrary viewpoint.

At the outset, we cannot overemphasize that, in our judgment, most situations where the State has a justifiable interest in regulating speech will fall within one or more of the various established exceptions, discussed above but not applicable here, to the usual rule that governmental bodies may not prescribe the form or content of individual expression. Equally important to our conclusion is the constitutional backdrop against which our decision must be made. The constitutional right of free expression is powerful medicine in a society as diverse and populous as ours. It is designed and intended to remove governmental restraints from the arena of public discussion, putting the decision as to what views shall be voiced largely into the hands of each of us, in the hope that use of such freedom will ultimately produce a more capable citizenry and more perfect polity and in the belief that no other approach would comport with the premise of individual dignity and choice upon which our political system rests. See Whitney v California, 274 US 357, 375–377 (1927) (Brandeis, J., concurring).

To many, the immediate consequence of this freedom may often appear to be only verbal tumult, discord, and even offensive utterance. These are, however, within established limits, in truth necessary side effects of the broader enduring values which the process of open debate permits us to achieve. That the air may at times seem filled with verbal cacophony is, in this sense not a sign of weakness but of strength. We cannot lose sight of the fact that, in what otherwise might seem a trifling and annoying instance of individual distasteful abuse of a privilege, these fundamental societal values are truly implicated.

Against this perception of the constitutional policies involved, we discern certain more particularized considerations that peculiarly call for reversal of this conviction. First, the principle contended for by the State seems inherently boundless. How is one to distinguish this from any other offensive word? Surely the State has no right to cleanse public debate to the point where it is grammatically palatable to the most squeamish among us. Yet no readily ascertainable general principle exists for stopping short of that result were we to affirm the judgment below. For, while the particular four-letter word being litigated here is perhaps more distasteful than most others of its genre, it is nevertheless often true that one man's vulgarity is another's lyric. Indeed, we think it is largely because governmental officials cannot make principled distinctions in this area that the Constitution leaves matters of taste and style so largely to the individual.

Additionally, we cannot overlook the fact, because it is well illustrated by the episode involved here, that much linguistic expression serves a dual communicative function: it conveys not only ideas capable of relatively precise, detached explication, but otherwise inexpressible emotions as well. In fact, words are often chosen as much for their emotive as their cognitive force. We cannot sanction the view that the Constitution, while solicitous of the cognitive content of individual speech, has little or no regard for that emotive function which, practically speaking, may often be the more important element of the overall message sought to be communicated. . . .

Finally, and in the same vein, we cannot indulge the facile assumption that one can forbid particular words without also running a substantial risk of suppressing ideas in the process. Indeed, governments might soon seize upon the censorship of particular words as a convenient guise for banning the expression of unpopular views. We have been able, as noted above, to discern little social benefit that might result from running the risk of opening the door to such grave results.

It is, in sum, our judgment that, absent a more particularized and compelling reason for its actions, the State may not, consistently with the First and Fourteenth Amendments, make the simple public display here involved of this single four-letter expletive a criminal offense. Because that is the only arguably sustainable rationale for the conviction here at issue, the judgment below must be

Reversed

Mr. Justice Blackmun, with whom The Chief Justice and Mr. Justice Black join.

I dissent. . . .

Cohen's absurd and immature antic, in my view, was mainly conduct and little speech. Further, the case appears to me to be well within the sphere of Chaplinsky v New Hampshire, 315 US 568 (1942), where Mr. Justice Murphy, a known champion of First Amendment freedoms, wrote for a unanimous bench. As a consequence, this Court's agonizing First Amendment values seems misplaced and unnecessary. . . .

## Notes and Questions

1. Cohen was not prosecuted for disagreeing with U.S. draft policies but rather for the particular way in which he expressed his disagreement. Would it be unreasonable to expect Cohen to choose other language to convey his sentiments?
2. What if, on being arrested, Cohen had directed some of the language emblazoned on his jacket at the arresting police officer? Could he rely on the same First Amendment principles discussed in Justice Harlan's opinion for immunity from prosecution? *See Robinson v. State,* 588 N.E.2d 533 (Ind. App. 1992) (holding, by 2–1 decision, that the First Amendment does not prevent the disorderly conduct conviction of a defendant who told a police officer to "get the fuck away," and that his investigation was "bullshit," and called the officer a "lying motherfucker," because such expression constitutes "fighting words" that do not enjoy constitutional protection).

(Continues)

(*Continued*)

3. In August 1998 Timothy Boomer fell out of his canoe into a Northern Michigan river and unleashed a string of curse words overheard by a married couple and their two young children. He was prosecuted for violating a state law dating back to 1897 that provides, "Any person who shall use any indecent, immoral, obscene, vulgar or insulting language in the presence or hearing of any woman or child shall be guilty of a misdemeanor." Mich. Stat. Ann. § 750.337. The statute had never been construed by the state courts. *See Prak v. Gregart*, 749 F. Supp. 825 (W.D. Mich. 1990). Boomer was convicted after a jury trial in June 1999.[4] On appeal, should the conviction in "the case of the cursing canoeist," as the episode commonly was called, be upheld? Is the statute unconstitutionally vague or overly broad? Does *Cohen v. California* recognize that curse words uttered under such circumstances are a protected form of speech under the First Amendment? Does the statute's inclusion of women as a "protected class" seem problematic? How would you rule?

## Constitutional Issues in Anti-Terrorism Laws

Since September 11, 2001 the United States has been engaged in a "war on terror." Much of the controversy surrounding the prosecution and detention of alleged terrorists has centered on issues related to searches and seizures, indefinite detention, torture, and other items associated with criminal procedure. These issues are beyond the scope of this book. However, several matters related to fighting terrorism fall squarely within the field of substantive criminal law.

Two of the most controversial statutes upon which anti-terrorism prosecutions have been based, 18 U.S.C.A. §§ 2339 A and B, involve providing material support to terrorists and terrorist organizations, respectively.

18 U.S.C. § 2339A. Providing material support to terrorists

(a) Whoever provides material support or resources or conceals or disguises the nature, location, source, or ownership of material support or resources, knowing or intending that they are to be used in preparation for, or in carrying out, a violation of [certain enumerated criminal offenses] or in preparation for, or in carrying out, the concealment of an escape from the commission of any such violation, or attempts or conspires to do such an act, shall be fined under this title or imprisoned.

18 U.S.C. § 2339B. Providing material support or resources to designated foreign terrorist organizations

(a)(1) Whoever knowingly provides material support or resources to a foreign terrorist organization, or attempts or conspires to do so, shall be fined under this title or imprisoned.

Although these statutes were part of the Antiterrorism and Effective Death Penalty Act of 1996, they have been increasingly used to prosecute individuals since the 9/11 attacks. Prosecution for providing material support to terrorists or terrorist organizations has raised a number of constitutional issues we discussed in this chapter, including the First Amendment rights of speech and association and Fifth Amendment due process matters related to their potential vagueness and overbreadth. *United States v. Hammoud*, presented below, provides a good example of how courts have considered these issues.

## *United States v. Hammoud*, 381 F.3d 316 (4th Cir. 2004), vacated on other grounds, 543 U.S. 1097, 125 S. Ct. 1051, 160 L. Ed. 2d 997 (2005)

William W. Wilkins, Chief Judge.

. . . The facts underlying Hammoud's convictions and sentence are largely undisputed. We therefore recount them briefly. . . .

Hizballah is an organization founded by Lebanese Shi'a Muslims in response to the 1982 invasion of Lebanon by Israel. Hizballah provides various forms of humanitarian aid to Shi'a Muslims in Lebanon. However, it is also a strong opponent of Western presence in the Middle East, and it advocates the use of terrorism in support of its agenda. Hizballah is particularly opposed to the existence of Israel and to the activities of the American government in the Middle East. Hizballah's general secretary is Hassan Nasserallah, and its spiritual leader is Sheikh Fadlallah. . . .

In 1992, Hammoud, a citizen of Lebanon, attempted to enter the United States on fraudulent documents. After being detained by the INS, Hammoud sought asylum. While the asylum application was pending, Hammoud moved to Charlotte, North Carolina, where his brothers and cousins were living. Hammoud ultimately obtained permanent resident status by marrying a United States citizen.

At some point in the mid-1990s, Hammoud, his wife, one of his brothers, and his cousins all became involved in a cigarette smuggling operation. The conspirators purchased large quantities of cigarettes in North Carolina, smuggled them to Michigan, and sold them without paying Michigan taxes. This scheme took advantage of the fact that Michigan imposes a tax of $7.50 per carton of cigarettes, while the North Carolina tax is only 50 cents. It is estimated that the conspiracy involved a quantity of cigarettes valued at roughly $7.5 million and that the state of Michigan was deprived of $3 million in tax revenues.

In 1996, Hammoud began leading weekly prayer services for Shi'a Muslims in Charlotte. These services were often conducted at Hammoud's home. At these meetings, Hammoud—who is acquainted with both Nasserallah and Fadlallah, as well as Sheikh Abbas Harake, a senior military commander for Hizballah—urged the attendees to donate money to Hizballah. Hammoud would then forward the money to Harake. The Government's evidence demonstrated that on one occasion, Hammoud donated $3,500 of his own money to Hizballah.

Based on these and other activities, Hammoud was charged with various immigration violations, sale of contraband cigarettes, money laundering, mail fraud, credit card fraud, and racketeering. Additionally, Hammoud was charged with conspiracy to provide material support to a designated FTO [Foreign Terrorist Organization] and with providing material support to a designated FTO, both in violation of 18 U.S.C.A. § 2339B. The latter § 2339B charge related specifically to Hammoud's personal donation of $3,500 to Hizballah.

At trial, one of the witnesses against Hammoud was Said Harb, who grew up in the same Lebanese neighborhood as Hammoud. Harb testified regarding his own involvement in the cigarette smuggling operation and also provided information regarding the provision of "dual use" equipment (such as global positioning systems, which can be used for both civilian and military activities) to Hizballah. The Government alleged that this conduct was part of the conspiracy to provide material support to Hizballah. Harb testified that Hammoud had declined to become involved in providing equipment because he was helping Hizballah in his own way. Harb also testified that when he traveled to Lebanon in September 1999, Hammoud gave him $3,500 for Hizballah.

[The jury convicted Hammoud of 14 offenses, and the judge imposed the maximum permissible sentence for each count, totaling 155 years in prison.] . . .

*IL Constitutionality of 18 U.S C.A. § 2339B*

Section 2339B, which was enacted as part of the Antiterrorism and Effective Death Penalty Act of 1996 (AEDPA), provides for a maximum penalty of 15 years imprisonment for any person who "knowingly provides material support or resources to a foreign terrorist organization, or attempts or conspires to do so." 18 U.S.C.A. § 2339B(a)(1). The term "material support" is defined as "currency or other financial securities, financial services, lodging, training, safehouses, false documentation or identification, communications equipment, facilities, weapons, lethal substances, explosives, personnel, transportation, and other physical assets, except medicine or religious materials."

Hammoud maintains that § 2339B is unconstitutional in a number of respects. . . .

*A. Freedom of Association*

Hammoud first contends that § 2339B impermissibly restricts the First Amendment **right of association**. Hammoud concedes (at least for purposes of this argument) that Hizballah engages in terrorist activity. But, he also notes the undisputed fact that Hizballah provides humanitarian aid to citizens of Lebanon. Hammoud argues that because Hizballah engages in both legal and illegal activities, he can be found criminally liable for providing material support to Hizballah only if he had a specific intent to further the organization's illegal aims. Because § 2339B lacks such a specific intent requirement, Hammoud argues that it unconstitutionally restricts the freedom of association.

It is well established that "[t]he First Amendment . . . restricts the ability of the State to impose liability on an individual solely because of his association with another." *NAACP v. Claiborne Hardware Co.*, 458 U.S. 886, 918-19 (1982); *see Scales v. United States*, 367 U.S. 203, 229 (1961) (noting that a

*(Continues)*

"blanket prohibition of association with a group having both legal and illegal aims . . . [would pose] a real danger that legitimate political expression or association would be impaired"). Therefore, it is a violation of the First Amendment to punish an individual for mere membership in an organization that has legal and illegal goals. Any statute prohibiting association with such an organization must require a showing that the defendant specifically intended to further the organization's unlawful goals. Hammoud maintains that because § 2339B does not contain such a specific intent requirement, his conviction violates the First Amendment.

Hammoud's argument fails because § 2339B does not prohibit mere association; it prohibits the *conduct* of providing material support to a designated FTO. Therefore, cases regarding mere association with an organization do not control. Rather, the governing standard is found in *United States v. O'Brien*, 391 U.S. 367 (1968), which applies when a facially neutral statute restricts some expressive conduct. Such a statute is valid if it is within the constitutional power of the Government; if it furthers an important or substantial governmental interest; if the governmental interest is unrelated to the suppression of free expression; and if the incidental restriction on alleged First Amendment freedoms is no greater than is essential to the furtherance of that interest.

Section 2339B satisfies all four prongs of the *O'Brien* test. First, § 2339B is clearly within the constitutional power of the government, in view of the government's authority to regulate interactions between citizens and foreign entities. *See Regan v. Wald*, 468 U.S. 222, 244 (1984) (holding that restrictions on travel to Cuba do not violate the Due Process Clause). Second, there can be no question that the government has a substantial interest in curbing the spread of international terrorism. *See Humanitarian Law Project v. Reno*, 205 F.3d 1130, 1135 (9th Cir.2000). Third, the Government's interest in curbing terrorism is unrelated to the suppression of free expression. Hammoud is free to advocate in favor of Hizballah or its political objectives—§ 2339B does not target such advocacy.

Fourth and finally, the incidental effect on expression caused by § 2339B is no greater than necessary. In enacting § 2339B and its sister statute, 18 U.S.C.A. § 2339A, Congress explicitly found that "foreign organizations that engage in terrorist activity are so tainted by their criminal conduct that any contribution to such an organization facilitates that conduct." AEDPA § 301(a)(7). As the Ninth Circuit reasoned,

> [i]t follows that all material support given to [foreign terrorist] organizations aids their unlawful goals. Indeed, . . . terrorist organizations do not maintain open books. Therefore, when someone makes a donation to them, there is no way to tell how the donation is used. Further, . . . even contributions earmarked for peaceful purposes can be used to give aid to the families of those killed while carrying out terrorist acts, thus making the decision to engage in terrorism more attractive. More fundamentally, money is fungible; giving support intended to aid an organization's peaceful activities frees up resources that can be used for terrorist acts. *Humanitarian Law Project*, 205 F.3d at 1136 (footnote omitted).

In light of this reasoning, the prohibition on material support is adequately tailored to the interest served and does not suppress more speech than is necessary to further the Government's legitimate goal. We therefore conclude that § 2339B does not infringe on the constitutionally protected right of free association.

### B. Overbreadth

Hammoud next argues that § 2339B is overbroad. A statute is overbroad only if it "punishes a substantial amount of protected free speech, judged in relation to the statute's plainly legitimate sweep." *Virginia v. Hicks*, 539 U.S. 113, 118-19 (2003) (internal quotation marks omitted). The overbreadth must be substantial "not only in an absolute sense, but also relative to the scope of the law's plainly legitimate applications." *Id.* at 120. It is also worth noting that when, as here, a statute is addressed to conduct rather than speech, an overbreadth challenge is less likely to succeed. *See id.* at 124 ("Rarely, if ever, will an overbreadth challenge succeed against a law or regulation that is not specifically addressed to speech or to conduct necessarily associated with speech (such as picketing or demonstrating).").

Hammoud argues that § 2339B is overbroad because (1) it prohibits mere association with an FTO, and (2) it prohibits such plainly legitimate activities as teaching members of an FTO how to apply for grants to further the organization's humanitarian aims. As discussed above, § 2339B does not prohibit mere association with an FTO and therefore is not overbroad on that basis. Regarding Hammoud's second overbreadth argument, it may be true that the material support prohibition of § 2339B encompasses some forms of expression that are entitled to First Amendment protection. *Cf. Humanitarian Law Project,* 205 F.3d at 1138 (holding that "training" prong of material support definition is vague because it covers such forms of protected expression as "instruct[ing] members of a designated group on how to petition the United Nations to give aid to their group"). Hammoud has utterly failed to demonstrate, however, that any overbreadth is substantial in relation to the legitimate reach of § 2339B. *See Hicks,* 539 U.S. at 122 ("The overbreadth claimant bears the burden of demonstrating, from the text of the law and from actual fact, that substantial overbreadth exists." (alteration & internal quotation marks omitted)).

*C. Vagueness*

Hammoud next argues that the term "material support" is unconstitutionally vague. "The void-for-vagueness doctrine requires that penal statutes define crimes so that ordinary people can understand the conduct prohibited and so that arbitrary and discriminatory enforcement is not encouraged." *United States v. McLamb,* 985 F.2d 1284, 1291 (4th Cir.1993). In evaluating whether a statute is vague, a court must consider both whether it provides notice to the public and whether it adequately curtails arbitrary enforcement. *See Kolender v. Lawson,* 461 U.S. 352, 357-58 (1983).

Section 2339B easily satisfies this standard. As noted above, the term "material support" is specifically defined as a number of enumerated actions. Hammoud relies on *Humanitarian Law Project,* in which the Ninth Circuit ruled that two components of the material support definition— "personnel" and "training"—were vague. *See Humanitarian Law Project,* 205 F.3d at 1137-38. The possible vagueness of these prongs of the material support definition does not affect Hammoud's conviction, however, because he was specifically charged with providing material support in the form of currency. *See United States v. Rahman,* 189 F.3d 88, 116 (2d Cir.1999) (per curiam) (rejecting vagueness challenge because allegedly vague term was not relevant to Appellant's conviction). There is nothing at all vague about the term "currency." . . .

*IX. Conclusion*

For the reasons set forth above, we reject each of Hammoud's challenges to his convictions and sentence. We therefore affirm the judgment of the district court in its entirety. . . .

### Notes and Questions

1. What exactly did Hammoud do (what was his conduct) that violated the law?
2. Would Hammoud have fared better if he were an American citizen?
3. Do you believe the statutes in question impose an incidental restriction on the freedom of speech that is "no greater than is essential"? Should this standard change during wartime?

## Equal Protection of the Laws

The Fourteenth Amendment provides, in relevant part, that no State shall "deny any person within its jurisdiction the equal protection of the laws." The U.S. Supreme Court has interpreted the Fifth Amendment's Due Process Clause as requiring the federal government to observe equal protection principles as well. *See Bolling v. Sharpe,* 347 U.S. 497 (1954). Although it makes sense that the criminal laws should apply equally throughout a jurisdiction, it was not always so during this country's history. For example, before the conclusion of the Civil War many southern states had dual systems of law for whites and blacks. Thus in Georgia black slaves who killed whites automatically were executed, even if they acted in self-defense. Punishments differed markedly according to whether a crime was committed by a slave, a "free person of color," or a white. For instance, a black man convicted of raping a white woman received a mandatory death sentence,

rape committed by anyone else against a white woman was punishable by a prison term of 2 to 20 years, and the rape of a black woman was punishable by imprisonment and a fine, at the court's discretion. *McCleskey v. Kemp,* 481 U.S. 279, 329–30 (1987) (Brennan, J., dissenting). Even after the Civil War ended, some southern states continued to maintain dual legislative standards for crime and punishment, known as "Black Codes," based on race.[5]

Under contemporary laws, equal protection challenges arise more frequently in the context of allegations that, *as applied,* a statute operates in an impermissibly discriminatory manner. On occasion, however, statutes still are written so that they have unequal effect. The normal justification offered is that real differences exist between the classes of people affected; because equal protection principles only require that people who are similar in relevant respects be treated similarly, there are no logical or legal problems associated with laws that have a different impact on people who truly are different.

## Michael M. v. Superior Court of Sonoma County, 450 U.S. 464, 101 S. Ct. 1200, 67 L. Ed. 2d 437 (1981)

Justice Rehnquist announced the judgment of the Court and delivered an opinion, in which The Chief Justice, Justice Stewart, and Justice Powell joined.

The question presented in this case is whether California's "statutory rape" law, § 261.5 of the Cal Penal Code Ann (West Supp 1981), violates the Equal Protection Clause of the Fourteenth Amendment. Section 261.5 defines unlawful sexual intercourse as "an act of sexual intercourse accomplished with a female not the wife of the perpetrator, where the female is under the age of 18 years." The statute thus makes men alone criminally liable for the act of sexual intercourse.

In July 1978, a complaint was filed in the Municipal Court of Sonoma County, Cal., alleging that petitioner, then a 17½-year-old male, had had unlawful sexual intercourse with a female under the age of 18, in violation of § 261.5. The evidence adduced at a preliminary hearing showed that at approximately midnight on June 3, 1978, petitioner and two friends approached Sharon, a 16½-year-old female, and her sister as they waited at a bus stop. Petitioner and Sharon, who had already been drinking, moved away from the others and began to kiss. After being struck in the face for rebuffing petitioner's initial advances, Sharon submitted to sexual intercourse with petitioner. Prior to trial, petitioner sought to set aside the information on both state and federal constitutional grounds, asserting that § 261.5 unlawfully discriminated on the basis of gender. The trial court and the California Court of Appeal denied petitioner's request for relief and petitioner sought review in the Supreme Court of California.

The Supreme Court held that "section 261.5 discriminates on the basis of sex because only females may be victims, and only males may violate the section." 601 P2d 572, 574. The court then subjected the classification to "strict scrutiny," stating that it must be justified by a compelling state interest. It found that the classification was "supported not by mere social convention but by the immutable physiological fact that it is the female exclusively who can become pregnant." Ibid. Canvassing "the tragic human costs of illegitimate teenage pregnancies," including the large number of teenage abortions, the increased medical risk associated with teenage pregnancies, and the social consequences of teenage child-bearing, the court concluded that the State has a compelling interest in preventing such pregnancies. Because males alone can "physiologically cause the result which the law properly seeks to avoid," the court further held that the gender classification was readily justified as a means of identifying offender and victim. For the reasons stated below, we affirm the judgment of the California Supreme Court. . . .

Unlike the California Supreme Court, we have not held that gender-based classifications are "inherently suspect" and thus we do not apply so-called "strict scrutiny" to those classifications. Our cases have held, however, that the traditional minimum rationality test takes on a somewhat "sharper focus" when gender-based classifications are challenged. In Reed v Reed, 404 US 71 (1971), for example, the Court stated that a gender-based classification will be upheld if it bears a "fair and substantial relationship" to legitimate state ends, while in Craig v Boren [429 U.S. 190], 197 [(1976)], the Court restated the test to require the classification to bear a "substantial relationship" to "important governmental objectives."

Underlying these decisions is the principle that a legislature may not "make overbroad generalizations based on sex which are entirely unrelated to any differences between men and women or which demean the ability or social status of the affected class."

But because the Equal Protection Clause does not "demand that a statute necessarily apply equally to all persons" or require "'things which are different in fact . . . to be treated in law as though they were the same,'" this Court has consistently upheld statutes where the gender classification is not invidious, but rather realistically reflects the fact that the sexes are not similarly situated in certain circumstances. As the Court has stated, a legislature may "provide for the special problems of women."

Applying those principles to this case, the fact that the California Legislature criminalized the act of illicit sexual intercourse with a minor female is a sure indication of its intent or purpose to discourage that conduct. Precisely why the legislature desired that result is of course somewhat less clear. This Court has long recognized that "[i]nquiries into congressional motives or purposes are a hazardous matter," and the search for the "actual" or "primary" purpose of a statute is likely to be elusive. Here, for example, the individual legislators may have voted for the statute for a variety of reasons. Some legislators may have been concerned about preventing teenage pregnancies, others about protecting young females from physical injury or from the loss of "chastity," and still others about promoting various religious and moral attitudes towards premarital sex.

The justification for the statute offered by the State, and accepted by the Supreme Court of California, is that the legislature sought to prevent illegitimate teenage pregnancies. That finding, of course, is entitled to great deference. And although our cases establish that the State's asserted reason for the enactment of a statute may be rejected, if it "could not have been a goal of the legislation," this is not such a case.

We are satisfied not only that the prevention of illegitimate pregnancy is at least one of the "purposes" of the statute, but also that the State has a strong interest in preventing such pregnancy. At the risk of stating the obvious, teenage pregnancies, which have increased dramatically over the last two decades, have significant social, medical, and economic consequences for both the mother and her child, and the State. Of particular concern to the State is that approximately half of all teenage pregnancies end in abortion. And of those children who are born, their illegitimacy makes them likely candidates to become wards of the State.

We need not be medical doctors to discern that young men and young women are not similarly situated with respect to the problems and the risks of sexual intercourse. Only women may become pregnant, and they suffer disproportionately the profound physical, emotional, and psychological consequences of sexual activity. The statute at issue here protects women from sexual intercourse at an age when those consequences are particularly severe.

The question thus boils down to whether a State may attack the problem of sexual intercourse and teenage pregnancy directly by prohibiting a male from having sexual intercourse with a minor female. We hold that such a statute is sufficiently related to the State's objectives to pass constitutional muster.

Because virtually all of the significant harmful and inescapably identifiable consequences of teenage pregnancy fall on the young female, a legislature acts well within its authority when it elects to punish only the participant who, by nature, suffers few of the consequences of his conduct. It is hardly unreasonable for a legislature acting to protect minor females to exclude them from punishment. Moreover, the risk of pregnancy itself constitutes a substantial deterrence to young females. No similar natural sanctions deter males. A criminal sanction imposed solely on males thus serves to roughly "equalize" the deterrents on the sexes.

We are unable to accept petitioner's contention that the statute is impermissibly underinclusive and must, in order to pass judicial scrutiny, be *broadened* so as to hold the female as criminally liable as the male. It is argued that this statute is not *necessary* to deter teenage pregnancy because a gender-neutral statute, where both male and female would be subject to prosecution, would serve that goal equally well. The relevant inquiry, however, is not whether the statute is drawn as precisely as it might have been, but whether the line chosen by the California Legislature is within constitutional limitations.

In any event, we cannot say that a gender-neutral statute would be as effective as the statute California has chosen to enact. The State persuasively contends that a gender-neutral statute would frustrate its interest in effective enforcement. Its view is that a female is surely less likely to report violations of the statute if she herself would be subject to criminal prosecution. In an area already fraught with prosecutorial difficulties, we decline to hold that the Equal Protection Clause requires a legislature to enact a statute so broad that it may well be incapable of enforcement. . . .

There remains only petitioner's contention that the statute is unconstitutional as it is applied to him because he, like Sharon, was under 18 at the time of sexual intercourse. Petitioner argues that the statute is flawed because it presumes that as between two persons under 18, the male is the culpable aggressor. We find petitioner's contentions unpersuasive. Contrary to his assertions, the statute does not rest on the assumption that males are generally the aggressors. It is instead an attempt by a legislature to prevent illegitimate teenage pregnancy by providing an additional deterrent for men. The age of the man is irrelevant since young men are as capable as older men of inflicting the harm sought to be prevented. . . .

Accordingly, the judgment of the California Supreme Court is
Affirmed. . . .
Justice Brennan, with whom Justices White and Marshall join, dissenting. . . .

After some uncertainty as to the proper framework for analyzing equal protection challenges to statutes containing gender-based classifications, this Court settled upon the proposition that a statute containing a

*(Continues)*

gender-based classification cannot withstand constitutional challenge unless the classification is substantially related to the achievement of an important governmental objective. This analysis applies whether the classification discriminates against males or against females. The burden is on the government to prove both the importance of its asserted objective and the substantial relationship between the classification and that objective. And the State cannot meet that burden without showing that a gender-neutral statute would be a less effective means of achieving that goal.

The State of California vigorously asserts that the "important governmental objective" to be served by § 261.5 is the prevention of teenage pregnancy. It claims that its statute furthers this goal by deterring sexual activity by males—the class of persons it considers more responsible for causing those pregnancies. But even assuming that prevention of teenage pregnancy is an important governmental objective and that it is in fact an objective of § 261.5, California still has the burden of proving that there are fewer teenage pregnancies under its gender-based statutory rape law than there would be if the law were gender neutral. To meet this burden, the State must show that because its statutory rape law punishes only males, and not females, it more effectively deters minor females from having sexual intercourse.

The plurality assumes that a gender-neutral statute would be less effective than § 261.5 in deterring sexual activity because a gender-neutral statute would create significant enforcement problems. . . .

However, a State's bare assertion that its gender-based statutory classification substantially furthers an important governmental interest is not enough to meet its burden of proof under Craig v Boren. Rather, the State must produce evidence that will persuade the Court that its assertion is true.

The State has not produced such evidence in this case. Moreover, there are at least two serious flaws in the State's assertion that law enforcement problems created by a gender-neutral statutory rape law would make such a statute less effective than a gender-based statute in deterring sexual activity.

First, the experience of other jurisdictions, and California itself, belies the plurality's conclusion that a gender-neutral statutory rape law "may well be incapable of enforcement." There are now at least 37 States that have enacted gender-neutral statutory rape laws. . . . California has introduced no evidence that those States have been handicapped by the enforcement problems the plurality finds so persuasive. Surely, if those States could provide such evidence, we might expect that California would have introduced it. . . .

The second flaw in the State's assertion is that even assuming that a gender-neutral statute would be more difficult to enforce, the State has still not shown that those enforcement problems would make such a statute less effective than a gender-based statute in deterring minor females from engaging in sexual intercourse. Common sense, however, suggests that a gender-neutral statutory rape law is potentially a *greater* deterrent of sexual activity than a gender-based law, for the simple reason that a gender-neutral law subjects both men and women to criminal sanctions and thus arguably has a deterrent effect on twice as many potential violators. Even if fewer persons were prosecuted under the gender-neutral law, as the State suggests, it would still be true that twice as many persons would be *subject* to arrest. The State's failure to prove that a gender-neutral law would be a less effective deterrent than a gender-based law, like the State's failure to prove that a gender-neutral law would be difficult to enforce, should have led this Court to invalidate § 261.5.

Until very recently, no California court or commentator had suggested that the purpose of California's statutory rape law was to protect young women from the risk of pregnancy. Indeed, the historical development of § 261.5 demonstrates that the law was initially enacted on the premise that young women, in contrast to young men, were to be deemed legally incapable of consenting to an act of sexual intercourse. Because their chastity was considered particularly precious, those young women were felt to be uniquely in need of the State's protection. In contrast, young men were assumed to be capable of making such decisions for themselves; the law therefore did not offer them any special protection.

It is perhaps because the gender classification in California's statutory rape law was initially designed to further these outmoded sexual stereotypes, rather than to reduce the incidence of teenage pregnancies, that the State has been unable to demonstrate a substantial relationship between the classification and its newly asserted goal. But whatever the reason, the State has not shown that Cal Penal Code § 261.5 is any more effective than a gender-neutral law would be in deterring minor females from engaging in sexual intercourse. It has therefore not met its burden of proving that the statutory classification is substantially related to the achievement of its asserted goal. . . .

*[handwritten note: seriousness: Harm done & degree of culpability.]*

# Constitutional Limits on the Government's Power to Punish

## Proportionality

### Andrew von Hirsch. "Doing Justice: The Choice of Punishments."

If one asks how severely a wrongdoer deserves to be punished, a familiar principle comes to mind: *Severity of punishment should be commensurate with the seriousness of the wrong.* Only grave wrongs merit severe penalties; minor misdeeds deserve lenient punishments. Disproportionate penalties are undeserved—severe sanctions for minor wrongs or vice versa. This principle has variously been called a principle of "**proportionality**" or "**just deserts**"; we prefer to call it *commensurate deserts,* a phrase that better suggests the concepts involved. In the most obvious cases, the principle seems a truism (who would wish to imprison shoplifters for life, or let murderers off with small fines?). Yet, whether and how it should be applied in allocating punishments has been in dispute. . . .

The principle looks retrospectively to the seriousness of the offender's past crime or crimes. "Seriousness" depends both on the harm done (or risked) by the act and on the degree of the actor's culpability. (When we speak of the seriousness of "the crime," we wish to stress that we are *not* looking exclusively to the act, but also to how much the actor can be held to blame for the act and its consequences.) If the offender had a prior criminal record at the time of conviction, the number and gravity of those prior crimes should be taken into account in assessing seriousness.

*Source:* Excerpts from "The Principle of Commensurate Deserts," from *Doing Justice: The Choice of Punishments* by Andrew von Hirsch. Copyright © 1976 by Andrew von Hirsch. Reprinted by permission of Hill and Wang, a division of Farrar, Straus and Giroux, LLC.

In Chapter 1 we introduced the "just deserts" principle and the related postulate that punishments should be calibrated to correspond to the seriousness of a crime. As the passage quoted above suggests, "seriousness" typically is measured both in terms of the harm caused (or risked) by the offender and the offender's culpability. Culpability normally is thought to have some relationship to *mens rea*; harms intentionally caused generally reflect a higher degree of culpability than unintentional harms. Perhaps more controversial, although doubtless reflected in sentencing practices, is the suggestion that an offender's prior criminal conduct, or record, should be considered when assessing the offender's culpability for a new offense.

In this section we are spared from analyzing the numerous policy decisions offered in support of particular legislative punishment schemes or scrutinizing what accounts for a judge's sentencing decision in an individual case. Our task is different, although not necessarily less complex. In keeping with our examination of *constitutional* limitations on legislative and judicial *power* to punish, we do not deal directly with what makes a punishment system or a particular criminal sentence seem more or less defensible, fair, or effective. The constitutional provision most relevant to our inquiry is the Eighth Amendment's ban on "**cruel and unusual punishments.**"

We first attempt to give more specific meaning to the concept of cruel and unusual punishments by examining death penalty cases. We then consider whether noncapital sentences can violate the Eighth Amendment's prohibition.

For nearly two centuries in this country's history, the power of the states and the federal government to impose capital punishment went virtually unchallenged. Although the death penalty was not used at all times in all states, several crimes were punishable by death well into the 20th century in a majority of jurisdictions. Murder was by far the most likely to result in a death sentence, but kidnapping, rape, burglary, treason, espionage, and a few other offenses remained capital crimes in many states and/or under federal law.

In 1972 in *Furman v. Georgia,* 408 U.S. 238 (1972), a 5–4 majority of the U.S. Supreme Court ruled that capital punishment, as then administered, violated the Eighth Amendment's cruel and unusual punishments clause. The *Furman* decision invalidated essentially all death penalty laws then in effect throughout the country. The justices did not rule that capital punishment was inherently unconstitutional. Three of the five justices in the tenuous *Furman* "majority" (each of the five wrote an individual opinion not joined by the others) identified *procedural* infirmities with how the death penalty was administered. Only Justices Brennan and Marshall were of the opinion that capital punishment was under all circumstances unconstitutional. Under the laws in effect when *Furman*

was decided, juries had unregulated discretion to choose between a death sentence and a prison sentence after an offender's conviction for a capital crime. They received no instructions or guidance about what factors were legally relevant to such a momentous decision, and they often had precious little information to help inform their sentencing decisions.

The states quickly went to work in the wake of *Furman* to revise their death penalty procedures. In 1976 the U.S. Supreme Court announced its approval of the capital punishment statutes from three states that had enacted two major procedural reforms. First, the death penalty was reserved for a narrower range of legislatively defined offenses. Second, a separate sentencing proceeding was held at the conclusion of the guilt-determination trial. This separate penalty trial allowed for a wide array of information relevant to punishment to be presented to the jury or judge, who retained the discretion to sentence the offender to death or life imprisonment. *See Gregg v. Georgia,* 428 U.S. 153 (1976); *Proffitt v. Florida,* 428 U.S. 242 (1976); *Jurek v. Texas,* 428 U.S. 262 (1976). At the same time the Court declared unconstitutional statutes from two states that removed all discretion from the sentencing authority and mandated capital punishment on conviction. *See Woodson v. North Carolina,* 428 U.S. 280 (1976); *Roberts v. Louisiana,* 428 U.S. 325 (1976).

A host of subsidiary issues made their way to the Supreme Court after the justices' approval in 1976 of the revised "guided discretion" death penalty statutes. Directly on the heels of the Court's 1976 death penalty decisions, all of which involved capital sentences imposed for murder, the justices were asked to decide whether capital punishment for the crime of rape was "cruel and unusual" under the Eighth Amendment. The offender, Ehrlich Anthony Coker, had raped a 16-year-old married woman. He argued that the death penalty was constitutionally excessive—or disproportionate—as punishment for this crime.

*[handwritten margin note: no guidelines]*

*[handwritten note: Furman v. Georgia: uninstructed capital punishment unconstit...]*

## Coker v. Georgia, 433 U.S. 584, 97 S. Ct. 2861, 53 L. Ed. 2d 982 (1977)

Mr. Justice White announced the judgment of the Court and filed an opinion in which Mr. Justice Stewart, Mr. Justice Blackmun, and Mr. Justice Stevens, joined. . . .

### I.

While serving various sentences for murder, rape, kidnapping, and aggravated assault, petitioner escaped from the Ware Correctional Institution near Waycross, Ga., on September 2, 1974. At approximately 11 o'clock that night, petitioner entered the house of Allen and Elnita Carver through an unlocked kitchen door. Threatening the couple with a "board," he tied up Mr. Carver in the bathroom, obtained a knife from the kitchen, and took Mr. Carver's money and the keys to the family car. Brandishing the knife and saying "you know what's going to happen to you if you try anything, don't you," Coker then raped Mrs. Carver. Soon thereafter, petitioner drove away in the Carver car, taking Mrs. Carver with him. Mr. Carver, freeing himself, notified the police; and not long thereafter petitioner was apprehended. Mrs. Carver was unharmed.

Petitioner was charged with escape, armed robbery, motor vehicle theft, kidnaping, and rape. . . . The jury returned a verdict of guilty, rejecting his general plea of insanity. A sentencing hearing was then conducted in accordance with the procedures dealt with at length in Gregg v Georgia, 428 US 153 (1976), where this Court sustained the death penalty for murder when imposed pursuant to the statutory procedures. The jury was instructed that it could consider as aggravating circumstances whether the rape had been committed by a person with a prior record of conviction for a capital felony and whether the rape had been committed in the course of committing another capital felony, namely, the armed robbery of Allen Carver. The court also instructed, pursuant to statute, that even if aggravating circumstances were present, the death penalty need not be imposed if the jury found they were outweighed by mitigating circumstances, that is, circumstances not constituting justification or excuse for the offense in question, "but which, in fairness and mercy, may be considered as extenuating or reducing the degree" of moral culpability or punishment. The jury's verdict on the rape count was death by electrocution. Both aggravating circumstances on which the court instructed were found to be present by the jury.

## II.

It is now settled that the death penalty is not invariably cruel and unusual punishment within the meaning of the Eighth Amendment. . . .

In sustaining the imposition of the death penalty in Gregg, however, the Court firmly embraced the holdings and dicta from prior cases, to the effect that the Eighth Amendment bars not only those punishments that are "barbaric" but also those that are "excessive" in relation to the crime committed. Under Gregg, a punishment is "excessive" and unconstitutional if it (1) makes no measurable contribution to acceptable goals of punishment and hence is nothing more than the purposeless and needless imposition of pain and suffering; or (2) is grossly out of proportion to the severity of the crime. A punishment might fail the test on either ground. Furthermore, these Eighth Amendment judgments should not be, or appear to be, merely the subjective views of individual Justices; judgment should be informed by objective factors to the maximum possible extent. To this end, attention must be given to the public attitudes concerning a particular sentence—history and precedent, legislative attitudes, and the response of juries reflected in their sentencing decisions are to be consulted. In Gregg, after giving due regard to such sources, the Court's judgment was that the death penalty for deliberate murder was neither the purposeless imposition of severe punishment nor a punishment grossly disproportionate to the crime. But the Court reserved the question of the constitutionality of the death penalty when imposed for other crimes.

## III.

That question, with respect to rape of an adult woman, is now before us. We have concluded that a sentence of death is grossly disproportionate and excessive punishment for the crime of rape and is therefore forbidden by the Eighth Amendment as cruel and unusual punishment. (*See* Footnote 4.)

### A.

As advised by recent cases, we seek guidance in history and from the objective evidence of the country's present judgment concerning the acceptability of death as a penalty for rape of an adult woman. At no time in the last 50 years has a majority of the States authorized death as a punishment for rape. In 1925, 18 States, the District of Columbia, and the Federal Government authorized capital punishment for the rape of an adult female. By 1971 just prior to the decision in Furman v Georgia, that number had declined, but not substantially, to 16 States plus the Federal Government. Furman then invalidated most of the capital punishment statutes in this country, including the rape statutes, because, among other reasons, of the manner in which the death penalty was imposed and utilized under those laws.

With their death penalty statutes for the most part invalidated, the States were faced with the choice of enacting modified capital punishment laws in an attempt to satisfy the requirements of Furman or of being satisfied with life imprisonment as the ultimate punishment for *any* offense. Thirty-five States immediately reinstituted the death penalty for at least limited kinds of crime. This public judgment as to the acceptability of capital punishment, evidenced by the immediate, post-Furman legislative reaction in a large majority of the States, heavily influenced the Court to sustain the death penalty for murder in Gregg v Georgia.

But if the "most marked indication of society's endorsement of the death penalty for murder is the legislative response to Furman," Gregg v Georgia, supra, at 179–180, it should also be a telling datum that the public judgment with respect to rape, as reflected in the statutes providing the punishment for that crime, has been dramatically different. In reviving death penalty laws to satisfy Furman's mandate, none of the States that had not previously authorized death for rape chose to include rape among capital felonies. Of the 16 States in which rape had been a capital offense, only three provided

*(Continues)*

the death penalty for rape of an adult woman in their revised statutes—Georgia, North Carolina, and Louisiana. In the latter two States, the death penalty was mandatory for those found guilty, and those laws were invalidated by Woodson and Roberts. When Louisiana and North Carolina, responding to those decisions, again revised their capital punishment laws, they reenacted the death penalty for murder but not for rape; none of the seven other legislatures that to our knowledge have amended or replaced their death penalty statutes since July 2, 1976, including four States (in addition to Louisiana and North Carolina) that had authorized the death sentence for rape prior to 1972 and had reacted to Furman with mandatory statutes, included rape among the crimes for which death was an authorized punishment. . . .

The upshot is that Georgia is the sole jurisdiction in the United States at the present time that authorizes a sentence of death when the rape victim is an adult woman, and only two other jurisdictions provide capital punishment when the victim is a child. The current judgment with respect to the death penalty for rape is not wholly unanimous among state legislatures, but it obviously weighs very heavily on the side of rejecting capital punishment as a suitable penalty for raping an adult woman.

## B.

It was also observed in Gregg that "[t]he jury . . . is a significant and reliable objective index of contemporary values because it is so directly involved," 428 US, at 181, and that it is thus important to look to the sentencing decisions that juries have made in the course of assessing whether capital punishment is an appropriate penalty for the crime being tried. . . .

According to the factual submissions in this Court, out of all rape convictions in Georgia since 1973—and that total number has not been tendered—63 cases had been reviewed by the Georgia Supreme Court as of the time of oral argument; and of these, six involved a death sentence, one of which was set aside, leaving five convicted rapists now under sentence of death in the State of Georgia. Georgia juries have thus sentenced rapists to death six times since 1973. This obviously is not a negligible number; and the State argues that as a practical matter juries simply reserve the extreme sanction for extreme cases of rape and that recent experience surely does not prove that jurors consider the death penalty to be a disproportionate punishment for every conceivable instance of rape, no matter how aggravated. Nevertheless, it is true that in the vast majority of cases, at least 9 out of 10, juries have not imposed the death sentence.

## IV.

These recent events evidencing the attitude of state legislatures and sentencing juries do not wholly determine this controversy, for the Constitution contemplates that in the end our own judgment will be brought to bear on the question of the acceptability of the death penalty under the Eighth Amendment. Nevertheless, the legislative rejection of capital punishment for rape strongly confirms our own judgment, which is that death is indeed a disproportionate penalty for the crime of raping an adult woman.

We do not discount the seriousness of rape as a crime. It is highly reprehensible, both in a moral sense and in its almost total contempt for the personal integrity and autonomy of the female victim and for the latter's privilege of choosing those with whom intimate relationships are to be established. Short of homicide, it is the "ultimate violation of self." It is also a violent crime because it normally involves force, or the threat of force or intimidation, to overcome the will and the capacity of the victim to resist. Rape is very often accompanied by physical injury to the female and can also inflict mental and psychological damage. Because it undermines the community's sense of security, there is public injury as well.

Rape is without doubt deserving of serious punishment; but in terms of moral depravity and of the injury to the person and to the public, it does not compare with murder, which does involve the unjustified taking of human life. Although it may be accompanied by another crime, rape by definition does not include the death of or even the serious injury to another person. The murderer kills; the rapist, if no more than that, does not. Life is over for the victim of the murderer; for the rape victim, life may not be nearly so happy as it was, but it is not over and normally is not beyond repair. We have the abiding conviction that the death penalty, which "is unique in its severity and irrevocability," is an excessive penalty for the rapist who, as such, does not take human life.

This does not end the matter, for under Georgia law, death may not be imposed for any capital offense, including rape, unless the jury or judge finds one of the statutory aggravating circumstances and then elects to impose that sentence. For the rapist to be executed in Georgia, it must therefore be found not only that he committed rape but also that one or more of the following aggravating circumstances were present; (1) that the rape was committed by a person with a prior record of conviction for a capital felony; (2) that the rape was committed while the offender was engaged in the commission of another capital felony, or aggravated battery; or (3) the rape "was outrageously or wantonly vile, horrible or inhuman in that it involved torture, depravity of mind, or aggravated battery to the victim." Here, the first two of these aggravating circumstances were alleged and found by the jury.

Neither of these circumstances, nor both of them together, change our conclusion that the death sentence imposed on Coker is a disproportionate punishment for rape. Coker had prior convictions for capital felonies—rape, murder, and kidnaping—but these prior convictions do not change the fact that the instant crime being punished is a rape not involving the taking of life. . . .

We note finally that in Georgia a person commits murder when he unlawfully and with malice aforethought, either express or implied, causes the death of another human being. He also commits that crime when in the commission of a felony he causes the death of another human being, irrespective of malice. But even where the killing is deliberate, it is not punishable by death absent proof of aggravating circumstances. It is difficult to accept the notion, and we do not, that the rapist, with or without aggravating circumstances, should be punished more heavily than the deliberate killer as long as the rapist does not himself take the life of his victim. The judgment of the Georgia Supreme Court upholding the death sentence is reversed, and the case is remanded to that court for further proceedings not inconsistent with this opinion.

So ordered. . . .

Mr. Justice Powell, concurring in the judgment in part and dissenting in part.

I concur in the judgment of the Court on the facts of this case, and also in the plurality's reasoning supporting the view that ordinarily death is disproportionate punishment for the crime of raping an adult woman. Although rape invariably is a reprehensible crime, there is no indication that petitioner's offense was committed with excessive brutality or that the victim sustained serious or lasting injury. The plurality, however, does not limit its holding to the case before us or to similar cases. Rather, in an opinion that ranges well beyond what is necessary, it holds that capital punishment *always*—regardless of the circumstances—is a disproportionate penalty for the crime of rape. . . . It is . . . quite unnecessary for the plurality to write in terms so sweeping as to foreclose each of the 50 state legislatures from creating a narrowly defined substantive crime of aggravated rape punishable by death. (*See* Footnote 1.) . . .

The deliberate viciousness of the rapist may be greater than that of the murderer. Rape is never an act committed accidentally. Rarely can it be said to be unpremeditated. There also is wide variation in the effect on the victim. The plurality opinion says that "[l]ife is over for the victim of the murderer; for the rape victim, life may not be nearly so happy as it was, but it is not over and normally is not beyond repair." But there is indeed "extreme variation" in the crime of rape. Some victims are so grievously injured physically or psychologically that life is beyond repair.

Thus, it may be that the death penalty is not disproportionate punishment for the crime of aggravated rape. Final resolution of the question must await careful inquiry into objective indicators of society's "evolving standards of decency," particularly legislative enactments and the responses of juries in capital cases. . . .

In a proper case a more discriminating inquiry than the plurality undertakes well might discover that both juries and legislatures have reserved the ultimate penalty for the case of an outrageous rape resulting in serious, lasting harm to the victim. I would not prejudge the issue. To this extent, I respectfully dissent.

Mr. Chief Justice Burger, with whom Mr. Justice Rehnquist joins, dissenting. . . .

## (1)

On December 5, 1971, the petitioner, Ehrlich Anthony Coker, raped and then stabbed to death a young woman. Less than eight months later Coker kidnaped and raped a second young woman. After twice raping this 16-year-old victim, he stripped her, severely beat her with a club, and dragged her into a wooded area where he left her for dead. He was apprehended and pleaded guilty to offenses stemming from these incidents. He was sentenced by three separate courts to three life terms, two 20-year terms, and one 8-year term of imprisonment. Each judgment specified that the sentences it imposed were to run consecutively rather than concurrently. Approximately 1 ½ years later, on

*(Continues)*

September 2, 1974, petitioner escaped from the state prison where he was serving these sentences. He promptly raped another 16-year-old woman in the presence of her husband, abducted her from her home, and threatened her with death and serious bodily harm. It is this crime for which the sentence now under review was imposed.

The Court today holds that the State of Georgia may not impose the death penalty on Coker. In so doing, it prevents the State from imposing any effective punishment upon Coker for his latest rape. The Court's holding, moreover, bars Georgia from guaranteeing its citizens that they will suffer no further attacks by this habitual rapist. In fact, given the lengthy sentences Coker must serve for the crimes he has already committed, the Court's holding assures that petitioner—and others in his position—will henceforth feel no compunction whatsoever about committing further rapes as frequently as he may be able to escape from confinement and indeed even within the walls of the prison itself. To what extent we have left States "elbow room" to protect innocent persons from depraved human beings like Coker remains in doubt.

## (2)

. . . Unlike the plurality, I would narrow the inquiry in this case to the question actually presented: Does the Eighth Amendment's ban against cruel and unusual punishment prohibit the State of Georgia from executing a person who has, within the space of three years, raped three separate women, killing one and attempting to kill another, who is serving prison terms exceeding his probable lifetime and who has not hesitated to escape confinement at the first available opportunity? Whatever one's view may be as to the State's constitutional power to impose the death penalty upon a rapist who stands before a court convicted for the first time, this case reveals a chronic rapist whose continuing danger to the community is abundantly clear. . . .

Surely recidivism, especially the repeated commission of heinous crimes, is a factor which may properly be weighed as an aggravating circumstance, permitting the imposition of a punishment more severe than for one isolated offense. . . .

As a factual matter, the plurality opinion is correct in stating that Coker's "prior convictions do not change the fact that the instant crime being punished is a rape not involving the taking of life"; however, it cannot be disputed that the existence of these prior convictions makes Coker a substantially more serious menace to society than a first-time offender. . . .

In my view, the Eighth Amendment does not prevent the State from taking an individual's "well-demonstrated propensity for life-endangering behavior" into account in devising punitive measures which will prevent inflicting further harm upon innocent victims. . . .

## (3)

The plurality acknowledges the gross nature of the crime of rape. A rapist not only violates a victim's privacy and personal integrity, but inevitably causes serious psychological as well as physical harm in the process. The long-range effect upon the victim's life and health is likely to be irreparable; it is impossible to measure the harm which results. . . . Rape is not a mere physical attack—it is destructive of the human personality. The remainder of the victim's life may be gravely affected, and this in turn may have a serious detrimental effect upon her husband and any children she may have. . . . To speak blandly, as the plurality does, of rape victims who are "unharmed," or to classify the human outrage of rape, as does Mr. Justice Powell, in terms of "excessive[ly] brutal," versus "moderately brutal," takes too little account of the profound suffering the crime imposes upon the victims and their loved ones.

Despite its strong condemnation of rape, the Court reaches the inexplicable conclusion that "the death penalty . . . is an excessive penalty" for the perpetrator of this heinous offense. This, the Court holds, is true even though in Georgia the death penalty may be imposed only where the rape is coupled with one or more aggravating circumstances. . . .

## (a)

The plurality opinion bases its analysis, in part, on the fact that "Georgia is the sole jurisdiction in the United States at the present time that authorizes a sentence of death when the rape victim is an adult woman." Surely, however, this statistic cannot be deemed determinative, or even particularly relevant. . . .

More to the point, however, it is myopic to base sweeping constitutional principles upon the narrow experience of the past five years. Considerable uncertainty was introduced into this area of the law by this Court's Furman decision. . . .

Failure of more States to enact statutes imposing death for rape of an adult woman may thus reflect hasty legislative compromise occasioned by time pressures following Furman, a desire to wait on the experience of those States which did enact such statutes, or simply an accurate forecast of today's holding.

In any case, when considered in light of the experience since the turn of this century, where more than one-third of American jurisdictions have consistently provided the death penalty for rape, the plurality's focus on the experience of the immediate past must be viewed as truly disingenuous. . . .

The Court has repeatedly pointed to the reserve strength of our federal system which allows state legislatures, within broad limits, to experiment with laws, both criminal and civil, in the effort to achieve socially desirable results. . . .

Statutory provisions in criminal justice applied in one part of the country can be carefully watched by other state legislatures, so that the experience of one State becomes available to all. Although human lives are in the balance, it must be remembered that failure to allow flexibility may also jeopardize human lives—those of the victims of undeterred criminal conduct. . . .

It is difficult to believe that Georgia would long remain alone in punishing rape by death if the next decade demonstrated a drastic reduction in its incidence of rape, an increased cooperation by rape victims in the apprehension and prosecution of rapists, and a greater confidence in the rule of law on the part of the populace. . . .

## (b)

The subjective judgment that the death penalty is simply disproportionate for the crime of rape is even more disturbing than the "objective" analysis discussed supra. The plurality's conclusion on this point is based upon the bare fact that murder necessarily results in the physical death of the victim, while rape does not. However, no Member of the Court explains why this distinction has relevance, much less constitutional significance. It is, after all, not irrational—nor constitutionally impermissible— for a legislature to make the penalty more severe than the criminal act it punishes in the hope it would deter wrongdoing. . . .

Until now, the issue under the Eighth Amendment has not been the state of any particular victim after the crime, but rather whether the punishment imposed is grossly disproportionate to the evil committed by the perpetrator. As a matter of constitutional principle, that test cannot have the primitive simplicity of "life for life, eye for eye, tooth for tooth." Rather States must be permitted to engage in a more sophisticated weighing of values in dealing with criminal activity which consistently poses serious danger of death or grave bodily harm. . . .

The clear implication of today's holding appears to be that the death penalty may be properly imposed only as to crimes resulting in death of the victim. This casts serious doubt upon the unconstitutional [sic] validity of statutes imposing the death penalty for a variety of conduct which, though dangerous, may not necessarily result in any immediate death, e.g., treason, airplane hijacking, and kidnaping. . . . We cannot avoid judicial notice that crimes such as airplane hijacking, kidnaping, and mass terrorist activity constitute a serious and increasing danger to the safety of the public. It would be unfortunate indeed if the effect of today's holding were to inhibit States and the Federal Government from experimenting with various remedies—including possibly imposition of the penalty of death—to prevent and deter such crimes.

Some sound observations, made only a few years ago, deserve repetition:

"Our task here, as must so frequently be emphasized and re-emphasized, is to pass upon the constitutionality of legislation that has been enacted and that is challenged. This is the sole task for judges. We should not allow our personal preferences as to the wisdom of legislative and congressional action, or our distaste for such action, to guide our judicial decision in cases such as these. The temptations to cross that policy line are very great. In fact, as today's decision reveals, they are almost irresistible." Furman v Georgia, 408 US, at 411 . . . (Blackmun, J., dissenting).

Whatever our individual views as to the wisdom of capital punishment, I cannot agree that it is constitutionally impermissible for a state legislature to make the "solemn judgment" to impose such penalty for the crime of rape. . . .

## Notes and Questions

1. Justice White's plurality opinion in *Coker* stresses that the Court's judgment about whether a punishment is constitutionally excessive or disproportionate to the crime for which it is imposed, "should not be, or appear to be, merely the subjective views of individual Justices; judgment should be informed by objective factors to the maximum possible extent." Return to the "objective" reasons

*(Continues)*

offered by Justice White in support of the conclusion that Coker's death sentence violated the Eighth Amendment. Do you find these reasons persuasive, or do you share Chief Justice Burger's skepticism about them? If reliance on "the subjective views of individual Justices" is to be avoided, why, in part IV of his opinion, does Justice White insist that "the Constitution contemplates that in the end our own judgment will be brought to bear" on the issue of disproportionate punishment?

2. How do Justice Powell's views differ from those of Justice White's? In light of the circumstances of Coker's crime, why should the rape he committed not be considered an "aggravated" one?

3. Chief Justice Burger charges that the Court's decision "prevents the State from imposing any effective punishment on Coker for his latest rape." If Georgia is denied using the death penalty, is Coker essentially immune from additional punishment? How might Justice White respond to this issue?

4. Between 1930 and 1967 approximately 407 of the 455 people (89.5 percent) executed in this country for rape were nonwhite.[7] This history of racial disproportionality in capital punishment for rape received no mention in *Coker*. (Coker was white.) Is this history irrelevant to a contemporary assessment of whether capital punishment is constitutionally unacceptable for rape?[8]

5. Although Coker's victim was just 16 years old, she was married and the justices characterized her several times as an "adult." Under *Coker*, is capital punishment necessarily disproportionate for the crime of raping a child? The Supreme Court considered this question in *Kennedy v. Louisiana*, 128 S. Ct. 2641 (2008). Patrick Kennedy was convicted of raping his 8-year-old stepdaughter and was sentenced to death under a Louisiana statute that authorized capital punishment for raping a child under the age of 13. Five other states—Georgia, Montana, Oklahoma, South Carolina, and Texas— similarly had adopted laws allowing the death penalty for raping a child, although Louisiana and Georgia were alone in not requiring that the defendant have a previous rape conviction before becoming death-penalty eligible. In a 5–4 decision authored by Justice Kennedy the Supreme Court relied on the same mode of analysis used in *Coker* to invalidate the statute and the death sentence under review. The majority opinion concluded that because 44 states and the federal death penalty law did not permit capital punishment for the crime of raping a child, along with the fact that only two individuals had been sentenced to death for this offense since 1964, was "evidence of a national consensus" against this practice. "[T]he history of the death penalty for this and other nonhomicide crimes" helped confirm this judgment. The justices then embarked on an independent inquiry, which confirmed their consultation of objective measures and reinforced the view that death is a constitutionally excessive punishment for a crime of this nature. Conceding that the rape of a child is an abhorrent offense, the majority opinion broadly concluded that, "[a]s it relates to crimes against individuals, . . . the death penalty should not be expanded to instances where the victim's life was not taken." Justice Kennedy noted that this holding "is limited to crimes against individual persons. We do not address, for example, crimes defining and punishing treason, espionage, terrorism, and drug kingpin activity, which are offenses against the State." The opinion additionally noted that "by in effect making the punishment for child rape and murder equivalent, a State that punishes child rape by death may remove a strong incentive for the rapist not to kill the victim. Assuming the offender behaves in a rational way, as one must to justify the penalty on grounds of deterrence, the penalty in some respects gives less protection, not more, to the victim, who is often the sole witness to the crime."

Justice Alito's dissenting opinion, which was joined by Chief Justice Roberts and Justices Scalia and Thomas, argued that the recent adoption by six states of the death penalty for this crime reflected a trend that the Court's opinion had prematurely stunted and that six was not an insignificant number of states that had chosen to punish child rape with death in light of *Coker*'s uncertain precedent, which might have discouraged other jurisdictions from enacting legislation. The dissent also took the majority opinion to task for invading the province of legislatures. "The Court's policy arguments concern matters that legislators should—and presumably do—take into account in deciding whether to enact a capital child-rape statute, but these arguments are irrelevant to the question that is before us in this case. Our cases have cautioned against using '"the aegis of the Cruel and Unusual Punishment Clause" to cut off the normal democratic processes,' but the Court forgets that warning here."

How would you vote on this issue if you were a legislator? And how would you vote if you were a member of the Supreme Court?

Proportionality issues also arise outside of the death penalty context. For example, if a man stole three golf clubs and, on his conviction for felony grand theft, was sentenced to life imprisonment with no parole eligibility for 25 years, could he successfully argue that his punishment is constitutionally excessive and hence violates the Eighth Amendment? Would it matter if he had a substantial prior criminal record, bringing him within the terms of the penalty-enhancement provisions of "three strikes" legislation?

## *Ewing v. California*, 538 U.S. 11, 123 S. Ct. 1179, 155 L. Ed. 2d 108 (2003)

[In March 2000, Gary Ewing stole three golf clubs worth $399 apiece from a Los Angeles County golf course pro shop. He had been paroled from prison the prior year. His additional criminal history included a 1984 theft conviction; a 1988 conviction for felony grand theft auto (subsequently dismissed on his completion of probation); a 1990 petty theft conviction; separate battery and theft convictions in 1992; a January 1993 burglary conviction; a February 1993 conviction for possession of drug paraphernalia; a July 1993 conviction for appropriating lost property; September 1993 convictions for possessing a firearm and trespassing; and December 1993 convictions for three burglaries and robbery for which he received the nine-year prison sentence from which he was paroled in 1999.

[California had adopted "three strikes" legislation in 1994. Under this law, "[i]f the defendant has one prior 'serious' or 'violent' felony conviction, he must be sentenced to 'twice the term otherwise provided as punishment for the current felony conviction.' If the defendant has two or more prior 'serious' or 'violent' felony convictions, he must receive 'an indeterminate term of life imprisonment'" with varying dates for parole eligibility. Prosecutors and trial courts are given the discretion to treat certain offenses, known as "wobblers," as either misdemeanors or felonies for purposes of the three strikes sentencing provisions.

[Following Ewing's conviction for felony grand theft in connection with the stolen golf clubs, the prosecutor alleged and the trial court found that Ewing had been convicted of four serious or violent felonies for the 1993 burglaries and robbery. He was sentenced to 25 years to life imprisonment following his felony grand theft conviction. The state courts upheld his sentence on appeal and the U.S. Supreme Court granted certiorari to "decide whether the Eighth Amendment prohibits the State of California from sentencing a repeat felon to a prison term of 25 years to life under the State's 'Three Strikes and You're Out' law."]

Justice O'Connor announced the judgment of the Court and delivered an opinion, in which the Chief Justice and Justice Kennedy join. . . .

### II.

### A.

The Eighth Amendment, which forbids cruel and unusual punishments, contains a "narrow proportionality principle" that "applies to noncapital sentences." *Harmelin v. Michigan*, 501 U.S. 957, 996-997 (1991) (Kennedy, J., concurring in part and concurring in judgment). We have most recently addressed the proportionality principle as applied to terms of years in a series of cases beginning with *Rummel v. Estelle*, [445 U.S. 263 (1980)].

In *Rummel*, we held that it did not violate the Eighth Amendment for a State to sentence a three-time offender to life in prison with the possibility of parole. . . . Like Ewing, Rummel was sentenced to a lengthy prison term under a recidivism statute. Rummel's two prior offenses were a 1964 felony for "fraudulent use of a credit card to obtain $80 worth of goods or services," and a 1969 felony conviction for "passing a forged check in the amount of $28.36." His triggering offense was a conviction for felony theft—"obtaining $120.75 by false pretenses." . . .

In *Hutto v. Davis*, 454 U.S. 370 (1982) (*per curiam*), the defendant was sentenced to two consecutive terms of 20 years in prison for possession with intent to distribute nine ounces of marijuana and distribution of marijuana. We held that such a sentence was constitutional . . . .

Three years after *Rummel*, in *Solem v. Helm*, 463 U.S. 277, 279 (1983), we held that the Eighth Amendment prohibited "a life sentence without possibility of parole for a seventh nonviolent felony." The triggering offense in *Solem* was "uttering a 'no account' check for $100." We specifically stated that the Eighth Amendment's ban on cruel and unusual punishments "prohibits . . . sentences that are disproportionate to the crime committed," and that the "constitutional principle of proportionality has been recognized explicitly in this Court for almost a century." The *Solem* Court then explained that three factors may be relevant to a determination of whether a sentence is so disproportionate that it violates the Eighth Amendment: "(i) the gravity of the offense and the harshness of the penalty; (ii) the sentences imposed on other criminals in the same jurisdiction; and (iii) the sentences imposed for commission of the same crime in other jurisdictions."

Applying these factors in *Solem*, we struck down the defendant's sentence of life without parole. We specifically noted the contrast between that sentence and the sentence in *Rummel*, pursuant to which the defendant was eligible for parole. . . .

*(Continues)*

*(Continued)*

Eight years after *Solem*, we grappled with the proportionality issue again in *Harmelin*. *Harmelin* was not a recidivism case, but rather involved a first-time offender convicted of possessing 672 grams of cocaine. He was sentenced to life in prison without possibility of parole. A majority of the Court rejected Harmelin's claim that his sentence was so grossly disproportionate that it violated the Eighth Amendment. The Court, however, could not agree on why his proportionality argument failed. . . .

Justice Kennedy, joined by two other Members of the Court, concurred in part and concurred in the judgment. Justice Kennedy specifically recognized that "[t]he Eighth Amendment proportionality principle also applies to noncapital sentences." He then identified four principles of proportionality review—"the primacy of the legislature, the variety of legitimate penological schemes, the nature of our federal system, and the requirement that proportionality review be guided by objective factors"—that "inform the final one: The Eighth Amendment does not require strict proportionality between crime and sentence. Rather, it forbids only extreme sentences that are 'grossly disproportionate' to the crime." Justice Kennedy's concurrence also stated that *Solem* "did not mandate" comparative analysis "within and between jurisdictions."

The proportionality principles in our cases distilled in Justice Kennedy's concurrence guide our application of the Eighth Amendment in the new context that we are called upon to consider.

## B.

For many years, most States have had laws providing for enhanced sentencing of repeat offenders. Yet between 1993 and 1995, three strikes laws effected a sea change in criminal sentencing throughout the Nation. These laws responded to widespread public concerns about crime by targeting the class of offenders who pose the greatest threat to public safety: career criminals. . . .

Throughout the States, legislatures enacting three strikes laws made a deliberate policy choice that individuals who have repeatedly engaged in serious or violent criminal behavior, and whose conduct has not been deterred by more conventional approaches to punishment, must be isolated from society in order to protect the public safety. Though three strikes laws may be relatively new, our tradition of deferring to state legislatures in making and implementing such important policy decisions is longstanding.

Our traditional deference to legislative policy choices finds a corollary in the principle that the Constitution "does not mandate adoption of any one penological theory." A sentence can have a variety of justifications, such as incapacitation, deterrence, retribution, or rehabilitation. Some or all of these justifications may play a role in a State's sentencing scheme. Selecting the sentencing rationales is generally a policy choice to be made by state legislatures, not federal courts.

When the California Legislature enacted the three strikes law, it made a judgment that protecting the public safety requires incapacitating criminals who have already been convicted of at least one serious or violent crime. Nothing in the Eighth Amendment prohibits California from making that choice. . . .

California's justification is no pretext. Recidivism is a serious public safety concern in California and throughout the Nation. According to a recent report, approximately 67 percent of former inmates released from state prisons were charged with at least one "serious" new crime within three years of their release. In particular, released property offenders like Ewing had higher recidivism rates than those released after committing violent, drug, or public-order offenses. . . .

The State's interest in deterring crime also lends some support to the three strikes law. We have long viewed both incapacitation and deterrence as rationales for recidivism statutes . . . .

To be sure, California's three strikes law has sparked controversy. Critics have doubted the law's wisdom, cost-efficiency, and effectiveness in reaching its goals. This criticism is appropriately directed at the legislature, which has primary responsibility for making the difficult policy choices that underlie any criminal sentencing scheme. We do not sit as a "superlegislature" to second-guess these policy choices. It is enough that the State of California has a reasonable basis for believing that dramatically enhanced sentences for habitual felons "advance[s] the goals of [its] criminal justice system in any substantial way."

## III.

Against this backdrop, we consider Ewing's claim that his three strikes sentence of 25 years to life is unconstitutionally disproportionate to his offense of "shoplifting three golf clubs." Brief for Petitioner 6. We first address the gravity of the offense compared to the harshness of the penalty. At the threshold, we note that Ewing incorrectly frames the issue. The gravity of his offense was not merely "shoplifting three golf clubs." Rather, Ewing was convicted of felony grand theft for stealing nearly $1,200 worth of merchandise after previously having been convicted of at least two "violent" or "serious" felonies. . . .

That grand theft is a "wobbler" under California law is of no moment. Though California courts have discretion to reduce a felony grand theft charge to a misdemeanor, it remains a felony for all purposes "unless and until the trial court imposes a misdemeanor sentence." . . .

In weighing the gravity of Ewing's offense, we must place on the scales not only his current felony, but also his long history of felony recidivism. Any other approach would fail to accord proper deference to the policy judgments that find expression in the legislature's choice of sanctions. . . . To give full effect to the State's choice of this legitimate penological goal, our proportionality review of Ewing's sentence must take that goal into account.

Ewing's sentence is justified by the State's public-safety interest in incapacitating and deterring recidivist felons, and amply supported by his own long, serious criminal record. Ewing has been convicted of numerous misdemeanor and felony offenses, served nine separate terms of incarceration, and committed most of his crimes while on probation or parole. His prior "strikes" were serious felonies including robbery and three residential burglaries. To be sure, Ewing's sentence is a long one. But it reflects a rational legislative judgment, entitled to deference, that offenders who have committed serious or violent felonies and who continue to commit felonies must be incapacitated. . . . Ewing's is not "the rare case in which a threshold comparison of the crime committed and the sentence imposed leads to an inference of gross disproportionality." *Harmelin*, 501 U.S., at 1005 (Kennedy, J., concurring in part and concurring in judgment).

We hold that Ewing's sentence of 25 years to life in prison, imposed for the offense of felony grand theft under the three strikes law, is not grossly disproportionate and therefore does not violate the Eighth Amendment's prohibition on cruel and unusual punishments. The judgment of the California Court of Appeal is affirmed. . . .

Justice Scalia, concurring in the judgment. . . .

Proportionality—the notion that the punishment should fit the crime—is inherently a concept tied to the penological goal of retribution. "[I]t becomes difficult even to speak intelligently of 'proportionality,' once deterrence and rehabilitation are given significant weight,"—not to mention giving weight to the purpose of California's three strikes law: incapacitation. In the present case, the game is up once the plurality has acknowledged that "the Constitution does not mandate adoption of any one penological theory," and that a "sentence can have a variety of justifications, such as incapacitation, deterrence, retribution, or rehabilitation." . . .

Perhaps the plurality should revise its terminology, so that what it reads into the Eighth Amendment is not the unstated proposition that all punishment should be reasonably proportionate to the gravity of the offense, but rather the unstated proposition that all punishment should reasonably pursue the multiple purposes of the criminal law. That formulation would make it clearer than ever, of course, that the plurality is not applying law but evaluating policy.

Because I agree that petitioner's sentence does not violate the Eighth Amendment's prohibition against cruel and unusual punishments, I concur in the judgment.

Justice Breyer, with whom Justice Stevens, Justice Souter, and Justice Ginsburg join, dissenting.

The constitutional question is whether the "three strikes" sentence imposed by California upon repeat-offender Gary Ewing is "grossly disproportionate" to his crime. The sentence amounts to a real prison term of at least 25 years. The sentence-triggering criminal conduct consists of the theft of three golf clubs priced at a total of $1,197. The offender has a criminal history that includes four felony convictions arising out of three separate burglaries (one armed). . . .

## I.

This Court's precedent sets forth a framework for analyzing Ewing's Eighth Amendment claim. The Eighth Amendment forbids, as "cruel and unusual punishments," prison terms (including terms of years) that are "grossly disproportionate." In applying the "gross disproportionality" principle, courts must keep in mind that "legislative policy" will primarily determine the appropriateness of a punishment's "severity," and hence defer to such legislative policy judgments. . . .

The plurality applies Justice Kennedy's analytical framework in *Harmelin* . . . To implement this approach, courts faced with a "gross disproportionality" claim must first make "a threshold comparison of the crime committed and the sentence imposed." If a claim crosses that threshold—itself a *rare* occurrence—then the court should compare the sentence at issue to other

*(Continues)*

*(Continued)*

sentences "imposed on other criminals" in the same, or in other, jurisdictions. The comparative analysis will "validate" or invalidate "an initial judgment that a sentence is grossly disproportionate to a crime."

I recognize the warnings implicit in the Court's frequent repetition of words such as "rare." Nonetheless I believe that the case before us is a "rare" case—one in which a court can say with reasonable confidence that the punishment is "grossly disproportionate" to the crime.

## II.

Ewing's claim crosses the gross disproportionality "threshold." . . .

Three kinds of sentence-related characteristics define the relevant comparative spectrum: (a) the length of the prison term in real time, *i.e.,* the time that the offender is likely actually to spend in prison; (b) the sentence-triggering criminal conduct, *i.e.,* the offender's actual behavior or other offense-related circumstances; and (c) the offender's criminal history.

In *Rummel,* the Court held constitutional (a) a sentence of life imprisonment *with parole available within 10 to 12 years,* (b) for the offense of obtaining $120 by false pretenses, (c) committed by an offender with two prior felony convictions (involving small amounts of money). In *Solem,* the Court held unconstitutional (a) a sentence of life imprisonment *without parole,* (b) for the crime of writing a $100 check on a nonexistent bank account, (c) committed by an offender with six prior felony convictions (including three for burglary). Which of the three pertinent comparative factors made the constitutional difference?

The third factor, prior record, cannot explain the difference. The offender's prior record was *worse* in *Solem,* where the Court found the sentence too long, than in *Rummel,* where the Court upheld the sentence. The second factor, offense conduct, cannot explain the difference. The nature of the triggering offense—viewed in terms of the actual monetary loss—in the two cases was about the same. The one critical factor that explains the difference in the outcome is the length of the likely prison term measured in real time. In *Rummel,* where the Court upheld the sentence, the state sentencing statute authorized parole for the offender, Rummel, after 10 or 12 years. In *Solem,* where the Court struck down the sentence, the sentence required the offender, Helm, to spend the rest of his life in prison.

Now consider the present case. The third factor, *offender characteristics*—*i.e.,* prior record—does not differ significantly here from that in *Solem.* Ewing's prior record consists of four prior felony convictions (involving three burglaries, one with a knife) contrasted with Helm's six prior felony convictions (including three burglaries, though none with weapons). The second factor, *offense behavior,* is worse than that in *Solem,* but only to a degree. It would be difficult to say that the actual behavior itself here (shoplifting) differs significantly from that at issue in *Solem* (passing a bad check) or in *Rummel* (obtaining money through false pretenses). Rather the difference lies in the *value* of the goods obtained. That difference, measured in terms of the most relevant feature (loss to the victim, *i.e.,* wholesale value) and adjusted for the irrelevant feature of inflation, comes down (in 1979 values) to about $379 here compared with $100 in *Solem,* or (in 1973 values) to $232 here compared with $120.75 in *Rummel.* . . .

The difference in *length* of the real prison term—the first, and critical, factor in *Solem* and *Rummel*—is considerably more important. Ewing's sentence here amounts, in real terms, to at least 25 years without parole or good-time credits. That sentence is considerably shorter than Helm's sentence in *Solem,* which amounted, in real terms, to life in prison. Nonetheless Ewing's real prison term is more than twice as long as the term at issue in *Rummel,* which amounted, in real terms, to at least 10 or 12 years. And, Ewing's sentence, unlike Rummel's (but like Helm's sentence in *Solem*), is long enough to consume the productive remainder of almost any offender's life. (It means that Ewing himself, seriously ill when sentenced at age 38, will likely die in prison.)

The upshot is that the length of the real prison term—the factor that explains the *Solem/Rummel* difference in outcome—places Ewing closer to *Solem* than to *Rummel,* though the greater value of the golf clubs that Ewing stole moves Ewing's case back slightly in *Rummel's* direction. Overall, the comparison places Ewing's sentence well within the twilight zone between *Solem* and *Rummel*—a zone where the argument for unconstitutionality is substantial, where the cases themselves cannot determine the constitutional outcome.

Second, Ewing's sentence on its face imposes one of the most severe punishments available upon a recidivist who subsequently engaged in one of the less serious forms of criminal conduct. I do not deny the seriousness of shoplifting, which an *amicus curiae* tells us costs retailers in the range of

$30 billion annually. But consider that conduct in terms of the factors that this Court mentioned in *Solem*—the "harm caused or threatened to the victim or society," the "absolute magnitude of the crime," and the offender's "culpability." In respect to all three criteria, the sentence-triggering behavior here ranks well toward the bottom of the criminal conduct scale. . . .

Third, some objective evidence suggests that many experienced judges would consider Ewing's sentence disproportionately harsh. The United States Sentencing Commission (having based the federal Sentencing Guidelines primarily upon its review of how judges had actually sentenced offenders) does not include shoplifting (or similar theft-related offenses) among the crimes that might trigger especially long sentences for recidivists . . . .

Taken together, these three circumstances make clear that Ewing's "gross disproportionality" argument is a strong one. That being so, his claim *must* pass the "threshold" test. If it did not, what would be the function of the test? A threshold test must permit *arguably* unconstitutional sentences, not only *actually* unconstitutional sentences, to pass the threshold—at least where the arguments for unconstitutionality are unusually strong ones. A threshold test that blocked every ultimately invalid constitutional claim—even strong ones—would not be a *threshold* test but a *determinative* test. And, it would be a *determinative* test that failed to take account of highly pertinent sentencing information, namely, comparison with other sentences. Sentencing comparisons are particularly important because they provide proportionality review with *objective* content. By way of contrast, a threshold test makes the assessment of constitutionality highly subjective. And, of course, so to transform that *threshold* test would violate this Court's earlier precedent.

## III.

Believing Ewing's argument a strong one, sufficient to pass the threshold, I turn to the comparative analysis. A comparison of Ewing's sentence with other sentences requires answers to two questions. First, how would other jurisdictions (or California at other times, *i.e.*, without the three strikes penalty) punish the *same offense conduct?* Second, upon what other conduct would other jurisdictions (or California) impose the *same prison term?* Moreover, since hypothetical punishment is beside the point, the relevant prison time, for comparative purposes, is *real* prison time, *i.e.*, the time that an offender must *actually serve.* . . .

The upshot is that comparison of other sentencing practices, both in other jurisdictions and in California at other times (or in respect to other crimes), validates what an initial threshold examination suggested. Given the information available, given the state and federal parties' ability to provide additional contrary data, and given their failure to do so, we can assume for constitutional purposes that the following statement is true: Outside the California three strikes context, Ewing's recidivist sentence is virtually unique in its harshness for his offense of conviction, and by a considerable degree. . . .

## V.

. . . Ewing's sentence is, at a minimum, 2 to 3 times the length of sentences that other jurisdictions would impose in similar circumstances. That sentence itself is sufficiently long to require a typical offender to spend virtually all the remainder of his active life in prison. These and the other factors that I have discussed, along with the questions that I have asked along the way, should help to identify "gross disproportionality" in a fairly objective way—at the outer bounds of sentencing.

In sum, even if I accept for present purposes the plurality's analytical framework, Ewing's sentence (life imprisonment with a minimum term of 25 years) is grossly disproportionate to the triggering offense conduct—stealing three golf clubs—Ewing's recidivism notwithstanding.

For these reasons, I dissent. . . .

## Notes and Questions

1. From a retributive perspective, is it defensible that Gary Ewing should be required to serve a minimum of 25 years, and perhaps the rest of his life, in prison for stealing three golf clubs? Is Ewing's prior record a legitimate consideration in evaluating the propriety of his sentence based on a just deserts rationale?

2. Do you agree with Justice Scalia that, "once deterrence and rehabilitation are given significant weight," the principle of proportionality between crime and punishment becomes exceedingly difficult, if not impossible to apply? On the other hand, if a legislature were to order life imprisonment for (recidivist) double-parking, would this punishment be immune from judicial review under Justice Scalia's analysis?

*(Continues)*

Should it be? *Should* a court ever "second guess" a penalty scheme enacted by a democratically elected legislature? Does Justice Breyer's opinion in *Ewing* more closely resemble judicial second guessing of a legislative judgment or constitutional analysis based on neutral principles of adjudication?

3. The three separate felonies for which Rummel was convicted (*Rummel v. Estelle*, 445 U.S. 263 (1980)) involved a total property value of just over $229. Under Texas's habitual felon statute, he received a sentence of life imprisonment (with parole eligibility). The Supreme Court rejected Rummel's claim that the life sentence amounted to a disproportionate punishment for his crimes. Do you agree? How would you analyze Rummel's claim?

4. *Solem v. Helm*, 463 U.S. 277 (1983), like *Rummel,* involved a life prison sentence imposed on a repeat felony offender. The defendant's crimes included three convictions for third-degree burglary, obtaining money by false pretenses, grand larceny, driving while intoxicated (third offense), and writing a bad check with intent to defraud. Unlike *Rummel,* the life sentence imposed in *Solem* did not allow for parole eligibility. A majority of the Court ruled that the sentence of life imprisonment without the possibility of parole was constitutionally excessive. Do you agree? Because the defendant in *Rummel* had no guarantee that he would be paroled, is a distinction based on parole *eligibility* sufficient to justify the different results in *Rummel* and *Solem*? In fact, Rummel was released approximately 8 months after the Court decided his case based on a different constitutional violation affecting his sentence. *Solem v. Helm,* 463 U.S., at 297 n. 25.

5. How "serious" is the crime of possessing greater than 650 grams of cocaine, the offense for which the offender in *Michigan v. Harmelin,* 501 U.S. 957 (1991) received a mandatory sentence of life imprisonment without parole? Is a legislature or a court in a better position to make that assessment? Following the U.S. Supreme Court's decision in *Harmelin,* the Michigan Supreme Court ruled that mandatory life imprisonment without parole eligibility for possessing more than 650 grams of cocaine violated the Michigan Constitution's prohibition against "cruel or unusual" punishments. The state court invalidated the "no-parole" feature of the law and directed that offenders already serving sentences under the statute would be eligible for parole consideration after serving 10 years of their sentences. *People v. Bullock,* 485 N.W.2d 866 (Mich. 1992).

6. Applying Justice Kennedy's analysis in *Harmelin,* which was adopted by the plurality in *Ewing,* the Maryland Court of Appeals (that state's highest court) ruled that a 20-year sentence imposed for common law battery was "grossly disproportionate" to the defendant's offense, which consisted of slapping his wife in the face. However, the court further ruled that a 30-year sentence imposed for a separate count of common law battery in which the defendant struck his wife with an iron was not "grossly disproportionate" and did not violate the defendant's rights under the Eighth Amendment or the state constitution. *Thomas v. State,* 634 A.2d 1 (Md. 1993).

## Conclusion

The crimes, defenses, and punishments comprising the substantive criminal law are primarily shaped by legislation and complementary judicial decisions. Still, these laws must operate within a constitutional framework. This chapter reviewed several important constitutional principles and their application to the criminal law.

We began by considering jurisdictional issues, focusing on *United States v. Lopez* and that case's implications for the appropriate division of responsibility between the federal and state governments in the administration of the criminal law. We next examined constitutional dimensions of the legality principle. The vagueness doctrine promotes fair notice and helps check the potential for arbitrary enforcement of the laws. Although stridently criticized when decided and rarely applied thereafter, *Lambert v. California* nevertheless represents one aspect of the fair notice requirement. The *Lambert* principle treads a fine line by attempting to ensure that the criminal law is not a trap for the unwary yet not endorsing ignorance of the law as an excuse in the normal case. The prohibition

against *ex post facto* laws also is directly related to the legality principle, as illustrated in the cases we considered.

The chapter then covered three major constitutional limitations on the government's power to define conduct as a crime: substantive due process, the First Amendment, and equal protection principles. *Coker v. Georgia* recognized a "proportionality" principle that curbs governmental authority to punish in death penalty cases, and in *Ewing v. California,* although upholding a 25-year to life prison sentence for the theft of golf clubs under "three strikes" legislation, a majority of the justices affirmed that a prison sentence also can represent a constitutionally excessive punishment for a crime.

It is important to recognize the constitutional boundaries within which the criminal law operates. Notwithstanding those limitations, the federal and state governments have tremendous latitude in defining crimes, creating defenses, and crafting punishments. The study of the criminal law focuses primarily on the legislative and judicial rules governing crimes, defenses, and punishments and on their underlying justifications. Those issues command our attention in the chapters that follow.

## Key Terms

| | |
|---|---|
| cruel and unusual punishment | overbreadth doctrine |
| equal protection of the laws | proportionality |
| *ex post facto* laws | right of association |
| fair notice | right of privacy |
| just deserts | *stare decisis* |
| the legality principle | substantive due process |
| offensive conduct | torts |
| offensive speech | vagueness |

## Review Questions

1. What is the relationship between "vagueness" and the *due process clause*?
2. What is the difference between the vagueness and overbreadth doctrines?
3. What are *ex post facto* laws and what is their relevance in criminal matters?
4. How do *substantive due process* principles help restrict the reach of the criminal law?
5. Discuss the circumstances under which "free speech" rights do and do not include the right to engage in "offensive speech."
6. How is offensive speech different from offensive conduct? Provide examples.
7. Where is the right of privacy found in the Constitution?
8. Anti-terrorism prosecutions have generally been based on three statutes. Identify those statutes.
9. What is the rationale behind the principle of "just deserts"?
10. Describe a criminal situation for which a convicted person would be given a sentence. Propose a sentence, providing justification that would balance the need for punishment while being considerate of potential claims that the punishment might be in violation of the Eighth Amendment.

## Notes

1. *See* Stephen JF. *A History of the Criminal Law of England.* New York: Burt Franklin; 1883.
2. *See* LaFave WR, Scott AW. *Criminal Law,* 2nd ed. St. Paul, MN: West Publishing Co.; 1986:246–247.
3. Hamilton A. *The Federalist Papers* (Clinton Rossiter, ed.) (No. 84). New York: The New American Library of World Literature; 1961:511–512.
4. *See* DeCaro F. The wavy borderline between free speech and foul. *New York Times,* June 20, 1999, sec. 9, p. 2; Bradsher K. Canoeist goes to court, fighting for right to curse. *New York Times,* June 3, 1999, p. A18.
5. *See* Kennedy R. *Race, Crime, and the Law.* New York: Vintage Books; 1997:84–85.
6. *See generally* Tribe LH. *American Constitutional Law,* 2nd ed. Mineola, NY: The Foundation Press; 1988:1439–1466, 1558–1565.
7. Wolfgang ME, Riedel M. Race, judicial discretion, and the death penalty. *Ann Am Acad Polit Soc Sci.* 1973;407:119.
8. *See generally* Dorin D. Two different worlds: Criminologists, justices and racial discrimination in the imposition of capital punishment in rape cases. *J Criminal Law Criminol.* 1981;72:1667.

## Footnote for *Papachristou v. City of Jacksonville*, 405 U.S. 156, 92 S. Ct. 839, 31 L. Ed. 2d 110 (1972)

1. Jacksonville Ordinance Code § 26-57 provided at the time of these arrests and convictions as follows:

> Rogues and vagabonds, or dissolute persons who go about begging, common gamblers, persons who use juggling or unlawful games or plays, common drunkards, common night walkers, thieves, pilferers or pickpockets, traders in stolen property, lewd, wanton and lascivious persons, keepers of gambling places, common railers and brawlers, persons wandering or strolling around from place to place without any lawful purpose or object, habitual loafers, disorderly persons, persons neglecting all lawful business and habitually spending their time by frequenting houses of ill fame, gaming houses, or places where alcoholic beverages are sold or served, persons able to work but habitually living upon the earnings of their wives or minor children shall be deemed vagrants and, upon conviction in the Municipal Court shall be punished as provided for Class D offenses.

## Footnote for *Cohen v. California*, 403 U.S. 15, 91 S. Ct. 1780, 29 L. Ed. 2d 284 (1971)

3. It is illuminating to note what transpired when Cohen entered a courtroom in the building. He removed his jacket and stood with it folded over his arm. Meanwhile, a policeman sent the presiding judge a note suggesting that Cohen be held in contempt of court. The judge declined to do so and Cohen was arrested by the officer only after he emerged from the courtroom.

## Footnotes for *Coker v. Georgia*, 433 U.S. 584, 97 S. Ct. 2861, 53 L. Ed. 2d 982 (1977)

4. Because the death sentence is a disproportionate punishment for rape, it is cruel and unusual punishment within the meaning of the Eighth Amendment even though it may measurably serve the legitimate ends of punishment and therefore is not invalid for its failure to do so. We observe that in the light of the legislative decisions in almost all of the States and in most of the countries around the world, it would be difficult to support a claim that the death penalty for rape is an indispensable part of the States' criminal justice system (concurring opinion of Justice Powell).
1. It is not this Court's function to formulate the relevant criteria that might distinguish aggravated rape from the more usual case, but perhaps a workable test would embrace the factors identified by Georgia: the cruelty or viciousness of the offender, the circumstances and manner in which the offense was committed, and the consequences suffered by the victim. . . .

# 3 | General Elements of Crimes

## Chapter Objectives

- Understand the concept of strict liability
- Learn the general elements of criminal behavior
- Consider and understand the different mental states giving rise to criminal liability
- Distinguish between *general intent crimes* and *specific intent crimes*
- Understand how the Model Penal Code views criminal culpability
- Understand the doctrine of *transferred intent*
- Understand when one's *failure to act* constitutes a crime
- Understand the distinction between an act versus a status
- Learn the meaning of the *concurrence principle*
- Understand the concept of "*harm*" and its importance to the criminal law
- Understand the importance of *causation* to the criminal law, including the difference between "*cause in fact*" and "*proximate cause*"

The criminal law often is divided conceptually into a *special part* and a *general part*. The special part consists of the specific rules that define crimes such as murder, manslaughter, burglary, robbery, and many others. The general part is made up of the foundational principles and doctrines that inform our judgments about culpability. The specific rules and the underlying premises of culpability are intimately related. Murder, for example, is a different and more serious crime than manslaughter because we consider **mens rea** to be one component of blameworthiness.

In this chapter we focus on the general elements of crimes: *mens rea*, the act, the concurrence principle, harm, and causation. These principles inevitably resurface in later chapters as we examine specific crimes. Similarly, we cannot avoid considering specific offenses in this chapter because they are the vehicle through which the general principles operate and are illustrated. This interrelationship between the general and special parts of the criminal law will become increasingly apparent as our studies continue. We begin by taking a closer look at the *mens rea* element of crimes.

*murder vs. manslaughter:*

## Mens Rea

### General Principles: "Strict Liability" Crimes?

Oliver Wendell Holmes once remarked that "even a dog distinguishes between being stumbled over and being kicked."[1] This intuitively satisfying observation captures the significance of *mens rea* to the criminal law. Surprisingly, however, very early in English history the criminal law took little account of the mental state accompanying the infliction of harm. Even today, as we shall see, certain categories of "strict liability" crimes exist that dispense with the requirement of an evil mind or a wrongful intent. Predictably, the role of *mens rea* in the criminal law is considerably more complex than Holmes' aphorism would suggest.

### Francis Bowes Sayre, "Mens Rea."

No problem of criminal law is of more fundamental importance or has proved more baffling through the centuries than the determination of the precise mental element or *mens rea* necessary for crime. For hundreds of years the books have repeated with unbroken cadence that *Actus non facit reum nisi mens sit rea.* "There can be no crime, large or small, without an evil mind," says Bishop. "It is therefore a principle of our legal system, as probably it is of every other, that the essence of an offence is the wrongful intent, without which it cannot exist."

But when it comes to attaching a precise meaning to *mens rea*, courts and writers are in hopeless disagreement. . . .

Up to the twelfth century the conception of *mens rea* in anything like its modern sense was non-existent. In certain cases at least criminal liability might attach irrespective of the actor's state of mind. But because the old records fail to set forth a *mens rea* as a general requisite of criminality one must not reach the conclusion that even in very early times the mental element was entirely disregarded. The very nature of the majority of the early offenses rendered them impossible of commission without a criminal intent. Waylaying and robbery are impossible without it; so is rape; and the same is roughly true of housebreaking. . . .

By the end of the twelfth century two influences were making themselves strongly felt. One was the Roman law which, resuscitated in the universities in the eleventh and twelfth centuries, was sweeping over Europe with new power. Scholars and legal writers were kindled with burning enthusiasm for old Roman law texts. The Roman notions of *dolus* and *culpa* were taken up with fresh interest, and in some cases the attempt was made to graft these onto English law. . . .

A second influence, even more powerful, was the canon law, whose insistence upon moral guilt emphasized still further the mental element in crime. In the determination of sin the mental element must be scrutinized quite as closely as the physical act. . . .

It was almost inevitable, therefore, that the emphasis placed by Bracton upon the mental element in criminality should take permanent root and become part of the established law. Under the pervasive influence of the Church, the teaching of the penitential books that punishment should be dependent upon moral guilt gave powerful impetus to this growth, for the very essence of moral guilt is a mental element.

Henceforth, the criminal law of England, developing in the general direction of moral blameworthiness, begins to insist upon a *mens rea* as an essential of criminality. . . .

By the second half of the seventeenth century, it was universally accepted law that an evil intent was as necessary for felony as the act itself. . . .

While the conception of a general *mens rea* as a requisite for felony was thus coming into prominence, the exigencies of the developing law made necessary at the same time a more and more clear cut conception of exactly what constituted this evil or felonious intent. At the outset when the *mens rea* necessary for criminality was based on general moral blameworthiness, the conception was an exceedingly vague one. As a result of the slow judicial process of discriminating one case from another and "taking of diversities," much sharper and more precise lines gradually came to be drawn as to the exact mental requisites for various crimes. Since each felony involved different social and public interests, the mental requisites for one almost inevitably came to differ from those of another. . . .

The growing insistence after Bracton's day upon moral blameworthiness as one of the factors underlying criminality resulted not only in the development of various technical mental requisites for different felonies but also in the slow formulation of recognized general defenses to criminal liability. The conception of blameworthiness or moral guilt is necessarily based upon a free mind voluntarily choosing evil rather than good; there can be no criminality in the sense of moral shortcoming if there is no freedom of choice or normality of will capable of exercising a free choice. After the twelfth century new general defenses begin to take shape such as insanity, infancy, compulsion or the like, based upon the lack of a guilty mind and thus negativing moral blameworthiness. . . .

A study of the historical development of the mental requisites of crime leads to certain inescapable conclusions. In the first place, it seems clear that *mens rea*, the mental factor necessary to prove criminality, has no fixed continuing meaning. The conception of *mens rea* has varied with the changing underlying conceptions and objectives of criminal justice. At the beginning when the object of criminal administration was to restrict and supplant the blood feud, the mental factor was of importance insofar as it determined the provocative nature of the offense; a malicious burning of another's dwelling house being far more provocative than an accidental one, judges must distinguish between malicious and accidental burnings. Under the dominating influence of the canon law and the penitential books the underlying objective of criminal justice gradually came to be the punishment of evil-doing; as a result the mental factors necessary for criminality were based upon a mind bent on evil-doing in the sense of moral wrong. Our modern objective tends more and more in the direction, not of awarding adequate punishment for moral wrongdoing, but of protecting social and public interests. To the extent that this objective prevails, the mental element requisite for criminality, if not altogether dispensed with, is coming to mean, not so much a mind bent on evil-doing as an intent to do that which unduly endangers social or public interests. As the underlying objective of criminal administration has almost unconsciously shifted, and is shifting, the basis of the requisite *mens rea* has imperceptibly shifted, lending a change to the flavor, if not to the actual content, of the criminal state of mind which must be proved to convict. . . .

Indeed, the strong current of modern decisions toward applying in the criminal law an objective standard, to which all must measure up at their peril, in place of the older subjective standard, under which defendants are punishable only for failing to measure up to their capacities, is only another manifestation of the same trend of the criminal law. Certain it is that in modern times we have moved far from the old fourteenth century conception of *mens rea* as a mind bent on moral wrongdoing.

In the second place, it is equally clear that whatever the early conception of *mens rea* may have been, as the law grew the requisite mental elements of the various felonies developed along different lines to meet exigencies and social needs which varied with each felony. The original requirement of an underlying evil motive derived from the canonists' conception of moral guilt came to be supplanted by the requirement of specific forms of intent evolved separately for each particular felony. In this way, the criminal law seems to have progressed from motive to intent. . . .

Since crimes involving a specific intent vary as widely with regard to the requisite intent as with the requisite act, it is hardly necessary to add that a study of the specific intent required for one such crime is of but little assistance in determining the precise mental element necessary for another. It is hopeless to find any general universal concept of *mens rea* applicable to all such crimes alike. Generalizations in this field are dangerous. Even though the statutory requirements for a specific intent are laid down for two crimes in the same words, not infrequently the meanings to be attached to the same word formulæ differ vastly.

In the third place, the strong tendency of the early days to link criminal liability with moral guilt made it necessary to free from punishment those who perhaps satisfied the requirements of specific intent for particular crimes but who, because of some personal mental defect or restraint, should not be convicted of any crime. The person who lacked a normal intelligence because of mental disease, or who lacked discretion because of tender years, or who through fear of death lacked the power to choose his conduct—all these must escape punishment if criminality was to be based upon moral guilt. Thus, there developed certain well-recognized defenses in criminal law, affecting one's general capacity to commit crime, originally based upon and springing from the same lack of a blameworthy mind but, as the law developed to meet new needs, becoming more and more differentiated and fixed in arbitrary molds. . . .

We have now reached a point where we can understand something of the realities of the underlying problem of *mens rea*. In view of the long development and widely differing forms of the mental element required for conviction in different groups of cases, it is manifestly futile to seek to attach to that much abused term any single precise meaning. Neither general moral blameworthiness, nor the intent to do that which provokes general social disapproval, nor the intent to commit some breach of contract or tort or crime, nor any other single form of intent will fit all cases.

Yet the great majority of the leading writers on criminal law have sought to analyze *mens rea* into some single intent or state of mind. Blackstone defines it as a "vicious will." Just what constitutes a "vicious will," however, no one can say. Bishop, so far as one can judge from his strikingly vague statements, makes *mens rea* mean the intent to do what is morally wrong. . . .

What too many have failed to realize is that the exact nature of *mens rea* cannot be determined until one has a clear understanding of just what the term covers. If *mens rea* is used in contrast with "specific intent" and its scope is narrowed so as to exclude questions of personal incapacity, crimes requiring specific intent, and crimes based upon negligence, it becomes possible to find in it a fairly precise single constituent element, as, for example, the intent to commit that which constitutes a crime other than a light police offense. On the other hand, if the term *mens rea* is given a broader meaning to designate whatever mental element is necessary to convict for *any* crime, thus including specific intent and questions raised by personal incapacities, it becomes at once clear that *mens rea* can never be analyzed into any single constituent element or group of elements because no single state of mind common to all crimes exists. Some of the most careful writers on criminal law have used the term to cover whatever mental element is necessary to convict for any particular crime; and to avoid the endless confusion which more restricted, ill-defined meanings are likely to produce, this seems sound.

The conclusion to which one is thus driven is by no means a negative one. It focuses our thinking upon the important fact that there is no single precise requisite state of mind common to all crime. The criminal state of mind of the child differs materially from that of the insane person. These both differ from that of the sane adult acting under a mistake of fact. All three differ from the state of mind which must be proved to convict for murder. An intensive study of the substantive law covering each separate group becomes necessary to reach an adequate understanding of the various states of mind requisite for criminality. The old conception of *mens rea* must be discarded, and in its place must be substituted the new conception of *mentes reae*.

*Source:* Sayre FB. *Mens Rea.* 45 *Harvard Law Review*. 1932; 45:974.

mens rea ⟶ mentes reae.

## *Morissette v. United States*, 342 U.S. 246, 72 S. Ct. 240, 96 L. Ed. 288 (1951)

Mr. Justice Jackson delivered the opinion of the Court.

This would have remained a profoundly insignificant case to all except its immediate parties had it not been so tried and submitted to the jury as to raise questions both fundamental and far-reaching in federal criminal law, for which reason we granted certiorari.

On a large tract of uninhabited and untilled land in a wooded and sparsely populated area of Michigan, the Government established a practice bombing range over which the Air Forces dropped simulated bombs at ground targets. These bombs consisted of a metal cylinder about forty inches long and eight inches across, filled with sand and enough black powder to cause a smoke puff by which the strike could be located. At various places about the range signs read "Danger—Keep Out—Bombing Range." Nevertheless, the range was known as good deer country and was extensively hunted.

Spent bomb casings were cleared from the targets and thrown into piles "so that they will be out of the way." They were not stacked or piled in any order but were dumped in heaps, some of which had been accumulating for four years or upwards, were exposed to the weather and rusting away.

Morissette, in December of 1948, went hunting in this area but did not get a deer. He thought to meet expenses of the trip by salvaging some of these casings. He loaded three tons of them on his truck and took them to a nearby farm, where they were flattened by driving a tractor over them. After expending this labor and trucking them to market in Flint, he realized $84. . . .

The loading, crushing and transporting of these casings were all in broad daylight, in full view of passersby, without the slightest effort at concealment. When an investigation was started, Morissette voluntarily, promptly and candidly told the whole story to the authorities, saying that he had no intention of stealing but thought the property was abandoned, unwanted and considered of no value to the Government. He was indicted, however, on the charge that he "did unlawfully, willfully and knowingly steal and convert" property of the United States of the value of $84, in violation of 18 U.S.C. § 641, which provides that "whoever embezzles, steals, purloins, or knowingly converts" government property is punishable by fine and imprisonment. Morissette was convicted and sentenced to imprisonment for two months or to pay a fine of $200. The Court of Appeals affirmed, one judge dissenting.

On his trial, Morissette, as he had at all times told investigating officers, testified that from appearances he believed the casings were castoff and abandoned, that he did not intend to steal the property, and took it with no wrongful or criminal intent. The trial court, however, was unimpressed and ruled: "He took it because he thought it was abandoned and he knew he was on government property. . . . That is no defense. . . . I don't think anybody can have the defense they thought the property was abandoned on another man's piece of property." The court stated: "I will not permit you to show this man thought it was abandoned. . . . I hold in this case that there is no question of abandoned property." The court refused to submit or to allow counsel to argue to the jury whether Morissette acted with innocent intention. It charged: "And I instruct you that if you believe the testimony of the government in this case, he intended to take it. . . . He had no right to take this property. . . . [A]nd it is no defense to claim that it was abandoned, because it was on private property. . . . And I instruct you to this effect: That if this young man took this property (and he says he did), without any permission (he says he did), that was on the property of the United States Government (he says it was), that it was of the value of one cent or more (and evidently it was), that he is guilty of the offense charged here. If you believe the government, he is guilty. . . . The question on intent is whether or not he intended to take the property. He says he did. Therefore, if you believe either side, he is guilty." Petitioner's counsel contended, "But the taking must have been with a felonious intent." The court ruled, however: "That is presumed by his own act."

The Court of Appeals' . . . construction of the statute is that it creates several separate and distinct offenses, one being knowing conversion of government property. The court ruled that this particular offense requires no element of criminal intent. This conclusion was thought to be required by the failure of Congress to express such a requisite and this Court's decisions in *United States v. Behrman*, 258 U.S. 280 and *United States v. Balint*, 258 U.S. 250.

## I.

In those cases this Court did construe mere omission from a criminal enactment of any mention of criminal intent as dispensing with it. If they be deemed precedents for principles of construction generally applicable to federal penal statutes, they authorize this conviction. Indeed, such adoption of the literal reasoning announced in those cases would do this and more—it would sweep out of all federal crimes, except when expressly preserved, the ancient requirement of a culpable state of mind. We think a resume of their historical background is convincing that an effect has been ascribed to them more comprehensive than was contemplated and one inconsistent with our philosophy of criminal law.

The contention that an injury can amount to a crime only when inflicted by intention is no provincial or transient notion. It is as universal and persistent in mature systems of law as belief in freedom of the human will and a consequent ability and duty of the normal individual to choose between good and evil. A relation between some mental element and punishment for a harmful act is almost as instinctive as the child's familiar exculpatory "But I didn't mean to," and has afforded the rational basis for a tardy and unfinished substitution of deterrence and reformation in place of retaliation and vengeance as the motivation for public prosecution. Unqualified acceptance of this doctrine by English common law in the Eighteenth Century was indicated by Blackstone's sweeping

*(Continues)*

statement that to constitute any crime there must first be a "vicious will." Common-law commentators of the Nineteenth Century early pronounced the same principle, although a few exceptions not relevant to our present problem came to be recognized.

Crime, as a compound concept, generally constituted only from concurrence of an evil-meaning mind with an evil-doing hand, was congenial to an intense individualism and took deep and early root in American soil. As the states codified the common law of crimes, even if their enactments were silent on the subject, their courts assumed that the omission did not signify disapproval of the principle but merely recognized that intent was so inherent in the idea of the offense that it required no statutory affirmation. Courts, with little hesitation or division, found an implication of the requirement as to offenses that were taken over from the common law. The unanimity with which they have adhered to the central thought that wrongdoing must be conscious to be criminal is emphasized by the variety, disparity and confusion of their definitions of the requisite but elusive mental element. However, courts of various jurisdictions, and for the purposes of different offenses, have devised working formulae, if not scientific ones, for the instruction of juries around such terms as "felonious intent," "criminal intent," "malice aforethought," "guilty knowledge," "fraudulent intent," "wilfulness," "scienter," to denote guilty knowledge, or "*mens rea*," to signify an evil purpose or mental culpability. By use or combination of these various tokens, they have sought to protect those who were not blameworthy in mind from conviction of infamous common-law crimes.

However, the Balint and Behrman offenses belong to a category of another character, with very different antecedents and origins. The crimes there involved depend on no mental element but consist only of forbidden acts or omissions. This, while not expressed by the Court, is made clear from examination of a century-old but accelerating tendency, discernible both here and in England, to call into existence new duties and crimes which disregard any ingredient of intent. The industrial revolution multiplied the number of workmen exposed to injury from increasingly powerful and complex mechanisms, driven by freshly discovered sources of energy, requiring higher precautions by employers. Traffic of velocities, volumes and varieties unheard of, came to subject the wayfarer to intolerable casualty risks if owners and drivers were not to observe new cares and uniformities of conduct. Congestion of cities and crowding of quarters called for health and welfare regulations undreamed of in simpler times. Wide distribution of goods became an instrument of wide distribution of harm when those who dispersed food, drink, drugs, and even securities, did not comply with reasonable standards of quality, integrity, disclosure and care. Such dangers have engendered increasingly numerous and detailed regulations which heighten the duties of those in control of particular industries, trades, properties or activities that affect public health, safety or welfare.

While many of these duties are sanctioned by a more strict civil liability, lawmakers, whether wisely or not, have sought to make such regulations more effective by invoking criminal sanctions to be applied by the familiar technique of criminal prosecutions and convictions. This has confronted the courts with a multitude of prosecutions, based on statutes or administrative regulations, for what have been aptly called "public welfare offenses." These cases do not fit neatly into any of such accepted classifications of common-law offenses, such as those against the state, the person, property, or public morals. Many of these offenses are not in the nature of positive aggressions or invasions, with which the common law so often dealt, but are in the nature of neglect where the law requires care, or inaction where it imposes a duty. Many violations of such regulations result in no direct or immediate injury to person or property but merely create the danger or probability of it which the law seeks to minimize. While such offenses do not threaten the security of the state in the manner of treason, they may be regarded as offenses against its authority, for their occurrence impairs the efficiency of controls deemed essential to the social order as presently constituted. In this respect, whatever the intent of the violator, the injury is the same, and the consequences are injurious or not according to fortuity. Hence, legislation applicable to such offenses, as a matter of policy, does not specify intent as a necessary element. The accused, if he does not will the violation, usually is in a position to prevent it with no more care than society might reasonably expect and no more exertion than it might reasonably exact from one who assumed his responsibilities. Also, penalties commonly are relatively small, and conviction does no grave damage to an offender's reputation. Under such considerations, courts have turned to construing statutes and regulations which make no mention of intent as dispensing with it and holding that the guilty act alone makes out the crime. This has not, however, been without expressions of misgiving. . . .

In overruling a contention that there can be no conviction on an indictment which makes no charge of criminal intent but alleges only making of a sale of a narcotic forbidden by law, Chief Justice Taft, wrote:

"While the general rule at common law was that the scienter was a necessary element in the indictment and proof of every crime, and this was followed in regard to statutory crimes even where the statutory definition did not in terms include it . . . , there has been a modification of this view in respect to prosecutions under statutes the purpose of which would be obstructed by such a requirement. It is a question of legislative intent to be construed by the court. . . ." *United States v. Balint, supra* (258 U.S. 251, 252).

He referred, however, to "regulatory measures in the exercise of what is called the police power, where the emphasis of the statute is evidently upon achievement of some social betterment rather than the punishment of crimes, as in cases of mala in se," and drew his citation of supporting authority chiefly from state court cases dealing with regulatory offenses.

On the same day, the Court determined that an offense under the Narcotic Drug Act does not require intent, saying, "If the offense be a statutory one, and intent or knowledge is not made an element of it, the indictment need not charge such knowledge or intent." *United States v. Behrman* (258 U.S. at 288).

Of course, the purpose of every statute would be "obstructed" by requiring a finding of intent, if we assume that it had a purpose to convict without it. Therefore, the obstruction rationale does not help us to learn the purpose of the omission by Congress. And since no federal crime can exist except by force of statute, the reasoning of the Behrman opinion, if read literally, would work far-reaching changes in the composition of all federal crimes. . . .

Neither this Court nor, so far as we are aware, any other has undertaken to delineate a precise line or set forth comprehensive criteria for distinguishing between crimes that require a mental element and crimes that do not. We attempt no closed definition, for the law on the subject is neither settled nor static. The conclusion reached in the *Balint* and *Behrman* Cases has our approval and adherence for the circumstances to which it was there applied. A quite different question here is whether we will expand the doctrine of crimes without intent to include those charged here.

Stealing, larceny, and its variants and equivalents, were among the earliest offenses known to the law that existed before legislation; they are invasions of rights of property which stir a sense of insecurity in the whole community and arouse public demand for retribution, the penalty is high and, when a sufficient amount is involved, the infamy is that of a felony, which, says Maitland, is ". . . as bad a word as you can give to man or thing." . . .

Congress, therefore, omitted any express prescription of criminal intent from the enactment before us in the light of an unbroken course of judicial decision in all constituent states of the Union holding intent inherent in this class of offense, even when not expressed in a statute. Congressional silence as to mental elements in an Act merely adopting into federal statutory law a concept of crime so well defined in common law and statutory interpretation by the states may warrant quite contrary inferences than the same silence in creating an offense new to general law, for whose definition the courts have no guidance except the Act. Because the offenses before this Court in the *Balint* and *Behrman* Cases were of this latter class, we cannot accept them as authority for eliminating intent from offenses incorporated from the common law. Nor do exhaustive studies of state court cases disclose any well-considered decisions applying the doctrine of crime without intent to such enacted common-law offenses. . . .

The Government asks us by a feat of construction radically to change the weights and balances in the scales of justice. The purpose and obvious effect of doing away with the requirement of a guilty intent is to ease the prosecution's path to conviction, to strip the defendant of such benefit as he derived at common law from innocence of evil purpose, and to circumscribe the freedom heretofore allowed juries. Such a manifest impairment of the immunities of the individual should not be extended to common-law crimes on judicial initiative. . . .

We hold that mere omission from § 641 of any mention of intent will not be construed as eliminating that element from the crime denounced.

## II.

It is suggested, however, that the history and purpose of § 641 imply something more affirmative as to elimination of intent from at least one of the offenses charged under it in this case. The

(Continues)

argument does not contest that criminal intent is retained in the offenses of embezzlement, stealing and purloining, as incorporated into this section. But it is urged that Congress joined with those, as a new, separate and distinct offense, knowingly to convert government property, under circumstances which imply that it is an offense in which the mental element of intent is not necessary. . . .

Congress, by the language of this section, has been at pains to incriminate only "knowing" conversions. But, at common law, there are unwitting acts which constitute conversions. In the civil tort, except for recovery of exemplary damages, the defendant's knowledge, intent, motive, mistake and good faith are generally irrelevant. If one takes property which turns out to belong to another, his innocent intent will not shield him from making restitution or indemnity, for his well-meaning may not be allowed to deprive another of his own.

Had the statute applied to conversions without qualification, it would have made crimes of all unwitting, inadvertent and unintended conversions. Knowledge, of course, is not identical with intent and may not have been the most apt words of limitation. But knowing conversion requires more than knowledge that defendant was taking the property into his possession. He must have had knowledge of the facts, though not necessarily the law, that made the taking a conversion. In the case before us, whether the mental element that Congress required be spoken of as knowledge or as intent, would not seem to alter its bearing on guilt. For it is not apparent how Morissette could have knowingly or intentionally converted property that he did not know could be converted, as would be the case if it was in fact abandoned or if he truly believed it to be abandoned and unwanted property. . . .

We find no grounds for inferring any affirmative instruction from Congress to eliminate intent from any offense with which this defendant was charged.

## III.

. . . Where intent of the accused is an ingredient of the crime charged, its existence is a question of fact which must be submitted to the jury. . . .

It follows that the trial court may not withdraw or prejudge the issue by instruction that the law raises a presumption of intent from an act. . . .

Moreover, the conclusion supplied by presumption in this instance was one of intent to steal the casings, and it was based on the mere fact that defendant took them. The court thought the only question was, "Did he intend to take the property?" That the removal of them was a conscious and intentional act was admitted. But that isolated fact is not an adequate basis on which the jury should find the criminal intent to steal or knowingly convert, that is, *wrongfully* to deprive another of possession of property. Whether that intent existed, the jury must determine, not only from the act of taking, but from that together with defendant's testimony and all of the surrounding circumstances. . . .

Of course, the jury, considering Morissette's awareness that these casings were on government property, his failure to seek any permission for their removal and his self-interest as a witness, might have disbelieved his profession of innocent intent and concluded that his assertion of a belief that the casings were abandoned was an afterthought. Had the jury convicted on proper instructions it would be the end of the matter. But juries are not bound by what seems inescapable logic to judges. They might have concluded that the heaps of spent casings left in the hinterland to rust away presented an appearance of unwanted and abandoned junk, and that lack of any conscious deprivation of property or intentional injury was indicated by Morissette's good character, the openness of the taking, crushing and transporting of the casings, and the candor with which it was all admitted. They might have refused to brand Morissette as a thief. Had they done so, that too would have been the end of the matter.

Reversed.

## A Note on Strict Liability

In *United States v. Freed*, 401 U.S. 601 (1971), the defendants were indicted for possessing unregistered hand grenades. The district court dismissed the indictment because it lacked a "scienter" allegation (*i.e.*, it did not charge that the defendants knew that the hand grenades had not been registered). The U.S. Supreme Court reversed.

The Act requires no specific intent or knowledge that the hand grenades were unregistered. It makes it unlawful for any person "to receive or possess a firearm which is not registered to him." . . .

The presence of a "vicious will" or *mens rea* was long a requirement of criminal responsibility. But the list of exceptions grew, especially in the expanding regulatory area involving activities affecting public health, safety, and welfare. . . .

The present case [involves] . . . a regulatory measure in the interest of the public safety, which may well be premised on the theory that one would hardly be surprised to learn that possession of hand grenades is not an innocent act. They are highly dangerous offensive weapons, no less dangerous than the narcotics involved in *United States v. Balint* where a defendant was convicted of sale of narcotics against his claim that he did not know the drugs were covered by a federal act. We say with Chief Justice Taft in that case:

"It is very evident from a reading of it that the emphasis of the section is in securing a close supervision of the business of dealing in these dangerous drugs by the taxing officers of the Government and that it merely uses a criminal penalty to secure recorded evidence of the disposition of such drugs as a means of taxing and restraining traffic. Its manifest purpose is to require every person dealing in drugs to ascertain at his peril whether that which he sells comes within the inhibition of the statute, and if he sells the inhibited drug in ignorance of its character, to penalize him. Congress weighed the possible injustice of subjecting an innocent seller to a penalty against the evil of exposing innocent purchasers to danger from the drug, and concluded that the latter was the result preferably to be avoided." *Id.*, at 253–254. . . .

If, in light of the Court's conclusion in *United States v. Freed*, the possession of unregistered hand grenades is a "strict liability" crime, what about a violation of the National Firearms Act which, in relevant part, makes it a felony punishable by up to 10 years in prison "for any person . . . to . . . possess a firearm" not properly registered under federal law (26 U.S.C.A. § 5861(d))? One type of firearm that must be registered is a "machine gun," defined as "any weapon which shoots, . . . or can be readily restored to shoot, automatically more than one shot, without manual reloading, by a single function of the trigger." 26 U.S.C.A. § 5845(b). In *Staples v. United States,* 511 U.S. 600 (1994), federal agents seized a rifle from the defendant's home that had been modified to be capable of fully automatic fire, bringing it within the statutory definition of a machine gun. The rifle was not registered. The defendant claimed that he did not know that the rifle would fire fully automatically and that absent proof of such knowledge he could not be convicted for violating the National Firearms Act. The trial court disagreed, ruling that the government was not required to prove the defendant "knew of the characteristics of his weapon that made it a 'firearm' under the Act."

The U.S. Supreme Court reversed. Justice Thomas's majority opinion rejected the argument that Congress's failure to include a specific *mens rea* element in the Act's registration provisions converted the measure into a strict liability crime. His opinion distinguished *Freed* as follows:

As the Government concedes, *Freed* did not address the issue presented here. In *Freed,* we decided only that § 5861(d) does not require proof of knowledge that a firearm is *unregistered.* The question presented by a defendant who possesses a weapon that is a "firearm" for purposes of the Act, but who knows only that he has a "firearm" in the general sense of the term, was not raised or considered. And our determination that a defendant need not know that his weapon is unregistered suggests no conclusion concerning whether § 5861(d) requires the defendant to know of the features that make his weapon a statutory "firearm"; different elements of the same offense can require different mental states.

Moreover, our analysis in *Freed* likening the Act to the public welfare statute in *Balint* rested entirely on the assumption that the defendant *knew* that he was dealing with hand grenades—that is, that he knew he possessed a particularly dangerous type of weapon (one within the statutory definition of a "firearm"), possession of which was not entirely "innocent" in and of itself. . . . The predicate for that analysis is eliminated when, as in this case, the very question to be decided is *whether* the defendant must know of the particular characteristics that make his weapon a statutory firearm. . . .

The potentially harsh penalty attached to violation of § 5861(d)—up to 10 years' imprisonment—confirms our reading of the Act. Historically, the penalty imposed under a statute has been a significant consideration in determining whether the statute should be construed as dispensing with *mens rea.* Certainly, the cases that first defined the concept of the public welfare offense almost uniformly involved statutes that provided for only light penalties such as fines or short jail sentences, not imprisonment in the state penitentiary. . . .

As commentators have pointed out, the small penalties attached to such offenses logically complemented the absence of a *mens rea* requirement: In a system that generally requires a "vicious will" to establish a

*(Continues)*

crime, 4 W. Blackstone, Commentaries, imposing severe punishments for offenses that require no *mens rea* would seem incongruous. . . .

If Congress had intended to make outlaws of gun owners who were wholly ignorant of the offending characteristics of their weapons, and to subject them to lengthy prison terms, it would have spoken more clearly to that effect.

## General Intent and Specific Intent

After *mens rea* is established as an element of a crime, further discriminations become necessary. As indicated in *Staples v. United States, supra,* "different elements of the same offense can require different mental states." In *Frey v.* State, presented below, we are introduced to the distinction between "**general intent**" and "**specific intent**" **crimes**. The lack of clarity of such terms has led many jurisdictions to abandon them in favor of more precisely defined mental states.

### *Frey v. State,* 708 So.2d 918 (Fla. 1998)

Shaw, Justice.

. . . Deputy Britt was on uniformed patrol at 11:30 P.M., April 20, 1994, when he saw Thomas Frey acting suspiciously near Earl's Trailer Park. Britt asked Frey for identification, and when a radio check showed an outstanding arrest warrant, Britt attempted to handcuff him. Frey, who was very drunk (his blood alcohol level was .388, or approximately four times the legal limit for driving), said, "I'm not going to jail," and grabbed Britt's throat with both hands, choking him. Britt tried to break free but could not. The deputy kicked and punched Frey, and in a final attempt to free himself, shot Frey in the legs. Both Britt and Frey were treated at the hospital for their injuries.

Frey was charged with aggravated battery on a law enforcement officer and resisting arrest with violence. He was tried before a jury and in closing argument defense counsel argued that Frey had been too drunk to form the specific intent to commit the crimes. The prosecutor, on the other hand, told the jury that while voluntary intoxication is a defense to aggravated battery, it is not a defense to resisting arrest with violence. The judge in his instructions to the jury echoed the prosecutor's statement of the law. Frey was convicted of battery and resisting arrest with violence. . . .

Frey argues that resisting arrest with violence is a specific intent crime and that his requested instruction on voluntary intoxication should have been given on this charge. . . .

We disagree.

Voluntary intoxication has long been recognized in Florida as a defense to specific intent crimes, as this Court noted in *Linehan v. State,* 476 So.2d 1262 (Fla.1985):

. . . In *Garner* [*v. State,* 9 So. 835 (Fla. 1891)] we stated that when a specific or particular intent is an essential or constituent element of the offense, intoxication, although voluntary, becomes a matter for consideration . . . with reference to the capacity or ability of the accused to form or entertain the particular intent, or . . . whether the accused was in such a condition of mind as to form a premeditated design. Where a party is too drunk to entertain or be capable of forming the essential particular intent, such intent can of course not exist, and no offense of which such intent is a necessary ingredient, [can] be perpetrated. 9 So. at 845.

*Linehan,* 476 So.2d at 1264. The defense, however, is unavailable for general intent crimes. *Id.*

Professor LaFave describes the general contours of specific intent, as opposed to general intent, crimes:

[T]he most common usage of "specific intent" is to designate a special mental element which is required above and beyond any mental state required with respect to the *actus reus* of the crime. Common law larceny, for example, requires the taking and carrying away of the property of another, and the defendant's mental state as to this act must be established, but in addition it must be shown that there was an "intent to steal" the property. Similarly, common law burglary requires a breaking and entry into the dwelling of another, but in addition to the mental state connected with these acts it must also be established that the defendant acted "with intent to commit a felony therein." The same situation prevails with many statutory crimes: assault "with intent to kill" as to certain aggravated assaults; confining another "for the purpose of ransom or reward" in kidnapping; making an untrue statement "designedly, with intent to defraud" in the crime of false pretenses; etc.

Wayne R. LaFave & Austin W. Scott, Jr., *Substantive Criminal Law* § 3.5(e)(1986) (footnotes omitted).

To determine whether resisting arrest with violence is a general intent or specific intent crime, we look to the plain language of the statute:

843.01 Resisting officer with violence to his person.—Whoever knowingly and willfully resists, obstructs, or opposes any officer . . . in the lawful execution of any legal duty, by offering or doing violence to the person of such officer . . . is guilty of a felony of the third degree. . . . § 843.01, Fla. Stat. (1993).

The statute's plain language reveals that no heightened or particularized, i.e., no specific, intent is required for the commission of this crime, only a general intent to "knowingly and willfully" impede an officer in the performance of his or her duties. . . .

Only if the present statute were to be recast to require a heightened or particularized intent would the crime of resisting arrest with violence be a specific intent crime. . . .

Anstead, Justice, concurring in part and dissenting in part. . . .

I believe that the artificial distinction we have established between general and specific intent, with only specific intent crimes warranting additional defenses such as voluntary intoxication, often leads to incongruous and harsh results. Countless commentators and courts have criticized the lack of a principled and useful basis for maintaining this distinction. As one commentator has noted: . . .

Since the terms do not clearly delineate for the jury (or anyone else) what blameworthy state of mind must exist in any given situation, it would seem senseless to instruct a jury in these amorphous terms. It would be much better to tell the jury that, for guilt, a defendant must have thought about (or have been reckless concerning) certain definite things. If he did, and also performed the requisite acts, he is to be found guilty. If he did not so contemplate and act, he is to be acquitted.

William Roth, *General vs. Specific Intent: A Time for Terminological Understanding in California*, 7 Pepp. L. Rev. 67, 77–78 (1979). . . .

Since this perplexing division between "general" and "specific" is judicially created, we should seriously consider whether now is the time to revise this ill-conceived framework. Rather than splitting hairs and attempting to draw a bright line through the murky and ill-defined netherworld that separates general from specific intent, our time would be better spent giving effect to the legislative intent behind a particular statute and focusing on the degree of culpability along the lines clearly delineated in the Model Penal Code. (*See* Footnote 5.) Other than the "nebulous distinction" separating general from specific intent crimes, no compelling policy reasons exist which support the availability of additional defenses in Florida to "specific" intent crimes such as first-degree murder, robbery, kidnapping, aggravated assault, battery, aggravated battery, burglary, escape, and theft, while denying the application of such defenses to "general" intent crimes such as resisting a police officer with violence or arson. The only difference I can see is that, for the most part, the statutes defining the former category have the magic words "with intent to," while the latter crimes do not.

## "GENERAL" vs. "SPECIFIC"

In *State v. Stasio*, 396 A.2d 1129 (N.J. 1979), the New Jersey Supreme Court grappled with the distinction between specific and general intent. Quoting Professor Hall's treatise, the court reasoned:

The current confusion resulting from diverse uses of "general intent" is aggravated by dubious efforts to differentiate that from "specific intent." Each crime . . . has its distinctive *mens rea, e.g.,* intending to have forced intercourse, intending to break and enter a dwelling-house and to commit a crime there, intending to inflict a battery, and so on. It is evident that there must be as many *mentes reae* as there are crimes. And whatever else may be said about an intention, an essential characteristic of it is that it is directed toward a definite end. To assert therefore that an intention is "specific" is to employ a superfluous term just as if one were to speak of a "voluntary act."

*(Continues)*

*(Continued)*

*Id.* at 1132–33 (quoting Jerome Hall, *General Principles of Criminal Law* 142 (2d ed. 1960)). The New Jersey high court went on to explain that:

[D]istinguishing between specific and general intent gives rise to incongruous results by irrationally allowing intoxication to excuse some crimes but not others. In some instances if the defendant is found incapable of formulating the specific intent necessary for the crime charged, such as assault with intent to rob, he may be convicted of a lesser included general intent crime, such as assault with a deadly weapon. In other cases there may be no related general intent offense so that intoxication would lead to acquittal. . . .

. . . [W]here the more serious offense requires only a general intent, such as rape, intoxication provides no defense, whereas it would be a defense to an attempt to rape, specific intent being an element of that offense. Yet the same logic and reasoning which impels exculpation due to the failure of specific intent to commit an offense would equally compel the same result when a general intent is an element of the offense.

*Stasio,* 396 A.2d at 1133–34 (citations omitted). . . .

Even the United States Supreme Court has recognized that "the mental element in criminal law encompasses more than the two possibilities of 'specific' and 'general' intent." *See Liparota v. United States,* 471 U.S. 419, 423 n. 5 (1985). Indeed, the Court has explained that:

This ambiguity [in the terms specific intent and general intent] has led to a movement away from the traditional dichotomy of intent and toward an alternative analysis of *mens rea.* This new approach [is] exemplified by the American Law Institute's Model Penal Code. . . .

. . . [T]here is [an] ambiguity inherent in the traditional distinction between specific intent and general intent. Generally, even time-honored and common-law crimes consist of several elements, and complex statutorily defined crimes exhibit this characteristic to an even greater degree. Is the same state of mind required of the actor for each element of the crime, or may some elements require one state of mind and some another? . . . "[C]lear analysis requires that the question of the kind of culpability required to establish the commission of an offense be faced separately with respect to each material element of the crime."

*United States v. Bailey,* 444 U.S. 394, 403–406 (1980) (quoting Model Penal Code § 2.02 comments at 123 (Tentative Draft No.4, 1955)).

Consistent with the views expressed above, Professors LaFave and Scott suggest an alternative method for evaluating the effect of voluntary intoxication on a defendant's ability to exhibit the requisite *mens rea* of a particular crime:

[I]t may be said that it is better, when considering the effect of the defendant's voluntary intoxication upon his criminal liability, to stay away from those misleading concepts of general intent and specific intent. Instead one should ask, first, what intent (or knowledge) if any does the crime in question require; and, then, if the crime requires some intent (knowledge), did the defendant in fact entertain such an intent (or, did he in fact know what the crime requires him to know).

Wayne R. LaFave & Austin W. Scott, Jr., *Substantive Criminal Law* § 4.10 at 554 (1986). . . .

## THIS CASE

The extreme facts of this case underscore the faulty rationale, if any, for maintaining the irrational division of criminal intent between "general" and "specific." As the majority opinion notes, Mr. Frey had a blood alcohol level of 0.388, approximately *four* times the legal limit for driving. . . . The arresting officer, Deputy Britt, testified that he believed Frey was drunk because he swayed, smelled of alcohol, babbled, could not stand still, and spoke in a loud, slurred and unintelligible voice. An emergency room physician testified that when persons have such a high level of alcohol in their systems, they may suffer blackouts, thus meaning they can do something and not remember it later. Finally, the trial judge commented that he had only seen one person with a higher blood alcohol content than Frey's in three to four years. He noted that emergency medical personnel usually take such severely intoxicated people directly to the hospital since normally "you're going to die on them. And I'm worried about that." By any measure, Mr. Frey was severely intoxicated and a serious question exists as to his capability of forming an intent, general or specific, to commit the crime at issue.

Against this factual backdrop, let us consider the criminal offense involved herein. Section 843.01, Florida Statutes (1993), provides:

Whoever *knowingly and willfully* resists, obstructs, or opposes any officer . . . in the lawful execution of any legal duty, by offering or doing violence to the person of such officer . . . is guilty of a felony of the third degree. . . .

(Emphasis added.) The statute defines the prohibited act and the requisite degree of blameworthiness to establish guilt. The statute's language requires that the offender's level of culpability be greater than negligence or recklessness by including a "knowledge" element. It logically follows that if a person is charged with "knowingly and willfully" restricting, obstructing, or opposing a law enforcement officer "in the lawful execution of any legal duty," one element of the crime is that the alleged offender knew that the person he was resisting was a law enforcement officer. . . .

## CONCLUSION

Consistent with the proposals of Professors Scott and LaFave discussed above, the American Law Institute committee has explained that when "purpose" or "knowledge" is an element of a crime, proof of intoxication may logically negate the existence of either. To violate section 843.01, it is evident that "knowledge" of the fact that one is obstructing an officer is an element of the crime of resisting arrest with violence. Therefore, under the sensible "element" approach to determining whether voluntary intoxication can negate the mental element of a crime, it is apparent that a defendant would be allowed to put on evidence that his level of intoxication rendered him unable to form the "knowledge" element of the crime of resisting arrest with violence under section 843.01.

I therefore conclude that the trial court should have granted petitioner's request for an instruction of the defense of voluntary intoxication. . . .

### Notes and Questions

1. How could the "specific intent" instruction that Frey requested have helped his defense to the charge of resisting arrest with violence? Does the attempt to clarify the "specific" vs. "general" intent distinction in footnote 2 of the majority opinion illustrate how Florida's resisting arrest with violence statute could be rewritten as a "specific intent" crime? Footnote 2 suggested, "For instance, the statute might be recast to read: 'Whoever knowingly and willfully resists . . . an officer . . . in the lawful execution of any legal duty, with the intent of doing violence to the person of such officer . . . is guilty of a felony of the third degree.'" Is the dissent's argument of assistance: "one element of the crime is that the alleged offender knew that the person he was resisting was a law enforcement officer"? Is the majority opinion's quotation from LaFave and Scott's *Substantive Criminal Law* treatise instructive, to the effect that "general intent" is the "mental state required with respect to the *actus reus* of the crime"? What is the *actus reus* of Frey's crime of resisting arrest with violence?

2. What could be characterized as the defendant's "general intent" in *Morissette*? What might be characterized as the "specific intent"—which the government failed to establish—in that case?

3. As the dissent in *Frey* makes clear, considerable confusion has been generated by the law's attempt to separate general intent from specific intent crimes. The distinction is important in contexts in addition to the defense of voluntary intoxication (which we consider in greater detail in Chapter 5). For example, the wrongful taking of another's property does not qualify as larceny unless the offender has the "specific intent" of permanently depriving the owner of possession. Thus if the person taking the property intends to return it at a later time, a larceny has not been committed under the traditional definition of that crime. Similarly, if burglary requires the breaking and entry of a dwelling *with the intent to commit a felony therein* (the "specific intent"), a wrongful breaking and entry, without more, fails to qualify. *See generally State v. Bridgeforth*, 750 P.2d 3 (Ariz. 1988); *State v. Orsello*, 554 N.W.2d 70 (Minn. 1996); *United States v. Blair*, 54 F.3d 639 (10th Cir. 1995).

*Four separate mental states in criminal culp.*

## Model Penal Code Approach

The Model Penal Code (MPC) has done away with the elusive distinction between general intent and specific intent. The MPC defines four separate mental states associated with criminal culpability: **purposely**, **knowingly**, **recklessly**, and **negligently**. MPC § 2.02(2). It further requires that the appropriate mental element be established "with respect to each material element of the offense." MPC § 2.02(1). The different forms of *mens rea* recognized by the MPC are defined and discussed in the following case.

## *United States v. Osguthorpe,* 13 F. Supp. 2d 1215 (D. Utah 1998)

Benson, District Judge.

. . . Defendant D.A. Osguthorpe is a 76 year old veterinarian who has been involved in the sheep ranching business for over 40 years. Defendant has routinely put his sheep out to summer pasture in Summit County on private ground located in the mountains above Park City, Utah. This private ground borders the Wasatch–Cache National Forest. There is no indication that, during the relevant time period, Osguthorpe ever held a permit to graze his sheep in the neighboring National Forest. In late 1994, the Forest Service issued a Notice of Violation against Osguthorpe for "placing or allowing unauthorized livestock to enter or be in the National Forest" pursuant to 36 C.F.R. § 261.7(a). Osguthorpe did not contest this Notice of Violation and was subsequently sentenced by the magistrate judge to one year unsupervised probation and fined $65.00.

In November and December of 1996, the Forest Service issued a total of three additional Notices of Violation pursuant to 36 C.F.R. § 261.7 alleging that Osguthorpe was guilty of allowing his livestock to enter or be in the National Forest and failing to remove the livestock when requested by a forest officer. . . .

Following a hearing on May 1, 1997, the magistrate judge found that *mens rea* is not required for a violation of 36 C.F.R. § 261.7(a). The magistrate judge determined that the Forest Service could prove its case by simply proving that the defendant's sheep were on Forest Service lands without authorization. Following the magistrate judge's ruling, . . . Osguthorpe changed his plea to guilty on one of the Notices of Violation and the Forest Service agreed to dismiss the other two Notices of Violation. However, the agreement was conditioned on defendant's retaining the right to appeal the magistrate judge's interpretation of 36 C.F.R. § 261.7(a). If successful on appeal, Osguthorpe may withdraw his guilty plea and proceed to trial. In that event, the Forest Service will be entitled to reassert the other Notices of Violation against Mr. Osguthorpe. On June 2, 1997, the magistrate judge accepted the conditional plea bargain and two months later sentenced Osguthorpe to five years supervised probation, conditioned upon spending 30 days in a halfway house, and imposed a fine of $5,000.00. . . .

The issue raised on appeal centers on the interpretation of one word: *allowing*. 36 C.F.R. § 261.7(a) prohibits "placing or allowing unauthorized livestock to enter or be in the National Forest System or other lands under Forest Service control." The Forest Service contends, and the magistrate judge agreed, that "placing or allowing" does not require any *mens rea* and the statute is one of strict liability. Defendant argues that the regulation in question does not eliminate a *mens rea* requirement. This Court finds that 36 C.F.R. § 261.7(a) is not one of strict liability and does require a showing of *mens rea*.

## I. Existence of a *Mens Rea* Requirement

It has long been true that "[t]he existence of a *mens rea* is the rule of, rather than the exception to, the principles of Anglo-American criminal jurisprudence." *Dennis v. United States,* 341 U.S. 494, 500 (1951). Despite this general rule, however, there are criminal statutes that have no intent requirement. Failure to stop at a stop sign is a common example of such a strict liability violation. The question in the instant case is whether 36 C.F.R. § 261.7(a) fits in the strict liability category.

As in all cases of statutory interpretation, we begin with the actual language employed by the regulation in question. . . .

The dictionary definitions of the terms *placing* and *allowing* are helpful in interpreting the meaning of § 261.7(a). According to The Random House Dictionary of the English Language (2d ed.1987) the term *placing* means "to put or set in a particular place, position, situation, or relation." *Allowing* is defined as "to give permission to or for . . . ; to permit by neglect; . . . to approve [or] sanction; . . . to permit something to happen or to exist." *Id.* The root word *allow* is a synonym of the word *permit* which denotes "granting or conceding the right of someone to do something." *Id.* The common and ordinary meanings of both *placing* and *allowing* indicate that some volition must be present. While *placing* signifies more active participation, *allowing* equally requires some permissiveness, acquiescence or approval. Thus, looking at the plain language of the Act, and applying ordinary definitions to the words employed, it is apparent that the statute does not altogether dispense with the *mens rea* element. To "allow" one's livestock to be on forest service property requires some level of involvement on the part of the owner. A sheep rancher has not "allowed" his sheep to be on forest service property, for example, if his sheep were released from the owner's locked pen by the action of a third party unknown to the owner and thereafter the sheep moved onto the government's property.

If the Department of Agriculture had wanted to write a strict liability regulation it could have easily done so. "Any person's sheep found on Forest land is guilty of an offense," would be a strict liability regulation. The present regulation is not.

Pointing to the legislative history, the Government argues that any *mens rea* requirement that may have existed in the regulation was removed in the 1977 amended version of 36 C.F.R. § 261.7(a). Prior to 1977, the regulation prohibited "[w]illfully allowing livestock to enter upon or to be upon such lands. . . ." 35 F.R. 3165 (1970). The amendment replaced "willfully allowing" with the words "placing or allowing." As drafters of the regulation, the Department of Agriculture did not, however, add words to indicate that the statute was to be one of strict liability. Therefore, the question is whether the elimination of the word *willfully* did away with any *mens rea*, or intent, requirement and made the statute one of strict liability. The Supreme Court has made it clear that the "mere omission from [a criminal statute] of any mention of intent will not be construed as eliminating that element from the crimes denounced." *Morissette v. United States*, 342 U.S. 246, 263 (1952). . . .

Accordingly, this Court is reluctant to dispense with a *mens rea* requirement without a clear indication that the drafters of the regulation intended such a result. The removal of the word *willfully* from 36 C.F.R. § 261.7(a) does not create strict liability. In order to prevail, the Forest Service must show that the defendant acted with the necessary *mens rea*.

## II. The Appropriate Level of *Mens Rea*

Having held that § 261.7(a) is not a strict liability statute, it is left for this Court to determine the appropriate level of *mens rea* required under the statute. As the Supreme Court has noted "[f]ew areas of criminal law pose more difficulty than the proper definition of the *mens rea* required for any particular crime." *United States v. Bailey*, 444 U.S. 394, 403 (1980). This is especially true where, as in the present case, "the language of the Act provides minimal assistance in determining what standard of intent is appropriate, and the sparse legislative history of the criminal provisions is similarly unhelpful." *United States v. United States Gypsum Co.*, 438 U.S. 422, 443–44 (1978). In order to arrive at the appropriate level of *mens rea*, this Court must "turn to more general sources and traditional understandings of the nature of the element of intent in the criminal law [and] try to avoid 'the variety, disparity and confusion' of judicial definitions of the 'requisite but elusive mental element' of criminal offenses." *Id.* at 444.

Much of the confusion regarding the requisite *mens rea* has stemmed from a historical shift in traditional *mens rea* analysis, for which courts have turned to various sources for guidance. The Supreme Court has consistently looked to the Model Penal Code as an avenue of resolving questions of this nature. . . . The Model Penal Code replaces the common law distinction between "specific intent" and "general intent" with four general levels of intent: purposely, knowingly, recklessly, and negligently. This "hierarchy of culpable states of mind" is listed in "descending order of culpability" with purposely being the most onerous standard and negligently the most lenient.

The question remaining, therefore, is which of the four levels of culpability should be the minimum required showing under § 261.7(a). The fact that the 1977 Amendment to the Act removed the word "willful" and added "placing and allowing" does not dispense with a *mens rea* requirement altogether, but it does serve to indicate that the drafters intended some lowering of the standard. Under these circumstances, it would be inappropriate to set the *mens rea* hurdle at "purposely." Particularly helpful in deciding which of the remaining three levels of culpability should be adopted is the Model Penal Code's interpretation of "willful." Section 2.02(8) of the Model Penal Code states that "[a] requirement that an offense be committed *wilfully is satisfied if a person acts knowingly* with respect to the material elements of the offense, unless a purpose to impose further requirements appears." (Emphasis added.) Because it is true that wilfully and knowingly are often used interchangeably, it would not make sense to adopt a knowingly standard in lieu of the elimination of its equivalent in the 1977 Amendment.

Of the remaining two levels, this Court finds that recklessness better fits the regulation in question and should be the minimum required showing in order to convict under § 261.7(a). While it is true there is no clear rationale in the regulation or legislative history which would indicate that reckless-ness, as opposed to negligence, should be the appropriate standard, this Court makes such ruling based, in part, upon the long held notion that "ambiguity concerning the ambit of criminal statutes

*(Continues)*

should be resolved in favor of lenity." *Rewis v. United States*, 401 U.S. 808, 812 (1971). Accordingly, the defendant may be convicted of a violation of § 261.7(a) only if the Government proves beyond a reasonable doubt that he recklessly, knowingly, or purposely *allowed* his livestock to be on Forest Service property.

The Magistrate Judge's decision is REVERSED. . . .

## Notes and Questions

1. Construct different fact scenarios to illustrate how Dr. Osguthorpe might have (a) purposely, (b) knowingly, (c) recklessly, and (d) negligently allowed his sheep to graze in the National Forest.
2. Dr. Osguthorpe was issued three Notices of Violation in November and December of 1996 alleging that he had unlawfully allowed his sheep to enter land within the National Forest. He did not contest a similar violation in late 1994. If the prosecution must prove that he "recklessly" allowed his sheep to enter the land, of what possible relevance is the 1994 violation? Assuming that the Notices of Violation were issued immediately after each incident, might the prosecution have a viable argument that Dr. Osguthorpe "knowingly" or "purposely" allowed his sheep to enter the National Forest property by the time of the last alleged violation in December of 1996? Note that if the prosecution did establish a "knowing" or "purposeful" violation, Dr. Osguthorpe could be convicted even though "recklessly" is the *mens rea* implicit in the definition of the offense. See MPC § 2.02(5).

## *Mens Rea* Requirements for Different Elements of Crimes

In *United States v. Bailey*, 444 U.S. 394 (1980), Justice Rehnquist raised the question, "Is the same state of mind required of the actor for each element of the crime, or may some elements require one state of mind and some another?" For example, if a person is charged with assaulting a police officer, is it sufficient to prove that the assault was committed "purposely" or "intentionally" or must the prosecution also prove that the defendant "knew" he or she was assaulting a police officer?

## *Commonwealth v. Flemings*, 652 A.2d 1282 (Pa. 1995)

Papadakos, Justice.

. . . On October 11, 1990, two members of the vice squad of the Erie Police Department were working undercover as drug purchasers. Appellee approached the vehicle in which the two officers were seated and essentially offered to sell them cocaine. Appellee left and returned a short time later accompanied by one Tisa Howard. Ms. Howard approached the passenger side of the vehicle where she conversed with Officer Yeaney. Meanwhile, Appellee approached the driver's side where he conversed with Officer Mioduszewski. As Officer Mioduszewski leaned toward the passenger window to consummate a drug transaction with Ms. Howard, his Smith and Wesson pistol was exposed to Appellee who then stole the pistol; and as Officer Mioduszewski turned around, Appellee had the gun pointed directly at him, as well as in the direction of Officer Yeaney and Ms. Howard. Appellee slowly backed off while pointing the gun at the officers. He then fled on foot. The officers gave chase and eventually caught Appellee exiting a nearby house. Prior to catching him, Appellee had stated, "Officers, I'll give you back your gun." Prior to that statement, the officers had not identified themselves as such. Appellee admitted that most of these events occurred, but testified that he did not know that Yeaney or Mioduszewski were police officers, but when he saw the firearm, he became frightened and took it so that no one would get injured. . . .

[T]he jury convicted Appellee of . . . two counts of aggravated assault. At the completion of the trial, Appellee requested that the jury be instructed that Appellee must have known that the undercover officers were police officers when he pointed the gun at them in order to be found guilty of aggravated assault on a police officer. The trial court refused this requested instruction concluding that knowledge of the fact that the victims were police officers is not an element of the crime under 18 P.S. § 2702(a)(3). During the course of their deliberations, the jury asked the trial court whether Appellee had to know whether the victims were police officers at the time of the assault. The trial court answered that it was not necessary that Appellee know they were police.

On appeal, the Superior Court reversed and remanded for a new trial holding that knowledge by Appellee that the victims were police officers was an element of the crime and must be proven. . . .

18 P.S. § 2702(a)(3) provides:

(a) Offense defined.—A person is guilty of aggravated assault if he:

(3) attempts to cause or intentionally or knowingly causes bodily injury to a police officer . . . in the performance of duty; . . .

The Superior Court first noted that attempted aggravated assault is a specific intent crime. . . .

Admittedly, this case raises an issue of first impression in Pennsylvania that as a matter of pure verbal logic could go either way. We are guided, however, by the United States Supreme Court's decision in *United States v. Feola,* 420 U.S. 671 (1975), where the court concluded, with respect to a comparable federal statute, that knowledge that a victim is a federal officer is not an element of the crime of assaulting a federal officer.

Prior to 1988, 18 U.S.C. § 111 provided; in part, as follows:

Whoever forcibly assaults, resists, opposes, impedes, intimidates, or interferes with any person designated in section 1114 of this title [that is, any federal officer] while engaged in or on account of the performance of his official duties, shall be fined not more than $5,000 or imprisoned not more than three years, or both. . . .

In *Feola,* the United States Supreme Court specifically held, by a 7 to 2 majority authored by Mr. Justice Blackmun, that criminal liability for the offense of assaulting a federal officer under 18 U.S.C. § 111 does not depend on whether or not the assailant harbored the specific intent to assault a federal officer. While conceding that either this conclusion, or its opposite, was "plausible" as a matter of determining legislative intent, the court concluded that Congress intended to protect federal officers, as well as federal functions, and that the rejection of a state scienter requirement was consistent with both purposes. 420 U.S. at 679.

Mr. Justice Blackmun went on to reason for the majority as follows:

We conclude, from all of this, that in order to effectuate the congressional purpose of according maximum protection to federal officers by making prosecution for assault upon them cognizable in the federal courts § 111 cannot be construed as embodying an unexpressed requirement that an assailant be aware that his victim is a federal officer. All the statute requires is an intent to assault, not an intent to assault a federal officer. A contrary conclusion would give insufficient protection to the agent enforcing an unpopular law, and none to the agent active under cover.

This interpretation poses no risk of unfairness to defendants. It is no snare for the unsuspecting. Although the perpetrator of a narcotics "rip-off," such as the one involved here, may be surprised to find that his intended victim is a federal officer in civilian apparel, he nevertheless knows from the very outset that his planned course of conduct is wrongful. The situation is not one where legitimate conduct becomes unlawful solely because of the identity of the individual or agency affected. In a case of this kind the offender takes his victim as he finds him. . . .

420 U.S. at 684–685.

Mr. Justice Blackmun's reasoning is equally applicable to the case at bar and our aggravated assault statute.

The liability imposed by 18 Pa.C.S. § 2702(a)(3) is not absolute, since to be convicted of aggravated assault under that section, the defendant must be shown to have possessed a criminal *mens rea, i.e.,* the intent to cause bodily injury. Thus, the defendant's ignorance of an officer's official status is relevant in those rare cases in which an officer fails to identify himself and then engages in a course of conduct which could reasonably be interpreted as the unlawful use of force directed either at the defendant or his property. Under such circumstances, a defendant would normally be justified in using reasonable force against his assailant. He could then be found to have exercised self defense, which would negate the existence of *mens rea.*

The same principle does not apply, however, where a defendant clearly intends to commit a crime and unwittingly chooses a police officer as his victim. Such a scenario scarcely argues the existence of an "honest mistake." Rather, under these circumstances the defendant's ignorance of the victim's official status is irrelevant since he knows from the outset that his planned course of conduct is unlawful. Once he chooses to engage in such conduct, he takes his victim as he finds him.

*(Continues)*

*(Continued)*

Similarly, the statute's requirement that an officer be "in the performance of duty" in no way implies that liability depends on whether the defendant is aware of his victim's official status. The duties of a police officer, like the officers in the instant case, frequently include undercover investigation in which the officer's official status is intentionally concealed. We do not interpret the language "in performance of duty" to require a defendant to have knowledge of the officer's official status since such a reading would all but strip the undercover officer of the protection the legislature intended to afford him. Rather, we hold that a defendant's lack of knowledge should only be considered in those cases in which a defendant acts with the *mistaken* belief that he is threatened with an intentional tort by a private citizen. That is not, however, quite the case here. Although appellant may in fact have believed that Officer Mioduszewski was a rival drug dealer, appellant's seizure of the officer's gun, which he then aimed at Mioduszewski, was a preemptive action, not one which is consistent with self-defense. His knowledge or otherwise of the officer's status is, thus, under the present facts, irrelevant—there is no honest mistake involved in the situation here.

In short, on the facts before us, the offender must take his victim as he finds him. Appellee here was clearly a wrongdoer. Knowledge that the victim is a police officer is not an element of the crime of aggravated assault under 18 P.S. § 2702(a)(3) and need not be proven. Proof of intent to assault is sufficient. . . .

## Notes and Questions

1. In *United States v. Feola*, 420 U.S. 671 (1975), which is discussed in *Flemings*, Justice Blackmun's opinion stressed that the Court's interpretation of the federal statute depended on "whether Congress intended to condition responsibility for violation of [18 U.S.C.A.] § 111 on the actor's awareness of the identity of his victim."

   > If the primary purpose is to protect federal law enforcement personnel, that purpose could well be frustrated by the imposition of a strict scienter requirement. On the other hand, if § 111 is seen primarily as an anti-obstruction statute, it is likely that Congress intended criminal liability to be imposed only when a person acted with the specific intent to impede enforcement activities. Otherwise, it has been said: "Were knowledge not required in obstruction of justice offenses described by these terms, wholly innocent (or even socially desirable) behavior could be transformed into a felony by the wholly fortuitous circumstance of the concealed identity of the person resisted." [*Citing United States v. Fernandez*, 497 F.2d 730, 744 (9th Cir. 1974) (Hufstedler, J., concurring).]

   The *Feola* Court concluded that "Congress intended to protect *both* federal officers and federal functions" and that "rejection of a strict scienter requirement is consistent with both purposes" (emphasis in original).

2. In light of the heavy emphasis in *Feola* on Congress's intent in enacting the federal legislation, how much weight should that decision have in *Flemings*, where the Pennsylvania Supreme Court was interpreting a state statute? What policy objectives are promoted by a statute that allows heightened punishment for assaulting a police officer? Is it "fair" to promote those objectives if the offender did not know that his or her victim in fact was a police officer?

3. The court in *Flemings* suggests that a different outcome would be likely if the defendant had a plausible claim that his conduct was motivated by self-defense. Why should ignorance of an alleged assault victim's status as a police officer matter when self-defense is reasonably at issue but not matter under other circumstances?

4. The rule adopted in *Flemings* and *Feola* is not universally followed. Some jurisdictions require proof that the defendant knew that his or her assault victim was a police officer before a conviction for that crime is allowed. *See, e.g., State v. Morey*, 427 A.2d 479 (Me. 1981); *Bundren v. State*, 274 S.E.2d 455 (Ga. 1981).

## Transferred Intent

What if, acting with the purpose or intent to harm a specific individual, A, a defendant's aim is bad with the result that B rather than A is injured? Can the defendant avoid criminal responsibility for the injuries to B on the ground that he or she had no intent to harm B? If the defendant misses both A and B and, contrary to his or her purpose or intent, instead shatters a window behind A's head, is the defendant guilty of malicious destruction of the window? We begin with the latter question, as we confront issues that occasionally are classified, somewhat misleadingly, under the rubric of "**transferred intent**."

## *Regina v. Pembliton,* 12 Cox Crim. Cases 607 (Court of Criminal Appeal 1874)

Lord Coleridge, C.J.—I am of opinion that this conviction must be quashed. The facts of the case are these. The prisoner and some other persons who had been drinking in a public house were turned out of it at about 11 P.M. for being disorderly, and then they began to fight in the street near the prosecutor's window. The prisoner separated himself from the others, and went to the other side of the street, and picked up a stone, and threw it at the persons he had been fighting with. The stone passed over their heads, and broke a large plate glass window in the prosecutor's house, doing damage to an amount exceeding 5/. The jury found that the prisoner threw the stone at the people he had been fighting with, intending to strike one or more of them with it, but not intending to break the window. The question is whether under an indictment for unlawfully and maliciously committing an injury to the window in the house of the prosecutor the proof of these facts alone, coupled with the finding of the jury, will do? Now I think that is not enough. The indictment is framed under the 24 & 25 Vict. c. 97, s. 51. The Act is an Act relating to malicious injuries to property, and sect. 51 enacts that whosoever shall unlawfully and maliciously commit any damage, &c., to or upon any real or personal property whatsoever of a public or a private nature, for which the punishment is hereinbefore provided, to an amount exceeding 5/ shall be guilty of a misdemeanor. There is also the 58th section which deserves attention. "Every punishment and forfeiture by this Act imposed on any person maliciously committing any offence, whether the same be punishable upon indictment or upon summary conviction, shall equally apply and be enforced whether the offence shall be committed from malice conceived against the owner of the property in respect of which it shall be committed, or otherwise." It seems to me on both these sections that what was intended to be provided against by the Act is the wilfully doing an unlawful Act, and that the Act must be wilfully and intentionally done on the part of the person doing it, to render him liable to be convicted. Without saying that, upon these facts, if the jury had found that the prisoner had been guilty of throwing the stone recklessly, knowing that there was a window near which it might probably hit, I should [not] have been disposed to interfere with the conviction, yet as they have found that he threw the stone at the people he had been fighting with intending to strike them and not intending to break the window, I think the conviction must be quashed. I do not intend to throw any doubt on the cases which have been cited and which show what is sufficient to constitute malice in the case of murder. They rest upon the principles of the common law, and have no application to a statutory offence created by an Act in which the words are carefully studied. . . .

## Notes and Questions

1. According to Lord Coleridge's interpretation, what is the precise nature of "the act" that "must be wilfully and intentionally done" to justify a conviction?
2. What does Lord Coleridge suggest about the propriety of a conviction based on the defendant's "recklessly" (as opposed to "intentionally") breaking the window?
3. What if Pembliton had aimed the stone at a specific person, A, against whom he bore ill will, but missed and struck and injured another person, B? Under Lord Coleridge's reasoning, should he be acquitted of an unlawful battery committed against B? In *Regina v. Latimer,* 17 Q.B.D. 359 (1886), Lord Coleridge stated the following:

   > It is common knowledge that a man who has an unlawful and malicious intent against another, and, in attempting to carry it out, injures a third person, is guilty of what the law deems malice against the person injured, because the offender is doing an unlawful act, and has that which the judges call general malice, and that is enough.

   Is the following distinction between *Pembliton* and *Latimer* helpful?

   > *Pembliton* . . . involved not only a transfer of intent from one victim to another, but also a change in the nature of the intent. In *Pembliton,* the defendant threw a rock at people standing in the street, but missed them and broke a window. The court held that transferred intent could not be applied to the offense of unlawful and malicious property damage because the defendant never intended to damage property.
   > Perkins and Boyce similarly distinguish between transfers of intent involving the "same mental pattern," *i.e.,* where only the object of the intent is shifted, and transfers of intent involving a

*(Continues)*

"different mental pattern," *i.e.*, where the crime intended differs from the crime committed. [Rollin M. Perkins & Ronald N. Boyce, *Criminal Law* 922-923 (Mineola, NY: The Foundation Press; 1982).] Only the first category can accurately be described as "transferred intent." *Poe v. State*, 671 A.2d 501, 507–508 n. 3 (Md. 1996) (Raker, J., concurring).

**4.** If we accept the general principle distinguishing *Latimer* and *Pembliton,* then the hypothetical we originally posed—where the defendant aims at A but misses and hits B—should be resolved fairly easily. The following case exposes additional issues regarding "transferred intent" and helps explain why that term may confuse the analysis rather than clarify it.

## *Harvey v. State,* 681 A.2d 628 (Md. App.), *cert. denied,* 686 A.2d 635 (Md. 1996)

[In June 1994 a fight broke out between two groups of young men in an apartment complex parking lot. The defendant's brother was involved in the fight. The defendant, Latrice "Kitty Cat" Harvey, passed a gun to a companion and instructed him to shoot a specified member of the rival gang. Her companion fired several shots, all of which missed the intended target, but one of which struck a bystander, Tiffany Evans, in the leg. Ms. Evans' wound was not fatal. Under Maryland law, the defendant was considered a (second-degree) principal, and thus equally responsible for the shooting as the actual shooter. She was charged with assault with intent to murder and reckless endangerment.

[The trial court ruled that the prosecution could rely on the doctrine of transferred intent to prove the *mens rea* required for assault with intent to murder. In relevant part, the trial judge instructed the jury:

[The doctrine of transferred intent means that the intent follows the bullet. . . .

[The doctrine of transferred intent applies to the specific intent to murder. Transferred intent means that if one specifically intends injury to another person, and in an effort to accomplish the injury or harm upon a person other than the one intended, he is guilty of the same kind of crime as if his aim had been more accurate.

[The fact that a person actually was killed instead of the intended victim, is immaterial, and the only question is what would have been the degree of guilt, if the result intended actually had been accomplished. The intent is transferred to the person whose death or harm has been caused.

[The defendant was convicted as charged. On appeal, she argued that "the doctrine of transferred intent is inapplicable to the crime of assault with intent to murder and erroneously relieved the State of its obligation to prove the required *mens rea* of a specific intent to kill directed at the actual assault victim, Tiffany Evans."]

Moylan, Judge. . . .

### Transferred Intent Generally

Suppose the intended victim in the cross-hairs of the gunsight is the President of the United States, Franklin Delano Roosevelt. Suppose the assassin's aim is unsure and the unintended recipient of the errant shot is Mayor Anton Cermak of Chicago. What is the guilt of the assassin with respect to Mayor Cermak, to whom the assassin bore no ill-will nor ever intended any harm? We encounter the issue of transferred intent, whereunder it is sometimes said that the intent follows the bullet.

As we attempt to follow the badly aimed or otherwise errant bullet that misses or is deflected from A (the intended target) and then hits or comes perilously close to B (the unintended target), a matrix of no less than nine combinations of criminal harms or *acti rei* presents itself. On the vertical axis, the intended target may have been (1) aimed at but missed, (2) hit but only wounded, or (3) hit and killed. With respect to each of those possibilities, there are then three further possibilities on the horizontal axis. Those are where the unintended target may have been (1) hit and killed, (2) hit but only wounded, or (3) endangered but missed. Which combinations are appropriate subjects for the application of the transferred intent doctrine?

|  | Unintended Target Hit and Killed | Unintended Target Hit But Not Killed | Unintended Target Missed |
|---|---|---|---|
| Intended Target Missed | ? | ? | ? |
| Intended Target Hit But Not Killed | ? | ? | ? |
| Intended Target Hit and Killed | ? | ? | ? |

## When the Intended Victim Is Missed and the Unintended Victim Is Killed

The classic transferred intent scenario was that in which lethal force was directed toward an intended victim, missed its target, and killed an unintended victim. That was the context in which the doctrine was hammered out as part of English common law. The doctrine was early recognized at common law. Sir Matthew Hale, in *History of the Pleas of the Crown* (published posthumously in 1736), said, at 466:

> To these may be added the cases abovementioned, *viz.* if A. by malice fore-thought strikes at B. and missing him strikes C. whereof he dies, tho he never bore any malice to C. yet it is murder, and the law transfers the malice to the party slain; the like of poisoning. . . .

In analyzing the development of the "transferred intent" doctrine at the common law, *Evans v. State*, 349 A.2d 300 (Md. App. 1975), *aff'd* 362 A.2d 629 (Md. 1976) pointed out that although the notion that the intent "transferred" from one victim to another was, in effect, a legal fiction in the course of the law's development, the doctrine today is eminently sound in application:

> In earlier evolutionary stages, a legal fiction or a procedural device may have been at work. It is now clearly recognized, however, that what is involved is simply a rule of substantive law that the *mens rea* of murder as to anyone coupled with the *actus reus* of a homicide is sufficient to constitute the crime of murder. As long as the *mens rea* and the *actus reus* correspond in time, there is no requirement that the *mens rea* be directed specifically at the actual victim. The modern and better explanations of the doctrine point out the inappropriateness of the word "transferred" in the earlier case law. . . .

To a similar effect is LaFave and Scott, *Criminal Law* (1972), at 253, pointing out that the right result is reached but that the earlier "sort of reasoning is, of course, pure fiction":

> These proper conclusions of law as to criminal liability in the bad-aim situation are sometimes said to rest upon the ground of "transferred intent" . . . This sort of reasoning is, of course, pure fiction. A never really intended to harm C; but it is not necessary, in order to impose criminal liability upon A, to pretend that he did. What is really meant, by this round-about method of explanation, is that when one person (A) acts (or omits to act) with intent to harm another person, (B), but because of a bad aim he harms a third person (C) whom he did not intend to harm, the law considers him (as it ought) just as guilty as if he had actually harmed the intended victim. In other words, criminal homicide, battery, arson and malicious mischief do not require that the defendant cause harm to the intended victim; an unintended victim will do just as well. (Footnotes omitted.)

In that classic scenario, one corner of the matrix was readily filled in:

|  | Unintended Target Hit and Killed | Unintended Target Hit But Not Killed | Unintended Target Missed |
|---|---|---|---|
| Intended Target Missed | TRANSFERRED INTENT | ? | ? |
| Intended Target Hit But Not Killed | ? | ? | ? |
| Intended Target Hit and Killed | ? | ? | ? |

## The Conceptual Problem

Some of the early, and simplistic, explanations of the transferred intent doctrine gave rise to some troubling conceptual problems. The classic formulation envisioned a single *actus reus*—the death of the unintended victim. If the single *mens rea*—the specific intent to kill the intended victim, *e.g.*—could then be "transferred" to the unintended victim, the unitary *mens rea* could combine with the unitary *actus reus* to produce one unitary and doctrinally tidy crime. *Q.E.D.*

*(Continues)*

(*Continued*)

The simple arithmetic explanation proved inadequate, however, when there was more than one *actus reus*. Suppose, in addition to the death of the unintended victim, the intended victim had also been killed or, at least, wounded by the bullet in its flight. If the *mens rea* had to be used to prove the crime against the intended victim, what was then left to be "transferred" to the case involving the unintended victim? The conceptual problem also arose even where the deadly force missed the intended victim completely but the State nonetheless sought to charge the assailant with the inchoate crime of attempted murder or assault with intent to murder. If the *mens rea* were in limited supply, to which of two crimes should it be allocated? How could a single *mens rea* be made to do double duty? It may now seem silly but this sort of anguishing was, in the course of the law's development, a doctrinal stumbling block. . . .

By thinking of the *mens rea* in such finite terms—as some discrete unit that must be either here or there—we have created a linguistic problem for ourselves where no real-life problem existed. Criminal *acts*, consummated or inchoate, are discrete events that can be both pinpointed and counted. A *mens rea*, by contrast, is an elastic thing of unlimited supply. It neither follows nor fails to follow the bullet. It does not go anywhere. It remains in the brain of the criminal actor and never moves. It may combine with a single *actus reus* to make a single crime. It may as readily combine with a hundred *acti rei*, intended and unintended, to make a hundred crimes, consummated and inchoate. Unforeseen circumstances may multiply the criminal acts for which the criminal agent is responsible. A single state of mind, however, will control the fact of guilt and the level of guilt of them all.

## The Fate of the Intended Victim Is Immaterial

The transferred intent doctrine assumed greater utility once we began to free ourselves from some of the constraints that were the unintended consequences of its metaphorically inapt label. Once we stopped conceptualizing a defendant's *mens rea* as a single finite unit that might be "transferred" from one *actus reus* to another, we were free to view it as a pervasive state of moral fault or criminal purpose, of unlimited supply, that could influence any number of expected or unexpected consequences that might flow from it. The arithmetic problem was finessed, and the guilt of the assailant (or his accomplice) *vis-a-vis* the unintended victim was unaffected by the fate of the intended victim. As far as the case with respect to the unintended victim was concerned, it made no difference whether the intended victim had been (1) aimed at and missed, (2) hit but only wounded, or (3) hit and killed. It similarly made no difference whether the assailant (and/or accomplice) had been charged with a crime against the intended victim or not. There was no danger of depleting the *mens rea*.

The "transferred" *mens rea* *vis-a-vis* the unintended victim or victims will not be affected in any way, therefore, by what happens to the intended victim. If B is hit or endangered by a bullet aimed at A, whatever crime may have occurred with respect to B is a constant regardless of whether A is missed, injured, or killed. . . . It will be of critical legal significance, however, whether B, the unintended victim, is (1) endangered but not hit, (2) hit and killed, or (3) hit but only wounded. What happens to the intended target does not matter. It is what happens to the unintended but actual victim that controls our analysis. . . .

Thus, the doctrine of transferred intent operates with full force whenever the unintended victim is hit and killed. It makes no difference whether the intended victim is (1) missed, (2) hit and killed, or (3) hit and only wounded. It makes no difference whether the defendant is charged with a crime against the intended victim or not. The entire left-hand column of the matrix is now complete:

|  | Unintended Target Hit and Killed | Unintended Target Hit But Not Killed | Unintended Target Missed |
| --- | --- | --- | --- |
| Intended Target Missed | TRANSFERRED INTENT | ? | ? |
| Intended Target Hit But Not Killed | TRANSFERRED INTENT | ? | ? |
| Intended Target Hit and Killed | TRANSFERRED INTENT | ? | ? |

## When the Unintended Victim Is Neither Killed Nor Injured

The business of "transferring" the *mens rea* of a specific intent to kill from an intended victim to an unintended victim (or, more properly, simply applying it to the unintended victim) becomes far more complex when dealing with inchoate criminal homicides such as assault with intent to murder, attempted murder (in either degree), and attempted voluntary manslaughter. . . . When what is being considered is a charge of inchoate homicide—attempted murder, attempted voluntary manslaughter, or assault with intent to murder—and the unintended victim has *not* been hit or injured in any way, there will be *no* "transfer" of intent from the intended victim to the unintended victim. The pioneering analysis in this regard was done by Judge Alpert in *Harrod v. State,* 499 A.2d 959 (Md. App. 1985). The defendant, throwing a hammer, missed his intended victim and almost hit an infant lying in a nearby port-a-crib. The State urged that the criminal intent aimed at the intended victim should be transferred to the threat posed to the unintended victim. After reviewing generally the law of transferred intent, Judge Alpert pointed out that all of the cases surveyed by *Gladden v. State,* 330 A.2d 176 (Md. 1974), were cases in which the unintended victim had actually been injured. . . . This Court then held squarely that when there is no harm to the unintended victim, the doctrine of transferred intent is inapplicable:

> To extend the doctrine of transferred intent to cases where the [un]intended victim is not harmed would be untenable. The absurd result would be to make one criminally culpable for each unintended victim who, although in harm's way, was in fact not harmed by a missed attempt towards a specific person. We refuse, therefore, to extend the doctrine of transferred intent to cases where a third person is not in fact harmed.

499 A.2d 959. . . .

Thus, the right-hand side of the matrix is also filled in:

|  | Unintended Target Hit and Killed | Unintended Target Hit But Not Killed | Unintended Target Missed |
|---|---|---|---|
| Intended Target Missed | TRANSFERRED INTENT | ? | No TRANSFERRED INTENT |
| Intended Target Hit But Not Killed | TRANSFERRED INTENT | ? | No TRANSFERRED INTENT |
| Intended Target Hit and Killed | TRANSFERRED INTENT | ? | No TRANSFERRED INTENT |

## When the Unintended Victim Is Injured But Not Killed

It is the intermediate situation—when the unintended victim is actually hit though not killed—that has divided the Court of Appeals. . . . It is, after all, only with respect to *consummated homicide* that the law necessarily must concern itself with a notion like transferred intent. There is a necessity principle at work that is not present when no death has resulted.

In cases involving the actual consummated homicide of an unintended victim, the necessity is that the homicidal agent can only be convicted of the homicide if the law can attribute to him one of the murderous *mentes reae*. It is frequently impossible to do that without resort to the transferred intent doctrine. Unless one strains to bring the death of an unintended victim under the coverage of the common law felony-murder doctrine (the assault with intent to murder the intended victim would, after all, be a felony involving a threat to human life), it might be impossible to convict the homicidal agent for the death he unquestionably caused of the unintended victim. . . .

Similarly, a depraved-heart theory would not be available to the prosecution unless the defendant were aware of the actual or probable presence of the unintended victim within the field of fire and also aware, perhaps, that his aim was bad. In the homicide cases, where the transferred intent doctrine historically developed, it is frequently a choice between that theory of guilt or nothing. . . .

In terms of punishment, moreover, only consummated criminal homicide has that profusion of levels of guilt—normal forms, aggravated forms, mitigated forms—that depend on subtle differences in the *mens rea* of the homicidal agent. The punishment for the unintended homicide will necessarily depend on the type of intent that is "transferred." It could in certain cases be life imprisonment. If the *mens rea* directed toward the intended victim would only have produced murder in the second degree, however, the intent that is transferred will limit the punishment for the unintended death to

*(Continues)*

a maximum of thirty years. If the *mens rea* directed toward the intended victim were mitigated, moreover, the maximum penalty for the unintended death would be limited to ten years. Homicide law needs the transferred intent doctrine.

There are, by contrast, no unsolvable problems in punishing the unintended battery of a chance or unintended victim. . . . Indeed, the majority opinion in *Ford v. State*, 625 A.2d 984, at n. 14 (Md. 1993), noted that the non-application of the transferred intent doctrine to cases of inchoate criminal homicide does not create the punishment vacuum that might be present in cases of consummated criminal homicide:

> We note that refusal to apply transferred intent to attempted murder by no means relieves a defendant of criminal liability for the harm caused to unintended victims. The defendant clearly can be convicted of attempted murder as to the primary victim and some other crime, such as criminal battery, as to other victims.

In the case of unintended victims who are simply in harm's way and are not actually injured, the crime of reckless endangerment is also available to pick up much of the slack and to make resort to the transferred intent doctrine less compelling. . . .

At a very fundamental level, there is an argument, based on internal consistency, against using the transferred intent doctrine in cases of inchoate homicide. We are concerned, after all, with a single crime—assault with intent to murder (or its common law analogue of attempted murder). It would be randomly haphazard to say with respect to that single crime that the transferred intent doctrine sometimes applies and sometimes does not. If transferred intent will not be used (as it is not) to elevate a simple assault into an assault with intent to murder when the victim is not touched, it makes no sense to say that precisely the same crime will be elevated into an assault with intent to murder when the victim is touched. . . .

There is one further argument based on logical *inconsistency*. If an assault with intent to murder misses its target and hits an unintended victim and the aggravating *mens rea* were then to be "transferred" so as to transform the otherwise simple battery of the unintended victim into a constructive assault with intent to murder, what are the limits of such logic? Suppose the assaultive force that misfired had instead been unleashed with the intent to rob the targeted victim; would the otherwise simple battery of the unintended victim thereby become a constructive assault with intent to rob? Suppose the assaultive force that misfired had been with the intent to rape the targeted victim; would the otherwise simple battery of the unintended victim thereby become a constructive assault with intent to rape? . . . Why should one aggravating *mens rea* be transferable but others not? . . .

Consummated criminal homicide is, in the last analysis, *sui generis*. Many of its complexities, such as the transferred intent doctrine, simply do not travel well to other criminal climes. There is, moreover, no reason of necessity for making the transferred intent doctrine travel to climes other than that of actual, consummated criminal homicides. For the rest, the actuality of the real *mens rea* properly combined with its precisely related *actus reus* is enough to establish guilt at the appropriate level without any necessary resort to an intention-shifting legal fiction. The inchoate criminal homicides are in no need of such a device. Thus, we hold, the completed matrix looks like this:

| | Unintended Target Hit and Killed | Unintended Target Hit But Not Killed | Unintended Target Missed |
| --- | --- | --- | --- |
| Intended Target Missed | TRANSFERRED INTENT | No TRANSFERRED INTENT | No TRANSFERRED INTENT |
| Intended Target Hit But Not Killed | TRANSFERRED INTENT | No TRANSFERRED INTENT | No TRANSFERRED INTENT |
| Intended Target Hit and Killed | TRANSFERRED INTENT | No TRANSFERRED INTENT | No TRANSFERRED INTENT |

## Reducing to the Lowest Common Denominator

The full matrix, although helpful perhaps to illustrate the evolutionary development of the transferred intent doctrine, is, in the last analysis, unnecessarily redundant. For more efficient reference, it is well-advised to reduce the doctrine's applicability to its lowest common denominator:

| Unintended Target Killed | Unintended Target Not Killed |
|---|---|
| TRANSFERRED INTENT | No TRANSFERRED INTENT |

In a nutshell:

THE FATE OF THE INTENDED TARGET IS IMMATERIAL. IF THE UNINTENDED VICTIM IS KILLED, THE TRANSFERRED INTENT DOCTRINE APPLIES. IF THE UNINTENDED VICTIM IS NOT KILLED, THE TRANSFERRED INTENT DOCTRINE DOES NOT APPLY.

## The Jury Instruction in This Case

The unintended victim, Tiffany Evans, was not killed. It was, therefore, error to have instructed the jury on the subject of transferred intent. The error was obviously prejudicial in that the State was thereby erroneously relieved of its obligation, on the charge of assault with intent to murder, to prove the required *mens rea* of a specific intent to kill Tiffany Evans. The conviction for assault with intent to murder must be reversed.

## The Reckless Endangerment

The erroneously given instruction on transferred intent, on the other hand, had no effect on the reckless endangerment conviction and the appellant, indeed, makes no claim in that regard. . . .

## Notes and Questions

1. The court rules that Harvey cannot be convicted of assaulting Tiffany Evans with the intent to murder her (a charge equivalent to attempted murder). Could Harvey be convicted of the attempted murder of anyone? If so, does this fact help make sense of the conclusion that she should not (also) be guilty of the attempted murder of Evans, who was inadvertently struck by the bullet? On the other hand, if Evans had been killed and not merely wounded by the bullet, could Harvey be convicted of the murder of anyone other than Evans? Does this fact help make sense of the different outcomes regarding murder, attempted murder, and "bad aim" cases that seek to rely on the doctrine of transferred intent?[2]

2. In *Harvey*, all four shots missed the intended victim but one of the errant bullets struck Tiffany Evans in the leg, wounding but not killing her. Assume that a second bullet struck and killed Harvey's brother, whom she was seeking to protect, not kill. Of what additional crime, if any, is Harvey guilty?

3. If Harvey had instructed her companion not to shoot a rival gang member but rather to shoot out the tires of his car, and a bullet went astray and struck and wounded a bystander, is Harvey guilty of attempted murder? Can a viable distinction be made between an act involving an unintended victim, involving the same type of harm contemplated, and an act involving a harm that is altogether different from the type contemplated? *Cf., Regina v. Pembliton, supra.*

4. Harvey's conviction for "reckless endangerment" was affirmed, even though her conviction for assault with the intent to commit murder was reversed. What justifies this result? Consult MPC § 2.02(c) for a definition of "recklessly."

## Criminal Negligence and Recklessness

We have considered several crimes that involve "intentional," "purposeful," or "knowing" conduct. The following case examines the remaining types of mental culpability recognized by the MPC: "recklessly" and "negligently."

> ### *People v. Haney*, 284 N.E.2d 564 (N.Y. 1972)
>
> Jasen, Judge.
>
> On this appeal, by the People, the question posed concerns the sufficiency of the evidence before a Grand Jury to support an indictment against the defendant for criminally negligent homicide in violation of section 125.10 of the Penal Law. The charge arose from an automobile accident which resulted in the death of Angela Palazzo, a pedestrian.
>
> The minutes of the Grand Jury contain the following pertinent testimony. The associate medical examiner read from the medical report of the autopsy which revealed the cause of Mrs. Palazzo's death. According to the medical report, it was due to multiple external contusions, fracture of the vertebra and lower extension bleeding in both chest cavities—conditions which, in the opinion of the medical examiner, were brought about by a "very considerable amount of force." An eyewitness to the accident testified that at approximately 6:30 A.M. on April 28, 1968, she observed the deceased step off a city bus that had stopped at the corner of Castleton and Bard Avenues in Staten Island. After the bus had continued on and the signal light turned green in her favor, the deceased started to cross Castleton Avenue at the intersection. While in the middle of the street, she was struck by an oncoming automobile driven by the defendant, Booker W. Haney. The witness added that the car was "some distance away" from the intersection and "coming fast" when she first saw it, and that prior to the moment of impact, she did not hear a horn or the screeching of brakes. Patrolman Thomas Roche of the automobile investigation squad, who conducted the investigation of the accident, stated that the automobile which struck the deceased eventually crashed into a utility pole, causing considerable damage to it. In addition, he said that Mrs. Palazzo's body was located 186 feet west of Bard Avenue, approximately 100 feet in front of the point where the defendant's automobile had finally stopped. His investigation further determined that, based on the lengthy skid marks from the defendant's car, the defendant's car was traveling at least 52 miles per hour at the time it struck the utility pole. Detective John Plohetski testified that after he placed the defendant under arrest, and informed him of his rights, the defendant kept repeating, "I didn't mean to hit her, I didn't mean to hit her."
>
> The indictment returned against the defendant charged him with criminally negligent homicide "in that, among other things, he drove a vehicle at a high, reckless, dangerous and unlawful rate of speed; in that he failed and neglected to stop said vehicle at the intersection . . . although the traffic signal situation at said intersection was red . . . and did thereby cause the death of the said Angela Palazzo."
>
> The defendant, after pleading not guilty, made a motion for inspection of the Grand Jury minutes. Supreme Court, Criminal Term, granted the motion and dismissed the indictment, holding that "[t]he evidence before the Grand Jury, even though unexplained and uncontraindicated, would not justify conviction by a trial jury. . . ." (298 N.Y.S.2d 415, 422). The Appellate Division unanimously affirmed . . . .
>
> Section 125.10 of the Penal Law provides that a person is guilty of the crime of criminally negligent homicide when, with **criminal negligence**, he causes the death of another person. Subdivision 4 of section 15.05 of the Penal Law defines "Criminal negligence" as follows: "A person acts with criminal negligence with respect to a result or to a circumstance described by a statute defining an offense when he fails to perceive a substantial and unjustifiable risk that such result will occur or that such circumstance exists. The risk must be of such nature and degree that the failure to perceive it constitutes a gross deviation from the standard of care that a reasonable person would observe in the situation." . . .
>
> A persistent problem, faced by the courts and legislatures alike, has been the formulation of the "extra" qualities that distinguish unintended homicides, which give rise to criminal liability, from those which, at most, produce civil liability for negligence. The Model Penal Code (Tent. Draft No.4 [April 25, 1955], Comments to § 2.02, at p. 128) observes, concerning the judicial and statutory definitions of conduct causing death which is criminal, although unintentional: "Thus under statutes, as at common law, the concept of criminal negligence has been left to judicial definition and the definitions vary greatly in their terms. As Jerome Hall has put it, the 'judicial essays run in terms of "wanton and wilful negligence," "gross negligence," and more illuminating yet, "that degree of

negligence that is more than the negligence required to impose tort liability." The apex of ambiguity is "wilful, wanton negligence" which suggests a triple contradiction—"negligence" implying inadvertence; "wilful," intention; and "wanton," recklessness.' (Citation omitted.) Much of this confusion is dispelled, in our view, by a clear-cut distinction between recklessness and negligence, in terms of the actor's awareness of the risk involved."

Cognizant of this problem, and in an endeavor to "crystallize [this] area of culpability and liability," the Legislature incorporated in the revised Penal Law the dichotomy proposed by the Model Penal Code. Thus, the revised Penal Law makes unintended homicide manslaughter in the second degree, when it is committed "recklessly" (Penal Law, § 125.15) and when committed "negligently," though not recklessly, it is criminally negligent homicide. (Penal Law, § 125.10.)

The distinction between these two crimes is provided in section 15.05 (subds. 3, 4) of the Penal Law, which specifically describes the mental state requisite for each. The reckless offender is aware of the proscribed risk and "consciously disregards" it, while the criminally negligent offender is not aware of the risk created and, hence, cannot be guilty of consciously disregarding it. Since the criminally negligent offender's liability arises only from a culpable failure to perceive the risk, his culpability is obviously less than that of the reckless offender who consciously disregards the risk. It is, however, "appreciably greater than that required for ordinary civil negligence by virtue of the 'substantial and unjustifiable' character of the risk involved and the factor of 'gross deviation' from the ordinary standard of care." Enactment of section 125.10 represents a marked change from prior law as the former Penal Law contained no crime truly equivalent to it. The present law lacks the moral implication of murder (Penal Law, § 125.25) or manslaughter in the first or second degree (Penal Law, §§ 125.15, 125.20), each of which involves awareness of the harm which will (or in some degrees probably will) result from the offender's conduct Criminally negligent homicide, in essence, involves the failure to perceive the risk in a situation where the offender has a legal duty of awareness. It, thus, serves to provide an offense applicable to conduct which is obviously socially undesirable. "[It proscribes] conduct which is inadvertent as to risk only because the actor is insensitive to the interests and claims of other persons in society." (Model Penal Code, Tent. Draft No. 9, *supra,* at p. 53.) The Legislature, in recognizing such conduct as criminal, endeavored to stimulate people towards awareness of the potential consequences of their conduct and influence them to avoid creating undesirable risks.

What amounts to a violation of this section depends, of course, entirely on the circumstances of the particular conduct. Whether in those circumstances the act or acts causing death involved a substantial and unjustifiable risk, and whether the failure to perceive it was such as to constitute a gross deviation from the standard of care which a reasonable man would have observed under the same circumstances, are questions that generally must be left directly to the trier of the facts. In other words, "[t]he tribunal must evaluate the actor's failure of perception and determine whether, under all the circumstances, it was serious enough to be condemned." (Model Penal Code, Tent. Draft No. 4, *supra,* at p. 126.) While it is difficult to clarify further these questions, it would seem sufficiently clear that for proper determination of these questions, two main considerations should be emphasized. Firstly, criminal liability cannot be predicated upon every careless act merely because its carelessness results in another's death; and, secondly, the elements of the crime "preclude the proper condemnation of inadvertent risk creation unless 'the significance of the circumstances of fact would be apparent to one who shares the community's general sense of right and wrong.'" (Model Penal Code, Tent. Draft No. 9, *supra,* at p. 53, citing Hart, The Aims of the Criminal Law, 23 Law & Contemp. Prob. 401, 417.)

Turning to the case before us, upon consideration of the totality of the circumstances surrounding the defendant's conduct, we conclude that the People put forward before the Grand Jury sufficient evidence to "warrant a conviction by the trial jury." In other words, the evidence presented to the Grand Jury was the equivalent of prima facie proof that the crime charged had been committed by the defendant.

The evidence discloses that Mrs. Palazzo was struck by the defendant's car while crossing Castleton Avenue at the intersection of Castleton and Bard Avenues with the traffic signal green in her favor. At the time she was struck, Mrs. Palazzo was half-way across Castleton Avenue. The evidence also indicates that the defendant failed to obey the red traffic signal at the intersection of Castleton and Bard Avenues and stop his motor vehicle, as required by law, before proceeding through the intersection. Additionally, there is evidence that he was traveling at a high rate of speed

*(Continues)*

*(Continued)*

(approximately 52 mph) just before and at impact. Furthermore, no visual obstruction to the sighting of Mrs. Palazzo, lawfully crossing at the intersection, was apparent. It should be abundantly clear that such conduct cannot be characterized as mere carelessness, sufficient only to establish liability for ordinary civil negligence. Rather, from this evidence, and the reasonable inferences to be drawn therefrom, a jury could find the defendant guilty of criminally negligent homicide. . . . To hold otherwise, and excuse the flagrant disregard manifested here, would sanction conduct at which the statute was clearly aimed, and, in effect, abolish the crime of criminally negligent homicide in all homicides resulting from a misuse of a motor vehicle.

Accordingly, the order appealed from should be reversed, and the indictment reinstated.

## Notes and Questions

1. What separates "reckless" conduct from criminally "negligent" conduct? What separates *criminal* negligence from *civil* negligence? Which mental state most accurately describes Haney's *mens rea*?
2. Why is "reckless" conduct generally subject to harsher punishment than criminally negligent conduct? Why (generally) is civilly negligent conduct not "punished" at all?
3. If a defendant is genuinely unaware that his or her behavior risks injuring another, what justification is there for imposing criminal punishment? We consider this question—along with many others related to *mens rea*—in greater detail in Chapter 6, when we study criminal homicide.

## The Act

The criminal law does not punish evil thoughts alone; *mens rea* that does not culminate in some sort of action (or, as we shall see, the failure to act when there is a legal duty to do so) is immune from criminal sanctions. This basic principle is tested in application in the so-called *inchoate crimes*—such as solicitation, conspiracy, and attempts—where one question involves *how much* action must be taken before the criminal law intervenes. We explore this and other issues involving inchoate crimes in Chapter 12. Here, we examine the act, or **actus reus**, principle from another perspective.

The first general principle of liability set forth in the MPC is the requirement for a voluntary act.

(1) A person is not guilty of an offense unless his liability is based on conduct which includes a voluntary act or the omission to perform an act of which he is physically capable.
(2) The following are not voluntary acts within the meaning of this Section:
    (a) a reflex or convulsion;
    (b) a bodily movement during unconsciousness or sleep;
    (c) conduct during hypnosis or resulting from hypnotic suggestion;
    (d) a bodily movement that otherwise is not a product or the effort or determination of the actor, either conscious or habitual.

## Voluntary Act Requirement

## *Martin v. State,* 17 So.2d 427 (Ala. App. 1944)

Simpson, Judge.

Appellant was convicted of being drunk on a public highway, and appeals. Officers of the law arrested him at his home and took him onto the highway, where he allegedly committed the proscribed acts, viz., manifested a drunken condition by using loud and profane language.

The pertinent provisions of our statute are: "Any person who, while intoxicated or drunk, appears in any public place where one or more persons are present, . . . and manifests a drunken condition by boisterous or indecent conduct, or loud and profane discourse, shall, on conviction, be fined," etc. Code 1940, Title 14, Section 120.

Under the plain terms of this statute, a voluntary appearance is presupposed. The rule has been declared, and we think it sound, that an accusation of drunkenness in a designated public place cannot be established by proof that the accused, while in an intoxicated condition, was involuntarily and forcibly carried to that place by the arresting officer.

Conviction of appellant was contrary to this announced principle and, in our view, erroneous. It appears that no legal conviction can be sustained under the evidence, so, consonant with the prevailing rule, the judgment of the trial court is reversed and one here rendered discharging appellant. . . .

## Notes and Questions

1. Martin was (a) intoxicated (b) in a public place (c) where people were present and (d) used loud and profane language, as prohibited by Title 14, § 120 of the Alabama Code. Is the court correct that "the plain terms of the statute" require reversal? Why would it be unfair to uphold Martin's conviction under "the plain terms of the statute"?
2. Is there another theory under which Martin's conviction might be challenged? Specifically, might it be argued that Martin lacked the *mens rea* necessary to support a conviction? Which is the more "basic" principle: that Martin did not appear intoxicated in public by his "voluntary acts" or that he did not possess the requisite *mens rea*?

## *People v. Decina*, 138 N.E.2d 799 (N.Y. 1956)

Froessel, Judge.

At about 3:30 P.M. on March 14, 1955, a bright, sunny day, defendant was driving, alone in his car, in a northerly direction on Delaware Avenue in the city of Buffalo. The portion of Delaware Avenue here involved is 60 feet wide. At a point south of an overhead viaduct of the Erie Railroad, defendant's car swerved to the left, across the center line in the street, so that it was completely in the south lane, traveling 35 to 40 miles per hour.

It then veered sharply to the right, crossing Delaware Avenue and mounting the easterly curb at a point beneath the viaduct and continued thereafter at a speed estimated to have been about 50 or 60 miles per hour or more. During this latter swerve, a pedestrian testified that he saw defendant's hand above his head; another witness said he saw defendant's left arm bent over the wheel, and his right hand extended towards the right door.

A group of six schoolgirls were walking north on the easterly sidewalk of Delaware Avenue, two in front and four slightly in the rear, when defendant's car struck them from behind. One of the girls escaped injury by jumping against the wall of the viaduct. The bodies of the children struck were propelled northward onto the street and the lawn in front of a coal company, located to the north of the Erie viaduct on Delaware Avenue. Three of the children, 6 to 12 years old, were found dead on arrival by the medical examiner, and a fourth child, 7 years old, died in a hospital two days later as a result of injuries sustained in the accident.

After striking the children, defendant's car continued on the easterly sidewalk, and then swerved back onto Delaware Avenue once more. It continued in a northerly direction, passing under a second viaduct before it again veered to the right and remounted the easterly curb, striking and breaking a metal lamppost. With its horn blowing steadily apparently because defendant was "stooped over" the steering wheel the car proceeded on the sidewalk until it finally crashed through a 7¼-inch brick wall of a grocery store, injuring at least one customer and causing considerable property damage.

When the car came to a halt in the store, with its horn still blowing, several fires had been ignited. Defendant was stooped over in the car and was "bobbing a little." To one witness he appeared dazed, to another unconscious, lying back with his hands off the wheel. Various people present shouted to defendant to turn off the ignition of his car, and "within a matter of seconds the horn stopped blowing and the car did shut off."

Defendant was pulled out of the car by a number of bystanders and laid down on the sidewalk. To a policeman who came on the scene shortly he appeared "injured, dazed"; another witness said that "he looked as though he was knocked out, and his arm seemed to be bleeding." An injured customer in the store, after receiving first aid, pressed defendant for an explanation of the accident and he told her: "I blacked out from the bridge." . . .

[The defendant was transported to a hospital and later interviewed by Dr. Wechter, whose testimony regarding the interview was admitted at trial. The court ruled that the doctor's testimony was erroneously admitted because it violated the physician-patient privilege. It then considered whether the indictment should have been dismissed.]

*(Continues)*

[Dr. Wechter] asked defendant how he felt and what had happened. Defendant, who still felt a little dizzy or blurry, said that as he was driving he noticed a jerking of his right hand, which warned him that he might develop a convulsion, and that as he tried to steer the car over to the curb he felt himself becoming unconscious, and he thought he had a convulsion. He was aware that children were in front of his car, but did not know whether he had struck them.

Defendant then proceeded to relate to Dr. Wechter his past medical history, namely, that at the age of 7 he was struck by an auto and suffered a marked loss of hearing. In 1946 he was treated in this same hospital for an illness during which he had some convulsions. Several burr holes were made in his skull and a brain abscess was drained. Following this operation defendant had no convulsions from 1946 through 1950. In 1950 he had four convulsions, caused by scar tissue on the brain. From 1950 to 1954 he experienced about 10 or 20 seizures a year, in which his right hand would jump although he remained fully conscious. In 1954, he had 4 or 5 generalized seizures with loss of consciousness, the last being in September, 1954, a few months before the accident. Thereafter he had more hospitalization, a spinal tap, consultation with a neurologist, and took medication daily to help prevent seizures.

On the basis of this medical history, Dr. Wechter made a diagnosis of Jacksonian epilepsy, and was of the opinion that defendant had a seizure at the time of the accident. . . .

Defendant was indicted and charged with violating section 1053-a of the Penal Law, Consol. Laws, c. 40. Following his conviction, after a demurrer to the indictment was overruled, the Appellate Division . . . granted a new trial upon the ground that the "transactions between the defendant and Dr. Wechter were between physician and patient for the purpose of treatment and that treatment was accomplished," and that evidence thereof should not have been admitted. . . .

We turn first to the subject of defendant's cross appeal, namely, that his demurrer should have been sustained, since the indictment here does not charge a crime. The indictment states essentially that defendant, knowing "that he was subject to epileptic attacks or other disorder rendering him likely to lose consciousness for a considerable period of time," was culpably negligent "in that he consciously undertook to and did operate his Buick sedan on a public highway" and "while so doing" suffered such an attack which caused said automobile "to travel at a fast and reckless rate of speed, jumping the curb and driving over the sidewalk" causing the death of 4 persons. In our opinion, this clearly states a violation of section 1053-a of the Penal Law. The statute does not require that a defendant must deliberately intend to kill a human being, for that would be murder. Nor does the statute require that he knowingly and consciously follow the precise path that leads to death and destruction. It is sufficient, we have said, when his conduct manifests a "disregard of the consequences which may ensue from the act, and indifference to the rights of others. No clearer definition, applicable to the hundreds of varying circumstances that may arise, can be given. Under a given state of facts, whether negligence is culpable is a question of judgment." *People v. Angelo,* 159 N.E. 394, 396 (N.Y.).

Assuming the truth of the indictment, as we must on a demurrer, this defendant knew he was subject to epileptic attacks and seizures that might strike at any time. He also knew that a moving motor vehicle uncontrolled on [a] public highway is a highly dangerous instrumentality capable of unrestrained destruction. With this knowledge, and without anyone accompanying him, he deliberately took a chance by making a conscious choice of a course of action, in disregard of the consequences which he knew might follow from his conscious act, and which in this case did ensue. How can we say as a matter of law that this did not amount to culpable negligence within the meaning of section 1053-a?

To hold otherwise would be to say that a man may freely indulge himself in liquor in the same hope that it will not affect his driving, and if it later develops that ensuing intoxication causes dangerous and reckless driving resulting in death, his unconsciousness or involuntariness at that time would relieve him from prosecution under the statute. His awareness of a condition which he knows may produce such consequences as here, and his disregard of the consequences, renders him liable for culpable negligence, as the courts below have properly held.

To have a sudden sleeping spell, an unexpected heart or other disabling attack, without any prior knowledge or warning thereof, is an altogether different situation, and there is simply no basis for comparing such cases with the flagrant disregard manifested here.

It is suggested in the dissenting opinion that a new approach to licensing would prevent such disastrous consequences upon our public highways. But would it and how and when? The mere possession of a driver's license is no defense to a prosecution under section 1053-a; nor does it assure continued ability to drive during the period of the license. . . . Section 1053-a places a personal responsibility on each driver of a vehicle whether licensed or not and not upon a licensing agency.

Accordingly, the Appellate Division properly sustained the lower court's order overruling the demurrer, as well as its denial of the motion in arrest of judgment on the same ground. . . .

### Notes and Questions

1. If the accident in which Decina was involved resulted from his suffering an epileptic seizure and losing consciousness, how was it (if at all) attributable to his voluntary actions?
2. If Decina had no previous history of seizures, should the charge against him stand?
3. Several cases have dealt with "somnambulism" (sleep-walking) and "automatism" or "unconsciousness" as defenses to criminal charges. *See State v. Mercer,* 165 S.E.2d 328 (N.C. 1969); *McClain v. State,* 678 N.E.2d 104 (Ind. 1997); *Polston v. State,* 685 P.2d 1 (Wyo. 1984); *Lewis v. State,* 27 S.E.2d 659 (Ga. 1943); *Fitzhugh v. State,* 43 So. 2d 831 (Ala. App.).[3]

## Act Versus Status

## *Robinson v. California,* 370 U.S. 660, 82 S. Ct. 1417, 8 L. Ed. 2d 758 (1962)

Mr. Justice Stewart delivered the opinion of the Court.

A California statute makes it a criminal offense for a person to "be addicted to the use of narcotics." This appeal draws into question the constitutionality of that provision of the state law, as construed by the California courts in the present case.

The appellant was convicted after a jury trial in the Municipal Court of Los Angeles. The evidence against him was given by two Los Angeles police officers. . . .

The trial judge instructed the jury that the statute made it a misdemeanor for a person "either to use narcotics, or to be addicted to the use of narcotics. . . . That portion of the statute referring to the 'use' of narcotics is based upon the 'act' of using. That portion of the statute referring to 'addicted to the use' of narcotics is based upon a condition or **status**. They are not identical. . . . To be addicted to the use of narcotics is said to be a status or condition and not an act. It is a continuing offense and differs from most other offenses in the fact that [it] is chronic rather than acute; that it continues after it is complete and subjects the offender to arrest at any time before he reforms. The existence of such a chronic condition may be ascertained from a single examination, if the characteristic reactions of that condition be found present."

The judge further instructed the jury that the appellant could be convicted under a general verdict if the jury agreed *either* that he was of the "status" *or* had committed the "act" denounced by the statute. "All that the People must show is either the defendant did use a narcotic in Los Angeles County, or that while in the City of Los Angeles he was addicted to the use of narcotics. . . ."

Under these instructions the jury returned a verdict finding the appellant "guilty of the offense charged." . . .

The broad power of a State to regulate the narcotic drugs traffic within its borders is not here in issue. . . .

Such regulation, it can be assumed, could take a variety of valid forms. A State might impose criminal sanctions, for example, against the unauthorized manufacture, prescription, sale, purchase, or possession of narcotics within its borders. In the interest of discouraging the violation of such laws, or in the interest of the general health or welfare of its inhabitants, a State might establish a program of compulsory treatment for those addicted to narcotics. Such a program of treatment might require periods of involuntary confinement. And penal sanctions might be imposed for failure to comply with established compulsory treatment procedures. *Cf. Jacobson v. Massachusetts,* 197 U.S. 11. Or a State might choose to attack the evils of narcotics traffic on broader fronts also—through

*(Continues)*

public health education, for example; or by efforts to ameliorate the economic and social conditions under which those evils might be thought to flourish. In short, the range of valid choice which a State might make in this area is undoubtedly a wide one, and the wisdom of any particular choice within the allowable spectrum is not for us to decide. Upon that premise we turn to the California law in issue here.

It would be possible to construe the statute under which the appellant was convicted as one which is operative only upon proof of the actual use of narcotics within the State's jurisdiction. But the California courts have not so construed this law. Although there was evidence in the present case that the appellant had used narcotics in Los Angeles, the jury were instructed that they could convict him even if they disbelieved that evidence. The appellant could be convicted, they were told, if they found simply that the appellant's "status" or "chronic condition" was that of being "addicted to the use of narcotics." And it is impossible to know from the jury's verdict that the defendant was not convicted upon precisely such a finding. . . .

This statute, therefore, is not one which punishes a person for the use of narcotics, for their purchase, sale or possession, or for antisocial or disorderly behavior resulting from their administration. It is not a law which even purports to provide or require medical treatment. Rather, we deal with a statute which makes the "status" of narcotic addiction a criminal offense, for which the offender may be prosecuted "at any time before he reforms." California has said that a person can be continuously guilty of this offense, whether or not he has ever used or possessed any narcotics within the State, and whether or not he has been guilty of any antisocial behavior there. . . .

It is unlikely that any State at this moment in history would attempt to make it a criminal offense for a person to be mentally ill, or a leper, or to be afflicted with a venereal disease. A State might determine that the general health and welfare require that the victims of these and other human afflictions be dealt with by compulsory treatment, involving quarantine, confinement, or sequestration. But, in the light of contemporary human knowledge, a law which made a criminal offense of such a disease would doubtless be universally thought to be an infliction of cruel and unusual punishment in violation of the Eighth and Fourteenth Amendments.

We cannot but consider the statute before us as of the same, category. In this Court counsel for the State recognized that narcotic addiction is an illness. Indeed, it is apparently an illness which may be contracted innocently or involuntarily. We hold that a state law which imprisons a person thus afflicted as a criminal, even though he has never touched any narcotic drug within the State or been guilty of any irregular behavior there, inflicts a cruel and unusual punishment in violation of the Fourteenth Amendment. To be sure, imprisonment for ninety days is not, in the abstract, a punishment which is either cruel or unusual. But the question cannot be considered in the abstract. Even one day in prison would be cruel and unusual punishment for the "crime" of having a common cold. . . .

Reversed. . . .

Mr. Justice Harlan, concurring.

I am not prepared to hold that on the present state of medical knowledge it is completely irrational and hence unconstitutional for a State to conclude that narcotics addiction is something other than an illness, nor that it amounts to cruel and unusual punishment for the State to subject narcotics addicts to its criminal law. . . . Since addiction alone cannot reasonably be thought to amount to more than a compelling propensity to use narcotics, the effect of this instruction was to authorize criminal punishment for a bare desire to commit a criminal act.

If the California statute reaches this type of conduct, and for present purposes we must accept the trial court's construction as binding, it is an arbitrary imposition which exceeds the power that a State may exercise in enacting its criminal law. . . .

Mr. Justice White, dissenting.

If appellant's conviction rested upon sheer status, condition or illness or if he was convicted for being an addict who had lost his power of self-control, I would have other thoughts about this case. But this record presents neither situation. . . .

I am not at all ready to place the use of narcotics beyond the reach of the States' criminal laws. I do not consider appellant's conviction to be a punishment for having an illness or for simply being in some status or condition, but rather a conviction for the regular, repeated or habitual use of narcotics immediately prior to his arrest and in violation of the California law. . . . The Court recognizes no degrees of addiction. The Fourteenth Amendment is today held to bar any prosecution for addiction regardless of the degree or frequency of use, and the Court's opinion bristles with indications of further consequences. If it is "cruel and unusual punishment" to convict appellant for addiction, it is difficult to understand why it would be any less offensive to the Fourteenth Amendment to convict him for use on the same evidence of use which proved he was an addict. It is significant that in purporting to reaffirm the power of the States to deal with the narcotics traffic, the Court does not include among the obvious powers of the State the power to punish for the use of narcotics. I cannot think that the omission was inadvertent. . .

## Notes and Questions

1. In *Robinson*, the majority opinion concludes that "[e]ven one day in prison would be a cruel and unusual punishment for the 'crime' of having a common cold." What justifies this assertion?
2. Is narcotics addiction an "illness"? Under what criteria? In the mythical society of Erewhon, Samuel Butler described a society in which pulmonary consumption (tuberculosis) is a crime and acts such as forgery, robbery, and arson stem from illness requiring hospitalization.[4]
3. Is narcotics addiction a "status"? If so, what conduct, presumably, leads to such a status? Does it matter how such a status is acquired?
4. If criminal punishment would be "cruel and unusual" for "the 'crime' of having a common cold," would it be permissible to punish one who has a cold for sneezing? For sneezing in public? Would it be permissible to punish a narcotics addict for using narcotics?

## *Powell v. Texas*, 392 U.S. 514, 88 S. Ct. 2145, 20 L. Ed. 2d 1254 (1968)

Mr. Justice Marshall announced the judgment of the Court and delivered an opinion in which the Chief Justice, Mr. Justice Black, and Mr. Justice Harlan join.

In late December 1966, appellant was arrested and charged with being found in a state of intoxication in a public place, in violation of Texas Penal Code, Art 477 (1952), which reads as follows:

"Whoever shall get drunk or be found in a state of intoxication in any public place, or at any private house except his own, shall be fined not exceeding one hundred dollars."

Appellant was tried in the Corporation Court of Austin, Texas, found guilty, and fined $20. He appealed to the County Court at Law No. 1 of Travis County, Texas, where a trial de novo was held. His counsel urged that appellant was "afflicted with the disease of chronic alcoholism," that "his appearance in public [while drunk was] . . . not of his own volition," and therefore that to punish him criminally for that conduct would be cruel and unusual, in violation of the Eighth and Fourteenth Amendments to the United States Constitution.

The trial judge in the county court, sitting without a jury, made certain findings of fact, but ruled as a matter of law that chronic alcoholism was not a defense to the charge. He found appellant guilty, and fined him $50. There being no further right to appeal within the Texas judicial system, appellant appealed to this Court. . . .

### I.

The principal testimony was that of Dr. David Wade, a Fellow of the American Medical Association, duly certificated in psychiatry. His testimony consumed a total of 17 pages in the trial transcript. . . . Dr. Wade sketched the outlines of the "disease" concept of alcoholism; noted that there is no generally accepted definition of "alcoholism"; alluded to the ongoing debate within the medical profession over whether alcohol is actually physically "addicting" or merely psychologically "habituating"; and concluded that in either case a "chronic alcoholic" is an "involuntary drinker," who is "powerless not to drink," and who "loses his self-control over his drinking." He testified that he had examined appellant, and that appellant is a "chronic alcoholic," who "by the time he has reached [the state of intoxication] . . . is not able to control his behavior, and [who] . . . has reached this point because he has an uncontrollable compulsion to drink." Dr. Wade also responded in the negative to the question whether appellant has "the willpower to resist the constant excessive consumption of alcohol." . . .

*(Continues)*

(*Continued*)

On cross-examination, Dr. Wade admitted that when appellant was sober he knew the difference between right and wrong, and he responded affirmatively to the question whether appellant's act of taking the first drink in any given instance when he was sober was a "voluntary exercise of his will." Qualifying his answer, Dr. Wade stated that "these individuals have a compulsion, and this compulsion, while not completely overpowering, is a very strong influence, an exceedingly strong influence, and this compulsion coupled with the firm belief in their mind that they are going to be able to handle it from now on causes their judgment to be somewhat clouded."

Appellant testified concerning the history of his drinking problem. He reviewed his many arrests for drunkenness; testified that he was unable to stop drinking; stated that when he was intoxicated he had no control over his actions and could not remember them later, but that he did not become violent; and admitted that he did not remember his arrest on the occasion for which he was being tried. On cross-examination, appellant admitted that he had had one drink on the morning of the trial and had been able to discontinue drinking. In relevant part, the cross-examination went as follows:

"Q. You took that one at eight o'clock because you wanted to drink?
"A. Yes, sir.
"Q. And you knew that if you drank it, you could keep on drinking and get drunk?
"A. Well, I was supposed to be here on trial, and I didn't take but that one drink.
"Q. You knew you had to be here this afternoon, but this morning you took one drink and then you knew that you couldn't afford to drink any more and come to court; is that right?
"A. Yes, sir, that's right.
"Q. So you exercised your will power and kept from drinking anything today except that one drink?
"A. Yes, sir, that's right". . . .

On redirect examination, appellant's lawyer elicited the following:

"Q. Leroy, isn't the real reason why you just had one drink today because you just had enough money to buy one drink?
"A. Well, that was just give to me.
"Q. In other words, you didn't have any money with which you could buy any drinks yourself?
"A. No, sir, that was give to me.
"Q. And that's really what controlled the amount you drank this morning, isn't it?
"A. Yes, sir.
"Q. Leroy, when you start drinking, do you have any control over how many drinks you can take?
"A. No, sir."

Evidence in the case then closed. The State made no effort to obtain expert psychiatric testimony of its own, or even to explore with appellant's witness the question of appellant's power to control the frequency, timing, and location of his drinking bouts, or the substantial disagreement within the medical profession concerning the nature of the disease, the efficacy of treatment and the prerequisites for effective treatment. . . .

Following this abbreviated exposition of the problem before it, the trial court indicated its intention to disallow appellant's claimed defense of "chronic alcoholism." Thereupon defense counsel submitted, and the trial court entered, the following "findings of fact":

"(1) That chronic alcoholism is a disease which destroys the afflicted person's will power to resist the constant, excessive consumption of alcohol.

"(2) That a chronic alcoholic does not appear in public by his own volition but under a compulsion symptomatic of the disease of chronic alcoholism.

"(3) That Leroy Powell, defendant herein, is a chronic alcoholic who is afflicted with the disease of chronic alcoholism."

Whatever else may be said of them, those are not "findings of fact" in any recognizable, traditional sense in which that term has been used in a court of law; they are the premises of a syllogism transparently designed to bring this case within the scope of this Court's opinion in *Robinson v. California*, 370 U.S. 660 (1962). Nonetheless, the dissent would have us adopt these "findings" without critical examination; it would use them as the basis for a constitutional holding that "a person may not be punished if the condition essential to constitute the defined crime is part of the pattern of his disease and is occasioned by a compulsion symptomatic of the disease."

The difficulty with that position . . . is that it goes much too far on the basis of too little knowledge. In the first place, the record in this case is utterly inadequate to permit the sort of informed and responsible adjudication which alone can support the announcement of an important and wide-ranging new constitutional principle. We know very little about the circumstances surrounding the drinking bout which resulted in this conviction, or about Leroy Powell's drinking problem, or indeed about alcoholism itself. The trial hardly reflects the sharp legal and evidentiary clash between fully prepared adversary litigants which is traditionally expected in major constitutional cases. The State put on only one witness, the arresting officer. The defense put on three—a policeman who testified to appellant's long history of arrests for public drunkenness, the psychiatrist, and appellant himself.

Furthermore, the inescapable fact is that there is no agreement among members of the medical profession about what it means to say that "alcoholism" is a "disease." . . .

The trial court's "finding" that Powell "is afflicted with the disease of chronic alcoholism," which "destroys the afflicted person's will power to resist the constant, excessive consumption of alcohol" covers a multitude of sins. Dr. Wade's testimony that appellant suffered from a compulsion which was an "exceedingly strong influence," but which was "not completely overpowering" is at least more carefully stated, if no less mystifying. Jellinek insists that conceptual clarity can only be achieved by distinguishing carefully between "loss of control" once an individual has commenced to drink and "inability to abstain" from drinking in the first place. Presumably a person would have to display both characteristics in order to make out a constitutional defense, should one be recognized. Yet the "findings" of the trial court utterly fail to make this crucial distinction, and there is serious question whether the record can be read to support a finding of either loss of control or inability to abstain. . . .

It is one thing to say that if a man is deprived of alcohol his hands will begin to shake, he will suffer agonizing pains and ultimately he will have hallucinations; it is quite another to say that a man has a "compulsion" to take a drink, but that he also retains a certain amount of "free will" with which to resist. It is simply impossible, in the present state of our knowledge, to ascribe a useful meaning to the latter statement. This definitional confusion reflects, of course, not merely the undeveloped state of the psychiatric art but also the conceptual difficulties inevitably attendant upon the importation of scientific and medical models into a legal system generally predicated upon a different set of assumptions.

## II.

Despite the comparatively primitive state of our knowledge on the subject, it cannot be denied that the destructive use of alcoholic beverages is one of our principal social and public health problems. The lowest current informed estimate places the number of "alcoholics" in America (definitional problems aside) at 4,000,000, and most authorities are inclined to put the figure considerably higher. The problem is compounded by the fact that a very large percentage of the alcoholics in this country are "invisible"—they possess the means to keep their drinking problems secret, and the traditionally uncharitable attitude of our society toward alcoholics causes many of them to refrain from seeking treatment from any source. . . .

There is as yet no known generally effective method for treating the vast number of alcoholics in our society. . . . Thus it is entirely possible that, even were the manpower and facilities available for a full-scale attack upon chronic alcoholism, we would find ourselves unable to help the vast bulk of our "visible"—let alone our "invisible"—alcoholic population.

However, facilities for the attempted treatment of indigent alcoholics are woefully lacking throughout the country. It would be tragic to return large numbers of helpless, sometimes dangerous and frequently unsanitary inebriates to the streets of our cities without even the opportunity to sober up adequately which a brief jail term provides. Presumably no State or city will tolerate such a state of affairs. Yet the medical profession cannot, and does not, tell us with any assurance that, even if the buildings, equipment and trained personnel were made available, it could provide anything more than slightly higher-class jails for our indigent habitual inebriates. Thus we run the grave risk that nothing will be accomplished beyond the hanging of a new sign—reading "hospital"—over one wing of the jailhouse.

One virtue of the criminal process is, at least, that the duration of penal incarceration typically has some outside statutory limit; this is universally true in the case of petty offenses, such as public

*(Continues)*

drunkenness, where jail terms are quite short on the whole. "Therapeutic civil commitment" lacks this feature; one is typically committed until one is "cured." Thus, to do otherwise than affirm might subject indigent alcoholics to the risk that they may be locked up for an indefinite period of time under the same conditions as before, with no more hope than before of receiving effective treatment and no prospect of periodic "freedom."

Faced with this unpleasant reality, we are unable to assert that the use of the criminal process as a means of dealing with the public aspects of problem drinking can never be defended as rational. The picture of the penniless drunk propelled aimlessly and endlessly through the law's "revolving door" of arrest, incarceration, release and re-arrest is not a pretty one. But before we condemn the present practice across-the-board, perhaps we ought to be able to point to some clear promise of a better world for these unfortunate people. Unfortunately, no such promise has yet been forthcoming. . . .

Ignorance likewise impedes our assessment of the deterrent effect of criminal sanctions for public drunkenness. . . .

Obviously, chronic alcoholics have not been deterred from drinking to excess by the existence of criminal sanctions against public drunkenness. But all those who violate penal laws of any kind are by definition undeterred. The longstanding and still raging debate over the validity of the deterrence justification for penal sanctions has not reached any sufficiently clear conclusions to permit it to be said that such sanctions are ineffective in any particular context or for any particular group of people who are able to appreciate the consequences of their acts. . . .

## III.

. . . Appellant, however, seeks to come within the application of the Cruel and Unusual Punishment Clause announced in *Robinson v. California*, which involved a state law making it a crime to "be addicted to the use of narcotics." . . .

On its face the present case does not fall within that holding, since appellant was convicted, not for being a chronic alcoholic, but for being in public while drunk on a particular occasion. The State of Texas thus has not sought to punish a mere status, as California did in *Robinson*; nor has it attempted to regulate appellant's behavior in the privacy of his own home. Rather, it has imposed upon appellant a criminal sanction for public behavior which may create substantial health and safety hazards, both for appellant and for members of the general public, and which offends the moral and esthetic sensibilities of a large segment of the community. This seems a far cry from convicting one for being an addict, being a chronic alcoholic, being "mentally ill, or a leper. . . ."

*Robinson* so viewed brings this Court but a very small way into the substantive criminal law. And unless *Robinson* is so viewed it is difficult to see any limiting principle that would serve to prevent this Court from becoming, under the aegis of the Cruel and Unusual Punishment Clause, the ultimate arbiter of the standards of criminal responsibility, in diverse areas of the criminal law, throughout the country.

It is suggested in dissent that *Robinson* stands for the "simple" but "subtle" principle that "[c]riminal penalties may not be inflicted upon a person for being in a condition he is powerless to change." In that view, appellant's "condition" of public intoxication was "occasioned by a compulsion symptomatic of the disease" of chronic alcoholism, and thus, apparently, his behavior lacked the critical element of *mens rea*. Whatever may be the merits of such a doctrine of criminal responsibility, it surely cannot be said to follow from *Robinson*. The entire thrust of *Robinson's* interpretation of the Cruel and Unusual Punishment Clause is that criminal penalties may be inflicted only if the accused has committed some act, has engaged in some behavior, which society has an interest in preventing, or perhaps in historical common law terms, has committed some *actus reus*. It thus does not deal with the question of whether certain conduct cannot constitutionally be punished because it is, in some sense, "involuntary" or "occasioned by a compulsion." . . .

Ultimately, then, the most troubling aspects of this case, were *Robinson* to be extended to meet it, would be the scope and content of what could only be a constitutional doctrine of criminal responsibility. In dissent it is urged that the decision could be limited to conduct which is "a characteristic and involuntary part of the pattern of the disease as it afflicts" the particular individual, and that "[i]t it not foreseeable" that it would be applied "in the case of offenses such as driving a car while intoxicated, assault, theft, or robbery." That is limitation by fiat. In the first place, nothing in the logic of the dissent would limit its application to chronic alcoholics. If Leroy Powell cannot be convicted of public intoxication, it is difficult to see how a State can convict an individual for murder, if that individual, while exhibiting normal behavior in all other respects, suffers from a "compulsion" to kill, which is an "exceedingly strong influence," but "not completely overpowering." Even if we limit our consideration to chronic alcoholics, it would seem impossible to confine the principle within the arbitrary bounds which the dissent seems to envision. . . .

Traditional common-law concepts of personal accountability and essential considerations of federalism lead us to disagree with appellant. We are unable to conclude, on the state of this record or on the current state of medical knowledge, that chronic alcoholics in general, and Leroy Powell in particular, suffer from such an irresistible compulsion to drink and to get drunk in public that they are utterly unable to control their performance of either or both of these acts and thus cannot be deterred at all from public intoxication. And in any event this Court has never articulated a general constitutional doctrine of *mens rea*.

We cannot cast aside the centuries-long evolution of the collection of interlocking and overlapping concepts which the common law has utilized to assess the moral accountability of an individual for his antisocial deeds. The doctrines of *actus reus, mens rea,* insanity, mistake, justification, and duress have historically provided the tools for a constantly shifting adjustment of the tension between the evolving aims of the criminal law and changing religious, moral, philosophical, and medical views of the nature of man. This process of adjustment has always been thought to be the province of the States. . . . It is simply not yet the time to write into the Constitution formulas cast in terms whose meaning, let alone relevance, is not yet clear either to doctors or to lawyers.

Affirmed.

Mr. Justice Black, whom Mr. Justice Harlan joins, concurring . . .

I agree with Mr. Justice Marshall that the findings of fact in this case are inadequate to justify the sweeping constitutional rule urged upon us. I could not, however, consider any findings that could be made with respect to "voluntariness" or "compulsion" controlling on the question whether a specific instance of human behavior should be immune from punishment as a constitutional matter. When we say that appellant's appearance in public is caused not by "his own" volition but rather by some other force, we are clearly thinking of a force that is nevertheless "his" except in some special sense. The accused undoubtedly commits the proscribed act and the only question is whether the act can be attributed to a part of "his" personality that should not be regarded as criminally responsible. . . .

[P]unishment of such a defendant can clearly be justified in terms of deterrence, isolation, and treatment. On the other hand, medical decisions concerning the use of a term such as "disease" or "volition," based as they are on the clinical problems of diagnosis and treatment, bear no necessary correspondence to the legal decision whether the overall objectives of the criminal law can be furthered by imposing punishment. For these reasons, much as I think that criminal sanctions should in many situations be applied only to those whose conduct is morally blameworthy, see *Morissette v. United States,* 342 U.S. 246 (1952), I cannot think the States should be held constitutionally required to make the inquiry as to what part of a defendant's personality is responsible for his actions and to excuse anyone whose action was, in some complex, psychological sense, the result of a "compulsion." . . .

The rule of constitutional law urged upon us by appellant would have a revolutionary impact on the criminal law, and any possible limits proposed for the rule would be wholly illusory. If the original boundaries of *Robinson* are to be discarded, any new limits too would soon fall by the wayside and the Court would be forced to hold the States powerless to punish any conduct that could be shown to result from a "compulsion," in the complex, psychological meaning of that term. The result, to choose just one illustration, would be to require recognition of "irresistible impulse" as a complete defense to any crime; this is probably contrary to present law in most American jurisdictions.

The real reach of any such decision, however, would be broader still, for the basic premise underlying the argument is that it is cruel and unusual to punish a person who is not morally blameworthy. I state the proposition in this sympathetic way because I feel there is much to be said for avoiding the use of criminal sanctions in many such situations. But the question here is one of constitutional law. The legislatures have always been allowed wide freedom to determine the extent to which moral culpability should be a prerequisite to conviction of a crime.

The criminal law is a social tool that is employed in seeking a wide variety of goals, and I cannot say the Eighth Amendment's limits on the use of criminal sanctions extend as far as this viewpoint would inevitably carry them. . . .

*(Continues)*

*(Continued)*

Mr. Justice White, concurring in the result.

If it cannot be a crime to have an irresistible compulsion to use narcotics, *Robinson v. California*, I do not see how it can constitutionally be a crime to yield to such a compulsion. Punishing an addict for using drugs convicts for addiction under a different name. Distinguishing between the two crimes is like forbidding criminal conviction for being sick with flu or epilepsy but permitting punishment for running a fever or having a convulsion. Unless *Robinson* is to be abandoned, the use of narcotics by an addict must be beyond the reach of the criminal law. Similarly, the chronic alcoholic with an irresistible urge to consume alcohol should not be punishable for drinking or for being drunk.

Powell's conviction was for the different crime of being drunk in a public place. Thus even if Powell was compelled to drink, and so could not constitutionally be convicted for drinking, his conviction in this case can be invalidated only if there is a constitutional basis for saying that he may not be punished for being in public while drunk. . . .

The trial court said that Powell was a chronic alcoholic with a compulsion not only to drink to excess but also to frequent public places when intoxicated. Nothing in the record before the trial court supports the latter conclusion, which is contrary to common sense and to common knowledge. . . . Before and after taking the first drink, and until he becomes so drunk that he loses the power to know where he is or to direct his movements, the chronic alcoholic with a home or financial resources is as capable as the nonchronic drinker of doing his drinking in private, of removing himself from public places, and, since he knows or ought to know that he will become intoxicated, of making plans to avoid his being found drunk in public. For these reasons, I cannot say that the chronic alcoholic who proves his disease and a compulsion to drink is shielded from conviction when he has knowingly failed to take feasible precautions against committing a criminal act, here the act of going to or remaining in a public place. On such facts the alcoholic is like a person with smallpox, who could be convicted for being on the street but not for being ill, or, like the epileptic, who could be punished for driving a car but not for his disease.

The fact remains that some chronic alcoholics must drink and hence must drink *somewhere*. Although many chronics have homes, many others do not. For all practical purposes the public streets may be home for these unfortunates, not because their disease compels them to be there, but because, drunk or sober, they have no place else to go and no place else to be when they are drinking. This is more a function of economic station than of disease, although the disease may lead to destitution and perpetuate that condition. For some of these alcoholics I would think a showing could be made that resisting drunkenness is impossible and that avoiding public places when intoxicated is also impossible. As applied to them this statute is in effect a law which bans a single act for which they may not be convicted under the Eighth Amendment—the act of getting drunk. . . .

These prerequisites to the possible invocation of the Eighth Amendment are not satisfied on the record before us. . . .

Mr. Justice Fortas, with whom Mr. Justice Douglas, Mr. Justice Brennan, and Mr. Justice Stewart join, dissenting. . . .

The sole question presented is whether a criminal penalty may be imposed upon a person suffering the disease of "chronic alcoholism" for a condition—being "in a state of intoxication" in public—which is a characteristic part of the pattern of his disease and which, the trial court found, was not the consequence of appellant's volition but of "a compulsion symptomatic of the disease of chronic alcoholism." . . .

This case does not raise any question as to the right of the police to stop and detain those who are intoxicated in public, whether as a result of the disease or otherwise; or as to the State's power to commit chronic alcoholics for treatment. Nor does it concern the responsibility of an alcoholic for criminal *acts*. We deal here with the mere *condition* of being intoxicated in public.

Although there is some problem in defining the concept, its core meaning, as agreed by authorities, is that alcoholism is caused and maintained by something other than the moral fault of the alcoholic, something that, to a greater or lesser extent depending upon the physiological or psychological makeup and history of the individual, cannot be controlled by him. . . .

*Robinson* stands upon a principle which, despite its subtlety, must be simply stated and respectfully applied because it is the foundation of individual liberty and the cornerstone of the relations between a civilized state and its citizens: Criminal penalties may not be inflicted upon a person for being in a condition he is powerless to change. . . .

In the present case, appellant is charged with a crime composed of two elements—being intoxicated and being found in a public place while in that condition. The crime, so defined, differs from that in *Robinson*. The statute covers more than a mere status. But the essential constitutional defect here is the same as in *Robinson*, for in both cases the particular defendant was accused of being in a condition which he had no capacity to change or avoid. The trial judge sitting as trier of fact found, upon the medical and other relevant testimony, that Powell is a "chronic alcoholic." He defined appellant's "chronic alcoholism" as "a disease which destroys the afflicted person's will power to resist the constant, excessive consumption of alcohol." He also found that "a chronic alcoholic does not appear in public by his own volition but under a compulsion symptomatic of the disease of chronic alcoholism." I read these findings to mean that appellant was powerless to avoid drinking; that having taken his first drink, he had "an uncontrollable compulsion to drink" to the point of intoxication; and that, once intoxicated, he could not prevent himself from appearing in public places. . . .

The findings in this case, read against the background of the medical and sociological data to which I have referred, compel the conclusion that the infliction upon appellant of a criminal penalty for being intoxicated in a public place would be "cruel and unusual punishment" within the prohibition of the Eighth Amendment. This conclusion follows because appellant is a "chronic alcoholic" who, according to the trier of fact, cannot resist the "constant excessive consumption of alcohol" and does not appear in public by his own volition but under a "compulsion" which is part of his condition. . . .

## Notes and Questions

1. Is alcoholism a "disease"? A "status"? For what specific offense was Leroy Powell convicted and punished?
2. Four justices dissented in *Powell*. Justice White's concurring opinion is crucial to the rejection of Powell's claim. Under what facts might Powell have secured Justice White's vote, and thus prevailed?
3. If the dissent had prevailed in *Powell*, would a chronic alcoholic have a defense to driving while intoxicated? What limits does the dissent place on the exculpatory rule it would adopt? Is it clear how those limits would be applied?
4. Justice Fortas's dissenting opinion in *Powell* states that *Robinson v. California* stands on the principle that: "Criminal penalties may not be inflicted upon a person for being in a condition he is powerless to change." Do you agree? Does Justice Black?

## Culpable Omissions

Can the *failure* to act be an "act" for the purposes of the criminal law? For example, assume that you are an excellent swimmer. While sitting near the shallow end of a swimming pool you notice a 2-year-old child slip into the water. She has no adult supervision. She quickly is in water over her head, and it is obvious that she cannot swim. You see her struggle and then go under water. No one else is aware of her plight. You could easily reach into the pool and pull her to safety. If you do not, she will certainly drown. If you do nothing (*i.e.*, if you fail to act) and she does drown, your behavior certainly is morally reprehensible. Have you committed a crime?

### *People v. Beardsley*, 113 N.W. 1128 (Mich. 1907)

McAlvay, C.J. Respondent was convicted of manslaughter before the circuit court for Oakland county, and was sentenced to the state prison at Jackson for a minimum term of one year and a maximum term not to exceed five years.

He was a married man living at Pontiac, and at the time the facts herein narrated occurred he was working as a bartender and clerk at the Columbia Hotel. He lived with his wife in Pontiac, occupying two rooms on the ground floor of a house. . . .

His wife being temporarily absent from the city, respondent arranged with a woman named Blanche Burns, who at the time was working at another hotel, to go to his apartment with him. He had been acquainted with her for some time. They knew each other's habits and character. They

*(Continues)*

had drunk liquor together, and had on two occasions been in Detroit and spent the night together in houses of assignation. On the evening of Saturday, March 18, 1905, he met her at the place where she worked, and they went together to his place of residence. They at once began to drink, and continued to drink steadily, and remained together, day and night, from that time until the afternoon of the Monday following, except when respondent went to his work on Sunday afternoon. There was liquor at these rooms, and when it was all used they were served with bottles of whisky and beer by a young man who worked at the Columbia Hotel, and who also attended respondent's fires at the house. He was the only person who saw them in the house during the time they were there together. Respondent gave orders for liquor by telephone. On Monday afternoon, about 1 o'clock, the young man went to the house to see if anything was wanted. At this time he heard respondent say they must fix up the rooms, and the woman must not be found there by his wife, who was likely to return at any time. During this visit to the house the woman sent the young man to a drug store to purchase, with money she gave him, camphor and morphine tablets. He procured both articles. There were six grains of morphine in quarter-grain tablets. She concealed the morphine from respondent's notice, and was discovered putting something into her mouth by him and the young man as they were returning from the other room after taking a drink of beer. She in fact was taking morphine. Respondent struck the box from her hand. Some of the tablets fell on the floor, and of these respondent crushed several with his foot. She picked up and swallowed two of them, and the young man put two of them in the spittoon. Altogether it is probable she took from three to four grains of morphine. The young man went away soon after this. Respondent called him by telephone about an hour later, and after he came to the house requested him to take the woman into the room in the basement which was occupied by a Mr. Skoba. She was in a stupor, and did not rouse when spoken to. Respondent was too intoxicated to be of any assistance, and the young man proceeded to take her downstairs. While doing this, Skoba arrived, and together they put her in his room on the bed. Respondent requested Skoba to look after her, and let her out the back way when she waked up. Between 9 and 10 o'clock in the evening, Skoba became alarmed at her condition. He at once called the city marshal and a doctor. An examination by them disclosed that she was dead.

Many errors are assigned by respondent, who asks to have his conviction set aside. The principal assignments of error are based upon the charge of the court and refusal to give certain requests to charge, and are upon the theory that under the undisputed evidence in the case, as claimed by the people and detailed by the people's witnesses, the respondent should have been acquitted and discharged. In the brief of the prosecutor, his position is stated as follows: "It is the theory of the prosecution that the facts and circumstances attending the death of Blanche Burns in the house of respondent were such as to lay upon him a duty to care for her, and the duty to take steps for her protection, the failure to take which was sufficient to constitute such an omission as would render him legally responsible for her death. . . . There is no claim on the part of the people that the respondent was in any way an active agent in bringing about the death of Blanche Burns, but simply that he owed her a duty which he failed to perform, and that in consequence of such failure on his part she came to her death." Upon this theory a conviction was asked and secured.

The law recognizes that under some circumstances the omission of a duty owed by one individual to another, where such omission results in the death of the one to whom the duty is owing, will make the other chargeable with manslaughter. This rule of law is always based upon the proposition that the duty neglected must be a legal duty, and not a mere moral obligation. It must be a duty imposed by law or by contract, and the omission to perform the duty must be the immediate and direct cause of death.

Although the literature upon the subject is quite meager and the cases few, nevertheless the authorities are in harmony as to the relationship which must exist between the parties to create the duty, the omission of which establishes legal responsibility. One authority has briefly and correctly stated the rule, which the prosecution claims should be applied to the case at bar, as follows: "If a person who sustains to another the legal relation of protector, as husband to wife, parent to child, master to seaman, etc., knowing such person to be in peril, willfully and negligently fails to make such reasonable and proper efforts to rescue him as he might have done, without jeopardizing his own life, or the lives of others, he is guilty of manslaughter, at least, if by reason of his omission of duty the dependent person dies. So one who from domestic relationship, public duty, voluntary choice, or otherwise, has the custody and care of a human being, helpless either from imprisonment, infancy, sickness, age, imbecility, or other incapacity of mind or body is bound to execute the charge with proper diligence and will be held guilty of manslaughter, if by culpable negligence he lets the helpless creature die." 21 Am. & Eng. Enc. of Law (2d Ed. p. 102). . . .

Seeking for a proper determination of the case at bar by the application of the legal principles involved, we must eliminate from the case all consideration of mere moral obligation, and discover whether respondent was under a legal duty towards Blanche Burns at the time of her death, knowing her to be in peril of her life, which required him to make all reasonable and proper effort to save her, the omission to perform which duty would make him responsible for her death. This is the important and determining question in this case. If we hold that such legal duty rested upon respondent, it must arise by implication from the facts and circumstances already recited. The record in this case discloses that the deceased was a woman past 30 years of age. She had been twice married. She was accustomed to visiting saloons and to the use of intoxicants. She previously had made assignations with this man in Detroit at least twice. There is no evidence or claim from this record that any duress, fraud, or deceit had been practiced upon her. On the contrary, it appears that she went upon this carouse with respondent voluntarily, and so continued to remain with him. Her entire conduct indicates that she had ample experience in such affairs.

It is urged by the prosecutor that the respondent "stood towards this woman for the time being in the place of her natural guardian and protector, and as such owed her a clear legal duty which he completely failed to perform." The cases cited and digested establish that no such legal duty is created based upon a mere moral obligation. The fact that this woman was in his house created no such legal duty as exists in law and is due from a husband towards his wife, as seems to be intimated by the prosecutor's brief. Such an inference would be very repugnant to our moral sense. Respondent had assumed either in fact or by implication no care or control over his companion. Had this been a case where two men under like circumstances had voluntarily gone on a debauch together, and one had attempted suicide, no one would claim that this doctrine of legal duty could be invoked to hold the other criminally responsible for omitting to make effort to rescue his companion. How can the fact that in this case one of the parties was a woman change the principle of law applicable to it? Deriving and applying the law in this case from the principle of decided cases, we do not find that such legal duty as is contended for existed in fact or by implication on the part of respondent towards the deceased, the omission of which involved criminal liability. We find no more apt words to apply to this case than those used by Mr. Justice Field, in *United States v. Knowles,* [4 Sawy. (U.S.) 517, Fed. Cas. No. 15,540]: "In the absence of such obligations, it is undoubtedly the moral duty of every person to extend to others assistance when in danger, . . . and, if such efforts should be omitted by anyone when they could be made without imperiling his own life, he would by his conduct draw upon himself the just censure and reproach of good men; but this is the only punishment to which he would be subjected by society." . . .

The conviction is set aside, and respondent is ordered discharged.

## Notes and Questions

1. On a late night in March 1964 Catherine ("Kitty") Genovese was walking from her car to her apartment complex in Queens, New York. An assailant grabbed and then stabbed her. She screamed loudly. Windows opened and lights were turned on in the apartment complex and one man yelled, "Let that girl alone!" The assailant left. Lights went out. As Ms. Genovese struggled to get closer to her building, the man returned and stabbed her again. As she screamed, windows again were illuminated and opened. The assailant drove away in his car. Ms. Genovese crawled to the foot of the stairs of the apartment building. The man returned and stabbed her a third time, shortly after 3:35 A.M. The police received their first call at 3:50 A.M. They were at the scene within 2 minutes. Kitty Genovese was dead. Approximately 38 people in the apartment complex had witnessed the stabbing.[5]

2. In March 1983 a young woman entered a tavern in New Bedford, Massachusetts, to buy cigarettes. Six patrons grabbed her and swung her onto a pool table. She was kicking and screaming. One of the men tried to force her to perform fellatio. Another raped her. Several other patrons looked on, some of whom reportedly yelled, "Go for it!" The woman eventually escaped and, clad only in a shirt and one shoe, fled into the street and flagged down a passing truck.[6] *The Accused*, a movie starring Jodie Foster, was made portraying the incident.

3. Should criminal liability be imposed for the failure of others to aid another person in peril in cases like *Beardsley,* or those involving Kitty Genovese, or the New Bedford, Massachusetts, pool-table rape? Are there specific circumstances that would present stronger or weaker cases for making the failure to act, or a **culpable omission**, a crime?

*(Continues)*

*(Continued)*

4. What explains the law's reluctance to require another to act, at the peril of criminal punishment, when there is simply a "moral" (and not a "legal") duty to come to another's aid? Does the difficulty in defining the precise nature and scope of the "duty to act" present a problem? Does the problem lie in expanding the web of criminal liability to unreasonable dimensions? For example, would all 38 people who witnessed Kitty Genovese's stabbing be guilty of a crime? And, of what crime? Some form of criminal homicide? Or, a special offense relating to "failure to come to another's aid"? Would requiring others to take action somehow threaten individual autonomy or compromise privacy by encouraging "snooping"? Are there meaningful moral distinctions between harm that follows from a failure to act (*e.g.*, not rescuing a drowning toddler) and harm that follows affirmative action (*e.g.*, pushing the toddler into the swimming pool, causing her to drown)? Are any of these possible objections sufficiently persuasive to avoid making the failure to act a crime when no more than a "moral" obligation impels action?

5. Consider the state **"Good Samaritan" laws,** which impose criminal sanctions for the failure to aid others under the designated circumstances, in Exhibit 3-1.

6. As *Beardsley* makes clear, the failure to act when there is a *legal* duty to do so may constitute a crime. In *Jones v. United States*, 308 F.2d 307 (D.C. Cir. 1972), the court explained as follows:

> There are at least four situations in which the failure to act may constitute breach of a legal duty. One can be held criminally liable: first, where a statute imposes a duty to care for another; second, where one stands in a certain status relationship to another; third, where one has assumed a contractual duty to care for another; and fourth, where one has voluntarily assumed the care of another and so secluded the helpless person as to prevent others from rendering aid. . . .

To that list, LaFave[7] added the following:

(5) *Duty based on creation of peril.*
(6) *Duty to control conduct of others* (*e.g.*, parent duty to control child, employer duty to control employee).
(7) *Duty of landowner* (to provide for the safety of others invited onto land).

7. Should there be criminal liability for the failure to act under the following circumstances?
   a. A parent neglects to feed or provide necessary medical care for his or her child, resulting in the child's death. *State v. Williams*, 484 P.2d 1167 (Wash. App. 1971); *Nozza v. State*, 288 So.2d 560 (Fla. App. 1974).
   b. An adult child neglects to feed or provide necessary medical care for his or her elderly parent, who is unable to tend to his or her own needs. *Billingslea v. State*, 780 S.W.3d 271 (Tex. Crim. App. 1989); *People v. Heitzman*, 886 P.2d 1229 (Cal 1994); *Davis v. Commonwealth*, 335 S.E.2d 375 (Va. 1985); *Sieniarecki v. State*, 724 So.2d 626 (Fla. App. 1998).
   c. A mother fails to protect her child from beatings or sexual assaults administered by the child's father, a sibling, or the mother's companion. *State v. Williquette*, 385 N.W.2d 145 (Wis. 1986); *Degren v. State*, 722 A.2d 887 (Md. 1999); *State v. Walden*, 293 S.E.2d 780 (N.C. 1982).
   d. A man takes no action to protect the child of his live-in female companion from abuse inflicted by the child's mother. *State v. Miranda*, 715 A.2d 680 (Ct. 1998). *Compare Pope v. State*, 396 A.2d 1054 (Md. 1979).
   e. A landlord, in violation of state law, fails to provide a fire escape from a multiple unit dwelling, when tenants of a rented apartment die after they are unable to escape from the burning building. *People v. Nelson*, 128 N.E.2d 391 (N.Y. 1955).
   f. A man rapes a woman. Distraught, the woman flees and falls or jumps into a river, as the man looks on and takes no action. The woman drowns. *Jones v. State*, 43 N.E.2d 1017 (Ind. 1942).

# Exhibit 3-1  State Good Samaritan Laws

## General Laws of Rhode Island Annotated

§ 11-56-1 Duty to Assist

Any person at the scene of an emergency who knows that another person is exposed to, or has suffered, grave physical harm shall, to the extent that he or she can do so without danger or peril to himself or herself or to others, give reasonable assistance to the exposed person. Any person violating the provisions of this section shall be guilty of a petty misdemeanor and shall be subject to imprisonment for a term not exceeding six (6) months, or by a fine of not more than five hundred dollars ($500), or both.

## Vermont Statutes Annotated

Art 12, § 519 Emergency Medical Care

(a) A person who knows that another is exposed to grave physical harm shall, to the extent that the same can be rendered without danger or peril to himself or without interference with important duties owed to others, give reasonable assistance to the exposed person unless that assistance or care is being provided by others.

(b) A person who provides reasonable assistance in compliance with subsection (a) of this section shall not be liable in civil damages unless his acts constitute gross negligence or unless he will receive or expects to receive remuneration. Nothing contained in this subsection shall alter existing law with respect to tort liability of a practitioner of the healing arts for acts committed in the ordinary course of his practice.

(c) A person who willfully violates subsection (a) of this section shall be fined not more than $100.00.

## Wisconsin Statutes Annotated

§ 940.34 Duty to Aid Victim or Report Crime

(1)(a) Whoever violates sub. (2)(a) is guilty of a Class C misdemeanor. . . .

(2)(a) Any person who knows that a crime is being committed and that a victim is exposed to bodily harm shall summon law enforcement officers or other assistance or shall provide assistance to the victim. . . .

(d) A person need not comply with this subsection if any of the following apply:
   1. Compliance would place him or her in danger.
   2. Compliance would interfere with duties the person owes to others.
   3. In the circumstances described under par. (a), assistance is being summoned or provided by others. . . .

(3) If a person renders emergency care for a victim, § 895.48(1) applies. Any person who provides other reasonable assistance under this section is immune from civil liability for his or her acts or omissions in providing the assistance. This immunity does not apply if the person receives or expects to receive compensation for providing the assistance.

## Concurrence Principle

*actus & mens must co-exist!*

"No proposition is more bedrock than that a crime consists of an *actus reus* and a *mens rea* and that the two must coexist. Where there is no coincidence in time between the guilty act and the guilty mind, there is, by definition, no crime." *Hall v. State,* 314 A.2d 704, 705 (Md. App. 1974).

The **concurrence principle** requires a logical connection or nexus between *mens rea* and an act before the conduct is labeled criminal. In particular, the act must be attributable to, or must have been motivated by, a guilty mind. For example, assume A enters a mountain cabin during a blizzard without the owner's permission to take refuge from the storm. After entering, he sees valuables and decides to take them with him when the weather calms. He is not guilty of burglary if that offense is defined as "breaking and entering a dwelling with the intention to commit a felony therein." This is so because at the time of his entry into the cabin A had no intention of committing felonious larceny or another felony; he sought only to escape the blizzard. His intention to steal, formed after his entry, did not concur with his breaking and entering the cabin.

In one widely cited case, the defendant Jackson was convicted of murder and sentenced to death in Kentucky.

*concurrence principle....*

He argued that the prosecution's proof did not establish that the crime was committed in Kentucky. With a companion, one Walling, Jackson had administered what he believed to be a lethal dose of cocaine to Pearl Bryan in Cincinnati, Ohio. Ms. Bryan was pregnant, presumably by Jackson. Jackson and Walling then transported Bryan to Campbell County, Kentucky, and decapitated her to prevent identification of the body. According to the prosecution's evidence, Ms. Bryan in fact was still alive when she was brought into Kentucky. She was killed in Kentucky by the decapitation. Absent such evidence, Kentucky would have no jurisdiction to try Jackson for murder, because the crime would have occurred in Ohio.

If Jackson's story is believed (*i.e.,* that he thought Ms. Bryan already was dead before she was decapitated in Kentucky), what possible problem related to the "concurrence" of *mens rea* and *actus reus* exists regarding Jackson's conviction for murder in Kentucky? Should this problem require the invalidation of his murder conviction and death sentence in that state? The Kentucky Court of Appeals thought not.

If [Pearl Bryan] was dead [before entering Campbell County, Kentucky], as might be supposed from her making no outcry, a verdict of guilty could not have been rendered; but if she was then alive, though appearing to be dead, and by the cutting of her throat she was killed while in Campbell County, then the jury might find a verdict of guilty, although the cutting off of the head was merely for the purpose of destroying the chance of identification, or for any other purpose. At best, the instruction does not authorize a verdict of conviction unless Jackson is shown to have cut off the head of his victim in Campbell County, and while she was in fact alive; and if he did this he was guilty of murder, though believing her already dead, if the act succeeded, and was but a part of the felonious attempt to kill her in Cincinnati. *Jackson* v. *Commonwealth*, 38 S.W. 422, 428 (Ky. App. 1896).

The result in *Jackson* has been criticized. Jerome Hall[8] maintains as follows:

> [Jackson] was certainly guilty of an attempt to murder in Ohio. But the mistake of fact in Kentucky excluded the *mens rea* of criminal homicide there; and the court's reliance upon the defendant's homicidal intention in Ohio was clearly violative of the principle of concurrence. . . . The Kentucky court apparently believed that since the defendant set out intentionally to kill and since he did, in fact, kill, the failure of the cocaine to do that in Ohio was unimportant. But the fact remains that the *mens rea* did not concur with the actual killing.

If the criticisms of *Jackson* have merit, of what offense, if any, could the defendant have been convicted in Kentucky? In Ohio? Would Jackson's avoiding a conviction for murder epitomize a classic "legal technicality," or do the jurisdictional issues and the concurrence principle render his Kentucky murder conviction manifestly unfair?

## Harm

Few philosophical arguments are as well known as John Stuart Mill's essay on social liberty, "On Liberty," in which Mill explores "the nature and limits of the power which can be legitimately exercised by society over the individual." Mill[9] describes the essence of his position as follows:

The object of this Essay is to assert one very simple principle, as entitled to govern absolutely the dealings of society with the individual in the way of compulsion and control, whether the means used be physical force in the form of legal penalties, or the moral coercion of public opinion. That principle is, that the sole end for which mankind are warranted, individually or collectively, in interfering with the liberty of action of any of their number, is self-protection. That the only purpose for which power can be rightfully exercised over any member of a civilised community, against his will, is to prevent harm to others. His own good, either physical or moral, is not a sufficient warrant. He cannot rightfully be compelled to do or forbear because it will be better for him to do so, because it will make him happier, because, in the opinions of others, to do so would be wise, or even right. These are good reasons for remonstrating with him, or reasoning with him, or persuading him, or entreating him, but not for compelling him, or visiting him with any evil in case he do otherwise. To justify that, the conduct from which it is desired to deter him, must be calculated to produce evil to someone else. The only part of the conduct of anyone, for which he is amenable to society, is that which concerns others. In the part which merely concerns himself, his independence is, of right, absolute. Over himself, over his own body and mind, the individual is sovereign.

When Mill asserts that "the only purpose for which power can be rightfully exercised over any member of a civilised community, against his will, is to prevent harm to others," what is the meaning of the crucial phrase, "harm to others"? And, which aspects of an individual's conduct "concern others" and which "merely concern himself"? Resolving those definitional and conceptual difficulties, of course, is crucial to give content to Mill's fundamental premise.

For example, Patrick Devlin[10] argued that "[w]ithout shared ideas on politics, morals, and ethics no society can exist." He further contends that "society may use the law to preserve morality in the same way as it uses it to safeguard anything else that is essential to its existence." How compatible are these views with Mill's thesis?

### *Picou v. Gillum*, 874 F.2d 1519 (11th Cir. 1989)

Powell, Associate Justice [(Retired, U.S. Supreme Court)]:

The question presented is whether the federal Constitution prohibits Florida from requiring riders of motorcycles to wear protective headgear. . . . Appellant contended that the statute violated federal constitutional rights to Due Process, Equal Protection, and privacy. . . . Helmet statutes have been the subject of numerous published opinions from state courts. Although a few courts in the late 1960's and early 1970's held motorcycle helmet laws unconstitutional, each of these cases has been reversed or overruled. Courts in subsequent cases have uniformly upheld the provisions. . . .

Appellant concedes that his case is not covered by existing precedents defining the right to privacy. He contends, however, that those precedents stand for a broader proposition: that the Constitution protects the "right to be let alone." See *Bowers v. Hardwick*, 478 U.S. 186, 199 (1986) (Blackmun, J., dissenting); *Olmstead v. United States*, 277 U.S. 438, 478 (Brandeis, J., dissenting). He further casts his argument in terms of a right to be free from "paternalistic" legislation. In other words, appellant argues that the Constitution forbids enforcement of any statute aimed only at protecting a State's citizens from the consequences of their own foolish behavior and not at protecting others. . . .

Whatever merit may exist in appellant's . . . contention that paternalistic legislation is necessarily invalid, this argument is inapplicable to Fla. Stat. § 316.211. The helmet requirement does not implicate appellant alone. Motorcyclists normally ride on public streets and roads that are maintained and policed by public authorities. Traffic is often heavy, and on highways proceeds at high rates of speed. The required helmet and faceshield may prevent a rider from becoming disabled by flying objects on the road, which might cause him to lose control and involve other vehicles in a serious accident.

It is true that a primary aim of the helmet law is prevention of unnecessary injury to the cyclist himself. But the costs of this injury may be borne by the public. A motorcyclist without a helmet is more likely to suffer serious head injury than one wearing the prescribed headgear. State and local governments provide police and ambulance services, and the injured cyclist may be hospitalized at public expense. If permanently disabled, the cyclist could require public assistance for many years. As Professor Tribe has expressed it, "[in] a society unwilling to abandon bleeding bodies on the highway, the motorcyclist or driver who endangers himself plainly imposes costs on others." L. Tribe, *American Constitutional Law* § 15-12, at 1372 (2d ed. 1988). Leaving aside the deference traditionally accorded to state highway safety regulation, we think Florida's helmet requirement a rational exercise of its police powers.

There is a strong tradition in this country of respect for individual autonomy and mistrust of paternalistic legislation. Appellant, like many of his predecessors in helmet law cases, cites John Stuart Mill for the proposition that "the only purpose for which power can rightfully be exercised over any member of a civilised community, against his will, is to prevent harm to others. His own good, either physical or moral, is not a sufficient warrant." J. Mill, *On Liberty* (1859). In fact, Thomas Jefferson presaged Mill by three quarters of century, writing in 1787 that "the legitimate powers of government extend to such acts only as are injurious to others." *Notes on the State of Virginia* in *Jefferson, Writings* 285 (Library of America ed. 1984). But the impressive pedigree of this political ideal does not readily translate into a constitutional right.

Legislatures and not courts have the primary responsibility for balancing conflicting interests in safety and individual autonomy. Indeed, the evidence suggests that arguments asserting the importance of individual autonomy may prevail in the political process. In the mid-1970's, opponents of helmet requirements successfully lobbied for amendment of a federal law that allowed withholding of federal highway funds from States without helmet statutes. More recently, Massachusetts' mandatory seat belt law was repealed by referendum after opponents attacked it as an infringement on personal liberties.

Subsequent studies suggest that repeal of these safety measures can have a substantial cost in lives and property. But it is no more our role to impose a helmet requirement on this ground than to invalidate Florida's helmet law on the grounds urged by appellant. Although a narrow range of privacy rights are shielded from the political process by the Constitution, the desirability of laws such as the Florida helmet requirement is a matter for citizens and their elected representatives to decide.

## Notes and Questions

1. As with motorcycle helmet laws, the courts have uniformly rejected challenges to compulsory seat belt laws. *See, e.g., Kelver v. State,* 808 N.E.2d 154 (Ind. App. 2004); *People v. Kuhrig,* 498 N.E.2d 1158 (Ill. 1986), *appeal dismissed,* 479 U.S. 1073 (1987); *State v. Hartog,* 440 N.W.2d 852 (Ia.), *cert. denied,* 493 U.S. 1005 (1989).

2. Consistent with the rationale of *Picou,* could a state make cigarette smoking a crime? Drinking alcohol? Ingesting caffeine? Granting that "[t]he helmet requirement does not implicate appellant alone" is a more refined analysis required that examines the magnitude of the government's interests and the degree to which individual liberties are constrained? Or, is the court's observation that the constitutional right of privacy does not extend to a claim of this nature sufficient to entrust the matter to the judgment of the legislature?

## State v. Brown, 364 A.2d 27 (N.J. Super. 1976), aff'd 381 A.2d 1231 (N.J. App. 1977)

Bachman, J.S.C.

. . . [D]efendant contends that he is not guilty of the alleged atrocious assault and battery because he and Mrs. Brown, the victim, had an understanding to the effect that if she consumed any alcoholic beverages (and/or became intoxicated), he would punish her by physically assaulting her. The testimony revealed that the victim was an alcoholic. On the day of the alleged crime she indulged in some spirits, apparently to Mr. Brown's dissatisfaction. As per their agreement, defendant sought to punish Mrs. Brown by severely beating her with his hands and other objects. . . .

Some courts have allowed the defense of consent in civil suits, while denying it in criminal prosecutions for battery. According to these courts, there are two different interests at stake. While criminal law is designed to protect the interests of society as a whole, the civil law is concerned with enforcing the rights of each individual within the society. So, while the consent of the victim may relieve defendant of liability in tort, this same consent has been held irrelevant in a criminal prosecution, where there is more at stake than a victim's rights. . . .

As recently as 1934 it was held in England that if an act is unlawful in the sense of it being in itself a criminal act, it is plain that it cannot be rendered lawful because the person to whose detriment it is done consents to it. No person can license another to commit a crime. *Regina v. Donovan,* 2 K.B. 498, 507 (1934).

The reasoning and public interest that is of concern and served by this rule is that of peace, health and good order. An individual or victim cannot consent to a wrong that is committed against the public peace. . . . It has been stated, and perhaps rightly so, that the only true consent to a criminal act is that of the community.

This is so because these acts (the physical assaults by defendant upon Mrs. Brown), even if done in private, have an impingement (whether direct or indirect) upon the community at large in that the very doing of them may tend to encourage their repetition and so to undermine public morals.

*State v. Fransua,* 85 N.M. 173 (App.Ct. 1973), bears further illustration and support for this court's holding, as it is a classic and recent case of an invitation and consent to an aggravated assault. There, as the result of an argument the victim, in compliance with defendant's wishes, produced a loaded pistol, laid it within defendant's reach and said: there's the gun, if you want to shoot me go ahead. The defendant picked up the pistol and shot the victim, wounding him seriously. In response to defendant's argument of consent on the part of the victim, the court wisely opined,

> We cannot agree. It is generally conceded that a state enacts criminal statutes making certain violent acts crimes for at least two reasons: One reason is to protect the persons of its citizens; the second, however, is to prevent a breach of the peace [citing] *State v. Seal,* 76 N.M. 461 (1966). While we entertain little sympathy for either the victim's absurd actions or the defendant's equally unjustified act of pulling the trigger, we will not permit the defense of consent to be raised in such cases. Whether the victims of crimes have so little regard for their own safety as to request injury, the public has a stronger and overriding interest in preventing and prohibiting acts such as these. [510 P.2d at 107] . . .

There are a few situations in which the consent of the victim (actual or implied) is a defense. These situations usually involve ordinary physical contact or blows incident to sports such as football, boxing, or wrestling. But this is expected and understood by the participants. The state cannot later be heard to charge a participant with criminal assault upon another participant if the injury complained of resulted from activity that is reasonably within the rules and purview of the sports activity.

However this is not to be confused with sports activities that are not sanctioned by the state. Thus, street fighting which is disordering and mischievous on many obvious grounds (even if for a purse and consented to), and encounters of that kind which tend to and have the specific objective of causing bodily harm, serve no useful purpose, but rather tend to breach the peace and thus are unlawful. No one is justified in striking another, except it be in self defense, and similarly whenever two persons go out to strike each other and do so, each is guilty of an assault. . . .

As stated by this court in its ruling and by the court in *People v. Samuels* it is a matter of common knowledge that a normal person in full possession of his or her mental faculties does not freely and seriously consent to the use upon his or herself of force likely to produce great bodily harm. Those persons that do freely consent to such force and bodily injury no doubt require the enforcement of the very laws that were enacted to protect them and other humans. . . .

This court concludes that, as a matter of law, no one has the right to beat another even though that person may ask for it. Assault and battery cannot be consented to by a victim, for the State makes it unlawful and is not a party to any such agreement between the victim and perpetrator. To allow an otherwise criminal act to go unpunished because of the victim's consent would not only threaten the security of our society but also might tend to detract from the force of the moral principles underlying the criminal law. A major dissent to the view that the victim's consent may be a valid defense to a charge of assault has been voiced by the noted English jurist, Sir Patrick Devlin, who stated,

> It is not a defense to any form of criminal assault that the victim thought his punishment well deserved and consented to it. To make a good defense the accused must prove that the law gave him the right to chastise and that he exercised it reasonably. There are certain standards of behavior or moral principles which society requires to be observed; and the breach of them is an offense not merely against the person who is injured, but against society as a whole. [*Devlin, The Enforcement of Morals*, 6 (1965)]

Thus, for the reasons given, the State has an interest in protecting those persons who invite, consent to and permit others to assault and batter them. Not to enforce these laws which are geared to protect such people would seriously threaten the dignity, peace, health and security of our society.

## Notes and Questions

1. How important to the court's decision in *Brown* is its assertion that "it is a matter of common knowledge that a normal person in full possession of his or her mental faculties does not freely and seriously consent to the use upon his or herself of force likely to produce great bodily harm"? Is the court suggesting that Mrs. Brown was not fully competent to consent to receive the beating? Is there evidence to that effect? Would it matter if in fact she were competent and considered the prospects of receiving a beating preferable, in the long run, to succumbing to her temptations to resume drinking?

2. Consider the following news item:

   > . . . Sheriff's detectives in Franklin County, Wash., say a man had his friend shoot him in the shoulder so he wouldn't have to go to work.
   >
   > When he first spoke with deputies [the man] said he'd been the victim of a drive-by shooting while he was jogging. . . . But detectives told KONA radio that [the man] later acknowledged he asked his friend to shoot him so he could get some time off work and avoid a drug test.
   >
   > The friend has been arrested for investigation of reckless endangerment. [The man who was shot] is expected to be charged with false reporting. . . .
   >
   > *Albany Times Union*, p. A5 (March 2, 2008).

3. How expansive is the principle recognized in *Picou* and *Brown*? In what activities that occasion an appreciable risk of injury should individuals be allowed to engage free of governmental sanction? Race car driving? Sky diving? Russian roulette? Can the (threatened) social harm inherent in such activities be offset by corresponding social benefits? What assessment of social benefits and costs *should* lead some risky conduct to be made criminal, whereas other risky conduct remains unregulated by the criminal laws?

4. When violence is condoned in a socially approved activity—such as boxing, football, or hockey—are there nevertheless limits to what is acceptable? Even if Mike Tyson and Evander Holyfield are free to punch one another within the rules of heavyweight boxing, would Tyson be guilty of criminal assault (or mayhem) for biting off a portion of Holyfield's ear? What amount of "harm" should the criminal law tolerate within an arena where violence is inherent in a socially approved activity?

5. Consult MPC § 2.11, which addresses the circumstances under which a victim's consent constitutes a defense to a criminal charge.

## People v. Schacker, 670 N.Y.S.2d 308 (Suffolk Co. Dist. Ct. 1998)

Lawrence Donohue, Judge.

. . . The defendant is charged with Assault in the Third Degree in violation of Penal Law 120.00.1. The factual portion of the information reads as follows:

> The defendant, [Robert Schacker,] at the Superior Ice Rink . . . during an ice hockey game and after a play was over and the whistle had blown, did come up behind Andrew Morenberg who was standing near the goal net and did strike him on the back of the neck and caused him to strike his head on the crossbar of the net, causing him to sustain a concussion, headaches, blurred vision, and memory loss. Injuries had been treated at St. John's Hospital, Smithtown. . . .

Both an X-ray and a CAT scan showed no damage. The final diagnosis was "Contusion Forehead." Thus the medical records show only minor injuries.

Since the defendant had pled "not guilty," the people are required to prove that the defendant possessed the conscious intent to cause physical injury to the complainant. The fact that the act occurred in the course of a sporting event is a defense that tends to deny that the requisite intent was present.

As Chief Judge Benjamin Cardozo said in *Murphy v. Steeplechase Amusement Co.*, 166 N.E. 173 (N.Y.): "One who takes part in such a sport accepts the dangers that inhere in it so far as they are obvious and necessary, just as a fencer accepts the risk of a thrust by his antagonist or a spectator at a ball game the chance of contact with the ball. . . . A different case would be here if the dangers inherent in the sport were obscure or unobserved . . . or so serious as to justify the belief that precautions of some kind must have been taken to avert them." Persons engaged in athletic competition are generally held to have legally assumed the risk of injuries which are known, apparent and reasonably foreseeable consequences of participation. Hockey players assume the risk of injury by voluntarily participating in a hockey game at an ice rink.

This tort rule states a policy "intended to facilitate free and vigorous participation in athletic activities." *Benitez. v. New York City Board of Education*, 541 N.E.2d 29 (N.Y.). This policy would be severely undermined if the usual criminal standards were applied to athletic competition, especially ice hockey. If cross checking, tripping and punching were criminal acts, the game of hockey could not continue in its present form.

The complainant does not assume the risk of reckless or intentional conduct. However, it must be recognized that athletic competition includes intentional conduct that has the appearance of criminal acts. In fact, in many sporting events, physical injuries are caused by contact with other players. However, the players are "legally deemed to have accepted personal responsibility for" the risks inherent in the nature of sport. This includes intentional acts which result in personal injury. Thus, in order to allege a criminal act which occurred in a hockey game, the factual portion of the information must allege acts that show that the intent was to inflict physical injury which was unrelated to the athletic competition. Although play may have terminated, the information herein does not show that the physical contact had no connection with the competition. Furthermore, the injuries must be so severe as to be unacceptable in normal competition, requiring a change in the nature of the game. That type of injury is not present in this case. Firstly, the physical injury resulted from hitting the net, not from direct contact with the defendant. Secondly, the hospital records do not indicate severe trauma to the complainant.

The idea that a hockey player should be prosecuted runs afoul of the policy to encourage free and fierce competition in athletic events. The people argued at the hearing that this was a non-checking hockey league. While the rules of the league may prohibit certain conduct, thereby reducing the potential injuries, nevertheless, the participant continues to assume the risk of a strenuous and competitive athletic endeavor. The normal conduct in a hockey game can not be the standard for criminal activity under the Penal Law, nor can the Penal Law be imposed on a hockey game without running afoul of the policy of encouraging athletic competition.

For the foregoing reasons, the interest of justice requires a dismissal of this charge pursuant to CPL 170.40.

## Notes and Questions

1. In *State v. Shelley*, 929 P.2d 489 (Wash. App.), *rev. denied*, 946 P.2d 402 (Wash. 1997), the defendant was convicted of second-degree assault after he punched an opponent during a pickup basketball

game played at a college facility. The court affirmed the conviction. It held that "the harm suffered"—in this case, a broken jaw—does not dictate "whether the [consent] defense is available or not." It reasoned that "[t]he correct inquiry is whether the conduct of defendant constituted foreseeable behavior in the play of the game." Furthermore, for the consent defense to be available to the defendant, "the injury must have occurred as a byproduct of the game itself." Based on these standards, did the court in *Schacker* err in dismissing the assault information? Or, is the dismissal in *Schacker* more the result of the particular allegations contained in the information? How strict must compliance be with the rules before aggressive contact is defined as a criminal assault?

2. Should the amount of harm tolerated in a hockey game under the general justification of consent, as in *Schacker,* be greater than that tolerated in the context of *State v. Brown*? Why or why not?

3. For cases involving criminal assault charges stemming from altercations during basketball games, *see State v. Floyd,* 466 N.W.2d 919 (Ia. App. 1990); *State v. Guidugli,* 811 N.E.2d 567 (Ohio App. 2004).

4. For more regarding the criminal law's application to conduct related to athletic competitions, *see* Harary[11] and Clarke.[12]

*cause in fact (but for)*

We encounter other issues relating to the construct of "harm" in later chapters. In particular, we confront the principle of harm in the commission of inchoate offenses, such as conspiracy and attempts, in Chapter 12. We also consider application of the harm principle when we examine whether causing the death of a fetus is tantamount to killing a "human being" for purposes of the law governing criminal homicide (see Chapter 6).

## Causation

An individual is not guilty of a crime unless his or her conduct "causes" the harm proscribed by law. It is helpful to begin analyzing **causation** issues by distinguishing between *cause-in-fact* (or "but for" causation) and the *proximate* (or "legal") *cause* of a harm. It is necessary, but generally not sufficient, to conclude that "but for" the defendant's conduct, the harm at issue would not have occurred. To impose criminal liability, more than simple cause-in-fact usually must be established; it further must be shown (in most cases) that the defendant's conduct is the legally operative, or proximate, cause of the harm.

The MPC addresses causation in section 2.03. It initially requires proof of cause-in-fact as follows:

§ 2.03(1) Conduct is the cause of a result when:
(a) It is an antecedent but for which the result in question would not have occurred. . . .

The MPC dispenses with the terminology of "proximate cause" but it makes clear that but for causation is not necessarily the sole inquiry:

§ 2.03(1)(b) [and] the relationship between the conduct and result satisfies any additional causal requirements imposed by the Code or by the law defining the offense.

The MPC's general approach, for crimes other than strict liability offenses, is to analyze whether the defendant caused a "particular result" with the *mens rea*—purposely, knowingly, recklessly, or negligently—required

*proximate (legal cause)*

by the law defining the offense (see § 2.03(2), (3)). The causation question is resolved by determining whether the injury or harm "is not too remote or accidental in its occurrence to have a [just] bearing on the actor's liability or on the gravity of his offense." For strict liability crimes causation "is not established unless the actual result is a probable consequence of the actor's conduct." MPC § 2.03(4).

Causation issues in the criminal law can be somewhat mystifying to unravel, although, as we shall see, the courts have developed general rules that cut at least part way through the thicket. Keep in mind that judgments about whether a defendant's conduct caused an alleged harm are, at bottom, *normative* in character. The MPC is explicit in recognizing this fact, with its focus on whether the particular result is "too remote or accidental in its occurrence to have a [just] bearing on the actor's liability or on the gravity of the offense." MPC §§ 2.03(2)(b), (3)(b).

The more specific rules associated with causation essentially are guideposts for arriving at a judgment about whether it is fair, or just, to hold an accused responsible for bringing about a specific harm.

\* \* \*

In September 1983 five teenagers were returning from a ZZ Top concert shortly before 4:00 A.M. in Glens Falls, New York-, when the van driven by one of them veered off of the road and slammed into a utility pole. None of the youths was seriously injured by the impact. However, the crash snapped the utility pole and brought electrical power lines down onto the vehicle. As the youths exited the van one of them, not the driver, received a fatal jolt of electricity. A police officer later explained, "The wires were hitting the top of the van, they were grounded when they got out and they were all (shocked)." Over two cases of empty beer bottles were in the van, and the teens admitted to drinking liquor after the concert. On hearing about the accident, the principal of the high school at which the youths were enrolled "lamented, 'With all of the inducements for alcohol now, all the kids have to do is turn on the television, and they get the message that with beer, you can be real popular, a great athlete or whatever.'"[13]

- Did the driver of the van "cause" the death of the electrocuted teen?
- Did the ZZ Top concert "cause" the young man's death?
- Did the media glamorization of alcohol "cause" the death?
- What are the antecedent but for causes of this tragic accident? What is (are) the legally relevant or proximate cause(s) of death?
- By what rules should we decide issues of causation in the criminal law?

---

### *Kusmider v. State,* 688 P.2d 957 (Alaska App. 1984)

Bryner, Chief Judge.

Thomas Kusmider was convicted, after a jury trial, of murder in the second degree, AS 11.41.110. He appeals, contending that Superior Court Judge Karl S. Johnstone erred in excluding evidence relating to the proximate cause of the victim's death. We affirm.

On November 15, 1982, Kusmider's girlfriend told Kusmider that an acquaintance, Arthur Villella, had sexually assaulted her. Kusmider went to Villella's home in Anchorage. A confrontation ensued, and Kusmider shot Villella. The bullet entered Villella's neck above the Adam's apple. Although the wound did not sever any major arteries, it damaged smaller vessels, causing blood to drain down Villella's windpipe.

Villella was unconscious by the time an ambulance arrived. He was attended by paramedics, who inserted a tube in his windpipe to help his breathing. En route to the hospital, however, Villella began flailing his arms and pulled the tube from his throat. Villella died approximately one hour after arriving at the hospital.

At Kusmider's trial, a pathologist testified that Villella's death was caused by the gunshot wound to his throat. However, the pathologist stated that the wound, while life-threatening, might have been survivable. Kusmider then asked the court for permission to present evidence on the issue of proximate cause. He argued that, if allowed to pursue the issue, he might be able to establish that Villella would have survived the gunshot wound if he had not been able to pull the tube from his windpipe. Kusmider maintained that the paramedics who transported Villella might have been negligent in failing to restrain Villella's arms. Kusmider insisted that he was entitled to have the jury consider whether possible negligence by the paramedics constituted an intervening or superseding cause of Villella's death, rendering the gunshot wound too remote to be considered the proximate cause of death.

Judge Johnstone precluded Kusmider from pursuing the issue of proximate cause before the jury. The judge ruled that negligent failure to provide appropriate medical assistance could not, under the circumstances, interrupt the chain of proximate causation and that, therefore, no jury issue of proximate cause had been raised by Kusmider's offer of proof.

On appeal, Kusmider renews his argument, contending that the jury should have been permitted to hear evidence on the issue of proximate cause. We believe that Kusmider's argument is flawed. Kusmider is correct in asserting that proximate cause is ordinarily an issue for the jury. Of course, in every criminal case the state must establish and the jury must find that the defendant's conduct was the actual cause, or cause-in-fact, of the crime charged in the indictment. Here, testimony that Villella actually died from the gunshot wound was undisputed, and the actual cause of death was not in issue. On appeal, Kusmider does not argue that the trial court's exclusion of evidence relating to proximate cause infringed in any way on the jury's ability to determine actual cause.

Case law and commentators agree that, when death is occasioned by negligent medical treatment of an assault victim, the original assailant ordinarily remains criminally liable for the death, even if it can be shown that the injuries inflicted in the assault were survivable; under such circumstances, proximate cause is not interrupted unless the medical treatment given to the injured person was grossly negligent and unforeseeable. (*See* footnote 1.)

In the present case, Kusmider offered to prove only that the paramedics who treated Villella might have been negligent in failing to restrain Villella's arms. Kusmider did not argue that he could demonstrate gross negligence or recklessness, nor did he contend that the circumstances surrounding Villella's death were unforeseeable. Since, as a matter of law, only grossly negligent and unforeseeable mistreatment would have constituted an intervening cause of death and interrupted the chain of proximate causation, we conclude that Judge Johnstone did not err in excluding evidence relating solely to the issue of negligence by the paramedics who treated Villella.

---

Even assuming Kusmider had offered to prove that the conduct of the paramedics was both unforeseeable and grossly negligent, we would still conclude that the trial court correctly excluded the evidence relating to proximate cause. In cases involving death from injuries inflicted in an assault, courts have uniformly held that the person who inflicted the injury will be liable for the death despite the failure of third persons to save the victim.

One commentator notes:

"The question is not what would have happened, but what did happen," and there can be no break in the legally-recognized chain of causation by reason of a possibility of intervention which did not take place, because a "negative act" is never superseding. Moreover, an injury is the proximate cause of resulting death although the deceased would have recovered had he been treated by the most approved surgical methods, or by more skillful methods, or "with more prudent care," or "with a different diet and better nursing," or "with proper caution and attention." The same is true even if the injured person did not take proper care of himself, or neglected to obtain medical treatment, or delayed too long in doing so, or refused to submit to a surgical operation despite medical advice as to its necessity. R. Perkins & R. Boyce, *Criminal Law* § 9 at 799–800 (footnotes omitted).

Here, Kusmider did not claim that the conduct of the paramedics inflicted any new injuries on Villella nor did he even assert that the paramedics aggravated the injuries inflicted by the gunshot wound. Rather, the gist of his claim was that negligence in failing to restrain Villella's arms enabled Villella to disrupt the apparently successful emergency treatment that he had begun to receive. Thus, in support of his argument that proximate cause had been interrupted, Kusmider attempted to rely exclusively on the proof of a negative act: that the paramedics negligently failed to take adequate precautions to restrain Villella. Since Kusmider never offered to prove that the paramedics engaged in any affirmative conduct that might have aggravated Villella's injuries and hastened his death, he could not, as a matter of law, have established a break in the chain of proximate cause, even if he could have shown that the paramedics committed gross and unforeseeable negligence by their failure to act. In short, no matter how negligent the paramedics may have been in failing to prevent Villella's death, it is manifest that the gunshot fired by Kusmider remained a substantial factor—if not the *only* substantial factor—in causing Villella's death. No more is required for purposes of establishing proximate cause.

Because the evidence proffered by Kusmider could not, as a matter of law, have established a break in the chain of proximate causation, we hold that Judge Johnstone did not err in excluding this evidence from trial. . . .

## Notes and Questions

1. Would you expect a different result in *Kusmider* if the shooting victim had arrived safely at the hospital, undergone an operation, and only thereafter died as a result of an infection attributable to physicians' negligent treatment of his wound? What if, while at the hospital, the victim contracted a fatal disease unrelated to the shooting and died as a result of the disease? What distinguishes these two examples? If the defendant intended to cause the victim's death and the death would not have occurred but for the defendant's conduct, should it matter precisely how that result eventually came about? Or, is it important that in the former scenario the infection seems to have a more direct link to the defendant's conduct, whereas in the latter scenario the disease seems to be more removed from, or independent of, the defendant's conduct? *See Bush v. Commonwealth*, 78 Ky. 268 (1880) (holding that defendant who wounded his victim, requiring medical treatment, was not responsible for causing the victim's death after the victim caught scarlet fever from the treating physician and died).

2. In footnote 1 of its opinion, the court in *Kusmider* suggests it would have been "groundless" for the defendant to have argued that the victim's affirmative act of pulling the tube from his windpipe would have affected the proximate cause analysis. For a case presenting similar circumstances, *see United States v. Hamilton*, 182 F. Supp. 548 (D. D.C. 1960).

3. In *Jones v. State*, 43 N.E.2d 1017 (Ind. 1942), the defendant raped a 12-year-old girl who, after the assault and while still in great distress, fell into a river and drowned. The defendant took no action to save her from the river. Is Jones guilty of "causing" her death? Is your answer to this question made any easier by applying principles of criminal liability based on the alleged failure to take action when there is a duty to act? (See the discussion under "Culpable Omissions," above.) A jury found Jones guilty of second-degree murder, and the Indiana Supreme Court affirmed.

*(Continues)*

(Continued)

If the 12-year-old victim of Jones' sexual assault had jumped into the river instead of falling into it, would that fact affect the causation analysis? What if, distraught over being assaulted, the girl jumped into the river the next day and drowned: Would Jones be responsible for causing her death and thus properly convicted of murder? What if the rape victim had endured the psychological trauma of the assault until her 21st birthday and then, leaving a note that she could no longer live with the grief the rape had caused, she jumped into the river and drowned? Would Jones be responsible for causing her death?

4. Kurt Bowman was convicted of driving while intoxicated and causing the death of the passenger in his car, Brenda Keyser, as a result of a one-car crash near Elkhart, Indiana. Ms. Keyser was not wearing her seat belt, and Bowman introduced evidence tending to show that she would not have suffered fatal injuries if she had been. He requested the trial judge to instruct the jury that, "if you find that [Ms. Keyser's] failure to wear the available safety belt was the direct cause of her death, you must find Kurt Bowman not guilty. . . ." Should the judge be required to give the requested instruction? The Indiana Court of Appeals ruled that the instruction was properly denied. *Bowman v. State*, 564 N.E.2d 309 (Ind. App. 1990), *affirmed in part and reversed in part on other grounds*, 577 N.E.2d 569 (Ind. 1991).

5. Assume that a defendant removes or bends a stop sign so that it is no longer readily visible to motorists approaching an intersection. If a motorist proceeds from the relevant direction into the intersection without stopping and her car is struck by another car, resulting in the death of one of the drivers, can the defendant be held responsible for causing the death and be convicted of manslaughter? Would it matter if the motorist who drove through the tampered-with stop sign happened to be speeding? See *State v. Hallett*, 619 P.2d 335 (Utah 1980) (affirming defendant's conviction).

# *People v. Schmies*, 51 Cal. Rptr. 2d 185 (Cal. App.), *rev. denied* (Cal. 1996)

Sparks, Acting Presiding Justice.

In this case we consider the question of causation in a criminal case involving police pursuit of a fleeing motorist. The principal question relates to the admissibility of evidence concerning the reasonableness of the pursuing officers' conduct.

Defendant Claude Alex Schmies fled from an attempted traffic stop and engaged in a high-speed vehicle chase with peace officers. During the chase, one of the pursuing patrol cars struck another car. The driver of the other car was killed and the police officer was injured. Charged with a variety of offenses, defendant was acquitted of second degree murder but convicted of vehicular manslaughter with gross negligence and reckless driving causing great bodily injury.

Sentenced to a total unstayed sentence of seven years four months, defendant appeals. . . .

[W]e . . . reject the claim that the trial court improperly precluded evidence relating to the reasonableness of the police officers' conduct during the pursuit.

## FACTUAL AND PROCEDURAL BACKGROUND

At 4:30 P.M. on October 24, 1992, California Highway Patrol (CHP) Officer Steven Petch was driving southbound on Interstate 5. Another CHP Officer, Christopher Homen, was approximately two-tenths of a mile behind him in another patrol car. Defendant, who was driving a motorcycle, entered the highway and accelerated to a speed of approximately 90 miles per hour. Officer Petch activated his lights for a traffic stop. Defendant slowed down, looked over his shoulder and then sped up.

Officer Petch turned on his flashing lights and siren and radioed to Officer Homen, "Here we go." Officer Petch also notified radio dispatch about the pursuit and, after getting behind the motorcycle and obtaining its license number, asked dispatch to check on the vehicle. The report came back clear: the motorcycle had not been reported stolen, nor were there any outstanding warrants.

The pursuit continued along Interstate 5 at speeds in excess of 90 miles per hour. Officer Petch drove alongside defendant to try to keep him on the freeway. He got a clear look at defendant and the motorcycle.

Defendant cut in front of Officer Petch and took an exit from the freeway. Both Officer Petch and Officer Homen followed in their cars. Officer Petch had turned on all of his emergency lights as well as his siren. Officer Homen did not have overhead lights, but had activated a light on the side of his car and also turned on his siren.

Defendant drove through stop signs and over double yellow lines to weave around cars. He drove through red lights as well. Defendant went at speeds of up to 95 miles per hour, and came perilously close to falling off his motorcycle on sharp turns.

Defendant drove through the intersection of Churncreek and Parsons. Officer Petch followed. A car on Parsons that had been stopped to allow Officer Petch to go through started into the intersection, apparently unaware of the second patrol car driven by Officer Homen. Officer Homen tried to avoid the car but hit it broadside. The driver, Jane Abbett, was killed and Officer Homen was injured. . . .

At trial, Officers Petch and Homen described their pursuit of defendant and the crash between Homen and Abbett. . . .

As discussed in detail below, defendant argued to the jury that Officer Homen's actions in the pursuit were a superseding intervening act, breaking the chain of causation and relieving defendant of liability for the death of Abbett and the injuries to Officer Homen. . . .

## DISCUSSION

### I. Causation Evidence

At trial, defendant attempted to demonstrate that Officer Homen's actions broke the chain of causation, absolving defendant of responsibility for Abbett's death and Officer Homen's injuries. To this end, defendant tried to obtain a complete copy of the CHP's pursuit policy to learn whether Officer Homen's actions violated CHP guidelines. . . . The trial court denied the motion, ruling that the policy was irrelevant to determining whether Officer Homen's actions were reasonably foreseeable.

Similarly, the court refused to permit defendant's expert witnesses to testify as to the reasonableness of the pursuit. The court distinguished the question of foreseeability from the question of reasonableness, and ruled that while defendant could argue that Officer Homen's conduct was not reasonably foreseeable, he could not introduce evidence relating to the reasonableness of Officer Homen's action, an area the court deemed irrelevant to the issues at hand. . . .

The reasonableness of the officers' conduct, as defendant proffered it, would have arguable relevance only with respect to the question of foreseeability of the harm that ensued from defendant's course of conduct. Foreseeability of harm is not itself an element of the offense, but is a recognized factor to be considered in determining whether the defendant acted with gross negligence and, if so, whether his conduct was the proximate or legal cause of the ensuing harm. Defendant focuses on the causation issue as the disputed issue upon which the reasonableness of the officers' conduct is claimed to have had probative value.

The principles of causation apply to crimes as well as torts. "Just as in tort law, the defendant's act must be the legally responsible cause (*'proximate cause'*) of the injury, death or other harm which constitutes the crime." Thus, in the language of the standard jury instruction, to constitute a homicide "there must be, in addition to the death of a human being, an unlawful act which was a cause of that death." . . . Thus, in homicide cases, a "cause of the [death of the decedent] is an act or omission that sets in motion a chain of events that produces as a direct, natural and probable consequence of the act or omission the [death] and without which the [death] would not occur." In general, "[p]roximate cause is clearly established where the act is directly connected with the resulting injury, with no intervening force operating." In this case there was an intervening force in operation, the pursuit by the CHP officers and the collision between the vehicle of one of the officers and the decedent's car. But that does not end the inquiry because "[d]efendant may also be criminally liable for a result directly caused by his act, even though there is another contributing cause."

Intervening causes in criminal cases are typically described as either "dependent" or "independent." A **dependent intervening cause** will not absolve a defendant of criminal liability while an **independent intervening cause** breaks the chain of causation and does absolve the defendant. "An intervening cause may be a normal or involuntary result of the defendant's original act. Such a cause is said to be 'dependent,' and does not supersede; *i.e.,* the defendant is liable just as in the direct causation case." An "independent" intervening "act may be so disconnected and unforeseeable as to be a superseding cause; *i.e.,* in such a case the defendant's act will be a remote, and not the proximate, cause." In the words of the Restatement Second of Torts, again in the context of, negligence, "[w]here the negligent conduct of the actor creates or increases the foreseeable risk of harm through

*(Continues)*

the intervention of another force, and is a substantial factor in causing the harm, such intervention is not a superseding cause." (Rest.2d Torts, § 442A, p. 468.) Stated another way, "[t]he intervention of a force which is a normal consequence of a situation created by the actor's negligent conduct is not a superseding cause of harm which such conduct has been a substantial factor in bringing about." (Rest.2d Torts, § 443, p. 472.)

These rules governing superseding cause are succinctly stated in BAJI No. 3.79, specifying when a third party's negligence is not a superseding cause. Under this instruction, a defendant is not relieved of liability by a third party's intervening negligence if at the time of his conduct he realized or reasonably should have realized that a third party might so act or the risk of harm was reasonably foreseeable; or a reasonable person knowing the situation existing at the time of the conduct of the third party would not have regarded it as highly extraordinary that the third party had so acted; or the conduct of the third person was not extraordinarily negligent and was a normal consequence of the situation created by defendant. Extraordinary, under this instruction on superseding causes, "means unforeseeable, unpredictable, and statistically extremely improbable." (Com. to BAJI No. 3.79 (8th ed. 1994 bound vol.) p. 104.)

The question whether defendant's acts caused Abbett's death and Officer Homen's injuries is to be determined by the trier of fact according to these general principles governing proximate causation.

As the Court of Appeal noted in [*People v.*] *Harris,* [52 Cal.App.3d 419, 427, (1975)], "[a] defendant may be criminally liable for a result directly caused by his act even if there is another contributing cause. If an intervening cause is a normal and reasonably foreseeable result of defendant's original act the intervening act is 'dependent' and not a superseding cause, and will not relieve defendant of liability." Moreover, as leading commentators have put it, "[i]f the intervening act or other cause is reasonably foreseeable, it will not supersede. As in tort law: (1) The consequence need not have been a strong probability; a possible consequence which might reasonably have been contemplated is enough. (2) The precise consequence need not have been foreseen; it is enough that the defendant should have foreseen the possibility of some harm of the kind which might result from his act." In other words, "it is only an unforeseeable intervening cause, an extraordinary and abnormal occurrence, which rises to the level of an exonerating, superseding cause."

Defendant claims his defense would have been that "the actions taken by the CHP officers and the decedent, especially Officer Homen, were so inappropriate as to be a superseding cause which would break the chain of proximate cause—*i.e.,* they were the sole proximate cause of the death." But the court's orders, he argues on appeal, "precluded defense counsel from presenting evidence and arguing the 'abnormality' or 'nature' of the conduct of the officers." . . .

If we were concerned with a tort action, a disciplinary hearing, or even a criminal action *against the officers,* then we would be focused upon the officers' conduct and the "reasonableness" thereof, from their point of view. In that respect we would consider the harm that was reasonably foreseeable to the officers, and whether their conduct created an unreasonable risk of harm to others. But, we reiterate, the negligence or other fault of the officers is not a defense to the charge against defendant. The fact that the officers may have shared responsibility or fault for the accident does nothing to exonerate defendant for his role. In short, whether the officers' conduct could be described with such labels as negligent, careless, tortious, cause for discipline, or even criminal, in an action against them, is not at issue with respect to the defendant here. In this sense the "reasonableness" of the officers' conduct, focused upon their point of view and their blameworthiness for the death, is not relevant.

The issue with respect to defendant focuses upon his point of view, that is, whether the harm that occurred was a reasonably foreseeable consequence of his conduct at the time he acted. Since the officers' conduct was a direct and specific response to defendant's conduct, the claim that their conduct was a superseding cause of the accident can be supported only through a showing that their conduct was so unusual, abnormal, or extraordinary that it could not have been foreseen. . . . In this case the trial court understood and was legally correct in making the distinction we have been discussing, that is, the distinction between reasonableness of the officers' conduct from their point of view and reasonable foreseeability of that conduct from the defendant's point of view. The court endeavored to explain that distinction as clearly as it could and repeatedly advised the defense that it would not be precluded from introducing evidence and argument with respect to the reasonable foreseeability of the officers' conduct. . . .

With respect to defendant's appellate assertion regarding the pursuit policy guidelines of the CHP, we find no error for several reasons. . . .

[W]hether the officers violated the CHP pursuit guidelines is immaterial. The question is whether defendant realized or should have realized that the CHP officers would pursue his fleeing motorcycle. In this case the evidence clearly shows that defendant knew, or at the very least, should have known that his flight would cause the officers to pursue him. This is true if for no other reason than the officers were in fact pursuing him and he nevertheless continued to flee. This illegal and dangerous act by defendant caused the officers to pursue him and ultimately caused the fatal accident. It adds not one whit to say that the officers violated the CHP pursuit guidelines. The test, as we have recounted, is not whether the officers acted reasonably but rather whether defendant realized or should have realized that the officers would respond as they did. Assume, for purposes of illustration, a bank has a written policy that its armed guards should not fire their weapons at an armed robber if the bank is full of customers. Nevertheless, in the course of an armed robbery during business hours the guard, fearing the robber might injure or kill someone, predictably fires at the robber but misses and kills a customer. Can the defendant robber charged with the murder of the customer establish that the shooting by the guard was a superseding cause because it violated the bank's rules? The answer to that hypothetical is the same as the answer to the identical claim in this case: no. Just as the robber has no knowledge of the bank's rules, so too defendant lacks any knowledge of the CHP guidelines. If it is reasonably foreseeable that the guard, in the heat of the frightening moment, will fire at the robber, or that the CHP officers will give chase to a fleeing motorcycle traveling over 90 miles per hour, it is no defense to prove a rule violation. The task of the jury is to determine whether the officers' response was so extraordinary that it was unforeseeable, unpredictable and statistically extremely improbable. A rule violation may give rise to civil liability or disciplinary action, but it has nothing to do with the foreseeability of the officers' conduct. . . .

In summary, in this case defendant, acting with gross negligence, created a situation to which CHP officers responded. As a result of defendant's conduct and the officers' response, a vehicle collision occurred in which a third party was killed and one of the officers was injured. The question whether defendant's conduct is legally responsible for the death and injury depends upon whether the officers' conduct can be regarded as a superseding cause. That issue depends upon whether the danger was reasonably foreseeable to defendant rather than upon the reasonableness of the officers' response. The court made that distinction, explained it carefully in its ruling, and repeatedly advised that evidence and argument with respect to reasonable foreseeability would not be excluded. The defense pointed to no specific relevant evidence that it was dissuaded from presenting by the court's ruling. The jury was expressly instructed that "an intervening act may be so disconnected and unforeseeable as to be a superseding cause," absolving defendant of liability. The jury simply rejected this claim. There is no cause for reversal on this record. . . .

## Notes and Questions

1. Why should the defendant be expected to "reasonably foresee" action by the police that may deviate substantially from what the police's own operating procedures define as reasonable or appropriate under the circumstances? Would it matter if the defendant in *Schmies* or the bank robber in the court's hypothetical had actual knowledge of the procedures in place for governing police behavior?

2. The court in *Schmies* distinguishes among a number of concepts relevant to causation analysis. For example, it discusses dependent and independent "intervening" causes and the circumstances under which those intervening causes are considered to be "superseding" or "supervening," so as to insulate the defendant's conduct from being identified as a legally relevant (proximate) cause of the harm. Consider the following definitions and rules.

> When . . . one dominant cause is found it is treated as the "sole cause" for the purposes of the criminal case. . . . A "sole cause" which intervenes between defendant's act and the result in question is spoken of a "superseding cause." . . .
>
> A cause which produces a result without the aid of any intervening cause is a direct cause. . . .
> An intervening cause is one which comes between an antecedent and a consequence.[14]

\*\*\*

> The criminal law tends to distinguish between "responsive" (or "dependent") and "coincidental" (or "independent") intervening causes . . . .

*(Continues)*

A responsive intervening cause is an act that occurs in reaction or response to the defendant's prior wrongful conduct. For example, suppose that D1 operates his boat at an unsafe speed, causing it to capsize. V1, his drunken passenger, drowns foolishly attempting to swim to shore. V1's actions constitute a responsive intervening cause in his own death, *i.e.,* his life-saving efforts were in response to D1's initial improper conduct. Or, suppose that D2 seriously wounds V2. V2 is taken to a hospital where he receives poor medical treatment by physician X and dies. In D2's prosecution for the death, X's negligent conduct constitutes a responsive intervening cause: X's medical actions were in response to D2's act of wounding V2.

Generally speaking, a responsive intervening cause does *not* relieve the initial wrongdoer of criminal responsibility, unless the response was not only unforeseeable, but highly abnormal or bizarre. This outcome is justifiable. The defendant's initial wrongdoing caused the response. Since he is responsible for the presence of the intervening force, the defendant should not escape liability unless the intervening force was so out-of-the-ordinary that it is no longer fair to hold him criminally responsible for the outcome. . . .

A coincidental intervening cause is a force that does not occur in response to the initial wrongdoer's conduct. The only relationship between the defendant's conduct and the intervening cause is that the defendant placed the victim in a situation where the intervening cause could independently act upon him. For example, suppose that D1 robs V1, a passenger in D1's car, and then abandons V1 on a rural road. Sometime later, driver X1 strikes and kills V1, who is standing in the middle of the road. X1's conduct is a coincidental intervening cause: D1's actions did not cause X1 to drive down that road on that particular occasion; D1 simply put V1 on the road where X1's independent conduct acted upon V1.

Or, suppose that D2 wounds V2. V2 is taken to a hospital for medical treatment, where he is killed by X2, a "knife-wielding maniac" who is running through the hospital killing everyone in sight. Again, X2 is a coincidental intervening cause: X2 was going to be running through that hospital killing victims whether or not V2 was there. This is a case in which V2 was in the wrong place at the wrong time, put there by D2's original wrongdoing.

The common law rule of thumb is that a coincidental intervening cause relieves the original wrongdoer of criminal responsibility, unless the intervention was foreseeable. In the present examples, therefore, it would be necessary to determine whether D1 and D2, as reasonable people, should have foreseen, respectively, that V1 would be struck by another car, and that V2 would be the victim of a criminal intermediary. In the first case, it may have been foreseeable that another car would drive down that road and strike V1. In the second hypothetical, X2's criminal activities were probably abnormal enough to relieve D2 of liability for the ensuing death, unless, for example, the events occurred in a high-security penal institution.[15]

*Source:* Reprinted from *Understanding Criminal Law* by Joshua Dressler with permission. Copyright © 2006 Matthew Bender & Company, Inc., a member of the LexisNexis Group. All rights reserved.

Do these guidelines help give content to the MPC's general directive, that the actor is not criminally responsible for a result or injury that is "too remote or accidental in its occurrence to have a [just] bearing on the actor's liability or the gravity of the offense"?

3. In *Schmies,* what is properly characterized as the "direct" cause of Ms. Abbett's death? Is that cause "intervening" with respect to the defendant's conduct? Is it a "dependent" intervening cause that is responsive to the defendant's prior conduct or an "independent" intervening cause that is merely coincidental? Based on this classification, what test should be applied to determine whether the intervening cause is "superseding" or whether the defendant's conduct fairly can be said to be the proximate cause of Ms. Abbett's death?

4. Assume that an intruder unlawfully enters another's house and awakens the homeowner while rummaging through his or her bedroom, or that an assailant displays a weapon and demands money from a person walking down a street. Terrified by the confrontation, the victim suffers a heart attack and dies. Should the burglar or armed robber, respectively, be held responsible for causing the death and thus convicted of a criminal homicide, perhaps even murder? We consider this type of causation issue in Chapter 6 when we discuss the felony-murder rule.

5. What if an assault victim suffers from hemophilia, a condition that impedes coagulation of the blood so that even a minor cut or wound can result in profuse bleeding, even resulting in death? Should the victim's "preexisting condition" be defined as the legally operative cause of death? If not, and the administered blow is the proximate cause of death, should the offender's culpability at least be diminished based on *mens rea* considerations (assuming the offender did not know or had no reason to believe that the victim suffered from hemophilia)? See *State v. Frazier,* 98 S.W.2d 707 (Mo. 1936).

### *State v. Rogers*, 992 S.W.2d 393 (Tenn. 1999)

[The defendant, Wilbert Rogers, stabbed James Bowdery in the heart with a butcher knife on May 6, 1994. During an operation to repair his heart Bowdery went into cardiac arrest, resulting in a loss of oxygen to the brain and severe brain damage. He remained in a coma roughly 15 months, and died on August 7, 1995, from kidney complications associated with being comatose for such an extensive length of time. Rogers was convicted of second-degree murder. On appeal he argued that the passage of 15 months between his stabbing Bowdery and Bowdery's death precluded a conviction for criminal homicide. In support of his argument, he relied on the continued vitality of the common law "**year-and-a-day rule**"—a rule that bars a criminal homicide conviction when more than a year lapses between the defendant's injurious act and the victim's death.]

## A. Historical Development

The year-and-a-day rule is deeply rooted in the common law. Its lineage is generally traced to the thirteenth century where the rule was originally utilized as a statute of limitations governing the time in which an individual might initiate a private action for murder known as "appeal of death." . . .

By the eighteenth century, however, the year-and-a-day rule had been extended to the law governing public prosecutions so that a homicide prosecution could not be brought unless the victim died within a year and one day of the injury. . . .

Three justifications are ordinarily given for the common law rule. The first and most often cited justification is that thirteenth century medical science was incapable of establishing causation beyond a reasonable doubt when a great deal of time had elapsed between the injury to the victim and the victim's death. Therefore, it was presumed that a death which occurred more than a year and one day from the assault or injury was due to natural causes rather than criminal conduct.

Second, it has often been said that the rule arose from the early function of the jury as a reporter of the happenings of the vicinage. Even if expert medical testimony had been adequate to establish causation at common law, it would not have been admissible. Unlike current procedure, in early English courts, jurors were required to rely upon their own knowledge to reach a verdict, and they could not rely upon the testimony of witnesses having personal knowledge of the facts or upon expert opinion testimony.

Finally, the rule has occasionally been characterized as an attempt to ameliorate the harshness of the common law practice of indiscriminately imposing the death penalty for all homicides—first degree murder and manslaughter alike.

## B. Modern Status

Despite its early common law recognition and near universal acceptance, the rule has fallen into disfavor and has been legislatively or judicially abrogated by the vast majority of jurisdictions which have recently considered the issue. Most courts describe the rule as outmoded and obsolete since the reasons justifying its recognition no longer exist. In characterizing the rule as an anachronism, courts have particularly emphasized the many advances of medical science. . . .

Although the rule has been judicially abrogated in a number of jurisdictions, none of those courts have adopted a new time limit to replace the year-and-a-day rule. In declining to do so, courts emphasize that no arbitrary time frame is needed because abolition of the year-and-a-day rule does not relieve the State of its burden of proving causation beyond a reasonable doubt. . . . These courts also point out that there is no statute of limitations for murder prosecutions and find implementation of an arbitrary time limit which bars murder prosecutions if the victim does not die within a specified period of time inconsistent with public policy. . . .

The State strongly urges this Court to judicially abrogate the common law rule. While the defendant concedes that this Court has the power to abrogate the rule, he nonetheless argues that we should defer action on this issue to the judgment of the General Assembly. Rogers contends that the General Assembly is in a better position to determine an appropriate substitute time limit to replace the year-and-a-day rule.

*(Continues)*

*(Continued)*

This Court has "not hesitated to abolish obsolete common-law doctrines," and we have recognized that "we have a special duty to do so where it is the Court, rather than the Legislature, which has recognized and nurtured" the common law rule. . . .

Since, as previously stated, the year-and-a-day rule has its roots in the common law, and has in fact never been a part of the statutory law of this State, we refuse the defendant's suggestion to defer this issue to the General Assembly's judgment. . . .

Without question, the reasons which prompted common law courts to recognize the rule no longer exist. Medical science can now sustain the critically wounded for months and even years beyond what might have been imagined only a few decades ago. Comparable progress has been made in the development of diagnostic skills, so that problems of medical causation are more readily resolved. Modern pathologists are able to determine the cause of death with much greater accuracy than was possible in earlier times. Moreover, jurors today may rely upon expert testimony, even when the testimony relates to an ultimate issue of fact such as causation. . . . Finally, the death penalty is no longer indiscriminately imposed for all homicides. . . . Accordingly, we hereby abolish the common law rule, and by doing so, join the majority of other jurisdictions which have recently considered the issue. We also reject the defendant's invitation to adopt a substitute time limit to replace the year-and-a-day rule. We agree with those courts which have held that no arbitrary time frame is needed because abolition of the year-and-a-day rule does not relieve the State of its burden proving causation beyond a reasonable doubt. . . .

## D. *Ex Post Facto* Violation

Having determined that the common law rule should be abolished, we must next consider the defendant's claim that the *ex post facto* clauses of the state and federal constitution require that our decision be prospectively applied. The State asserts that abrogation of the rule can be retrospectively applied since it is not an unforeseeable judicial enlargement of a criminal statute. . . . Given the fact that the rule has been abolished by every court which has squarely faced the issue, and given the fact that the validity of the rule has been questioned in this State in light of the passage of the 1989 Act, we conclude that our decision abrogating the rule is not an unexpected and unforeseen judicial construction of a principle of criminal law. Moreover, we emphasize that abolition of the rule does not allow the State to obtain a conviction upon less proof, nor does its abolition impose criminal sanctions for conduct that was heretofore innocent. Accordingly, we apply our decision abolishing the rule retroactively to the facts of this case and affirm the defendant's conviction. . . .

## Notes and Questions

1. What is the relationship between the year-and-a-day rule and the issue of proximate cause?
2. Many other courts that have abrogated the year-and-a-day rule have announced that their decisions would apply prospectively only to avoid the *ex post facto* issue discussed in *Rogers*. *See, e.g., Commonwealth v. Lewis,* 409 N.E.2d 771 (Mass. 1980); *State v. Young,* 390 A.2d 556 (N.J. 1978); *State v. Pine,* 524 A.2d 1104 (R.I. 1987); *State v. Vance,* 403 S.E.2d 495 (N.C. 1991); *People v. Stevenson,* 331 N.W. 2d 143 (Mich. 1982). In 2001 the U.S. Supreme Court reviewed the Tennessee Supreme Court's decision in *Rogers* to decide whether the state court's abrogation of the year-and-a-day rule could be applied retrospectively without violating U.S. Constitutional principles. The justices affirmed the state court ruling in a 5–4 decision. *Rogers v. Tennessee,* 532 U.S. 451 (2001). Justice O'Connor's majority opinion distinguished between retroactive application being given to rules of criminal law through *legislative* action—appropriately analyzed pursuant to *Ex Post Facto* principles—and rules that are modified *judicially,* or through common law development—which are analyzed under Due Process principles.

> In the context of common law doctrines (such as the year and a day rule), there often arises a need to clarify or even to reevaluate prior opinions as new circumstances and fact patterns present themselves. Such judicial acts, whether they be characterized as "making" or "finding" the law, are a necessary part of the judicial business in States in which the criminal law retains some of its common law elements. Strict application of *ex post facto* principles in that context would unduly impair the incremental and reasoned development of precedent that is the foundation of the common law system. The common law, in short, presupposes a measure of evolution that is incompatible with stringent application of *ex post facto* principles. It was on account of concerns

such as these that *Bouie v. City of Columbia*, 378 U.S. 347 (1964) restricted due process limitations on the retroactive application of judicial interpretations of criminal statutes to those that are "unexpected and indefensible by reference to the law which had been expressed prior to the conduct in issue." *Bouie,* 378 U.S., at 354 (internal quotation marks omitted).

We believe this limitation adequately serves the common law context as well. It accords common law courts the substantial leeway they must enjoy as they engage in the daily task of formulating and passing upon criminal defenses and interpreting such doctrines as causation and intent, reevaluating and refining them as may be necessary to bring the common law into conformity with logic and common sense. It also adequately respects the due process concern with fundamental fairness and protects against vindictive or arbitrary judicial lawmaking by safeguarding defendants against unjustified and unpredictable breaks with prior law. Accordingly, we conclude that a judicial alteration of a common law doctrine of criminal law violates the principle of fair warning, and hence must not be given retroactive effect, only where it is "unexpected and indefensible by reference to the law which had been expressed prior to the conduct in issue." *Ibid.* . . .

There is . . . nothing to indicate that the Tennessee court's abolition of the rule in petitioner's case represented an exercise of the sort of unfair and arbitrary judicial action against which the Due Process Clause aims to protect. Far from a marked and unpredictable departure from prior precedent, the court's decision was a routine exercise of common law decisionmaking in which the court brought the law into conformity with reason and common sense. It did so by laying to rest an archaic and outdated rule that had never been relied upon as a ground of decision in any reported Tennessee case. . . .

Justices Stevens, Scalia, Thomas, and Breyer dissented.

Can you distinguish the Court's ruling in *Rogers* from its ruling in *Stogner v. California*, 539 U.S. 607 (2003), which prohibited retroactive application of an extended statute of limitations to allow prosecution of a suspect for crimes allegedly committed during a period when prosecution would have been barred under the previously operative statute of limitations? We considered *Stogner* in connection with our discussion of *ex post facto* principles in Chapter 2.

**3.** The *Rogers* court declined to substitute a different fixed period of time for the year-and-a-day rule. Through legislation, Washington used to require that a death occur within 3 years and 1 day of a defendant's harmful conduct to sustain a criminal homicide conviction. However, a 1997 amendment eliminated that time limit and specified that death could occur "at any time." Wash. Rev. Stat. Ann. § 9A.32.010 (West Supp. 1999). As *Rogers* indicates, no courts have adopted a specific time limit to replace the year-and-a-day rule.

**4.** Consider the following news article:

William J. Barnes shot and partly paralyzed a Philadelphia police officer in 1966, and he served 20 years for it and related offenses.

But last month, 41 years after the shooting, the district attorney filed new charges of murder after the officer . . . died of an infection she says stems from the shooting. Mr. Barnes, now 71, was sent back to prison.

"The law is that when you set in motion a chain of events," District Attorney Lynne M. Abraham said, "a perpetrator of a crime is responsible for every single thing that flows from that chain of events, no matter how distant, as long as we can prove the chain is unbroken."

She plans to prove that the bullet that lodged near [the officer's] spine in 1966 led to the urinary tract infection that led to his death last month.[16]

*Five general elements of crime.*

## Conclusion

This chapter examined the five general elements of crime: *mens rea,* the act, the concurrence principle, harm, and causation. We are far from finished with these issues. Not surprisingly, they surface repeatedly as we consider specific crimes and how the rules defining those crimes are applied to distinct facts. Before we move to the "special" part of the criminal law, in Chapters 4 and 5 we consider defenses to crimes that are based on the doctrines of justification and excuse, respectively.

## Key Terms

*actus reus*
causation
cause-in-fact
concurrence principle
criminal negligence
culpable omissions
dependent intervening cause
general intent crimes
Good Samaritan laws
independent intervening
 cause

knowingly
*mens rea*
negligently
proximate cause
purposely
recklessly
specific intent crimes
status
transferred intent
year-and-a-day rule

## Review Questions

1. Define *mens rea*. Give an example of *mens rea* with regard to a crime.
2. Make a distinction between *general intent crimes* and *specific intent crimes*. Give an example of each in your response.
3. How does the Model Penal Code view the distinction of *general* and *specific* intent regarding criminal culpability?
4. List the four levels of criminal culpability as posited by the Model Penal Code.
5. Explain the "doctrine of transferred intent" and give at least two examples of how this doctrine is used in criminal law.
6. Why is the distinction between an act and a status important in considering whether a crime has occurred?
7. Explain what is meant by the *voluntary act* requirement.
8. What is meant by a *culpable omission*? Give an example.
9. What is meant by the *concurrence principle*?
10. What is the *year-and-a-day rule*?

## Notes

1. Howe MD, ed. *The Common Law*. Boston: Little, Brown; 1963:7.
2. *See generally* Dressler J. *Understanding Criminal Law*, 4th ed. Newark, NJ: LexisNexis; 2006:134–136.
3. *See generally* Saunders KW. Voluntary acts and the criminal law: justifying culpability based on the existence of volition. *Univ Pitt Law Rev.* 1988;49:443; Moore MS. More on act and crime. *Univ Penn Law Rev.* 1994;142:1749; Corrado M. Is there an act requirement in the criminal law? *Univ Penn Law Rev.* 1994;142:1529.
4. Butler S. *Erewhon and Erewhon Revisited*. New York: The Modern Library; 1955:88, 106–111 (orig. 1872); *See generally* Moore MS. Causation and the excuses. *Calif Law Rev.* 1985;73:1091, 1112.
5. *See* Gansberg M. "37 Who Saw Murder Didn't Call The Police," *New York Times* p. 1, March 27, 1964; Rosenthal AM. *Thirty-Eight Witnesses*. New York: McGraw-Hill; 1964.
6. *See Commonwealth v. Vieira*, 519 N.E.2d 1320 (Mass. 1988); Yeager D. A radical community of aid: a rejoinder to opponents of affirmative duties to help strangers. *Wash Univ Law Q.* 1993;71:1, 21.
7. LaFave WR. *Criminal Law*, 3rd ed. West Group 2000:218–219.
8. Hall J. *General Principles of Criminal Law*, 2nd ed. Indianapolis, IN: Bobbs-Merrill; 1960:189; *See also* Dressler, p. 213 (suggesting that the court in *Jackson* "ignore[d] the concurrence principle").
9. Mill JS. *On Liberty*. S. Collini, ed. Cambridge: Cambridge University Press; 1989:13. Reprinted with the permission of Cambridge University Press.
10. Devlin P. *The Enforcement of Morals*. New York: Oxford University Press; 1965.
11. Harary C. Aggressive play or criminal assault? An in-depth look at sports violence and criminal liability. *Columb J Law Arts.* 2002;25:197.
12. Clarke CA. Law and order on the courts: the application of criminal liability for intentional fouls during sporting events. *Ariz State Law J.* 2000;32:1149.
13. Mahoney J. "Freak Crash Electrocutes 18-Year-Old," *Albany Times Union* A1, A14, Sept. 23, 1983.
14. Perkins RM, Boyce RN. *Criminal Law*. Mineola, NY: Foundation Press; 1982:781, 787, 790.
15. Dressler J. *Understanding Criminal Law*, 4th ed. Newark, NJ: LexisNexis; 2006:204–206.
16. Urbina I. "41 Years After Crime, Prosecutor Says Assault Victim Is Now Murder Victim," *New York Times*, p. A16, Sept. 19, 2007.

## Footnote for *Frey v. State,* 708 So.2d 918 (Fla. 1998)

5. The Model Penal Code does not distinguish between general and specific intent, instead establishing the following degrees of culpability as applied to each statutory element of the alleged crime: purposely, knowingly, recklessly, or negligently. Model Penal Code § 2.02 (Proposed Official Draft 1962). . . .

## Footnote for *Kusmider v. State,* 688 P.2d 957 (Alaska App. 1984)

1. At the time of trial, Kusmider argued only that the paramedics who treated Villella might have been negligent in failing to restrain Villella's arms and that this negligence might have amounted to an intervening factor interrupting proximate cause. Kusmider did not argue that Villella's conduct in pulling the tube from his own windpipe might have affected proximate cause. It is clear that such an argument would have been groundless, since it is well-settled that proximate cause is not affected when death results after the victim of an assault fails to obtain prompt medical treatment or engages in conduct that interferes with effective medical treatment. See, *e.g., Commonwealth v. Cheeks*, 223 A.2d 291, 294 (Pa.1966); W. LaFave & A. Scott, Criminal Law § 35 at 256–59 (1972).

# Justification Defenses

## Chapter Objectives

- Understand what is meant by an affirmative defense
- Differentiate between *justification defenses* and *excuse defenses*
- Understand requirements relating to the *deadly force* doctrine
- Understand what is required for a successful self-defense claim
- Understand the limitations on using force to defend property, a home, or a third person
- Understand the implications of resisting lawful arrest
- Understand the defenses of necessity and choice of evils

In criminal prosecutions the government is required to prove beyond a reasonable doubt that the defendant is guilty of the crime charged. The prosecution must accordingly prove the existence of each element of the crime charged (*mens rea, actus reus*, causation, etc.) pursuant to this standard. However, even if the government establishes all elements of the charged offense, the defendant might nevertheless be acquitted.

The criminal law permits a number of *affirmative defenses* that can either absolve the defendant of criminal liability or greatly reduce the seriousness of the offense. It is important to understand the distinction between affirmative defenses and defenses based on the prosecution's asserted inability to prove one or more elements of the charged crime. Defenses of the latter variety are triggered automatically on a defendant's plea of not guilty. They may be supplemented by the defendant's offering evidence in denial of the charges. For example, a defendant relying on an alibi defense might present evidence that he or she was somewhere else when a crime was committed. A rape defendant might present evidence that the victim consented to intercourse. A defendant charged with possession of marijuana may present evidence that the substance was actually oregano. These arguments attack the sufficiency of the government's proof rather than raising an affirmative defense. If they cause the jury to have a reasonable doubt about whether the prosecution has proven all elements of the crime, the defense has been successful and the defendant will be acquitted.

Affirmative defenses are different. With an **affirmative defense** the defendant essentially admits the essential elements of the crime charged but offers an excuse or justification that negates criminal responsibility. Before a jury can consider evidence relating to an affirmative defense, the defendant must present sufficient evidence to the trial judge to establish the possible existence of the defense. This *burden of production,* or burden of "going forward," does not require a great deal of evidence. As long as the defendant presents enough evidence to fairly put the defense at issue, the burden of production has been met. Once the defendant meets this burden of production, the burden may shift to the prosecution to prove beyond a reasonable doubt that the defendant's actions were not justified or excused under the law. However, in many jurisdictions the defendant also has the ultimate burden of persuasion on at least some affirmative defenses, although that burden might be by a preponderance of the evidence, or by clear and convincing evidence, instead of the more demanding standard of beyond a reasonable doubt.

Affirmative defenses include both "excuse" and "justification" rationales. **Excuse defenses** involve instances where the defendant admits committing the accused act, and even that the act was wrong and the ensuing harm regrettable, but claims that his or her conduct should be excused because of extenuating circumstances such as insanity or duress. We deal with excuse defenses in Chapter 5. Justification defenses are different. They involve instances where the defendant admits committing the accused act but claims that the actions were appropriate under the circumstances. With **justification defenses** the defendant essentially says, "Yes I did it, and I would do it again if placed in the same situation." For example, although the law normally punishes killing, if an individual shoots and kills a knife-wielding assailant who is threatening his or her life, we are likely to have little hesitation in concluding that the person under attack was justified in using deadly force in self-defense. It would be inequitable to punish a person who is confronted with the choice of killing an unlawful assailant or instead suffering death or serious injury at the assailant's hands. The law would authorize the individual to protect him- or herself, and we would conclude that, under the circumstances, the homicide was fully justified.

We consider self-defense and other justification defenses in this chapter.

## Self-Defense

By far, the most commonly used justification defense is self-defense. **Self-defense** is the use of force against another person to repel an unprovoked attack. Although it may seem axiomatic that a person is allowed to use force to defend him- or herself, the law surrounding self-defense is anything but simple.

Several factors must be considered in determining whether a person acted in lawful self-defense. Successful self-defense claims typically include the following factors:

1. The defendant is the victim of an unlawful attack.
2. The defendant did not provoke the attack.
3. The defendant actually and reasonably believes that he or she is in imminent danger of bodily harm.
4. The defendant uses only the amount of force needed to repel the attack.
5. In some jurisdictions, the defendant does not have the opportunity to retreat before resorting to deadly defensive force.

The key to understanding the law of self-defense is to remember that it is centered on the concepts of reasonableness, necessity, and proportionality. A person is allowed to do what a reasonable person would do under the same circumstances. It must be necessary (or at least reasonably perceived as necessary) to use defensive force, and the force used must be proportionate to the threatened unlawful aggressive force. As you read the following cases that deal with claims of self-defense, ask yourself whether the defendant acted reasonably (*i.e.*, complied with the "necessity" and "proportionality" requirements for using defensive force) under the circumstances that (reasonably) were perceived to exist.

## Provocation and the "Aggressor"

As stated above, a person may not claim self-defense if he or she is the unlawful aggressor at the time of the conflict. In this context one court has defined an aggressor as a person "who commits an unlawful act reasonably calculated to produce an affray foreboding injurious or fatal consequences." *United States v. Peterson*, 483 F.2d 1222, 1233 (D.C. Cir. 1973). Keep this definition in mind as you read *Brown v. State*.

---

### *Brown v. State*, 698 P.2d 671 (Alaska App. 1985)

Bryner, Chief Judge. . . .

Elijah J. Brown was convicted of assaulting his friend of thirty years, E.M. Miller. Their friendship had apparently disintegrated, due in part to Miller's friendship with Brown's wife. Brown and his wife had been separated for three years when Brown went to his wife's apartment at around 6:00 P.M. on November 4, 1982, to deliver some mail to his daughter. Brown became upset when he discovered his wife and Miller together in his wife's living room. A short argument ensued.

There was conflicting testimony as to whether Miller threatened Brown. Brown claimed that Miller threatened him with a .44 magnum handgun; Miller denied this claim. In any event, Brown left the apartment and Miller went to a private club known as "the party house." It is undisputed that Miller had a .44 magnum in his pocket at the party house. Brown testified that a short time after arguing with Miller, he returned to his wife's apartment and argued with her until she asked him to leave. Brown went home, decided that he had to go talk to Miller again and, after arming himself with a .22 rifle, took a cab to the party house.

Although Brown and Miller gave different versions of what happened inside the party house, they agreed that Brown entered carrying his rifle while Miller was standing at the bar and that Brown, after saying something to Miller, fired his rifle, shooting Miller in the chest. Miller responded, firing one round from his handgun before it jammed, and Brown fired a second shot into the ceiling. Brown and Miller then struggled with one another until they were separated.

At trial, Miller testified that Brown shot him without **provocation**. In contrast, Brown testified that he entered the party house, approached Miller without pointing the rifle at him, and told Miller he wanted to talk. According to Brown, after he and Miller had spoken for about a minute, Miller turned as if to walk away, then suddenly turned back. When Miller turned back toward Brown, his coat was open and he held the .44 magnum, lowering it at Brown. In response to Miller's action, Brown raised his rifle and managed to fire a shot at Miller just before Miller fired at Brown. Brown repeatedly testified that he had only intended to talk to Miller and that he fired only in self-defense.

In addressing this defense during closing arguments to the jury, the prosecution stated:

> If you find that there is self-defense in this case and it's a meritorious defense and you find so by the preponderance of the evidence, that it is a meritorious defense, then [Brown]'s not guilty of anything because self-defense . . . applies to all four [degrees of assault].

Brown did not object to this argument. The trial court then gave two jury instructions dealing with the issue of self-defense. In its initial instruction the court told the jury that the defendant bears the burden of proof on the issue of self-defense. Jury Instruction 13 provided, in relevant part:

I have previously instructed you that the State always bears the burden of proving the defendant's guilt beyond a reasonable doubt. However, in this case the defendant has asserted the defense of self-defense, which is an affirmative defense. As to that defense only the defendant must bear the burden of proof, about which I will instruct you more fully.

The court's next instruction, Jury Instruction 14, dealt with the use of deadly force and concluded that, "unless the State has proven beyond a reasonable doubt that the defendant did not act in self-defense, you must find the defendant not guilty." Although Jury Instruction 13 improperly shifted the burden of proof on the issue of self-defense and conflicted with Jury Instruction 14, Brown failed to object to it.

On appeal, Brown argues that Instruction 13 constituted plain error. The state acknowledges that the instruction was an incorrect statement of the law but maintains that it did not amount to plain error because, as a matter of law, Brown was not entitled to assert self-defense. Relying on *Bangs v. State,* 608 P.2d 1, 5 (Alaska 1980), the state claims that Brown became an initial aggressor and forfeited his right to claim self-defense when he armed himself and sought to confront Miller. The state maintains that, since as a matter of law Brown was not entitled to any self-defense instruction, he could not have suffered prejudice from the incorrect self-defense instruction.

In *Bangs,* a heated exchange took place between Bangs and his eventual victim, Troyer, during which Troyer grabbed Bangs and attempted to choke him. Bangs escaped, walked rapidly to his nearby trailer, grabbed a loaded revolver and returned to the scene of the struggle. Bangs testified that he pointed his gun at Troyer, cocked it, and challenged Troyer to "come on." When Troyer lunged at him, Bangs shot. The Alaska Supreme Court, relying on *State v. Millett,* 273 A.2d 504, 510 (Me.1971), held that even when the evidence was viewed in the light most favorable to Bangs, he was not entitled to a self-defense instruction, because he had been the initial aggressor.

We do not believe the holding in *Bangs* to be dispositive in the present case. A defendant in a criminal case is entitled to a jury instruction as to his theory of the case if there is some evidence to support it. In the present case, if there was some evidence presented at trial to support the conclusion that Brown was not an initial aggressor, then he was entitled to have the jury instructed on self-defense. In both *Bangs* and *Millett* undisputed evidence established that the defendants, anticipating resistance, procured guns for the sole purpose of armed confrontation with their intended victims. Both defendants, after arming themselves, challenged their victims to physical combat with the apparent purpose of provoking a response. Under these circumstances both were found to be initial aggressors as a matter of law. By contrast, in this case there was evidence that Brown was not an initial aggressor—that he did not confront Miller to seek combat or challenge Miller for the purpose of provoking a physical response. Brown testified that when he entered the party house he intended only to talk to Miller, that he communicated that intent, and that he first raised his .22 rifle when Miller drew his own gun.

*Bangs* does not deprive a defendant of the right generally recognized at common law to "seek his adversary for the purpose of a peaceful solution as to their differences." As noted in *State v. Bristol,* 84 P.2d 757, 765 (Wyo. 1938),

[N]either the fact of arming himself, nor the fact of going to the restaurant, even if he knew that the [victim] was there, was sufficient to deprive the defendant of the right of self-defense. The criterion is as to what he then did or said when he found the [victim], and whether what he said or did was reasonably calculated to cause the [victim] to be provoked into attacking the defendant.

Thus, in determining whether Brown was a first aggressor, the crucial inquiry is not whether he was armed when he went to meet with Miller; rather, it is whether his assault occurred "in the course of a dispute provoked by the defendant at a time when he [knew] or ought reasonably to [have known] that the encounter [would] result in mortal combat." *State v. Millett,* 273 A.2d at 510. This is an inquiry that must be resolved in light of the totality of the evidence presented at trial.

While it is only Brown's own testimony which supports his theory that he entered the party house in a non-aggressive manner with the sole intention of conversing with Miller, the burden to produce some evidence is not a heavy one. Viewing the evidence in the light most favorable to Brown, we believe that there is some evidence that Brown did not provoke a dispute with Miller under circumstances that he knew or should have known would result in mortal combat. Brown was therefore entitled to have his self-defense claim—weak as it may have been—properly determined by the jury.

*(Continues)*

*(Continued)*

The law is well settled in Alaska that once some evidence places self-defense in issue, the state has the burden of disproving the existence of the defense beyond a reasonable doubt. In Jury Instruction 13 the trial court incorrectly placed the burden of proof for self-defense upon Brown. As the state acknowledges, this instruction is erroneous, since it raises the possibility that the jury might view it as shifting the burden of proof on a closely contested issue. . . .

The conviction is REVERSED.

## Notes and Questions

1. Apply the definition of an aggressor stated in *United States v. Peterson* to Brown's conduct. Was Brown the aggressor? Should he have been allowed to argue self-defense?

2. Note that in *Brown* the trial court's "Jury Instruction 13" erroneously shifted the burden of persuasion to the defendant on the issue of self-defense. It correctly charged, in "Jury Instruction 14," that the prosecution was required to prove beyond a reasonable doubt that the defendant did not act in lawful self-defense. Alaska follows the majority rule in requiring the State to prove the absence of self-defense beyond a reasonable doubt, once the defense has met its burden of production and has placed self-defense at issue. However, the U.S. Supreme Court has ruled that Due Process principles do not prohibit the States from shifting the burden of persuasion to the defense to establish self-defense, if they choose to do so. *Martin v. Ohio,* 480 U.S. 228 (1987).

3. A person who is the initial aggressor in an altercation normally cannot claim self-defense in response to the victim's retaliatory blows. However, does the initial aggressor forfeit the right to defend him- or herself from an attack after the initial altercation has ended? Assume D punches his archenemy, V, for no apparent reason while they are at a bar and knocks V unconscious. Quite satisfied with himself, D leaves the bar and heads toward his car in the parking lot. Moments later the infuriated V wakes up and runs after D. V catches up with D in the parking lot and takes a swing at him. Seeing V out of the corner of his eye, D ducks just in time and then punches V as V again advances toward him. Can D claim self-defense if charged with assault for the second altercation? The general rule is yes. As long as the initial aggressor, D, has clearly communicated that he has withdrawn from the situation and ended the initial aggression, the right to self-defense may be asserted.

## Reasonableness of Belief

A defendant can prevail on a self-defense claim only if his or her actions were reasonable. The concept of reasonableness is used throughout the law, although it is not easily defined. As a general proposition, it means what an average person would do under similar circumstances. The difficulty of determining whether an act was reasonable is compounded by the need to decide whose state of mind should control. Should we use a **subjective** criterion (*i.e.,* whether the act seemed reasonable to the actor at the time)? Should we employ an **objective** standard, based on the perspectives of the average person on the street? Or should we invoke some kind of hybrid consideration, based partly on the actor's unique qualities and partly on a more detached, objective standard regarding what would have been "reasonable" under the circumstances? These issues had to be addressed by the New York Court of Appeals in *People v. Goetz,* widely known as the "subway vigilante" case. As you read *Goetz,* bear in mind that this is an appeal by the prosecution challenging the trial court's dismissal of an indictment.

## *People v. Goetz,* 497 N.E.2d 41, 506 N.Y.S.2d 18 (1986)

Chief Judge Wachtler.

A Grand Jury has indicted defendant on attempted murder, assault, and other charges for having shot and wounded four youths on a New York City subway train after one or two of the youths approached him and asked for $5. The lower courts, concluding that the prosecutor's charge to the Grand Jury on the defense of justification was erroneous, have dismissed the attempted murder, assault and weapons possession charges. We now reverse and reinstate all counts of the indictment.

### I.

The precise circumstances of the incident giving rise to the charges against defendant are disputed, and ultimately it will be for a trial jury to determine what occurred. We feel it necessary, however, to provide some factual background to properly frame the legal issues before us. Accordingly, we have summarized the facts as they appear from the evidence before the Grand Jury. We stress, however, that we do not purport to reach any conclusions or holding as to exactly what transpired or whether defendant is blameworthy. The credibility of witnesses and the reasonableness of defendant's conduct are to be resolved by the trial jury.

On Saturday afternoon, December 22, 1984, Troy Canty, Darryl Cabey, James Ramseur, and Barry Allen boarded an IRT express subway train in The Bronx and headed south toward lower Manhattan. The four youths rode together in the rear portion of the seventh car of the train. Two of the four, Ramseur and Cabey, had screwdrivers inside their coats, which they said were to be used to break into the coin boxes of video machines.

Defendant Bernhard Goetz boarded this subway train at 14th Street in Manhattan and sat down on a bench toward the rear section of the same car occupied by the four youths. Goetz was carrying an unlicensed .38 caliber pistol loaded with five rounds of ammunition in a waistband holster. The train left the 14th Street station and headed towards Chambers Street.

It appears from the evidence before the Grand Jury that Canty approached Goetz, possibly with Allen beside him, and stated "give me five dollars." Neither Canty nor any of the other youths displayed a weapon. Goetz responded by standing up, pulling out his handgun and firing four shots in rapid succession. The first shot hit Canty in the chest, the second struck Allen in the back, the third went through Ramseur's arm and into his left side; the fourth was fired at Cabey, who apparently was then standing in the corner of the car, but missed, deflecting instead off of a wall of the conductor's cab. After Goetz briefly surveyed the scene around him, he fired another shot at Cabey, who then was sitting on the end bench of the car. The bullet entered the rear of Cabey's side and severed his spinal cord.

All but two of the other passengers fled the car when, or immediately after, the shots were fired. The conductor, who had been in the next car, heard the shots and instructed the motorman to radio for emergency assistance. The conductor then went into the car where the shooting occurred and saw Goetz sitting on a bench, the injured youths lying on the floor or slumped against a seat, and two women who had apparently taken cover, also lying on the floor. Goetz told the conductor that the four youths had tried to rob him.

While the conductor was aiding the youths, Goetz headed towards the front of the car. The train had stopped just before the Chambers Street station and Goetz went between two of the cars, jumped onto the tracks and fled. Police and ambulance crews arrived at the scene shortly thereafter. Ramseur and Canty, initially listed in critical condition, have fully recovered. Cabey remains paralyzed, and has suffered some degree of brain damage.

On December 31, 1984, Goetz surrendered to police in Concord, New Hampshire, identifying himself as the gunman being sought for the subway shootings in New York nine days earlier. Later that day, after receiving *Miranda* warnings, he made two lengthy statements, both of which were tape recorded with his permission. In the statements, which are substantially similar, Goetz admitted that he had been illegally carrying a handgun in New York City for three years. He stated that he had first purchased a gun in 1981 after he had been injured in a mugging. Goetz also revealed that twice between 1981 and 1984 he had successfully warded off assailants simply by displaying the pistol.

According to Goetz's statement, the first contact he had with the four youths came when Canty, sitting or lying on the bench across from him, asked "how are you," to which he replied "fine." Shortly thereafter, Canty, followed by one of the other youths, walked over to the defendant and stood to his left, while the other two youths remained to his right in the corner of the subway car. Canty then said "give me five dollars." Goetz stated that he knew from the smile on Canty's face that they wanted to "play with me." Although he was certain that none of the youths had a gun, he had a fear, based on prior experiences, of being "maimed."

Goetz then established "a pattern of fire," deciding specifically to fire from left to right. His stated intention at that point was to "murder [the four youths], to hurt them, to make them suffer as much as possible." When Canty again requested money, Goetz stood up, drew his weapon, and began firing, aiming for the center of the body of each of the four. Goetz recalled that the first two he shot "tried to run through the crowd [but] they had nowhere to run." Goetz then turned to his right to "go after the other two." One of these two "tried to run through the wall of the train, but . . . he had nowhere to go." The other youth (Cabey) "tried pretending that he wasn't with [the others]" by standing still, holding on to one of the subway hand straps, and not looking at Goetz. Goetz nonetheless fired his fourth shot at him. He then ran back to the first two youths to make sure they had been "taken care of." Seeing that they had both been shot, he spun back to check on the latter two. Goetz noticed that the youth who had been standing still was now sitting on a bench and seemed unhurt. As Goetz told the police, "I said '[y]ou seem to be all right, here's another,'" and he then fired the shot which severed

*(Continues)*

*(Continued)*

Cabey's spinal cord. Goetz added that "if I was a little more under self-control . . . I would have put the barrel against his forehead and fired." He also admitted that "if I had had more [bullets], I would have shot them again, and again, and again."

## II.

After waiving extradition, Goetz was brought back to New York and arraigned on a felony complaint charging him with attempted murder and criminal possession of a weapon.

. . . On March 27, 1985, the second Grand Jury filed a 10-count indictment, containing four charges of attempted murder (Penal Law §§ 110.00, 125.25[1]), four charges of assault in the first degree (Penal Law § 120.10[1]), one charge of reckless endangerment in the first degree (Penal Law § 120.25), and one charge of criminal possession of a weapon in the second degree (Penal Law § 265.03 [possession of loaded firearm with intent to use it unlawfully against another]). Goetz was arraigned on this indictment on March 28, 1985.

. . . On October 14, 1985, Goetz moved to dismiss the charges contained in the second indictment alleging, among other things, that the evidence before the second Grand Jury was not legally sufficient to establish the offenses charged (*see* CPL 210.20(1[b])), and that the prosecutor's instructions to the Grand Jury on the defense of justification were erroneous and prejudicial to the defendant so as to render its proceedings defective (*see* CPL 210.20[1][c] 210.35[5]). . . .

In an order dated January 21, 1986, the Criminal Term granted Goetz's motion to the extent that it dismissed all counts of the second indictment other than the reckless endangerment charge, with leave to resubmit these charges to a third Grand Jury. The court, after inspection of the Grand Jury minutes, first rejected Goetz's contention that there was not legally sufficient evidence to support the charges. It held, however, that the prosecutor, in a supplemental charge elaborating upon the justification defense, had erroneously introduced an objective element into this defense by instructing the grand jurors to consider whether Goetz's conduct was that of a "reasonable man in [Goetz's] situation." The court citing prior decisions from both the First and Second Departments (*see, e.g., People v. Santiago,* 488 N.Y.S.2d 4 [1st Dept]; *People v. Wagman,* 471 N.Y.S.2d 147 [2d Dept.]), concluded that the statutory test for whether the use of deadly force is justified to protect a person should be wholly subjective, focusing entirely on the defendant's state of mind when he used such force. It concluded that dismissal was required for this error because the justification issue was at the heart of the case. . . .

On appeal by the People, a divided Appellate Division, 501 N.Y.S.2d 326, affirmed Criminal Term's dismissal of the charges. . . .

## III.

Penal Law article 35 recognizes the defense of justification, which "permits the use of force under certain circumstances" (*see, People v. McManus,* 67 N.Y.2d 541, 545). One such set of circumstances pertains to the use of force in defense of a person, encompassing both self-defense and defense of a third person (Penal Law § 35.15). Penal Law § 35.15(1) sets forth the general principles governing all such uses of force: "[a] person may . . . use physical force upon another person when and to the extent he *reasonably believes* such to be necessary to defend himself or a third person from what he *reasonably believes* to be the use or imminent use of unlawful physical force by such other person" (emphasis added).

Section 35.15(2) sets forth further limitations on these general principles with respect to the use of "deadly physical force": "A person may not use deadly physical force upon another person under circumstances specified in subdivision one unless (a) He *reasonably believes* that such other person is using or about to use deadly physical force . . . or (b) He *reasonably believes* that such other person is committing or attempting to commit a kidnapping, forcible rape, forcible sodomy or robbery" (emphasis added).

Thus, consistent with most justification provisions, Penal Law § 35.15 permits the use of deadly physical force only where requirements as to triggering conditions and the necessity of a particular response are met. As to the triggering conditions, the statute requires that the actor "reasonably believes" that another person either is using or about to use deadly physical force or is committing or attempting to commit one of certain enumerated felonies, including robbery. As to the need for the use of deadly physical force as a response, the statute requires that the actor "reasonably believes" that such force is necessary to avert the perceived threat.

Because the evidence before the second Grand Jury included statements by Goetz that he acted to protect himself from being maimed or to avert a robbery, the prosecutor correctly chose to charge the justification defense in section 35.15 to the Grand Jury (*see* CPL 190.25[6]; *People v. Valles,* 62 N.Y.2d 36, 38). The prosecutor properly instructed the grand jurors to consider whether the use of deadly physical force was justified to prevent either serious physical injury or a robbery, and, in doing so, to separately analyze the defense with respect to each of the charges. He elaborated upon the prerequisites for the use of deadly physical force essentially by reading or paraphrasing the language in Penal Law § 35.15. . . .

When the prosecutor had completed his charge, one of the grand jurors asked for clarification of the term "reasonably believes." The prosecutor responded by instructing the grand jurors that they were to consider the circumstances of the incident and determine "whether the defendant's conduct was that of a reasonable man in the defendant's situation." It is this response by the prosecutor— and specifically his use of "a reasonable man"—which is the basis for the dismissal of the charges by the lower courts. As expressed repeatedly in the Appellate Division's plurality opinion, because section 35.15 uses the term "*he* reasonably believes," the appropriate test, according to that court, is whether a defendant's beliefs and reactions were "reasonable to *him.*" Under that reading of the statute, a jury which believed a defendant's testimony that he felt that his own actions were warranted and were reasonable would have to acquit him, regardless of what anyone else in defendant's situation might have concluded. Such an interpretation defies the ordinary meaning and significance of the term "reasonably" in a statute, and misconstrues the clear intent of the Legislature, in enacting section 35.15, to retain an objective element as part of any provision authorizing the use of deadly physical force.

Penal statutes in New York have long codified the right recognized at common law to use deadly physical force, under appropriate circumstances, in self-defense. These provisions have never required that an actor's belief as to the intention of another person to inflict serious injury be correct in order for the use of deadly force to be justified, but they have uniformly required that the belief comport with an objective notion of reasonableness. The 1829 statute, using language which was followed almost in its entirety until the 1965 recodification of the Penal Law, provided that the use of deadly force was justified in self-defense or in the defense of specified third-persons "when there shall be a reasonable ground to apprehend a design to commit a felony, or to do some great personal injury, and there shall be imminent danger of such design being accomplished."

In *Shorter v. People,* 2 N.Y. 193, we emphasized that deadly force could be justified under the statute even if the actor's beliefs as to the intentions of another turned out to be wrong, but noted there had to be a reasonable basis, viewed objectively, for the beliefs. We explicitly rejected the position that the defendant's own belief that the use of deadly force was necessary sufficed to justify such force regardless of the reasonableness of the beliefs (*id.,* at pp. 200–201 ). . . .

In 1961 the Legislature established a Commission to undertake a complete revision of the Penal Law and the Criminal Code. The impetus for the decision to update the Penal Law came in part from the drafting of the Model Penal Code by the American Law Institute, as well as from the fact that the existing law was poorly organized and in many aspects antiquated. . . . Following the submission by the Commission of several reports and proposals, the Legislature approved the present Penal Law in 1965 (L. 1965, ch. 1030), and it became effective on September 1, 1967. . . . While using the Model Penal Code provisions on justification as general guidelines, however, the drafters of the new Penal Law did not simply adopt them verbatim.

The provisions of the Model Penal Code with respect to the use of deadly force in self-defense reflect the position of its drafters that any culpability which arises from a mistaken belief in the need to use such force should be no greater than the culpability such a mistake would give rise to if it were made with respect to an element of a crime (*see,* ALI, Model Penal Code and Commentaries, part I, at 32, 34 [hereafter cited as MPC Commentaries]; Robinson, Criminal Law Defenses, *op. cit.,* at 410). Accordingly, under Model Penal Code § 3.04(2)(b), a defendant charged with murder (or attempted murder) need only show that he "*believe[d]* that [the use of deadly force] was necessary to protect himself against death, serious bodily injury, kidnapping or [forcible] sexual intercourse" to prevail on a self-defense claim (emphasis added). If the defendant's belief was wrong, and was recklessly, or negligently formed, however, he may be convicted of the type of homicide charge requiring only a reckless or negligent, as the case may be, criminal intent (*see,* Model Penal Code § 309[2]; MPC Commentaries, *op. cit.,* part I, at 32, 150). . . .

*(Continues)*

(*Continued*)

New York did not follow the Model Penal Code's equation of a mistake as to the need to use deadly force with a mistake negating an element of a crime, choosing instead to use a single statutory section which would provide either a complete defense or no defense at all to a defendant charged with any crime involving the use of deadly force. The drafters of the new Penal Law adopted in large part the structure and content of Model Penal Code § 3.04, but, crucially, inserted the word "reasonably" before "believes."

The plurality below agreed with defendant's argument that the change in the statutory language from "reasonable ground," used prior to 1965, to "he reasonably believes" in Penal Law § 35.15 evinced a legislative intent to conform to the subjective standard contained in Model Penal Code § 3.04. This argument, however, ignores the plain significance of the insertion of "reasonably." Had the drafters of section 35.15 wanted to adopt a subjective standard, they could have simply used the language of section 3.04. "Believes" by itself requires an honest or genuine belief by a defendant as to the need to use deadly force. Interpreting the statute to require only that the defendant's belief was "reasonable to *him*," as done by the plurality below, would hardly be different from requiring only a genuine belief; in either case, the defendant's own perceptions could completely exonerate him from any criminal liability.

We cannot lightly impute to the Legislature an intent to fundamentally alter the principles of justification to allow the perpetrator of a serious crime to go free simply because that person believed his actions were reasonable and necessary to prevent some perceived harm. To completely exonerate such an individual, no matter how aberrational or bizarre his thought patterns, would allow citizens to set their own standards for the permissible use of force. It would also allow a legally competent defendant suffering from delusions to kill or perform acts of violence with impunity, contrary to fundamental principles of justice and criminal law.

We can only conclude that the Legislature retained a reasonableness requirement to avoid giving a license for such actions. . . .

The conclusion that section 35.15 retains an objective element to justify the use of deadly force is buttressed by the statements of its drafters. The executive director and counsel to the Commission which revised the Penal Law have stated that the provisions of the statute with respect to the use of deadly physical force largely conformed with the prior law, with the only changes they noted not being relevant here. . . . [I]n *People v. Miller*, 39 N.Y.2d 543 . . . , we held that a defendant charged with homicide could introduce, in support of a claim of self-defense, evidence of prior acts of violence committed by the deceased of which the defendant had knowledge. The defense, as well as the plurality below, place great emphasis on the statement in *Miller* that "the crucial fact at issue [is] the state of mind of the defendant." This language, however, in no way indicates that a wholly subjective test is appropriate. To begin, it is undisputed that section 35.15 does contain a subjective element, namely that the defendant believed that deadly force was necessary to avert the imminent use of deadly force or the commission of certain felonies. Evidence that the defendant knew of prior acts of violence by the deceased could help establish his requisite beliefs. Moreover, such knowledge would also be relevant on the issue of reasonableness, as the jury must consider the circumstances a defendant found himself in, which would include any relevant knowledge of the nature of persons confronting him. . . .

Goetz also argues that the introduction of an objective element will preclude a jury from considering factors such as the prior experiences of a given actor and thus, require it to make a determination of "reasonableness" without regard to the actual circumstances of a particular incident. This argument, however, falsely presupposes that an objective standard means that the background and other relevant characteristics of a particular actor must be ignored. To the contrary, we have frequently noted that a determination of reasonableness must be based on the "circumstances" facing a defendant or his "situation." Such terms encompass more than the physical movements of the potential assailant. As just discussed, these terms include any relevant knowledge the defendant had about that person. They also necessarily bring in the physical attributes of all persons involved, including the defendant. Furthermore, the defendant's circumstances encompass any prior experiences he had which could provide a reasonable basis for a belief that another person's intentions were to injure or rob him or that the use of deadly force was necessary under the circumstances.

Accordingly, a jury should be instructed to consider this type of evidence in weighing the defendant's actions. The jury must first determine whether the defendant had the requisite beliefs under section 35.15, that is, whether he believed **deadly force** was necessary to avert the imminent use of deadly force or the commission of one of the felonies enumerated therein. If the People do not prove beyond a reasonable doubt that he did not have such beliefs, then the jury must also consider whether these beliefs were reasonable. The jury would have to determine, in light of all the "circumstances," as explicated above, if a reasonable person could have had these beliefs.

The prosecutor's instruction to the second Grand Jury that it had to determine whether, under the circumstances, Goetz's conduct was that of a reasonable man in his situation was thus essentially an accurate charge. . . .

The Grand Jury has indicted Goetz. It will now be for the petit jury to decide whether the prosecutor can prove beyond a reasonable doubt that Goetz's reactions were unreasonable and therefore excessive. . . .

## Notes and Questions

1. Goetz was eventually tried and acquitted of attempted murder and assault. He was convicted of illegal possession of a firearm. Based on the facts stated in the opinion, do you believe the jury was right in acquitting Goetz of the more serious charges? Does the jury's verdict mean that the jurors would have done what Goetz did had they been placed in similar circumstances?

2. A jury in a civil action brought by the victims of the shooting found Goetz liable and awarded the victims several million dollars in damages.

3. Justice Holmes wrote, "Detached reflection cannot be demanded in the presence of an uplifted knife." *Brown v. United States*, 256 U.S. 335, 343 (1921). This observation has been interpreted to mean that one must consider the heat of the moment and that individuals often must make split-second judgments before using force in apparent self-defense. Does the difference between what Goetz may have known and believed while standing in the subway car, as opposed to what may have been apparent based on evidence presented months later in a court of law, affect how you view the jury's verdict in *Goetz*?

4. Assume that Bob and Bill get into an argument over who was a better baseball pitcher, Randy Johnson or Greg Maddux. Bob, fed up with Bill's obstinacy, yells, "That's it you lame-brain, I'll show you!" He then reaches into his pocket. Thinking Bob is going to pull a gun, Bill draws out his own gun and kills Bob. Inside Bob's pocket was a baseball encyclopedia, not a gun. Can Bill claim self-defense even if he was mistaken in his belief that Bob had a weapon? If his belief that he was in imminent danger was reasonable, he can claim self-defense even if no danger actually existed. The reasonableness of an individual's belief is normally a question for a jury to decide. In making this decision, a jury could consider whether Bill knew that Bob had a history of violence or knew that Bob owned a gun or normally carried a weapon. Remember, the standard to be used by the jury is what a reasonable person in Bill's circumstances, with Bill's level of knowledge and experience, would have believed.

## Deadly Force

A person may use only that amount of force which is reasonably necessary to repel his or her attacker. This principle of **"proportionality"** between the aggressive and defensive force means that a person may use deadly force in self-defense only if he or she reasonably believes that it is necessary to prevent imminent death or serious bodily injury. Only non-deadly defensive force may be used if an attack does not involve the threat of death or serious bodily harm.

What constitutes a reasonably perceived threat of death or serious bodily injury can be a difficult question. For example, can a person ever use deadly defensive force if the assailant is unarmed? Would it matter if the assailant has the strength and pugilistic prowess of a heavyweight boxer such as Mike Tyson in his prime? Consider the dilemma faced by the court in *Howard v. State*.

## *Howard v. State,* 390 So.2d 1070 (Ala. App. 1980)

Clark, Retired Circuit Judge.

This is an appeal from a judgment of conviction and a sentence to ten years' imprisonment for manslaughter in the first degree of Barbara Young. . . .

The undisputed evidence shows that Barbara Young was killed by one or both of two shots of a pistol fired by defendant, that Betty Jo Pettaway was hit by another shot from the pistol fired by defendant, and that the three shots occurred in rapid succession.

The only seriously contested issue in the homicide case was as to the asserted defense of self-defense, or defense of defendant's mother, or both, included within the defendant's plea of not guilty. . . .

The tragedy occurred on a Sunday evening, April 22, 1979, in the backyard of a "hit house," a term employed by the witnesses as a house, a residence, at which liquor was sold without a license and gambling by cards was promoted.

*(Continues)*

(*Continued*)

The two alleged victims and defendant and her mother, Rosena Howard, engaged in a game of cards in a back room of the house, at which at the time were approximately twenty-five persons, most of whom were imbibing freely. At some point in the card game, defendant's mother accused the alleged victim of the homicide of cheating. After the game broke up, defendant's mother and Barbara Young started fighting inside the house. The fight was broken up by the intervention of others; defendant, her mother and a friend then went outside. Betty Jo Pettaway and Barbara Young followed them outside and a fight between defendant and Barbara broke out again. While the fight was going on outside, defendant's mother obtained a pistol from her purse and struck Barbara over her eye with it. The pistol was taken from defendant's mother by a man and given to a female from whom defendant obtained the pistol and fired the three shots while her mother and Barbara Young were fighting and while Barbara had defendant's mother by the hair. . . .

A part of the transcript of the court's oral charge is as follows:

There is one other thing in the [question] of self-defense that the law says a person in self-defense is justified only in using such force as is necessary to repel an attack on him and that in the eyes of the law, if you attacked me with your fist, I don't have the right to shoot you to stop the attack.

The transcript also shows that at the conclusion of the court's oral charge the following occurred:

" . . . Are there any exceptions to the oral charge? . . .

"Mr. Hughes (Counsel for Defendant): In addition to those, we would accept (sic) one portion of the Court's charge, and I possibly could have misunderstood it when the Court said that if one were attacked by somebody with fists, that certainly there would be no right to retaliate with a gun or shoot him. I would ask the Court to add to that, that it would depend on the circumstances under which there was an attack by fists or whatever. . . .

The transcript shows that in continuing to charge the jury, the court . . . did not again discuss the point made by defendant's counsel, as quoted above.

The portion of the court's oral charge quoted above as to the right of a person to shoot another who has attacked him with his fist is not a correct statement of the law, although it can be a very powerful and persuasive statement that the State would have the right to make in argument to a jury in its plea that a defendant should be found guilty of a homicide notwithstanding the defense of self-defense.

A correct statement of the precise applicable principle is found in the opinion of Chief Justice Stone in *Scales v. State,* 11 So. 121 (Ala. 1891-92):

. . . Fisticuff blows do not, as a rule, inflict the grievous bodily harm, which, other means of escape being cut off, will excuse the slaying of the assailant. "When a man is struck with the naked hand, and has no reason to apprehend a design to do him great bodily harm, he must not return the blow with a dangerous weapon."

Appellant relies upon *George v. State,* 40 So. 961 (Ala. 1906), which is substantially in point as to what the trial court incorrectly stated in this case, as well as in that case. In *George,* the trial court had stated "that a blow from the hand or fist never justified the use of a deadly weapon." The Supreme Court said, "The law is that a blow from the hand or fist under ordinary circumstances, neither justifies nor excuses the use of a deadly weapon." It reversed the trial court for its failure to give defendant's requested charge 1, "Gentlemen of the jury, I charge you that an assault with the hand or fist never justifies or excuses a homicide under ordinary circumstances, and it is for you to decide whether the facts in this case are within the ordinary reason or not."

It appears from *George, supra,* that there was no exception to the pertinent part of the court's oral charge but that the erroneous instruction quoted therefrom was a material factor in the reversal by reason of the failure to give defendant's requested charge 1.

As to the question under consideration, *Dilburn v. State,* 77 So. 983 (Ala.App. 1918), is directly in point. It states in less than a one column opinion:

The trial judge in his oral charge charged the jury:

"He (the defendant) would not be justified in using a deadly weapon if struck by the fist, or any other assault which would not likely cause serious bodily harm."

This was in effect charging the jury that under the evidence the defendant was not justified in using a deadly weapon, and that the blow struck by the fists was not likely to cause serious bodily harm, which was the very question then being submitted to the jury. The rule is that the killing of one who is the assailant must be under a reasonable apprehension of loss of life or of great bodily harm, and the danger must appear to be so imminent at the moment of the assault as to present no alternative of escaping its consequences except by resisting. *Scales v. State*, 11 So. 121. It was said in Shorters' case "When a man is struck with the naked hand, and has no reason to apprehend a design to do him great bodily harm, he must not return the blow with a dangerous weapon," and this expression was quoted with approval in the *Scales Case, supra.*

But it is a question for the jury to satisfy itself of all the evidence in the case whether or not the defendant was in imminent and manifest danger either of losing his own life or of suffering grievous bodily harm, or that it appeared so to the mind of a reasonable man. That part of the oral charge of the court excepted to was in conflict with the foregoing views, and for that error the judgment must be reversed.

The part of the court's oral charge in the instant case to which an exception was taken contains the same vice as that found in the oral charge in *Dilburn* and requires a like result, a reversal. . . .

For the error indicated, the judgment of the trial court must be reversed and the cause remanded. . . .

### Notes and Questions

1. Do you believe Howard acted reasonably? Should she at least be permitted to place the question of self-defense before the jury?
2. What if a person honestly but unreasonably believes she is in imminent danger of death or serious injury and kills her assailant? In many states such circumstances would present a case of "imperfect" self-defense, resulting in a conviction for manslaughter instead of murder. Does this doctrine seem fair? We give further consideration to such issues in Chapter 6.

## Retreat

The amount of force used in claimed self-defense implicates the proportionality principle. The question of whether a person must retreat from the unlawful aggressor, if that is a viable option, before using force in claimed self-defense implicates the necessity principle. The near-universal rule is that a person need not retreat before using non-deadly force in self-defense. Jurisdictions are split, however, on whether a person must retreat before using deadly defensive force. These issues arise in *Cooper v. United States*.

### *Cooper v. United States*, 512 A.2d 1002 (D.C. 1986)

Belson, Associate Judge: . . .

Leon Cooper and his brother Robert Parker lived with their mother, Alice Cooper. In the early part of August 1981, Parker unexpectedly left home for 10 days. Early on the morning of August 12th, he returned. He did not tell his mother or brother where he had been.

Parker stayed home for much of the day. Mrs. Cooper returned from work in the evening, and Cooper returned from his job shortly afterward. Cooper was carrying a pistol when he returned. The three were sitting in Mrs. Cooper's small living room when the two brothers began to quarrel after Cooper asked Parker where he had been during the past 10 days.

Suddenly, the quarrel escalated, and the two brothers found themselves standing in the middle of the living room, shouting at each other. Parker hit Cooper in the head with a small radio; Mrs. Cooper ran upstairs to call for help. She then heard a "pop." She went downstairs and saw Parker lying on the floor, Cooper said "I have shot my brother" and "Mama, I am so sorry. I mean——." Cooper later told the police that he had just shot his brother, that his brother was hitting him with the radio and "I couldn't take it anymore and I just shot him."

At trial, Cooper's counsel objected to instruction 5.16B, the standard instruction given when the defendant raises a claim of self-defense. The court instructed the jury, in pertinent part:

> Now, if the defendant—If the defendant could have safely retreated but did not do so, his failure to retreat is a circumstance which you may consider together with all the other circumstances in determining whether he went further in repelling the danger, real or apparent, than he was justified in doing so under the circumstances.
>
> Before a person can avail himself [of] the plea of self-defense against a charge of homicide, he must do everything in his power, consistent with his own safety, to avoid the danger and avoid the necessity of taking life. However, if the defendant actually believed that he was in imminent danger of death or serious bodily harm, and that deadly force was necessary to repel such danger, he was not required to retreat or

*(Continues)*

consider whether he could safely retreat. He was entitled to stand his ground and use such force as was reasonably necessary under the circumstances to save his life or protect himself from serious bodily harm.

Appellant took the position that the second sentence of the instruction which begins, "Before a person can avail himself [of] the plea of self-defense," inappropriately imposed a duty to retreat in the face of an attack. The trial court overruled counsel's objections. Defense counsel then asked the trial court for a **"castle doctrine"** instruction, *i.e.,* that a person has no duty whatsoever to retreat when attacked in his own home. The trial judge denied the request, stating that, in his opinion, the castle doctrine applies when a person in his home is attacked by a stranger or one who comes onto the premises without permission, but not when a fight occurs between two co-occupants. The jury found Cooper guilty of voluntary manslaughter while armed and carrying a pistol without a license, and the trial court sentenced him to a jail term of 8 to 24 years for the armed manslaughter conviction, and a consecutive term of 1 year for carrying a pistol without a license. This appeal followed.

# I.

We consider first whether, under the law of this jurisdiction, a person generally has the duty to retreat in the face of an assault by another, when retreat is a feasible alternative.

In *Gillis v. United States,* 400 A.2d 311 (D.C.1979), this court considered whether a person threatened with death or serious bodily harm has a duty to retreat, if it can be done safely, before using deadly force in defense. Gillis claimed that he had acted in self-defense when a man named Smith approached him on a deserted street late at night and accused him of being with Smith's girlfriend. Gillis claimed that Smith reached in Smith's pocket, and pulled out a shiny object. Gillis then pulled out a pistol, shot, and mortally wounded Smith. Gillis was convicted of second-degree murder while armed.

On appeal, Gillis asserted that the trial court's jury instruction erroneously implied the existence of a duty to retreat, that instruction being essentially the instruction set out above except for its omission of the second sentence. We reviewed the conflicting precedents in the law of the District of Columbia regarding the duty to retreat, noting first that in *Marshall v. United States*, 45 App.D.C. 373 (1916), the court held "[t]he right of a defendant when in imminent danger to take life does not depend upon whether there was an opportunity to escape. One under such circumstances is not compelled to stand aside, or to flee." *Gillis, supra,* 400 A.2d at 312, quoting *Marshall, supra,* 45 App.D.C. at 384. This formulation expressed an emphasis consistent with the so-called "American rule," which holds that one is not required to retreat whether he is attacked in his home or elsewhere, but may stand his ground and defend himself.

We also noted in *Gillis,* however, that later in *Laney v. United States,* 54 App.D.C. 56, 294 F. 412 (1923), the Circuit Court of Appeals had written:

> "It is well-settled rule that, before a person can avail himself of the plea of self-defense against the charge of homicide, he must do everything in his power, consistent with his safety, to avoid the danger and avoid the necessity of taking life. . . . In other words, no necessity for killing an assailant can exist, so long as there is a safe way open to escape the conflict." *Gillis, supra,* 400 A.2d at 312 (quoting Laney, supra, 54 App.D.C. at 54, 294 F.2d at 414).

*Laney* used language consistent with what is known as the common law, "retreat to the wall," rule.

Faced with these apparently conflicting precedents, the *Gillis* court reconciled them in what it termed a "middle ground" approach to self-defense. The middle ground approach imposes no duty to retreat, but it "permit[s] the jury to consider whether a defendant, if he safely could have avoided further encounter by stepping back or walking away, was actually or apparently in imminent danger of bodily harm. In short, this rule permits the jury to determine if the defendant acted too hastily, was too quick to pull the trigger." Id. [at 313]. We affirmed in *Gillis,* holding that the instruction given did not impose a duty to retreat, but allowed a failure to retreat, together with all the other circumstances, to be considered by the jury in determining whether the case was truly one of self-defense.

The unique question presented in this appeal is whether, when one is assaulted in one's home by a co-occupant, the availability of a means of retreat is as much a consideration as it otherwise is under the middle ground approach. No cases in this jurisdiction have been called to our attention which address the question regarding an occupant's duty or ability to retreat in the face of an assault by a co-occupant, nor have we been able to identify any. We look, then, to see how other courts have addressed this issue.

We begin by noting that the question whether an occupant has a duty to retreat when assaulted by a co-occupant will not arise in those jurisdictions which follow the American rule, for in those jurisdictions one can stand one's ground regardless of where one is assaulted, or by whom. Therefore, whatever guidance is available is provided by the courts of those jurisdictions which follow the common law, "retreat to the wall," rule.

Those jurisdictions following the common law rule have almost universally adopted the "castle doctrine" that one who through no fault of his own is attacked in one's own home is under no duty to retreat. While the status of the castle doctrine in the District of Columbia has never been squarely decided, we will assume for purposes of this discussion that appellant is correct in maintaining that the doctrine is applicable here.

Courts following the common law rule have split, however, regarding whether a defendant is entitled to a castle doctrine instruction when the defendant is assaulted by a co-occupant. An early case addressing this question was *People v. Tomlins*, 107 N.E. 496 (N.Y. 1914). In *Tomlins*, a father shot and killed his son in their cottage. The New York Court of Appeals held that an instruction which informed the jury that the father had a duty to retreat was erroneous. The court first noted that if a man is assaulted in his home, "he may stand his ground and resist the attack. He is under no duty to take to the fields and the highways, a fugitive from his own home." The court then held that the rule is the same whether the attack is initiated by an intruder or a co-occupant; "why . . . should one retreat from his own house, when assailed by a partner or co-tenant, any more than when assailed by a stranger who is lawfully on the premises? Whither shall he flee, and how far, and when may he be permitted to return?"

As other courts grappled with this question, they often returned to the questions posed by the *Tomlins* court, although frequently reaching a different result. A majority of courts favor giving a castle doctrine instruction when a defendant claims self-defense when attacked in his home by a co-occupant, while a substantial minority holds that the castle doctrine does not apply in this special circumstance. Those decisions which favor giving a castle doctrine instruction stress the occupant's interest in remaining in the home, while those that oppose giving the instruction focus on the entitlement of both combatants to occupy the house and the fact that they usually are related, and reason that the parties have some obligation to attempt to defuse the situation.

Having examined these authorities, we are convinced that the reasoning of those jurisdictions holding that a castle doctrine instruction should not be given in instances of co-occupant attacks is the more compelling. As the Florida Supreme Court noted in *[State v.] Bobbitt*, [415 So.2d 724 (Fla. 1982)], both the decedent husband and accused wife in the case before it "had equal rights to be in the 'castle' and neither had the legal right to eject the other." The court further observed:

> We see no reason why a mother should not retreat from her own son, even in her own kitchen. Such a view does not render her defenseless against a member of her family gone berserk, because . . . a person placed in the position of imminent danger of death or great bodily harm to himself by the wrongful attack of another has no duty to retreat if to do so would increase his own danger of death or great bodily harm.

Although in *Bobbitt*, the Florida Supreme Court used the analogy of a mother attacked by her son, that court's reasoning is also applicable to situations where a daughter attacks her father, a husband attacks his wife, or as here, a brother attacks his brother. Indeed, all co-occupants, even those unrelated by blood or marriage, have a heightened obligation to treat each other with a degree of tolerance and respect. That obligation does not evaporate when one co-occupant disregards it and attacks another. We are satisfied, moreover, that an instruction that embraces the middle ground approach appropriately permits the jury to consider the truly relevant question, *i.e.*, whether a defendant, "if he safely could have avoided further encounter by stepping back or walking away, was actually or apparently in imminent danger of bodily harm." We hold that evidence that the defendant was attacked in his home by a co-occupant did not entitle him to an instruction that he had no duty whatsoever to retreat. The trial court did not err in refusing to give a castle doctrine instruction under the circumstances of this case.

## II.

Cooper also contends that instruction 5.16B is an incorrect and inadequate expression of this jurisdiction's law when a defendant claims that he acted in self-defense to ward off an attack occurring in his own home. We have already rejected the main thrust of his argument, *i.e.*, that the jury must be given a castle doctrine instruction when a person is attacked in his home by a co-occupant. Cooper, however, argues also that the instruction actually encourages the jury to draw an adverse inference from the defendant's failure to retreat by including the following sentence: "Before a person can avail himself of the plea of self-defense against a charge of homicide, he must do everything in his power,

*(Continues)*

*(Continued)*

consistent with his own safety, to avoid the danger and avoid the necessity of taking life." This court has previously considered, and rejected, that contention.

In *Carter v. United States* 475 A.2d 1118 (D.C. 1984), the defendant had killed a man during a fight at his mother's house. Among other contentions, *Carter* asserted that the same language in instruction 5.16B to which *Cooper* objects inappropriately imposed a duty to retreat. The *Carter* court noted that when viewed in isolation, the sentence appears to impose a duty to retreat, and we agree that this sentence is not an ideal formulation to explain the tension between a defendant's duty to avoid danger and his right to defend himself when that danger is imminent. The *Carter* court went on to hold, however, that the propriety of a trial court's instruction is determined from the context of the overall charge and, taken as a whole, the instruction correctly stated the law in this jurisdiction. The court wrote, "[T]he content of the [instruction to which Cooper now objects] did not impose upon appellant a mandatory duty to retreat. Rather, it served to clarify a factor which the jury was privileged to consider when evaluating a claim of self-defense." *Carter* directly rejects Cooper's contention that because the second sentence in instruction 5.16B can be read to imply that the defendant has a duty to retreat, the entire instruction is fatally defective, and we are bound by that panel's decision.

In view of the foregoing, we affirm Cooper's convictions.

## Notes and Questions

1. Most jurisdictions do not require a person who is attacked with deadly force to retreat before using deadly force in self-defense. This rule is based on the theory that right should not yield to wrong; to require retreat would reward wrongdoers and embolden aggressors.

   A significant minority of jurisdictions require that a person must retreat before using deadly defensive force. It is important to note that retreat is required only if the person knows she or he can do so with complete safety. Otherwise, even in "retreat" jurisdictions, a claim of self-defense involving deadly force will not be barred.

2. An exception to the retreat rule in jurisdictions recognizing that duty is the castle doctrine. As noted in *Cooper*, the castle exception to the retreat requirement permits a person attacked in his or her home to use deadly force in self-defense rather than retreat, even if retreat is available. Some courts have expanded the castle doctrine to include yards and buildings around a home, whereas others have limited it by excluding common hallways and entryways in apartment buildings.

3. Courts are divided over how to handle the castle doctrine and the duty to retreat when cohabitants are involved in a physical altercation. Some courts agree with the position taken by the court in *Cooper* based on the belief that because co-occupants must live together, it is better to require retreat and have a cooling off period if a safe retreat is available rather than have bloodshed and the relationship ruined. Other courts have held that it is never reasonable to require a person to retreat within his or her own home. As stated in *People v. Tomlins*, 213 N.Y. 240, 244, (1914), "whither shall he flee, and how far, and when may he be permitted to return?"

## Imminent Harm and the Battered Woman Syndrome

A person may only use force in self-defense if he or she is threatened with ***imminent* harm**. Imminent means "near at hand" or "ready to take place." *State v. Janes*, 850 P.2d 495 (Wash. 1993). A person who makes a preemptive strike against another who threatens to assault him or her at some future time may not be entitled to rely on self-defense. The issue of when an assault is imminent may be especially relevant in cases where a battered woman assaults or kills her abusive husband.

---

### *State v. Stewart,* 763 P.2d 572 (Kan. 1988)

Lockett, Justice.

. . . Peggy Stewart fatally shot her husband, Mike Stewart, while he was sleeping. She was charged with murder in the first degree. Defendant pled not guilty, contending that she shot her husband in self-defense. Expert evidence showed that Peggy Stewart suffered from the **battered woman syndrome**. Based upon the battered woman syndrome, the trial judge instructed the jury on self-defense. The jury found Peggy Stewart not guilty.

The State stipulates that Stewart "suffered considerable abuse at the hands of her husband," but contends that the trial court erred in giving a self-defense instruction since Peggy Stewart was in no imminent danger when she shot her sleeping husband. We agree that under the facts of this case the giving of the self-defense instruction was erroneous. . . .

. . . Peggy Stewart married Mike Stewart in 1974. Evidence at trial disclosed a long history of abuse by Mike against Peggy and her two daughters from one of her prior marriages. Laura, one of Peggy's daughters, testified that early in the marriage Mike hit and kicked Peggy, and that after the first year of the marriage Peggy exhibited signs of severe psychological problems. . . .

In 1977, two social workers informed Peggy that they had received reports that Mike was taking indecent liberties with her daughters. Because the social workers did not want Mike to be left alone with the girls, Peggy quit her job. In 1978, Mike began to taunt Peggy by stating that Carla, her 12-year-old daughter, was "more of a wife" to him than Peggy.

Later, Carla was placed in a detention center, and Mike forbade Peggy and Laura to visit her. When Mike finally allowed Carla to return home in the middle of summer, he forced her to sleep in an un-air conditioned room with the windows nailed shut, to wear a heavy flannel nightgown, and to cover herself with heavy blankets. Mike would then wake Carla at 5:30 A.M. and force her to do all the housework. Peggy and Laura were not allowed to help Carla or speak to her.

When Peggy confronted Mike and demanded that the situation cease, Mike responded by holding a shotgun to Peggy's head and threatening to kill her. Mike once kicked Peggy so violently in the chest and ribs that she required hospitalization. Finally, when Mike ordered Peggy to kill and bury Carla, she filed for divorce. Peggy's attorney in the divorce action testified in the murder trial that Peggy was afraid for both her and her children's lives. . . .

Mike's intimidation of Peggy continued to escalate. One morning, Laura found her mother hiding on the school bus, terrified and begging the driver to take her to a neighbor's home. That Christmas, Mike threw the turkey dinner to the floor, chased Peggy outside, grabbed her by the hair, rubbed her face in the dirt, and then kicked and beat her.

After Laura moved away, Peggy's life became even more isolated. Once, when Peggy was working at a cafe, Mike came in and ran all the customers off with a gun because he wanted Peggy to go home and have sex with him right that minute. He abused both drugs and alcohol, and amused himself by terrifying Peggy, once waking her from a sound sleep by beating her with a baseball bat. He shot one of Peggy's pet cats, and then held the gun against her head and threatened to pull the trigger. Peggy told friends that Mike would hold a shotgun to her head and threaten to blow it off, and indicated that one day he would probably do it.

In May 1986, Peggy left Mike and ran away to Laura's home in Oklahoma. It was the first time Peggy had left Mike without telling him. Because Peggy was suicidal, Laura had her admitted to a hospital. There, she was diagnosed as having toxic psychosis as a result of an overdose of her medication. On May 30, 1986, Mike called to say he was coming to get her. . . .

. . . Peggy testified that Mike threatened to kill her if she ever ran away again. As soon as they arrived at the house, Mike forced Peggy into the house and forced her to have oral sex several times.

The next morning, Peggy discovered a loaded .357 magnum. She testified she was afraid of the gun. She hid the gun under the mattress of the bed in a spare room. Later that morning, as she cleaned house, Mike kept making remarks that she should not bother because she would not be there long, or that she should not bother with her things because she could not take them with her. She testified she was afraid Mike was going to kill her.

Mike's parents visited Mike and Peggy that afternoon. Mike's father testified that Peggy and Mike were affectionate with each other during the visit. Later, after Mike's parents had left, Mike forced Peggy to perform oral sex. After watching television, Mike and Peggy went to bed at 8:00 P.M. As Mike slept, Peggy thought about suicide and heard voices in her head repeating over and over, "kill or be killed." At this time, there were two vehicles in the driveway and Peggy had access to the car keys. About 10:00 P.M., Peggy went to the spare bedroom and removed the gun from under the mattress, walked back to the bedroom, and killed her husband while he slept. She then ran to the home of a neighbor, who called the police.

When the police questioned Peggy regarding the events leading up to the shooting, Peggy stated that things had not gone quite right that day, and that when she got the chance she hid the gun under the mattress. She stated that she shot Mike to "get this over with, this misery and this torment." . . .

Two expert witnesses testified during the trial. The expert for the defense, psychologist Marilyn Hutchinson, diagnosed Peggy as suffering from "battered woman syndrome," or post-traumatic stress syndrome. Dr. Hutchinson testified that Mike was preparing to escalate the violence in retaliation for Peggy's running away. She testified that loaded guns, veiled threats, and increased sexual demands are indicators of the escalation of the cycle. Dr. Hutchinson believed Peggy had a repressed knowledge that she was in a "really grave lethal situation."

(Continues)

The State's expert, psychiatrist Herbert Modlin, neither subscribed to a belief in the battered woman syndrome nor to a theory of learned helplessness as an explanation for why women do not leave an abusive relationship. . . .

At defense counsel's request, the trial judge gave an instruction on self-defense to the jury. The jury found Peggy not guilty.

[Under Kansas law, although the prosecution may not appeal an acquittal in a criminal case, the state supreme court can consider questions of law arising from acquittals if it deems their resolution "essential to the just administration of criminal law."]

. . . The question reserved is whether the trial judge erred in instructing on self-defense when there was no imminent threat to the defendant and no evidence of any argument or altercation between the defendant and the victim contemporaneous with the killing. . . .

The State claims that under the facts the instruction should not have been given because there was no lethal threat to defendant contemporaneous with the killing. . . . Under the common law, the excuse for killing in self-defense is founded upon necessity, be it real or apparent. . . .

These common-law principles were codified in K.S.A. 21-3211, which provides:

> "A person is justified in the use of force against an aggressor when and to the extent it appears to him and he reasonably believes that such conduct is necessary to defend himself or another against such aggressor's imminent use of unlawful force."

The traditional concept of self-defense has posited one-time conflicts between persons of somewhat equal size and strength. When the defendant claiming self-defense is a victim of long-term domestic violence, such as a battered spouse, such traditional concepts may not apply. Because of the prior history of abuse, and the difference in strength and size between the abused and the abuser, the accused in such cases may choose to defend during a momentary lull in the abuse, rather than during a conflict. However, in order to warrant the giving of a self-defense instruction, the facts of the case must still show that the spouse was in imminent danger close to the time of the killing.

A person is justified in using force against an aggressor when it appears to that person and he or she reasonably believes such force to be necessary. A reasonable belief implies both an honest belief and the existence of facts which would persuade a reasonable person to that belief. . . .

Where self-defense is asserted, evidence of the deceased's long-term cruelty and violence towards the defendant is admissible. In cases involving battered spouses, expert evidence of the battered woman syndrome is relevant to a determination of the reasonableness of the defendant's perception of danger. . . . However, no jurisdictions have held that the existence of the battered woman syndrome in and of itself operates as a defense to murder.

In order to instruct a jury on self-defense, there must be some showing of an imminent threat or a confrontational circumstance involving an overt act by an aggressor. There is no exception to this requirement where the defendant has suffered long-term domestic abuse and the victim is the abuser. In such cases, the issue is not whether the defendant believes homicide is the solution to past or future problems with the batterer, but rather whether circumstances surrounding the killing were sufficient to create a reasonable belief in the defendant that the use of deadly force was necessary. . . .

Here, . . . there is an absence of imminent danger to defendant: Peggy told a nurse at the Oklahoma hospital of her desire to kill Mike. She later voluntarily agreed to return home with Mike when he telephoned her. She stated that after leaving the hospital Mike threatened to kill her if she left him again. Peggy showed no inclination to leave. In fact, immediately after the shooting, Peggy told the police that she was upset because she thought Mike would leave her. Prior to the shooting, Peggy hid the loaded gun. The cars were in the driveway and Peggy had access to the car keys. After being abused, Peggy went to bed with Mike at 8 P.M. Peggy lay there for two hours, then retrieved the gun from where she had hidden it and shot Mike while he slept.

Under these facts, the giving of the self-defense instruction was erroneous. Under such circumstances, a battered woman cannot reasonably fear imminent life-threatening danger from her sleeping spouse. . . .

To hold otherwise in this case would in effect allow the execution of the abuser for past or future acts and conduct. . . .

The appeal is sustained.

Herd, Justice, dissenting:

The sole issue before us on the question reserved is whether the trial court erred in giving a jury instruction on self-defense. We have a well-established rule that a defendant is entitled to a self-defense instruction if there is any evidence to support it, even though the evidence consists solely of the defendant's testimony. It is for the jury to determine the sincerity of the defendant's belief she needed to act in self-defense, and the reasonableness of that belief in light of all the circumstances. . . .

[A]ppellee met her burden of showing some competent evidence that she acted in self-defense, thus making her defense a jury question. She testified she acted in fear for her life, and Dr. Hutchinson corroborated this testimony. The evidence of Mike's past abuse, the escalation of violence, his threat of killing her should she attempt to leave him, and Dr. Hutchinson's testimony that appellee was indeed in a "lethal situation" more than met the minimal standard of "any evidence" to allow an instruction to be given to the jury. . . .

. . . [Dr.] Hutchinson qualified as an expert on the battered woman syndrome and analyzed the uncontroverted facts for the jury. She concluded appellee was a victim of the syndrome and reasonably believed she was in imminent danger. . . .

The majority implies its decision is necessary to keep the battered woman syndrome from operating as a defense in and of itself. It has always been clear the syndrome is not a defense itself. Evidence of the syndrome is admissible only because of its relevance to the issue of self-defense. The majority of jurisdictions have held it beyond the ordinary jury's understanding why a battered woman may feel she cannot escape, and have held evidence of the battered woman syndrome proper to explain it. The expert testimony explains how people react to circumstances in which the average juror has not been involved. It assists the jury in evaluating the sincerity of the defendant's belief she was in imminent danger requiring self-defense and whether she was in fact in imminent danger.

Dr. Hutchinson explained to the jury at appellee's trial the "cycle of violence" which induces a state of "learned helplessness" and keeps a battered woman in the relationship. She testified appellee was caught in such a cycle. The cycle begins with an initial building of tension and violence, culminates in an explosion, and ends with a "honeymoon." The woman becomes conditioned to trying to make it through one more violent explosion with its battering in order to be rewarded by the "honeymoon phase," with its expressions of remorse and eternal love and the standard promise of "never again." After all promises are broken time after time and she is beaten again and again, the battered woman falls into a state of learned helplessness where she gives up trying to extract herself from the cycle of violence. She learns fighting back only delays the honeymoon and escalates the violence. If she tries to leave the relationship, she is located and returned and the violence increases. She is a captive. She begins to believe her husband is omnipotent, and resistance will be futile at best.

It is a jury question to determine if the battered woman who kills her husband as he sleeps fears he will find and kill her if she leaves, as is usually claimed. Under such circumstances the battered woman is not under actual physical attack when she kills but such attack is imminent, and as a result she believes her life is in imminent danger. She may kill during the tension-building stage when the abuse is apparently not as severe as it sometimes has been, but nevertheless has escalated so that she is afraid the acute stage to come will be fatal to her. She only acts on such fear if she has some survival instinct remaining after the husband-induced "learned helplessness."

Dr. Hutchinson testified the typical batterer has a dichotomous personality, in which he only shows his violent side to his wife or his family. A batterer's major characteristic is the need to blame all frustration on someone else. In a typical battering relationship, she said, the husband and wife are in traditional sex roles, the wife has low self-esteem, and the husband abuses drugs or alcohol. The husband believes the wife is his property and what he does to her is no one's business. There is usually a sense of isolation, with the woman not allowed to speak with friends or children. Overlying the violence is the intimation of death, often created by threats with weapons.

It was Dr. Hutchinson's opinion Mike was planning to escalate his violence in retaliation against appellee for running away. She testified that Mike's threats against appellee's life, his brutal sexual acts, and appellee's discovery of the loaded gun were all indicators to appellee the violence had escalated and she was in danger. Dr. Hutchinson believed appellee had a repressed knowledge she was in what was really a gravely lethal situation. She testified appellee was convinced she must "kill or be killed." . . .

The majority claims permitting a jury to consider self-defense under these facts would permit anarchy. This underestimates the jury's ability to recognize an invalid claim of self-defense. . . .

The majority bases its opinion on its conclusion appellee was not in imminent danger, usurping the right of the jury to make that determination of fact. The majority believes a person could not be in imminent danger from an aggressor merely because the aggressor dropped off to sleep. This is a fallacious conclusion. For instance, picture a hostage situation where the armed guard inadvertently

*(Continues)*

drops off to sleep and the hostage grabs his gun and shoots him. The majority opinion would preclude the use of self-defense in such a case.

The majority attempts to buttress its conclusion appellee was not in imminent danger by citing 19th Century law. The old requirement of "immediate" danger is not in accord with our statute on self-defense, K.S.A. 21-3211, and has been emphatically overruled by case law. Yet this standard permeates the majority's reasoning. A review of the law in this state on the requirement of imminent rather than immediate danger to justify self-defense is therefore required. . . .

In *State v. Hundley*, [693 P.2d 475 (Kan. 1985)], we joined other enlightened jurisdictions in recognizing that the jury in homicide cases where a battered woman ultimately kills her batterer is entitled to all the facts about the battering relationship in rendering its verdict. The jury also needs to know about the nature of the cumulative terror under which a battered woman exists and that a batterer's threats and brutality can make life-threatening danger imminent to the victim of that brutality even though, at the moment, the batterer is passive. Where a person believes she must kill or be killed, and there is the slightest basis in fact for this belief, it is a question for the jury as to whether the danger was imminent. . . .

## Notes and Questions

1. What is the importance of the distinction made in the dissenting opinion between danger being "immediate" and "imminent"? Consider the following portion of Justice Martin's dissenting opinion in *State v. Norman*, 378 S.E.2d 8 (N.C. 1989), a case resembling *Stewart* in which a majority of the North Carolina Supreme Court ruled that a battered woman who shoots her sleeping husband is not entitled to a jury instruction on the issue of self-defense. "[The proper] interpretation of the meaning of 'imminent' is reflected in the Comments to the Model Penal Code: 'The actor must believe that his defensive action is immediately necessary and the unlawful force against which he defends must be force that he apprehends will be used on the present occasion, but he need not apprehend that it will be immediately used.'" 378 S.E.2d at 19, n. 1. Does the hypothetical used in the dissenting opinion in *Stewart*, about a captive who takes advantage of the opportunity to kill his guard while the guard is sleeping, help illustrate the conceptual difference between danger being "imminent" as opposed to "immediate"?

2. Of what relevance is the battered woman syndrome to Peggy Stewart's claim of self-defense? Does the "learned helplessness" aspect of the battered woman syndrome arguably help explain why Stewart simply didn't leave her abusive husband while he slept, instead of killing him? If that is the point, is the implication that battered women need not act as "reasonable" people and instead are entitled to have their self-defense claims evaluated using a standard of the "reasonable battered woman"?

3. For a case holding that a defendant who claims to be a battered woman and admits to killing her abusive husband in his sleep is entitled to have the jury consider whether she acted in lawful self-defense, *see State v. Leidholm*, 334 N.W.2d 811 (N.D. 1983).

4. *See generally,* Sheryl McCarthy, "Injustice After All," *U.S.A. Today,* p. 15A, May 9, 2006 (discussing killings committed by children and women against their abusers and the subsequent processing of those cases within the criminal justice system).

## Use of Force in Resisting Unlawful Arrest

One issue confronting the criminal justice system is determining what amount of force, if any, a citizen may use in resisting an unlawful arrest. Whereas a citizen has a right to freedom, does he or she have the right to resort to force to assert that freedom against a police officer who is attempting to make an arrest without just cause? In some jurisdictions a defendant has the right to use non-deadly force under such circumstances. In other jurisdictions, if the person making the arrest is clearly a police officer, the defendant has no right to resist even if the arrest is plainly wrongful. In *Commonwealth v. Moreira,* the Supreme Judicial Court of Massachusetts discusses the pros and cons of the respective approaches.

### *Commonwealth v. Moreira,* 447 N.E.2d 1224 (Mass. 1983)

Nolan, Justice.

After a jury trial in the Superior Court, the defendant was convicted of assault and battery on one Joseph P. Munroe, a Somerville police officer, and he appealed. The Appeals Court reversed the judgment and ordered a new trial on the basis of the trial judge's erroneous instructions with respect to the defense of justification. We granted further appellate review to resolve the issue of a person's right to resist an unlawful arrest.

There was evidence which would allow the jury to find that the police had probable cause to stop the defendant and his companions to inquire into their possession of a handgun on a public street late at night. This would have occurred outside the defendant's home if the defendant had not pushed Officer Munroe into the house, where he assaulted the officer. This evidence was sufficient to support a conclusion that the officer's entry into the defendant's house was involuntary and, therefore, lawful, and that the defendant was without justification to resist, or aid another in resisting a police investigation. This evidence would support a conviction of assault and battery.

There was also evidence presented by the defendant which would warrant the jury in finding that the defendant was with his brother and other companions at a street corner when his brother observed the police approaching. The defendant, his brother, and a friend drove in the brother's van to the defendant's home a few blocks away. The defendant entered his house, went to the kitchen, took some ice out of the freezer and placed it on the table. He then returned to the front door where his friend handed him a bottle of vodka which he put in the kitchen sink. When the defendant returned to the front door, the two police officers pushed the defendant's brother and friend aside, and Officer Munroe burst into the house, striking the defendant and sending him violently to the floor. The defendant retaliated, striking Munroe twice with his fists. When Munroe kicked the defendant in the knee, the defendant punched the officer again, hit him with a wrench, and was pushed into the living room and struck with a billy club. The defendant denied striking Munroe with a wrench. The gun which the defendant allegedly had in his possession was not found. This evidence, if believed, would warrant the jury in concluding that the police made an unlawful entry into the defendant's home.

Although the defendant requested that the judge instruct the jury as to the defendant's right to resist an unlawful police intrusion into his home, the judge did not so instruct the jury. The defendant, likewise, requested that the judge instruct the jury that they should consider whether the police had a reasonable basis for believing that the defendant was engaged in the commission of a felony and, thus, whether the police officers' conduct was lawful. The judge declined to give such an instruction.

The Appeals Court concluded that the judge's instructions "had the effect of usurping the jury's fact-finding function on the issue of the lawfulness of the police conduct and, therefore, served to deprive the defendant of a substantial ground of his justification defense." *Commonwealth v. Moreira,* 436 N.E.2d 423 (Mass.App. 1982). On this basis, the Appeals Court reversed the conviction. We agree with that disposition.

We granted the Commonwealth's application for further appellate review to resolve the issue of a person's right to use force to resist an unlawful arrest. In *Regina v. Tooley,* 2 Ld. Raymond Rep. 1296, 1299–1301 (Q.B.1709), it was held that a person had the right to use force to resist an unlawful arrest. The court declared, "[I]f one be imprisoned upon an unlawful authority, it is a sufficient provocation to all people out of compassion; much more where it is done under a colour of justice, and where the liberty of the subject is invaded, it is a provocation to all the subjects of England." Id. at 1301. In 1865, we held that a person had the right to resist forcibly an unlawful arrest. Upon review of this area of the law, we conclude that this rule is no longer consistent with the needs of modern society and should be abrogated.

We note that the trend in this country has been away from the old rule and toward the resolution of disputes in court. Since 1709, society has changed. In this era of constantly expanding legal protection of the rights of the accused in criminal proceedings, an arrestee may be reasonably required to submit to a possibly unlawful arrest and to take recourse in the legal processes available to restore his liberty. *State v. Koonce,* 214 A.2d 428 (N.J. App.Div. 1965). An arrestee has the benefit of liberal bail laws, appointed counsel, the right to remain silent and to cut off questioning, speedy arraignment, and speedy trial. As a result of these rights and procedural safeguards, the need for the common law rule disappears—self-help by an arrestee has become anachronistic. As the New Jersey court wrote, self-help "is antisocial in an urbanized society." Thus, the modern view, adopted by eleven States by judicial decision, and by nineteen States by legislative enactment, emerges: a person may not resist an unlawful arrest which is accomplished without excessive force. If a police officer is making an illegal arrest, but without excessive force, the remedy is to be found in the courts. *Miller v. State,* 462 P.2d 421, 426 (Alaska 1969). The legality of an arrest may often be a close question as to which even lawyers and judges may disagree. Such a close question is more properly decided by a detached magistrate rather than by the participants in what may well be a highly volatile imbroglio. As the Alaska court wrote in the *Miller* case, "We feel that the legality of a peaceful arrest should be determined by courts of law and not through a trial by battle in the streets." Id. at 427.

Accordingly, we conclude that in the absence of excessive or unnecessary force by an arresting officer, a person may not use force to resist an arrest by one who he knows or has good reason to believe is an

*(Continues)*

*(Continued)*

authorized police officer, engaged in the performance of his duties, regardless of whether the arrest was unlawful in the circumstances. . . .

Our conclusion does not apply to cases in which the police officer uses excessive force in his attempt to subdue the arrestee. In such a situation, the disposition of the case depends on the application of the rules pertaining to self-defense. Thus, we conclude that where the officer uses excessive or unnecessary force to subdue the arrestee, regardless of whether the arrest is lawful or unlawful, the arrestee may defend himself by employing such force as reasonably appears to be necessary. Moreover, once the arrestee knows or reasonably should know that if he desists from using force in self-defense, the officer will cease using force, the arrestee must desist. Otherwise, he will forfeit his defense.

The questions whether the officer used excessive force and whether the arrestee used reasonable force to resist the excessive force are questions of fact to be resolved by the jury on proper instruction by the trial judge. Application of this rule will not require that the arrestee act in a manner commensurate with such action as would follow detached and reasoned reflection. Some recognition must be given to the frailty of human nature. The rule merely requires the finder of fact to determine whether the arrestee's conduct was reasonable in light of all the circumstances.

In summary, we conclude that in the absence of excessive force by the arresting officer, an arrestee is not privileged to use force to resist an unlawful arrest if the arrestee knows or reasonably believes that the arresting officer is an authorized police officer. However, where the arresting officer uses excessive force in his attempt to subdue the arrestee, the arrestee has the right to use such force as is reasonably necessary to repel such excessive force. . . .

Judgment of the Superior Court reversed.

## Notes and Questions

1. A leading reason for a change from the common law rule permitting individuals to physically resist an unlawful arrest is the increased sophistication of firepower and the portability of weapons. Does it seem sensible that rules of law should be modified in response to social developments of this nature?

2. Do you agree with the court's distinction between when an unlawful arrest is made with unnecessary force and without such force? Might resisting arrest when a police officer is using any force merely cause the incident to escalate and increase the likelihood of serious injury?

3. The position taken by the Massachusetts Supreme Judicial Court is just one of several options that courts and legislatures have adopted. Other rules include never allowing citizens to use force against the police, always allowing citizens to use non-deadly force when unlawfully arrested, and allowing the use of force only against plain-clothed police officers. What do you believe is the appropriate policy? Does it depend on whose interest is considered paramount?

## Defense of Another

Although not invariably true in earlier times, contemporary law now recognizes, in most circumstances, that an individual is justified in using reasonable force in defense of another person when she or he reasonably believes that the other person is in imminent danger of unlawful bodily harm. As with self-defense, the amount of force used in defense of others is limited only to such force as is necessary to quell the danger. Thus deadly defensive force may be used only in instances where it reasonably appears that the third person is under deadly attack.

In most jurisdictions the right to use force to protect a third person closely parallels the justifiable use of force in self-defense. For the defense to be successful, the defender's actions must reasonably be perceived to be necessary and must be proportionate to the threatened harm. However, special difficulties can arise when a person comes to the aid of a third person who is being attacked by another who, unbeknownst to the defender, is a non-uniformed police officer discharging official duties. In some jurisdictions, if

the defender reasonably failed to perceive that the apparent attacker was a police officer, his or her actions will be justified. Other jurisdictions, conversely, follow what is known as the *alter ego rule*, meaning that the defender only has the right to use defensive force if the third person lawfully could have used force to defend him- or herself. Thus a good Samaritan who unwittingly intervenes to help a third party unlawfully resist an arrest, for example, will not be able to assert that she was attempting to defend that individual from what (reasonably) appeared to be unlawful violence, and she will be criminally liable for her actions.

The Model Penal Code (MPC) contains a four-part test to determine whether a person's actions in defending a third person are justified. Section 3.05 of the MPC provides that a person may use force to protect a third person if

1. He would be justified in using force to protect himself if the facts were as he believed them to be;
2. The person whom he seeks to protect would be justified in using force to protect himself;
3. The actor believes intervention is necessary to protect the third person; and

**4.** If the third person would be required to retreat under the rules of self-defense, the actor must try to facilitate such retreat by the third person before using deadly force.

This approach thus rejects the alter ego rule discussed above. What policy reasons might justify, respectively, the "reasonable belief" and the "alter ego" rules regarding defense of a third person?

## Defense of Home and Property

It makes sense that a person is allowed to use reasonable force to protect him- or herself from a person who is threatening to cause imminent bodily harm. Should the same be true if a person is not in physical danger but rather the person's property is in danger of being stolen, damaged, or destroyed?

Under the common law a person is justified in using a reasonable amount of non-deadly force to defend personal or real property. Before using force, the owner of the property must reasonably believe force is needed to prevent the imminent, unlawful dispossession of his or her property. In addition, some jurisdictions require the property owner to ask the person taking the property to cease and desist before force may be used. Once the property has been taken, however, the owner may not use force to recapture it unless in fresh pursuit of the taker.

The MPC provides owners with somewhat more leeway in using force to protect or reclaim their property. Under § 3.06 (1)(b), a person may use force to recapture personal property, even if not in fresh pursuit, if he or she believes that the individual taking the property has no claim of right to it. Pursuant to § 3.06 (3)(a), "[t]he use of force is justifiable . . . only if the actor first requests the person against whom such force is used to desist from his interference with the property, unless the actor believes that: (i) such request would be useless; or (ii) it would be dangerous . . . ; or (iii) substantial harm will be done to the physical condition of the property which is sought to be protected before the request can effectively be made." Consistent with the proportionality principle, under § 3.06 (3)(d), deadly force cannot be used solely to protect personal property. Section 3.09 of the Code governs situations where the actor's beliefs relevant to the justifiable use of force are mistaken.

One especially controversial aspect involving the defense of property is the use of deadly force to protect one's home against invasion by intruders. Jurisdictions have adopted different rules related to the so-called castle doctrine, including how closely associated the defense of habitat must be with defense of self in order to make the use of deadly force justified against a home intruder. These issues figure into the court's decision in *Warrington v. State*. castle doctrine . . . .

---

### *Warrington v. State,* 840 A.2d 590 (Del. 2003)

Berger, Justice:

. . . Robert Wesley Warrington ("Wes"), then 22, and Andrew Warrington ("Drew"), then 18, are brothers who lived with their father at 100 Port Lewes in Sussex County. Wes owed an acquaintance, Jesse Pecco, approximately $800 for drugs that Wes had consumed instead of selling. In order to partially repay the debt, Wes forged a check from his father's bank account, making it out to himself in the amount of $700. Wes gave the check to Pecco on Friday, August 11, 2000, and the two men agreed to meet on Monday to cash the check.

Pecco did not go to the meeting place. Instead, he drove to 100 Port Lewes, and parked his car directly behind Wes's car so as to immobilize it. Pecco then entered the dwelling through its unlocked front door. Drew, who was upstairs watching television, heard shouts coming from the first floor. When he went downstairs to see what was happening, he found Pecco involved in a physical struggle with Wes. Drew soon realized that the two were fighting over control of a knife that Pecco was holding. Drew struck Pecco from behind, causing him to release the knife. According to Wes, Pecco then had the opportunity to leave the house, but instead chased Drew, who fled up the stairs. Both brothers maintain that Pecco was the aggressor in the fight, and that they believed he posed a threat.

The two brothers testified that they gained the upper hand as Wes stabbed Pecco repeatedly with the knife and Drew struck him repeatedly with a fireplace poker. . . .

During the altercation, a 911 call was made from the Warrington residence. DNA from blood marks found on the telephone used to make the call matched Pecco's DNA. . . . Drew [claimed] . . . that it was he who dialed the number, only to have Pecco knock the phone from his hands. The jury listened to the sounds of the fight, as recorded on the 911 tape, before reaching its conclusion regarding self-defense. The tape revealed that, towards the end of the fight, Pecco was pleading with the brothers to stop attacking him. He asked, "Why are you guys trying to kill me?" to which one of the brothers responded, "Good reasons." As he died, Pecco said, "Wes, show me some love. Give me a hug before I die. Give me a hug." Testimony demonstrated that Drew responded by kicking him in the face and telling him to shut up.

*(Continues)*

. . . Wes and Drew's (collectively "defendants") defense to the Pecco murder charges was self-defense within a dwelling. Accordingly, the trial judge instructed the jury:

A defense raised in this case is justification. . . . The defense stems from the defendants' assertion that, at the time in question, their actions were justified. The elements of the defense of justification, in this case, are as follows: (1) That the defendants were in their own dwelling at the time of the incident. (2) That Mr. Pecco was [an] intruder unlawfully in defendants' dwelling at the time of the incident. . . . (3(b)) That the defendants reasonably believed that Mr. Pecco would inflict [personal] injury upon them. . . .

. . . At another point in the instruction, the judge clarified,

As to the justification defense of a person unlawfully in a dwelling, if you find Mr. Pecco an intruder unlawfully in the defendants' dwelling and if the defendant overcame Jesse Pecco so that the defendant no longer believed he was in danger of physical injury or personal injury, and therefore, the defendant knew the use of deadly force was no longer necessary, then the continued use of deadly force was not justified. In other words, *if a person is initially justified in defending himself, but then knows that the danger to him has passed then the subsequent use of deadly force is not justified.* (Emphasis added.)

The jury found both Drew and Wes guilty of murder in the first degree, possession of a deadly weapon during the commission of a felony, and conspiracy in the first degree. . . .

Delaware's statute defining self-defense within a dwelling provides:

In the prosecution of an occupant of a dwelling charged with killing or injuring an intruder who was unlawfully in said dwelling, it shall be a defense that the occupant was in the occupant's own dwelling at the time of the offense, and:

(1) The encounter between the occupant and intruder was sudden and unexpected, compelling the occupant to act instantly; or

(2) The occupant reasonably believed that the intruder would inflict personal injury upon the occupant or others in the dwelling; or

(3) The occupant demanded that the intruder disarm or surrender, and the intruder refused to do so.

The question presented in this case is whether the reasonable belief that the intruder would inflict injury must be contemporaneous with the forceful actions taken against the intruder. Defendants contend that, once both intrusion and reasonable belief of danger have occurred, the rightful occupant of the dwelling has "license to slay the intruder" without considering whether the intruder has been disabled so as to no longer pose a threat.

The doctrine of self-defense within a dwelling differs from self-defense in other contexts. A defendant attacked outside her home must retreat rather than use deadly force, but inside her own dwelling she has no duty to retreat, even if she could do so with "complete safety." . . .

In Delaware, the added protection granted to the occupant of a dwelling is not absolute. The statute provides a defense only in three circumstances: when the occupant 1) encounters the intruder suddenly and must act "instantly;" 2) reasonably believes that the intruder will inflict injury; or 3) demands that the intruder disarm or surrender and the intruder refuses. These three circumstances share a common element—the occupant is being placed in immediate peril. The plain meaning of this immediacy requirement is that the statute only affords protection if the occupant is confronted with one of the three circumstances *at the time* the occupant uses deadly force on the intruder.

Nothing in the statutory language or Delaware case law supports defendants' contention that what begins as self-defense can be turned into a license to kill. Defendants rely on several extremely broad self-defense within a dwelling statutes from other jurisdictions in arguing their position. For example, the California statute provides: "[a]ny person using force intended or likely to cause death . . . within his or her residence shall be presumed to have held a reasonable fear of imminent peril of death or great bodily injury . . . when that force is used. . . ." [Cal.Penal Code § 198.5]. Colorado's so-called "make my day" statute provides the occupant of a dwelling immunity from prosecution:

[A]ny occupant of a dwelling is justified in using . . . deadly force against another person when that other person has made an unlawful entry into the dwelling and when the occupant has a reasonable belief that such other person has committed a crime in the dwelling in addition to the uninvited entry, or is committing or intends to commit a crime against a person or property in addition to the uninvited entry and when the occupant reasonably believes that such other person might use any physical force, no matter how slight, against any occupant. [Colo. Rev. Stat. § 18-1-704.5.]

Delaware's statute, however, is not comparable to either of these statutes. Moreover, defendants have failed to identify any jurisdiction, including California and Colorado, where an occupant of a dwelling is justified in using deadly force after the intruder has been totally subdued.

. . . Based on the foregoing, we conclude that the "added" jury instruction explaining the limitation on the defense of self-defense within a dwelling correctly stated the law. Accordingly, the judgments of the Superior Court are hereby AFFIRMED.

## Notes and Questions

1. How does the Delaware statute addressing the justifiable use of force against an intruder within a dwelling differ from the rules governing the justifiable self-defense outside of the home?
2. How does the Delaware statute differ from the analogous laws in California and Colorado, which are discussed within the opinion?
3. Are laws authorizing the more permissive use of deadly force by homeowners against intruders reasonable in light of the special nature of the home as a sanctuary, or do they invite citizens to become "judge, jury, and executioner," to "shoot first and ask questions later," and unnecessarily encourage bloodshed?[1]

## *People v. Ceballos*, 526 P.2d 241 (Cal. 1974)

Burke, Justice. . . .

Defendant lived alone in a home in San Anselmo. The regular living quarters were above the garage, but defendant sometimes slept in the garage and had about $2,000 worth of property there.

In March 1970 some tools were stolen from defendant's home. On May 12, 1970, he noticed the lock on his garage doors was bent and pry marks were on one of the doors. The next day he mounted a loaded .22 caliber pistol in the garage. The pistol was aimed at the center of the garage doors and was connected by a wire to one of the doors so that the pistol would discharge if the door was opened several inches.

The damage to defendant's lock had been done by a 16-year-old boy named Stephen and a 15-year-old boy named Robert. On the afternoon of May 15, 1970, the boys returned to defendant's house while he was away. Neither boy was armed with a gun or knife. After looking in the windows and seeing no one, Stephen succeeded in removing the lock on the garage doors with a crowbar, and, as he pulled the door outward, he was hit in the face with a bullet from the pistol.

Stephen testified: He intended to go into the garage "(f)or musical equipment" because he had a debt to pay to a friend. His "way of paying that debt would be to take (defendant's) property and sell it" and use the proceeds to pay the debt. He "wasn't going to do it (*i.e.*, steal) for sure, necessarily." He was there "to look around," and "getting in, I don't know if I would have actually stolen."

Defendant, testifying in his own behalf, admitted having set up the trap gun. He stated that after noticing the pry marks on his garage door on May 12, he felt he should "set up some kind of a trap, something to keep the burglar out of my home." When asked why he was trying to keep the burglar out, he replied, " . . . Because somebody was trying to steal my property . . . and I don't want to come home some night and have the thief in there . . . usually a thief is pretty desperate . . . and . . . they just pick up a weapon . . . if they don't have one . . . and do the best they can."

When asked by the police shortly after the shooting why he assembled the trap gun, defendant stated that "he didn't have much and he wanted to protect what he did have." . . .

The jury found defendant guilty of assault with a deadly weapon. . . .

Defendant contends that had he been present he would have been justified in shooting Stephen since Stephen was attempting to commit burglary, that under cases such as *United States v. Gilliam*, [25 F.Cas. 1319 (D.C.D.C. 1882)], defendant had a right to do indirectly what he could have done directly, and that therefore any attempt by him to commit a violent injury upon Stephen was not "unlawful" and hence not an assault. The People argue that the rule in Gilliam is unsound, that as a matter of law a trap gun constitutes excessive force, and that in any event the circumstances were not in fact such as to warrant the use of deadly force.

The issue of criminal liability under statutes such as Penal Code section 245 where the instrument employed is a trap gun or other deadly mechanical device appears to be one of first impression in this

*(Continues)*

state, but in other jurisdictions courts have considered the question of criminal and civil liability for death or injuries inflicted by such a device. . . .

In the United States, courts have concluded that a person may be held criminally liable under statutes proscribing homicides and shooting with intent to injure, or civilly liable, if he sets upon his premises a deadly mechanical device and that device kills or injures another. However, an exception to the rule that there may be criminal and civil liability for death or injuries caused by such a device has been recognized where the intrusion is, in fact, such that the person, were he present, would be justified in taking the life or inflicting the bodily harm with his own hands. The phrase "were he present" does not hypothesize the actual presence of the person, but is used in setting forth in an indirect manner the principle that a person may do indirectly that which he is privileged to do directly.

Allowing persons, at their own risk, to employ deadly mechanical devices imperils the lives of children, firemen and policemen acting within the scope of their employment, and others. Where the actor is present, there is always the possibility he will realize that deadly force is not necessary, but deadly mechanical devices are without mercy or discretion. Such devices "are silent instrumentalities of death. They deal death and destruction to the innocent as well as the criminal intruder without the slightest warning. The taking of human life (or infliction of great bodily injury) by such means is brutally savage and inhuman." [*State v. Plumlee,* 149 So. 425, 430 (La. 1933).]

It seems clear that the use of such devices should not be encouraged. Moreover, whatever may be thought in torts, the foregoing rule setting forth an exception to liability for death or injuries inflicted by such devices "is inappropriate in penal law for it is obvious that it does not prescribe a workable standard of conduct; liability depends upon fortuitous results." [Model Penal Code (Tent. Draft No. 8, § 3.06, com. 15).] We therefore decline to adopt that rule in criminal cases.

Furthermore, even if that rule were applied here, as we shall see, defendant was not justified in shooting Stephen. Penal Code section 197 provides: "Homicide is . . . justifiable . . . 1. When resisting any attempt to murder any person, or to commit a felony, or to do some great bodily injury upon any person; or, 2. When committed in defense of habitation, property, or person, against one who manifestly intends or endeavors, by violence or surprise, to commit a felony. . . ." Since a homicide is justifiable under the circumstances specified in section 197, an attempt to commit a violent injury upon another under those circumstances is justifiable.

By its terms subdivision 1 of Penal Code section 197 appears to permit killing to prevent any "felony," but in view of the large number of felonies today and the inclusion of many that do not involve a danger of serious bodily harm, a literal reading of the section is undesirable. *People v. Jones,* 12 Cal. Rptr. 777 (Cal. App. 1961), in rejecting the defendant's theory that her husband was about to commit the felony of beating her and that therefore her killing him to prevent him from doing so was justifiable, stated that Penal Code section 197 "does no more than codify the common law and should be read in light of it." *Jones* read into section 197, subdivision 1, the limitation that the felony be "some atrocious crime attempted to be committed by force." *Jones* further stated, "the punishment provided by a statute is not necessarily an adequate test as to whether life may be taken for in some situations it is too artificial and unrealistic. We must look further into the character of the crime, and the manner of its perpetration. When these do not reasonably create a fear of great bodily harm, as they could not if defendant apprehended only a misdemeanor assault, there is no cause for the exaction of a human life." [12 Cal.Rptr., at 780.]

*Jones* involved subdivision 1 of Penal Code section 197, but subdivision 2 of that section is likewise so limited. The term "violence of surprise" in subdivision 2 is found in common law authorities and, . . . developed [so] . . . that killing or use of deadly force to prevent a felony was justified only if the offense was a forcible and atrocious crime. . . .

Examples of forcible and atrocious crimes are murder, mayhem, rape and robbery. In such crimes "from their atrocity and violence human life (or personal safety from great harm) either is, or is presumed to be, in peril."

Burglary has been included in the list of such crimes. However, in view of the wide scope of burglary under Penal Code section 459, as compared with the common law definition of that offense, in our opinion it cannot be said that under all circumstances burglary under section 459 constitutes a forcible and atrocious crime.

Where the character and manner of the burglary do not reasonably create a fear of great bodily harm, there is no cause for exaction of human life, or for the use of deadly force. The character and manner of the burglary could not reasonably create such a fear unless the burglary threatened, or was reasonably believed to threaten, death or serious bodily harm.

In the instant case the asserted burglary did not threaten death or serious bodily harm, since no one but Stephen and Robert was then on the premises. A defendant is not protected from liability merely by the fact that the intruder's conduct is such as would justify the defendant, were he present, in believing that the intrusion threatened death or serious bodily injury. There is ordinarily the possibility that the defendant, were he present, would realize the true state of affairs and recognize the intruder as one whom he would not be justified in killing or wounding.

We thus conclude that defendant was not justified under Penal Code section 197, subdivisions 1 or 2, in shooting Stephen to prevent him from committing burglary.

## Notes and Questions

1. What seems to be the principal reason behind the court's ruling? Do you agree with it?
2. In 1984, California added the following provision to its Penal Code:

   > Any person using force intended or likely to cause death or great bodily injury within his or her residence shall be presumed to have held a reasonable fear of imminent peril of death or great bodily injury to self, family, or a member of the household when that force is used against another person, not a member of the family or household, who unlawfully and forcibly enters or has unlawfully and forcibly entered the residence and the person using the force knew or had reason to believe that an unlawful and forcible entry occurred.
   >
   > As used in this section, great bodily injury means a significant or substantial physical injury.
   > Cal. Penal Code § 198.5.

   If Ceballos had been home at the time of the intrusion, would he have been authorized to shoot the intruders under this statutory provision? Under the Delaware law discussed in *Warrington*? Under the Colorado statute discussed in that case?
3. What if, instead of mounting a pistol on his garage door, Ceballos had left his large, vicious guard dog in the garage while he was gone? Would the dog's propensity to bark on hearing noises make a difference? *See generally State v. Cook*, 594 S.E.2d 819 (N.C. App.), *aff'd*, 606 S.E.2d 118 (N.C. 2004), holding that defendant was properly convicted of "assault with a deadly weapon" after he ran from a pursuing police officer into the back yard of his sister's home, and there issued a "bite him" command to his sister's dog, whereupon the dog bit the officer in the ankle. The dissenting opinion reviews cases from several jurisdictions involving dogs as alleged "deadly weapons." *See also Muller v. Boyd*, 855 N.Y.S.2d 651 (App. Div. 2008), granting plaintiff's motion for a new trial in a civil case in which the jury had refused to impose liability for bite injuries inflicted on a house guest by a dog that "had undergone 'guard dog' training, in which it had been trained to 'go after' intruders," and had previously bitten other guests.

## Public Duty

A public officer is generally immune from prosecution for actions stemming from exercising his or her duty as a public servant. Such immunity extends to the use of force, trespass, and the taking of property. A defense will be available as long as the public official does not abuse or exceed his or her authority.

The issue of a person carrying out a public duty by committing an act that would be criminal if not performed as part of a public duty arises most frequently in the use of force by police to effectuate an arrest. Under the common law a police officer would be justified in using any force necessary to capture and arrest a fleeing felon. Deadly force could not be used to arrest a person suspected of committing a misdemeanor. The fleeing felon rule, as the common law rule is commonly known, has been criticized for a number of reasons. In the heat of the moment it is not always possible for a police officer to know exactly what crime a suspect may have committed. Maybe it was a felony. Maybe it was a misdemeanor, or even no crime at all. The uncertainty caused problems for police officers who were called upon to take quick action. In addition, although some misdemeanors are of a violent nature (*e.g.*, assault), some felonies are nonviolent (*e.g.*, tax evasion). It would be a dubious policy to permit a police officer to shoot a tax cheat and not to shoot a violent offender.

In the 20th century states and police departments in the United States moved away from the fleeing felon rule and adopted standards that limited the use of deadly force by a police officer. In *Tennessee v. Garner*, the U.S. Supreme Court set forth standards to determine whether police may use deadly force to arrest a suspect.

## *Tennessee v. Garner*, 471 U.S. 1, 105 S. Ct. 1694, 85 L. Ed. 2d 1 (1985)

Justice White delivered the opinion of the Court. . . .

At about 10:45 P.M. on October 3, 1974, Memphis Police Officers Elton Hymon and Leslie Wright were dispatched to answer a "prowler inside call." Upon arriving at the scene they saw a woman standing on her porch and gesturing toward the adjacent house. She told them she had heard glass breaking and that "they" or "someone" was breaking in next door. While Wright radioed the dispatcher to say that they were on the scene, Hymon went behind the house. He heard a door slam and saw someone run across the backyard. The fleeing suspect, who was appellee-respondent's decedent, Edward Garner, stopped at a 6-feet-high chain link fence at the edge of the yard. With the aid of a flashlight, Hymon was able to see Garner's face and hands. He saw no sign of a weapon, and, though not certain, was "reasonably sure" and "figured" that Garner was unarmed. He thought Garner was 17 or 18 years old and about 5'5" or 5'7" tall. (*See* footnote 2.) While Garner was crouched at the base of the fence, Hymon called out "police, halt" and took a few steps toward him. Garner then began to climb over the fence. Convinced that if Garner made it over the fence he would elude capture, Hymon shot him. The bullet hit Garner in the back of the head. Garner was taken by ambulance to a hospital, where he died on the operating table. Ten dollars and a purse taken from the house were found on his body.

In using deadly force to prevent the escape, Hymon was acting under the authority of a Tennessee statute and pursuant to Police Department policy. The statute provides that "[i]f, after notice of the intention to arrest the defendant, he either flee or forcibly resist, the officer may use all the necessary means to effect the arrest." Tenn.Code Ann. § 40-7-108 (1982). The Department policy was slightly more restrictive than the statute, but still allowed the use of deadly force in cases of burglary. The incident was reviewed by the Memphis Police Firearm's Review Board and presented to a grand jury. Neither took any action.

Garner's father then brought this action in the Federal District Court for the Western District of Tennessee, seeking damages under 42 U.S.C. § 1983 for asserted violations of Garner's constitutional rights. The complaint alleged that the shooting violated the Fourth, Fifth, Sixth, Eighth, and Fourteenth Amendments of the United States Constitution. It named as defendants Officer Hymon, the Police Department, its Director, and the Mayor and city of Memphis. After a 3-day bench trial, the District Court entered judgment for all defendants. It dismissed the claims against the Mayor and the Director for lack of evidence. It then concluded that Hymon's actions were authorized by the Tennessee statute, which in turn was constitutional. Hymon had employed the only reasonable and practicable means of preventing Garner's escape. . . .

The Court of Appeals reversed and remanded. . . .

Whenever an officer restrains the freedom of a person to walk away, he has seized that person. While it is not always clear just when minimal police interference becomes a seizure, there can be no question that apprehension by the use of deadly force is a seizure subject to the reasonableness requirement of the Fourth Amendment.

### A.

A police officer may arrest a person if he has probable cause to believe that person committed a crime. Petitioners and appellant argue that if this requirement is satisfied the Fourth Amendment has nothing to say about how that seizure is made. This submission ignores the many cases in which this Court, by balancing the extent of the intrusion against the need for it, has examined the reasonableness of the manner in which a search or seizure is conducted. To determine the constitutionality of a seizure "[w]e must balance the nature and quality of the intrusion on the individual's Fourth Amendment interests against the importance of the governmental interests alleged to justify the intrusion." *United States v. Place,* 462 U.S. 696, 703 (1983). We have described "the balancing of competing interests" as "the key principle of the Fourth Amendment." *Michigan v. Summers,* 452 U.S. 692, 700 n. 12 (1981). Because one of the factors is the extent of the intrusion, it is plain that reasonableness depends on not only when a seizure is made, but also how it is carried out. . . .

### B.

. . . [N]otwithstanding probable cause to seize a suspect, an officer may not always do so by killing him. The intrusiveness of a seizure by means of deadly force is unmatched. The suspect's fundamental interest in his own life need not be elaborated upon. The use of deadly force also frustrates the interest of the individual, and of society, in judicial determination of guilt and punishment. Against these interests are ranged governmental interests in effective law enforcement. It is argued that overall violence will be reduced by encouraging the peaceful submission of suspects who know that they may be shot if they flee. Effectiveness in making arrests requires the resort to deadly force, or at least the meaningful threat thereof. "Being able to arrest such individuals is a condition precedent to the state's entire system of law enforcement." Brief for Petitioners 14.

Without in any way disparaging the importance of these goals, we are not convinced that the use of deadly force is a sufficiently productive means of accomplishing them to justify the killing of nonviolent suspects. The use of deadly force is a self-defeating way of apprehending a suspect and so setting the criminal justice mechanism in motion. If successful, it guarantees that that mechanism will not be set in motion. And while the meaningful threat of deadly force might be thought to lead to the arrest of more live suspects by discouraging escape attempts, the presently available evidence does not support this thesis. The fact is that a majority of police departments in this country have forbidden the use of deadly force against nonviolent suspects. If those charged with the enforcement of the criminal law have abjured the use of deadly force in arresting nondangerous felons, there is a substantial basis for doubting that the use of such force is an essential attribute of the arrest power in all felony cases. Petitioners and appellant have not persuaded us that shooting nondangerous fleeing suspects is so vital as to outweigh the suspect's interest in his own life.

The use of deadly force to prevent the escape of all felony suspects, whatever the circumstances, is constitutionally unreasonable. It is not better that all felony suspects die than that they escape. Where the suspect poses no immediate threat to the officer and no threat to others, the harm resulting from failing to apprehend him does not justify the use of deadly force to do so. It is no doubt unfortunate when a suspect who is in sight escapes, but the fact that the police arrive a little late or are a little slower afoot does not always justify killing the suspect. A police officer may not seize an unarmed, nondangerous suspect by shooting him dead. The Tennessee statute is unconstitutional insofar as it authorizes the use of deadly force against such fleeing suspects.

It is not, however, unconstitutional on its face. Where the officer has probable cause to believe that the suspect poses a threat of serious physical harm, either to the officer or to others, it is not constitutionally unreasonable to prevent escape by using deadly force. Thus, if the suspect threatens the officer with a weapon or there is probable cause to believe that he has committed a crime involving the infliction or threatened infliction of serious physical harm, deadly force may be used if necessary to prevent escape, and if, where feasible, some warning has been given. As applied in such circumstances, the Tennessee statute would pass constitutional muster. . . .

The District Court concluded that Hymon was justified in shooting Garner because state law allows, and the Federal Constitution does not forbid, the use of deadly force to prevent the escape of a fleeing felony suspect if no alternative means of apprehension is available. This conclusion made a determination of Garner's apparent dangerousness unnecessary. The court did find, however, that Garner appeared to be unarmed, though Hymon could not be certain that was the case. . . .

In reversing, the Court of Appeals accepted the District Court's factual conclusions and held that "the facts, as found, did not justify the use of deadly force." We agree. Officer Hymon could not reasonably have believed that Garner—young, slight, and unarmed—posed any threat. Indeed, Hymon never attempted to justify his actions on any basis other than the need to prevent an escape. The District Court stated in passing that "[t]he facts of this case did not indicate to Officer Hymon that Garner was 'non-dangerous.'" This conclusion is not explained, and seems to be based solely on the fact that Garner had broken into a house at night. However, the fact that Garner was a suspected burglar could not, without regard to the other circumstances, automatically justify the use of deadly force. Hymon did not have probable cause to believe that Garner, whom he correctly believed to be unarmed, posed any physical danger to himself or others.

The dissent argues that the shooting was justified by the fact that Officer Hymon had probable cause to believe that Garner had committed a nighttime burglary. While we agree that burglary is a serious crime, we cannot agree that it is so dangerous as automatically to justify the use of deadly force. . . .

The judgment of the Court of Appeals is affirmed, and the case is remanded for further proceedings consistent with this opinion.

Justice O'Connor, with whom The Chief Justice and Justice Rehnquist join, dissenting.

The Court today holds that the Fourth Amendment prohibits a police officer from using deadly force as a last resort to apprehend a criminal suspect who refuses to halt when fleeing the scene of a

*(Continues)*

nighttime burglary. This conclusion rests on the majority's balancing of the interests of the suspect and the public interest in effective law enforcement. Notwithstanding the venerable common-law rule authorizing the use of deadly force if necessary to apprehend a fleeing felon, and continued acceptance of this rule by nearly half the States, the majority concludes that Tennessee's statute is unconstitutional inasmuch as it allows the use of such force to apprehend a burglary suspect who is not obviously armed or otherwise dangerous. Although the circumstances of this case are unquestionably tragic and unfortunate, our constitutional holdings must be sensitive both to the history of the Fourth Amendment and to the general implications of the Court's reasoning. By disregarding the serious and dangerous nature of residential burglaries and the longstanding practice of many States, the Court effectively creates a Fourth Amendment right allowing a burglary suspect to flee unimpeded from a police officer who has probable cause to arrest, who has ordered the suspect to halt, and who has no means short of firing his weapon to prevent escape. I do not believe that the Fourth amendment supports such a right, and I accordingly dissent. . . .

Because burglary is a serious and dangerous felony, the public interest in the prevention and detection of the crime is of compelling importance. Where a police officer has probable cause to arrest a suspected burglar, the use of deadly force as a last resort might well be the only means of apprehending the suspect. With respect to a particular burglary, subsequent investigation simply cannot represent a substitute for immediate apprehension of the criminal suspect at the scene. Indeed, the Captain of the Memphis Police Department testified that in his city, if apprehension is not immediate, it is likely that the suspect will not be caught. Although some law enforcement agencies may choose to assume the risk that a criminal will remain at large, the Tennessee statute reflects a legislative determination that the use of deadly force in prescribed circumstances will serve generally to protect the public. Such statutes assist the police in apprehending suspected perpetrators of serious crimes and provide notice that a lawful police order to stop and submit to arrest may not be ignored with impunity. . . .

Against the strong public interests justifying the conduct at issue here must be weighed the individual interests implicated in the use of deadly force by police officers. The majority declares that "[t]he suspect's fundamental interest in his own life need not be elaborated upon." This blithe assertion hardly provides an adequate substitute for the majority's failure to acknowledge the distinctive manner in which the suspect's interest in his life is even exposed to risk. For purposes of this case, we must recall that the police officer, in the course of investigating a nighttime burglary, had reasonable cause to arrest the suspect and ordered him to halt. The officer's use of force resulted because the suspected burglar refused to heed this command and the officer reasonably believed that there was no means short of firing his weapon to apprehend the suspect. Without questioning the importance of a person's interest in his life, I do not think this interest encompasses a right to flee unimpeded from the scene of a burglary. The legitimate interests of the suspect in these circumstances are adequately accommodated by the Tennessee statute: to avoid the use of deadly force and the consequent risk to his life, the suspect need merely obey the valid order to halt. . . .

Whatever the constitutional limits on police use of deadly force in order to apprehend a fleeing felon, I do not believe they are exceeded in a case in which a police officer has probable cause to arrest a suspect at the scene of a residential burglary, orders the suspect to halt, and then fires his weapon as a last resort to prevent the suspect's escape into the night. I respectfully dissent.

## Notes and Questions

1. Can you succinctly articulate the standard announced by the Court that governs whether police are allowed to use deadly force to make an arrest? Do you believe this standard is appropriate? What changes, if any, would you make?
2. Note that *Garner* involves a civil action, not a criminal prosecution. Officer Hymon was following state law and departmental guidelines when he shot Mr. Garner. Although the departmental policy was ruled unconstitutional, it arguably would be unfair to punish Hymon for doing what he was trained and instructed to do. However, if an officer now violates the strictures of *Garner*, should he or she be criminally prosecuted? Should it matter how gross the violation is, or should officers who wrongfully use deadly force routinely be prosecuted?
3. Do you believe limiting the power of the police to use deadly force to apprehend a suspect makes their job more difficult? Do you believe suspects are more likely to run from police if they believe they will not be shot for doing so?

## Necessity and Choice of Evils

The defense of **necessity**, sometimes called "**choice of evils**," is a justification defense. It is raised when a person commits what under normal circumstances would be a crime to prevent a greater harm from occurring. Under the early common law, the harm sought to be avoided had to be of a natural origin (such as a storm, earthquake, or flood), as opposed to arising from human origin. This requirement has largely disappeared through case law and legislative action. Today, defendants generally are allowed to raise a necessity defense regardless of the source of the peril, as long as the other requirements of the defense are present.

Two classic examples of necessity are a person using another's dock to secure his boat during a sudden storm or burning down another's house to create a firebreak against a raging fire that is threatening to engulf an entire town. Although both acts ordinarily would be criminal—the first involving a trespass and the second arson—they would be justified if committed to prevent a greater social harm from occurring.

The necessity defense is frequently raised by protesters. For example, people who block access to abortion clinics might argue they are doing so to prevent the murder of unborn children. People who free animals from laboratories might argue they do so to protect the lives of rats and rabbits. In the 1970s and 1980s, in an effort to stop the proliferation of nuclear power plants, people often attempted to block access to them. *State v. Warshow* deals with the prosecution of a number of such protesters for trespassing at Vermont's Yankee Nuclear Power Station.

---

### *State v. Warshow*, 410 A.2d 1000 (Vt. 1979)

Barney, Chief Justice.

The defendants were part of a group of demonstrators that traveled to Vernon, Vermont, to protest at the main gate of a nuclear power plant known as Vermont Yankee. The plant had been shut down for repairs and refueling, and these protestors had joined a rally designed to prevent workers from gaining access to the plant and placing it on-line.

They were requested to leave the private premises of the power plant by representatives of Vermont Yankee and officers of the law. The defendants were among those who refused, and they were arrested and charged with unlawful trespass.

The issue with which this appeal of their convictions is concerned relates to a doctrine referred to as the defense of necessity. At trial the defendants sought to present evidence relating to the hazards of nuclear power plant operation which, they argued, would establish that defense. After hearing the defendants' offer of proof the trial court excluded the proffered evidence and refused to grant compulsory process for the witnesses required to present the defense. The jury instruction requested on the issue of necessity was also refused, and properly preserved for appellate review.

In ruling below, the trial court determined that the defense was not available. It is on this basis that we must test the issue.

The defense of necessity is one that partakes of the classic defense of "confession and avoidance." It admits the criminal act, but claims justification. . . .

The doctrine is one of specific application insofar as it is a defense to criminal behavior. This is clear because if the qualifications for the defense of necessity are not closely delineated, the definition of criminal activity becomes uncertain and even whimsical. The difficulty arises when words of general and broad qualification are used to describe the special scope of this defense.

In the various definitions and examples recited as incorporating the concept of necessity, certain fundamental requirements stand out:

(1) there must be a situation of emergency arising without fault on the part of the actor concerned;
(2) this emergency must be so imminent and compelling as to raise a reasonable expectation of harm, either directly to the actor or upon those he was protecting;
(3) this emergency must present no reasonable opportunity to avoid the injury without doing the criminal act; and
(4) the injury impending from the emergency must be of sufficient seriousness to out measure the criminal wrong.

It is the defendants' position that they made a sufficient offer of proof to establish the elements of the necessity defense to raise a jury question. The trial court rejected this contention on the ground, among others, that the offer did not sufficiently demonstrate the existence of an emergency or imminent danger.

This ruling was sound, considering the offer. The defendants wished to subpoena witnesses to testify to the dangers of nuclear accidents and the effect of low-level radiation. It was conceded that

*(Continues)*

there had been no serious accident at Vermont Yankee, but defendants contended that the consequences could be so serious that the mere possibility should suffice. This is not the law.

There is no doubt that the defendants wished to call attention to the dangers of low-level radiation, nuclear waste, and nuclear accident. But low-level radiation and nuclear waste are not the types of imminent danger classified as an emergency sufficient to justify criminal activity. To be imminent, a danger must be, or must reasonably appear to be, threatening to occur immediately, near at hand, and impending. We do not understand the defendants to have taken the position in their offer of proof that the hazards of low-level radiation and nuclear waste buildup are immediate in nature. On the contrary, they cite long-range risks and dangers that do not presently threaten health and safety. Where the hazards are long term, the danger is not imminent, because the defendants have time to exercise options other than breaking the law.

Nor does the specter of nuclear accident as presented by these defendants fulfill the imminent and compelling harm element of the defense. The offer does not take the position that they acted to prevent an impending accident. Rather, they claimed that they acted to foreclose the "chance" or "possibility" of accident. This defense cannot lightly be allowed to justify acts taken to foreclose speculative and uncertain dangers. Its application must be limited to acts directed to the prevention of harm that is reasonably certain to occur. Therefore the offer fails to satisfy the imminent danger element. The facts offered would not have established the defense.

These acts may be a method of making public statements about nuclear power and its dangers, but they are not a legal basis for invoking the defense of necessity. Nor can the defendants' sincerity of purpose excuse the criminal nature of their acts. . . .

Judgment affirmed.

Hill, Justice, concurring.

While I agree with the result reached by the majority, I am unable to agree with their reasoning. As I see it, the sole issue raised by this appeal is whether the trial court erred in not allowing the defendants to present the defense of necessity.

For the purposes of this concurrence I briefly restate the relevant factual background. The convictions under review arose out of a peaceful anti-nuclear demonstration conducted on the property of Vermont Yankee Nuclear Power Station in Vernon on October 8, 1977. Although the power station had been shut down for six to eight weeks prior to the demonstration, it was about to be refueled and recommence operation. Based upon their belief that nuclear power presented real and substantial dangers, defendants blocked the entrance to the power station to prevent its further operation.

At trial defendants sought to raise the affirmative defense of necessity, arguing that they were faced with a choice of evils—either violate the literal terms of the law or comply with the law and allow the commission of a more egregious wrong, *i.e.*, the proliferation of nuclear power—and that they chose the course which would result in the least harm to the public, even though it meant violating the criminal law. . . .

The defense of necessity proceeds from the appreciation that, as a matter of public policy, there are circumstances where the value protected by the law is eclipsed by a superseding value, and that it would be inappropriate and unjust to apply the usual criminal rule. The balancing of competing values cannot, of course, be committed to the private judgment of the actor, but must, in most cases, be determined at trial with due regard being given for the crime charged and the higher value sought to be achieved.

Determination of the issue of competing values and, therefore, the availability of the defense of necessity is precluded, however, when there has been a deliberate legislative choice as to the values at issue. The common law defense of necessity deals with imminent dangers from obvious and generally recognized harms. It does not deal with non-imminent or debatable harms, nor does it deal with activities that the legislative branch has expressly sanctioned and found not to be harms.

Both the state of Vermont and the federal government have given their imprimatur to the development and normal operation of nuclear energy and have established mechanisms for the regulation of nuclear power. Implicit within these statutory enactments is the policy choice that the benefits of nuclear energy outweigh its dangers.

If we were to allow defendants to present the necessity defense in this case we would, in effect, be allowing a jury to redetermine questions of policy already decided by the legislative branches of the federal and state governments. This is not how our system of government was meant to operate. . . .

Since defendants' defense of necessity was foreclosed by a deliberate legislative policy choice, there was no error on the trial court's part in not allowing the defense to be presented.

Billings, Justice, dissenting.

This is an appeal from convictions for unlawful trespass. The sole issue before the Court is whether the trial court erred in excluding evidence on the defense of necessity at trial on the offer of proof made by the defendants at the preliminary hearing and at the trial.

The defendants, appearing pro se, made an offer of proof detailing the elements of the defense of necessity. The trial judge did not rule on the sufficiency of the offer, but excluded the offered evidence, stating that "the Court does not view it as a justification or a defense as the act as alleged (sic) in Vermont at this time." . . .

Here the trial court prevented the introduction of any evidence on the issue of necessity in spite of the offer of proof. The test is whether the offer of proof was sufficient in form and substance to show that the tendered evidence was relevant and admissible, and material. The offer must be only specific and concrete enough to make apparent to the trial court the existence of facts, which, if proved, would make the evidence relevant to an issue in the case and otherwise admissible.

The majority states that the danger of low-level radiation and nuclear waste, which the defendants offered to prove, are "not the types of imminent danger classified as an emergency sufficient to justify criminal activity." Furthermore, the majority dismisses those portions of the proof dealing with the threat of a nuclear accident by characterizing them as mere "speculative and uncertain dangers." In doing so the majority has decided to so read the evidence as to give credibility only to that evidence offered on the effects of low-level radiation. This approach is clearly inconsistent with the case of State v. Fernie, [285 A.2d 726, 727 (Vt. 1971)], in which then Justice Barney held that it is reversible prejudicial error for the trial court to exclude a whole line of material evidence tending to support a defense even though "(d)eficiencies in the admissibility of such evidence might . . . have later appeared." It is not for this Court to weigh the credibility of the evidence in this manner where there is evidence offered on the elements of the defense. While this case might stand in a different posture if, at the close of the defendants' evidence, they had failed to introduce evidence, as offered, on each and every element of the defense sufficient to make out a prima facie case, that situation is not before us. Furthermore, it is not for the trial judge to rule on the ultimate credibility and weight of the evidence. Where there is evidence offered which supports the elements of the defense, the questions of reasonableness and credibility are for the jury to decide.

The defendants offered evidence on all the requisite elements of the defense of necessity. They stated as follows:

[They had] a feeling that there was a situation of an emergency or imminent danger that would have occurred with the start up of the reactor on October 8th the time of (their) alleged crime . . . the chance . . . of the nuclear power plant having a serious accident which would cause . . . great untold damage to property and lives and health for many generations.

The defendants also stated that "there was reasonable belief that it would have been an emergency had they started that reactor up . . . there was a very good chance of an accident there for which there is no insurance coverage or very little." Specifically, the defendants offered to show by expert testimony that there were defects in the cooling system and other aspects of the power plant which they believed could and would result in a meltdown within seven seconds of failure on the start up of the plant. In addition, the defendants went to great lengths to base their defense on the imminent danger that would result from the hazardous radiation emitted from the plant and its wastes when the plant resumed operations.

While the offer made by the defendants was laced with statements about the dangers they saw in nuclear power generally, it is clear that they offered to show that the Vermont Yankee facility at which they were arrested was an imminent danger to the community on the day of the arrests; that, if it commenced operation, there was a danger of meltdown and severe radiation damage to persons and property. In support of this contention, the defendants stated that they would call experts familiar with the Vermont Yankee facility and the dangerous manner of its construction, as well as other experts who would testify on the effects of meltdown and radiation leakage, on the results of governmental testing, and on the regulation of the Vermont Yankee facility. These witnesses were highly qualified to testify about the dangers at the Vermont Yankee facility based either on personal knowledge or on conditions the defendants offered to show existed at the time of the trespass.

Furthermore, the defendants offered to show that, in light of the imminent danger of an accident, they had exhausted all alternative means of preventing the start up of the plant and the immediate catastrophe it would bring. Under the circumstances of imminent danger arising from the start up of

*(Continues)*

*(Continued)*

the plant, coupled with the resistance of Vermont Yankee and government officials, which the defendants offered to prove, nothing short of preventing the workers access to start up the plant would have averted the accident that the defendants expected.

Through this offer, it cannot be said, without prejudgment, that the defendants failed to set forth specific and concrete evidence, which, if proven, would establish the existence of an imminent danger of serious proportions through no fault of the defendants which could not be averted without the trespass. Whether the defendants' expectations and opportunities were reasonable under the circumstances of this case is not for the trial court to decide without hearing the evidence. From a review of the record, I am of the opinion that the offer here measured up to the standard required and that the trial court struck too soon in excluding the offered evidence. . . .

I am of the opinion that the defendants are entitled to present evidence on the defense of necessity as it exists at common law. To deny them this opportunity is to deny them a fair trial merely because they express unpopular political views. I would hold that they are entitled to present evidence on this defense even though they may well fail to establish a sufficient case to send the issue to the jury for deliberation, because I find the offer of proof sufficient for this purpose under the law of Vermont.

## Notes and Questions

1. Note the two limitations on the use of a necessity defense discussed by the majority and concurring opinions. The harm sought to be prevented must be imminent. Concern about a future harm is insufficient to support a necessity defense. In addition, if the legislature has voiced its preference for the continuation or legality of the activity that is the impetus for the protest, a necessity defense will be precluded. Do you believe these requirements are mere technicalities designed to prevent civil protest or do you believe they are based on legitimate policy considerations?

2. The MPC contains a choice of evils defense that differs to some degree from common law necessity. Most noticeable is the lack of an imminent danger requirement in the MPC. Rather than requiring that the actor act to prevent an imminent harm as required under the common law, § 3.02 of the MPC only requires that the defendant believes that the act is "necessary to avoid a harm or evil . . . ." Does this approach seem more reasonable? Consider a situation where a person is hiking in a park when he hears over the radio that there is a tornado warning for the area. The warning says the tornado should be passing over the park in 20 minutes. The hiker immediately takes off running toward a maintenance facility he just passed. He finds the facility locked. Rather than staying out in the dangerous conditions, he breaks the door down and goes into the underground workstation. Is he guilty of an illegal entry? Under the common law, if the storm were still 15 minutes off a prosecutor could argue that he was not in imminent danger. Although such a result might seem unfair and absurd, it is legally plausible.

3. Under the necessity doctrine, the harm sought to be avoided must be greater than the harm committed. This being said, should there be limits on what a person can do to avoid or prevent a harm? Should a person be able to kill another to save his or her life? Should he or she be able to kill a person to save three other lives? See *Queen v. Dudley and Stephens* in Chapter 1 for a discussion of this conundrum.

## Conclusion

Justification defenses involve the defendant admitting that what he or she did would ordinarily violate the law but arguing that the conduct and resulting harm were justified under the circumstances. These defenses involve a societal balancing of evils. The law has determined that some actions are justified to prevent a greater harm, including the use of force to protect self or another from being injured by an unlawful aggressor or causing property damage or personal injury to prevent more serious harms from occurring. If a judge decides that such defenses lawfully may be presented, the jury ultimately will determine whether an act was justified or criminal.

## Key Terms

affirmative defense
battered woman syndrome
castle doctrine
choice of evils
deadly force
excuse defenses
imminent harm

justification defenses
necessity
objective standard
proportionality
provocation
self-defense
subjective criterion

## Review Questions

1. Using the Model Penal Code as a guide, describe how a "justification defense" might be established in a homicide case.
2. Explain the meaning of the *deadly force* doctrine.

3. Is a person ever "justified" in resisting a lawful arrest? Why or why not?
4. In the case of self-defense or the defense of another, what does the term *reasonable force* mean?
5. How does the battered woman syndrome relate to justification defenses?
6. How does the doctrine of *reasonable force* differ between cases involving human life and those involving only real property?
7. How much force can a person use to defend property? Are there different standards for defending one's home? Why?
8. Under what circumstances must a person retreat from an altercation instead of "standing his or her ground" and using defensive force?
9. What is meant by a necessity or "choice of evils" defense?

## Notes

1. *See* Young ME. "'Castle Law' Arms Texas Homeowners With Right to Shoot: Does New Law Make Them Quicker to Pull the Trigger?" *Dallas Morning News*, Jan. 20, 2008; Blumenthal R. "Fatal Shootings Test Limits of New Self-Defense Law in Texas," *New York Times*, p. 32, Dec. 13, 2007.

## Footnote for *Tennessee v. Garner*, 471 U.S. 1, 105 S. Ct. 1694, 85 L. Ed. 2d 1 (1985)

2. In fact, Garner, an eighth-grader, was 15. He was 5'4" tall and weighed somewhere around 100 or 110 pounds.

# CHAPTER

# 5 Excuse Defenses

## Chapter Objectives

- Distinguish between *excuse defenses* and *justification defenses*
- Understand the different tests used for the defense of insanity
- Distinguish *mistakes of law* from *mistakes of fact* as excuse defenses
- Understand *duress* as a defense
- Understand the difference between the subjective and objective tests of entrapment
- Know the limitations of *entrapment* as a defense
- Understand the role *intoxication* plays in excuse defenses

One function of the criminal law is to define and enforce the minimal standards of conduct that citizens living within an organized society are expected to observe. These fundamental expectations are codified by statute, and their violation is subject to punishment. As we have seen, punishing people for violating the criminal law serves multiple purposes, including retribution, deterrence, and incapacitation.

However, not all people who violate the prohibitions of the criminal law are fully responsible for their conduct and hence may not be sufficiently blameworthy to merit punishment. And, depending on the extent of and reasons for their diminished responsibility, punishing them may not promote the utilitarian objectives of the criminal law, including governing their future behavior (specific deterrence) or serving as an example to others (general deterrence). Incarcerating (or executing) individuals who cause harm but lack the ability to make rational choices or control their behavior would certainly incapacitate them, but civil commitment seems more appropriate than criminal sanctions to protect society from dangerous people who are not blameworthy if we remain true to the retributive or "just deserts" aspect of punishment.

Some conduct that ordinarily would result in criminal punishment occurs because a person is coerced into doing it. Some conduct is attributable to mistakes. Some conduct occurs because the individual committing it lacks the capacity to understand in a meaningful way what he or she is doing or that the conduct is wrong. In situations such as these, it is possible that the actor will be excused from criminal liability notwithstanding the considerable harm that his or her conduct has caused.

Excuse defenses involve instances where the defendant does not dispute committing an act that caused harm, or even that the conduct was wrong, but claims that he or she should be excused from conviction and punishment because of extenuating circumstances. They stand in contrast to justification defenses, which we considered in Chapter 4, which presume that under the circumstances the individual's otherwise unlawful behavior represented a socially approved *act* or a correct choice; the conduct should not be punished because it was fully justified. Excuse defenses focus not on the act—which is not condoned—but rather on the *actor*, who is operating under some sort of disabling condition that leads us to conclude that he or she is not to be blamed for committing the act in question. We are thus willing to excuse the individual, even though we may greatly lament the harm his or her conduct occasioned.

In this chapter we consider several excuse defenses. We examine their elements, their limitations, and the policy reasons that explain their recognition.

## Insanity

One of the most controversial areas of the criminal law is the **insanity defense**. It may seem unjust if a person who has committed an atrocious act is found not guilty of committing a crime. In June 2001 the nation reacted with horror at the news that Andrea Yates had drowned her five children, ages 7 months to 7 years, in a bathtub in her Houston-area home. Charged with murder, Yates originally was convicted and sentenced to life imprisonment. Her convictions were reversed on appeal, *Yates v. State*, 171 S.W.3d 215 (Tex. App. 2005), and a different jury found her not guilty by reason of insanity after a new trial. When we consider the goals of the criminal law and the reasons behind the insanity defense, the use of this defense becomes somewhat more comprehensible.

Punishment is designed for blameworthy individuals who violate the criminal law. Yet it is not the purpose of the criminal law to punish people who lack responsibility for their conduct and consequently are not blameworthy.

For example, we may refrain from punishing a 3-year-old child for misbehavior because the child does not understand that she has done something wrong. It seems similarly inappropriate to punish a mentally ill person whose disease robs him of the ability to comprehend or control his actions or to appreciate that what he did was wrong. Mental illness, however, is a much more nebulous condition than being 3 years old. Moreover, mental illness (a medical concept) is not synonymous with insanity (a legal concept). We will find that it has been quite challenging to (1) conceptualize and (2) identify in a specific individual the precise degree and consequences of mental illness that should result in

the legal judgment that a person should not be held criminally responsible for his or her conduct.

## M'Naghten Rule

The standard or test used to determine if a person is legally insane varies among the states and under federal law. A leading test that historically has been used throughout much of the United States is known as the **M'Naghten rule**. This rule derived from a 19th century English case and has been both lauded and criticized, as explained in *State v. Johnson*, below.

*excuse vs. Justification defenses.*

### *State v. Johnson*, 399 A.2d 469 (R.I. 1979)

Doris, Justice.

The sole issue presented by this appeal is whether this court should abandon the M'Naghten test in favor of a new standard for determining the criminal responsibility of those who claim they are blameless by reason of mental illness. For the reasons stated herein, we have concluded that the time has arrived to modernize our rule governing this subject.

Before punishing one who has invaded a protected interest, the criminal law generally requires some showing of culpability in the offender. The requirement of a *mens rea*, or guilty mind, is the most notable example of the concept that before punishment may be exacted, blameworthiness must be demonstrated. That some deterrent, restraint, or rehabilitative purpose may be served is alone insufficient. It has been stated that the criminal law reflects the moral sense of the community. "The fact that the law has, for centuries, regarded certain wrongdoers as improper subjects for punishment is a testament to the extent to which that moral sense has developed. Thus, society has recognized over the years that none of the three asserted purposes of the criminal law rehabilitation, deterrence and retribution is satisfied when the truly irresponsible, those who lack substantial capacity to control their actions, are punished." *United States v. Freeman,* 357 F.2d 606, 615 (2d Cir. 1966). The law appreciates that those who are substantially unable to restrain their conduct are, by definition, incapable of being deterred and their punishment in a correctional institution provides no example for others.

The law of criminal responsibility has its roots in the concept of free will. As Mr. Justice Jackson stated:

"How far one by an exercise of free will may determine his general destiny or his course in a particular matter and how far he is the toy of circumstance has been debated through the ages by theologians, philosophers, and scientists.

Whatever doubts they have entertained as to the matter, the practical business of government and administration of the law is obliged to proceed on more or less rough and ready judgments based on the assumption that mature and rational persons are in control of their own conduct." *Gregg Cartage & Storage Co. v. United States*, 316 U.S. 74, 79–80 (1942).

Our law proceeds from this postulate and seeks to fashion a standard by which criminal offenders whose free will has been sufficiently impaired can be identified and treated in a manner that is both humane and beneficial to society at large. The problem has been aptly described as distinguishing between those cases for which a correctional-punitive disposition is appropriate and those in which a medical-custodial disposition is the only kind that is legally permissible.

Because language is inherently imprecise and there is a wide divergence of opinion within the medical profession, no exact definition of "insanity" is possible. Every legal definition comprehends elements of abstraction and approximation that are particularly difficult to apply in marginal cases. Our inability to guarantee that a new rule will always be infallible, however, cannot justify unyielding adherence to an outmoded standard, solely at variance with contemporary medical and legal knowledge. Any legal standard designed to assess criminal responsibility must satisfy several objectives. It must accurately reflect the underlying principles of substantive law and community values while comporting with the realities of scientific understanding. The standard must be phrased in order to make fully

*(Continues)*

(*Continued*)

available to the jury such psychiatric information as medical science has to offer regarding the individual defendant, yet be comprehensible to the experts, lawyers, and jury alike. Finally, the definition must preserve to the trier of facts, be it judge or jury, its full authority to render a final decision. These considerations are paramount in our consideration of the rule to be applied in this jurisdiction in cases in which the defense of lack of criminal responsibility due to a mental illness is raised.

# I.

The historical evolution of the law of criminal responsibility is a fascinating, complex story. For purposes of this opinion, however, an exhaustive historical discussion is unnecessary; a brief sketch will therefore suffice. The renowned "right-wrong" test had antecedents in England as early as 1582. In that year the Eirenarcha, written by William Lambard of the Office of the Justices of Peace, laid down as the test of criminal responsibility "knowledge of good or evil." During the 1700's the language of the test shifted its emphasis from "good or evil" to "know." During the eighteenth century, when these tests and their progeny were evolving, psychiatry was hardly a profession, let alone a science. Belief in demonology and witchcraft was widespread and became intertwined with the law of responsibility. So eminent a legal scholar as Blackstone adamantly insisted upon the existence of witches and wizards as late as the latter half of the eighteenth century. The psychological theories of phrenology and monomania thrived and influenced the development of the "right and wrong" test. Both of these compartmentalized concepts have been soundly rejected by modern medical science which views the human personality as a fully integrated system. By historical accident, however, the celebrated case of Daniel M'Naghten froze these concepts into the common law just at the time when they were beginning to come into disrepute.

Daniel M'Naghten attempted to assassinate Sir Robert Peel, Prime Minister of England, but mistakenly shot Peel's private secretary instead. This assassination had been preceded by several attempts on the lives of members of the English Royal House, including Queen Victoria herself. When M'Naghten was tried in 1843 the jury was charged with a test heavily influenced by the enlightened work of Dr. Isaac Ray who was severely critical of the "right and wrong" rule. After the jury acquitted M'Naghten the public indignation, spearheaded by the Queen, was so pronounced that the Judges of England were summoned before the House of Lords to justify their actions. In an extraordinary advisory opinion, issued in a pressure-charged atmosphere, Lord Chief Justice Tindal, speaking for all but one of the 15 judges, reversed the charge used at trial and articulated what has become known as the M'Naghten rule. The principal rule in M'Naghten's Case, 8 Eng.Rep. 718 (1843) states:

> "To establish a defense on the ground of insanity it must be clearly proved that, at the time of committing the act, the party accused was laboring under such a defect of reason, from disease of the mind, as not to know the nature and quality of the act he was doing, or if he did know it, that he did not know that what he was doing was wrong." 8 Eng. Rep. at 722.

This dual-pronged test, issued in response to the outrage of a frightened Queen, rapidly became the predominant rule in the United States.

This jurisdiction has long adhered to the M'Naghten standard for determining criminal responsibility. . . .

The M'Naghten rule has been the subject of considerable criticism and controversy for over a century. The test's emphasis upon knowledge of right or wrong abstracts a single element of personality as the sole symptom or manifestation of mental illness. M'Naghten refuses to recognize volitional or emotional impairments, viewing the cognitive element as the singular cause of conduct. . . .

M'Naghten has been further criticized for being predicated upon an outmoded psychological concept because modern science recognizes that "insanity" affects the whole personality of the defendant, including the will and emotions. One of the most frequent criticisms of M'Naghten has been directed at its all-or-nothing approach, requiring total incapacity of cognition. We agree that:

> "Nothing makes the inquiry into responsibility more unreal for the psychiatrist than limitation of the issue to some ultimate extreme of total incapacity, when clinical experience reveals only a graded scale with marks along the way. . . .
>
> "The law must recognize that when there is no black and white it must content itself with different shades of gray." Model Penal Code, § 4.01, Comment at 158 (Tent Draft No. 4, 1955).

By focusing upon total cognitive incapacity, the M'Naghten rule compels the psychiatrist to testify in terms of unrealistic concepts having no medical meaning. Instead of scientific opinions, the rule calls for a moral or ethical judgment from the expert which judgment contributes to usurpation of the jury's function as decision maker. . . .

That these criticisms have had a pronounced effect is evidenced by the large and growing number of jurisdictions that have abandoned their former allegiance to M'Naghten in favor of the Model Penal Code formulation. We also find these criticisms persuasive and agree that M'Naghten's serious deficiencies necessitate a new approach.

Responding to criticism of M'Naghten as a narrow and harsh rule, several courts supplemented it with the "irresistible impulse" test. Under this combined approach, courts inquire into both the cognitive and volitional components of the defendant's behavior. Although a theoretical advance over the stringent right and wrong test, the irresistible impulse doctrine has also been the subject of widespread criticism. Similar to M'Naghten's absolutist view of capacity to know, the irresistible impulse is considered in terms of a complete destruction of the governing power of the mind. A more fundamental objection is that the test produces the misleading notion that a crime impulsively committed must have been perpetrated in a sudden and explosive fit. Thus, the irresistible impulse test excludes those "far more numerous instances of crimes committed after excessive brooding and melancholy by one who is unable to resist sustained psychic compulsion or to make any real attempt to control his conduct."

The most significant break in the century-old stranglehold of M'Naghten came in 1954 when the Court of Appeals for the District of Columbia declared that, "an accused is not criminally responsible if his unlawful act was the product of mental disease or mental defect." *Durham v. United States*, 214 F.2d 862, 874–75 (D.C. Cir. 1954). The **"product" test**, first pioneered by the Supreme Court of New Hampshire in *State v. Pike*, 49 N.H. 399, 402 (1869), was designed to facilitate full and complete expert testimony and to permit the jury to consider all relevant information, rather than restrict its inquiry to data relating to a sole symptom or manifestation of mental illness. *Durham* generated voluminous commentary and made a major contribution in recasting the law of criminal responsibility. In application, however, the test was plagued by significant deficiencies. The elusive, undefined concept of productivity posed serious problems of causation and gave the jury inadequate guidance. Most troublesome was the test's tendency to result in expert witnesses' usurpation of the jury function. As a result, in *Washington v. United States*, 390 F.2d 444, 455–56 (D.C. Cir. 1967), the court took the extreme step of proscribing experts from testifying concerning productivity altogether. Finally, in *United States v. Brawner*, 471 F.2d 969 (D.C. Cir. 1972), the court abandoned *Durham*, decrying the "trial by label" that had resulted. The author of *Durham*, Chief Judge Bazelon, stated that testimony couched in terms of the legal conclusion that an act was or was not the product of mental disease invited the jury to abdicate its responsibility as ultimate decision maker, and acquiesce in the experts' conclusions.

Several commentators have advocated abolition of the separate defense of lack of criminal responsibility due to a mental illness. Proponents contend that abolition would result in the responsibility issue being more properly considered as the existence vel non of the *mens rea*. Under a common proposal the criminal process would be bifurcated: first, the jury would resolve the question of guilt, and second, a panel of experts would determine the appropriate disposition. Arguably, abolition of the separate defense is subject to constitutional objections because it potentially abrogates the right to trial by jury and offends the guarantee of due process. We believe that such a drastic measure, if advisable at all, is appropriately left to the legislative process.

Responding to the criticism of the M'Naghten and irresistible impulse rules, the American Law Institute incorporated a new test of criminal responsibility into its Model Penal Code. (*See* Footnote 5.) The **Model Penal Code test** has received widespread and ever growing acceptance. It has been adopted with varying degrees of modification in 26 states and by every federal court of appeals that has addressed the issue. Although no definition can be accurately described as the perfect or ultimate pronouncement, we believe that the Model Penal Code standard represents a significant, positive improvement over our existing rule. Most importantly, it acknowledges that volitional as well as cognitive impairments must be considered by the jury in its resolution of the responsibility issue. The test replaces M'Naghten's unrealistic all-or-nothing approach with the concept of "substantial" capacity. Additionally, the test employs vocabulary sufficiently in the common ken that its use at trial will permit a reasonable three-way dialogue between the law-trained judges and lawyers, the medical-trained experts, and the jury.

*(Continues)*

(*Continued*)

Without question the essential dilemma in formulating any standard of criminal responsibility is encouraging a maximum informational input from the expert witnesses while preserving to the jury its role as trier of fact and ultimate decision maker. As one court has aptly observed:

"At bottom, the determination whether a man is or is not held responsible for his conduct is not a medical but a legal, social or moral judgment. Ideally, psychiatrists much like experts in other fields should provide grist for the legal mill, should furnish the raw data upon which the legal judgment is based. It is the psychiatrist who informs as to the mental state of the accused his characteristics, his potentialities, his capabilities. But once this information is disclosed, it is society as a whole, represented by judge or jury, which decides whether a man with the characteristics described should or should not be held accountable for his acts." *United States v. Freeman*, 357 F.2d at 619–20.

Because of our overriding concern that the jury's function remain inviolate, we today adopt the following formulation of the Model Penal Code test:

A person is not responsible for criminal conduct if at the time of such conduct, as a result of mental disease or defect, his capacity either to appreciate the wrongfulness of his conduct or to conform his conduct to the requirements of law is so substantially impaired that he cannot justly be held responsible.

The terms "mental disease or defect" do not include an abnormality manifested only by repeated criminal or otherwise antisocial conduct. (*See* Footnote 8.)

There are several important reasons why we prefer this formulation. The greatest strength of our test is that it clearly delegates the issue of criminal responsibility to the jury, thus precluding possible usurpation of the ultimate decision by the expert witnesses. Under the test we have adopted, the jury's attention is appropriately focused upon the legal and moral aspects of responsibility because it must evaluate the defendant's blameworthiness in light of prevailing community standards. Far from setting the jury at large, as in the majority Model Penal Code test the defendant must demonstrate a certain form of incapacity. That is, the jury must find that a mental disease or defect caused a substantial impairment of the defendant's capacity to appreciate the wrongfulness of his act or to conform his conduct to legal requirements. Our new test emphasizes that the degree of "substantial" impairment required is essentially a legal rather than a medical question. Where formerly under M'Naghten total incapacity was necessary for exculpation, the new standard allows the jury to find that incapacity less than total is sufficient. Because impairment is a matter of degree, the precise degree demanded is necessarily governed by the community sense of justice as represented by the trier of fact.

Several other components of our new test require elucidation. Our test consciously employs the more expansive term "appreciate" rather than "know." Implicit in this choice is the recognition that mere theoretical awareness that a certain course of conduct is wrong, when divorced from appreciation or understanding of the moral or legal impact of behavior, is of little import. A significant difference from our former rule is inclusion in the new test of the concept that a defendant is not criminally responsible if he lacked substantial capacity to conform his conduct to the requirements of law. As we noted at the outset, our law assumes that a normal individual has the capacity to control his behavior; should an individual manifest free will in the commission of a criminal act, he must be held responsible for that conduct. Mental illness, however, can effectively destroy an individual's capacity for choice and impair behavioral controls. . . .

The second paragraph of our test is designed to exclude from the concept of "mental disease or defect" the so-called psychopathic or sociopathic personality. We have included this language in our test to make clear that mere recidivism alone does not justify acquittal. We recognize that this paragraph has been the source of considerable controversy. Nevertheless, we believe that its inclusion in our test is necessary to minimize the likelihood of the improper exculpation of defendants who are free of mental disease but who knowingly and deliberately pursue a life of crime.

As we have emphasized previously, preserving the respective provinces of the jury and experts is an important concern. Consonant with modern medical understanding, our test is intended to allow the psychiatrist to place before the jury all of the relevant information that it must consider in reaching its decision. We adhere to Dean Wigmore's statement that when criminal responsibility is in issue, "Any and all conduct of the person is admissible in evidence." Nevertheless, the charge to the jury must include unambiguous instructions stressing that regardless of the nature and extent of the experts' testimony, the issue of exculpation remains at all times a legal and not a medical question. In determining the issue of responsibility the jury has two important tasks.

First, it must measure the extent to which the defendant's mental and emotional processes were impaired at the time of the unlawful conduct. The answer to that inquiry is a difficult and elusive one, but no more so than numerous other facts that a jury must find in a criminal trial. Second, the jury must assess that impairment in light of community standards of blameworthiness. The jury's unique qualifications for making that determination justify our unusual deference to the jury's resolution of the issue of responsibility. For it has been stated that the essential feature of a jury "lies in the interposition between the accused and his accuser of the commonsense judgment of a group of laymen, and in the community participation and shared responsibility that results from that group's determination of guilt or innocence." *Williams v. Florida*, 399 U.S. 78, 100 (1970). Therefore, the charge should leave no doubt that it is for the jury to determine: (1) the existence of a cognizable mental disease or defect, (2) whether such a disability resulted in a substantial impairment at the time of the unlawful conduct of the accused's capacity either to appreciate the wrongfulness of his conduct or to conform his conduct to the requirements of the law, and consequently, (3) whether there existed a sufficient relationship between the mental abnormality and the condemned behavior to warrant the conclusion that the defendant cannot justly be held responsible for his acts.

So there will be no misunderstanding of the thrust of this opinion, mention should be made of the treatment to be afforded individuals found lacking criminal responsibility due to a mental illness under the test we have adopted. Unquestionably the security of the community must be the paramount interest. Society withholds criminal sanctions out of a sense of compassion and understanding when the defendant is found to lack capacity. It would be an intolerable situation if those suffering from a mental disease or defect of such a nature as to relieve them from criminal responsibility were to be released to continue to pose a threat to life and property. The General Laws provide that a person found not guilty because he was "insane" at the time of the commission of a crime shall be committed to the Director of the State Department of Mental Health for observation. At a subsequent judicial hearing if he is found to be dangerous, the person must be committed to a public institution for care and treatment. This procedure insures society's protection and affords the incompetent criminal offender necessary medical attention.

Our test as enunciated in this opinion shall apply to all trials commenced after the date of this opinion. The defendant in the instant case is entitled to a new trial solely on the issue of criminal responsibility. The defendant's appeal is sustained and the case is remanded to the Superior Court for a new trial in accordance with the opinions expressed herein.

## Retreat from M'Naghten

The M'Naghten test has been criticized on several fronts. One theme is the unforgiving, absolutist nature of the test. The M'Naghten rule can be interpreted to require total impairment—e.g., "did not know"—as opposed to excusing an individual whose capacities were "substantially impaired"—the standard preferred by the Rhode Island Supreme Court in *Johnson*.

Another criticism of the M'Naghten rule is its purely cognitive perspective. Under the rule the insanity defense fails if a person knows the nature and quality of the act committed and knows that it is wrong. Although the term "know" is ambiguous, courts have treated it as meaning intellectual awareness or understanding. It has been argued that this meaning is too restrictive for determining a person's "sanity" or legal responsibility for a crime. This is because even if a person knows what he or she is doing, the person may not be able to control himself or herself from doing it. The Model Penal Code (MPC) test (and the Rhode Island variation) includes an "irresistible impulse" component that accounts for this possibility. This so-called "volitional prong" permits a defense of insanity to succeed based on disease-related impediments to behavioral control, even if the defendant "knew" (the "cognitive prong") the nature and consequences of his or her action and that it was wrong.

Another item to be considered involves deciding whether the defendant knew his or her act was "wrong"; should this term refer to the defendant's understanding that the act violated the law (was "legally" wrong), or should it mean that the defendant knew that the act was "morally" wrong, which arguably connotes a deeper level of appreciation. *See United States v. Ewing*, 494 F.3d 607 (7th Cir. 2007), *cert. den.*, 128 S. Ct. 925 (2008). How might it be possible for a mentally ill defendant to know, for example, that there is a law against murder yet still maintain that killing another person is not morally wrong?

In response to the harshness of the M'Naghten rule, for a period of time the Circuit Court for the District of Columbia followed what is known as the *Durham* rule or the "disease-product" test for insanity. Under this test, as discussed in *Johnson*, the jury would simply be asked whether the accused's act "was the product of mental disease or defect." *Durham v. United States*, 214 F.2d 862, 874–875 (D.C. Cir. 1954). If it was, a valid insanity defense existed. A problem soon revealed with this "but-for" test was that expert testimony by psychiatrists regarding whether the defendant's conduct was a product of a mental disease or defect threatened to supplant the jury's role in determining criminal responsibility. As we have noted, a verdict of not guilty by reason of insanity is a legal judgment that involves an assessment of the defendant's blameworthiness; it is not a determination properly made by medical professionals even if they are trained in psychiatry. The D.C. Circuit used this test for nearly 20 years before abandoning it in 1972. *United States v. Brawner*, 471 F.2d 969 (D.C. Cir. 1972). New Hampshire is the only state that currently follows a version of the "disease-product" test for insanity.

* * *

In 1982 a jury in Washington, D.C., found John Hinckley, Jr. not guilty by reason of insanity of all charges related to the assassination attempt on President Ronald Reagan. The MPC test for insanity was used at Hinckley's trial. The verdict was incomprehensible to many people who had viewed television replays of the assassination attempt and was roundly denounced in many quarters. The backlash from this verdict led to Congress's enactment of the **Federal Insanity Defense Reform Act** in 1984, 18 U.S.C. § 17. This statute makes it more difficult for a defendant to be found insane than the tests under the MPC and M'Naghten. The statute governs all trials in federal court that involve the insanity defense. It does not apply to state trials, although several states now have modeled their insanity tests after the federal standard. Under the statute, the defense must prove by clear and convincing evidence that "at the time of the commission of the acts constituting the offense, the defendant, as a result of a severe mental disease or defect, was unable to appreciate the nature and quality or the wrongfulness of his acts."

Several factors make this a difficult standard to meet. First, only *severe* mental diseases or defects qualify as predicates for the insanity defense; less disabling infirmities do not suffice. Second, under the test the defendant must have been *unable* to appreciate right from wrong, which is a repudiation of the more forgiving "lacks the substantial capacity" approach reflected in the MPC standard. Third, again deviating from the MPC, the federal test eliminates the "volitional" prong and is *exclusively cognitive*. Finally, the *defendant* must prove each element of the test by *clear and convincing evidence*. In many jurisdictions the government is required to prove beyond a reasonable doubt that the defendant should not be acquitted by reason of insanity (as was true at Hinckley's trial) after the defendant puts insanity at issue. And, where the burden of persuasion on insanity is shifted to the defendant, in many jurisdictions that burden can be met by a preponderance of the evidence rather than the federal test's more demanding "clear and convincing evidence" standard. Who bears the burden of persuasion, and what that burden is, can be critically important to the outcome of insanity trials. Conflicting testimony is often presented about the defendant's mental status where sanity is contested, and when the jury is uncertain or confused about which evidence to believe, the party with the burden of persuasion will lose.

We learn more about the federal test for insanity in *United States v. Waagner*.

---

## United States v. Waagner, 319 F.3d 962 (7th Cir. 2003)

Terence T. Evans, Circuit Judge.

Clayton L. Waagner says that after his daughter suffered a miscarriage, he heard a voice ask "how could [he] grieve so hard over this one when millions are killed, or murdered, every year." Waagner said the voice, which only he could hear, belonged to God and that it went on to say "I have called you to be my warrior and I want you to go to war against the abortion industry." Describing himself as a "warrior for pre-born children," Waagner embarked on what ultimately became a 2-year, cross-country crime spree. The spree included staking out abortion clinics, stealing a 4-wheel drive Yukon on which he logged 30,000 miles, stealing a Winnebago motor home, robbing gas stations, burglarizing residences, and even evading Pennsylvania state troopers after a high-speed chase. Waagner, a convicted felon, stole firearms during his burglaries, and he went to great lengths to avoid apprehension, going so far as downloading police frequencies from the Internet and storing them on a CD-ROM so he could monitor police movements on a scanner.

Waagner was eventually apprehended by an Illinois state trooper in September of 1999 and subsequently charged with a couple of federal offenses—possession of a firearm by a felon and possession of a stolen motor vehicle which crossed state lines. *See* 18 U.S.C. § 922(g)(1) and 18 U.S.C. § 2313(a).

It is, of course, certainly not surprising that someone who claims to hear bizarre commands from God and then embarks on a massive crime spree has more than a few mental problems. And Waagner did. This became clear when he filed a notice of intent to raise an insanity defense to the charges against him. After filing his notice, the government requested a psychiatric examination, and Waagner was evaluated by Dr. Daniel S. Greenstein, a clinical psychologist whose diagnosis was adjustment disorder, delusional disorder grandiose type, and antisocial personality disorder. Dr. Greenstein testified at Waagner's trial that the diagnoses of adjustment disorder and antisocial personality disorder were not severe mental diseases or defects. However, he testified that Waagner's "delusional disorder grandiose type" is a severe mental disease and that he would not necessarily be able to appreciate the wrongfulness of his actions. The insanity defense primarily rested on this opinion, but the jury didn't buy it, as Waagner was convicted on both counts.

Despite his loss at the trial, Waagner proved to be a tough nut to crack: he escaped from the DeWitt County jail in Clinton, Illinois, where he was in custody awaiting sentencing. The escape led to 9 more months of freedom before he was arrested again in December of 2001.

Back in court, Waagner pled guilty to a charge of escape under 18 U.S.C. § 751(a). Later, he was sentenced on all three counts of conviction. The district judge found that Waagner was an armed career criminal who possessed firearms in connection with crimes of violence, that he obstructed justice by escaping and, finally, that he had not shown acceptance of responsibility for either the original charges or the escape. Based on these findings, Waagner was sentenced to 327 months imprisonment on the gun charge, 120 months on the stolen vehicle charge, and 37 months on the escape charge. The gun and stolen vehicle sentences were ordered to run concurrent, but the escape sentence was consecutive—resulting in a total sentence of 364 months.

On this appeal, Waagner seeks a new trial on the original charges. His argument rests on a claim that the jury clearly erred in concluding that he failed to meet his burden of proof on the insanity defense and that the court erred in failing to give the jury a requested instruction on the consequences that would flow from a finding of not guilty by reason of mental defect. Failing that, Waagner tacks on a trio of challenges to the sentence imposed by the district judge.

Waagner's first claim, that the jury clearly erred in failing to find him not guilty by reason of insanity, is an offshoot of a routine challenge to the sufficiency of the evidence argument. As we have noted, a defendant making an ordinary sufficiency challenge "faces a nearly insurmountable hurdle [in that we will] consider the evidence in the light most favorable to the Government, defer to the credibility determination of the jury, and overturn a verdict only when the record contains no evidence, regardless of how it is weighed, from which the jury could find guilt beyond a reasonable doubt." *United States v. Szarwark*, 168 F.3d 993, 995 (7th Cir. 1999) (quoting *United States v. Moore*, 115 F.3d 1348, 1363 (7th Cir. 1997)).

In an insanity case, unlike a typical challenge to the sufficiency of the evidence, the defendant's burden is even greater. At trial, it is the defendant, not the government, that must carry the burden of proving insanity (which is an affirmative defense) by clear and convincing evidence. 18 U.S.C. § 17(b). Because, under 704(b) of the Federal Rules of Evidence, legal insanity is a question to be decided by the trier of fact, the finding here by the jury may not be disturbed unless it is clearly erroneous. *United States v. Reed*, 997 F.2d 332, 334 (7th Cir. 1993).

To succeed on an insanity defense, a defendant must prove that as the result of a severe mental disease or defect he was unable to appreciate the nature and quality or wrongfulness of his acts. 18 U.S.C. § 17(a). So the question becomes, what was the evidence the jury considered, and was it so one-sided that any decision except a finding of not guilty by reason of mental defect must be cast aside? We think not.

Dr. Greenstein offered the only expert opinion evidence in this case. But on cross-examination, the doctor testified that whether Waagner even had a delusional disorder was a very close call, made more difficult because it involved religious beliefs. He noted that a person would not necessarily be delusional simply because the person believed, based on strongly held religious beliefs, that killing abortion doctors was morally justified. Because it was a close call, Dr. Greenstein testified that he felt ethically obliged to err in favor of diagnosing Waagner with a delusional disorder. Plus, the doctor candidly acknowledged that on the same facts a different psychologist might come to a different conclusion.

Dr. Greenstein's equivocal testimony is not surprising given that none of the objective evidence pointed very strongly toward a finding that Waagner had a severe mental disorder. Likewise, the diagnosis was further drawn into question because Waagner, despite an extensive criminal record, had no recorded history of psychiatric illness.

Dr. Greenstein's diagnosis of delusional disorder is even more dubious given that his evaluation was based mainly on Waagner's own self-reported, and we think self-serving, statements. Dr. Greenstein did not interview any other persons regarding Waagner's mental condition. Indeed, the evidence showed that Waagner failed to mention the "voices" he heard to others, particularly his wife, son, and a partner in crime who testified for the government and said he never observed Waagner acting irrationally. Likewise, various officers testified that they had no difficulty communicating with Waagner, that he acted rationally, responded to their questions, never appeared distracted, and never mentioned abortion or abortion clinics to them. And so, the evidence that Waagner suffered from a severe mental

*(Continues)*

disease or defect when he committed his crimes was far from clear and convincing. In fact, it seems to us that the jury reached the correct result, not only one that was not clearly erroneous.

But even if the evidence of a disqualifying mental disorder was stronger, Waagner would still be a long way away from steering his boat into the acquittal harbor. That's because the mental disorder question is the lesser of two things to be proved to carry the affirmative defense of insanity. He had to prove that his criminal conduct was the result of his mental disorder, not the result of something which seems more likely in this case, an antisocial personality disorder. Even if he truly believed that he was "God's warrior" against the abortion industry, that does not mean that his criminal conduct was a result of his delusions. Regarding his possession of the stolen motor vehicle, Waagner admitted that God never told him to steal a vehicle, let alone a Winnebago motor home. He further admitted that he needed a big vehicle because of his large family, that the Winnebago fulfilled that purpose, and that he was not going to drive to an abortion clinic with his wife in the Winnebago. Regarding his possession of the firearms, he also admitted that God did not specifically tell him that he needed to steal a Beretta .22 caliber pistol during a burglary.

Waagner also admitted that he had committed the crimes of unlawful possession of a firearm by a felon and possession of stolen goods in 1993, at least a half dozen years before his alleged delusions struck home. Therefore, a jury could reasonably conclude that Waagner, who admitted committing these very crimes without God's urging, possessed the firearms and the stolen Winnebago as a result of either a desire to remain a fugitive, or as a result of his antisocial personality disorder, or both, and not because he was in the grip of some bizarre delusions.

Finally, it seems to us that the evidence was overwhelming that Waagner appreciated the wrongfulness of his conduct. First, he gave the arresting Illinois state trooper false identification, and he lied about how he happened to be driving the Winnebago. Later, after confessing, he even asked the trooper if he could work out a deal. He was also selective in what he would discuss, declining to talk about his involvement in a Kentucky armed robbery. This all points to a man who knew he was in a jam and wanted to avoid, or minimize if possible, his criminal liability. This is not the stuff of a man whose actions are uncontrollable because of a severe delusional disorder.

In the face of all this evidence, and a lot of additional evidence we have not discussed (for instance, in 1999 he chose not to try to murder a doctor when a police officer arrived at an abortion clinic), Waagner's primary argument seems to be that he should win on the issue because the government "failed to present any expert testimony." Given the nature of Dr. Greenstein's testimony and the other evidence of Waagner's sanity at the time of the offense, the government was not required to present any expert testimony of its own. *United States v. Bennett*, 908 F.2d 189, 195 (7th Cir. 1990) ("The government is not required to rebut expert testimony with its own expert as it may accomplish the same result by presenting lay witnesses and other evidence and by undermining the defense expert's credibility through cross examination."). There was, in other words, no need for the government to get its own expert and bog the trial down to what often happens in cases like this—a "battle of experts" before a jury of lay people. Reviewing the entire record in this case, we conclude that the jury did not err, let alone clearly err, in finding that Waagner failed to prove the affirmative defense of insanity by clear and convincing evidence.

Waagner next argues that the district court erred in refusing his proffered insanity instruction which would have told the jury he would be committed to a "suitable facility" if he was found not guilty by reason of insanity. This instruction was necessary, he contends, because his prior bad acts and mental state would have caused the jury to believe he might be released and return to his anti-abortion mission, if his insanity defense was accepted. We review *de novo* a district court's decision to not give a jury instruction. *See United States v. Andreas*, 216 F.3d 645, 668–69 (7th Cir. 2000).

The Seventh Circuit pattern jury instructions for federal criminal trials address this issue. The committee comment to Instruction 6.02 states, "In *Shannon v. United States*, 512 U.S. 573 (1994), the Supreme Court held that a jury may be instructed on [the] automatic commitment requirement of § 4243, but only to counteract inaccurate or misleading information presented to the jury during trial." Waagner does not argue that the government presented inaccurate or misleading information to the jury, so there was no reason for the district judge to give his proposed instruction. *See, e.g., United States v. Fisher*, 10 F.3d 115, 122–23 (3d Cir. 1993) (consequences instruction unnecessary when prosecutor did not suggest that defendant would be a danger to the community if found insane); *United States v. Thigpen*, 4 F.3d 1573, 1578 (11th Cir. 1993) (consequences instruction appropriate if government presents inadmissible evidence, argument, or questions implying that defendant will be released back into society if found insane). And this leads us to the sentencing issues, which require little comment.

\* \* \*

For these reasons, the judgment of the district court is Affirmed.

## Notes and Questions

1. A handful of states, including Idaho, Kansas, Montana, and Utah, have essentially eliminated the insanity defense by making evidence of mental illness relevant only if it interferes with a criminal defendant's ability to formulate the *mens rea* requirement of the charged offense. Although not squarely confronting the constitutionality of abolishing the insanity defense, the Supreme Court has ruled that the states have wide latitude regarding the specifics of this defense. *Clark v. Arizona,* 548 U.S. 735 (2006). What do you think about the advisability of narrowly limiting the availability of a defense based on mental illness to the accused's ability to form the *mens rea* comprising an element of the charged crime?[1]

2. Several states have enacted laws authorizing a verdict of **guilty but mentally ill (GBMI)** as an alternative to not guilty by reason of insanity. A GBMI verdict signifies that the defendant was mentally ill at the time of the crime yet still criminally responsible for his or her conduct. Unlike a verdict of not guilty by reason of insanity, GBMI represents a guilty verdict and entails a conviction, so the defendant is subject to the full punishment authorized by the criminal law. The defendant may or may not qualify for treatment for mental illness after a GBMI verdict. Critics complain that GBMI statutes mislead juries into believing they are issuing "compromise verdicts" that are not materially different from regular guilty verdicts and result in punishing individuals who are not truly responsible for their conduct and should be acquitted by reason of insanity. Proponents argue that GBMI verdicts help clarify that mental illness is not synonymous with insanity and help safeguard society from dangerous people while allowing juries appropriately to hold people criminally responsible for their conduct. *See, e.g., State v. Neely,* 876 P.2d 222 (N.M. 1994).[2]

3. It is important to understand that competence to stand trial is different from insanity. A defendant is considered competent to stand trial if, at the time of the trial, he or she is able to understand the charges filed and to assist counsel in preparing and presenting a defense. Due process prohibits the trial of a person who lacks such minimal competency. *See Cooper v. Oklahoma,* 517 U.S. 348 (1996); *Pate v. Robinson,* 383 U.S. 375 (1966). Insanity relates to a different time period—when the act was committed, as opposed to when the trial is to begin—and, as we have seen, involves very different criteria. Because a trial cannot take place if the defendant is incompetent to be tried, no judge or jury can even consider an insanity defense under those circumstances. Thus, somewhat confusingly perhaps, only "competent" defendants can be found "insane."

## Mistake

People make mistakes all the time. Luckily, it is rare that a mistake leads to a criminal act, let alone a criminal prosecution. From time to time it does happen. In this section we consider the circumstances under which a person's mistake qualifies as a defense to a charged crime.

We consider different kinds of mistakes: mistakes of law and mistakes of fact. A mistake of law is presented when a person does not know the proper meaning of a law or otherwise misinterprets it. A mistake of fact involves a misunderstanding about the true nature of events (other than the law) that, in this context, results in the commission of an act that ordinarily would be a crime. Although neither type of mistake guarantees that a defendant will be acquitted, we will see that the law generally is much more reluctant to excuse conduct that results from a mistake about the meaning of the law than a mistake based on a misunderstanding of the facts.

### Mistake of Law

There is an old saying that ignorance of the law is no excuse. As we saw in Chapter 2, and particularly in our consideration of *Lambert v. California*, that statement is not always accurate. There are times, rare as they may be, when a person's ignorance or mistake about a law excuses otherwise criminal conduct. The primary reasons for limiting the use of a **mistake of law** defense are the dubious policy implications of "rewarding" people who make little effort to understand the law by recognizing such a defense and because people can easily fabricate such an explanation. If mistake of law were routinely recognized as a defense, people would have little incentive to understand legal requirements and anybody charged with a crime could maintain, "But I did not know it was against the law," and thus be relieved of liability.

Although rare, under limited instances mistake of law may qualify as a defense. If a person relies on information provided by a responsible government official or law enforcement officer and it turns out that the information was incorrect, mistake of law may be recognized as a defense.

A defendant may also be excused from a criminal violation if he or she took reasonable steps to learn about the specific law before acting. What comprises reasonable steps in this context is debatable. Consider the circumstances presented in *People v. Marrero.*

## People v. Marrero, 507 N.E.2d 1068 (N.Y. 1987)

Bellacosa, Judge.

[Julio Marrero was employed as a corrections officer at a federal prison in Connecticut. He was arrested in New York City for carrying a concealed weapon without a permit. New York Penal Law § 265.20(A)(1)(a) exempted "peace officers" from this permit requirement. Under New York Criminal Procedure Law § 1.20(33)(h), "peace officer" includes: "An attendant, or an official, or guard of any state prison or of any penal correctional institution." Marrero argued that the statutory exemption provided to a "guard . . . of *any penal correctional institution*" reasonably could be understood to apply to his status as a guard in a federal prison.]

The defense of mistake of law is not available to a Federal corrections officer arrested in a Manhattan social club for possession of a loaded .38 caliber automatic pistol who claimed he mistakenly believed he was entitled, pursuant to the interplay of CPL 2.10, 1.20 and Penal Law § 265.20, to carry a handgun without a permit as a peace officer.

In a prior phase of this criminal proceeding, defendant's motion to dismiss the indictment upon which he now stands convicted was granted; then it was reversed and the indictment reinstated by a divided Appellate Division. . . .

On the trial of the case, the court rejected the defendant's argument that his personal misunderstanding of the statutory definition of a peace officer is enough to excuse him from criminal liability under New York's mistake of law statute (Penal Law § 15.20). The court refused to charge the jury on this issue and defendant was convicted of criminal possession of a weapon in the third degree. We affirm the Appellate Division order, upholding the conviction.

Defendant was a Federal corrections officer in Danbury, Connecticut, and asserted that status at the time of his arrest in 1977. He claimed at trial that there were various interpretations of fellow officers and teachers, as well as the peace officer statute itself, upon which he relied for his mistaken belief that he could carry a weapon with legal impunity.

The starting point for our analysis is the New York mistake statute as an outgrowth of the dogmatic common-law maxim that ignorance of the law is no excuse. The central issue is whether defendant's personal misreading or misunderstanding of a statute may excuse criminal conduct in the circumstances of this case.

The common-law rule on mistake of law was clearly articulated in *Gardner v. People*, 62 N.Y. 299. In *Gardner*, the defendants misread a statute and mistakenly believed that their conduct was legal. The court insisted, however, that the "mistake of law" did not relieve the defendants of criminal liability. The statute at issue, relating to the removal of election officers, required that prior to removal, written notice must be given to the officer sought to be removed. The statute provided one exception to the notice requirement: "removal . . . shall only be made after notice in writing . . . unless made while the inspector is actually on duty on a day of registration, revision of registration, or election, and for improper conduct." The defendants construed the statute to mean that an election officer could be removed without notice for improper conduct at any time. The court ruled that removal without notice could only occur for improper conduct on a day of registration, revision of registration or election.

In ruling that the defendant's misinterpretation of the statute was no defense, the court said: "The defendants made a mistake of law. Such mistakes do not excuse the commission of prohibited acts. 'The rule on the subject appears to be, that in acts mala in se, the intent governs, but in those mala prohibita, the only inquiry is, has the law been violated?' The act prohibited must be intentionally done. A mistake as to the fact of doing the act will excuse the party, but if the act is intentionally done, the statute declares it a misdemeanor, irrespective of the motive or intent. . . . The evidence offered [showed] that the defendants were of [the] opinion that the statute did not require notice to be given before removal. This opinion, if entertained in good faith, mitigated the character of the act, but was not a defence [sic]" (*Gardner v. People*, 62 N.Y. 299, 304, *supra*). . . .

The desirability of the Gardner-type outcome, which was to encourage the societal benefit of individuals' knowledge of and respect for the law, is underscored by Justice Holmes' statement: "It is no doubt true that there are many cases in which the criminal could not have known that he was breaking the law, but to admit the excuse at all would be to encourage ignorance where the law-maker has determined to make men know and obey, and justice to the individual is rightly outweighed by the larger interests on the other side of the scales" (Holmes, The Common Law, at 48 [1881]).

The revisors of New York's Penal Law intended no fundamental departure from this common-law rule in Penal Law § 15.20, which provides in pertinent part:

"§ 15.20. Effect of ignorance or mistake upon liability.

\* \* \*

"2. A person is not relieved of criminal liability for conduct because he engages in such conduct under a mistaken belief that it does not, as a matter of law, constitute an offense, unless such mistaken belief is founded upon an official statement of the law contained in (a) a statute or other enactment . . . (d) an interpretation of the statute or law relating to the offense, officially made or issued by a public servant, agency, or body legally charged or empowered with the responsibility or privilege of administering, enforcing or interpreting such statute or law.". . .

The defendant claims as a first prong of his defense that he is entitled to raise the defense of mistake of law under section 15.20(2)(a) because his mistaken belief that his conduct was legal was founded upon an official statement of the law contained in the statute itself. Defendant argues that his mistaken interpretation of the statute was reasonable in view of the alleged ambiguous wording of the peace officer exemption statute, and that his "reasonable" interpretation of an "official statement" is enough to satisfy the requirements of subdivision (2)(a). However, the whole thrust of this exceptional exculpatory concept, in derogation of the traditional and common-law principle, was intended to be a very narrow escape valve. Application in this case would invert that thrust and make mistake of law a generally applied or available defense instead of an unusual exception which the very opening words of the mistake statute make so clear, i.e., "A person is not relieved of criminal liability for conduct . . . unless" (Penal Law § 15.20). The momentarily enticing argument by defendant that his view of the statute would only allow a defendant to get the issue generally before a jury further supports the contrary view because that consequence is precisely what would give the defense the unintended broad practical application.

The prosecution further counters defendant's argument by asserting that one cannot claim the protection of mistake of law under section 15.20(2)(a) simply by misconstruing the meaning of a statute but must instead establish that the statute relied on actually permitted the conduct in question and was only later found to be erroneous. To buttress that argument, the People analogize New York's official statement defense to the approach taken by the Model Penal Code (MPC). Section 2.04 of the MPC provides:

"Section 2.04. Ignorance or Mistake.

\* \* \*

"(3) A belief that conduct does not legally constitute an offense is a defense to a prosecution for that offense based upon such conduct when . . . (b) he acts in reasonable reliance upon an official statement of the law, afterward determined to be invalid or erroneous, contained in (i) a statute or other enactment."

Although the drafters of the New York statute did not adopt the precise language of the Model Penal Code provision with the emphasized clause, it is evident and has long been believed that the Legislature intended the New York statute to be similarly construed. In fact, the legislative history of section 15.20 is replete with references to the influence of the Model Penal Code provision. . . .

It was early recognized that the "official statement" mistake of law defense was a statutory protection against prosecution based on reliance of a statute that did in fact authorize certain conduct. "It seems obvious that society must rely on some statement of the law, and that conduct which is in fact 'authorized' . . . should not be subsequently condemned. The threat of punishment under these circumstances can have no deterrent effect unless the actor doubts the validity of the official pronouncement—a questioning of authority that is itself undesirable." While providing a narrow escape hatch, the idea was simultaneously to encourage the public to read and rely on official statements of the law, not to have individuals conveniently and personally question the validity and interpretation of the law and act on that basis. If later the statute was invalidated, one who mistakenly acted in reliance on the authorizing statute would be relieved of criminal liability. That makes sense and is fair. To go further does not make sense and would create a legal chaos based on individual selectivity.

*(Continues)*

*(Continued)*

In the case before us, the underlying statute never in fact authorized the defendant's conduct; the defendant only thought that the statutory exemptions permitted his conduct when, in fact, the primary statute clearly forbade his conduct. Moreover, by adjudication of the final court to speak on the subject in this very case, it turned out that even the exemption statute did not permit this defendant to possess the weapon. It would be ironic at best and an odd perversion at worst for this court now to declare that the same defendant is nevertheless free of criminal responsibility.

The "official statement" component in the mistake of law defense in both paragraphs (a) and (d) adds yet another element of support for our interpretation and holding. Defendant tried to establish a defense under Penal Law § 15.20(2)(d) as a second prong. But the interpretation of the statute relied upon must be "officially made or issued by a public servant, agency or body legally charged or empowered with the responsibility or privilege of administering, enforcing or interpreting such statute or law." We agree with the People that the trial court also properly rejected the defense under Penal Law § 15.20(2)(d) since none of the interpretations which defendant proffered meets the requirements of the statute. . . .

It must also be emphasized that, while our construction of Penal Law § 15.20 provides for narrow application of the mistake of law defense, it does not, as the dissenters contend, "rule out any defense based on mistake of law." To the contrary, mistake of law is a viable exemption in those instances where an individual demonstrates an effort to learn what the law is, relies on the validity of that law and, later, it is determined that there was a mistake in the law itself.

The modern availability of this defense is based on the theory that where the government has affirmatively, albeit unintentionally, misled an individual as to what may or may not be legally permissible conduct, the individual should not be punished as a result. This is salutary and enlightened and should be firmly supported in appropriate cases. However, it also follows that where, as here, the government is not responsible for the error (for there is none except in the defendant's own mind), mistake of law should not be available as an excuse.

We recognize that some legal scholars urge that the mistake of law defense should be available more broadly where a defendant misinterprets a potentially ambiguous statute not previously clarified by judicial decision and reasonably believes in good faith that the acts were legal. Professor Perkins, a leading supporter of this view, has said: "[i]f the meaning of a statute is not clear, and has not been judicially determined, one who has acted 'in good faith' should not be held guilty of crime if his conduct would have been proper had the statute meant what he 'reasonably believed' it to mean, even if the court should decide later that the proper construction is otherwise." (Perkins, Ignorance and Mistake in Criminal Law, 88 U.Pa.L.Rev. 35, 45.) . . .

We conclude that the better and correctly construed view is that the defense should not be recognized, except where specific intent is an element of the offense or where the misrelied-upon law has later been properly adjudicated as wrong. Any broader view fosters lawlessness. It has been said in support of our preferred view in relation to other available procedural protections: "A statute . . . which is so indefinite that it 'either forbids or requires the doing of an act in terms so vague that men of common intelligence must necessarily guess at its meaning and differ as to its application, violates the first essential of due process of law' and is unconstitutional. If the court feels that a statute is sufficiently definite to meet this test, it is hard to see why a defense of mistake of law is needed. Such a statute could hardly mislead the defendant into believing that his acts were not criminal, if they do in fact come under its ban. . . . [I]f the defense of mistake of law based on indefiniteness is raised, the court is . . . going to require proof . . . that the act was sufficiently definite to guide the conduct of reasonable men. Thus, the need for such a defense is largely supplied by the constitutional guarantee."

Strong public policy reasons underlie the legislative mandate and intent which we perceive in rejecting defendant's construction of New York's mistake of law defense statute. If defendant's argument were accepted, the exception would swallow the rule. Mistakes about the law would be encouraged, rather than respect for and adherence to law. There would be an infinite number of mistake of law defenses which could be devised from a good-faith, perhaps reasonable but mistaken, interpretation of criminal statutes, many of which are concededly complex. Even more troublesome are the opportunities for wrong minded individuals to contrive in bad faith solely to get an exculpatory notion before the jury. These are not in terrorem arguments disrespectful of appropriate adjudicative procedures; rather, they are the realistic and practical consequences were the dissenters' views to prevail. Our holding comports with a statutory scheme which was not designed to allow false and diversionary stratagems to be provided for many more cases than the statutes contemplated.

This would not serve the ends of justice but rather would serve game playing and evasion from properly imposed criminal responsibility.

Accordingly, the order of the Appellate Division should be affirmed.

Hancock, Judge (dissenting). . . .

The basic difference which divides the court may be simply put. Suppose the case of a man who has committed an act which is criminal not because it is inherently wrong or immoral but solely because it violates a criminal statute. He has committed the act in complete good faith under the mistaken but entirely reasonable assumption that the act does not constitute an offense because it is permitted by the wording of the statute. Does the law require that this man be punished? The majority says that it does and holds that (1) Penal Law § 15.20(2)(a) must be construed so that the man is precluded from offering a defense based on his mistake of law and (2) such construction is compelled by prevailing considerations of public policy and criminal jurisprudence. We take issue with the majority on both propositions.

There can be no question that under the view that the purpose of the criminal justice system is to punish blameworthiness or "choosing freely to do wrong," our supposed man who has acted innocently and without any intent to do wrong should not be punished. Indeed, under some standards of morality he has done no wrong at all. Since he has not knowingly committed a wrong there can be no reason for society to exact retribution. Because the man is law-abiding and would not have acted but for his mistaken assumption as to the law, there is no need for punishment to deter him from further unlawful conduct. Traditionally, however, under the ancient rule of Anglo-American common law that ignorance or mistake of law is no excuse, our supposed man would be punished.

The maxim "ignorantia legis neminem excusat" finds its roots in Medieval law when the "actor's intent was irrelevant since the law punished the act itself" and when, for example, the law recognized no difference between an intentional killing and one that was accidental. Although the common law has gradually evolved from its origins in Anglo-Germanic tribal law (adding the element of intent [*mens rea*] and recognizing defenses based on the actor's mental state—*e.g.*, justification, insanity and intoxication) the dogmatic rule that ignorance or mistake of law is no excuse has remained unaltered. Various justifications have been offered for the rule, but all are frankly pragmatic and utilitarian—preferring the interests of society (*e.g.*, in deterring criminal conduct, fostering orderly judicial administration, and preserving the primacy of the rule of law) to the interest of the individual in being free from punishment except for intentionally engaging in conduct which he knows is criminal.

Today there is widespread criticism of the common-law rule mandating categorical preclusion of the mistake of law defense. The utilitarian arguments for retaining the rule have been drawn into serious question, but the fundamental objection is that it is simply wrong to punish someone who, in good-faith reliance on the wording of a statute, believed that what he was doing was lawful. It is contrary to "the notion that punishment should be conditioned on a showing of subjective moral blameworthiness." This basic objection to the maxim "ignorantia legis neminem excusat" may have had less force in ancient times when most crimes consisted of acts which by their very nature were recognized as evil (malum in se). In modern times, however, with the profusion of legislation making otherwise lawful conduct criminal (malum prohibitum), the "common law fiction that every man is presumed to know the law has become indefensible in fact or logic."

With this background we proceed to a discussion of our disagreement with the majority's construction of Penal Law § 15.20(2)(a) and the policy and jurisprudential arguments made in support of that construction. . . .

Defendant stands convicted after a jury trial of criminal possession of a weapon in the third degree for carrying a loaded firearm without a license (Penal Law § 265.02). He concedes that he possessed the unlicensed weapon but maintains that he did so under the mistaken assumption that his conduct was permitted by law. Although at the time of his arrest he protested that he was a Federal corrections officer and exempt from prosecution under the statute, defendant was charged with criminal possession of a weapon in the third degree. On defendant's motion before trial the court dismissed the indictment, holding that he was a peace officer as defined by CPL 2.10(26) and, therefore, exempted by Penal Law § 265.20 from prosecution under Penal Law § 265.02. The People appealed and the Appellate Division reversed and reinstated the indictment by a 3–2 vote. . . . The trial court rejected defendant's efforts to establish a defense of mistake of law under Penal Law § 15.20(2)(a). He was convicted and the Appellate Division has affirmed.

Defendant's mistaken belief that, as a Federal corrections officer, he could legally carry a loaded weapon without a license was based on the express exemption from criminal liability under Penal Law § 265.02 accorded in Penal Law § 265.20(a)(1)(a) to "peace officers" as defined in the Criminal

(Continues)

Procedure Law and on his reading of the statutory definition for "peace officer" in CPL 2.10(26) as meaning a correction officer "of any penal correctional institution," including an institution not operated by New York State. Thus, he concluded erroneously that, as a corrections officer in a Federal prison, he was a "peace officer" and, as such, exempt by the express terms of Penal Law § 265.20(a)(1)(a).

This mistaken belief, based in good faith on the statute defining "peace officer" (CPL 2.10[26]), is, defendant contends, the precise sort of "mistaken belief . . . founded upon an official statement of the law contained in . . . a statute or other enactment" which gives rise to a mistake of law defense under Penal Law §15.20(2)(a). He points out, of course, that when he acted in reliance on his belief he had no way of foreseeing that a court would eventually resolve the question of the statute's meaning against him and rule that his belief had been mistaken, as three of the five-member panel at the Appellate Division ultimately did in the first appeal.

The majority, however, has accepted the People's argument that to have a defense under Penal Law § 15.20(2)(a) "a defendant must show that the statute permitted his conduct, not merely that he believed it did." Here, of course, defendant cannot show that the statute permitted his conduct. To the contrary, the question has now been decided by the Appellate Division and it is settled that defendant was not exempt under Penal Law § 265.20(a)(1)(a). Therefore, the argument goes, defendant can have no mistake of law defense. While conceding that reliance on a statutory provision which is later found to be invalid would constitute a mistake of law defense (see, Model Penal Code § 2.04 [3][b][i]), the People's flat position is that "one's mistaken reading of a statute, no matter how reasonable or well intentioned, is not a defense."

Nothing in the statutory language suggests the interpretation urged by the People and adopted by the majority: that Penal Law § 15.20(2)(a) is available to a defendant not when he has mistakenly read a statute but only when he has correctly read and relied on a statute which is later invalidated. Such a construction contravenes the general rule that penal statutes should be construed against the State and in favor of the accused and the Legislature's specific directive that the revised Penal Law should not be strictly construed but "must be construed according to the fair import of [its] terms to promote justice and effect the objects of the law" (Penal Law § 5.00).

More importantly, the construction leads to an anomaly: only a defendant who is not mistaken about the law when he acts has a mistake of law defense. In other words, a defendant can assert a defense under Penal Law § 15.20(2)(a) only when his reading of the statute is correct—not mistaken. Such construction is obviously illogical; it strips the statute of the very effect intended by the Legislature in adopting the mistake of law defense. The statute is of no benefit to a defendant who has proceeded in good faith on an erroneous but concededly reasonable interpretation of a statute, as defendant presumably has. An interpretation of a statute which produces an unreasonable or incongruous result and one which defeats the obvious purpose of the legislation and renders it ineffective should be rejected.

Finally, the majority's disregard of the natural and obvious meaning of Penal Law § 15.20(2)(a) so that a defendant mistaken about the law is deprived of a defense under the statute amounts, we submit, to a rejection of the obvious legislative purposes and policies favoring jurisprudential reform underlying the statute's enactment. It is self-evident that in enacting Penal Law § 15.20(2) as part of the revision and modernization of the Penal Law the Legislature intended to effect a needed reform by abolishing what had long been considered the unjust archaic common-law rule totally prohibiting mistake of law as a defense. Had it not so intended it would simply have left the common-law rule intact. In place of the abandoned "ignorantia legis" common-law maxim the Legislature enacted a rule which permits no defense for ignorance of law but allows a mistake of law defense in specific instances, including the one presented here: when the defendant's erroneous belief is founded on an "official statement of the law." . . .

Any fair reading of the majority opinion, we submit, demonstrates that the decision to reject a mistake of law defense is based on considerations of public policy and on the conviction that such a defense would be bad, rather than on an analysis of CPL 15.20(2)(a) under the usual principles of statutory construction. The majority warns, for example, that if the defense were permitted "the exception would swallow the rule"; that "[m]istakes about the law would be encouraged"; that an "infinite number of mistake of law defenses . . . could be devised"; and that "wrong minded individuals [could] contrive in bad faith solely to get an exculpatory notion before the jury."

These considerations, like the People's argument that the mistake of law defense "'would encourage ignorance where knowledge is socially desired,'" are the very considerations which have been consistently offered as justifications for the maxim "ignorantia legis." That these justifications are unabashedly utilitarian cannot be questioned. It could not be put more candidly than by Justice Holmes in defending the common-law maxim more than 100 years ago: "*Public policy sacrifices the individual to the general good.* . . . It is no doubt true that there are many cases in which the criminal could not have known that he was breaking the law, but to admit the excuse at all would be to encourage ignorance where the law-maker has determined to make men know and obey, and *justice to the individual is rightly outweighed by the larger interests on the other side of the scales*" (Holmes, The Common Law, at 48 [1881]; emphasis added). Regardless of one's attitude toward the acceptability of these views in the 1980's, the fact remains that the Legislature in abandoning the strict "ignorantia legis" maxim must be deemed to have rejected them.

We believe that the concerns expressed by the majority are matters which properly should be and have been addressed by the Legislature. We note only our conviction that a statute which recognizes a defense based on a man's good-faith mistaken belief founded on a well-grounded interpretation of an official statement of the law contained in a statute is a just law. The law embodies the ideal of contemporary criminal jurisprudence "that punishment should be conditioned on a showing of subjective moral blameworthiness."

We do not believe that permitting a defense in this case will produce the grievous consequences the majority predicts. The unusual facts of this case seem unlikely to be repeated. Indeed, although the majority foresees "an infinite number of mistake of law defenses" New Jersey, which adopted a more liberal mistake of law statute in 1978, has apparently experienced no such adversity. Nor is there any reason to believe that courts will have more difficulty separating valid claims from "diversionary stratagem[s]" in making preliminary legal determinations as to the validity of the mistake of law defense than of justification or any other defense.

But these questions are now beside the point, for the Legislature has given its answer by providing that someone in defendant's circumstances should have a mistake of law defense (Penal Law § 15.20[2][a]). Because this decision deprives defendant of what, we submit, the Legislature intended that he should have, we dissent.

## Notes and Questions

1. Did Marrero take reasonable steps to learn the law? What more could he have done?
2. What if Marrero had asked his lawyer if he could carry his gun and his lawyer said yes? Would that make his defense stronger? Under the law, in most, if not all, jurisdictions relying on a lawyer's advice does not excuse otherwise criminal conduct.
3. Could an argument be made that the prosecution of Marrero violated any constitutional provisions discussed in Chapter 2? What does the majority opinion suggest along these lines?
4. Note the tension between the "utilitarian" principles embodied in Justice Holmes' writing and endorsed by the majority opinion and the "individual blameworthiness" approach favored by the dissent. Which perspective is more persuasive?

## Mistake of Fact

A defendant is not guilty of a crime if a **mistake of fact** negates the *mens rea* of the charged offense. The classic example is where Bob goes to a coat rack and picks up a coat believing it to be his. In fact, it is not his coat, but John's. Bob is not guilty of theft, or larceny, because he did not intend to take a coat belonging to John. He simply made a mistake of fact.

On the other hand, what if Bob points a gun at John, believing it to be unloaded, and pulls the trigger? In fact, the gun was loaded and Bob kills the unfortunate John. Is he guilty of a crime? If so, of what specific crime? Consider *State v. Sexton.*

## *State v. Sexton,* 733 A.2d 1125 (N.J. 1999)

O'Hern, J. . . .

On May 10, 1993, Shakirah Jones, a seventeen-year-old friend of defendant [15-year-old Ronald Sexton] and decedent [17-year-old Alquadir Matthews], overheard the two young men having what she described as a "typical argument." The two young men walked from a sidewalk into a vacant lot. Jones saw defendant with a gun in his hand, but she did not see defendant shoot Matthews.

*(Continues)*

Jones heard Matthews tell defendant, "there are no bullets in that gun," and then walk away. Defendant called Matthews back and said, "you think there are no bullets in this gun?" Matthews replied, "yeah." Jones heard the gun go off. A single bullet killed Matthews.

Acting on information received from Jones, police recovered a small caliber automatic pistol near the crime scene. The police did not trace the ownership of the gun, which may have been owned by Matthews's grandmother.

A ballistics expert testified that there was a spring missing from the gun's magazine, which prevented the other bullets from going into the chamber after the first bullet was discharged. In this condition, the gun would have to be loaded manually by feeding the live cartridge into the chamber prior to firing.

The expert later clarified that, if the magazine had been removed after one round had been inserted into the chamber, it would be impossible to see whether the gun was loaded without pulling the slide that covered the chamber to the rear. The expert agreed that, for someone unfamiliar with guns, once the magazine was removed, it was "probably a possible assumption" that the gun was unloaded.

Defendant's version was that when the two young men were in the lot, Matthews showed defendant a gun and "told me the gun was empty." Defendant "asked him was he sure," and "he said yes." When Matthews asked if defendant would like to see the gun, defendant said "yes." Defendant "took the gun and was looking at it," and "it just went off." He never unloaded the gun or checked to see if there were any bullets in the gun. He had never before owned or shot a gun.

A grand jury indicted defendant for purposeful or knowing murder, possession of a handgun without a permit, and possession of a handgun for an unlawful purpose. At the close of the State's case, defendant moved to dismiss the murder charge because the victim had told him that the gun was not loaded. The court denied the motion.

The court charged murder and the lesser-included offenses of aggravated manslaughter and reckless manslaughter. Concerning defendant's version of the facts, the court said:

> Defense contends this was a tragic accident. That Alquadir [Matthews], says the defense, handed the gun to Ronald [defendant]. Alquadir told Ronald, you know, the gun was not loaded. Ronald believed the gun was not loaded. Ronald did not think the gun was pointed at Alquadir when it went off. But the gun went off accidentally and, says the defense, that is a very tragic and sad accident but it is not a crime.
>
> If, after considering all the evidence in this case, including the evidence presented by the defense as well as the evidence presented by the State, if you have a reasonable doubt in your mind as to whether the State has proven all the elements of any of these crimes: murder, aggravated manslaughter, or reckless manslaughter, you must find the defendant not guilty of those crimes.

The jury found defendant not guilty of murder, aggravated manslaughter, or possession of a handgun for an unlawful purpose, but guilty of reckless manslaughter and unlawful possession of a handgun without a permit. . . .

On appeal, the Appellate Division reversed defendant's conviction on multiple grounds. The court found that the trial judge erroneously charged the jury on first degree murder, despite the absence of any credible evidence that defendant intended to kill or seriously injure Matthews. The court concluded that the unwarranted charge had the potential of leading the jury to a compromise verdict on reckless manslaughter instead of acquitting him entirely.

The Appellate Division also held that the trial court should have charged the jury that the State bore the burden of disproving beyond a reasonable doubt defendant's mistake-of-fact defense, and that the failure to do so was plain error. The Appellate Division relied on *Wilson v. Tard*, 593 F. Supp. 1091 (D.N.J. 1984), in which Judge Stern, applying New Jersey law, reversed a conviction of aggravated manslaughter because the trial court had charged the jury that defendant carried the burden of establishing his mistake-of-fact defense by a preponderance of the evidence. The Appellate Division noted that "[t]he critical holding of *Wilson* is that once the defendant, as here, presents evidence of a reasonable mistake of fact that would refute an essential element of the crime charged, the State's burden of proving each element beyond a reasonable doubt includes disproving the reasonable mistake of fact." . . .

The 1979 New Jersey Code of Criminal Justice, N.J.S.A. 2C:1-1 to -104-9 (the Code), followed the mental-state formulation of the MPC. The Code provides generally that no person should be guilty of an offense unless the person "acted purposely, knowingly, recklessly or negligently, as the law may require, with respect to each material element of the offense." N.J.S.A. 2C:2-2a. The precise delineation of these four states of criminal culpability, each drawn from the MPC and each defined in N.J.S.A. 2C:2-2b, represented an effort, as a framer of the Code described it, "to achieve greater individual justice through a closer relation between guilt and culpability, requiring workable definitions of the various culpability factors. These factors must be related precisely to each element of an offense, defense, or mitigation, and all unnecessary limitations upon individual culpability should be eliminated." . . .

The MPC also contains an express provision for mistake-of-fact defenses. "Its mistake of fact provision, while creating potential for conceptual confusion by continuing the common law characterization of the doctrine as a 'defense,' in fact sought to clarify the common law." The MPC expressly recognized that the doctrine did not sanction a true defense, but rather was an attack on the prosecution's ability to prove the requisite culpable mental state for at least one objective element of the crime. Hence, unlike enactments in many pre-MPC states, "the MPC expressly recognizes that the mistake of an accused need not be a reasonable mistake unless the Legislature has expressly decided that the requisite culpable mental state was minimal—'negligence' or perhaps, 'recklessness.'"

The MPC provides that, "Ignorance or mistake as to a matter of fact or law is a defense if: (a) the ignorance or mistake negatives the purpose, knowledge, belief, recklessness or negligence required to establish a material element of the offense; or (b) the law provides that the state of mind established by such ignorance or mistake constitutes a defense." Model Penal Code § 2.04 (1962). . . .

The Commentary to the Hawaii Criminal Code gives an easy example of how, under the MPC, a mistake of fact may negate culpability.

> [I]f a person is ignorant or mistaken as to a matter of fact . . . the person's ignorance or mistake will, in appropriate circumstances, prevent the person from having the requisite culpability with respect to the fact . . . as it actually exists. For example, a person who is mistaken (either reasonably, negligently, or recklessly) as to which one of a number of similar umbrellas on a rack is the person's and who takes another's umbrella should be afforded a defense to a charge of theft predicated on either intentionally or knowingly taking the property of another. . . . A reckless mistake would afford a defense to a charge requiring intent or knowledge—but not to an offense which required only recklessness or negligence. Similarly, a negligent mistake would afford a defense to a charge predicated on intent, knowledge, or recklessness—but not to an offense based on negligence. [State v. Cavness, 911 P.2d 95, 99–100 (Haw.Ct.App. 1996).] . . .

States have restricted the mistake-of-fact doctrine by imposing a reasonableness requirement. By thinking in terms of the reasonable person while failing to appreciate that the MPC's mistake-of-fact and culpability provisions are interrelated, these states have undermined the structure of the MPC.

To explain, we may consider again the case of the absent-minded umbrella thief. If only a reasonable mistake will provide a defense to the charge of theft, the absent-minded but careless restaurant patron will have no defense to a charge of theft. *People v. Navarro*, 160 Cal.Rptr. 692, 99 Cal.App.3d Supp. 1 (1979), explains how a mistake of fact, even though it is unreasonable, may constitute a defense to a crime requiring a culpable mental state higher than recklessness or negligence. In *Navarro*, defendant took some wooden beams from a construction site. He was charged with theft. He claimed that he thought the owner had abandoned the beams. The trial court instructed the jury that this would be a valid defense only if the scavenger's belief was reasonable. The reviewing court held that such an instruction was erroneous because if the jury "concluded that defendant in good faith believed that he had the right to take the beams, even though such belief was unreasonable . . . , defendant was entitled to an acquittal since the specific intent required to be proved as an element of the offense had not been established." Id. at 697 (footnote omitted). Otherwise, one would end up imposing liability for theft on a lesser basis than knowledge or purpose to steal the property of another.

The issue posed by our grant of certification was whether a mistake of fact was a defense to the charge of reckless manslaughter. The short answer to that question is: "It depends." The longer answer requires that we relate the type of mistake involved to the essential elements of the offense, the conduct proscribed, and the state of mind required to establish liability for the offense. Defendant insists that the State is required to disprove, beyond a reasonable doubt, his mistake-of-fact defense. Most states would agree with that statement. In *State v. Savoie*, 67 N.J. 439, 463 n. 8, 341 A.2d 598 (1975), the Court similarly said that once the defense of mistake of fact is raised, the burden of

(Continues)

persuasion is on the State. Just what does that mean? Does it mean that the State must prove that the mistake was unreasonable? We must begin by examining the language of the statute.

N.J.S.A 2C:2-4a allows a defense of ignorance or mistake "if the defendant reasonably arrived at the conclusion underlying the mistake" and the mistake either "negatives the culpable mental state required to establish the offense" or "[t]he law provides that the state of mind established by such ignorance or mistake constitutes a defense." The crime of manslaughter is a form of criminal homicide. "A person is guilty of criminal homicide if [the actor] purposely, knowingly [or] recklessly . . . causes the death of another human being." N.J.S.A. 2C:11-2. Criminal homicide constitutes aggravated manslaughter, a first-degree offense with special sentencing provisions, when the actor "recklessly causes death under circumstances manifesting extreme indifference to human life." N.J.S.A 2C:11-4a. Criminal homicide constitutes manslaughter, a second-degree crime, "when . . . [i]t is committed recklessly," that is, when the actor has recklessly caused death. N.J.S.A.2C:11-4b.

In this case, the jury has acquitted defendant of aggravated manslaughter. He can be retried only for reckless manslaughter. The culpable mental state of the offense is recklessness. N.J.S.A. 2C:2-2b(3) states:

> A person acts recklessly with respect to a material element of an offense when [the actor] consciously disregards a substantial and unjustifiable risk that the material element exists or will result from [the actor's] conduct. The risk must be of such a nature and degree that, considering the nature and purpose of the actor's conduct and the circumstances known to [the actor], its disregard involves a gross deviation from the standard of conduct that a reasonable person would observe in the actor's situation. . . .

The State argues that "[i]t is obvious that the firing of a gun at another human being without checking to see if it is loaded disregards a substantial risk." The State argues that at a minimum there must be some proof establishing that defendant "reasonably arrived at the conclusion underlying the mistake." To return to the language of N.J.S.A 2C:2-4a, does the mistake about whether the gun was loaded "negative the culpable mental state required to establish the offense," or does "the law provide that the state of mind established by such ignorance or mistake constitutes a defense"? Of itself, a belief that the gun is loaded or unloaded does not negate the culpable mental state for the crime of manslaughter. Thus, one who discharges a gun, believing it to be unloaded, is not necessarily innocent of manslaughter. . . .

The material elements of manslaughter are the killing of another human being with a reckless state of mind. The culpable mental state is recklessness—the conscious disregard of a substantial and unjustified risk that death will result from the conduct. What mistaken belief will negate this state of mind?

> [T]he translation is uncertain at its most critical point: in determining the kind of mistake that provides a defense when recklessness, the most common culpability level, as to a circumstance is required. Recall that a negligent or faultless mistake negates . . . recklessness. While a "negligent mistake" may be said to be an "unreasonable mistake," all "unreasonable mistakes" are not "negligent mistakes." A mistake may also be unreasonable because it is reckless. Reckless mistakes, although unreasonable, will not negate recklessness. Thus, when offense definitions require recklessness as to circumstance elements, as they commonly do, the reasonable-unreasonable mistake language inadequately describes the mistakes that will provide a defense because of the imprecision of the term "unreasonable mistake." Reckless-negligent-faultless mistake language is necessary for a full and accurate description. [Robinson & Grall, [Element Analysis in Defining Criminal Liability: The Model Penal Code and Beyond, 35 Stan. L.Rev. [726,] 729 (1983)].]

Thus, to disprove a reasonable mistake by proving that it is unreasonable, will turn out to be a mixed blessing for defendant. If the State may disprove a reasonable mistake by proving that the mistake was unreasonable, defendant may be convicted because he was negligent, as opposed to reckless, in forming the belief that the gun was unloaded. If recklessness is required as an element of the offense, "a merely negligent or faultless mistake as to that circumstance provides a defense," Id. at 728.

Correctly understood, there is no difference between a positive and negative statement on the issue—what is required for liability versus what will provide a defense to liability. Id. at 732. What is required in order to establish liability for manslaughter is recklessness (as defined by the Code) about whether death will result from the conduct. A faultless or merely careless mistake may negate that reckless state of mind and provide a defense.

How can we explain these concepts to a jury? We believe that the better way to explain the concepts is to explain what is required for liability to be established. The charge should be tailored to the factual circumstances of the case. The court should explain precisely how the offered defense plays into the element of recklessness. Something along the following lines will help to convey to the jury the concepts relevant to a reckless manslaughter charge:

> In this case, ladies and gentlemen of the jury, the defendant contends that he mistakenly believed that the gun was not loaded. If you find that the State has not proven beyond a reasonable doubt that the defendant was reckless in forming his belief that the gun was not loaded, defendant should be acquitted of the offense of manslaughter. On the other hand, if you find that the State has proven beyond a reasonable doubt that the defendant was reckless in forming the belief that the gun was not loaded, and consciously disregarded a substantial and unjustifiable risk that a killing would result from his conduct, then you should convict him of manslaughter.

Undoubtedly, our Committee on Model Criminal Charges can improve the formulation.

To sum up, evidence of an actor's mistaken belief relates to whether the State has failed to prove an essential element of the charged offense beyond a reasonable doubt. As a practical matter, lawyers and judges will undoubtedly continue to consider a mistake of fact as a defense. When we do so, we must carefully analyze the nature of the mistake in relationship to the culpable mental state required to establish liability for the offense charged. Despite the complexities perceived by scholars, the limited number of appeals on this subject suggests to us that juries have very little difficulty in applying the concepts involved. We may assume that juries relate the instructions to the context of the charge. For example, in the case of the carelessly purloined umbrella, we are certain that juries would have no difficulty in understanding that it would have been a reasonable mistake (although perhaps a negligent mistake) for the customer to believe that he or she was picking up the right umbrella.

To require the State to disprove beyond a reasonable doubt defendant's reasonable mistake of fact introduces an unnecessary and perhaps unhelpful degree of complexity into the fairly straightforward inquiry of whether defendant "consciously disregard[ed] a substantial and unjustifiable risk" that death would result from his conduct and that the risk was "of such a nature and degree that, considering the nature and purpose of the actor's conduct and the circumstances known to him, its disregard involve[d] a gross deviation from the standard of conduct that a reasonable person would observe in the actor's situation." N.J.S.A 2C:2-2b(3); N.J.S.A 2C:11-4b.

The judgment of the Appellate Division is affirmed. The matter is remanded to the Law Division for further proceedings in accordance with this opinion.

## Notes and Questions

1. Assume that Ronald Sexton is on trial exclusively for the "purposeful murder" of Alquadir Matthews. Assume further that Sexton claims that he did not know the gun was loaded—*i.e.*, that he made a "mistake of fact." In what sense is the claimed mistake a "defense" when the prosecution is required to establish, beyond a reasonable doubt and as an element of the crime, that the homicide was committed "purposefully"?

2. As the court's opinion in *Sexton* indicates, mistake of fact issues may be considerably more complicated for crimes that entail a *mens rea* of recklessness or negligence rather than intent (purposefulness) or knowledge. If Sexton makes the claim that he did not know the gun pointed at Matthews was loaded, is such a mistake incompatible with his conviction for a homicide committed recklessly or negligently? How might his mistake be relevant in this context?

3. Mistake of fact is not a defense to strict liability offenses. For example, as we consider in Chapter 7, in some states statutory rape is a strict liability crime. As a consequence, even a good faith and reasonable mistake about the age of a sexual partner is not a defense.

## Duress

Conduct that otherwise would be criminal may be excused if committed under **duress**. Some argue that duress can equally well be considered a form of justification, at least in most cases. A defendant raising the defense of duress essentially maintains that he or she engaged in the conduct to avoid harm to him or herself or another that was being threatened by a third party. The law may excuse an individual for succumbing to such pressure. We likely would not blame an individual, for example, who stole a loaf of bread from a grocery store or forged a signature on a check to avoid death or serious injury. We might be less willing to excuse the individual who claimed duress under other circumstances, for example, where the threatened harm was speculative or less serious, or where his or her conduct resulted in equally serious injury to another.

At common law, the defendant was required to establish the following elements to qualify for the defense of duress.

1. A reasonable belief or fear
2. Of a present, imminent, and impending threat of
3. Death or serious injury
4. To him- or herself or another (sometimes limited to a close relative)

5. From which there was no reasonable escape or alternative course of action
6. And the danger or threat was not brought about by the defendant's own conduct.[3]

Although many jurisdictions continue to require these elements of the common law defense, some have relaxed the common law requirements in various particulars. In *State v. Toscano* the New Jersey Supreme Court discusses modern trends in the defense of duress.

## *State v. Toscano*, 378 A.2d 755 (N.J. 1977)

Pashman, J. . . .

On April 20, 1972, the Essex County Grand Jury returned a 48-count indictment alleging that eleven named defendants and two unindicted co-conspirators had defrauded various insurance companies by staging accidents in public places and obtaining payments in settlement of fictitious injuries. The First Count of the indictment alleged a single conspiracy involving twelve different "staged" accidents over a span of almost three years. In the remaining counts, the participants were charged with separate offenses of conspiracy, obtaining money by false pretenses and receiving fraudulently obtained money.

Dr. Joseph Toscano, a chiropractor, was named as a defendant in the First Count and in two counts alleging a conspiracy to defraud the Kemper Insurance Company (Kemper). Prior to trial, seven of the eleven defendants pleaded guilty to various charges, leaving defendant as the sole remaining defendant charged with the conspiracy to defraud Kemper. Among those who pleaded guilty was William Leonardo, the architect of the alleged general conspiracy and the organizer of each of the separate incidents. Although the First Count was dismissed by the trial judge at the conclusion of the State's case, the evidence did reveal a characteristic modus operandi by Leonardo and his cohorts which is helpful in understanding the fraudulent scheme against Kemper. Typically, they would stage an accident or feign a fall in a public place. A false medical report for the "injured" person, together with a false verification of employment and lost wages, would then be submitted to the insurer of the premises. The same two doctors were used to secure the medical reports in every instance except that involving the claim against Kemper. Likewise, the confirmations of employment and lost wages were secured from the same pool of friendly employers. The insurance companies made cash payments to resolve the claims under their "quick settlement" programs, usually within a few weeks after the purported accidents. Leonardo took responsibility for dividing the funds to the "victims" of the accidents, to the doctors and employers, taking a substantial portion for himself.

Michael Hanaway, an unindicted co-conspirator who acted as the victim in a number of these staged accidents, testified that defendant was drawn into this scheme largely by happenstance. On January 6, 1970, Hanaway staged a fall at E.J. Korvette's in Woodridge, New Jersey, under the direction of Leonardo and Frank Neri, another defendant who pleaded guilty prior to trial. Dr. Miele, one of the two doctors repeatedly called upon by Leonardo to provide fraudulent medical reports, attested to Hanaway's claimed injuries on a form supplied by the insurer. Hanaway was subsequently paid $975 in settlement of his claim by the Underwriters Adjusting Company on behalf of Korvette's insurer.

In the meantime, however, the same trio performed a similar charade at the R.K.O. Wellmont Theater in Montclair, New Jersey. Kemper, which insured the R.K.O. Theater, was immediately notified of Hanaway's claim, and Dr. Miele was again enlisted to verify Hanaway's injuries on a medical report. However, because the R.K.O. accident occurred on January 8, 1970, only two days after the Korvette's incident Dr. Miele confused the two claims and mistakenly told Kemper's adjuster that he was treating Hanaway for injuries sustained at Korvette's. When Hanaway learned of the claims adjuster's suspicions, he informed William Leonardo who, in turn, contacted his brother Richard (a co-defendant at trial) to determine whether Toscano would agree to verify the treatments.

The State attempted to show that Toscano agreed to fill out the false medical report because he owed money to Richard Leonardo for gambling debts. It also suggested that Toscano subsequently sought to cover up the crime by fabricating office records of non-existent office visits by Hanaway. Defendant sharply disputed these assertions and maintained that he capitulated to William Leonardo's demands only because he was fearful for his wife's and his own bodily safety. Since it is not our function here to assess these conflicting versions, we shall summarize only those facts which, if believed by the jury, would support defendant's claim of duress.

Defendant first met Richard Leonardo in 1953 as a patient and subsequently knew him as a friend. Defendant briefly encountered the brother, William, in the late 1950's at Caldwell Penitentiary when Toscano served as a prison guard. Although William was an inmate, the doctor did not know him personally. Through conversations with some police officers and William's brother and father, however, he did learn enough about William to know of his criminal record. In particular, Richard told him many times that William was "on junk," that he had a gang, that "they can't keep up with the amount of money that they need for this habit," and that he himself stayed away from William.

Thus, when William first called the defendant at his office, asking for a favor, he immediately cut off the conversation on the pretext that he was with a patient. Although William had not specifically mentioned the medical form at that time, defendant testified that he was "nauseated" by "just his name." A few days later, on a Thursday evening, he received another call in his office. This time Leonardo asked defendant to make out a report for a friend in order to submit a bill to a claims adjuster. He was more insistent, stating that defendant was "going to do it," but defendant replied that he would not and could not provide the report. Once again the doctor ended the conversation abruptly by claiming, falsely, that he was with other persons.

The third and final call occurred on Friday evening. Leonardo was "boisterous and loud" repeating, "You're going to make this bill out for me." Then he said: "Remember, you just moved into a place that has a very dark entrance and you leave there with your wife. . . . You and your wife are going to jump at shadows when you leave that dark entrance." Leonardo sounded "vicious" and "desperate" and defendant felt that he "just had to do it" to protect himself and his wife. He thought about calling the police, but failed to do so in the hope that "it would go away and wouldn't bother me any more."

In accordance with Leonardo's instructions, defendant left a form in his mailbox on Saturday morning for Leonardo to fill in with the necessary information about the fictitious injuries. It was returned that evening and defendant completed it. On Sunday morning he met Hanaway at a pre-arranged spot and delivered a medical bill and the completed medical report. He received no compensation for his services, either in the form of cash from William Leonardo or forgiven gambling debts from Richard Leonardo. He heard nothing more from Leonardo after that Sunday.

Shortly thereafter, still frightened by the entire episode, defendant moved to a new address and had his telephone number changed to an unlisted number in an effort to avoid future contacts with Leonardo. He also applied for a gun permit but was unsuccessful. His superior at his daytime job with the Newark Housing Authority confirmed that the quality of defendant's work dropped so markedly that he was forced to question defendant about his attitude. After some conversation, defendant explained that he had been upset by threats against him and his wife. He also revealed the threats to a co-worker at the Newark Housing Authority.

After defendant testified, the trial judge granted the State's motion to exclude any further testimony in connection with defendant's claim of duress, and announced his decision not to charge the jury on that defense. He based his ruling on two decisions by the former Court of Errors and Appeals, *State v. Palmieri*, [107 A. 407 (N.J. 1919)] and *State v. Churchill*, [143 A. 330 (N.J. 1928)], which referred to the common law rule that a successful claim of duress required a showing of a "present, imminent and impending" threat of harm. As he interpreted these decisions, the defendant could not satisfy this standard by establishing his own subjective estimate of the immediacy of the harm. Rather, the defendant was obliged to prove its immediacy by an objective standard which included a reasonable explanation of why he did not report the threats to the police. Since Toscano's only excuse for failing to make such a report was his doubts that the police would be willing or able to protect him, the court ruled that his subjective fears were irrelevant.

After stating that the defense of duress is applicable only where there is an allegation that an act was committed in response to a threat of present, imminent and impending death or serious bodily harm, the trial judge charged the jury:

Now, one who is standing and receiving instructions from someone at the point of a gun is, of course, in such peril. One can describe such threat as being imminent, present and pending, and a crime committed under those circumstances, or rather conduct engaged in under those circumstances, even though criminal in nature, would be excused by reason of the circumstances in which it was committed.

(Continues)

Now, where the peril is not imminent, present and pending to the extent that the defendant has the opportunity to seek police assistance for himself and his wife as well, the law places upon such a person the duty not to acquiesce in the unlawful demand and any criminal conduct in which he may thereafter engage may not be excused. Now, this principle prevails regardless of the subjective estimate he may have made as to the degree of danger with which he or his wife may have been confronted. Under the facts of this case, I instruct you, as members of the jury, that the circumstances described by Dr. Toscano leading to his implication in whatever criminal activities in which you may find he participated are not sufficient to constitute the defense of duress. . . .

# II.

At common law the defense of duress was recognized only when the alleged coercion involved a use or threat of harm which is "present, imminent and pending" and "of such a nature as to induce a well grounded apprehension of death or serious bodily harm if the act is not done."

It was commonly said that duress does not excuse the killing of an innocent person even if the accused acted in response to immediate threats. Aside from this exception, however, duress was permitted as a defense to prosecution for a range of serious offenses, and many lesser crimes.

To excuse a crime, the threatened injury must induce "such a fear as a man of ordinary fortitude and courage might justly yield to." Although there are scattered suggestions in early cases that only a fear of death meets this test, an apprehension of immediate serious bodily harm has been considered sufficient to excuse capitulation to threats. Thus, the courts have assumed as a matter of law that neither threats of slight injury nor threats of destruction to property are coercive enough to overcome the will of a person of ordinary courage. A "generalized fear of retaliation" by an accomplice, unrelated to any specific threat, is also insufficient.

More commonly, the defense of duress has not been allowed because of the lack of immediate danger to the threatened person. When the alleged source of coercion is a threat of "future" harm, courts have generally found that the defendant had a duty to escape from the control of the threatening person or to seek assistance from law enforcement authorities.

Assuming a "present, imminent and impending" danger, however, there is no requirement that the threatened person be the accused. Although not explicitly resolved by the early cases, recent decisions have assumed that concern for the well-being of another, particularly a near relative, can support a defense of duress if the other requirements are satisfied. . . .

The insistence under the common law on a danger of immediate force causing death or serious bodily injury may be ascribed to its origins in early cases dealing with treason, to the proclivities of a "tougher-minded age," or simply to judicial fears of perjury and fabrication of baseless defenses. We do not discount the latter concern as a reason for caution in modifying this accepted rule, but we are concerned by its obvious shortcomings and potential for injustice. Under some circumstances, the commission of a minor criminal offense should be excusable even if the coercive agent does not use or threaten force which is likely to result in death or "serious" bodily injury. Similarly, it is possible that authorities might not be able to prevent a threat of future harm from eventually being carried out. As shown by *Commonwealth v. Reffitt*, [148 S.W. 48 (Ky. 1912)], and *Hall v. State*, [187 So. 392 (Fla. 1939)], the courts have not wholly disregarded the predicament of an individual who reasonably believes that appeals for assistance from law enforcement officials will be unavailing, but there has been no widespread acknowledgment of such an exception. Warnings of future injury or death will be all the more powerful if the prospective victim is another person, such as a spouse or child, whose safety means more to the threatened person than his own well-being. Finally, as the drafters of the Model Penal Code observed, "long and wasting pressure may break down resistance more effectively than a threat of immediate destruction." § 2.09, Comment at 8 (Tent.Draft No. 10, 1960).

Commentators have expressed dissatisfaction with the common law standard of duress. Stephen viewed the defense as a threat to the deterrent function of the criminal law, and argued that "it is at the moment when temptation is strongest that the law should speak most clearly and emphatically to the contrary." Stephen, 2 History of the Criminal Law in England 107 (1883). A modern refinement of this position is that the defense should be designed to encourage persons to act against their self-interest if a substantial percentage of persons in such a situation would do so. This standard would limit its applicability to relatively minor crimes and exclude virtually all serious crimes unless committed under threat of imminent death.

Others have been more skeptical about the deterrent effects of a strict rule. As the Alabama Supreme Court observed in an early case:

> That persons have exposed themselves to imminent peril and death for their fellow man, and that there are instances where innocent persons have submitted to murderous assaults, and suffered death, rather than take life, is well established; but such self-sacrifice emanated from other motives than the fear of legal punishment. *Arp v. State*, [12 So. 301, 303 (Ala. 1893)].

Building on this premise, some commentators have advocated a flexible rule which would allow a jury to consider whether the accused actually lost his capacity to act in accordance with "his own desire, or motivation, or will" under the pressure of real or imagined forces. The inquiry here would focus on the weaknesses and strengths of a particular defendant, and his subjective reaction to unlawful demands. Thus, the "standard of heroism" of the common law would give way, not to a "reasonable person" standard, but to a set of expectations based on the defendant's character and situation.

The drafters of the Model Penal Code and the New Jersey Penal Code sought to steer a middle course between these two positions by focusing on whether the standard imposed upon the accused was one with which "normal members of the community will be able to comply. . . ." They stated:

> . . . law is ineffective in the deepest sense, indeed it is hypocritical, if it imposes on the actor who has the misfortune to confront a dilemmatic choice, a standard that his judges are not prepared to affirm that they should and could comply with if their turn to face the problem should arise. Condemnation in such case is bound to be an ineffective threat; what is, however, more significant is that it is divorced from any moral base and is unjust. Where it would be both 'personally and socially debilitating' to accept the actor's cowardice as a defense, it would be equally debilitating to demand that heroism be the standard of legality. (Model Penal Code § 2.09, Comment at 7 (Tent.Draft No. 10, 1960), quoting Hart, "The Aims of the Criminal Law," 23 Law & Contemp. Prob. 401, 414 and n. 31 (1958); New Jersey Model Penal Code § 2C:2-9, Commentary at 71 (1971).)

Thus, they proposed that a court limit its consideration of an accused's "situation" to "stark, tangible factors which differentiate the actor from another, like his size or strength or age or health," excluding matters of temperament. They substantially departed from the existing statutory and common law limitations requiring that the result be death or serious bodily harm, that the threat be immediate and aimed at the accused, or that the crime committed be a non-capital offense. While these factors would be given evidential weight, the failure to satisfy one or more of these conditions would not justify the trial judge's withholding the defense from the jury.

Both the Prosecutor and the Attorney General substantially approve of the modifications suggested by the drafters of the model codes. However, they would allow the issue to be submitted to the jury only where the trial judge has made a threshold determination that the harm threatened was "imminent." Defendant, in a rather cryptic fashion, refers us to New York's statutory definition of duress, New York Penal Code § 40.00 (1970), which requires a showing of coercion by the use or threatened imminent use of unlawful force. However, he advocates leaving the question of immediacy to the jury.

For reasons suggested above, a per se rule based on immediate injury may exclude valid claims of duress by persons for whom resistance to threats or resort to official protection was not realistic. While we are hesitant to approve a rule which would reward citizens who fail to make such efforts, we are not persuaded that capitulation to unlawful demands is excusable only when there is a "gun at the head" of the defendant. We believe that the better course is to leave the issue to the jury with appropriate instructions from the judge.

Although they are not entirely identical, under both model codes defendant would have had his claim of duress submitted to the jury. Defendant's testimony provided a factual basis for a finding that Leonardo threatened him and his wife with physical violence if he refused to assist in the fraudulent scheme. Moreover, a jury might have found from other testimony adduced at trial that Leonardo's threats induced a reasonable fear in the defendant. Since he asserted that he agreed to complete the false documents only because of this apprehension, the requisite elements of the defense were established. Under the model code provisions, it would have been solely for the jury to determine whether a "person of reasonable firmness in his situation" would have failed to seek police assistance or refused to cooperate, or whether such a person would have been, unlike defendant, able to resist.

*(Continues)*

Exercising our authority to revise the common law, we have decided to adopt this approach as the law of New Jersey. Henceforth, duress shall be a defense to a crime other than murder if the defendant engaged in conduct because he was coerced to do so by the use of, or threat to use, unlawful force against his person or the person of another, which a person of reasonable firmness in his situation would have been unable to resist.

We have deliberately followed the language of the proposed New Jersey Penal Code in stating our holding and we expect trial judges to frame their jury charges in the same terms. The defendant shall have the burden of producing sufficient evidence to satisfy the trial judge that the fact of duress is in issue. Such evidence may appear in the State's case or that of the defendant. No longer will there be a preliminary judicial determination that the threats posed a danger of "present, imminent and impending" harm to the defendant or to another. In charging the jury, however, the trial judge should advert to this factor of immediacy, as well as the gravity of the harm threatened, the seriousness of the crime committed, the identity of the person endangered, the possibilities for escape or resistance and the opportunities for seeking official assistance. He should also emphasize that the applicable standard for judging the defendant's excuse is the "person of reasonable firmness in (the accused's) situation."

Finally, the trial judge will instruct the jury that the defendant has the burden of persuasion on the issue of duress and that he must establish the defense by a preponderance of the evidence in order to win an acquittal. To avoid confusion, the judge should stress that the prosecution must prove all other elements of the crime and disprove any other defenses beyond a reasonable doubt. Only if the prosecution has met that burden should the jury consider the affirmative defense of duress; at that point alone will the defendant have the burden of proving the defense by a preponderance of the evidence in order to prevail.

We recognize that in other instances where the initial burden of producing evidence in support of an affirmative defense has been placed on the defendant, the burden of disproving the defense beyond a reasonable doubt has remained with the State. In this case, however, we think it more appropriate as a matter of public policy to follow the practice utilized in insanity cases and to require the defendant to prove the existence of duress by a preponderance of the evidence.

The peculiar nature of duress, which focuses on the reasonableness of the accused's fear and his actual ability to resist unlawful demands, is not completely offset by the "person of reasonable firmness" standard. While the idiosyncrasies of an individual's temperament cannot excuse an inability to withstand such demands, his attributes (age, health, etc.) are part of the "situation" which the jury is admonished to consider. We think that the admittedly open-ended nature of this standard, with the possibility for abuse and uneven treatment, justifies placing the onus on the defendant to convince the jury.

## IV.

Defendant's conviction of conspiracy to obtain money by false pretenses is hereby reversed and remanded for a new trial.

## Notes and Questions

1. Note that the defendant's claim of duress is to be judged pursuant to the standard enunciated in § 2.09(1) of the MPC: the threatened unlawful force must be such "that a person of reasonable firmness in [the actor's] situation would have been unable to resist." Would you anticipate that what a "person of reasonable firmness" should be expected to resist depends, in large part, on the particulars of his or her "situation"? What aspects of the defendant, or of his or her "situation," should be considered? The nature and magnitude of the threat? The defendant's personal characteristics? The immediacy of the threat? The magnitude of harm committed by the defendant under the claimed duress? Alternative courses of action that might have been available to avoid succumbing to the threat and committing the harmful conduct?

2. Considering the above standards and factors, if you were on the jury in Toscano's case, would you be sympathetic to his claim that he should be excused for his role in the alleged insurance bilking scheme because he participated in it under duress?

3. Is the more flexible standard adopted by the New Jersey Supreme Court for evaluating claims of duress preferable to the more demanding common law test? What are the potential virtues, and the potential drawbacks, of this enhanced flexibility?

**4.** Should duress be available as a defense when the crime charged is a homicide? Under the common law, killing another person to save one's own life, or the life of a family member, was considered to be "the unacceptable choice." Making duress unavailable under such circumstances was premised on the belief that it is better for an individual to give up his or her own life than to kill an innocent person. The MPC does not expressly prohibit the defense of duress for a homicide. Why do you suppose that is? What should be expected of a "person of reasonable firmness in the actor's situation" when that situation involves the choice of "kill or be killed"?

**5.** In *United States v. Dixon*, 548 U.S. 1, 126 S. Ct. 2437, 165 L. Ed. 2d 299 (2006), the Supreme Court ruled that Due Process is not offended by making duress an affirmative defense that must be established by the accused by a preponderance of the evidence.

## Entrapment : Two tests to prove it...

**Entrapment** is one of the most controversial affirmative defenses permitted by the criminal law. The entrapment defense arguably is based more on public policy considerations—most importantly, imposing limits on how far law enforcement should be allowed to go in encouraging or inducing citizens to commit crimes or in participating in criminal activity—than fitting within the traditional framework of "excuse" defenses.[4] While asserting the defense of entrapment, defendants typically claim that law enforcement officers improperly induced them into committing a crime. These situations often arise in the context of sting operations where police "set up" defendants by facilitating or encouraging the criminal act. The fine line the police must walk is between providing a person with the opportunity to commit a crime that he or she was already inclined to commit and manufacturing or inducing criminal conduct that a person who "takes the bait" otherwise would not have committed.

Strong feelings often are elicited regarding law enforcement's use of sting operations as well as whether entrapment should be recognized as a defense. Opponents of the entrapment defense argue that it allows criminals caught red-handed to be acquitted for no legitimate reason, and particularly hampers law enforcement in crimes such as drug transactions, bribery, and prostitution. Such offenses typically do not involve a complaining victim, making proactive enforcement efforts by the police particularly important. Proponents of the defense maintain that it is necessary to keep the police from manufacturing crimes. They argue that sting operations waste scarce police resources and that there are enough crimes to be

investigated and solved without the police helping to create more.

Two dominant tests are used by courts in determining whether a defendant should benefit from the entrapment defense. The **subjective test**, used in the federal system and in most state courts, requires a defendant to establish that (1) government conduct induced the commission of the crime, which (2) the defendant was not predisposed to commit. If a defendant was predisposed to commit the crime, the entrapment defense is not available. This approach shifts much of the emphasis away from the conduct of the police and toward the character and criminal history of the defendant. The subjective test is designed to allow the police to capture unwary criminals but not ensnare unwary innocents. It may be difficult for a defendant with a criminal history to raise an entrapment defense in jurisdictions that use the subjective test because the prosecution can argue that the defendant's prior record shows that he or she was predisposed to commit the crime presently charged. Opponents of this approach argue that it gives the police free rein to use unsavory tactics in an attempt to ensnare known criminals in the commission of a crime.

The second approach relies on an **objective test**, which has been adopted by a significant minority of jurisdictions and is the basis of the relevant MPC provision, § 2.13. The objective test focuses on the conduct of the police and essentially considers whether the police's actions would likely have induced a normal, law-abiding citizen to commit the crime. Because the emphasis is on imposing general limits on law enforcement, an individual defendant's predisposition to commit a crime is irrelevant. In *State v. Doran*, the Ohio Supreme Court explains its preference for the subjective test for entrapment.

*objective vs. subjective*

## *State v. Doran*, 449 N.E.2d 1295 (Ohio 1983)

Frank D. Celebrezze, Chief Justice.

As a result of a series of drug transactions, appellant, William S. Doran, was indicted on six counts of aggravated trafficking under R.C. 2925.03(A)(1) and (5) and one count of permitting drug abuse under R.C. 2925.13(A). The particular circumstances of these transactions are as follows.

In October 1980, appellant picked up a hitchhiker named Nona F. Wilson. Appellant did not know Wilson prior to this occasion. Unknown to appellant, Wilson was an agent of Medway Enforcement Group, a multi-county undercover drug enforcement group. Wilson was not a law enforcement officer,

*(Continues)*

but was paid by Medway to introduce undercover police agents to prospective drug dealers. Medway paid Wilson $50 for arranging a drug buy with a first offender and $100 for arranging a buy with a repeat offender.

While appellant drove Wilson home, the two struck up a conversation and became friendly. During their conversation, Wilson asked appellant if he dealt in drugs. Appellant responded that he did not. Over the next several weeks appellant frequently spoke with Wilson over the telephone and occasionally saw her in person. Invariably Wilson relayed to appellant her desperate need for money. Wilson explained to appellant that her ex-husband had custody of her children and that she needed money to hire a lawyer to regain her children's custody. Wilson repeatedly suggested to appellant that if he could obtain drugs for her to sell, she could make the money she needed. Appellant declined and counselled Wilson to find a job.

Approximately two weeks after appellant and Wilson met, she requested that appellant meet her at a Wadsworth bar. There, Wilson introduced appellant to David High. High was introduced simply as a friend of Wilson, but was in reality an undercover narcotics agent with Medway.

In the days that followed, Wilson continued to press appellant into obtaining drugs for her to sell. Wilson's pleas became increasingly emotional and she would often break down and cry. Wilson even confessed to appellant that she was contemplating kidnapping her children in order to regain their custody. Appellant continuously resisted Wilson's pleas until finally he gave in and agreed to attempt to locate a supplier of drugs.

On November 18, 1980, appellant informed Wilson that he may have found a supplier. On that date appellant met with Wilson and High. Wilson explained to appellant that High was assisting her with the purchase of the drugs. Appellant received $200 from High, and Wilson and High then left. Shortly thereafter, Wilson and High saw appellant, at which time appellant delivered two tinfoil packets to them which contained phencyclidine (PCP).

Sometime later, Wilson told appellant that she had found an apartment but needed money for a deposit in order to move in. On November 21, 1980, under circumstances similar to the earlier sale, another drug transaction was completed between appellant and High, in Wilson's presence.

A third sale took place on November 25, 1980, after Wilson told appellant that she was despondent over her inability to adequately clothe her children. This sale took place in the same manner as the earlier two sales with appellant delivering the drugs to High in Wilson's presence.

Over the next three to four months, three additional sales were completed. These last three sales were arranged and carried out between appellant and High alone. None of the latter sales involved the physical presence of Wilson, even though she continued to contact appellant and maintain that these drug transactions were necessary to satisfy a considerable gambling debt, the payment of which would prevent her from turning to prostitution. Prior to the final sale, High told appellant that only one more sale would be necessary in order for Wilson and High to get married and finance a new beginning for themselves. After the sixth sale, appellant was arrested and indicted.

Appellant was tried before a jury and raised the defense of entrapment. Wilson testified as a *defense* witness. The trial court instructed the jury on the elements of the offenses and the requirement that the state had the burden to prove those elements beyond a reasonable doubt. . . . The trial court's definition of the defense of entrapment was, as set forth in his instruction:

"The defendant denies that he formed a purpose to commit a crime. He claims that he is excused because he was unlawfully entrapped by the undercover agent.

"Unlawful entrapment occurs when a police officer, informant or agent plants in the mind of the defendant the original idea and purpose inducing the defendant to commit a crime that he had not considered and which otherwise he had no intention of committing or would not have committed but for the inducement of the police officer, undercover agent or informant.

"If the defendant did not, himself, conceive the idea of committing a crime and it was suggested to him by the officer for the purpose of causing his arrest, the defendant must be found not guilty. However, if the defendant commits a crime while acting even in part in carrying out his own purpose or plan to violate the law, an entrapment is not unlawful and is not a defense even if the officer suggested the crime and provided the opportunity or facility or aided or encouraged its commission. A person is not entrapped when officers for the purpose of detecting crime merely present a defendant with an opportunity to commit an offense. Under such circumstances, craft and pretense may be used by law officers or agents to accomplish such purpose.

"If you find by credible evidence that the defendant had the predisposition and criminal design to commit the act into which he claims he was entrapped and that he was merely provided with opportunity to commit those acts for which he was both apt and willing, then he has not been unlawfully entrapped." . . .

Appellant was acquitted of the aggravated trafficking charges which related to the first three buys and of the charge of permitting drug abuse. Appellant was found guilty of the remaining three counts of aggravated trafficking which arose from the buys conducted solely between appellant and High, without the physical presence of Wilson. . . .

This appeal poses several previously unanswered questions significant to the administration of criminal justice in this state. . . .

We must initially choose between defining entrapment under the "subjective" or "objective" test. Succinctly stated, the subjective test of entrapment focuses upon the predisposition of the accused to commit an offense whereas the objective or "hypothetical-person" test focuses upon the degree of inducement utilized by law enforcement officials and whether an ordinary law-abiding citizen would have been induced to commit an offense.

The United States Supreme Court adopted the subjective test of entrapment for federal prosecutions in *Sorrells v. United States* (1932), 287 U.S. 435. That test has withstood several challenges to its continued viability. See *Sherman v. United States* (1958), 356 U.S. 369 and *United States v. Russell* (1973), 411 U.S. 423. However, the objective test has won favor with a minority of United States Supreme Court justices and has been adopted by several states.

This court has not yet defined which test is applicable in this state. Since defining the entrapment defense under either of the above standards does not implicate federal constitutional principles, we are not bound by *Sorrells* and its progeny and are free to adopt either standard.

Appellant advocates adoption of the objective test. The approach advanced by appellant would examine the conduct of the police officer or agent and require a determination of whether the police conduct would induce an ordinary law-abiding citizen to commit a crime. Appellant's position is that the conduct of the police or their agent in this case was compelling and outrageous in continuing to induce appellant into committing a crime after appellant had repeatedly refused to succumb to these inducements.

The state urges adoption of the subjective test. . . . The state suggests that the emphasis should be on the predisposition of the accused to commit a crime and that law enforcement should be free to use "artifice and stratagem" to apprehend those engaged in criminal activity.

For the reasons to follow, we hold that the defense of entrapment in Ohio will be defined under the subjective test.

We are constrained to reject the objective or hypothetical-person test because of the dangers inherent in the application of that standard. First, there is the danger that the objective approach will operate to convict those persons who should be acquitted. This is true because the objective test focuses upon the nature and degree of the inducement by the government agent and not upon the predisposition of the accused. Thus, even though the accused may not be individually predisposed to commit the crime, the inducement may not be of the type to induce a reasonably law-abiding citizen, and thus lead to the conviction of an otherwise innocent citizen.

Equally oppressive is the danger that the objective test may lead to acquittals of those who should be convicted. Again, the objective test emphasizes the effect of the inducement upon an ordinary law-abiding citizen and renders the predisposition of the accused irrelevant. If that is the case, a "career" criminal, or one who leaves little or no doubt as to his predisposition to commit a crime, will avoid conviction if the police conduct satisfies the objective test for entrapment.

A final danger is that adoption of the objective test may adversely impact on the accuracy of the fact-finding process. The determinative question of fact under the objective standard is what inducements were actually offered. Since most of these inducements will be offered in secrecy, the trial will more than likely be reduced to a swearing contest between an accused claiming that improper inducements were used and a police officer denying the accused's exhortations. Regardless of whether this situation favors the prosecution or defense, the fact-finder is still in the position of having to decide the truth solely upon the testimony of two diametrically opposed witnesses.

Conversely, the subjective test presents relatively few problems. By focusing on the predisposition of the accused to commit an offense, the subjective test properly emphasizes the accused's criminal culpability and not the culpability of the police officer. Indeed, the subjective test reduces the dangers of

*(Continues)*

convicting an otherwise innocent person and acquitting one deserving of conviction. In addition, the fact-finding process is enhanced because evidence of predisposition may come from objective sources.

Our sole reservation concerning the subjective test involves the scope of admissible evidence on the issue of an accused's predisposition. While evidence relevant to predisposition should be freely admitted, judges should be hesitant to allow evidence of the accused's bad reputation, without more, on the issue of predisposition. Rather, while by no means an exhaustive list, the following matters would certainly be relevant on the issue of predisposition: (1) the accused's previous involvement in criminal activity of the nature charged, (2) the accused's ready acquiescence to the inducements offered by the police, (3) the accused's expert knowledge in the area of the criminal activity charged, (4) the accused's ready access to contraband, and (5) the accused's willingness to involve himself in criminal activity. Under this approach, the evidence on the issue of an accused's predisposition is more reliable than the evidence of the nature of inducement by police agents under the objective test.

Consequently, where the criminal design originates with the officials of the government, and they implant in the mind of an innocent person the disposition to commit the alleged offense and induce its commission in order to prosecute, the defense of entrapment is established and the accused is entitled to acquittal. However, entrapment is not established when government officials "merely afford opportunities or facilities for the commission of the offense" and it is shown that the accused was predisposed to commit the offense.

In the case at bar, the trial court's instruction on entrapment, as previously set forth, is consonant with the definition of entrapment announced in this case. . . .

[The court reversed the defendant's conviction on other grounds, owing to the trial court's failure to clarify burden of proof issues concerning the entrapment issues.]

## Notes and Questions

1. No one made Doran sell drugs. He made a conscious decision to violate the law. Why should he be able to argue that he should be acquitted, under any statement of the entrapment defense?
2. Using the subjective test for entrapment, would you have voted for a guilty or a not guilty verdict if you were on Doran's jury? Would you find any additional information about Doran or the circumstances surrounding the drug transactions useful in arriving at a decision?
3. How, in particular, might the jury's assessment of an entrapment claim in Doran's case using the subjective, or predisposition, test be different depending on whether Doran had previous convictions for drug-related activity? Would Doran's prior record be relevant under the objective test of entrapment? Should it be?
4. How persuasive are the Ohio Supreme Court's reasons for preferring the subjective test for entrapment over the objective test? In particular, how likely is it that "the objective approach will operate to convict those persons who should be acquitted"? Will adoption of the subjective test minimize the perceived risk of "a swearing contest between an accused claiming that improper inducements were used and a police officer denying the accused's exhortations"? In many jurisdictions that make use of the objective test, a judge considers the evidence and makes a ruling as a matter of law on the entrapment defense. Would such an approach help obviate the *Doran* court's concerns about the "adverse impact on the accuracy of the fact-finding process"?
5. In Ohio's neighboring state, the Pennsylvania Supreme Court arrived at precisely the opposite conclusion and, after weighing the pros and cons of the respective approaches, opted for the objective test of entrapment instead of the subjective test. *See Commonwealth v. Weiskerger*, 554 A.2d 10 (Penn. 1989). As noted by the court in *Doran*, the entrapment test is not constitutionally based. Jurisdictions need not recognize this defense at all (although all states and the federal courts do), and they are at liberty to choose what test or tests for entrapment to use.
6. Some states have adopted a hybrid test for the entrapment defense that combines both an objective and a subjective inquiry. *See, e.g., State v. Buendia*, 912 P.2d 284 (N.M. 1996); *Cruz v. State*, 465 So. 2d 516 (Fla. 1985).
7. Although the subjective test arguably is more generous to law enforcement, it is not without limitations. *Jacobson v. United States*, 503 U.S. 540 (1992), discussed the length to which law enforcement agencies could go in targeting suspects. Jacobson subscribed to a magazine featuring pictures of nude boys before the passage of a law that would make possession of the material illegal. After enactment of the pornography statute, government agents continually sent him material in the mail, including subscription information, and information about fictitious lobbying organizations against government censorship. Twenty-six months after the mailings began, Jacobson placed an order for pornographic magazines. After the magazines were delivered he was arrested, charged with,

and convicted of possessing child pornography. The Supreme Court ruled that under the subjective test of entrapment the government was required to prove that the defendant's "predisposition" was not the product of government conduct. Based in part on the fact that Jacobson had never ordered illegal material in the past, the Court found that the government had created the predisposition and overturned Jacobson's conviction.

## Infancy

Part of being a child is getting into mischief. While the level and seriousness of mischief may vary from child to child, some mischievous activity by children is expected and generally accepted by society. As the mischief becomes more serious, and the child gets older, the justice system must determine when the child's actions should be considered criminal.

The criminal law requires more than harm-causing conduct before it ascribes blame and imposes punishment. It also requires that offenders have a culpable state of mind. Young children may lack the experience, cognitive ability, and moral capacity to make fully rational choices about their conduct. They also tend to be more impulsive in their behavior. Accordingly, the criminal law may refrain from holding young children responsible for their actions by recognizing the defense of **infancy**.

Under the common law a child under 7 years of age was considered incapable of having the mental capacity to commit a crime. As a result, the child could not be prosecuted for his or her actions. Children between the ages of 7 and 14 were presumed to lack the capacity to commit a crime, but the presumption was rebuttable. If the prosecutor could prove that the child in fact did have the ability to understand what he or she was doing and that it was wrong, a prosecution could proceed. Children over the age of 14 were presumed to have the capacity to understand what they were doing and therefore could be criminally prosecuted.

Today, all states have a juvenile court system to handle cases of children who commit acts that would be considered criminal if they were adults. Cases processed in the juvenile courts are not considered "criminal" matters but rather juvenile delinquency proceedings. Juveniles are entitled to certain procedural safeguards in juvenile delinquency adjudications, including the right to counsel, the right to confront accusing witnesses, notice of the charges against them, the right against compelled self-incrimination, the right against double jeopardy, and to have the charges against them proved beyond a reasonable doubt. *In re Gault*, 387 U.S. 1 (1967); *In re Winship*, 397 U.S. 358 (1970); *Breed v. Jones*, 421 U.S. 519 (1975). Although some jurisdictions also provide a right to trial by jury, the federal Constitution does not require jury trials in juvenile delinquency cases. Juvenile courts traditionally have operated under a philosophy of rehabilitation rather than punishment, although children nevertheless can be deprived of their liberty in juvenile correctional institutions and suffer the stigma of being branded a juvenile delinquent.

In *State v. Q.D. and M.S.* the Washington Supreme Court considered a somewhat counterintuitive question: Does the infancy defense apply to juvenile court proceedings?

### *State v. Q.D. and M.S.*, 685 P.2d 557 (Wash. 1984)

Dimmick, Justice.

Two juveniles appeal from separate adjudications which found that they had committed offenses which if committed by an adult would be crimes. The Court of Appeals, in these consolidated appeals, certified to this court the questions whether the statutory presumption of infant incapacity, RCW 9A.04.050 (*See* Footnote 1), applies to juvenile adjudications, and if it does, what standard of proof is required to rebut the presumption. Each defendant argues that the trial court's determinations of capacity were erroneous under any standard. Appellant Q.D. additionally argues that there was insufficient evidence to convict him of trespass in the first degree. Appellant M.S. contends that the imposition of a penalty under the crime victims' compensation act is either inapplicable or discretionary in juvenile court dispositions. . . .

Appellant Q.D. was found to have capacity per RCW 9A.04.050 in a pretrial hearing. He was 11½ years old at the time of the alleged offense. At trial a different judge determined he had committed trespass in the first degree. The evidence introduced to show capacity consisted of testimony from a case worker and a detective who had worked with him in connection with his plea of guilty to a burglary committed at age 10 years. The case worker testified that Q.D. was familiar with the justice system, was street wise, and that he used his age as a shield. The detective told the court that Q.D. was cooperative in the burglary investigation, and he appeared to know his rights. The evidence in the guilt phase consisted of testimony from the principal and a custodial engineer of

(*Continues*)

the school in which Q.D. was charged with trespass. The engineer testified that he saw Q.D. sitting on the school grounds about 2 P.M. playing with some keys that looked like the set belonging to the night custodian. When the engineer checked his desk which was in an unlocked office, he found that the keys were missing as was the burglar alarm key. The engineer could not be certain that he had seen the keys since the morning. He called the principal and they brought Q.D. into the office. When Q.D. arose from the chair he had been sitting on in the office, the burglar alarm key was discovered on a radiator behind the chair.

Appellant M.S., in a single proceeding, was found to have capacity and to have committed indecent liberties on a 4½-year-old child for whom she was babysitting. Evidence included the testimony of the victim, the victim's mother, a physician who had examined the victim, and a social worker who had interviewed the victim. M.S. was less than 3 months from 12 years old at the time of the offense. The issue of capacity was first raised by the defendant in a motion to dismiss at the close of the State's evidence. The State argued that defendant's proximity to the age when capacity is assumed, the defendant's threats to the victim not to tell what had happened, and her secrecy in carrying out the act were ample proof of capacity. The trial judge, in his oral ruling finding capacity, stated that the responsibility entrusted to the defendant by the victim's mother and her own parents in permitting her to babysit showed a recognition of the defendant's maturity.

# I.

## APPLICABILITY OF RCW 9A.04.050 TO JUVENILE COURTS . . .

At common law, children below the age of 7 were conclusively presumed to be incapable of committing crime, and children over the age of 14 were presumed capable and treated as adults. Children between these ages were rebuttably presumed incapable of committing crime. Washington codified these presumptions amending the age of conclusive incapacity to 7, and presumed capacity to 12 years of age. As recently as 1975, the Legislature again included the infancy defense in the criminal code. The purpose of the presumption is to protect from the criminal justice system those individuals of tender years who are less capable than adults of appreciating the wrongfulness of their behavior.

The infancy defense fell into disuse during the early part of the century with the advent of reforms intended to substitute treatment and rehabilitation for punishment of juvenile offenders. This *parens patriae* system, believed not to be a criminal one, had no need of the infancy defense.

The juvenile justice system in recent years has evolved from *parens patriae* scheme to one more akin to adult criminal proceedings. The United States Supreme Court has been critical of the *parens patriae* scheme as failing to provide safeguards due an adult criminal defendant, while subjecting the juvenile defendant to similar stigma, and possible loss of liberty. This court has acknowledged Washington's departure from a strictly *parens patriae* scheme to a more criminal one, involving both rehabilitation and punishment. Being a criminal defense, RCW 9A.04.050 should be available to juvenile proceedings that are criminal in nature.

The principles of construction of criminal statutes, made necessary by our recognition of the criminal nature of juvenile court proceedings, also compel us to conclude that RCW 9A.04.050 applies to proceedings in juvenile courts.

A finding that RCW 9A.04.050 does not apply to juvenile courts would render that statute meaningless or superfluous contrary to rules of construction. Juvenile courts have exclusive jurisdiction over all individuals under the chronological age of 18 who have committed acts designated criminal if committed by an adult. Declination of jurisdiction and transfer to adult court is limited to instances where it is in the best interest of the juvenile or the public. Thus, all juveniles who can avail themselves of the infancy defense will come under the jurisdiction of the juvenile court, and most will remain there. Implied statutory repeals are found not to exist where the two statutes can be reconciled and given effect. Goals of the Juvenile Justice Act of 1977 include accountability for criminal behavior and punishment commensurate with age and crime. A goal of the criminal code is to safeguard conduct that is not culpable. The infancy defense which excludes from criminal condemnation persons not capable of culpable, criminal acts, is consistent with the overlapping goals of the Juvenile Justice Act of 1977 and the Washington Criminal Code.

## II.

## STANDARD OF PROOF UNDER RCW 9A.04.050

The State has the burden of rebutting the statutory presumption of incapacity of juveniles age 8 and less than 12 years. Capacity must be found to exist separate from the specific mental element of the crime charged. While capacity is similar to the mental element of a specific crime or offense, it is not an element of the offense, but is rather a general determination that the individual understood the act and its wrongfulness. Both defendants liken the incapacity presumption to a jurisdictional presumption. Were capacity an element of the crime, proof beyond a reasonable doubt would be required. But capacity, not being an element of the crime, does not require as stringent a standard of proof.

Few jurisdictions have ruled on the appropriate standard of proof necessary to rebut the presumption of incapacity, and fewer still have discussed their reasoning for preferring one standard over another. It appears that other states have split between requiring proof beyond a reasonable doubt, and clear and convincing proof.

Our recent discussion of the standard of proof to be applied in involuntary commitment proceedings offers guidance. In *Dunner v. McLaughlin*, 676 P.2d 444 (Wash. 1984), we held that the burden of proof should be by clear, cogent and convincing evidence. In so holding, we recognized that the preponderance of the evidence standard was inadequate, but the proof beyond a reasonable doubt standard imposed a burden which, as a practical matter, was unreasonably difficult, thus undercutting the State's legitimate interests.

The Legislature, by requiring the State to rebut the presumption of incapacity, has assumed a greater burden than the minimal proof imposed by the preponderance of the evidence standard. On the other hand, to require the State to prove capacity beyond a reasonable doubt when the State must also prove the specific mental element of the charged offense by the same standard, is unnecessarily duplicative. Frequently, the same facts required to prove *mens rea* will be probative of capacity, yet the overlap is not complete. Capacity to be culpable must exist in order to maintain the specific mental element of the charged offense. Once the generalized determination of capacity is found, the State must prove beyond a reasonable doubt that the juvenile defendant possessed the specific mental element. The clear and convincing standard reflects the State's assumption of a greater burden than does the preponderance of the evidence standard. At the same time, the liberty interest of the juvenile is fully protected by the requirement of proof beyond a reasonable doubt of the specific mental element. We therefore require the State to rebut the presumption of incapacity by clear and convincing evidence.

## III.

## EVIDENCE OF CAPACITY

We do not need to reach the question of whether there was substantial evidence to show that Q.D. understood the act of trespass or understood it to be wrong, as we reverse on other grounds. Nevertheless, a discussion of capacity in this case may prove instructive to trial courts. Q.D. argues that the evidence showed only that he was familiar with the juvenile system through his previous plea of guilty to a burglary charge, but did not show he understood the act and wrongfulness of trespass. The language of RCW 9A.04.050 clearly indicates that a capacity determination must be made in reference to the specific act charged: "understand the act . . . and to know that it was wrong." If Q.D. is correct that the evidence showed no more than a general understanding of the justice system, he would be correct in concluding that the State did not show an understanding and knowing wrongfulness of trespass. In addition, an understanding of the wrongfulness of burglary does not alone establish capacity in regard to trespass. While both offenses include entry or unlawfully remaining in a building, burglary also requires an intent to commit a crime against a person or property therein. Defendant may well understand that it is wrong to enter a locked building with the intention of committing a crime, but not know that entering an unlocked school building is wrong.

The issue of capacity was first raised on M.S.'s motion to dismiss at the end of the trial. The judge stated in response to arguments of counsel that he was persuaded by the confidence in defendant's maturity held by the mother of the victim and her own parents in permitting her to assume the responsibility for babysitting. Contrary to defendant's arguments that the trial judge created a prima facie proof of capacity based solely on babysitting, there was other evidence to support his finding of capacity. The defendant waited until she and the victim were alone evidencing a desire for secrecy.

*(Continues)*

The defendant later admonished the victim not to tell what happened, further supporting the finding that the defendant knew the act was wrong. Lastly, the defendant was less than 3 months from the age at which capacity is presumed to exist. There was clear and convincing circumstantial evidence that M.S. understood the act of indecent liberties and knew it to be wrong.

Finally, in response to the parties' requests for guidelines concerning the forum of the capacity hearing, we find the separate hearing in Q.D.'s case, and the single hearing of capacity and the substantive charge in M.S.'s case to be appropriate under the different circumstances in each. In Q.D.'s case, prior criminal history was the basis for attempting to prove capacity, and thus a separate hearing avoided prejudice. In M.S.'s case, the facts of the offense were offered to show capacity, and a separate hearing would be unduly repetitive. Rather than delineating a rigid rule, the circumstances should dictate whether a separate hearing is appropriate. . . . In the event that it is necessary to show capacity by proof of both criminal history and the particular facts of the offense charged, caution should be employed to prevent the introduction of evidence of prior history from prejudicing the determination on the merits. . . .

### CONCLUSION

The infancy defense, codified in RCW 9A.04.050 applies to juvenile proceedings. The proof that a juvenile of 8 years and less than 12 years understood the charged act and knew it to be wrong is by clear and convincing evidence. Because the State met this burden in M.S.'s case, the finding that she committed indecent liberties is affirmed. M.S.'s disposition . . . is likewise affirmed. Because the State failed to prove Q.D.'s entry into the school building, his conviction is reversed.

### Notes and Questions

1. Not all jurisdictions are in accord with Washington. For example, in *Gammon v. Berlat*, 696 P.2d 700 (Ariz. 1985), the Arizona Supreme Court explicitly held that the statutory presumptions regarding infancy do not apply to juvenile proceedings.
2. New York has a statute that prohibits the operation of a boat by a person under the age of 18. If a 14-year-old boy is charged with illegally driving a boat, can he use infancy as a defense? No, according to the court in *People v. Ullman*, 609 N.Y.2d 750 (1993). What reasons might justify such a ruling?

## Intoxication

Under the early common law intoxication was not a defense to a crime. As time progressed the common law began to permit evidence of intoxication to reduce criminal responsibility for some crimes. Moreover, the common law also began to make a sharp distinction between voluntary and involuntary intoxication.

Today, **involuntary intoxication** is universally recognized as a defense to a crime. To establish involuntary intoxication as a defense, the defendant generally must show that he or she unknowingly or involuntarily ingested an intoxicating substance that prevented him or her from forming the *mens rea* required for conviction or understanding the difference between right and wrong. Because the defendant was not at fault in becoming intoxicated, the law recognizes that criminal responsibility should not be imposed.

The law regarding **voluntary intoxication**, however, is far from uniform. Some jurisdictions permit the jury to consider a defendant's voluntary intoxication if the charged crime requires a specific intent. Some states limit consideration of voluntary intoxication to first-degree murder prosecutions and the issue of premeditation. Some states do not permit the jury to consider evidence of voluntary intoxication under any circumstances, and there is no constitutional obligation for them to do so. *Montana v. Egelhoff*, 518 U.S. 37 (1996).

In *State v. Cameron*, the New Jersey Supreme Court grapples with the proper role of a defendant's voluntary intoxication in a criminal prosecution.

### *State v. Cameron*, 514 A.2d 1302 (N.J. 1986)

Clifford, J. . . .

Defendant, Michele Cameron, age 22 at the time of trial, was indicted for second degree aggravated assault, possession of a weapon, a broken bottle, with a purpose to use it unlawfully, and fourth degree resisting arrest. A jury convicted defendant of all charges. [The Appellate Division reversed, "holding that it was error not to have given an intoxication charge."] . . .

The charges arose out of an incident of June 6, 1981, on a vacant lot in Trenton. The unreported opinion of the Appellate Division depicts the following tableau of significant events:

> The victim, Joseph McKinney, was playing cards with four other men. Defendant approached and disrupted the game with her conduct. The participants moved their card table to a new location within the lot. Defendant followed them, however, and overturned the table. The table was righted and the game resumed. Shortly thereafter, defendant attacked McKinney with a broken bottle. As a result of that attack he sustained an injury to his hand, which necessitated 36 stitches and caused permanent injury.
>
> Defendant reacted with violence to the arrival of the police. She threw a bottle at their vehicle, shouted obscenities, and tried to fight them off. She had to be restrained and handcuffed in the police wagon. . . .

The heart of the Appellate Division's reversal of defendant's conviction is found in its determination that voluntary intoxication is a defense when it negates an essential element of the offense—here, purposeful conduct. We agree with that proposition. Likewise are we in accord with the determinations of the court below that all three of the charges of which this defendant was convicted—aggravated assault, the possession offense, and resisting arrest—have purposeful conduct as an element of the offense; and that a person acts purposely "with respect to the nature of his conduct or a result thereof if it is his conscious object to engage in conduct of that nature or to cause such a result" (quoting *N.J.S.A.* 2C:2-2(b)(1)). We part company with the Appellate Division, however, in its conclusion that the circumstances disclosed by the evidence in this case required that the issue of defendant's intoxication be submitted to the jury.

The court below noted that every witness who testified gave some appraisal of defendant's condition. On the basis of that evidence the Appellate Division concluded that defendant's conduct was both bizarre and violent. She had been drinking and could not be reasoned with. The victim thought she was intoxicated and two police officers thought she was under the influence of something. Not one witness who testified thought that her conduct was normal. Therefore, it was for the jury to determine if she was intoxicated, and if so, whether the element of purposefulness was negated thereby.

The quoted passage reflects a misapprehension of the level of proof required to demonstrate intoxication for purposes of demonstrating an inability to engage in purposeful conduct.

Under the common law intoxication was not a defense to a criminal charge. Rather than being denominated a defense, intoxication was viewed as a "condition of fact," or, in a homicide case, as "a mere circumstance to be considered in determining whether premeditation was present or absent."

Notwithstanding the general proposition that voluntary intoxication is no defense, the early cases nevertheless held that in some circumstances intoxication could be resorted to for defensive purposes—specifically, to show the absence of a specific intent.

> The exceptional immunity extended to the drunkard is limited to those instances where the crime involves a specific, actual intent. When the degree of intoxication is such as to render the person incapable of entertaining such intent, it is an effective defence. If it falls short of this it is worthless. [*Warner v. State*, 29 A. 505 (N.J. 1894)]

The principle that developed from the foregoing approach—that intoxication formed the basis for a defense to a "specific intent" crime but not to one involving only "general" intent—persisted for about three-quarters of a century, or until this Court's decision in *State v. Maik*, 287 A.2d 715 (N.J. 1972).

Eventually the problems inherent in the application of the specific-general intent dichotomy surfaced. In *State v. Maik, supra*, this Court dwelt on the elusiveness of the distinction between "specific" and "general" intent crimes, particularly as that distinction determined what role voluntary intoxication played in a criminal prosecution. Chief Justice Weintraub's opinion for the Court restated the original proposition that "a defendant will not be relieved of criminal responsibility because he was under the influence of intoxicants or drugs voluntarily taken," and then set forth four exceptions to that rule: (1) the ingestion of drugs for medication, producing unexpected or bizarre results; (2) impairment of mental faculties negating only premeditation or deliberation, to preclude elevation to first degree murder; (3) reduction of felony homicide to second degree murder when the felonious intent is negated; and (4) when insanity results.

*Maik*, a murder prosecution, was not given a uniform reading. As later pointed out in *State v. Stasio*, 396 A.2d 1129 (N.J. 1979), the Appellate Division in *State v. Del Vecchio*, 361 A.2d 579

(*Continues*)

(*Continued*)

(N.J.Super. 1976), limited *Maik's* sweep to the proposition that voluntary intoxication is relevant only to the determination of whether a murder may be raised to first degree, whereas Judge Allcorn, dissenting in *State v. Atkins*, 377 A.2d 718 (N.J. Super. App. Div. 1977), rev'd, 396 A.2d 1122 (N.J. 1979), read *Maik* to rule out voluntary intoxication as a defense to any criminal prosecution, irrespective of whether a specific or general intent was an element of the offense. It thus fell to the Court in *Stasio* to resolve the difference in interpretations that had been accorded *Maik*, an undertaking that we recognize met with but limited success.

A majority of the Court in *Stasio* found the difference between general and specific intent to be "not readily ascertainable," and concluded that honoring the distinction would give rise to "incongruous results by irrationally allowing intoxication to excuse some crimes but not others." 396 A.2d 1129. On the one hand, therefore, the Court brought some stability to the area, with its holding that absent one or more of the exceptions stated in *Maik*, the principle that voluntary intoxication will not excuse criminal conduct was applicable to all crimes. But on the other hand, because of critically important legislation that was looming in the background at the time of the *Stasio* decision, the opinion may have posed more questions than it answered. . . .

Which brings us to the Code. . . .

As originally enacted . . . N.J.S.A. 2C:2-8 provided:

a. Except as provided in subsection d. of this section, intoxication of the actor is not a defense unless it negatives an element of the offense.

b. When recklessness establishes an element of the offense, if the actor, due to self-induced intoxication, is unaware of a risk of which he would have been aware had he been sober, such unawareness is immaterial. . . .

c. Intoxication which (1) is not self-induced or (2) is pathological is an affirmative defense if by reason of such intoxication the actor at the time of his conduct lacks substantial and adequate capacity either to appreciate its wrongfulness or to conform his conduct to the requirement of law.

d. Definitions. In this section unless a different meaning plainly is required:

(1) "Intoxication" means a disturbance of mental or physical capacities resulting from the introduction of substances into the body;

(2) "self-induced intoxication" means intoxication caused by substances which the actor knowingly introduces into his body, the tendency of which to cause intoxication he knows or ought to know, unless he introduces them pursuant to medical advice or under such circumstances as would afford a defense to a charge of crime. . . .

As is readily apparent, self-induced intoxication is not a defense unless it negatives an element of the offense. Under the common-law intoxication defense, as construed by the Commission, intoxication could either exculpate or mitigate guilt "if the defendant's intoxication, in fact, prevents his having formed a mental state which is an element of the offense and if the law will recognize the proof of the lack of that mental state." Thus, the Commission recognized that under pre-Code law, intoxication was admissible as a defense to a "specific" intent, but not a "general" intent, crime.

The original proposed Code rejected the specific/general intent distinction, choosing to rely instead on the reference to the four states of culpability for offenses under the Code: negligent, reckless, knowing, and purposeful conduct, N.J.S.A. 2C:2-2(b). Although the Code employs terminology that differs from that used to articulate the common-law principles referable to intoxication, the Commission concluded that the ultimately-enacted statutory intoxication defense would achieve the same result as that reached under the common law. In essence, "[t]hat which the cases now describe as a 'specific intent' can be equated, for this purpose, with that which the Code defines as 'purpose' and 'knowledge.' A 'general intent' can be equated with that which the Code defines as 'recklessness,' or criminal 'negligence.'" Therefore, according to the Commissioners, N.J.S.A. 2C:2-8(a) and (b) would serve much the same end as was achieved by the common-law approach. Specifically, N.J.S.A. 2C:2-8(a) permits evidence of intoxication as a defense to crimes requiring either "purposeful" or "knowing" mental states but it excludes evidence of intoxication as a defense to crimes requiring mental states of only recklessness or negligence.

N.J.S.A. 2C:2-8 was modeled after the Model Penal Code (MPC) § 2.08. The drafters of the MPC, as did the New Jersey Commission, criticized the specific-general intent distinction, and adopted instead the same four states of culpability eventually enacted in the Code. In the commentary, the drafters of the MPC expressly stated their intention that intoxication be admissible to disprove the culpability factors of purpose or knowledge, but that for crimes requiring only recklessness or negligence, exculpation based on intoxication should be excluded as a matter of law.

The drafters explicitly determined that intoxication ought to be accorded a significance that is entirely co-extensive with its relevance to disprove purpose or knowledge, when they are the requisite mental elements of a specific crime. . . . [W]hen the definition of a crime or a degree thereof requires proof of such a state of mind, the legal policy involved will almost certainly obtain whether or not the absence of purpose or knowledge is due to the actor's self-induced intoxication or to some other cause.

The policy reasons for requiring purpose or knowledge as a requisite element of some crimes are that in the absence of those states of mind, the criminal conduct would not present a comparable danger, or the actor would not pose as significant a threat. Moreover, the ends of legal policy are better served by subjecting to graver sanctions those who consciously defy legal norms. It was those policy reasons that dictated the result that the intoxication defense should be available when it negatives purpose or knowledge. The drafters concluded: "If the mental state which is the basis of the law's concern does not exist, the reason for its nonexistence is quite plainly immaterial."

Thus, when the requisite culpability for a crime is that the person act "purposely" or "knowingly," evidence of voluntary intoxication is admissible to disprove that requisite mental state. The language of N.J.S.A. 2C:2-8 and its legislative history make this unmistakably clear and lend support to Stasio's and Atkins' minority opinions.

The foregoing discussion establishes that proof of voluntary intoxication would negate the culpability elements in the offenses of which this defendant was convicted. The charges—aggravated assault, possession of a weapon with a purpose to use it unlawfully, and resisting arrest—all require purposeful conduct (aggravated assault uses "purposely" or "knowingly" in the alternative). The question is what level of intoxication must be demonstrated before a trial court is required to submit the issue to a jury. What quantum of proof is required?

The guiding principle is simple enough of articulation. We need not here repeat the citations to authorities already referred to in this opinion that use the language of "prostration of faculties such that defendant was rendered incapable of forming an intent." Justice Depue's instruction to a jury over a century ago, quoted with approval in State v. Treficanto, 146 A. 313 (N.J. 1929) remains good law:

> You should carefully discriminate between that excitable condition of the mind produced by drink, which is not incapable of forming an intent, but determines to act on a slight provocation, and such prostration of the faculties by intoxication as puts the accused in such a state that he is incapable of forming an intention from which he shall act. . . .

So firmly fixed in our case law is the requirement of "prostration of faculties" as the minimum requirement for an intoxication defense that we feel secure in our assumption that the legislature intended nothing different in its statutory definition of intoxication: "a disturbance of mental or physical capacities resulting from the introduction of substances into the body." In order to satisfy the statutory condition that to qualify as a defense intoxication must negative an element of the offense, the intoxication must be of an extremely high level. Therefore, consistency between the definition of intoxication and the effect given it by the legislature require that the standard be "prostration of faculties." Less certain is how that standard is to be satisfied. . . .

[W]e conclude that some of the factors pertinent to the determination of intoxication sufficient to satisfy the test of "prostration of faculties"—a shorthand expression used here to indicate a condition of intoxication that renders the actor incapable of purposeful or knowing conduct—are the following: the quantity of intoxicant consumed, the period of time involved, the actor's conduct as perceived by others (what he said, how he said it, how he appeared, how he acted, how his coordination or lack thereof manifested itself), any odor of alcohol or other intoxicating substance, the results of any tests to determine blood-alcohol content, and the actor's ability to recall significant events.

Measured by the foregoing standard and evidence relevant thereto, it is apparent that the record in this case is insufficient to have required the trial court to grant defendant's request to charge intoxication. . . .

True, the victim testified that defendant was drunk, and defendant herself said she felt "pretty intoxicated," "pretty bad," and "very intoxicated." But these are no more than conclusory labels, of little assistance in determining whether any drinking produced a prostration of faculties.

More to the point is the fact that defendant carried a quart of wine, that she was drinking (we are not told over what period of time) with other people on the vacant lot, that about a pint of the wine

*(Continues)*

was consumed, and that defendant did not drink this alone but rather "gave most of it out, gave some of it out." Defendant's conduct was violent, abusive, and threatening. But with it all there is not the slightest suggestion that she did not know what she was doing or that her faculties were so beclouded by the wine that she was incapable of engaging in purposeful conduct. That the purpose of the conduct may have been bizarre, even violent, is not the test. The critical question is whether defendant was capable of forming that bizarre or violent purpose, and we do not find sufficient evidence to permit a jury to say she was not.

Defendant's own testimony, if believed, would furnish a basis for her actions. She said she acted in self-defense, to ward off a sexual attack by McKinney and others. She recited the details of that attack and of her reaction to it with full recall and in explicit detail, explaining that her abuse of the police officers was sparked by her being upset by their unfairness in locking her up rather than apprehending McKinney.

Ordinarily, of course, the question of whether a defendant's asserted intoxication satisfies the standards enunciated in this opinion should be resolved by the jury. But here, viewing the evidence and the legitimate inferences to be drawn therefrom in the light most favorable to defendant, the best that can be made of the proof of intoxication is that defendant may have been extremely agitated and distraught. It may even be that a fact-finder could conclude that her powers of rational thought and deductive reasoning had been affected. But there is no suggestion in the evidence that defendant's faculties were so prostrated by her consumption of something less than a pint of wine as to render her incapable of purposeful or knowing conduct. The trial court correctly refused defendant's request. . . .

To recapitulate: (1) under the Code voluntary intoxication is a defense to a criminal charge that contains as an essential element proof that a defendant acted purposely or knowingly; (2) the Code definition of "intoxication" contemplates a condition by which the mental or physical capacities of the actor, because of the introduction of intoxicating substances into the body, are so prostrated as to render him incapable of purposeful or knowing conduct. . . .

The judgment below is reversed and the cause is remanded to the Appellate Division for further proceedings consistent with this opinion.

## Notes and Questions

1. Why should a person who voluntarily becomes intoxicated be allowed to escape criminal liability based on a lack of capacity to act "purposely" or "knowingly" or to understand what he or she is doing? Does recognizing a defense of voluntary intoxication have dubious policy implications? Or, should we be sympathetic to the oft-heard complaint, "It wasn't 'me,' it was the alcohol"?
2. What evidence of intoxication must a defendant offer before an intoxication defense is required to be submitted to a jury? Is this a different question than whether the defendant is entitled to prevail on such a defense? Should the jury at least have been required to consider the defense of involuntary intoxication in *Cameron*, even if it ultimately chose to reject the defense?

## Conclusion

Excuse defenses are based on the premise that an actor should not be held criminally responsible for conduct resulting from a "disabling condition" that renders the individual not truly blameworthy and hence not a good candidate for punishment. In this chapter we considered various disabling conditions that well might interfere with individuals' ability to exercise free will, make rational decisions, or control conduct, including mental illness, the threat of injury from another person, tender years, and intoxication. However, we have seen that "simply" being mentally ill, or subjected to a threat, or being young, or intoxicated is not necessarily sufficient to excuse criminal behavior. The disabling condition must have demonstrable consequences. Thus, for example, not all mentally ill defendants are "insane" and not all acts committed to avoid threatened harm qualify the actor for the defense of duress. The specific tests used in connection with excuse defenses and the circumstances under which they are available are critically important.

We likewise considered different approaches to the entrapment defense, which reflect policy decisions about how far law enforcement should be able to go in facilitating or encouraging individuals to engage in criminal behavior and how resilient those who "take the bait" are expected to be. The "objective" test for entrapment centers on society's interest in guarding against overly aggressive law enforcement tactics and makes the specific defendant's state of mind or willingness to commit a crime irrelevant. Conversely, the "subjective" test focuses on whether the defendant on trial was "predisposed" to commit the crime and seized the chance to do so when presented with an opportunity or whether the government instead induced or enticed an individual who would not otherwise have committed a crime to do so.

Unlike the justification defenses we considered in Chapter 4, a decision to excuse an individual for conduct that ordinarily would be a crime does not represent a societal judgment that the actor's behavior is condoned or was appropriate under the circumstances. Rather, we focus on the actor and his or her peculiar infirmities. Although we deeply lament a tragedy such as Andrea Yates' drowning her small children in a bathtub, we might nevertheless be willing to excuse her and others like her from criminal responsibility because they are not sufficiently blameworthy to be convicted and punished.

## Key Terms

| | |
|---|---|
| duress | mistake of law |
| entrapment | M'Naghten rule |
| Federal Insanity Defense Reform Act | Model Penal Code test for insanity |
| guilty but mentally ill (GBMI) | objective test for entrapment |
| infancy | product test |
| insanity defense | subjective test for entrapment |
| involuntary intoxication | voluntary intoxication |
| mistake of fact | |

## Review Questions

1. Briefly describe what is meant by *excuse defenses* and how they are different from *justification defenses*.
2. Explain the M'Naghten rule and its place in modern criminal law.
3. What is meant by *mistake of law*? Give an example in which a person might claim "mistake" as an excuse defense.
4. May a person claim *mistake of law* as an excuse defense if the "mistake" was that she or he did not understand the law? Why or why not?
5. Define *mistake of fact*. How does *mistake of fact* differ from *mistake of law*?
6. Does the the existence of a threat of harm automatically entitle a criminal defendant to rely on *duress* as an excuse defense? Why or why not?
7. List the elements of the *subjective test* used by courts in cases involving *entrapment*.
8. Describe the *objective test* used by courts in cases involving *entrapment*.
9. Differentiate between *voluntary* and *involuntary intoxication* and give an example of how each might be used as a defense.

## Notes

1. *See generally* Morse SJ, Hoffman MB. The uneasy entente between legal insanity and *mens rea*: beyond *Clark v. Arizona*. J. Criminal Law Criminol. 2007;97:1071.
2. Gundach-Evans AD. "*State v. Calin*": the paradox of the insanity defense and guilty but mentally ill statute, recognizing impairment without affording treatment. *S. Dak. Law Rev.* 2006;51:122.
3. *See* Dressler J. *Understanding Criminal Law*, 4th ed. Newark, NJ: LexisNexis, 2006:323–324.
4. *See* Robinson PH, Cahill MT. *Law Without Justice: Why Criminal Law Doesn't Give People What They Deserve*. Oxford, UK: Oxford University Press, 2006:137–139, 180–183.

## Footnotes for *State v. Johnson*, 399 A.2d 469 (RI 1979)

5. "(1) A person is not responsible for criminal conduct if at the time of such conduct, as a result of mental disease or defect, he lacks substantial capacity either to appreciate the criminality (wrongfulness) of his conduct or to conform his conduct to the requirements of law.

   "(2) As used in this article, the terms 'mental disease or defect' do not include an abnormality manifested only by repeated criminal or otherwise antisocial conduct" Model Penal Code, § 4.01 (Final Draft, 1962).

8. This test was proposed as an alternative by the minority of the American Law Institute Council that drafted the Model Penal Code. One of its most forceful advocates was Professor Herbert Wechsler, the reporter for the Model Penal Code. . . . The majority of the A.L.I. Council rejected this alternative because they deemed it unwise to present questions of justice to the jury. As this opinion indicates, we believe that the jury's resolution of the responsibility issue is in the final analysis always predicated upon the community sense of justice and it is preferable to be forthright about the basis of that decision.

## Footnote to *State v. Q.D. and M.S.*, 685 P.2d 557 (Wash. 1984)

1. RCW 9A.04.050 provides in part:
   "Children under the age of eight years are incapable of committing crime. Children of eight and under twelve years of age are presumed to be incapable of committing crime, but this presumption may be removed by proof that they have sufficient capacity to understand the act or neglect, and to know that it was wrong."

# 6 Criminal Homicide

## Chapter Objectives

- Understand the general elements of criminal homicide
- Identify the different elements of murder and manslaughter under the common law, Model Penal Code, and graded homicide jurisdictions
- Understand the felony murder rule
- Understand the concepts of "life" and "death" as important to the law of criminal homicide.

"**Homicide** obviously means the killing of a human being by a human being."[1] Not all homicides are criminal, however. Some killings are accidental and are not punished criminally. Others, as we have seen, may be justified (*e.g.,* in self-defense, in war, in lawful executions) or excused (*e.g.,* resulting from insanity or infancy) even though committed intentionally. Although all criminal homicides involve the same harm—the death of a human being—the law does not necessarily treat them as being equally reprehensible. For example, first-degree murder may be punishable by death or life imprisonment, whereas involuntary manslaughter or criminally negligent homicide may result in a short prison or jail sentence or even in probation.

Our primary objective in this chapter is to analyze the sorting principles used to distinguish various forms of criminal homicide. This task requires a foray into the distinctions between murder and voluntary manslaughter and between first- and second-degree murder and an examination of unintentional killings. We also probe more deeply into the requirement that a homicide entail the "killing of a human being." In particular, we consider whether prenatal life (*e.g.,* a fetus or an embryo) comes within the ambit of the law of criminal homicide. We finally explore the meaning of "death" in the criminal law.

## Sorting Principles

### Murder and Voluntary Manslaughter: The Common Law Approach

During early English history homicides (like other wrongs that today are considered crimes) were treated as private, civil matters between the killer and the victim's family.

The crown did not intervene to impose punishment. Instead, the slayer was obligated to pay compensation or damages to the deceased's family. Retaliatory killings and cycles of blood feuding between families were common. Homicides gradually became defined as crimes against the greater social order, and the state assumed responsibility for punishing them.

Two categories of homicide were recognized by the 12th century. Nonfelonious homicides—those that were justifiable or excused—became immunized from criminal punishment. In contrast, felonious or murderous homicides were punished by death. By the 1400s all culpable homicides were classified as **murder**. No further distinctions were made between unlawful killings, and all were punished capitally.

The indiscriminate application of the death penalty against murderers (as well as on many other kinds of felons) gave rise to a judicial practice called "benefit of clergy" that was used to alleviate the harsh punishment regime. Ecclesiastical courts had jurisdiction over members of the clergy who were accused of crimes. These courts did not impose capital punishment. A cleric charged with a crime thus was given the "**benefit of clergy**" and transferred from a royal court to an ecclesiastical court, where punishment customarily entailed up to a year's confinement and the branding of a thumb.

Illiteracy was widespread in medieval England, so readers frequently were enlisted to make the text of religious tracts accessible to the laity. To enable a broad class of offenders to qualify for benefit of clergy, judges began using a fiction that if an accused felon could read, he (women were never granted this status because they were not eligible for holy orders) was presumed to be a member of the clergy. Offenders who could not read typically memorized a passage from the Bible and recited it as proof of their literacy. The royal courts eventually retained jurisdiction over offenders who were given benefit of clergy and ceased transferring them to ecclesiastical courts. However, those offenders remained exempt from capital punishment. A statute passed in 1489 provided that benefit of clergy could be claimed just once. Thus the practice of thumb-branding represented a primitive (and painful) form of record keeping.

The judicial maneuvering related to conferring benefit of clergy to spare felons from the gallows "reduced the administration of justice to a sort of farce."[2] Parliament responded to this state of affairs. Through a series of statutes enacted between 1496 and 1547, benefit of clergy was denied in all cases of "murder of malice prepensed" or "malice aforethought."

Denying benefit of clergy for murder committed with malice aforethought directly led to the creation of a new category of criminal homicide: manslaughter. *The crucial distinguishing factor between murder and manslaughter was the presence or absence of malice aforethought.* Murder—an unlawful killing committed with malice—was for that reason a nonclergyable offense; unless granted a royal pardon a convicted murderer faced capital punishment. Manslaughter—an unlawful killing without malice—remained a felony punishable by death, but offenders convicted of that crime were still eligible for benefit of clergy.

Because "**malice**" distinguished murder and manslaughter, it was imperative for the courts to define that term. This proved to be a daunting task. Consider the definitions provided in *State v. Jensen*, 417 P.2d 273, 282–283 (Kan. 1966):

We turn to the early writers of the common law to ascertain their definition of malice as applied to common-law murder. Blackstone defines the term as:

". . . malice prepense, *malitia praecogitata,* is not so properly spite or malevolence to the deceased in particular, as any evil design in general; the dictate of a wicked, depraved, and malignant heart; *un disposition a faire un male chose:* and it may be either *express,* or *implied* in law. Express malice is when one, with a sedate deliberate mind and formed design, doth kill another: which formed design is evidenced by external circumstances discovering that inward intention; as lying in wait, antecedent menaces, former grudges, and concerted schemes to do him some bodily harm. . . .

". . . in many cases where no malice is expressed, the law will imply it: as, where a man willfully poisons another, in such a deliberate act the law presumes malice, though no particular enmity can be proved. And if a man kills another suddenly, without any, or without a considerable provocation, the law implies malice; for no person, unless of an abandoned heart, would be guilty of such an act, upon a slight or no apparent cause" (4 Comm. pp. 198–200). . . .

Definitions of malice proved unwieldy and often case-specific. As we shall see, the framers of the Model Penal Code (MPC) considered "malice" to be so ill defined that they abandoned the concept altogether in distinguishing between murder and manslaughter[3]:

At common law, murder was defined as the unlawful killing of another human being with "malice aforethought." Whatever the original meaning of that phrase, it became over time an "arbitrary symbol" used by judges to signify any of a number of mental states deemed sufficient to support liability for murder. Successive generations added new content to "malice aforethought" until it encompassed a variety of mental attitudes bearing no predictable relationship to the ordinary sense of the two words.

The trend for courts in jurisdictions that continue to rely on malice as the separating factor between murder and manslaughter is to dispense with the more ambiguous definitional language and substitute the underlying states of mind or conduct that cause an unlawful killing to be murder instead of manslaughter[4]:

[I]t will not solve modern homicide cases to say simply that murder is the unlawful killing of another with malice aforethought, that manslaughter is the unlawful killing of another without malice aforethought, and that no crime is committed if the killing is lawful. For an understanding of the crime of murder, it is necessary to consider, one by one, the various types of murder which the judges created and which, in general, remain to this day. . . .

To sum up these various modern types of murder, they are: (1) intent-to-kill murder; (2) intent-to-do-serious-bodily-injury murder; (3) felony murder; (4) depraved-heart murder.

We start by examining the category of killings known at common law as voluntary manslaughter. These are *intentional* homicides, which, although unlawful, do not qualify as murder. We initially consider killings committed "in the heat of passion on adequate provocation."

### Negating Malice: The Heat of Passion on Adequate Provocation

---

## Girouard v. State, 583 A.2d 718 (Md. 1991)

Cole, Judge.

In this case we are asked to reconsider whether the types of provocation sufficient to mitigate the crime of murder to manslaughter should be limited to the categories we have heretofore recognized, or whether the sufficiency of the provocation should be decided by the factfinder on a case-by-case basis. Specifically, we must determine whether words alone are provocation adequate to justify a conviction of manslaughter rather than one of second degree murder.

The Petitioner, Steven S. Girouard, and the deceased, Joyce M. Girouard, had been married for about two months on October 28, 1987, the night of Joyce's death. Both parties, who met while working in the same building, were in the army. They married after having known each other for

*(Continues)*

approximately three months. The evidence at trial indicated that the marriage was often tense and strained, and there was some evidence that after marrying Steven, Joyce had resumed a relationship with her old boyfriend, Wayne.

On the night of Joyce's death, Steven overheard her talking on the telephone to her friend, whereupon she told the friend that she had asked her first sergeant for a hardship discharge because her husband did not love her anymore. Steven went into the living room where Joyce was on the phone and asked her what she meant by her comments; she responded, "nothing." Angered by her lack of response, Steven kicked away the plate of food Joyce had in front at her. He then went to lie down in the bedroom.

Joyce followed him into the bedroom, stepped up onto the bed and onto Steven's back, pulled his hair and said, "What are you going to do, hit me?" She continued to taunt him by saying, "I never did want to marry you and you are a lousy fuck and you remind me of my dad." The barrage of insults continued with her telling Steven that she wanted a divorce, that the marriage had been a mistake and that she had never wanted to marry him. She also told him she had seen his commanding officer and filed charges against him for abuse. She then asked Steven, "What are you going to do?" Receiving no response, she continued her verbal attack. She added that she had filed charges against him in the Judge Advocate General's Office (JAG) and that he would probably be court martialed.

When she was through, Steven asked her if she had really done all those things, and she responded in the affirmative. He left the bedroom with his pillow in his arms and proceeded to the kitchen where he procured a long handled kitchen knife. He returned to Joyce in the bedroom with the knife behind the pillow. He testified that he was enraged and that he kept waiting for Joyce to say she was kidding, but Joyce continued talking. . . . Joyce reiterated that the marriage was a big mistake, that she did not love him and that the divorce would be better for her.

After pausing a moment, Joyce asked what Steven was going to do. What he did was lunge at her with the kitchen knife he had hidden behind the pillow and stab her 19 times. Realizing what he had done, he dropped the knife and went to the bathroom to shower off Joyce's blood. Feeling like he wanted to die, Steven went back to the kitchen and found two steak knives with which he slit his own wrists. He lay down on the bed waiting to die, but when he realized that he would not die from his self-inflicted wounds, he got up and called the police, telling the dispatcher that he had just murdered his wife.

When the police arrived they found Steven wandering around outside his apartment building. Steven was despondent and tearful and seemed detached, according to police officers who had been at the scene. He was unconcerned about his own wounds, talking only about how much he loved his wife and how he could not believe what he had done. Joyce Girouard was pronounced dead at the scene. . . .

Steven Girouard was convicted, at a court trial in the Circuit Court for Montgomery County, of second degree murder and was sentenced to 22 years incarceration, 10 of which were suspended. . . . The Court of Special Appeals affirmed the judgment of the circuit court in an unreported opinion. We granted certiorari to determine whether the circumstances of the case presented provocation adequate to mitigate the second degree murder charge to manslaughter.

Petitioner relies primarily on out of state cases to provide support for his argument that the provocation to mitigate murder to manslaughter should not be limited only to the traditional circumstances of: extreme assault or battery upon the defendant; mutual combat; defendant's illegal arrest; injury or serious abuse of a close relative of the defendant's; or the sudden discovery of a spouse's adultery. Petitioner argues that manslaughter is a catchall for homicides which are criminal but that lack the malice essential for a conviction of murder. Steven argues that the trial judge did find provocation (although he held it inadequate to mitigate murder) and that the categories of provocation adequate to mitigate should be broadened to include factual situations such as this one.

The State counters by stating that although there is no finite list of legally adequate provocations, the common law has developed to a point at which it may be said there are some concededly provocative acts that society is not prepared to recognize as reasonable. Words spoken by the victim, no matter how abusive or taunting, fall into a category society should not accept as adequate provocation. According to the State, if abusive words alone could mitigate murder to manslaughter, nearly every domestic argument ending in the death of one party could be mitigated to manslaughter. . . .

Initially, we note that the difference between murder and manslaughter is the presence or absence of malice. **Voluntary manslaughter** has been defined as "an *intentional* homicide, done in a sudden heat of passion, caused by adequate provocation, before there has been a reasonable opportunity for the passion to cool" (Emphasis in original). *Cox v. State*, 311 Md. 326, 331, 534 A.2d 1333 (1988).

There are certain facts that may mitigate what would normally be murder to manslaughter. For example, we have recognized as falling into that group: (1) discovering one's spouse in the act of sexual intercourse with another; (2) mutual combat; (3) assault and battery. There is also authority recognizing injury to one of the defendant's relatives or to a third party, and death resulting from resistance of an illegal arrest as adequate provocation for mitigation to manslaughter. Those acts mitigate homicide to manslaughter because they create passion in the defendant and are not considered the product of free will.

In order to determine whether murder should be mitigated to manslaughter we look to the circumstances surrounding the homicide and try to discover if it was provoked by the victim. Over the facts of the case we lay the template of the so-called "Rule of Provocation." The courts of this State have repeatedly set forth the requirements of the Rule of Provocation:

1. There must have been adequate provocation;
2. The killing must have been in the heat of passion;
3. If must have been a sudden heat of passion—that is, the killing must have followed the provocation before there had been a reasonable opportunity for the passion to cool;
4. There must have been a causal connection between the provocation, the passion, and the fatal act.

We shall assume without deciding that the second, third, and fourth of the criteria listed above were met in this case. We focus our attention on an examination of the ultimate issue in this case, that is, whether the provocation of Steven by Joyce was enough in the eyes of the law so that the murder charge against Steven should have been mitigated to voluntary manslaughter. For provocation to be "adequate," it must be "'calculated to inflame the passion of a reasonable man and tend to cause him to act for the moment from passion rather than reason.'" *Carter v. State,* 66 Md.App. at 572, 505 A.2d 545 quoting R. Perkins, *Perkins on Criminal Law* at p. 56 (2d ed. 1969). The issue we must resolve, then, is whether the taunting words uttered by Joyce were enough to inflame the passion of a *reasonable* man so that that man would be sufficiently infuriated so as to strike out in hot-blooded blind passion to kill her. Although we agree with the trial judge that there was needless provocation by Joyce, we also agree with him that the provocation was not adequate to mitigate second degree murder to voluntary manslaughter.

Although there are few Maryland cases discussing the issue at bar, those that do hold that words alone are not adequate provocation. . . .

In *Lang v. State,* 250 A.2d 276 (Md. App.), *cert. denied,* 396 U.S. 971 (1969), the Court of Special Appeals . . . did note, however, that words can constitute adequate provocation if they are accompanied by conduct indicating a present intention and ability to cause the defendant bodily harm. Clearly, no such conduct was exhibited by Joyce in this case. While Joyce did step on Steven's back and pull his hair, he could not reasonably have feared bodily harm at her hands. This, to us, is certain based on Steven's testimony at trial that Joyce was about 5'1" tall and weighed 115 pounds, while he was 6'2" tall, weighing over 200 pounds. Joyce simply did not have the size or strength to cause Steven to fear for his bodily safety. . . .

Other jurisdictions overwhelmingly agree with our cases and hold that words alone are not adequate provocation. . . . Aside from the cases, recognized legal authority in the form of treatises supports our holding. *Perkins on Criminal Law,* at p. 62, states that it is "with remarkable uniformity that even words generally regarded as 'fighting words' in the community have no recognition as adequate provocation in the eyes of the law." It is noted that

> mere words or gestures, however offensive, insulting, or abusive they may be, are not, according to the great weight of authority, adequate to reduce a homicide, although committed in a passion provoked by them, from murder to manslaughter, especially when the homicide was intentionally committed with a deadly weapon[.] (Footnotes omitted.) 40 C.J.S. *Homicide* § 47, at 909 (1944). *See also,* 40 Am.Jur.2d *Homicide* § 64, at 357 (1968).

Thus, with no reservation, we hold that the provocation in this case was not enough to cause a reasonable man to stab his provoker 19 times. Although a psychologist testified to Steven's mental problems and his need for acceptance and love, . . . [t]he standard is one of reasonableness; it does not and should not focus on the peculiar frailties of mind of the Petitioner. That standard of reasonableness has not been met here. We cannot in good conscience countenance holding that a verbal domestic argument ending in the death of one spouse can result in a conviction of manslaughter. We agree with the trial judge that social necessity dictates our holding. Domestic arguments easily escalate into furious fights. We perceive no reason for a holding in favor of those who find the easiest way to end a domestic dispute is by killing the offending spouse. . . .

*(Continues)*

## Notes and Questions

1. As the court notes in *Girouard,* the "mere words" doctrine was firmly established at common law. If we accept the premise that words can be hurtful and inspire great anger—inflaming "passion"—what policy reasons can be offered in support of the "mere words" doctrine?
2. In *Girouard* the defendant was not arguing that he was entitled to be acquitted of murder but only that the finder of fact should be allowed to *consider* his claim of provocation. What would be wrong with a rule that allowed the judge or jury to consider the adequacy of provocation on a case-by-case basis?
3. The *Girouard* court ruled that "the provocation in this case was not enough to cause a reasonable man to stab his provoker 19 times." Is that a fair statement of the issue? Note that manslaughter is a serious crime and that "reasonable" people, by definition, do not unlawfully kill others. What, precisely, is it that the law requires to be "reasonable" when a crime that otherwise would be considered murder is reduced to manslaughter?

## Burden of Proof and Provocation

Which party should be required to establish whether the defendant killed in the heat of passion on adequate provocation to distinguish murder from voluntary manslaughter? Should the defense be burdened with proving that the killing occurred in the heat of passion on adequate provocation, thus negating malice, or should the prosecution have to prove that the defendant did kill with malice, thus affirmatively refuting the heat of passion defense? The U.S. Supreme Court addressed this issue in *Mullaney v. Wilbur,* 421 U.S. 684, 95 S. Ct. 1881, 44 L. Ed. 2d 508 (1975).

Wilbur had beaten another man to death in a Maine hotel room after the victim allegedly made an unwanted homosexual advance. He was convicted of murder. The trial judge had charged the jury that

"malice aforethought is an essential and indispensable element of the crime of murder," without which the homicide would be manslaughter. The jury was further instructed, however, that if the prosecution established that the homicide was both intentional and unlawful, malice aforethought was to be conclusively implied unless the defendant proved by a fair preponderance of the evidence that he acted in the heat of passion on sudden provocation.

The Court ruled that this instruction unconstitutionally shifted the burden to the defendant to disprove an element of the crime, in violation of the rule announced in *In re Winship,* 397 U.S. 358, 90 S. Ct. 1068, 25 L. Ed. 2d 368 (1970) (see Chapter 1).

In Winship the ultimate burden of persuasion remained with the prosecution, although the standard had been reduced to proof by a fair preponderance of the evidence. In this case, by contrast, the State has affirmatively shifted the burden of proof to the defendant. The result, in a case such as this one where the defendant is required to prove the critical fact in dispute, is to increase further the likelihood of an erroneous murder conviction. . . .

We therefore hold that the Due Process Clause requires the prosecution to prove beyond a reasonable doubt the absence of the heat of passion on sudden provocation when the issue is properly presented in a homicide case. . . .

Under Maine law "malice" was (functionally) an element of the crime of murder. Would it be possible to define murder and manslaughter in such a way that the defendant lawfully could be required to prove the essential facts that distinguish an intentional, unlawful killing as manslaughter to avoid a conviction for murder? We encounter an important limitation on the Court's ruling in *Mullaney* when we consider *Patterson v. New York* later in this chapter.

## "Reasonable Person" Requirement

### *Bedder v. Director of Public Prosecutions,* 2 All E.R. 801 (H.L. 1954)

Lord Simonds L.C.: My Lords, this appeal raises once more a question of importance in the criminal law. . . .

The appellant, a youth of eighteen years, was convicted on May 27, 1954, at Leicester Assizes of the murder of Doreen Mary Redding, a prostitute. He appealed to the Court of Criminal Law Appeal on the substantial ground of misdirection, claiming that the learned judge who tried the case had wrongly directed the jury on the test of provocation and that, had they been rightly directed, they might have found him guilty not of murder but of manslaughter only. The Court of Criminal Appeal dismissed his appeal, holding that the jury had been rightly directed.

The relevant facts, so far as they bear on the question of provocation, can be shortly stated. The appellant has the misfortune to be sexually impotent, a fact which he naturally well knew, and, according to his own evidence, had allowed to prey on his mind. On the night of the crime he saw the prostitute with another man and, when they had parted, went and spoke to her and was led by her to a quiet court off a street in Leicester. There he attempted in vain to have intercourse with her whereupon—and I summarise the evidence in the way most favourable to him—she jeered at him and attempted to get away. He tried still to hold her and then she slapped him in the face and punched him in the stomach. He grabbed her shoulders and pushed her back from him whereat (I use his words): "She kicked me in the privates. Whether it was her knee or foot, I do not know. After that I do not know what happened till she fell." She fell, because he had taken a knife from his pocket and stabbed her with it twice, the second blow inflicting a mortal injury.

It was in these circumstances that the appellant pleaded that there had been such provocation by the deceased as to reduce the crime from murder to manslaughter, and the question is whether the learned judge rightly directed the jury on this issue. In my opinion, the summing-up of the learned judge was impeccable. Adapting the language used in this House in the cases of *Mancini v. Director of Public Prosecutions*, 28 Cr.App.R. 65; [1942] A.C. 1, and *Holmes v. Director of Public Prosecutions*, 31 Cr.App.R. 123; [1946] A.C. 588, to which I shall later refer, he thus directed the jury: "Provocation would arise if the conduct of the deceased woman, Mrs. Redding, to the prisoner was such as would cause a reasonable person, and actually caused the person, to lose his self-control suddenly and to drive him into such a passion and lack of self-control that he might use violence of the degree and nature which the prisoner used here. The provocation must be such as would reasonably justify the violence used, the use of a knife," and a little later he addressed them thus: "The reasonable person, the ordinary person, is the person you must consider when you are considering the effect which any acts, any conduct, any words, might have to justify the steps which were taken in response thereto, so that an unusually excitable or pugnacious individual, or a drunken one or a man who is sexually impotent is not entitled to rely on provocation which would not have led an ordinary person to have acted in the way which was in fact carried out. There may be, members of the jury, infirmity of mind and instability of character, but if it does not amount to insanity, it is no defence. Likewise, infirmity of body or infliction of the mind of the assailant is not material in testing whether there has been provocation by the deceased to justify the violence used so as to reduce the act of killing to manslaughter. They must be tested throughout this case by the reactions of a reasonable man to the acts, or series of acts, done by the deceased woman." . . .

The court thereupon approved and reiterated the proposition that the question for the jury was whether, on the facts as they found them from the evidence, the provocation was, in fact, enough to lead a reasonable person to do what the accused did. . . .

The argument, as I understood it, for the appellant was that the jury, in considering the reaction of the hypothetical reasonable man to the acts of provocation, must not only place him in the circumstances in which the accused was placed, but must also invest him with the personal physical peculiarities of the accused. Learned counsel, who argued the case for the appellant with great ability, did not, I think, venture to say that he should be invested with mental or temperamental qualities which distinguished him from the reasonable man. . . . But he urged that the reasonable man should be invested with the peculiar physical qualities of the accused, as in the present case with the characteristic of impotence, and the question should be asked: What would be the reaction of the impotent reasonable man in the circumstances? For that proposition I know of no authority; nor can I see any reason in it. It would be plainly illogical not to recognise an unusually excitable or pugnacious temperament in the accused as a matter to be taken into account but yet to recognise for that purpose some unusual physical characteristic, be it impotence or another. Moreover, the proposed distinction appears to me to ignore the fundamental fact that the temper of a man which leads him to react in such and such a way to provocation, is, or may be, itself conditioned by some physical defect. It is too subtle a refinement for my mind or, I think, for that of a jury to grasp that the temper may be ignored but the physical defect taken into account.

It was urged on your Lordships that the hypothetical reasonable man must be confronted with all the same circumstances as the accused, and that this could not be fairly done unless he was also invested with the peculiar characteristics of the accused. But this makes nonsense of the test. Its purpose is to invite the jury to consider the act of the accused by reference to a certain standard or norm of conduct and with this object the "reasonable" or the "average" or the "normal" man is invoked. If the reasonable man is then deprived in whole or in part of his reason, or the normal man endowed with abnormal characteristics, the test ceases to have any value. This is precisely the consideration which led this House in *Mancini's* case to say that an unusually excitable or pugnacious

*(Continues)*

person is not entitled to rely on provocation which would not have led an ordinary person to act as he did. In my opinion, then, the Court of Criminal Appeal was right in approving the direction given to the jury by the learned judge, and this appeal must fail. . . .

## Notes and Questions

1. In light of Bedder's particular infirmity, is it understandable how he might have been more readily provoked into a heat of passion by the conduct of his victim than would an "average" or "normal" "reasonable man"? If so, and if Bedder is not in fact an average, reasonable man, why should he be held to that standard?

2. What emerges as the dominant function of a legal test based on a standard of reasonableness: to describe the characteristics and reaction of a particular defendant and to assess individual culpability accordingly, or to prescribe a standard of conduct with which all are expected to comply, regardless of their idiosyncrasies? Which function *should* prevail: to measure criminal wrongdoing and assign punishment according to subjective or individual standards, or to impose and enforce objective standards that apply universally?

3. Even if legal standards should not be adjusted for "an unusually excitable or pugnacious individual, or a drunken one," is Bedder's infirmity of that same type?

## Cooling Time

As the Maryland Court of Appeals noted in *Girouard v. State, supra,* the "rule of provocation" requires that to qualify as manslaughter a killing must have been committed in "a sudden heat of passion—that is, the killing must have followed the provocation before there had been a reasonable opportunity for the passion to cool." We now consider the issue of "cooling time."

In *State v. Gounagias,* 153 P. 9 (Wash. 1915), the question of reasonable cooling time arose in a challenge to the defendant's conviction for first-degree murder. The evidence suggested that the homicide victim, Dan George, had sodomized the defendant, John Gounagias, on April 19, 1914, after Gounagias had become helplessly intoxicated. The next day Gounagias confronted George, and George dismissed the incident by retorting that, "You're all right, it did not hurt you." Gounagias implored George not to let on to their coworkers that the assault had occurred. Word of the incident nevertheless circulated, and Gounagias received occasional taunts and verbal jabs from his coworkers.

On May 6, over 2 weeks after George's assault, Gounagias went to a coffeehouse, where several of his coworkers greeted him with "laughing remarks and suggestive gestures." On receiving this reception Gounagias became highly agitated and "rushed from the coffeehouse, ran to his own house, made a necessary visit to the toilet, went to his mattress, took out the revolver and loaded it, went rapidly up the hill to the house where George lived, entered the house, and by the light of a match found George asleep in his bed, did not awaken him, but immediately shot him through the head, firing five shots, all that he had in the revolver. . . ."

The trial judge refused to allow the jury to hear about George's prior act of sodomizing Gounagias, ruling that the event was sufficiently removed in time from the homicide that it could not reasonably have been offered as provocation. In the court's view, a reasonable cooling time

had passed between the provoking incident and the killing, making testimony about the sodomy irrelevant. Gounagias appealed. The Washington Supreme Court discussed "cooling time" in the context of Washington's criminal homicide statute, which made the issue relevant to discriminating between first-degree and second-degree murder. This discussion is equally relevant to our consideration of the provocation rule and manslaughter.

The appellant contends that what would be such reasonable provocation as to be competent evidence in mitigation, and what would be a reasonable cooling time after such provocation, are always questions for the jury.

The doctrine of mitigation is briefly this: That if the act of killing, though intentional, be committed under the influence of sudden, intense anger, or heat of blood, obscuring the reason, produced by an adequate or reasonable provocation, and before sufficient time has elapsed for the blood to cool and reason to reassert itself, so that the killing is the result of temporary excitement rather than of wickedness of heart or innate recklessness of disposition, then the law, recognizing the standard of human conduct as that of the ordinary or average man, regards the offense so committed as of less heinous character than premeditated or deliberate murder. Measured as it must be by the conduct of the average man, what constitutes adequate cause is incapable of exact definition. . . .

There can be no doubt that the original outrage committed by the deceased would have been a sufficient provocation to take the case to the jury, if the appellant, immediately upon realizing its perpetration, had sought out and slain the deceased. There can be little doubt that, had the appellant slain the deceased when, on meeting him the next day, the deceased impudently treated the outrage as inconsequential, the question of provocation would have been for the jury. No court would be warranted in saying that such callous conduct, while the original wrong was but a day old, would have no reasonable tendency to produce immediate, uncontrollable anger, destroying the capacity for cool reflection in the average man. In such a case evidence of both the previous conduct and the insolent behavior of deceased on the subsequent meeting would have been admissible.

The appellant, however, did neither of these. On the meeting the next day, notwithstanding the insolence of the deceased, he admittedly condoned the offense, requesting that deceased preserve silence. It may even be conceded that, had the appellant met the deceased immediately after first discovering from words and gestures of others that deceased had circulated the story of the outrage and then, smarting under this added injury, had killed him, evidence of the whole transaction should have been submitted to the jury to determine the adequacy of the provocation. We have, however, been cited to no case, independent of a governing statute, in which it has been held that the immediate provocation, when referable for its provocative force to some antecedent outrage known to the accused from the beginning, was held sufficient to take the case to the jury, when not the act of or participated in by the deceased. At least one court has asserted that provocative words or acts, to have a reasonable tendency to produce a mitigating degree of anger and excitement in the ordinary man, must be the words or acts of the victim at the time and place of the killing. . . .

We are not prepared to go so far, since it would seem but natural that, on first seeing the gestures and hearing the words of others indicating that the story had been circulated, the appellant would certainly know that the deceased was responsible for its circulation as if he had been present and participating in the demonstrations, and that such knowledge would be as suddenly exasperating when *at first acquired* the one way as the other. But even this assumption does not meet the case in hand. According to the offered evidence the appellant let these things pass repeatedly for many days without molesting the deceased, even to the extent of a remonstrance. The offered evidence makes it clear that the appellant knew and appreciated for days before the killing the full meaning of the words, signs, and vulgar gestures of his countrymen, which, as the offer shows, he had encountered from day to day for about three weeks following the original outrage, wherever he went. The final demonstration in the coffeehouse was nothing new. It was exactly what the appellant, from his experience for the prior three weeks, must have anticipated. To say that it alone tended to create the sudden passion and heat of blood essential to mitigation is to ignore the admitted fact that the same thing had created no such condition on its repeated occurrence during the prior three weeks. To say that these repeated demonstrations, coupled with the original outrage, *culminated* in a sudden passion and heat of blood when he encountered the same character of demonstration in the coffeehouse on the night of the killing, is to say that sudden passion and heat of blood in the mitigative sense may be a cumulative result of repeated reminders of a single act of provocation occurring weeks before, and this, whether that provocation be regarded as the original outrage or the spreading of the story among appellant's associates, both of which he knew and fully realized for three weeks before the fatal

night. This theory of the cumulative effect of reminders of former wrongs, not of new acts of provocation by the deceased, is contrary to the idea of sudden anger as understood in the doctrine of mitigation. In the nature of the thing *sudden* anger cannot be cumulative. A provocation which does not cause instant resentment, but which is only resented after being thought upon and brooded over, is not a provocation sufficient in law to reduce intentional killing from murder to manslaughter, or under our statute to second degree murder, which includes every inexcusable, unjustifiable, unpremeditated, intentional killing. . . .

The evidence offered had no tendency to prove sudden anger and resentment. On the contrary, it did tend to prove brooding thought, resulting in the design to kill. It was therefore properly excluded. . . .

## Notes and Questions

1. How clear is the line between acting in the heat of passion on "sudden" provocation, and acting in an enraged manner after "brooding" and reflecting about an incident? What does the court mean in *Gounagias* when it says, "In the nature of the thing *sudden* anger cannot be cumulative"?

2. The *Gounagias* court suggests that a jury question might have been presented about reasonable cooling time if Gounagias had killed in response to various "triggering" incidents—such as George's initial derisive comments, or his coworkers' first taunts—even though those incidents were removed from the original provocation. Does this rule seem consistent with the requirement for "sudden" provocation? Why did Gounagias not benefit from this rule?

3. For a case presenting the issue of reasonable cooling time involving the trial judge's refusal to allow the jury to consider the charge of manslaughter, as opposed to murder, see *Allen v. State*, 647 P.2d 389 (Nev. 1982).

## MPC Approach: Murder Versus Manslaughter (for Purposeful Killings)

The MPC "undertakes a major restructuring of the law of homicide."[5] We already observed that the MPC does not rely on "malice" to distinguish murder and manslaughter. Other concepts familiar to the common law approach also are gone, including such venerable ones as "heat of passion," "provocation," and the "reasonable man." We focus in this section on purposeful (or intentional) killings that nevertheless may qualify as manslaughter under legislative schemes that make use of MPC principles. We consider unintentional forms of manslaughter—what traditionally (although not under the MPC) has been called "involuntary manslaughter"—later in this chapter.

Murder and manslaughter are defined under the MPC as follows:

§ 210.2 Murder

(1) Except as provided in Section 210.3(1)(b), criminal homicide constitutes murder when:

(a) it is committed purposely or knowingly; or (b) it is committed recklessly under circumstances manifesting extreme indifference to the value of human life. Such recklessness and indifference are presumed if the actor is engaged or is an accomplice in the commission of, or an attempt to commit, or flight after committing or attempting to commit robbery, rape or deviate sexual intercourse by force or threat of force, arson, burglary, kidnapping or felonious escape.

(2) Murder is a felony of the first degree [but a person convicted of murder may be sentenced to death, as provided in Section 210.6].

§ 210.3 Manslaughter

(1) Criminal homicide constitutes manslaughter when:

(a) it is committed recklessly; or

(b) a homicide which would otherwise be murder is committed under the influence of extreme mental or emotional disturbance for which there is reasonable explanation or excuse. The reasonableness of such explanation or excuse shall be determined from the viewpoint of a person in the actor's situation under the circumstances as he believes them to be.

(2) Manslaughter is a felony of the second degree.

For present purposes we are most interested in the form of killing described in § 210.3(1)(b) (*i.e.*, "a homicide which would otherwise be murder" that "is committed under the influence of extreme mental or emotional disturbance for which there is reasonable explanation or excuse"). It is necessary to refer to the definition of murder provided in § 210.2(1) to make sense of this description. In this context, we focus on the type of murder described in § 210.2(1)(a) (*i.e.*, a criminal homicide committed "purposely or knowingly").

The MPC's distinction between murder and manslaughter in some respects resembles the common law approach. In other respects it is significantly different. The official commentary accompanying the MPC explains that one major departure[6]

concerns the statement of the rule of provocation in Subsection (1)(b). The formulation in Subsection (1)(b) represents a substantial enlargement of the class of cases which would otherwise be murder but which could be reduced to manslaughter under then existing law because the homicidal act occurred in the "heat of passion" upon "adequate provocation." The decisive question is reframed to ask whether the homicide was committed "under the influence of extreme mental or emotional disturbance for which there is a reasonable explanation or excuse." The Model Code further provides that the "reasonableness of such explanation or excuse shall be determined from the viewpoint of a person in the actor's situation under the circumstances as he believes them to be."

This formulation treats on a parity with classic provocation cases situations where the provocative circumstance is something other than an injury inflicted by the deceased on the actor but nonetheless is an event that arouses extreme mental or emotional disturbance. There is a larger element of subjectivity in the standard than there was under prevailing law, though it is only the actor's "situation" and "the circumstances as he believed them to be," not his scheme of moral values, that are thus to be considered. The ultimate test, however, is objective; there must be a "reasonable" explanation or excuse for the actor's disturbance. This is to state in fair and realistic terms the criteria by which the mitigating import of mental or emotional distress should be appraised when it is a factor in so grave a crime as homicide. . . .

We learn more about the differing approaches between the MPC and the common law rules governing murder and manslaughter in the following case.

## State v. Dumlao, 715 P.2d 822 (Haw. App. 1986)

Heen, Judge.

Defendant Vidado B. Dumlao (Dumlao) appeals from his conviction of murder. Hawaii Revised Statutes (HRS) § 707-701 (1976). [§ 707-701(1) defines murder as: "Except as provided in section 707-702, a person commits the offense of murder if he intentionally or knowingly causes the death of another person."] He argues on appeal that the trial court erred in refusing to give his requested manslaughter instruction. Relying on *State v. O'Daniel*, 616 P.2d 1383 (Haw. 1980), he contends there was sufficient evidence that he shot his mother-in-law, Pacita M. Reyes (Pacita), while "under the influence of extreme mental or emotional disturbance for which there [was] a reasonable explanation" to support an instruction under HRS § 707–702(2) (1976) [which provides as follows: "In a prosecution for murder it is a defense, which reduces the offense to manslaughter, that the defendant was, at the time he caused the death of the other person, under the influence of extreme mental or emotional disturbance for which there is a reasonable explanation. The reasonableness of the explanation shall be determined from the viewpoint of a person in the defendant's situation under the circumstances as he believed them to be. . . .]. We agree and reverse. . . .

### HISTORY OF THE MITIGATING FACTOR IN MANSLAUGHTER

The principle that the presence of an extreme mental or emotional disturbance will reduce the offense of murder to manslaughter is a modification of the ancient distinction between slaying in cold blood and slaying in the heat of passion existing in Anglo-Saxon criminal law prior to the Norman conquest of 1066. The "Doctrine of Provocation" became firmly established in the common law in 1628 and the distinction between murder and manslaughter turned on the presence of heat of passion caused by adequate provocation. Donovan and Wildman, *Is the Reasonable Man Obsolete? A Critical Perspective on Self-Defense and Provocation,* 14 Loyola L. Rev. 435, 440 and 446 (1981) (hereafter *Donovan and Wildman*).

In the United States mutual combat, assault and adultery were gradually recognized as having been legally adequate provocation at common law to reduce murder to manslaughter. In some jurisdictions illegal arrest, injuries to third parties, and even words tending to give rise to heat of passion are sufficient provocation. . . .

The determination of the adequacy of the provocation gradually became a jury prerogative in marginal cases, and the reasonable person test was devised to assist the jury. Today the test has four elements: (1) provocation that would rouse a reasonable person to the heat of passion; (2) actual provocation of the defendant; (3) a reasonable person would not have cooled off in the time between the provocation and the offense; and (4) the defendant did not cool off. The reasonable person yardstick is strictly objective; neither the mental nor physical peculiarities of the accused are evaluated in determining whether the loss of self-control was "reasonable."

## CRITICISM OF THE "REASONABLE PERSON" TEST

As originally developed the provocation defense focused on the mental state of the accused as the test for moral culpability; however, under the objective or "reasonable person" test the individual's mental state is not the determinative factor.

Some commentators have remarked on the inconsistency of the reasonable person test.

The reasonable man test, being objective in nature, is antithetical to the concept of *mens rea*. Like all objective standards, it is an external standard of general application that does not focus on an individual accused's mental state. Thus, from the point of view of traditional Anglo-American jurisprudence, a paradox is inherent in the use of the reasonable man standard to test criminal responsibility: the presence or absence of criminal intent is determined by a standard which ignores the mental state of the individual accused. *Donovan and Wildman, supra,* at 451 (footnotes omitted).

The objective test placed the jury in the conceptually awkward, almost impossible, position of having to determine when it is reasonable for a reasonable person to act unreasonably.

In the law of contract and tort, and elsewhere in the criminal law, the test of the reasonable man indicates an ethical standard; but it seems absurd to say that the reasonable man will commit a felony the possible punishment for which is imprisonment for life. To say that the "ordinary" man will commit this felony is hardly less absurd. The reason why provoked homicide is punished is to deter people from committing the offence; and it is a curious confession of failure on the part of the law to suppose that, notwithstanding the possibility of heavy punishment, an ordinary person will commit it. If the assertion were correct, it would raise serious doubts whether the offence should continue to be punished.

Surely the true view of provocation is that it is a concession to "the frailty of human nature" in those exceptional cases where the legal prohibition fails of effect. It is a compromise, neither conceding the propriety of the act nor exacting the full penalty for it. This being so, how can it be that that paragon of virtue, the reasonable man, gives way to provocation? Williams, *Provocation and the Reasonable Man,* 1954 Crim.L.Rev. 740, 742.

The MPC's response to this criticism is discussed below. . . .
HRS § 707–702(2)
. . . Since HRS § 707–702(2) is derived from MPC § 210.3, we may look to the commentaries and cases from other jurisdictions explaining and construing that section for insight into the meaning of the language of our statute. . . .

## A.

"Extreme mental or emotional disturbance" sometimes is, but should not be, confused with the "insanity" defense. The point of the extreme emotional disturbance defense is to provide a basis for mitigation that differs from a finding of mental defect or disease precluding criminal responsibility. The disturbance was meant to be understood in relative terms as referring to a loss of self control due to intense feelings.

*(Continues)*

The extreme mental or emotional disturbance concept of the MPC must also be distinguished from the so-called "diminished capacity" defense.

> The doctrine of diminished capacity provides that evidence of an abnormal mental condition not amounting to legal insanity but tending to prove that the defendant could not or did not entertain the specific intent or state of mind essential to the offense should be considered for the purpose of determining whether the crime charged or a lesser degree thereof was in fact committed. *State v. Baker,* 691 P.2d 1166, 1168 (Haw. 1984).

Although the MPC does *not* recognize diminished capacity as a distinct category of mitigation, by placing more emphasis than does the common law on the actor's subjective mental state, it also may allow inquiry into areas which have traditionally been treated as part of the law of diminished responsibility or the insanity defense. . . .

An explanation of the term "extreme emotional disturbance" which reflects the situational or relative character of the concept was given in *People v. Shelton,* 385 N.Y.S.2d 708, 717 (Misc. 1976), as follows:

> [T]hat extreme emotional disturbance is the emotional state of an individual, who: (a) has no mental disease or defect that rises to the level [of insanity]; and (b) is exposed to an extremely unusual and overwhelming stress; and (c) has an extreme emotional reaction to it, as a result of which there is a loss of self-control and reason is overborne by intense feelings, such as passion, anger, distress, grief, excessive agitation or other similar emotions.

It is clear that in adopting the "extreme mental or emotional disturbance" concept, the MPC intended to define the provocation element of manslaughter in broader terms than had previously been done. It is equally clear that our legislature also intended the same result when it adopted the language of the MPC.

We turn then to the second prong of our analysis, the test to determine the reasonableness of the explanation for the mental or emotional disturbance. It is here that the most significant change has been made in the law of manslaughter.

## B.

The anomaly of the reasonable person test was corrected by the drafters of the MPC through the development of an objective/subjective test of reasonableness. Dressler, [*Rethinking Heat of Passion: A Defense in Search of a Rationale,* 73 J. Crim. L. and Criminology 421,431 (1982)] explains that,

> it makes the test more, although not entirely, subjective, by requiring the jury to test the reasonableness of the actor's conduct, "from the viewpoint of a person in the actor's situation." Thus, the actor's sex, sexual preference, pregnancy, physical deformities, and similar characteristics are apt to be taken into consideration in evaluating the reasonableness of the defendant's behavior. (Footnotes omitted.)

This more subjective version of the provocation defense goes substantially beyond the common law by abandoning preconceptions of what constitutes adequate provocation, and giving the jury wider scope. *Id.*

Under the prior law of provocation, personal characteristics of the defendant were not to be considered. Under the MPC a change from the old provocation law and the reasonable person standard has been effected by requiring the fact-finder to focus on a person in the defendant's situation. Thus, the MPC, while requiring that the explanation for the disturbance must be reasonable, provides that the reasonableness is determined from the defendant's viewpoint. The phrase "actor's situation," as used in § 210.3(b) of the MPC, is designedly ambiguous and is plainly flexible enough to allow the law to grow in the direction of taking account of mental abnormalities that have been recognized in the developing law of diminished responsibility. 1980 MPC Commentary, *supra,* at 72.

Moreover, the MPC does not require the provocation to emanate from the victim as was argued by the State here.

In light of the foregoing discussion and the necessity of articulating the defense in comprehensible terms, we adopt the test enunciated by the New York Court of Appeals in *People v. Casassa,* 404 N.E.2d 1310, 1316 (N.Y.), *cert. denied,* 449 U.S. 842 (1980):

> [W]e conclude that the determination whether there was reasonable explanation or excuse for a particular emotional disturbance should be made by viewing the subjective, internal situation in which the defendant

found himself and the external circumstances as he perceived them at the time, however inaccurate that perception may have been, and assessing from that standpoint whether the explanation . . . for his emotional disturbance was reasonable, so as to entitle him to a reduction of the crime charged from murder . . . to manslaughter. . . . [Footnote omitted.]

The language of HAS § 707–702(2) indicates that the legislature intended to effect the same change in the test for manslaughter in Hawaii's law as was made by the MPC. Therefore, we hold that under HAS § 707–702(2) the broader sweep of the emotional disturbance defense applies when considering whether an offense should be reduced from murder to manslaughter. To hold that the pre-penal code law of provocation continues to hold sway would be to render the language of § 707–702(2) meaningless.

Thus, we hold in the instant case that the trial court was required to instruct the jury as requested by Dumlao, if there was any evidence to support a finding that at the time of the offense he suffered an "extreme mental or emotional disturbance" for which there was a "reasonable explanation" when the totality of circumstances was judged from his personal viewpoint.

We turn now to the question of whether there was evidence to support the proffered instruction.

## II.

In the instant case, there was evidence of the following:

Arthur Golden, M.D., Dumlao's expert witness, stated at trial that his diagnosis of Dumlao was one of "paranoid personality disorder," which is a "long range, almost lifetime or certainly over many, many years, emotional or mental disorder. It is almost a way of functioning."

Dr. Golden diagnosed Dumlao as having unwarranted suspiciousness, one of the basic indicators of the "paranoid personality disorder." That unwarranted suspiciousness included pathological jealousy, which Dumlao suffered throughout his ten-year marriage. Dumlao harbored the belief that other males, including his wife's relatives, were somehow sexually involved with her. He could never figure out exactly who or where or how, yet this extreme suspiciousness persisted.

Dr. Golden described the second major sign of Dumlao's paranoid personality disorder as hypersensitivity, characterized by being easily slighted or quick to take offense, and a readiness to counterattack when a threat was perceived. For example, when somebody glanced or gazed at his wife, he would consider it a personal affront and could well believe that a sexual overture had been made to her. Dr. Golden believed that at the time of the offense Dumlao felt the need to counterattack because Dumlao perceived a very substantial threat. Dr. Golden stated that "[p]erhaps an ordinary individual in his situation would not have. He did."

Dr. Golden gave other examples from Dumlao's history, illustrating his extreme and irrational reactions to outwardly normal events. . . .

Dumlao's extreme and irrational jealousy concerning his wife was known to all the family members. According to Agapito, they couldn't even talk to their sister in Dumlao's presence. "If we have to talk to her, we have to talk from a distance because he suspects us." Furthermore, Dumlao "never allowed us to talk in a group."

Dumlao's testimony, describing his own perceptions of the night in question, further confirms the nature of his extreme jealousy. Dumlao described how he became suspicious and jealous of his wife that night because of the way that Agapito looked at him. . . .

Dumlao went on to describe how, after his father had counseled him and he went back in the room, he could hear the voices of the others talking in the living room. He thought that they were talking about him. He then came out to "investigate" with his gun in his waistband. He saw "Pedrito with his eyes at me, burning eye, angry eye, angry," and he saw Eduardo and Agapito alongside Pedrito. Dumlao testified that Pedrito rushed at him, holding a knife in his hand, saying, "My sister suffer ten years. You going to pay. Now you going to pay." He testified that when he pulled his gun to "try and scare [Pedrito]," the gun went off, firing the bullet that struck and killed Pacita.

## CONCLUSION

Reviewing the evidence within the context of the meaning of HRS § 707-702(2), we conclude that it was sufficient to require the trial court to give Dumlao's requested instruction on manslaughter. There was evidence, "no matter how weak, inconclusive or unsatisfactory," that Dumlao killed Pacita while under the influence of "extreme emotional disturbance." Whether a jury will agree that there was

*(Continues)*

such a disturbance or that the explanation for it was reasonable we cannot say. However, Dumlao was entitled to have the jury make that decision using the objective/subjective test. . . .

Reversed and remanded for new trial.

## Notes and Questions

1. How, if at all, is Dumlao likely to benefit by having his conduct judged "from the viewpoint of a person in the actor's situation," and not under the "reasonable man" standard?

2. What, precisely, is Dumlao's "situation"? Note that the Model Penal Code's standard is not entirely subjective: there must be a "reasonable explanation" for the extreme mental or emotional disturbance. However, "[t]he reasonableness of such explanation or excuse shall be determined from the viewpoint of a person in the actor's situation under the circumstances as he believes them to be."

> The critical element in the Model Code formulation is the clause requiring that reasonableness be assessed "from the viewpoint of a person in the actor's situation." The word "situation" is designedly ambiguous. On the one hand, it is clear that personal handicaps and some external circumstances must be taken into account. Thus, blindness, shock from traumatic injury, and extreme grief are all easily read into the term "situation." This result is sound, for it would be morally obtuse to appraise a crime for mitigation of punishment without reference to these factors. On the other hand, it is equally plain that idiosyncratic moral values are not part of the actor's situation. An assassin who kills a political leader because he believes it is right to do so cannot ask that he be judged by the standard of a reasonable extremist. Any other result would undermine the normative message of the criminal law. In between these two extremes, however, there are matters neither as clearly distinct from individual blameworthiness as blindness or handicap nor as integral a part of moral depravity as a belief in the rightness of killing. Perhaps the classic illustration is the unusual sensitivity to the epithet "bastard" of a person born illegitimate. An exceptionally punctilious sense of personal honor or an abnormally fearful temperament may also serve to differentiate an individual actor from the hypothetical reasonable man, yet none of these factors is wholly irrelevant to the ultimate issue of culpability. The proper role of such factors cannot be resolved satisfactorily by abstract definition of what may constitute adequate provocation. The Model Code endorses a formulation that affords sufficient flexibility to differentiate in particular cases between those special aspects of the actor's situation that should be deemed material for purpose of grading and those that should be ignored. There thus will be room for interpretation of the word "situation," and that is precisely the flexibility desired. There will be opportunity for argument about the reasonableness of explanation or excuse, and that too is a ground on which argument is required. In the end, the question is whether the actor's loss of self-control can be understood in terms that arouse sympathy in the ordinary citizen. . . .
> *Model Penal Code and Commentaries (Official Draft and Revised Comments)*, Part II, vol. 1, art. 210.3, pp. 62–63 (Philadelphia: The American Law Institute 1980).

3. In *State v. Seguritan*, 766 P.2d 128 (Haw. 1988), the Hawaii Supreme Court disapproved of the *Dumlao* court's reliance on the explanation of "extreme emotional disturbance" contained in *People v. Shelton*, 385 N.Y.S.2d 708 (Misc. 1976). In particular, the Hawaii Supreme Court concluded there was no support for the suggestion in *Shelton* that the defendant must be exposed to "extremely unusual and overwhelming stress" in order to benefit from the extreme mental or emotional disturbance standard.

4. How would Bedder likely have fared under the Model Penal Code's approach to differentiating between murder and manslaughter?

## Malice and the Burden of Proof

In *Mullaney v. Wilbur*, 421 U.S. 684 (1975), the U.S. Supreme Court ruled that when malice is an element of the crime of murder (or the functional equivalent), due process forbids a state from placing the burden of proof on the defendant to negate malice by establishing that a killing was committed in the heat of passion on sudden provocation. We have seen that the MPC does not rely on malice as an element of murder. In a jurisdiction that follows the MPC approach, may a state require the defendant to prove that he

or she acted "under the influence of extreme mental or emotional disturbance" to avoid a murder conviction (instead of manslaughter)? The Court considered this question just 2 years after it decided *Mullaney*, in *Patterson v. New York*, 432 U.S. 197 (1977).

Gordon Patterson was convicted of second-degree murder in a New York trial after he shot and killed John Northrup. The defendant had observed Northrup with his estranged wife while the latter was "in a state of semi-undress." Under New York law, the elements of murder in the second degree are (1) causing the death of a third person (2) with the intent to kill. It is an "affirmative defense" that the defendant "acted under the influence of extreme emotional disturbance for which there was a reasonable explanation or excuse, the reasonableness of which is to be determined from the viewpoint of a person in the defendant's situation under the circumstances as the defendant believed them to be." N.Y. Penal Law § 125.25(1). A person is guilty of first-degree manslaughter when

> With intent to cause the death of another person, he causes the death of such person or of a third person under circumstances which do not constitute murder because he acts under the influence of extreme emotional disturbance. . . . The fact that homicide was committed under the influence of extreme emotional disturbance constitutes a mitigating circumstance reducing murder to manslaughter in the first degree and need not be proved in any prosecution initiated under this subdivision. N.Y. Penal Law § 125.20(2).

In accordance with those laws, the trial court instructed the jury that the prosecution was required to prove beyond a reasonable doubt that Patterson intentionally caused Northrup's death to secure a second-degree murder conviction. The judge further charged the jury that

> the defendant had the burden of proving his affirmative defense by a preponderance of the evidence. The jury was told that if it found beyond a reasonable doubt that appellant had intentionally killed Northrup but that appellant had demonstrated by a preponderance of the evidence that he had acted under the influence of extreme emotional disturbance, it had to find appellant guilty of manslaughter instead of murder.

Patterson alleged that burdening him with proving the "affirmative defense" of extreme emotional disturbance violated the rule of *Mullaney v. Wilbur.* The New York Court of Appeals had distinguished *Mullaney* "on the ground that the New York statute involved no shifting of the burden to the defendant to disprove any fact essential to the offense charged since the New York affirmative defense of extreme emotional disturbance bears no direct relationship to any element of murder." The U.S. Supreme Court agreed (6–3). Justice White's majority opinion observed as follows:

> We cannot conclude that Patterson's conviction under the New York law deprived him of due process of law. The crime of murder is defined by the statute, which represents a recent revision of the state criminal code, as causing the death of another person with intent to do so. The death, the intent to kill, and causation are the facts that the State is required to prove beyond a reasonable doubt if a person is to be convicted of murder. No further facts are either presumed or inferred in order to constitute the crime. The statute does provide an affirmative defense—that the defendant acted under the influence of extreme emotional disturbance for which there was a reasonable explanation— which, if proved by a preponderance of the evidence,

would reduce the crime to manslaughter, an offense defined in a separate section of the statute. It is plain enough that if the intentional killing is shown, the State intends to deal with the defendant as a murderer unless he demonstrates the mitigating circumstances. . . .

> It seems to us that the State satisfied the mandate of *Winship* that it prove beyond a reasonable doubt "every fact necessary to constitute the crime with which [Patterson was] charged." 397 U.S., at 364. . . . [I]n revising its criminal code, New York provided the affirmative defense of extreme emotional disturbance, a substantially expanded version of the older heat-of-passion concept; but it was willing to do so only if the facts making out the defense were established by the defendant with sufficient certainty. The State was itself unwilling to undertake to establish the absence of those facts beyond a reasonable doubt, perhaps fearing that proof would be too difficult and that too many persons deserving treatment as murderers would escape that punishment if the evidence need merely raise a reasonable doubt about the defendant's emotional state. It has been said that the new criminal code of New York contains some 25 affirmative defenses which exculpate or mitigate but which must be established by the defendant to be operative. The Due Process Clause, as we see it, does not put New York to the choice of abandoning those defenses or undertaking to disprove their existence in order to convict of a crime which otherwise is within its constitutional powers to sanction by substantial punishment. . . .

Justice Powell's dissenting opinion charged that the Court's ruling in *Patterson* had reduced the rule of *Mullaney v. Wilbur* to a mere formalism:

> The test the Court today establishes allows a legislature to shift, virtually at will, the burden of persuasion with respect to any factor in a criminal case, so long as it is careful not to mention the nonexistence of that factor in the statutory language that defines the crime. The sole requirement is that any references to the factor be confined to those sections that provide for an affirmative defense. . . .

> With all respect, this type of constitutional adjudication is indefensibly formalistic. A limited but significant check on possible abuses in the criminal law now becomes an exercise in arid formalities. What *Winship* and *Mullaney* had sought to teach about the limits a free society places on its procedures to safeguard the liberty of its citizens becomes a rather simplistic lesson in statutory draftsmanship. Nothing in the Court's opinion prevents a legislature from applying this new learning to many of the classical elements of the crimes it punishes. It would be preferable, if the Court has found reason to reject the rationale of *Winship* and *Mullaney*, simply and straightforwardly to overrule those precedents. . . .

In Justice Powell's view, under *Patterson* a state could define murder

> as mere physical contact between the defendant and the victim leading to the victim's death, but then set up an affirmative defense leaving it to the defendant to prove that

he acted without culpable *mens rea*. The State, in other words, could be relieved altogether of responsibility for proving *anything* regarding the defendant's state of mind, provided only that the face of the statute meets the Court's drafting formulas.

## Different Grades of Murder

Initially, no discrimination was made between different kinds of murder, and that crime automatically was punished by death. Pennsylvania legislators thought this approach to be excessively harsh and inflexible. In 1794 they passed an innovative statute dividing murder into first degree, which remained a capital crime, and second degree, which was not punished capitally. Their intent was to identify especially heinous killings—those deserving punishment by death—as first-degree murder and reserve **second-degree murder** for other killings committed with malice. In relevant part, the statute provided[7]

> [A]ll murder, which shall be perpetrated by means of poison, or by lying in wait or by any other kind of wilful,

deliberate and premeditated killing, or which shall be committed in the perpetration or attempt to perpetrate any arson, rape, robbery, or burglary, shall be deemed murder of the first degree; and all other kinds of murder shall be deemed murder in the second degree. . . . Pa. Laws 1794, c. 257, sec. 2.

The "Pennsylvania formula" for distinguishing between first-degree and second-degree murder was quickly adopted in many other states. Certain defining characteristics of first-degree murder are more straightforward than others. It is comparatively easy to identify a killing perpetrated by poison or one committed in connection with an arson, rape, robbery, or burglary. What, however, is the apparent meaning of the important category of "any other kind of wilful, deliberate and premeditated killing"? Once defined, how can those mental states be proven? Conceptually, does it make sense to distinguish between capital and noncapital murder based on whether an offender killed in a "wilful, deliberate, and premeditated" manner?

### *State v. Brown*, 836 S.W.2d 530 (Tenn. 1992)

[Mack Brown was convicted of the first-degree murder of his 4-year-old son, Eddie. Evajean Brown, Mack's wife and Eddie's mother, who was also charged in the case, called for an ambulance, claiming that Eddie had fallen down some steps and thereafter had stopped breathing. An autopsy revealed multiple skull fractures; swelling to the brain; several injuries to internal organs; and bruises about the child's head, chest, legs, arm, buttocks, and scrotum. He also had suffered cuts, apparent cigarette burns, and a broken arm. Mack Brown admitted that he and his wife had spanked Eddie the morning of his death because Eddie had urinated and defecated on the floor. He recalled Eddie telling him, "I hate you," and stated that he "went blank" and "his next memory was of going downstairs and hearing Eddie behind him, falling onto the landing and into the door."

[Mack Brown was evaluated as being borderline mentally retarded. He also was diagnosed as having "recurrent major depression and a dependent personality, a condition characterized by inadequacy in decision-making and a tendency to allow another person to accept the major responsibilities for his life."

[The statute in effect at the time of the homicide defined first-degree murder as follows:

> Every murder perpetrated by means of poison, lying in wait, or by other kind of willful, deliberate, malicious, and premeditated killing, or committed in the perpetration of, or attempt to perpetrate, any murder in the first degree, arson, rape, robbery, burglary, larceny, kidnapping, aircraft piracy, or the unlawful throwing, placing or discharging of a destructive device or bomb, is murder in the first degree. TCA § 39-2-292(a)(1982).

[Brown appealed his conviction on the ground that there was insufficient evidence to support a verdict of first-degree murder, and in particular that evidence was lacking to support a finding of deliberation and premeditation.]

Daugherty, Justice. . . .

Based upon our review of the record, we conclude that the evidence in this case is insufficient to establish deliberation and premeditation. Hence, the defendant's conviction for first-degree murder cannot stand. However, we do find the evidence sufficient to sustain a conviction of second-degree murder.

At common law, there were no degrees of murder, but the tendency to establish a subdivision by statute took root relatively early in the development of American law. The pattern was set by a 1794 Pennsylvania statute that identified the more heinous kinds of murder as murder in the first degree, with all other murders deemed to be murder in the second degree. Some states have subdivided the offense into three or even four degrees of murder, but since the enactment of the first such statute in 1829, Tennessee has maintained the distinction at two. It is one which this Court has found to be "not only founded in mercy and humanity, but . . . well fortified by reason." *Poole v. State*, 61 Tenn. 289, 290 (1872).

From the beginning, the statutory definition of first-degree murder required the state to prove that "the killing [was] done *willfully*, that is, of purpose, with intent that the act by which the life of a party is taken should have that effect; *deliberately*, that is, with cool purpose; *maliciously*, that is, with malice aforethought; and *with premeditation*, that is, a design must be formed to kill, before the act, by which the death is produced, is performed." *Dale v. State*, 18 Tenn. (10 Yer.) 551, 552 (1837) (emphasis added). Because conviction of second-degree murder also requires proof of intent and malice, the two distinctive elements of first-degree murder are deliberation and premeditation. . . .

Intent to kill had long been the hallmark of common-law murder, and in distinguishing manslaughter from murder on the basis of intent, the courts recognized, in the words of an early Tennessee Supreme Court decision, that

> [t]he law knows of no specific time within which an intent to kill must be formed so as to make it murder [rather than manslaughter]. If the will accompanies the act, a moment antecedent to the act itself which causes death, it seems to be as completely sufficient to make the offence murder, as if it were a day or any other time. *Anderson v. State*, 2 Tenn. (2 Overt.) 6, 9 (1804).

Of course, the *Anderson* opinion predates the statutory subdivision of murder into first and second degrees. But the temporal concept initially associated in that case with intent, *i.e.*, that no definite period of time is required for the formation of intent, was eventually carried over and applied to the analysis of premeditation. . . .

Hence, perhaps the two most oft-repeated propositions with regard to the law of first-degree murder, that the essential ingredient of first-degree murder is premeditation and that premeditation may be formed in an instant, are only partially accurate, because they are rarely quoted in context. In order to establish first-degree murder, the premeditated killing must also have been done deliberately, that is, with coolness and reflection. . . . [E]ven if intent (or "purpose to kill") and premeditation ("design") may be formed in an instant, deliberation requires some period of reflection, during which the mind is "free from the influence of excitement, or passion." *Clarke v. State*, 402 S.W.2d 863, 868 (Tenn. 1966). . . .

This trend toward a confusion of premeditation and deliberation has not been unique to Tennessee. It was for a time reflected by the commentators. In *Clarke v. State, supra*, 402 S.W.2d at 868, the Court quoted from the 1957 edition of Wharton's Criminal Law and Procedure as follows:

> "Deliberation and premeditation involve a prior intention or design to do the act in question. It is not necessary, however, that this intention should have been conceived at any particular period of time, and it is sufficient that only a moment elapsed between the plan and its execution. . . ."

A more recent version of Wharton's Criminal Law, however, returns the discussion of premeditation and deliberation to its roots:

> Although an intent to kill, without more, may support a prosecution for common law murder, such a murder ordinarily constitutes murder in the first degree only if the intent to kill is accompanied by premeditation and deliberation. '**Premeditation**' is the process simply of thinking about a proposed killing before engaging in the homicidal conduct; and '*deliberation*' is the process of *carefully weighing such matters as the wisdom of going ahead with the proposed killing, the manner in which the killing will be accomplished, and the consequences which may be visited upon the killer if and when apprehended. 'Deliberation' is present if the thinking, i.e., the 'premeditation,' is being done in such a cool mental state, under such circumstances, and for such a period of time as to permit a 'careful weighing' of the proposed decision.* C. Torcia, *Wharton's Criminal Law* § 140 (14th ed. 1979) (emphasis added).

To the same effect is this analysis of the distinction between first- and second-degree murder found in 2 W. LaFave and A. Scott, *Substantive Criminal Law* § 7.7 (1986):

> Almost all American jurisdictions which divide murder into degrees include the following two murder situations in the category of **first-degree murder**: (1) intent-to-kill murder where there exists (in addition to the intent to kill) the elements of premeditation and deliberation, and (2) felony murder where the felony in question is one of five or six listed felonies, generally including rape, robbery, kidnapping, arson and burglary. Some states instead or in addition have other kinds of first-degree murder.
>
> (a) Premeditated, Deliberate, Intentional Killing. To be guilty of this form of first-degree murder the defendant must not only intend to kill but in addition he must premeditate the killing and deliberate about it. It is not easy to give a meaningful definition of the words 'premeditate' and 'deliberate' as they are used in

*(Continues)*

(*Continued*)

connection with first-degree murder. Perhaps the best that can be said of 'deliberation' is that it requires a cool mind that is capable of reflection, and of 'premeditation' that it requires that the one with the cool mind did in fact reflect, at least for a short period of time before his act of killing.

It is often said that premeditation and deliberation require only a 'brief moment of thought' or a 'matter of seconds,' and convictions for first-degree murder have frequently been affirmed where such short periods of time were involved. *The better view, however, is that to 'speak of premeditation and deliberation which are instantaneous, or which take no appreciable time, . . . destroys the statutory distinction between first and second-degree murder,'* and (in much the same fashion that the felony-murder rule is being increasingly limited) this view is growing in popularity. This is not to say, however, that premeditation and deliberation cannot exist when the act of killing follows immediately after the formation of the intent. The intention may be finally formed only as a conclusion of prior premeditation and deliberation, while in other cases the intention may be formed without prior thought so that premeditation and deliberation occurs only with the passage of additional time for 'further thought, and a turning over in the mind.' (Footnotes omitted; emphasis added.)

One further development in Tennessee law has tended to blur the distinction between the essential elements of first- and second-degree murder, and that is the matter of evidence of "repeated blows" being used as circumstantial evidence of premeditation. Obviously, there may be legitimate first-degree murder cases in which there is no direct evidence of the perpetrator's state of mind. Since that state of mind is crucial to the establishment of the elements of the offense, the cases have long recognized that the necessary elements of first-degree murder may be shown by circumstantial evidence. Relevant circumstances recognized by other courts around the country have included the fact "that a deadly weapon was used upon an unarmed victim; that the homicidal act was part of a conspiracy to kill persons of a particular class; that the killing was particularly cruel; that weapons with which to commit the homicide were procured; that the defendant made declarations of his intent to kill the victim; or that preparations were made before the homicide for concealment of the crime, as by the digging of a grave." *Wharton's Criminal Law, supra,* at § 140. This list, although obviously not intended to be exclusive, is notable for the omission of "repeated blows" as circumstantial evidence of premeditation or deliberation. . . .

Logically, of course, the fact that repeated blows (or shots) were inflicted on the victim is not sufficient, by itself, to establish first-degree murder. Repeated blows can be delivered in the heat of passion, with no design or reflection. Only if such blows are inflicted as the result of premeditation and deliberation can they be said to prove first-degree murder. . . .

It is consistent with the murder statute and with case law in Tennessee to instruct the jury in a first-degree murder case that no specific period of time need elapse between the defendant's formulation of the design to kill and the execution of that plan, but we conclude that it is prudent to abandon an instruction that tells the jury that "premeditation may be formed in an instant." Such an instruction can only result in confusion, given the fact that the jury must also be charged on the law of deliberation. If it was not clear from the opinions emanating from this Court within the last half-century, it is now abundantly clear that the deliberation necessary to establish first-degree murder *cannot* be formed in an instant. It requires proof, as the Sentencing Commission Comment to § 39-13-201(b) further provides, that the homicide was "committed with 'a cool purpose' and without passion or provocation," which would reduce the offense either to second-degree murder or to manslaughter, respectively.

This discussion leads us inevitably to the conclusion that Mack Brown's conviction for first-degree murder in this case cannot be sustained. . . .

Here, there simply is no evidence in the record that in causing his son's death, Mack Brown acted with the premeditation and deliberation required to establish first-degree murder. There is proof, circumstantial in nature, that the defendant acted maliciously toward the child, in the heat of passion or anger, and without adequate provocation—all of which would make him guilty of second-degree murder. The only possible legal basis upon which the state might argue that a first-degree conviction can be upheld in this case is the proof in the record that the victim had sustained "repeated blows." It was on this basis, and virtually no other, that we upheld a similar first-degree murder conviction for the death of a victim of prolonged child abuse in *State v. LaChance,* 524 S.W.2d 933 (Tenn. 1975). In view of our foregoing discussion concerning the shortcomings of such an analysis, we find it necessary to depart from much of the rationale underlying that decision.

In abandoning *LaChance,* we are following the lead of a sister state. In *Midgett v. State,* 729 S.W.2d 410 (Ark. 1987), the Arkansas Supreme Court was asked to affirm the first-degree murder conviction of a father who had killed his eight-year-old son by repeated blows of his fist. As was the case here, there was a shocking history of physical abuse to the child, established both by eyewitness testimony and by proof of old bruises and healed fractures. . . . [T]he *Midgett* court noted:

> The appellant argues, and we must agree, that in a case of child abuse of long duration the jury could well infer that the perpetrator comes not to expect death of the child from his action, but rather that the child will live so that the abuse may be administered again and again. Had the appellant planned his son's death, he could have accomplished it in a previous beating. . . .
>
> The evidence in this case supports only the conclusion that the appellant intended not to kill his son but to further abuse him or that his intent, if it was to kill the child, was developed in a drunken, heated, rage while disciplining the child. Neither of those supports a finding of premeditation or deliberation. *Midgett,* 729 S.W.2d at 413–414.

The Arkansas court, in strengthening the requirements for proof of premeditation and deliberation in a first-degree murder case involving a victim of child abuse, found it necessary to overrule prior case law to the extent that it was inconsistent with the opinion in *Midgett.* We do the same here. Like the *Midgett* court, we do not condone the homicide in this case, or the sustained abuse of the defenseless victim, Eddie Brown. We simply hold that in order to sustain the defendant's conviction, the proof must conform to the statute. Because the state has failed to establish sufficient evidence of first-degree murder, we reduce the defendant's conviction to second-degree murder and remand the case for resentencing. . . .

## Notes and Questions

1. Even if the killing of Eddie Brown was not "deliberate" and "premeditated," was it nevertheless sufficiently abhorrent to be considered as the most serious form of criminal homicide (*i.e.,* first-degree murder)? *Should* such a killing be defined as first-degree murder? How would legislation be drafted to accomplish this end?
2. "In homespun terminology, intentional murder is in the first degree if committed in cold blood, and is murder in the second degree if committed on impulse or in the sudden heat of passion." *Austin v. United States,* 382 F.2d 129, 137 (D.C. Cir. 1967).

## *State v. Forrest,* 362 S.E.2d 252 (N.C. 1987)

Meyer, Justice.

Defendant was convicted of the first-degree murder of his father, Clyde Forrest. The State having stipulated before trial to the absence of any statutory aggravating factors under N.C.G.S. § 15A-2000, the case was tried as a noncapital case, and defendant was sentenced accordingly to life imprisonment. . . . The facts of this case are essentially uncontested, and the evidence presented at trial tended to show the following series of events. On 22 December 1985, defendant John Forrest admitted his critically ill father, Clyde Forrest, Sr., to Moore Memorial Hospital. Defendant's father, who had previously been hospitalized, was suffering from numerous serious ailments, including severe heart disease, hypertension, a thoracic aneurysm, numerous pulmonary emboli, and a peptic ulcer. By the morning of 23 December 1985, his medical condition was determined to be untreatable and terminal. Accordingly, he was classified as "No Code," meaning that no extraordinary measures would be used to save his life, and he was moved to a more comfortable room.

On 24 December 1985, defendant went to the hospital to visit his ailing father. No other family members were present in his father's room when he arrived. While one of the nurse's assistants was tending to his father, defendant told her, "There is no need in doing that. He's dying." She responded, "Well, I think he's better." The nurse's assistant noticed that defendant was sniffing as though crying and that he kept his hand in his pocket during their conversation. She subsequently went to get the nurse.

When the nurse's assistant returned with the nurse, defendant once again stated his belief that his father was dying. The nurse tried to comfort defendant, telling him, "I don't think your father is as sick as you think he is." Defendant, very upset, responded, "Go to hell. I've been taking care of him for years. I'll take care of him." Defendant was then left alone in the room with his father.

*(Continues)*

Alone at his father's bedside, defendant began to cry and to tell his father how much he loved him. His father began to cough, emitting a gurgling and rattling noise. Extremely upset, defendant pulled a small pistol from his pants pocket, put it to his father's temple, and fired. He subsequently fired three more times and walked out into the hospital corridor, dropping the gun to the floor just outside his father's room.

Following the shooting, defendant, who was crying and upset, neither ran nor threatened anyone. Moreover, he never denied shooting his father and talked openly with law enforcement officials. Specifically, defendant made the following oral statements: "You can't do anything to him now. He's out of his suffering." "I killed my daddy." "He won't have to suffer anymore." "I know they can burn me for it, but my dad will not have to suffer anymore." "I know the doctors couldn't do it, but I could." "I promised my dad I wouldn't let him suffer." . . .

Though defendant's father had been near death as a result of his medical condition, the exact cause of the decedent's death was determined to be the four point-blank bullet wounds to his head. Defendant's pistol was a single-action .22-caliber five-shot revolver. The weapon, which had to be cocked each time it was fired, contained four empty shells and one live round.

At the close of the evidence, defendant's case was submitted to the jury for one of four possible verdicts: first-degree murder, second-degree murder, voluntary manslaughter, or not guilty. After a lengthy deliberation, the jury found defendant guilty of first-degree murder. Judge Cornelius accordingly sentenced defendant to the mandatory life term. . . .

In his second assignment of error, defendant asserts that the trial court committed reversible error in denying his motion for directed verdict as to the first-degree murder charge. Specifically, defendant argues that the trial court's submission of the first-degree murder charge was improper because there was insufficient evidence of premeditation and deliberation presented at trial. We do not agree, and we therefore overrule defendant's assignment of error. . . .

First-degree murder is the intentional and unlawful killing of a human being with malice and with premeditation and deliberation. Premeditation means that the act was thought out beforehand for some length of time, however short, but no particular amount of time is necessary for the mental process of premeditation.

**Deliberation** means an intent to kill, carried out in a cool state of blood, in furtherance of a fixed design for revenge or to accomplish an unlawful purpose and not under the influence of a violent passion, suddenly aroused by lawful or just cause or legal provocation.

The phrase "cool state of blood" means that the defendant's anger or emotion must not have been such as to overcome his reason.

Premeditation and deliberation relate to mental processes and ordinarily are not readily susceptible to proof by direct evidence. Instead, they usually must be proved by circumstantial evidence. Among other circumstances to be considered in determining whether a killing was with premeditation and deliberation are: (1) want of provocation on the part of the deceased; (2) the conduct and statements of the defendant before and after the killing; (3) threats and declarations of the defendant before and during the course of the occurrence giving rise to the death of the deceased; (4) ill-will or previous difficulty between the parties; (5) the dealing of lethal blows after the deceased has been felled and rendered helpless; and (6) evidence that the killing was done in a brutal manner. We have also held that the nature and number of the victim's wounds is a circumstance from which premeditation and deliberation can be inferred. . . .

Here, many of the circumstances that we have held to establish a factual basis for a finding of premeditation and deliberation are present. It is clear, for example, that the seriously ill deceased did nothing to provoke defendant's action. Moreover, the deceased was lying helpless in a hospital bed when defendant shot him four separate times. In addition, defendant's revolver was a five-shot single-action gun which had to be cocked each time before it could be fired. Interestingly, although defendant testified that he always carried the gun in his job as a truck driver, he was not working on the day in question but carried the gun to the hospital nonetheless.

Most persuasive of all on the issue of premeditation and deliberation, however, are defendant's own statements following the incident. Among other things, defendant stated that he had thought about putting his father out of his misery because he knew he was suffering. He stated further that he had promised his father that he would not let him suffer and that, though he did not think he could do it, he just could not stand to see his father suffer any more. These statements, together with the other circumstances mentioned above, make it clear that the trial court did not err in submitting to the jury the issue of first-degree murder based upon premeditation and deliberation. Accordingly, defendant's second assignment of error is overruled. . . .

Exum, Chief Justice, dissenting.

Almost all would agree that someone who kills because of a desire to end a loved one's physical suffering caused by an illness which is both terminal and incurable should not be deemed in law as culpable and deserving of the same punishment as one who kills because of unmitigated spite, hatred or ill will. Yet the Court's decision in this case essentially says there is no legal distinction between the two kinds of killing. Our law of homicide should not be so roughly hewn as to be incapable of recognizing the difference. I believe there are legal principles which, when properly applied, draw the desirable distinction and that both the trial court and this Court have failed to recognize and apply them. . . .

## Notes and Questions

1. Should the law recognize a distinction between the two kinds of killings described in Chief Justice Exum's dissent in *Forrest*? Do you believe Forrest is more or less deserving of being convicted of first-degree murder than Brown (*State v. Brown, supra*)? Would Forrest likely have fared better in a jurisdiction that relied on the MPC's approach to discriminating between criminal homicides?

2. How satisfactory are the concepts of premeditation and deliberation in distinguishing the most culpable forms of murder (*i.e.*, first degree) from less culpable ones in the context of a case like *Forrest*? Note that Forrest received a mandatory sentence of life imprisonment. Under present North Carolina law, if he had been convicted of second-degree murder, the judge would have had the authority to sentence him to prison for a term ranging from 94 months (just under 8 years) to 392 months (or over 32 years). N.C. Gen. Stat. §§ 14–17, 15A-1340.17.

## Diminished Capacity and Homicide

First-degree murder that requires proof of premeditation and deliberation is a "specific intent" crime (see Chapter 3). Defendants sometimes offer evidence that because of some disability, such as mental illness or mental retardation, they are incapable of formulating the specific intent necessary for first-degree murder. They argue, accordingly, that they can be convicted only of a lower grade of criminal homicide, typically second-degree murder but sometimes manslaughter. This type of defense to first-degree murder commonly is known as "diminished capacity" or "partial responsibility." For a variety of reasons the diminished capacity defense is not recognized in all jurisdictions. These issues are discussed in the following case.

## *Chestnut v. State*, 538 So.2d 820 (Fla. 1989)

Grimes, Justice.

This is a petition to review the First District Court of Appeal's decision in *Chestnut v. State*, 505 So. 2d 1352 (Fla. 1st DCA 1987), which held that evidence of diminished mental capacity was inadmissible to negate the specific intent required to convict of first-degree premeditated murder. The district court certified the following question as one of great public importance:

Is evidence of an abnormal mental condition not constituting legal insanity admissible for the purpose of proving either that the accused could not or did not entertain the specific intent or state of mind essential to proof of the offense, in order to determine whether the crime charged, or a lesser degree thereof, was in fact committed?

*Id.* at 1356. . . . We answer the question in the negative.

Chestnut and two codefendants, Jackie Bolesta and Gary German, robbed and killed the victim, Carl Brown, as the result of a robbery/murder scheme. German, the state's chief witness, was granted immunity and testified that Bolesta and Chestnut planned to rob and kill Brown on a trip to look at horses ostensibly being offered to Brown for sale. He testified that Bolesta demanded the victim's money while Chestnut held an axe handle over the victim and then struck him across the forehead. German further testified that Bolesta, armed with a machete, and Chestnut then took the victim "down the road." After the victim was killed, German and Chestnut hid the body in the woods. Chestnut was charged with both first-degree premeditated murder and felony murder. . . .

The state filed a pretrial motion seeking to prohibit anticipated testimony by defense witnesses who would present evidence concerning appellant's mental condition below that standard recognized by

*(Continues)*

the *M'Naghten* rule. The trial court granted the state's motion finding that "'absent an insanity plea, expert testimony as to mental status, especially when offered to bolster an affirmative defense would be improper in and of itself since it would only tend to confuse the jury.'"

At trial, counsel proffered expert testimony seeking to establish that Chestnut did not have the mental state required for premeditated first-degree murder. The proffered testimony revealed that Chestnut's intelligence was in the lowest five percent of the general population. Further, it showed that, several years earlier, appellant was kicked in the head by a bull, sustaining a fractured skull and brain damage which caused a posttraumatic seizure disorder that required medication. Chestnut also proffered evidence that he has diminished mental capacity with moderate impairment of verbal memory and has a passive personality which causes him to avoid physical confrontation; as a result, he is easily led and manipulated.

At the instruction conference, the trial court denied appellant's request for a special verdict form that would separate premeditated murder from felony murder. Chestnut was convicted of first-degree murder and sentenced to life imprisonment without parole for twenty-five years, the state declining to seek the death penalty. . . .

The issue presented by this case is not a new one. In his article entitled "Psychiatric Evidence in Criminal Cases for Purposes Other Than the Defense of Insanity," Professor Lewin explains:

> Partial responsibility has been recognized for at least 100 years but it was not until the late 1950's that the Supreme Court of California in a series of decisions promulgated the modern concept and excited the imaginations of forensic psychiatrists, behavioral scientists and related scholars. Simply stated, the theory is that if because of mental disease or defect a defendant cannot form the specific state of mind required as an essential element of a crime, he may be convicted only of a lower grade of the offense not requiring that particular mental element. For example, if D is charged with the premeditated slaying of V, partial responsibility would enable a psychiatrist to testify that a mental disease interfered with D's capacity to formulate a plan. Thus a jury could find that D acted impulsively and without premeditation and therefore find D guilty of a lesser grade of homicide. The defense is thus available to reduce first degree murder requiring the specific intent elements of deliberation, premeditation and intent to kill to second degree murder, or even to manslaughter. Although generally applied to first degree murder cases, it is in theory applicable to any crime requiring proof of a specific intent, such as larceny or robbery. 26 Syracuse L Rev. 1051, 1054–55 (1975) (footnotes omitted).

Differing terminology has sometimes clouded an understanding of the defense, as explained in *Muench v. Israel*, 715 F.2d 1124, 1142–43 (7th Cir. 1983), *cert. denied sub nom, Worthing v. Israel*, 467 U.S. 1228 (1984):

> Petitioners, of course, claim they are not attempting to impose upon Wisconsin what they call a "diminished responsibility defense," thereby attempting to capitalize on the somewhat misleading nature of that particular label for the doctrine. A distinction can be drawn between the theory advanced by petitioners—admitting evidence of mental illness which is explicitly tied (at least grammatically) to the specific *mens rea* at issue, and a doctrine which might accurately be called diminished responsibility—admitting evidence of mental illness as a vague and general mitigating factor.
>
> However, the courts have used the labels diminished responsibility, diminished capacity, and other nomenclature merely as a shorthand for the proposition that expert evidence of mental abnormalities is admissible on the question of whether the defendant in fact possessed a particular mental state which is an element of the charged offense. When a court rejects the doctrine of diminished capacity, it is saying that psychiatric evidence is inadmissible on the *mens rea* issue, as recent cases rejecting the doctrine explain.

Following the lead of California, approximately one-half the states and federal jurisdictions now approve the defense. Yet, it was recently noted in *State v. Wilcox*, 436 N.E.2d 523, 525 (Ohio 1982):

> At this juncture, however, it appears that enthusiasm for the diminished capacity defense is on the wane and that there is, if anything, a developing movement away from diminished capacity although the authorities at this point are still quite mixed in their views.

It is also clear that a state is not constitutionally compelled to recognize the doctrine of diminished capacity. *Muench v. Israel. See also Fisher v. United States*, 328 U.S. 463 (1946).

The adoption of the principle of diminished capacity has usually been justified on the premise that mentally deficient persons should be treated in the same manner as intoxicated persons. *See United*

*States v. Brawner,* 471 F.2d 969 (D.C.Cir.1972). However, several recent cases have sharply rejected this analogy. Thus, in *Bethea v. United States,* 365 A.2d 64, 88 (D.C. 1976), *cert. denied,* 433 U.S. 911 (1977), the court said:

> We recognize that there are exceptions to the basic principle that all individuals are presumed to have a similar capacity for *mens rea.* The rule that evidence of intoxication may be employed to demonstrate the absence of specific intent figured prominently in the *Brawner* court's advocacy of consistency in the treatment of expert evidence of mental impairment. The asserted analogy is flawed, however, by the fact that there are significant evidentiary distinctions between psychiatric abnormality and the recognized incapacitating circumstances. Unlike the notion of partial or relative insanity, conditions such as intoxication, medication, epilepsy, infancy, or senility are, in varying degrees, susceptible to quantification or objective demonstration and to lay understanding. As the Ninth Circuit observed in *Wahrlich v. Arizona,* 479 F.2d 1137, 1138 (9th Cir.), *cert. denied,* 414 U.S. 1011 (1973):
>
>> Exposure to the effects of age and of intoxicants upon state of mind is a part of common human experience which fact finders can understand and apply; indeed, they would apply them even if the state did not tell them they could. The esoterics of psychiatry are not within the ordinary ken. (Footnotes omitted.)

In the same vein, the court in *State v. Wilcox* said:

> It takes no great expertise for jurors to determine whether an accused was "'so intoxicated as to be mentally unable to intend anything (unconscious),'" . . . whereas the ability to assimilate and apply the finely differentiated psychiatric concepts associated with diminished capacity demands a sophistication (or as critics would maintain a sophistic bent) that jurors (and officers of the court) ordinarily have not developed. We are convinced as was the *Bethea* court, that these "significant evidentiary distinctions" preclude treating diminished capacity and voluntary intoxication as functional equivalents for purposes of partial exculpation from criminal responsibility. 436 N.E.2d at 530 (citation omitted).

The adverse consequences of adopting the defense of diminished capacity were recognized by the court in *Bethea v. United States:*

> Under the present statutory scheme, a successful plea of insanity avoids a conviction, but confronts the accused with the very real possibility of prolonged therapeutic confinement. If, however, psychiatric testimony were generally admissible to cast a reasonable doubt upon whatever degree of *mens rea* was necessary for the charged offense, thus resulting in outright acquittal, there would be scant reason indeed for a defendant to risk such confinement by arguing the greater form of mental deficiency. Thus, quite apart from the argument that the diminished capacity doctrine would result in a considerably greater likelihood of acquittal for those who by traditional standards would be held responsible, the future safety of the offender as well as the community would be jeopardized by the possibility that one who is genuinely dangerous might obtain his complete freedom merely by applying his psychiatric evidence to the threshold issue of intent. 365 A.2d at 90–91 (footnotes omitted).

To permit the defense of diminished capacity would invite arbitrary applications of the law because of the nebulous distinction between specific and general intent crimes. Moreover, a recognition of the defense would open the door to consequences which could seriously affect our society. In a case of first-degree premeditated murder, a finding of diminished mental capacity would serve to reduce the conviction to a lesser homicide. However, in the case of robbery, which was held to be a specific intent crime in *Bell v. State,* 394 So.2d 979 (Fla. 1981), the application of diminished capacity could result in an absolute acquittal of any crime whatsoever. This is so because the only necessarily lesser included offense of robbery is petit theft and that, too, is a specific intent crime. Apparently, the same would be true for battery. Since burglary is also a specific intent crime, one acquitted of that offense could only be convicted, if at all, of trespass. Unlike the case where one is found not guilty by reason of insanity, there would be no authority to commit these persons for treatment except through the use of civil remedies and its concomitant burdens.

In criticizing the principle of diminished capacity, Abraham Goldstein, in his comprehensive book entitled *The Insanity Defense,* stated:

> There are several reasons for limiting the subjective inquiry to instances when the insanity defense is pleaded. One is the reluctance to believe it is possible to know what passes through a man's mind. From this comes the feeling that acts are the only reliable indices, and with it the suspicion that the person who claims he was not "responsible" for his acts is lying. If he should win his freedom by virtue of what are

*(Continues)*

*(Continued)*

probably lies, the security of society would be threatened. A second is that the use of a subjective theory would probably result in more acquittals. Yet such acquittals might bring freedom for defendants who have proved themselves less able than most men to control their conduct, thereby increasing the threat to society. The insanity defense takes both these problems into account by allowing a full presentation of the defendant's mental life while, at the same time, providing a way of keeping him in custody if it should seem necessary. A. Goldstein, *The Insanity Defense* 192 (1967).

We acknowledge the cogent reasons expressed in *Bethea v. United States* for declining to adopt the defense of diminished capacity:

> The concept of insanity is simply a device the law employs to define the outer limits of that segment of the general population to whom these presumptions concerning the capacity for criminal intent shall not be applied. The line between the sane and the insane for the purposes of criminal adjudication is not drawn because for one group the actual existence of the necessary mental state (or lack thereof) can be determined with any greater certainty, but rather because those whom the law declares insane are demonstrably so aberrational in their psychiatric characteristics that we choose to make the assumption that they are incapable of possessing the specified state of mind. Within the range of individuals who are not "insane," the law does not recognize the readily demonstrable fact that as between individual criminal defendants the nature and development of their mental capabilities may vary greatly. . . . By contradicting the presumptions inherent in the doctrine of *mens rea*, the theory of diminished capacity inevitably opens the door to variable or sliding scales of criminal responsibility. We should not lightly undertake such a revolutionary change in our criminal justice system. 365 A.2d at 87–88 (footnotes omitted).

Finally, we note the pertinent comment in *State v. Wilcox*:

> [T]he effect of adopting a diminished capacity model transcends the doctrine's potential to transform criminal trials into psychiatric shouting matches. Rather, the diminished capacity theory forcefully challenges conventional concepts of culpability and "involve[s] a fundamental change in the common law theory of responsibility." *Fisher, supra,* 328 U.S. at 476. Echoing *Bethea,* "[w]e conclude that the potential impact of concepts such as diminished capacity or partial insanity—however labeled—is of a scope and magnitude which precludes their proper adoption by an expedient modification of the rules of evidence. If such principles are to be incorporated into our law of criminal responsibility, the change should lie within the province of the legislature." *Bethea, supra,* [365 A.2d] at page 92. See *Fisher, supra,* 328 U.S. at page 476.

. . . 436 N.E.2d at 533. . . .

It could be said that many, if not most, crimes are committed by persons with mental aberrations. If such mental deficiencies are sufficient to meet the definition of insanity, these persons should be acquitted on that ground and treated for their disease. Persons with less serious mental deficiencies should be held accountable for their crimes just as everyone else. If mitigation is appropriate, it may be accomplished through sentencing, but to adopt a rule which creates an opportunity for such persons to obtain immediate freedom to prey on the public once again is unwise. . . .

Overton, Justice, dissenting.

In my view, it is totally unreasonable and illogical to allow a defendant to present evidence of *voluntary* intoxication and drug use as a defense to the element of specific intent to commit first-degree premeditated murder, but then to prohibit another defendant from presenting objective evidence of *involuntary* organic brain damage on the same issue. This results in a clear injustice. . . .

## Notes and Questions

1. Reconsider *Frey v. State*, 708 So.2d 918 (Fla. 1998), another Florida case we considered in Chapter 3. Who seems like a better candidate to benefit from a legal doctrine that would allow a jury to conclude that he lacked the "specific intent" to commit a crime: Frey or Chestnut?

2. For several years the California courts occupied a leading role in recognizing diminished capacity in first-degree murder cases. *See People v. Wolff,* 394 P.2d 959 (Cal. 1964). Under California law the diminished capacity defense could even be used to reduce murder to manslaughter. *See People v. Conley,* 411 P.2d 911 (Cal. 1966). More recently, however, the California legislature has abolished the diminished capacity defense. Cal. Penal Code §§ 25(a), 28(a), (b). *See People v. Avena,* 916 P.2d 1000 (Cal. 1996). What dangers are lurking if evidence is admitted in a case like *Chestnut* to attempt to

establish that a defendant in a first-degree murder trial lacked the capacity to premeditate and deliberate? On the other hand, what problems arise if such evidence is not admissible or is only admissible in certain contexts (*e.g.*, voluntary intoxication) but not others (*e.g.*, mental retardation or mental illness)?

**3.** The terms "diminished capacity" and "partial responsibility" are potentially misleading because they imply that a defendant "is partially responsible for the commission of some offense. This, of course, is not true. If the defendant's mental condition was such that he did not premeditate and deliberate, then there is *no* responsibility for first degree murder but *full* responsibility for second degree murder" (emphasis in original).[8]

**4.** How does diminished capacity differ, in theory and in consequence, from the insanity defense?

## Felony Murder

The 1794 Pennsylvania legislation that created first-degree and second-degree murder included within first-degree murder those killings "committed in the perpetration or attempt to perpetrate any arson, rape, robbery or burglary." Unlike the other major type of first-degree murder we just considered—a "wilful, deliberate and premeditated killing"—the legislation is conspicuously silent about a *mens rea* requirement for killings committed in connection with the named felonies. This omission was not inadvertent. The felony-murder doctrine recognized under the common law, which applied to killings associated with felonies generally and not just the four named in the Pennsylvania statute, "operated to impose liability for murder based on the culpability required for the underlying felony without separate proof of culpability with regard to the death. The homicide, as distinct from the underlying felony, was thus an offense of strict liability."[9]

The felony-murder rule worked no unusual hardships in an era when *all* felonies were punished by death. It made little difference, for example, whether a robber, a burglar, or a rapist who killed during the commission of his or her crime was executed for the underlying felony or was deemed a murderer and executed for the accompanying homicide. Indeed, the felony-murder rule was used historically primarily to punish offenders who killed accidentally while *attempting* to commit a felony that was not successfully completed. At common law the attempt to commit a felony was only a misdemeanor. Accidental killings committed during an attempted felony could be defined as murder because the felony-murder doctrine includes killings committed during the *attempted* perpetration of a felony.[10]

As the law evolved, most felonies that did not involve the taking of life ceased to be punished capitally. The number of felonies greatly expanded over the ones traditionally recognized at common law. These changes caused many to question the continuing justification for a rule that automatically defined as murder any killing committed by the offender during the perpetration or attempted perpetration of a felony. A number of limitations sprouted to contain the application of the felony-murder rule. Some jurisdictions—consistent with the approach taken by the MPC—chose to abandon the rule altogether. We consider more about the rationale, application, and limitations on the felony-murder rule in the cases that follow.

### *People v. Stamp,* 82 Cal. Rptr. 598 (Cal. App. 1969)

Cobey, Associate Justice.

These are appeals by Jonathan Earl Stamp, Michael John Koory and Billy Dean Lehman, following jury verdicts of guilty of robbery and murder, both of the first degree. Each man was given a life sentence on the murder charge together with the time prescribed by law on the robbery count.

Defendants appeal their conviction of the murder of Carl Honeyman who, suffering from a heart disease, died between 15 and 20 minutes after Koory and Stamp held up his business, the General Amusement Company, on October 26, 1965, at 10:45 A.M. Lehman, the driver of the getaway car, was apprehended a few minutes after the robbery; several weeks later Stamp was arrested in Ohio and Koory in Nebraska. . . .

On this appeal appellants primarily rely upon their position that the felony-murder doctrine should not have been applied in this case due to the unforeseeability of Honeyman's death.

#### THE FACTS

Defendants Koory and Stamp, armed with a gun and a blackjack, entered the rear of the building housing the offices of General Amusement Company, ordered the employees they found there to go to the front of the premises, where the two secretaries were working. Stamp, the one with the gun, then

*(Continues)*

went into the office of Carl Honeyman, the owner and manager. Thereupon Honeyman, looking very frightened and pale, emerged from the office in a "kind of hurry." He was apparently propelled by Stamp who had hold of him by an elbow.

The robbery victims were required to lie down on the floor while the robbers took the money and fled out the back door. As the robbers, who had been on the premises 10 to 15 minutes, were leaving, they told the victims to remain on the floor for five minutes so that no one would "get hurt."

Honeyman, who had been lying next to the counter, had to use it to steady himself in getting up off the floor. Still pale, he was short of breath, sucking air, and pounding and rubbing his chest. As he walked down the hall, in an unsteady manner, still breathing hard and rubbing his chest, he said he was having trouble "keeping the pounding down inside" and that his heart was "pumping too fast for him." A few minutes later, although still looking very upset, shaking, wiping his forehead and rubbing his chest, he was able to walk in a steady manner into an employee's office. When the police arrived, almost immediately thereafter, he told them he was not feeling well and that he had a pain in his chest. About two minutes later, which was 15 to 20 minutes after the robbery had occurred, he collapsed on the floor. At 11:25 he was pronounced dead on arrival at the hospital. The coroner's report listed the immediate cause of death as heart attack.

The employees noted that during the hours before the robbery Honeyman had appeared to be in normal health and good spirits. The victim was an obese, sixty-year-old man, with a history of heart disease, who was under a great deal of pressure due to the intensely competitive nature of his business. Additionally, he did not take good care of his heart.

Three doctors, including the autopsy surgeon, Honeyman's physician, and a professor of cardiology from U.C.L.A., testified that although Honeyman had an advanced case of atherosclerosis, a progressive and ultimately fatal disease, there must have been some immediate upset to his system which precipitated the attack. It was their conclusion in response to a hypothetical question that but for the robbery there would have been no fatal seizure at that time. The fright induced by the robbery was too much of a shock to Honeyman's system. There was opposing expert testimony to the effect that it could not be said with reasonable medical certainty that fright could ever be fatal.

## SUFFICIENCY OF THE EVIDENCE RE CAUSATION

Appellants' contention that the evidence was insufficient to prove that the robbery factually caused Honeyman's death is without merit. . . .

A review of the facts as outlined above shows that there was substantial evidence of the robbery itself, that appellants were the robbers, and that but for the robbery the victim would not have experienced the fright which brought on the fatal heart attack.

## APPLICATION OF THE FELONY-MURDER RULE

Appellants' contention that the **felony-murder rule** is inapplicable to the facts of this case is also without merit. Under the felony-murder rule of section 189 of the [California] Penal Code, a killing committed in either the perpetration of or an attempt to perpetrate robbery is murder of the first degree. This is true whether the killing is willful, deliberate and premeditated, or merely accidental or unintentional, and whether or not the killing is planned as a part of the commission of the robbery. . . .

The doctrine presumes malice aforethought on the basis of the commission of a felony inherently dangerous to human life.

This rule is a rule of substantive law in California and not merely an evidentiary shortcut to finding malice as it withdraws from the jury the requirement that they find either express malice or the implied malice which is manifested in an intent to kill. Under this rule no intentional act is necessary other than the attempt to or the actual commission of the robbery itself. When a robber enters a place with a deadly weapon with the intent to commit robbery, malice is shown by the nature of the crime. There is no requirement that the killing occur, "while committing" or "while engaged in" the felony, or that the killing be "a part of" the felony, other than that the few acts be a part of one continuous transaction. Thus the homicide need not have been committed "to perpetrate" the felony. There need be no technical inquiry as to whether there has been a completion or abandonment of or desistence from the robbery before the homicide itself was completed.

The doctrine is not limited to those deaths which are foreseeable. Rather a felon is held strictly liable for *all* killings committed by him or his accomplices in the course of the felony. As long as the homicide is the direct causal result of the robbery the felony-murder rule applies whether or not the death was a natural or probable consequence of the robbery. So long as a victim's predisposing physical condition, regardless of its cause, is not the *only* substantial factor bringing about his death, that condition, and the robber's ignorance of it, in no way destroys the robber's criminal responsibility for the death. So long as life is shortened as a result of the felonious act, it does not matter that the victim might have died soon anyway. In this respect, the robber takes his victim as he finds him. . . .

## Notes and Questions

1. Are Stamp and his codefendants any *more* culpable, in a meaningful way, than robbers who, after gathering up valuables and making their getaway, do not have the misfortune of having one of their victims suffer a heart attack and die? Should they be guilty of first-degree murder, the most serious form of criminal homicide? Are they *less* culpable than robbers, who, after gathering up valuables, place a gun to the temple of their victims and deliberately shoot and kill them?

2. What if, instead of committing an armed robbery, Stamp had entered Mr. Honeyman's place of business and shoplifted merchandise worth several hundreds of dollars, making the crime a felonious larceny? If Mr. Honeyman observed the theft, became agitated, and suffered a heart attack and died, could/should Stamp be guilty of murder under the felony-murder rule?

3. What if Stamp and his codefendants had made a clean and uneventful getaway, and Mr. Honeyman suffered no immediate adverse effects? If, a few days later, Honeyman had a chance encounter with Stamp, became agitated, and suffered a heart attack and died, could/should Stamp be guilty of murder under a felony-murder rationale?

4. One of Stamp's codefendants, Lehman, did not even enter the store and had no contact with Mr. Honeyman. Should he be held responsible for murder?

5. What if, having been robbed before, Mr. Honeyman was armed when Stamp and Koory entered the store, and he (Honeyman) shot and killed Koory? Would/should Stamp be guilty of murdering Koory by application of the felony-murder rule?

6. For other cases involving application of the felony-murder rule where the cause of death was a heart attack, *see State v. Dixon,* 387 N.W.2d 682 (Neb. 1986) and decisions there cited.

## Nature of the Underlying Felony

If a killing occurs during the perpetration of a felony, should it matter what kind of felony was being committed for purposes of applying the felony-murder rule?

### *State v. Stewart,* 663 A.2d 912 (R.I. 1995)

Weisberger, Chief Justice.

This case comes before us on the appeal of the defendant, Tracy Stewart, from a judgment of conviction entered in the Superior Court on one count of second-degree murder in violation of G.L.1956 (1981 Reenactment) § 11-23-1. We affirm the judgment of conviction. The facts insofar as pertinent to this appeal are as follows.

On August 31, 1988, twenty-year-old Tracy Stewart (Stewart or defendant) gave birth to a son, Travis Young (Travis). Travis's father was Edward Young, Sr. (Young). Stewart and Young, who had two other children together, were not married at the time of Travis's birth. Travis lived for only fifty-two days, dying on October 21, 1988, from dehydration.

During the week prior to Travis's death, Stewart, Young, and a friend, Patricia McMasters (McMasters), continually and repeatedly ingested cocaine over a two- to three-consecutive-day period at the apartment shared by Stewart and Young. The baby, Travis, was also present at the apartment while Stewart, Young, and McMasters engaged in this cocaine marathon. . . .

The cocaine binge continued uninterrupted for two to three days. McMasters testified that during this time neither McMasters nor Stewart slept at all. McMasters testified that defendant was never far from her during this entire two- to three-day period except for the occasions when McMasters left the apartment to buy more cocaine. During this entire time, McMasters saw defendant feed Travis only once. Travis was in a walker, and defendant propped a bottle of formula up on the walker, using a blanket, for the baby to feed himself. McMasters testified that she did not see defendant hold the baby to feed him nor did she see defendant change Travis's diaper or clothes during this period.

*(Continues)*

(*Continued*)

Ten months after Travis's death defendant was indicted on charges of second-degree murder, wrongfully causing or permitting a child under the age of eighteen to be a habitual sufferer for want of food and proper care (hereinafter sometimes referred to as "wrongfully permitting a child to be a habitual sufferer"), and manslaughter. The second-degree-murder charge was based on a theory of felony murder. The prosecution did not allege that defendant intentionally killed her son but rather that he had been killed during the commission of an inherently dangerous felony, specifically, wrongfully permitting a child to be a habitual sufferer. Moreover, the prosecution did not allege that defendant intentionally withheld food or care from her son. Rather the state alleged that because of defendant's chronic state of cocaine intoxication, she may have realized what her responsibilities were but simply could not remember whether she had fed her son, when in fact she had not. . . .

The defendant was found guilty of both second-degree murder and wrongfully permitting a child to be a habitual sufferer. A subsequent motion for new trial was denied. This appeal followed. . . .

The defendant moved for judgment of acquittal . . . at the close of the state's case and again at the close of all the evidence. In regard to the felony-murder charge defendant claimed that the evidence was insufficient to prove that the crime of wrongfully permitting a child to be a habitual sufferer is an inherently dangerous felony. . . .

*Whether Wrongfully Permitting a Child To Be a Habitual Sufferer Is an Inherently Dangerous Felony.*

Rhode Island's murder statute, § 11-23-1, enumerates certain crimes that may serve as predicate felonies to a charge of first-degree murder. A felony that is not enumerated in § 11-23-1 can, however, serve as a predicate felony to a charge of second-degree murder. Thus the fact that the crime of wrongfully permitting a child to be a habitual sufferer is not specified in § 11-23-1 as a predicate felony to support a charge of first-degree murder does not preclude such crime from serving as a predicate to support a charge of second-degree murder.

In Rhode Island second-degree murder has been equated with common-law murder. At common law, where the rule is unchanged by statute, "[h]omicide is murder if the death results from the perpetration or attempted perpetration of an inherently dangerous felony." *Id.* (quoting Perkins, *Criminal Law* 44 (2d ed. 1969)). To serve as a predicate felony to a charge of second-degree murder, a felony that is not specifically enumerated in § 11-23-1 must therefore be an inherently dangerous felony.

The defendant contends that wrongfully permitting a child to be a habitual sufferer is not an inherently dangerous felony and cannot therefore serve as the predicate felony to a charge of second-degree murder. In advancing her argument, defendant urges this court to adopt the approach used by California courts to determine if a felony is inherently dangerous. This approach requires that the court consider the elements of the felony "in the abstract" rather than look at the particular facts of the case under consideration. With such an approach, if a statute can be violated in a manner that does not endanger human life, then the felony is not inherently dangerous to human life. *People v. Burroughs*, 678 P.2d 894, 898–900 (Cal. 1984); *People v. Caffero*, 207 Cal.App. 3d 678, 683–84 (1989). Moreover, the California Supreme Court has defined an act as "inherently dangerous to human life when there is 'a *high probability* that it will result in death.'"

In *Caffero, supra,* a two-and-one-half-week-old baby died of a massive bacterial infection caused by lack of proper hygiene that was due to parental neglect. The parents were charged with second-degree felony murder and felony-child abuse, with the felony-child-abuse charge serving as the predicate felony to the second-degree-murder charge. Examining California's felony-child-abuse statute in the abstract, instead of looking at the particular facts of the case, the court held that because the statute could be violated in ways that did not endanger human life, felony-child abuse was not inherently dangerous to human life. By way of example, the court noted that a fractured limb, which comes within the ambit of the felony-child-abuse statute, is unlikely to endanger the life of an infant, much less of a seventeen-year-old (the statute applied to all minors below the age of eighteen years, not only to young children). Because felony-child abuse was not inherently dangerous to human life, it could not properly serve as a predicate felony to a charge of second-degree felony murder.

The defendant urges this court to adopt the method of analysis employed by California courts to determine if a felony is inherently dangerous to life. Aside from California, it appears that Kansas is the only other state which looks at the elements of a felony in the abstract to determine if such felony is inherently dangerous to life. *See, e.g., State v. Wesson,* 802 P.2d 574, 581 (Kan. 1990) (holding that the sale of crack cocaine when viewed in the abstract is not inherently dangerous to human life); *State v. Underwood,* 615 P.2d 153, 161 (Kan. 1980) (holding that the unlawful possession of a firearm by an ex-felon when viewed in the abstract is not inherently dangerous to human life). The case of *Ford v. State,* 423 S.E.2d 255 (Ga. 1992), cited in defendant's brief for the proposition that possession of a firearm by an ex-felon is not an inherently dangerous felony which can support a felony-murder conviction, actually holds that the attendant circumstances of the particular case should be considered in determining whether the underlying felony "create[d] a foreseeable risk of death." In *Ford* the defendant (Ford) had previously been convicted of the felony of possession of cocaine with intent to distribute. Ford was visiting the home of his girlfriend's mother and had brought with him a semiautomatic pistol. While there he attempted to unload the pistol, but in so doing, he discharged the weapon, sending a bullet both through the floor and through the ceiling of a basement apartment located in the house. The bullet struck and killed the occupant of the basement apartment. There was no evidence that at the time of the shooting the defendant was aware of the existence of the apartment or of the victim's presence in it. Ford was charged with and convicted of felony murder, with the underlying felony being the possession of a firearm by a convicted felon.

The Georgia Supreme Court reversed the conviction for felony murder holding that a status felony, including the possession of a firearm by a previously-convicted felon, is not inherently dangerous. The court explained that there could indeed be circumstances in which such a felony could be considered dangerous (for example when the possession of a firearm was coupled with an aggravated assault or other dangerous felony) but that such circumstances were absent in that case. It held that in *that particular case,* which did not involve an assault or other criminal conduct, the underlying felony of possession of a firearm by a previously convicted felon was not inherently dangerous and thus could not serve as a predicate to the charge of felony murder.

We decline defendant's invitation to adopt the California approach in determining whether a felony is inherently dangerous to life and thus capable of serving as a predicate to a charge of second-degree felony murder. We believe that the better approach is for the trier of fact to consider the facts and circumstances of the particular case to determine if such felony was inherently dangerous in the manner and the circumstances in which it was committed, rather than have a court make the determination by viewing the elements of a felony in the abstract. We now join a number of states that have adopted this approach. *See, e.g., Jenkins v. State,* 230 A.2d 262 (Del. 1967); *State v. Wallace,* 333 A.2d 72 (Me. 1975); *Commonwealth v. Ortiz,* 560 N.E.2d 698 (Mass. 1990); *State v. Harrison,* 564 P.2d 1321 (N.M. 1977); *State v. Nunn,* 297 N.W.2d 752 (Minn. 1980).

A number of felonies at first glance would not appear to present an inherent danger to human life but may in fact be committed in such a manner as to be inherently dangerous to life. The crime of escape from a penal facility is an example of such a crime. On its face, the crime of escape is not inherently dangerous to human life. But escape may be committed or attempted to be committed in a manner wherein human life is put in danger. Indeed, in *State v. Miller,* [161 A. 222 (R.I. 1932)] this court upheld the defendant's conviction of second-degree murder on the basis of the underlying felony of escape when a prison guard was killed by an accomplice of the defendant during an attempted escape from the Rhode Island State prison. By way of contrast, the California Supreme Court has held that the crime of escape, viewed in the abstract, is an offense that is not inherently dangerous to human life and thus cannot support a second-degree felony-murder conviction. *People v. Lopez,* 489 P.2d 1372, 1376 (Cal. 1971) (In Bank).

The amendment of our murder statute to include any unlawful killing "committed during the course of the perpetration, or attempted perpetration, of felony manufacture, sale, delivery, or other distribution of a controlled substance otherwise prohibited by the provisions of chapter 28 of title 21" lends further support for not following California's approach to determining the inherent dangerousness of a felony. According to the statute a person who delivers phencyclidine (PCP), a controlled substance under section (e)(5) of schedule II of G.L. 1956 (1989 Reenactment) § 21-28-2.08, as amended by P.L.1991, ch. 211, § 1, to another person who then dies either as a result of an overdose or as a result of behavior precipitated by the drug use (such as jumping off a building because of the loss of spacial perception) could be charged with first-degree murder under § 11-23-1. Conversely, the California Court of Appeal has held that when viewed in the abstract, the standard used by California courts to determine whether a felony is inherently dangerous, the furnishing or selling of PCP is not a felony that carries a high

*(Continues)*

probability that death will result. *People v. Taylor*, 6 Cal.App.4th 1084, 1100 (1992). Consequently, the California Court of Appeal held that the felony of furnishing PCP could not serve as a predicate to a charge of second-degree felony murder. It is clear that there is a profound ideological difference in the approach of the Rhode Island Legislature from the holdings of the courts of the State of California concerning appropriate criminal charges to be preferred against one who furnishes PCP (and presumably a host of other controlled substances) to another person with death resulting therefrom. The lawmakers of the State of Rhode Island have deemed it appropriate to charge such a person with the most serious felony in our criminal statutes—first-degree murder. It appears that the appellate court of California, however, would hold that the most serious charge against one who furnishes PCP to another person with death resulting therefrom would be involuntary manslaughter.

The Legislature's recent amendment to our murder statute as well as this court's prior jurisprudence concerning second-degree felony murder reinforces our belief that we should not adopt the California approach to determine whether a felony is inherently dangerous. The proper procedure for making such a determination is to present the facts and circumstances of the particular case to the trier of fact and for the trier of fact to determine if a felony is inherently dangerous in the manner and the circumstances in which it was committed. This is exactly what happened in the case at bar. The trial justice instructed the jury that before it could find defendant guilty of second-degree murder, it must first find that wrongfully causing or permitting a child to be a habitual sufferer for want of food or proper care was inherently dangerous to human life "in its manner of commission." This was a proper charge. By its guilty verdict on the charge of second-degree murder, the jury obviously found that wrongfully permitting a child to be a habitual sufferer for want of food or proper care was indeed a felony inherently dangerous to human life in the circumstances of this particular case. . . .

The defendant's motions for judgment of acquittal on the felony-murder charge on the ground that wrongfully permitting a child to be a habitual sufferer is not an inherently dangerous felony were properly denied. . . .

## Notes and Questions

1. *People v. Burroughs*, 678 P.2d 894 (Cal. 1984), is cited in *Stewart* as a case involving application of California's requirement that a felony be inherently dangerous "in the abstract" to support a conviction for murder under the felony-murder doctrine. In *Burroughs* the defendant used aggressive massage and other nontraditional techniques in an effort to treat a young man's leukemia. The young man died of a massive hemorrhage after a massage session. The defendant was convicted of felony murder. The underlying felony supporting the murder conviction was practicing medicine without a license. The California Supreme Court invalidated the felony-murder conviction. The court relied on the fact that practicing medicine without a license "is not inherently so dangerous that by its very nature, it cannot be committed without creating a substantial risk that someone will be killed. . . ."

   Does it make more sense to evaluate the dangerousness of the felony used to support a felony-murder conviction in the abstract, as under the California rule, or "to consider the facts and circumstances of the particular case," as in *Stewart*? Which approach permits a broader application of the felony-murder rule?

   If Burroughs had feloniously practiced medicine without a license in Rhode Island as he had in California and thus helped cause the death of the man suffering from leukemia in the former state, would he be at risk of being convicted of felony murder? If Stewart had feloniously "wrongfully permitted a child to be a habitual sufferer" in California and thus helped cause the death of Travis Young in that state, would she be at risk of being convicted of felony murder? Which jurisdiction follows the better rule for determining whether the predicate felony supporting a felony-murder conviction is inherently dangerous?

2. Why should there be *any* limitation on the types of felonies that support a conviction for felony murder? What supplies the element of "malice" that is a part of the common law definition of murder in the context of the felony-murder rule?

3. If Burroughs is not guilty of murder under California's application of the felony-murder rule, might he be found guilty of another form of criminal homicide? See the discussions of "depraved heart" murder and manslaughter, below.

## Killing During the Perpetration of the Felony

### *State v. Hearron*, 619 P.2d 1157 (Kan. 1980)

Prager, Justice:

This is a direct appeal in a criminal action involving convictions of one count of felony murder, one count of conspiracy to commit burglary, two counts of felony theft, two counts of burglary, and one count of attempted burglary. The defendant, William R. Hearron, appeals only from the conviction of felony murder and does not challenge the other convictions.

The facts in the case are undisputed and are essentially as follows: On the evening of January 22, 1979, Ann Terry, who resided on North 70th Street in Kansas City, Kansas, was looking out a window of her home and saw three black youths walk up the driveway. She saw and heard them try to open the garage doors. She advised her husband, Delmer Terry, who immediately turned on an outside light over the garage. The three youths then fled. After notifying the police, the Terrys decided to look for the youths. They got into their automobile and followed a white van which was driving slowly in front of their house. After two or three blocks, the van turned left. The Terrys turned right, spotting the youths who had attempted to break into their garage. Delmer Terry stopped the car, got out, and accused them of attempting to break into his house. The boys pulled firearms. Three shots were fired, wounding Terry. Terry fell back into the car and collapsed. A nearby resident came to the scene and then called an ambulance and the police. While Mrs. Terry remained with her husband, the white van again appeared. The van stopped across from the Terry vehicle. Ann Terry was able to see defendant, William Hearron, looking at their car, before slowly driving on. Defendant was the driver and sole occupant of the van. The whole episode lasted approximately five minutes. Delmer Terry later died from his wounds.

On February 1, 1979, Kansas City police officers obtained a search warrant and searched defendant's house. The police found two black youths hiding in the attic. One of the boys, James Scaife, was identified as the person who shot and killed Mr. Terry. The police also found goods stolen in other January 22, 1979, burglaries. Scaife and defendant were tried together. Scaife was convicted of felony murder. Defendant was also convicted of felony murder as an aider and abettor.

On appeal, the defendant challenges the sufficiency of the evidence to establish felony murder. The defendant maintains, in substance, that the attempted burglary had been completed at the time the shooting occurred and, hence, the felony-murder rule should not be applied. We have concluded from the undisputed factual circumstances that this point is without merit. . . .

K.S.A. 21-3401 includes as murder the killing of a human being "committed in the perpetration or attempt to perpetrate any felony." Although that statute does not specifically include, within the felony-murder rule, the killing of another during flight from the scene of the crime, it is the established law of this state that flight from the scene of the crime may be considered as a part of the res gestae of the crime and a killing during flight may constitute felony murder. . . .

Other jurisdictions, likewise, hold flight from the scene of the crime to be a part of the res gestae and that a killing during the escape or flight may justify application of the felony-murder rule.

Some jurisdictions hold the killing to be within the res gestae of the underlying felony if committed during escape or attempt to escape and the accused has not yet reached a point of temporary safety. Other jurisdictions hold a killing to be within the felony-murder rule if the killing and felony are so "inextricably woven" that they may be considered as "one continuous transaction" or so connected that there is no break in the chain of events. Time, distance, and the causal relationship between the underlying felony and the killing are factors to be considered in determining whether the killing is a part of the felony and, therefore, subject to the felony-murder rule. Whether the underlying felony had been abandoned or completed prior to the killing so as to remove it from the ambit of the felony-murder rule is ordinarily a question of fact for the jury to decide. When we apply the factors of time, distance, and causal relationship to the facts of this case, we have no hesitancy in holding that it was a factual issue for the jury to determine whether the killing of Delmer Terry occurred during the commission of the attempted burglary. . . .

### Notes and Questions

1. Courts sometimes ask whether a killing was committed during an "immediate pursuit" of fleeing felons, or whether instead the killing occurred after the felons had reached "a place of temporary safety," or a "safe haven." *See People v. Salas*, 500 P.2d 7 (Cal. 1972).

*(Continues)*

2. If the defendants in *Hearron* were in their van and several blocks from the Terrys' home, was the attempted burglary still being perpetrated? Did the defendants appear to be "in flight" after the attempted burglary? Under what facts would the killing have been sufficiently removed from the attempted burglary to make the felony-murder rule inapplicable?
3. Given the facts on *Hearron*, why do you suppose the prosecution relied on a theory of felony murder instead of murder based on the intent to kill?

## Vicarious Liability: Killings by Cofelons

In *People v. Stamp* we noted that one of the defendants (Lehman) served as the getaway car driver and did not even enter the store. Nevertheless, Lehman was convicted of felony murder after a robbery victim suffered a heart attack and died. In *State v. Hearron* a codefendant (James Scaife) shot and killed the victim, yet Hearron was convicted of felony murder. We consider the issue of "**vicarious liability,**" or the responsibility of one felon for acts committed by his or her cofelons, in greater detail in Chapter 11. The issue of vicarious liability is important to the scope of the felony-murder doctrine. Is the rule so broad that all cofelons are equally guilty of murder committed by any one of the felons?

## United States v. Heinlein, 490 F.2d 725 (D.C. Cir. 1973)

McGowan, Circuit Judge:

Appellants were charged with felony-murder, murder in the second degree, rape while armed, and rape. They were convicted of felony-murder and the lesser included offense of assault with intent to commit rape while armed. When the jury was unable to agree as to punishment on the felony-murder count, the District Court sentenced appellant Heinlein to death, and both of the Walker brothers to prison sentences of twenty years to life. . . .

All of the participants in the events giving rise to these appeals appear to have lived in the nether world of chronic alcoholism, and the events themselves are of a singularly squalid nature. Because of this, as well as the difficulties of reconstructing—through the imperfect instrument of a chronic alcoholic—what happened in this instance in that confused and cloudy environment, this was obviously a difficult and distasteful case to try, both for judge and jury.

We have, accordingly, examined the record in this case with special care. We have concluded that, with the exception of what we believe to have been a misconception by the court of the law of felony-murder in its application to accomplices, appellants had a fair trial, and that no unacceptable risk of a miscarriage of justice resides in affirming Heinlein's convictions on both counts, and the convictions of the Walker brothers for assault with intent to commit rape while armed. The convictions of the latter for felony-murder are reversed.

### I.

Appellants chose not to testify at trial. Accordingly, the only purportedly eyewitness version of the events in question was given by Mr. James Harding, a chronic alcoholic. On the morning of April 13, 1968, so Harding testified, he and Marie McQueen, the murder victim, were released after overnight incarceration for drunkenness. After buying some wine, they met Bernard Heinlein and the Walker brothers, David and Frank, on the street. The five of them then went to an apartment occupied by the Walkers to drink the wine. Heinlein told McQueen that he wanted to have sexual relations with her, and the Walkers both voiced support of this proposal. When McQueen refused, the three appellants seized her, held her down, and began to remove her clothing. During the struggle, McQueen slapped Heinlein in the face. His response was to take a knife from his pocket and stab her, inflicting what proved to be a fatal wound. . . .

### II.

Appellants Frank and David Walker seek reversal of their felony-murder convictions on the ground that the jury was improperly instructed. Their counsel unsuccessfully requested use of the felony-murder instruction on accomplices contained in the Junior Bar Association Criminal Jury Instructions. The words in that instruction which appellants regard as of critical importance are contained in the phrase "in the course of the felony and in furtherance of the common purpose to commit the felony." These words are, in their submission, to be compared with the qualifying phrases in the instruction as given, namely, "in the course of," and "as a part of" the rape or attempted rape, and "within the scope of the rape which one or more of the defendants undertook to commit."

It may be true, as appellants assert, that the words "in furtherance of the common purpose" in the requested instruction would have focused the jury's attention more directly on the causational element of felony-murder than did the instruction given by the District Court. Those words emphasized that the slaying must be causally related to the objects of the felony, whereas under the latter mere coincidence of time and place between the felony and the murder might be thought to suffice. To that extent, the Junior Bar instruction conforms more nearly to the requirements of our statute as it has been interpreted in respect of accomplices, and is therefore preferable to the instruction given.

It may also be true, however, that the verbal differences between these two formulations, in terms of practical impact upon the jury, are not very great. A slaying which may be said to have occurred "within the scope of" a particular felony might also be regarded as "in furtherance of the common purpose to commit the felony." Under either instruction defense counsel could presumably argue to the jury that the slaying was so unrelated to the object of the felony as to be beyond its scope, in the one case, or not in furtherance of the common purpose to commit it, in the other. On our facts, the argument would be that Heinlein's stabbing of McQueen was an unexpected response to his being slapped in the face and was independent of any common purpose to rape. Indeed, any sudden killing of McQueen by one of the appellants could be regarded as frustrating and defeating that common purpose, and therefore alien to it.

If such argument had been permitted in this case, the differences between the two instructions might not be such as to compel reversal. The District Court, however, appears to have construed the statutory definition of felony-murder to preclude such a defense, and to have restricted argument to the jury accordingly. . . .

[T]he trial judge foreclosed counsel from arguing to the jury matters not going to intent as such, but which do relate to a concept which may, without significant alteration of meaning, be variously phrased as either (1) the scope of the felony which the participants undertook to commit, or (2) the furtherance of the common purpose which the participants entertained in embarking upon it.

We deal with the crime of felony-murder as it is defined in our statute. Our central task is, thus, one of statutory interpretation. It would perhaps be possible to read the relevant statute as contemplating only a coincidence in point of time and place between the commission of the felony and the occurrence of the slaying. This is what the trial judge did, but this does not, however, seem to be what the formulators of jury instructions in this jurisdiction have done. Both the Junior Bar instruction refused by the trial judge, and the one given by him, include phrases which seem to look beyond this concept of mere coincidence.

If these inclusions reflect a proper interpretation of the statute, then the trial judge erred in ruling that, on the evidence adduced, the slaying was clearly within the scope of the felony of rape which the defendants undertook to commit. By so doing, he prevented defense counsel from suggesting to the jurors that the evidence warranted a finding by them that, insofar as the Walker brothers were concerned, the scope of the common undertaking had been exceeded. They would of course, not have been obliged to make that finding, but it would appear to have been open to them even under the terms of the court's instruction as given.

The D.C. felony-murder statute is addressed in terms only to the person who kills while perpetrating a felony. Accomplices, like the Walker brothers, are exposed to first degree murder accountability by reason of the aiding and abetting statute. It is true that that exposure does not depend upon proof of an intent to kill on the part of the accomplice; that intent is supplied by the fact of participation in the felony giving rise to the killing. But, certainly as to accomplices, the matter does not end there. The common law concepts of causation and vicarious responsibility are operative. The accomplice who aids and abets the commission of a felony is legally responsible as a principal for all acts of the other person which are in furtherance of the common design or plan to commit the felony, or are the natural and probable consequence of acts done in the perpetration of the felony.

It was upon these terms that the common law encompassed the imposition of liability for first degree murder upon one who neither killed nor had any intent to kill, but only took part knowingly in the commission of a felony. . . .

Under the common law concept it seems clear that there was room for jury issues relating to whether the slaying occurred within the scope of the felony which the parties undertook to commit,

(Continues)

(Continued)

or in furtherance of their common plan or purpose to commit it. The Supreme Court of Oregon recently pointed to what appears to have been the common law concept. Quoting with approval Wharton's Criminal Law & Procedure, Vol. I, at 544, the court stated the rule in these terms:

> Something more than a mere coincidence of time and place between the wrongful act and the death is necessary. It must appear that there was such actual legal relation between the killing and the crime committed or attempted that the killing can be said to have occurred as a part of the perpetration of the crime, or in furtherance of an attempt or purpose to commit it. *State v. Schwensen*, 392 P.2d 328, 334 (Or. 1964).

It does not appear to be true, as some have supposed, that the current felony-murder statutes embody a legislative purpose to deter the commission of felonies to the point of embracing the coincidence rationale. . . .

Thus it would seem that the felony-murder instruction which has been used in this jurisdiction, as well as the one proposed for use by the Junior Bar Association, reflect an understanding that the statute embraces occasions when the jury may properly be urged to find that the homicidal act fell outside the scope of the felonious crime which the parties undertook to commit. It was error for the trial court to forbid counsel to argue this to the jury; and the felony-murder convictions of the Walker brothers must be reversed. . . .

## Notes and Questions

1. If the Walker brothers and Heinlein had set out to rob Marie McQueen, instead of rape her, and Heinlein had killed her in response to her resistance, would the prosecution have a stronger case regarding the Walker brothers' guilt for felony murder under a vicarious liability theory? Why or why not?
2. Does there appear to be a significant difference between the instructions requested by defense counsel in *Heinlein* and the instructions actually given? What does the court rule in this regard? How, precisely, did the trial judge err in this case?

## Note on the Death Penalty and Felony Murder

The defendant in *Enmund v. Florida,* 458 U.S. 782 (1982), was convicted of felony murder and sentenced to death. As the getaway car driver, he remained outside of the home entered by his two codefendants as the cofelons robbed and then killed the elderly couple residing in the home. The U.S. Supreme Court did not disturb Enmund's conviction for felony murder, but the justices ruled that the Eighth Amendment's cruel and unusual punishments clause prohibited his execution. The Court framed and analyzed the issue as follows:

[I]t is for us ultimately to judge whether the Eighth Amendment permits imposition of the death penalty on one such as Enmund who aids and abets a felony in the course of which a murder is committed by others but who does not himself kill, attempt to kill, or intend that a killing take place or that lethal force will be employed. We have concluded . . . that it does not. . . .

. . . It is fundamental that "causing harm intentionally must be punished more severely than causing the same harm unintentionally." H. Hart, Punishment and Responsibility 162 (1968). Enmund did not kill or intend to kill and thus his culpability is plainly different from that of the robbers who killed; yet the state treated them alike and attributed to Enmund the culpability of those who killed the [victims]. This was impermissible under the Eighth Amendment.

Five years after deciding *Enmund,* the Supreme Court revisited the constitutionality of imposing the death penalty for felony murder on defendants who did not personally kill the victims nor intend that the victims be killed. In *Tison v. Arizona,* 481 U.S. 137 (1987), Ricky and Raymond Tison helped their father and his cellmate, both of whom were convicted murderers, escape from prison. The party thereafter hijacked a car occupied by a family, with Ricky and Raymond actively participating in these events. Later, while the brothers were getting water for the family whose car had been commandeered, their father and his cellmate shot and killed the family members. The elder Tison subsequently was killed, and Ricky and Raymond were apprehended, convicted of felony murder, and sentenced to death. The Court, by a vote of 5–4, rejected their argument that *Enmund* required that their death sentences be invalidated. The majority opinion purported to distinguish *Enmund,* and announced a more limited Eighth Amendment rule in this context.

A narrow focus on the question of whether or not a given defendant "intended to kill," . . . is a highly unsatisfactory means of definitively distinguishing the most culpable and dangerous of murderers. Many who intend to, and do, kill are not criminally liable at all—those who act in self-defense or with other justification or excuse. Other intentional homicides, though criminal, are often felt undeserving of the death penalty—those that are the result of provocation. On the other hand, some nonintentional murderers may be among the most dangerous and inhumane of all—the person who tortures another not caring whether the victim lives or dies, or the robber who shoots someone in the course of the robbery, utterly indifferent to the fact that the desire to rob may have the unintended consequence of killing the victim as well as taking the victim's property. This reckless indifference to the value of human life may be every bit as shocking to the moral sense as an "intent to kill." Indeed it is for this very reason that the common law and modern criminal codes alike have classified behavior such as occurred in this case along with intentional murders. *See, e.g.,* G. Fletcher, Rethinking Criminal Law § 6.5, pp 447–448 (1978) ("[I]n the common law, intentional killing is not the only basis for establishing the most egregious form of criminal homicide. . . . For example, the Model Penal Code treats reckless killing, 'manifesting extreme indifference to the value of human life,' as equivalent to purposeful and knowing killing."). Enmund held that when "intent to kill" results in its logical though not inevitable consequence—the taking of human life—the Eighth Amendment permits the State to exact the death penalty after a careful weighing of the aggravating and mitigating circumstances. Similarly, we hold that the reckless disregard for human life implicit in knowingly engaging in criminal activities known to carry a grave risk of death represents a highly culpable mental state, a mental state that may be taken into account in making a capital sentencing judgment when that conduct causes its natural, though also not inevitable, lethal result.

The petitioners own personal involvement in the crimes was not minor, but rather, as specifically found by the trial court, "substantial." Far from merely sitting in a car away from the actual scene of the murders acting as the getaway driver to a robbery, each petitioner was actively involved in every element of the kidnapping-robbery and was physically present during the entire sequence of criminal activity culminating in the murder of the Lyons family and the subsequent flight. The Tisons' high level of participation in these crimes further implicates them in the resulting deaths. . . .

We will not attempt to precisely delineate the particular types of conduct and states of mind warranting imposition of the death penalty here. Rather, we simply hold that major participation in the felony committed, combined with reckless indifference to human life, is sufficient to satisfy the Enmund culpability requirement.

We consider the "reckless indifference" category of murder discussed in *Tison* later in this chapter.

## Killings Committed by Third Parties

In *United States v. Heinlein, Enmund v. Florida,* and *Tison v. Arizona,* defendants who did not personally kill homicide victims were prosecuted for felony murder on the theory that they were vicariously liable for killings committed by their *cofelons.* We now consider whether felony murder convictions can be sustained when a killing is committed by *a third party*—such as the intended victim of a robbery, or a police officer who responds to the scene, or a bystander—and not by one of the felons. Should the felons be convicted of murder when a third party kills in response to the felons' crime or attempted crime on the ground that they set in motion a chain of events that was reasonably likely to result in a homicide? Or, is a killing committed by a third party too remote from the felons' own conduct to hold the felons responsible for murder, which is the most serious type of criminal homicide? Should it matter whether the person killed is an innocent bystander, the intended victim of the felony, or a police officer, as opposed to one of the felons? We consider these issues in the following case.

### *People v. Hernandez*, 624 N.E.2d 661 (N.Y. 1993)

Simons, Judge.

This appeal raises the question whether a conviction of felony murder under Penal Law § 125.25(3) should be sustained where the homicide victim, a police officer, was shot not by one of the defendants but by a fellow officer during a gun battle following defendants' attempted robbery. Under the circumstances presented, we conclude that it should, and we therefore affirm.

### I.

Defendants Santana and Hernandez conspired to ambush and rob a man who was coming to a New York City apartment building to buy drugs. The plan was to have Santana lure him into the building stairwell where Hernandez waited with a gun. In fact, the man was an undercover State Trooper, wearing a transmitter, and backed up by fellow officers.

Once the Trooper was inside the building, Hernandez accosted him and pointed a gun at his head. A fight ensued during which the officer announced that he was a policeman, pulled out his service revolver and began firing. In the confusion, Hernandez, still armed, ran from the building into a courtyard where he encountered members of the police back-up unit. They ordered him to halt. Instead, he aimed his gun at one of the officers and moved toward him. The officers began firing, and one, Trooper Joseph Aversa, was fatally shot in the head. His body was found near the

(Continues)

*(Continued)*

area where Hernandez was apprehended after being wounded. Santana was arrested inside the building.

The evidence at trial did not establish who killed Aversa, but the People conceded that it effectively eliminated the possibility that either defendant was the shooter. Separate juries were empaneled for the two cases, and both defendants were convicted of felony murder and other charges.

On appeal, defendants contend that the felony murder charges should have been dismissed because neither one of them fired the fatal shot. The Appellate Division rejected that argument. . . .

# II.

Some 30 years ago, this Court affirmed the dismissal of a felony murder charge on the grounds that neither the defendant nor a cofelon had fired the weapon that caused the deaths (*People v. Wood*, 201 N.Y.S.2d 328, 167 N.E.2d 736). In *Wood*, the defendant and his companions were escaping from a fight outside a tavern when the tavern owner, attempting to aid police, fatally shot a bystander and one of defendant's companions. Defendant was charged with assault and felony murder. At the time, the relevant provision of section 1044 of the former Penal Law defined murder in the first degree as "[t]he killing of a human being . . . without a design to effect death, by a person engaged in the commission of, or in an attempt to commit a felony" (§ 1044[2]). We concluded that by the plain terms of the statute defendant could not be liable for murder, for the killing of the two men was not committed by a person "engaged in the commission of" a felony or a felony attempt. Relying on the statute's "peculiar wording," we decided the case without addressing whether a similar result would be required as a matter of common law. The *Wood* case acknowledged that other jurisdictions differed on whether to apply a proximate cause theory under which felons could be held responsible for homicides committed by nonparticipants or an agency theory under which felons would be responsible only if they committed the final, fatal act.

In 1965, the Legislature revised the felony murder statute by removing the language that had been dispositive in *Wood* and replacing it with a provision holding a person culpable for felony murder when, during the commission of an enumerated felony or attempt, either the defendant or an accomplice "causes the death of a person other than one of the participants" (Penal Law § 125.25[3]). Thus, this appeal raises the question of whether *Wood* remains good law despite the recasting of the Penal Law. . . .

The People . . . premise their argument on the established construction of the term "causes the death," which is now the operative language in the Penal Law. That term is used consistently throughout article 125 and has been construed to mean that homicide is properly charged when the defendant's culpable act is "a sufficiently direct cause" of the death so that the fatal result was reasonably foreseeable. In the People's view the evidence here meets that standard. They contend that it was highly foreseeable that someone would be killed in a shootout when Hernandez refused to put down his gun and instead persisted in threatening the life of one of the back-up officers. Thus, under the People's theory, Hernandez "caused the death" of Aversa. Because his attempt to avoid arrest was in furtherance of a common criminal objective shared with Santana, the People contend that the murder was properly attributed to Santana as well as under principles of accomplice liability.

In response, defendants assert that *People v. Wood*, though decided on narrow statutory grounds, states a rule that was followed for centuries at common law and one that has been embraced by a significant number of jurisdictions. The rationale for requiring that one of the cofelons be the shooter (or, more broadly, the person who commits the final, fatal act) has been framed in several ways. Some courts have held that when the victim or a police officer or a bystander shoots and kills, it cannot be said that the killing was in furtherance of a common criminal objective. Others have concluded that under such circumstances the necessary malice or intent is missing (*Wooden v. Commonwealth*, 222 Va. 758, 284 S.E.2d 811). Under the traditional felony murder doctrine, the malice necessary to make the killing murder was constructively imputed from the *mens rea* incidental to perpetration of the underlying felony. Thus, in *Wooden*, the Virginia Supreme Court concluded that where a nonparticipant in the felony is the shooter, there can be no imputation of the necessary malice to him, and no party in the causal chain has both the requisite *mens rea* and culpability for the *actus reus*. Still other courts have expressed policy concerns about extending felony murder liability. They have asserted that no deterrence value attaches when the felon is not the person immediately responsible for the death, or have contended that an expansive felony murder rule might unreasonably hold the felons responsible for the acts of others—for instance, when an unarmed felon is fleeing the scene and a bystander is hit by the bad aim of the armed victim.

## III.

Analysis begins with the statute. The causal language used in our felony murder provision and elsewhere in the homicide statutes has consistently been construed by this Court according to the rule in *People v. Kibbe* (321 N.E.2d 773 (N.Y.)), where we held that the accused need not commit the final, fatal act to be culpable for causing death. . . .

In light of the statutory language and the case law prior to the revision, we conclude that the Legislature intended what appears obvious from the face of the statute: that "causes" in the felony murder provision should be accorded the same meaning it is given in subdivisions (1) and (2) of section 125.25 of the Penal Law.

Unlike defendants and those courts adopting the so-called agency theory, we believe New York's view of causality, based on a proximate cause theory, to be consistent with fundamental principles of criminal law. Advocates of the agency theory suggest that no culpable party has the requisite *mens rea* when a nonparticipant is the shooter. We disagree. The basic tenet of felony murder liability is that the *mens rea* of the underlying felony is imputed to the participant responsible for the killing. By operation of that legal fiction, the transferred intent allows the law to characterize a homicide, though unintended and not in the common design of the felons, as an intentional killing. Thus, the presence or absence of the requisite *mens rea* is an issue turning on whether the felon is acting in furtherance of the underlying crime at the time of the homicide, not on the proximity or attenuation of the death resulting from the felon's acts. Whether the death is an immediate result or an attenuated one, the necessary *mens rea* is present if the causal act is part of the felonious conduct.

No more persuasive is the argument that the proximate cause view will extend criminal liability unreasonably. First, New York law is clear that felony murder does not embrace any killing that is coincidental with the felony but instead is limited to those deaths caused by one of the felons in furtherance of their crime. More than civil tort liability must be established; criminal liability will adhere only when the felons' acts are a sufficiently direct cause of the death. When the intervening acts of another party are supervening or unforeseeable, the necessary causal chain is broken, and there is no liability for the felons. Where a victim, a police officer or other third party shoots and kills, the prosecution faces a significant obstacle in proving beyond a reasonable doubt to a jury that the felons should be held responsible for causing the death.

Second, the New York felony murder statute spells out the affirmative defense available to the accomplice who does not cause the death (*see*, Penal Law § 125.25[3][a]–[d])). Defendants assert that our construction of the statute's causality language will mean that an accomplice whose partner is the shooter will have a defense but one whose unarmed partner causes the death will not. The plain language of the statute does not support that proposition. The statutory defense is available to the accomplice who (a) does not cause the death, (b) is unarmed, (c) has no reason to believe that the cofelon is armed and (d) has no reason to believe that the cofelon will "engage in conduct likely to result in death or serious physical injury." Thus, by its terms, the defense is not limited to situations where the cofelon kills with a weapon; it applies as well to instances where some other "conduct likely to result in death" is not within the contemplation of the accomplice.

In short, our established common-law rules governing determinations of causality and the availability of the statutory defense provide adequate boundaries to felony murder liability. The language of Penal Law § 125.25(3) evinces the Legislature's desire to extend liability broadly to those who commit serious crimes in ways that endanger the lives of others. That other States choose more narrow approaches is of no moment to our statutory scheme. Our Legislature has chosen not to write those limitations into our law, and we are bound by that legislative determination. . . .

## Notes and Questions

1. The court in *Hernandez* relied heavily on New York's murder statute in ruling as it did. Several other courts have refused to impose felony-murder liability on felons when a third party directly causes another's death. After extensive consideration of the issue over a period of several years, the Pennsylvania Supreme Court announced in *Commonwealth ex rel. Smith v. Myers*, 261 A.2d 550 (Pa. 1970), that it would no longer use the felony-murder doctrine to impute murder liability for third-party killings. The Minnesota Supreme Court concurred with that result in *State v. Branson*, 487 N.W.2d 880 (Minn. 1992), observing the following: "In considering whether one committing or attempting to commit a felony can be held criminally responsible for a death committed by a third

*(Continues)*

*(Continued)*

party during the commission of the felony, courts in most jurisdictions hold that the 'doctrine of felony murder does not extend to a killing, although growing out of the commission of the felony, if directly attributable to the act of one other than the defendant or those associated with him in the unlawful enterprise'" (citing cases). However, even in jurisdictions rejecting felony-murder liability when killings are committed by a third party, the courts sometimes conclude that a murder conviction can be supported on alternative grounds, such as under a "depraved heart" or recklessness rationale (discussed later in this chapter). *See, e.g., Taylor v. Superior Court of Alameda County*, 477 P.2d 131 (Cal. 1970).

2. Note the defense discussed in *Hernandez* that is made available by statute in New York to an accomplice in a felony who does not personally kill the homicide victim. Does this defense seem more forgiving than the common law felony-murder rule? Should Santana benefit by the statutory defense?

3. What if instead of a police officer being killed, Hernandez had been shot and killed by a bullet fired by one of the officers? Would Santana be guilty of murdering Hernandez under New York's felony-murder rule?

   Applying the "proximate cause" theory discussed in *Hernandez*, the Illinois Supreme Court ruled that "a defendant may be charged with first degree murder, on a felony murder theory," when the decedent is a cofelon who is killed by an intended victim of the defendant and cofelon." *People v. Dekens*, 695 N.E.2d 474 (Ill. 1998).

   > Consistent with the proximate cause theory, liability should lie for any death proximately related to the defendant's criminal conduct. Thus, the key question here is whether the decedent's death is the direct and proximate result of the defendant's felony. As our cases make clear, application of the felony-murder doctrine does not depend on the guilt or innocence of the person killed during the felony or on the identity of the person whose act causes the decedent's death.

4. Although the Illinois Supreme Court has reaffirmed the rule it adopted in *Dekens*, *People v. Klebanowski*, 852 N.E.2d 813 (Ill. 2006), not all courts are in agreement with it. *See, e.g., Campbell v. State*, 444 A.2d 1034 (Md. 1982) (using "agency" theory to reject felony-murder liability for surviving felon whose cofelon was shot and killed during an armed robbery attempt by either a police officer or the intended victim of the attempted robbery); *State v. Bonner*, 411 S.E.2d 598 (N.C. 1992) (holding that cofelons may not be charged with felony murder as a result of the death of an accomplice at the hands of an adversary to their crime).

## Should the Felony-Murder Rule Be Abolished?

England abolished the felony-murder rule in 1957. The drafters of the MPC did away with the felony-murder doctrine, although they did not make irrelevant the fact that a killing was committed during the perpetration of a dangerous felony. Recall that one type of murder liability under the MPC is predicated on killing "committed recklessly under circumstances manifesting extreme indifference to the value of human life." MPC § 210.2(1)(b). Under that same section: "Such recklessness and indifference are presumed if the actor is engaged or is an accomplice in the commission of, or an attempt to commit, or flight after committing or attempting to commit robbery, rape or deviate sexual intercourse by force or threat of force, arson, burglary, kidnapping or felonious escape." The drafters offered the reasons for adopting this different approach. They are as follows:[11]

> [U]nder the Model Code, as at common law, murder occurs if a person kills purposely, knowingly, or with extreme recklessness. Lesser culpability yields lesser liability, and a person who inadvertently kills another under circumstances not amounting to negligence is guilty of no crime at all. The felony-murder rule contradicts this scheme. It bases conviction of murder not on any proven culpability with respect to homicide but on liability for another crime. The underlying felony carries its own penalty and the additional punishment for murder is therefore gratuitous—gratuitous, at least in terms of what must have been proved at trial in a court of law.

> It is true, of course, that the felony-murder rule is often invoked where liability for murder exists on another ground. One who kills in the course of armed robbery is almost certainly guilty of murder in the form of intentional or extremely reckless homicide without any need of special doctrine. Similarly, a man who burns another's house will scarcely be heard to complain that he lacks the culpability for murder if the blaze kills a sleeping occupant. For the vast majority of cases it is probably true that homicide occurring during the commission or attempted commission of a felony is murder independent of the felony-murder rule. At bottom, continued adherence to the doctrine may rest on assessments of this sort.

> The problem is that criminal liability attaches to individuals, not generalities. It is a weak rejoinder to a complaint of unjust conviction to say that for most persons in the defendant's situation the result would have been appropriate. . . .

It remains indefensible in principle to use the sanctions that the law employs to deal with murder unless there is at least a finding that the actor's conduct manifested an extreme indifference to the value of human life. The fact that the actor was engaged in a crime of the kind that is included in the usual first-degree felony-murder enumeration or was an accomplice in such crime, as has been observed, will frequently justify such a finding. Indeed, the probability that such a finding will be justified seems high enough to warrant the presumption of extreme indifference that Subsection (1)(b) creates. But liability depends, as plainly it should, upon the crucial finding. The result may not differ often under such a formulation from that which would be reached under some form of the felony-murder rule. But what is more important is that a conviction on this basis rests solidly upon principle. . . .

Echoing this rationale, the Michigan Supreme Court abrogated the common law felony-murder doctrine in that state in *People v. Aaron*, 299 N.W.2d 304 (Mich. 1980). *See also People v. Burroughs*, 678 P.2d 894, 902–915 (Cal. 1984) (Bird, C.J., concurring) (advocating that the California Supreme Court should abolish the judicially created offense of second-degree felony murder). The rule has been abolished by statute in Hawaii and Kentucky and significantly limited by legislation and/or judicial decision in several other states.[12]

In repudiating the common law doctrine of felony murder in Michigan, that state's supreme court concluded as follows:

> Whatever reasons can be gleaned from the dubious origin of the felony-murder rule to explain its existence, those reasons no longer exist today. Indeed, most states . . . have recognized the harshness and inequity of the rule as is evidenced by the numerous restrictions placed on it. The felony-murder doctrine is unnecessary and in many cases unjust in that it violates the basic premises of individual moral culpability upon which our criminal law is based. *People v. Aaron*, 299 N.W.2d 304, 328 (Mich. 1980).

In what sense is the felony-murder rule "unnecessary"? In what sense is it "unjust"? Does it have countervailing uses or benefits that make its continuing existence justifiable? What practical consequences would ensue if the felony-murder rule were abolished? Scholarly commentary has been heavily critical of the felony-murder rule, yet the rule nevertheless endures, even if in somewhat modified form, under the law in most states.

## "Depraved Heart" Murder

Outside of the context of felony murder, can (should) a conviction for murder be sustained absent a showing that the offender *intentionally* killed his or her victim?

### Alston v. State, 643 A.2d 468 (Md. App. 1994), *affirmed*, 662 A.2d 247 (Md. 1995)

[Members of two gangs engaged in a shoot-out in a street in downtown Baltimore, exchanging 40 to 50 gun shots. An innocent bystander, 15-year-old Adrian Edmonds, was struck by a bullet and killed. The defendant, David L. "Diesel" Alston, was one of the gang members participating in the shoot-out, but no evidence identified him as the source of the bullet that killed the victim. He was convicted of second-degree murder. On appeal, he urged that the evidence was insufficient to support his conviction.]

Moylan, Judge. . . .

Murder, of course, requires that there be a homicide. Adrian Edmonds was indisputably the homicide victim. Murder further requires that the criminal agent perpetrate the homicide with malice. We now recognize that there are at least four forms of murderous malice. One of these requires neither a specific intent to kill nor a specific intent to do any grievous bodily harm. Neither does it involve the perpetration or attempted perpetration of a felony. This type of murder, which does not have an aggravated first-degree form, is referred to as **"depraved-heart" murder**. The particular murderous *mens rea* that supports it was described in *DeBettencourt v. State*, 48 Md.App. 522, 530 (1981):

> In homicide law, [the] classic form of malice is referred to as "express malice." In its vaster experience with infinite nuances, however, the law of homicide has recognized variant forms of malice. It refers to these as the various types of "implied malice" (more sophisticated modern analysis recognizes them as forms of "equivalent malice"). One of these variant forms of malice—the analogue of the hour—is that of "the depraved heart." It is the form that establishes that *the wilful doing of a dangerous and reckless act with wanton indifference to the consequences and perils involved,* is just as blameworthy, and just as worthy of punishment,

*(Continues)*

when the harmful result ensues, as is the express intent to kill itself. This highly blameworthy state of mind is not one of mere negligence (even enough to serve as the predicate for civil tort liability). It is not merely one even of gross criminal negligence (even enough to serve as the predicate for guilt of manslaughter). *It involves rather the deliberate perpetration of a knowingly dangerous act with reckless and wanton unconcern and indifference as to whether anyone is harmed or not.* The common law treats such a state of mind as just as blameworthy, just as antisocial and, therefore, just as truly murderous as the specific intents to kill and to harm (emphasis supplied). . . .

Rollin M. Perkins & Ronald N. Boyce, 60 *Criminal Law* (3d ed. 1982), describes depraved-heart murder:

> The standard has been codified in terms of activity which manifests "extreme indifference to the value of human life" and which "creates a grave risk of death," or "where all the circumstances of the killing show an abandoned and malignant heart."
>
> In other words, the intent to do an act in wanton and wilful disregard of the obvious likelihood of causing death or great bodily injury is a malicious intent. The word "wanton" is the key word here. . . . The difference is that in the act of the shooter there is an element of viciousness—an extreme indifference to the value of human life—that is not found in the act of the motorist. And it is the viciousness which makes the act "wanton" as well as "wilful" (footnotes omitted). . . .

In terms of the *actus reus* of depraved-heart murder, the question is whether the defendant engaged in conduct that created a very high risk of death or serious bodily injury to others. Wayne R. LaFave & Austin W. Scott, *Substantive Criminal Law* § 7.4(a) at 200 (1986), describes this necessary *actus reus*:

> For murder the degree of risk of death or serious bodily injury must be more than a mere unreasonable risk, more even than a high degree of risk. Perhaps the required danger may be designated a "very high degree" of risk to distinguish it from those lesser degrees of risk which will suffice for other crimes. Such a designation of conduct at all events is more accurately descriptive than that flowery expression found in the old cases and occasionally incorporated into some modern statutes—*i.e.*, conduct "evincing a depraved heart, devoid of social duty, and fatally bent on mischief." Although "very high degree of risk" means something quite substantial, it is still something far less than certainty or substantial certainty (footnotes omitted).

We have no difficulty in concluding that for approximately ten men to engage in an extended firefight on an urban street in a residential neighborhood was conduct that created a very high degree of risk of death or serious bodily injury to others. . . . Our conclusion in this regard that a very high risk was created is strengthened by the fact that we are talking about 11 P.M. on a hot July evening, when various persons, according to the evidence, were still sitting out on the front steps of rowhouses, quite aside from any question of whether there were persons moving in the street or on the sidewalks.

The nub of the problem is in identifying the lethal conduct in this case. From the point of view of the endangered neighbors and bystanders there on Presstman Street, from the point of view of the general public outraged at the proliferation of this type of mindless barbarism, and from the point of view ultimately of the criminal law itself, the lethal conduct in this case was the shoot-out itself. We are not going to segment it into forty or fifty discrete life-endangering acts, one per bullet, and then seek to determine which were propelled from the appellant's gun and which from the guns of others. How many discrete life-endangering acts are there, for instance, in a single burst from an automatic weapon? We consider, rather, the collective shoot-out itself to have been a single explosion of bullets, even as we might consider a series of exploding dynamite sticks to be a single explosion. There could have been no "gunfight at the OK Corral" without both the Clantons and the Earps. To pursue the metaphor, we would have to transfer the gunfight from the OK Corral to the schoolyard at recess time to approximate more closely the pagan unconcern with the rights and lives of others exhibited by both the appellant's gang and the "New York boys" on the night of July 14 on Presstman Street. Each of the participants contributed to the accumulation of critical mass. The critical mass triggered a chain reaction, which then took on a life of its own.

In terms of the *mens rea* of depraved-heart murder, Wayne R. LaFave & Austin W. Scott, *Substantive Criminal Law* § 7.4 at 200–201 (1986), is again instructive:

The distinctions between an unreasonable risk and a high degree of risk and a very high degree of risk are, of course, matters of degree, and there is no exact boundary line between each category; they shade gradually like a spectrum from one group to another. Some have thus questioned whether this is a sound basis upon which to make the important distinction between murder and manslaughter. More appealing is the Model Penal Code approach, whereunder a reckless killing is murder only if done "under circumstances manifesting extreme indifference to the value of human life." This language, which better serves the "purpose of communicating to jurors in ordinary language the task expected of them," has been substantially followed in many but not all of the modern codes (footnotes omitted).

In the words of the Model Penal Code, the conduct of the appellant, as well as the conduct of all of the other combatants, was conduct "manifesting extreme indifference to the value of human life." In terms of its unconcern or disregard of the lives of others, it was wanton and it was depraved. We could easily add other adjectives to the lexicon of depravity: inhuman, amoral, animalistic, savage. It was conduct, on the part of everyone involved, that civilized society is simply not going to tolerate.

We hold that the evidence was legally sufficient to support the conviction of the appellant for depraved-heart murder, to wit, murder in the second degree. . . .

## Notes and Questions

1. As the court's opinion in *Alston* notes, the drafters of the MPC sought to capture the essence of "depraved heart" murder by using somewhat less colorful and more precise terminology. The MPC analogue is a killing "committed recklessly under circumstances manifesting extreme indifference to the value of human life." MPC § 210.2(1)(b). Note the important language qualifying "recklessly." A criminal homicide committed "recklessly"—without more—is manslaughter under the MPC. MPC § 210.3(1)(a). A murder conviction must be supported by more than simple recklessness (sometimes referred to as "recklessness-plus"): conduct exhibiting "extreme indifference to the value of human life."

2. "Depraved heart" killings are said to evidence malice, thus qualifying them as a form of murder, because they are considered "just as blameworthy, and just as worthy of punishment . . . as is the express intent to kill itself." *DeBettencourt v. State*, 428 A.2d 479 (Md. App.), *cert. denied*, 290 Md. 713 (Md. 1981) (quoted in *Alston*). Do you agree with this assertion? Can you offer examples?

3. Some authorities suggest that depraved heart murder requires that the defendant's conduct put more than a single person at risk. The modern trend, however, is squarely contrary. *See, e.g., Windham v. State*, 602 So.2d 798 (Miss. 1992); *Robinson v. State*, 517 A.2d 94 (Md. 1986). *See generally Commonwealth v. Malone*, 47 A.2d 445 (Pa. 1946) (upholding murder conviction based on depraved heart rationale for death caused during a game of "Russian Poker," or Russian roulette).

4. As a practical matter, how is it possible to distinguish between "reckless" killings (which amount to manslaughter under statutes that adopt the MPC approach) and reckless killings that are marked by the offender's "extreme indifference to the value of human life"? *See generally People v. Poplis*, 281 N.E.2d 167 (N.Y. 1972) (upholding murder conviction under the latter standard for a death caused by "repeated" and "continuing" beatings of a child, which involved "something more serious than mere recklessness alone which has had an incidental tragic result"); *Brown v. Commonwealth*, 975 S.W.2d 922 (Ky. 1998); *Neitzel v. State*, 655 P.2d 325 (Alaska App. 1982).

## Driving While Intoxicated Fatality as Murder

Can a homicide resulting from a car crash caused by a drunk driver justify a conviction for murder, under the "depraved heart" or recklessness "manifesting extreme indifference to the value of human life" standards discussed above? This issue has been considered by a number of courts in recent years. Consider the following case.

### *State v. Doub*, 95 P.3d 116 (Kan. App. 2004)

Greene, J.

John P. Doub, III, appeals his conviction of second-degree murder pursuant to K.S.A. 21-3402(b), claiming insufficiency of evidence. . . .

*(Continues)*

*(Continued)*

Following a party for his softball team at a club where he admitted drinking six beers, Doub admitted that his pickup struck two parked vehicles and that he left the scene because he was concerned that he had been drinking. Doub ultimately admitted that, approximately 2 hours after striking the parked cars, he drove his pickup into the rear of a Cadillac in which 9-year-old Jamika Smith was a passenger. According to the State's accident investigator, the collision occurred as Doub's pickup, "going tremendously faster," drove "up on top of [the Cadillac]," initially driving it down into the pavement, and ultimately propelling it off the street and into a tree. Doub offered no aid to the victims, left the scene of the accident, and initially denied any involvement in the collision, suggesting that his pickup had been stolen. Some 15 hours after the collision, Smith died as a result of blunt traumatic injuries caused by the collision.

Approximately 6 months after these events, Doub admitted to a former girlfriend that he had a confrontation with his second ex-wife the evening of the collision, had been drinking alcohol and smoking crack, and had subsequently caused the collision. The girlfriend approached the authorities with Doub's statements, which suggested that Doub left the softball party, caused the collisions with the parked vehicles, left that scene, subsequently consumed the additional alcohol and crack cocaine, and then caused the collision resulting in Smith's death, all within a 2- to 3-hour period.

Doub was charged with . . . second-degree depraved heart murder, [and] with lesser included offenses of involuntary manslaughter and vehicular homicide . . . .

K.S.A.2003 Supp. 21-3402 defines second-degree murder as follows:

> "Murder in the second-degree is the killing of a human being committed:
> (a) Intentionally; or
> (b) unintentionally but recklessly under circumstances manifesting extreme indifference to the value of human life."

When the offense is committed pursuant to subsection (b), our courts have employed the common-law nomenclature of "depraved heart" second-degree murder.

In *State v. Robinson*, 261 Kan. 865, 876-78 (1997), our Supreme Court discussed the requirements for depraved heart murder:

> . . . "[D]epraved heart second-degree murder requires a conscious disregard of the risk, sufficient under the circumstances, to manifest extreme indifference to the value of human life. *Recklessness that can be assimilated to purpose or knowledge is treated as depraved heart second-degree murder*, and less extreme recklessness is punished as manslaughter. Conviction of depraved heart second-degree murder requires proof that the defendant acted recklessly under circumstances manifesting extreme indifference to the value of human life. This language describes a kind of culpability that differs in degree but not in kind from the ordinary recklessness required for manslaughter." (Emphasis added.)

In *Robinson*, the court specifically rejected the argument that the offense required general indifference to the value of *all* human life and concluded that the elements could be met if the defendant manifested an extreme indifference to the value of one specific human life. . . .

We find no reported decision in Kansas construing and applying K.S.A. 21-3402(b) in the context of a vehicular collision and therefore approach this appeal as a case of first impression. . . .

[O]ur focus is the statutory language adopted in Kansas that apparently had its genesis in the Model Penal Code first proposed in 1962, which required killing "recklessly under circumstances manifesting extreme indifference to the value of human life." A.L.I., Model Penal Code § 210.2 (Proposed Official Draft 1962).

Since 1975 the appellate courts of many states have acknowledged that the required state of mind for depraved heart murder can be attributed to the driver of an automobile. Our review of such cases reveals that most jurisdictions with statutory provisions patterned after the Model Penal Code have acknowledged that the offense may be committed by automobile. . . .

One commentator surveyed 20 cases between 1975 and 1986 and found the following factors as persuasive of the requisite state of mind:

> "1. *Intoxication*. The driver was using alcohol, illegal drugs, or both.
> "2. *Speeding*. Usually excessive rates are recorded.
> "3. *Near or nonfatal collisions shortly before the fatal accident*. Courts believe that collisions should serve as a warning to defendants that their conduct is highly likely to cause an accident. Failure to modify their driving is viewed as a conscious indifference to human life.

"4. *Driving on the wrong side of the road.* Many cases involve head-on collisions. Included here is illegally passing or veering into oncoming traffic.

"5. *Failure to aid the victim.* The driver left the scene of the accident and/or never attempted to seek aid for the victim.

"6. *Failure to heed traffic signs.* Usually more than once prior to the fatal accident, the driver ran a red light and/or stop sign.

"7. *Failure to heed warnings about reckless driving.* In *Pears v. State*, for example, the court cited as proof of Pears' extreme indifference to life the fact that he continued driving after he had been warned by police officers not to drive because he was intoxicated. In other cases a police pursuit of the driver for earlier traffic violations was an implicit warning that the defendant's driving was dangerous.

"8. *Prior record of driving offenses (drunk or reckless driving or both).* The relevance of a defendant's prior record for reckless or intoxicated driving is, as *United States v. Fleming* pointed out, not to show a propensity to drive while drunk but 'to establish that defendant had grounds to be aware of the risk his drinking and driving while intoxicated presented to others.'" Luria, *Death on the Highway: Reckless Driving as Murder,* 67 Or. L.Rev. 799, 823 (1988).

Application of these factors seems appropriate to determine whether evidence in a particular case meets the requisite state of mind, but we are mindful that no precise universal definition or exclusive criteria is appropriate. The comments to the Model Penal Code declare that "recklessness" must be of such an extreme nature that it demonstrates an indifference to human life similar to that held by one who commits murder purposely or knowingly, but precise definition is impossible.

"The significance of purpose or knowledge as a standard of culpability is that, cases of provocation or other mitigation apart, purposeful or knowing homicide demonstrates precisely such indifference to the value of human life. Whether recklessness is so extreme that it demonstrates similar indifference is not a question, it is submitted, that can be further clarified. It must be left directly to the trier of fact under instructions which make it clear that recklessness that can fairly be assimilated to purpose or knowledge should be treated as murder and that less extreme recklessness should be punished as manslaughter." A.L.I., Model Penal Code & Commentaries Part II § 210.2, Comment. 4, pp. 21–22 (1980).

. . . Doub argues that his conduct was not even sufficiently egregious to constitute vehicular homicide, citing *State v. Krovvidi*, 58 P.3d 687 (Kan. 2002), which reversed a conviction for vehicular homicide. The following language defines the crime of vehicular homicide, which is quite different from the language defining depraved heart murder:

"Vehicular homicide is the unintentional killing of a human being committed by the operation of an automobile, airplane, motor boat or other motor vehicle in a manner which creates an unreasonable risk of injury to the person or property of another and which constitutes a material deviation from the standard of care which a reasonable person would observe under the same circumstances." K.S.A. 21-3405.

The facts in *Krovvidi*, however, differed greatly from those before us. In *Krovvidi*, the State pointed exclusively to inattentive driving and the fact that the driver drove through a red light as factors showing conduct that rose to the level of culpability required under the vehicular homicide statute. Here, the facts are far more egregious. Moreover, depraved heart murder requires an entirely different level of culpability from that required for vehicular homicide. The following language from *Krovvidi* is instructive:

"In this case, there are no aggravating factors present. Krovvidi had not been drinking and was not under the influence of any drug, both factors which may provide the additional evidence to establish a material deviation. None of the passengers in his vehicle warned him as he was about to enter the intersection; none were concerned that his driving appeared reckless or that he was accelerating or speeding as he approached the intersection. Krovvidi was not speeding and proceeded through the intersection thinking his light was green. Absent additional aggravating factors, we conclude that his conduct does not amount to the material deviation required under the provisions [of] K.S.A 21-3405." 274 Kan. at 1075, 58 P.3d 687. . . .

Considering the presence of many of those factors significant to other courts, we are convinced that a rational factfinder could have found Doub guilty of depraved heart second-degree murder beyond a reasonable doubt. The evidence against Doub is particularly damning considering that (a) he admits that his driving was preceded by drinking; (b) he admits that he struck two parked cars and ignored commands to stop because he was concerned that he had been drinking; (c) he then consumed

(*Continues*)

additional alcohol and used crack cocaine; (d) he then resumed driving and caused a fatal collision, due in part to excessive speed; (e) he failed to render aid to the victims; and (f) he fled the scene in order to avoid criminal liability. We conclude that these facts clearly demonstrate an extreme indifference to human life.

Affirmed.

## Notes and Questions

1. Do the factors relied on in the court's opinion succeed in distinguishing Doub's conduct from the thousands of drunk driving-related fatalities that occur annually on America's highways and are not prosecuted as murder?
2. Should driving while intoxicated or under the influence of illegal drugs, without more, represent recklessness "manifesting extreme indifference to the value of human life"?
3. A number of other courts have sustained murder convictions for drunk-driving fatalities where the driver's conduct is sufficiently aggravated. *See Hamilton v. Commonwealth*, 560 S.W.2d 539 (Ky. 1977); *People v. Watson*, 637 P.2d 279 (Cal. 1981); *Pears v. State*, 672 P.2d 903 (Alaska App. 1983), *sentence modified on appeal*, 698 P.2d 1198 (Alaska 1985); *United States v. Fleming*, 739 F.2d 945 (4th Cir. 1984).

## Manslaughter Based on Unintentional Killings

We earlier considered manslaughter (traditionally known as *voluntary* manslaughter) based on intentional killings. Under the common law approach, a killing that otherwise would be murder is reduced to manslaughter if committed in the heat of passion on adequate provocation. Under the MPC's standard, a purposeful killing is manslaughter when committed under the influence of extreme mental or emotional disturbance for which there is reasonable explanation or excuse.

We now consider a different variety of manslaughter: killings committed unintentionally yet accompanied by a state of mind typically labeled as recklessness or gross and culpable negligence. The MPC defines reckless homicide as manslaughter and creates a lower order offense, negligent homicide, for killings committed with criminal negligence. See MPC §§ 210.3(1)(a), 210.4. Under statutes based on the common law, both reckless and criminally negligent homicides traditionally have been called *involuntary* manslaughter.

Keep in mind that a conviction for manslaughter requires more than the type of negligence or recklessness that would result in civil liability for a death resulting from, for example, an auto accident or medical malpractice. To help distinguish *criminal* recklessness and negligence from their civil counterparts, as well as from each other, consult the important definitions in MPC §§ 2.02(2)(c) and (d). We now take up the issue of imposing criminal responsibility for unintentional killings, where these heightened forms of recklessness and/or negligence characterize conduct that results in a death.

## *Commonwealth v. Feinberg*, 253 A.2d 636 (Pa. 1969)

Jones, Justice.

Appellant Max Feinberg owned and operated a cigar store in the skid-row section of Philadelphia. One of the products he sold was Sterno, a jelly-like substance composed primarily of methanol and ethanol and designed for cooking and heating purposes. Sterno was manufactured and sold in two types of containers, one for home use and one for industrial use. Before September 1963 both types of Sterno contained approximately 3.75% methanol, or wood alcohol, and 71% ethanol, or grain alcohol; of the two types of alcohols, methanol is far more toxic if consumed internally. Beginning in September of 1963, the Sterno company began manufacturing a new type of industrial Sterno which was 54% methanol. The cans containing the new industrial Sterno were identical to the cans containing the old industrial Sterno except in one crucial aspect: on the lids of the new 54% methanol Sterno were imprinted the words "Institutional Sterno. Danger. Poison. For use only as a Fuel. Not for consumer use. For industrial and commercial use. Not for home use." A skull and crossbones were also lithographed on the lid. . . .

[B]etween December 21 and December 28, appellant had sold approximately 400 cans of the new industrial Sterno. Between December 23 and December 30, thirty-one persons died in the skid-row area as a result of methanol poisoning. In many of the cases the source of the methanol was traced to the new industrial Sterno. Since appellant was the only retail outlet of this type of Sterno in Philadelphia, he was arrested and indicted on thirty-one counts charging involuntary manslaughter and on companion bills charging violation of the Pharmacy Act (Act of September 27, 1961, P.L. 1700, § 1 et seq., 63 P.S. § 390-1 et seq.).

Appellant was convicted on seventeen counts of involuntary manslaughter and on twenty-five counts of violating the Pharmacy Act by Judge Charles L. Guerin, sitting without a jury. Judge Guerin held that appellant had violated the Pharmacy Act and that, therefore, he was guilty of a misdemeanor–manslaughter in each of the seventeen cases. Five of the manslaughter convictions were appealed to the Superior Court which affirmed four of them, although on a different theory. *Commonwealth v. Feinberg*, 234 A.2d 913 (Pa. Super. 1967). In writing for a six-judge majority, Judge Montgomery held that appellant had not violated the Pharmacy Act and, therefore, was not guilty of a misdemeanor-manslaughter, but that the evidence justified the conclusion that appellant was guilty of involuntary manslaughter. . . .

The first question we must answer is whether appellant violated the Pharmacy Act. The Act defines any product containing more than one per cent methanol as a poison and provides that any person selling such a product must properly label the container, warn the purchaser of the dangerous propensities of the product and satisfy himself that the purchaser will use the product for a legitimate purpose.

Certain facts are clear in this case. First, the Sterno sold by appellant is a poison as defined by the Act. Second, appellant did not comply with the requirements outlined above. Judge Guerin held that this was enough to justify a conviction under the Act; the Superior Court unanimously disagreed. Judge Montgomery wrote, "Our close study of the 1961 Pharmacy Act leads us to the conclusion that it was not intended to cover general commercial products but was limited to drugs and devices as defined in the act and that the provisions respecting poison are to be followed only when poisonous drugs or poisonous devices are sold in connection with the practice of pharmacy or incident thereto." (234 A.2d at 916.) We agree with this conclusion. . . .

The second issue in this case is whether appellant is guilty of involuntary manslaughter in each or any of the four appeals presently before us. The Penal Code defines **involuntary manslaughter** as a death "happening in consequence of an unlawful act, or the doing of a lawful act in an unlawful way. . . ." (Act of June 24, 1939, P.L. 872, § 703, 18 P.S. § 4703). Since we have determined that appellant did not violate the Pharmacy Act in selling the new industrial Sterno, the second portion of this statutory definition must be controlling. When a death results from the doing of an act lawful in itself but done in an unlawful manner, in order to sustain a conviction for manslaughter the Commonwealth must present evidence to prove that the defendant acted in a rash or reckless manner.

The conduct of the defendant resulting in the death must be such a departure from the behavior of an ordinary and prudent man as to evidence a disregard of human life or an indifference to the consequences. Furthermore, there must be a direct causal relationship between the defendant's act and the deceased's death.

We have searched in vain for cases from this Commonwealth involving factual situations similar to the one now before us. We have, however, found four cases from other jurisdictions which are on point. In the leading case, *Thiede v. State*, 182 N.W. 570 (Neb. 1921), the defendant gave the deceased moonshine containing methanol, the drinking of which resulted in his death. While noting that the defendant had violated the state prohibition laws, the court refused to rest the manslaughter conviction on this statutory violation, holding that the manufacturing and distribution of moonshine was merely malum prohibitum and not malum per se. The court continued, "We cannot go so far as to say that [dispensing moonshine], prompted perhaps by the spirit of good-fellowship, though prohibited by law, could ever, by any resulting consequence, be converted into the crime of manslaughter; *but, where the liquor by reason of its extreme potency or poisonous ingredients*, is dangerous to use as an intoxicating beverage, where the drinking of it is capable of producing direct physical injury, other than as an ordinary intoxicant, and of perhaps endangering life itself, *the case is different, and the question of negligence enters; for, if the party furnishing the liquor knows, or was apprised of such facts that*

(Continues)

*he should have known of the danger, there then appears from his act a recklessness which is indifferent to results.* Such recklessness in the furnishing of intoxicating liquors, in violation of law, may constitute such an unlawful act as, if it results in causing death, will constitute manslaughter." (182 N.W. at 573.) (Emphasis added.)

We conclude, after studying the record, that appellant fits within the black-letter rule laid down in *Thiede* and that the Commonwealth has made out all the elements necessary to warrant a conviction for involuntary manslaughter. First, the record establishes that appellant sold the Sterno with the knowledge that at least some of his customers would extract the alcohol for drinking purposes. . . .

Second, appellant was aware, or should have been aware, that the Sterno he was selling was toxic if consumed. The new industrial Sterno was clearly marked as being poisonous. Even the regular Sterno is marked "Caution. Flammable. For Use only as a Fuel" and if consumed internally may have serious consequences. Furthermore, when appellant was informed about the first deaths from methanol poisoning, he told the boy who worked in his shop to tell any police who came around that there was no Sterno in the store. Appellant also told the police that he had never purchased any Sterno from the Richter Paper Company. This evidence indicates to us that appellant was aware that he was selling the Sterno for an illicit purpose.

Appellant presses several contentions for our consideration. First, he claims that the Commonwealth has not established the necessary causal link between the sale of the Sterno and the deaths. . . . The court in *Thiede*, in answering an argument similar to the one now made by appellant, stated: "Defendant contends that the drinking of liquor, by deceased was his voluntary act and served as an intervening cause, breaking the causal connection between the giving of the liquor by defendant and the resulting death. The drinking of the liquor, in consequence of defendant's act, was, however, what the defendant contemplated. Deceased, it is true, may have been negligent in drinking, but, where the defendant was negligent, then the contributory negligence of the deceased will be no defense in a criminal action." (182 N.W. at 574).

Appellant next criticizes the following sentence in Judge Montgomery's opinion: "In the light of the *recognized weaknesses* of the purchasers of the product, and appellant's greater concern for profit than with the results of his actions, he was grossly negligent and demonstrated a wanton and reckless disregard for the welfare of those *whom he might reasonably have expected* to use the product for drinking purposes." (234 A.2d at 917.) (Emphasis added.) Appellant urges that the Superior Court is here imposing an inequitable burden on sellers of Sterno by requiring them to recognize the "weaknesses" of their customers. Appellant has exaggerated the import of this sentence. The Superior Court was not imposing a duty on all sellers of Sterno to determine how their customers will use the product. The Court was merely saying that if a seller of Sterno is aware that the purchaser is an alcoholic and will use Sterno as a source of alcohol, then the seller is grossly negligent and wantonly reckless in selling Sterno to him. We do not think this imposes an intolerable burden on sellers of Sterno. . . .

In conclusion, then, we find that Judge Guerin did, in fact, consider the issue of involuntary manslaughter and did find appellant guilty of involuntary manslaughter. . . .

Orders affirmed.

## Notes and Questions

1. Feinberg sold a lawful commodity that was clearly marked as poison, which his adult customers used for a purpose for which the product was not intended. Why is Feinberg guilty of a crime?

2. If Feinberg was properly convicted for causing the death of his customers who purchased and then consumed the Sterno, of what specific type of criminal homicide is he guilty? The court cited evidence that Feinberg "sold the Sterno with the knowledge that at least some of his customers would extract the alcohol for drinking purposes." If he possessed such knowledge, could an argument be made that murder, and not manslaughter, is the more appropriate charge?

3. Does the court's ruling impose a duty on merchants to refrain from selling lawful wares to some customers who are willing to purchase them? Does it impose a duty on merchants to know or anticipate how customers will use (or misuse) the commodities they purchase? What action should Feinberg have taken when customers approached his cash register to pay for a can of Sterno?

# Criminal Negligence and Recklessness

## *Commonwealth v. Nixon,* 718 A.2d 311 (Pa. Super. 1998), *affirmed,* 761 A.2d 1151 (Pa. 2000), *cert. den.,* 532 U.S. 1008 (2001)

Del Sole, Judge:

Appellants, Dennis and Lorie Nixon, were convicted of involuntary manslaughter and endangering the welfare of a child. They were sentenced to two and one-half to five years incarceration and a fine of $1,000. This direct appeal followed. We affirm. . . .

Appellants were the parents of Shannon Nixon ("Shannon"). The Nixon family are members of the Faith Tabernacle Church, a religion in which illnesses are addressed through spiritual treatment rather than by medicine. Thus, when Shannon began to feel ill in June of 1997, Appellants took her to be "anointed" at the church and prayed for her recovery. Shannon initially felt better and told her parents she had received her "victory," a recovery in answer to prayer. However, soon Shannon was ill again. As Shannon became increasingly weak and fell into a coma, Appellants continued to pray for her recovery. Shannon died hours after she fell into a coma. During the autopsy, it was determined that Shannon died of complications related to the onset of diabetes acidosis. . . .

Appellants claim that Shannon had a right to refuse medical treatment pursuant to her constitutional right to privacy which would eliminate Appellants' duty to provide treatment. . . . Appellants additionally assert Shannon's ability to refuse medical treatment as a mature minor abrogated Appellants' duty of care. Because both of these arguments necessitate an abrogation of Appellants' parental duty of care, we will address them together.

Although Shannon, as a mature minor, had a right to refuse medical treatment pursuant to her constitutional right to privacy, this right does not discharge her parents' duty to override her decision when her life is in immediate danger. Our Supreme Court, in *Green Appeal,* 292 A.2d 387 (Pa. 1972), permitted a sixteen-year-old boy to refuse to undergo an operation based upon religious beliefs. However, the permission to refuse medical treatment extended to minors in *Green* was strictly limited to situations in which the minor's life was not threatened. Thus, *Green* did not provide Shannon with the legal means to refuse medical treatment at a time when her life was in danger. . . .

In *Commonwealth v. Barnhart,* 497 A.2d 616 (Pa. Super. 1985), another set of parents from Appellants' church were convicted of involuntary manslaughter and endangering the welfare of a child after relying solely on spiritual healing to treat their two-year-old son's cancer. This court held that every parent in the Commonwealth of Pennsylvania had a duty of care to their child, at the very least, "to avert the child's untimely death." *Barnhart,* at 621. The court elaborated on the nature of the parents' duty:

> A parent has the legal duty to protect her child, and the discharge of this duty requires affirmative performance. The inherent dependency of a child upon his parent to obtain medical aid, *i.e.,* the incapacity of a child to evaluate his condition and summon aid by himself, supports imposition of such a duty upon the parent.

*Id.* (citations omitted).

Appellants had a duty to their minor child Shannon to override her own religious beliefs and obtain medical treatment for her when her condition became life-threatening. Neither Shannon's right of privacy, nor her status as a mature minor abrogated that duty. Thus, Appellants' arguments fail on their merits.

Appellants argue that their prosecution violated notice requirements of due process where spiritual treatment was authorized by statute. Appellants' argument refers to a possible conflict between criminal statutes and child abuse statutes. The Child Protective Services Act ("The CPSA") provides:

> If, upon investigation, the county agency determines that a child has not been provided needed medical or surgical care because of seriously held religious beliefs of the child's parents, guardian or person responsible for the child's welfare, which beliefs are consistent with those of a bona fide religion, the child shall not be deemed to be physically or mentally abused. . . . 23 Pa.C.SA § 6303(b)(3).

The involuntary manslaughter statute provides:

> A person is guilty of involuntary manslaughter when as a direct result of doing an unlawful act in a reckless or grossly negligent manner or the doing of a lawful act in a reckless or grossly negligent manner, he causes the death of another person. 18 Pa.C.S.A. § 2504.

*(Continues)*

Appellants argue that because the CPSA exempts parents who utilize spiritual treatment for their children from characterization as child abusers, they did not have sufficient notice that their spiritual treatment of Shannon could be criminal under the involuntary manslaughter statute. We disagree.

We find that the CPSA and the involuntary manslaughter statutes are not in conflict in their plain meaning, as well as under a constitutional analysis. A plain reading of the statutes shows that an act which does not qualify as child abuse may still be done in a manner which causes death and thus qualifies as involuntary manslaughter. This precise situation occurred in this case. While the Nixons were not considered child abusers for treating their children through spiritual healing, when their otherwise lawful course of conduct led to a child's death, they were guilty of involuntary manslaughter. . . .

Appellants argue the trial court abused its discretion in imposing a sentence above the aggravated range outlined in the sentencing guidelines. . . . Appellants argue the trial court improperly considered their prior record in deciding their sentence. Appellants' prior records stem from the death of their child, Clayton, in 1991. Clayton was nine years old and died from complications arising from an ear infection. Because Appellants did not take Clayton for medical treatment, they were prosecuted and pled no contest to involuntary manslaughter and endangering the welfare of a child. In that instance, Appellants received two years of probation.

The trial court based its departure from the sentencing guidelines on the repetitive nature of Appellants' crime. The trial court adequately explained its rationale and justification for the sentence in its June 10, 1997, order. The court gave great emphasis to the lack of opportunity for rehabilitation of Appellants and possibility of a recurrence of these criminal events. We will not disturb the trial court's well-justified sentence. . . .

## Notes and Questions

1. Assume the Nixons loved their daughter and very much wanted her to recover from her illness. In trying to help their daughter they resorted to prayer, which was consistent with their religious faith and accepted by other adherents of the Faith Tabernacle Church. They apparently acted in accordance with their 16-year-old daughter's wishes, and she shared the same religious views. By whose standards, and what standards, have they committed a crime? Do they deserve criminal punishment?

2. Several other courts have found parents who repudiate medical assistance in reliance on spiritual or religious methods to attempt to alleviate their children's illnesses to be properly convicted of manslaughter. *See Walker v. Superior Court of Sacramento County,* 763 P.2d 852 (Cal. 1988), *cert. denied,* 491 U.S. 905 (1989); *Commonwealth v. Twitchell,* 617 N.E.2d 609 (Mass. 1993) (however, reversing conviction based on parents' reliance on Attorney General's opinion that statute authorized treatment by spiritual means); *State v. Hays,* 964 P.2d 1042 (Or. App.), *rev. denied,* 977 P.2d 1170 (Or. 1998), *cert. denied,* 527 U.S. 1006 (1999).

## Negligent Good Samaritans

Does it matter, in prosecutions for manslaughter based on criminal negligence or recklessness, that the defendant may have been motivated to help the homicide victim and had no intention to kill or even injure the deceased?

## *Commonwealth v. Pierce,* 138 Mass. 165 (1884)

Holmes, J.

The defendant has been found guilty of manslaughter, on evidence that he publicly practised as a physician, and, being called to attend a sick woman, caused her, with her consent, to be kept in flannels saturated with kerosene for three days, more or less, by reason of which she died. There was evidence that he had made similar applications with favorable results in other cases, but that in one the effect had been to blister and burn the flesh as in the present case.

The main questions which have been argued before us are raised by the fifth and sixth rulings requested on behalf of the defendant, but refused by the court, and by the instructions given upon the same matter. The fifth request was, shortly, that the defendant must have "so much knowledge or probable information of the fatal tendency of the prescription that [the death] may be reasonably presumed by the jury to be the effect of obstinate, wilful rashness, and not of an honest intent and expectation to cure." The seventh request assumes the law to be as thus stated. The sixth request was as follows: "If the defendant made the prescription with an honest purpose and intent to cure the deceased, he is not guilty of this offence, however gross his ignorance of the quality and tendency of the remedy prescribed, or of the nature of the disease, or of both." The eleventh request was substantially similar, except that it was confined to this indictment.

The court instructed the jury, that "it is not necessary to show an evil intent;" that, "if by gross and reckless negligence he caused the death, he is guilty of culpable homicide;" that "the question is whether the kerosene (if it was the cause of the death), either in its original application, renewal, or continuance, was applied as the result of foolhardy presumption or gross negligence on the part of the defendant;" and that the defendant was "to be tried by no other or higher standard of skill or learning than that which he necessarily assumed in treating her; that is, that he was able to do so without gross recklessness or foolhardy presumption in undertaking it." In other words, that the defendant's duty was not enhanced by any express or implied contract, but that he was bound at his peril to do no grossly reckless act when in the absence of any emergency or other exceptional circumstances he intermeddled with the person of another. . . .

We have to determine whether recklessness in this sense was necessary to make the defendant guilty of felonious homicide, or whether his acts are to be judged by the external standard of what would be morally reckless, under the circumstances known to him, in a man of reasonable prudence.

More specifically, the questions raised by the foregoing requests and rulings are whether an actual good intent and the expectation of good results are an absolute justification of acts, however foolhardy they may be if judged by the external standard supposed, and whether the defendant's ignorance of the tendencies of kerosene administered as it was will excuse the administration of it.

So far as civil liability is concerned, at least, it is very clear that what we have called the external standard would be applied, and that, if a man's conduct is such as would be reckless in a man of ordinary prudence, it is reckless in him. Unless he can bring himself within some broadly defined exception to general rules, the law deliberately leaves his idiosyncrasies out of account, and peremptorily assumes that he has as much capacity to judge and to foresee consequences as a man of ordinary prudence would have in the same situation. . . .

If this is the rule adopted in regard to the redistribution of losses, which sound policy allows to rest where they fall in the absence of a clear reason to the contrary, there would seem to be at least equal reason for adopting it in the criminal law, which has for its immediate object and task to establish a general standard, or at least general negative limits, of conduct for the community, in the interest of the safety of all. . . .

If a physician is not less liable for reckless conduct than other people, it is clear, in the light of admitted principle and the later Massachusetts cases, that the recklessness of the criminal no less than that of the civil law must be tested by what we have called an external standard. In dealing with a man who has no special training, the question whether his act would be reckless in a man of ordinary prudence is evidently equivalent to an inquiry into the degree of danger which common experience shows to attend the act under the circumstances known to the actor. The only difference is, that the latter inquiry is still more obviously external to the estimate formed by the actor personally than the former. But it is familiar law that an act causing death may be murder, manslaughter, or misadventure, according to the degree of danger attending it. If the danger is very great, as in the case of an assault with a weapon found by the jury to be deadly, or an assault with hands and feet upon a woman known to be exhausted by illness, it is murder.

The very meaning of the fiction of implied malice in such cases at common law was, that a man might have to answer with his life for consequences which he neither intended nor foresaw. To say that he was presumed to have intended them, is merely to adopt another fiction, and to disguise the truth. The truth was, that his failure or inability to predict them was immaterial, if, under the circumstances known to him, the court or jury, as the case might be, thought them obvious.

As implied malice signifies the highest degree of danger, and makes the act murder; so, if the danger is less, but still not so remote that it can be disregarded, the act will be called reckless, and will be manslaughter, as in the case of an ordinary assault with feet and hands, or a weapon not deadly, upon a well person. . . .

*(Continues)*

(*Continued*)

If the principle which has thus been established both for murder and manslaughter is adhered to, the defendant's intention to produce the opposite result from that which came to pass leaves him in the same position with regard to the present charge that he would have been in if he had had no intention at all in the matter. We think that the principle must be adhered to, where, as here, the assumption to act as a physician was uncalled for by any sudden emergency, and no exceptional circumstances are shown; and that we cannot recognize a privilege to do acts manifestly endangering human life, on the ground of good intentions alone.

We have implied, however, in what we have said, and it is undoubtedly true, as a general proposition, that a man's liability for his acts is determined by their tendency under the circumstances known to him, and not by their tendency under all the circumstances actually affecting the result, whether known or unknown. And it may be asked why the dangerous character of kerosene, or "the fatal tendency of the prescription," as it was put in the fifth request, is not one of the circumstances the defendant's knowledge or ignorance of which might have a most important bearing on his guilt or innocence.

But knowledge of the dangerous character of a thing is only the equivalent of foresight of the way in which it will act. We admit that, if the thing is generally supposed to be universally harmless, and only a specialist would foresee that in a given case it would do damage, a person who did not foresee it, and who had no warning, would not be held liable for the harm. If men were held answerable for everything they did which was dangerous in fact, they would be held for all their acts from which harm in fact ensued. The use of the thing must be dangerous according to common experience, at least to the extent that there is a manifest and appreciable chance of harm from what is done, in view either of the actor's knowledge or of his conscious ignorance. And therefore, again, if the danger is due to the specific tendencies of the individual thing, and is not characteristic of the class to which it belongs, which seems to have been the view of the common law with regard to bulls, for instance, a person to be made liable must have notice of some past experience, or, as is commonly said, "of the quality of his beast." But if the dangers are characteristic of the class according to common experience, then he who uses an article of the class upon another cannot escape on the ground that he had less than the common experience. Common experience is necessary to the man of ordinary prudence, and a man who assumes to act as the defendant did must have it at his peril. When the jury are asked whether a stick of a certain size was a deadly weapon, they are not asked further whether the defendant knew that it was so. . . The defendant knew that he was using kerosene. The jury have found that it was applied as the result of foolhardy presumption or gross negligence, and that is enough. Indeed, if the defendant had known the fatal tendency of the prescription, he would have been perilously near the line of murder. It will not be necessary to invoke the authority of those exceptional decisions in which it has been held, with regard to knowledge of the circumstances, as distinguished from foresight of the consequences of an act, that, when certain of the circumstances were known, the party was bound at his peril to inquire as to the others, although not of a nature to be necessarily inferred from what were known.

The remaining questions may be disposed of more shortly. When the defendant applied kerosene to the person of the deceased in a way which the jury have found to have been reckless, or, in other words, seriously and unreasonably endangering life according to common experience, he did an act which his patient could not justify by her consent, and which therefore was an assault notwithstanding that consent.

*Exceptions overruled.*

## Notes and Questions

1. For a more contemporary case finding a physician guilty of criminally negligent homicide, *see State v. Warden*, 813 P.2d 1146 (Utah 1991). Dr. Warden helped deliver a prematurely born infant, who weighed approximately 4 pounds, at the mother's home. Despite signs that the baby was having difficulty breathing, the doctor took no action to hospitalize the child and repeatedly assured the mother and her family that nothing was wrong with the baby. The child died the next morning. Several expert witnesses testified that the baby's symptoms were such that prompt hospitalization should have been ordered.

## Misdemeanor-Manslaughter

### *Todd v. State,* 594 So.2d 802 (Fla. App. 1992)

Griffin, Judge.

On March 18, 1990, appellant entered the Lighthouse Church and stole $110 from the collection plate. The theft was witnessed by several members of the congregation, one of whom, Richard Voegltin, took off in his car in pursuit of appellant. During the pursuit, Mr. Voegltin, who had a preexisting heart condition, began to experience cardiac dysrhythmia. He lost control of his vehicle, collided with a tree at low speed and died of cardiac arrest.

The state charged appellant with manslaughter, alleging that he caused the death of Mr. Voegltin by committing the misdemeanor offense of petty theft which caused Voegltin to pursue him in order to recover the stolen property. . . .

The issue, as presented to us, is whether Florida recognizes the misdemeanor manslaughter rule. Reduced to basics, the misdemeanor manslaughter rule is that an unintended homicide which occurs during the commission of an unlawful act not amounting to a felony constitutes the crime of involuntary manslaughter. It is sometimes referred to more broadly as "unlawful act manslaughter." . . .

The misdemeanor manslaughter rule has been the subject to surprisingly little analysis, although in their Handbook on Criminal Law, LaFave and Scott have included a detailed discussion and critique of this theory of criminal responsibility. They suggest that "[t]he trend today, barely underway, is to abolish altogether this type of involuntary manslaughter. . . ." The authors posit that to punish as homicide the result of an unlawful act that is unintended and produced without any consciousness of the risk of producing it is "too harsh" and "illogical." W. LaFave and H. [sic] Scott, *supra* at 602. . . .

Because of the facial simplicity of the misdemeanor manslaughter rule, its application by courts has led to some rather extraordinary findings of criminal liability for homicide. For example, a Texas court found liability for manslaughter on the following facts: The victim discovered the defendant committing adultery with the victim's wife. Adultery was a misdemeanor in Texas. The victim made a murderous attack on the defendant. In defending himself against the murderous attack, the defendant killed the victim. The court decided that since the victim's murderous attack was a foreseeable reaction to the defendant's criminal misconduct, the defendant was guilty of manslaughter. *Reed v. State,* 11 Tex.Ct.App. 509 (1882), discussed in Wilner, [Unintentional Homicide in the Commission of an Unlawful Act, 87, Pa.L.Rev. 811, 834–835 (1939)]. In *Commonwealth v. Mink,* 123 Mass. 422, 425 (1877), the defendant was attempting to commit suicide, but her fiancee intervened to try to stop her and was accidentally killed by the defendant. Because suicide was an unlawful act *malum in se,* the court found defendant guilty of manslaughter.

Over time, this theory of criminal responsibility has developed many complexities. Courts differ about whether the unlawful act must amount to a criminal offense and whether different standards should apply for *malum in se* or *malum prohibitum* offenses. In this case, neither of these issues is of concern. The offense in this case is a *malum in se* misdemeanor offense under the criminal law of Florida. However, the other principal point of divergence in the development of the misdemeanor manslaughter rule—the issue of causation—is critical to this case.

The views on the requirement of causation in unlawful act manslaughter differ widely among the various jurisdictions. In some instances, no causal relationship at all has been required. At the other extreme is the requirement that there be not only a direct causal relationship between the unlawful act and the death, but that the death must be a natural and probable consequence of the offense. An example . . . is the case of *Votre v. State,* 138 N.E. 257 (Ind. 1923) where, contrary to statute, the defendant gave whiskey to the victim, who was a minor. Consumption of the alcohol caused the victim to suffer a heart attack of which he died. The Indiana court held that the defendant was not guilty of manslaughter because the homicide must follow both as a part of the perpetration of the unlawful act and as a natural and probable consequence of it. Wilner, *supra* at 836. As LaFave and Scott and Wilner point out, application of this view of causation essentially converts the unlawful act type of manslaughter into culpable negligence manslaughter—a development which these commentators applaud. . . .

Florida courts, by simply interpreting the statutory definition of manslaughter ("[t]he killing of a human being by the act, procurement, or culpable negligence of another, without lawful justification . . ."), appear always to have understood the importance of causation as an element of this type of homicide. Our

*(Continues)*

*(Continued)*

courts also have appreciated the foreseeability element of causation. . . . If it is true that imposition of criminal responsibility for death requires a causative link at least equivalent to that required for tort liability, the analysis of this case is considerably simplified. Under Florida law, in a tort context, appellant's petty theft may have been a cause-in-fact of the victim's heart attack, but it was not the legal cause.

In this case, even if it were assumed that the stress of pursuit brought on the heart attack, it cannot be said that the petty theft was the legal cause of Mr. Voegltin's death. The crime itself was a minor property offense. There is no suggestion of any touching or any threat to anyone's person. This is not even a case, like a purse snatching, where violence was necessary to produce the theft. Nor is it asserted that Mr. Voegltin died from fright or horror at witnessing the crime. The state's traverse specifically asserts that it was the *pursuit* that caused the fatal heart attack. Although the petty theft did trigger a series of events that concluded in the death of Mr. Voegltin and was, in that sense, a "cause" of the death, the petty theft did not encompass the kind of direct, foreseeable risk of physical harm that would support a conviction of manslaughter. The relationship between the unlawful act committed (petty theft) and the result effected (death by heart attack during pursuit in an automobile) does not meet the test of causation historically or currently required in Florida for conviction of manslaughter.

REVERSED.

## Notes and Questions

1. Can *Todd* be reconciled with *People v. Stamp*? Do the *Todd* and *Stamp* courts differ in how they define proximate cause or just in the application of causation principles to the facts presented in the respective cases?

2. What is the distinction between offenses that are *malum in se* and *malum prohibitum*?

3. In *Commonwealth v. Williams*, 1 A.2d 812 (Pa. Super. 1938), the defendant, who had a valid driver's license for several years, unlawfully failed to renew it in 1936. While driving with his expired license in a prudent manner, he was forced to veer off the road by an approaching vehicle. His car collided with a telephone pole and his passenger was killed. He was convicted of unlawful act–manslaughter based on his driving with an expired driver's license. The appellate court reversed, holding that the death was not connected with, or caused by, the unlawful act but merely coincidental with it.

    Would it matter if Williams had been 15 (*i.e.*, an underage driver), whose unlawful act consisted of operating a motor vehicle without a driver's license?

4. The courts also have applied the misdemeanor-manslaughter rule in the context of blows delivered (*i.e.*, an assault and battery) that result in death. *See, e.g., People v. Datema*, 533 N.W.2d 272 (Mich. 1995); *Comber v. United States*, 584 A.2d 26 (D.C. App. 1990) (granting the defendants new trials based on erroneous instructions); *State v. Pray*, 378 A.2d 1322 (Me. 1977) (repudiating common law concept of unlawful act manslaughter).

## "Life" and "Death"

We established that a homicide is the killing of a human being by another person. We have yet to ascertain when human life begins and ends for purposes of the law of criminal homicide. That is, at what developmental point is neonatal life considered a "human being" under the criminal law? Conversely, by what criteria should it be determined that a death has occurred? We now turn to these issues.

## Life: The Killing of a "Human Being"

### *Hughes v. State*, 868 P.2d 730 (Okla. Crim. App. 1994)

Chapel, Judge:

On August 2, 1990, Appellant, Treva La-Nan Hughes, while intoxicated, drove her vehicle into oncoming traffic and collided with another vehicle. The driver of the other vehicle, Reesa Poole, was nine months pregnant and expected to deliver in four days. The collision caused Poole's stomach to hit the steering wheel of her car with such force that the steering wheel broke. Poole was taken to the hospital where an emergency cesarean section was performed. When the baby was delivered, the only sign of life was an extremely slow heartbeat. A pediatrician immediately began resuscitation efforts, which were unsuccessful.

After a jury trial in the District Court of Oklahoma County before the Honorable Eugene Mathews, District Judge, Hughes was convicted of First Degree Manslaughter under 21 O.S. 1981, § 711(1) ("[h]omicide is manslaughter in the first degree . . . [w]hen perpetrated without a design to effect death by a person while engaged in the commission of a misdemeanor"). The jury also convicted Hughes of Driving Under the Influence While Involved in a Personal Injury Accident, in violation of 47 O.S.Supp.1985, § 11-904. Hughes received an eight-year prison sentence for the manslaughter conviction and a six-month suspended sentence for the driving under the influence conviction. . . .

Hughes argues as her sole proposition of error that her manslaughter conviction should be reversed on the basis of the common law "born alive" rule. Because Oklahoma has neither altered nor abolished this remnant of the common law, it remains in effect pursuant to 22 O.S.1981, § 9. Under the **"born alive" rule,** "[a] child can not be the subject of homicide until its complete expulsion from the body of the mother, and must be alive and have independent existence." O. Warren, *Warren on Homicide* § 55 (1938). In this case of first impression, Hughes claims that the fetus Mrs. Poole was carrying was not born alive and that its death cannot be considered a homicide. We now abandon the common law approach and hold that whether or not it is ultimately born alive, an unborn fetus that was viable at the time of injury is a "human being" which may be the subject of a homicide under 21 O.S.1981, § 691 ("Homicide is the killing of one human being by another").

The dissent adopts the State's position that the "born alive" rule would not prohibit a manslaughter conviction in this case because the fetus in question was born alive. It is asserted that the fetus had a heartbeat at birth and the Legislature has determined that any fetus born with a heartbeat is in fact born alive. While this argument is superficially appealing, it has no merit. The evidence presented at trial established that the baby was dead when delivered.

The fetus did have a weak heartbeat. It was, however, brain dead according to the doctor to whom it was handed immediately upon delivery. Additionally, according to the testimony, the fetus had no blood pressure, no respiration and did not respond to any resuscitative efforts. We are not prepared to hold that a brain dead fetus was alive when born simply because its heart was beating weakly. . . .

Today's decision abandoning the "born alive" rule is based in significant part upon its origins, history and purpose. Common law authorities refer to the born alive rule as early as the 1300's. The born alive rule was necessitated by the state of medical technology in earlier centuries.

In fact, as late as the nineteenth century, prior to quickening "it was virtually impossible for either the woman, a midwife, or a physician to confidently know that the woman was pregnant, or, it follows, that the child *in utero* was alive." Hence, there was *no evidence of life* until quickening.

Yet, the quickening of a fetus did not constitute proof that the fetus was alive in the womb at any particular moment thereafter. The health of a child within the womb could not be determined until the child was observed after birth. "As a result, live birth was required to prove that the unborn child was alive and that the material acts were the proximate cause of death, because it could not otherwise be established if the child was alive in the womb at the time of the material acts."

Advances in medical and scientific knowledge and technology have abolished the need for the born alive rule. Specifically, the medical and scientific evidence before us establishes that the child within Poole's womb was a living, viable fetus at the time of the collision and that this child died as a result of the placental abruption which occurred when Poole's stomach hit and broke the steering wheel of her car.

Although in the minority, two states have expressly rejected the born alive rule: Massachusetts and South Carolina. The Massachusetts court addressed the question of whether a viable fetus is a "person" under that state's vehicular homicide statute in *Commonwealth v. Cass,* 467 N.E.2d 1324 (Mass. 1984). In reaching its decision, the *Cass* court . . . rejected the born alive rule, stating:

> The rule has been accepted as the established common law in every American jurisdiction that has considered the question. But the antiquity of a rule is no measure of its soundness. 'It is revolting to have no better reason for a rule of law than that so it was laid down in the time of Henry IV. It is still more revolting if the grounds upon which it was laid down have vanished long since, and the rule simply persists from blind imitation of the past.' . . . It is time to reexamine the grounds upon which this ancient rule was laid down.

*(Continues)*

The rationale offered for the rule since 1348 is that 'it is difficult to know whether [the defendant] killed the child or not. . . .' . . . That is, one could never be sure that the fetus was alive when the accused committed his act. However, difficulty of proving causation is no sound reason for denying criminal liability. Medical science now may provide competent proof as to whether the fetus was alive at the time of a defendant's conduct and whether his conduct was the cause of death. . . . We do not consider [fear of speculation] a sufficient reason for refusing to consider the killing of a fetus a homicide.

We think that the better rule is that infliction of prenatal injuries resulting in the death of a viable fetus, before or after it is born, is homicide. If a person were to commit violence against a pregnant woman and destroy the fetus within her, we would not want the death of the fetus to go unpunished. We believe that our criminal law should extend its protection to viable fetuses. *Id.* 467 N.E.2d at 1328–29 (citations omitted) (footnotes omitted).

One day later, South Carolina also rejected the born alive rule in *State v. Horne,* 319 S.E.2d 703 (S.C. 1984). As in *Cass,* the *Horne* court acknowledged that it had previously held that a viable fetus was a person in the context of a wrongful death action. The *Horne* court then stated that "[i]t would be grossly inconsistent for us to construe a viable fetus as a 'person' for the purposes of imposing civil liability while refusing to give it a similar classification in the criminal context." *Id.* 319 S.E.2d at 704. Accordingly, the court held that:

This Court has the right and the duty to develop the common law of South Carolina to better serve an ever-changing society as a whole. In this regard, the criminal law has been the subject of change. . . . The fact this particular issue has not been raised or ruled on before does not mean we are prevented from declaring the common law as it should be. Therefore, we hold an action for homicide may be maintained in the future when the state can prove beyond a reasonable doubt the fetus involved was viable, *i.e.,* able to live separate and apart from its mother without the aid of artificial support. *Id.*

This Court also has the right and duty to develop the common law of Oklahoma to serve the evolving needs of our citizens.

To that end, we believe that the citizens of Oklahoma would be best served by a definition of "human being" in the context of Section 691 which does not rely upon an obsolete, antiquated common law rule. Rather, the meaning of that term should be determined by the plain language and purpose of the statute. The purpose of Section 691 is, ultimately, to protect human life. A viable human fetus is nothing less than human life. As stated by the court in *Cass,* "[a]n offspring of human parents cannot reasonably be considered to be other than a human being . . . first within, and then in normal course outside, the womb." *Cass, supra,* 467 N.E.2d at 1325. Thus, the term "human being" in Section 691—according to its plain and ordinary meaning—includes a viable human fetus. . . .

We wish to make it absolutely clear that our holding shall not affect a woman's constitutional right to choose a lawful abortion based upon her constitutional right to privacy or a physician's right to perform one. *Roe v. Wade,* 410 U.S. 113, 153–66 (1973); *Planned Parenthood v. Casey,* 505 U.S. 833 (1992). Neither state statute nor caselaw can render a constitutionally protected abortion unlawful. Accordingly, today's decision to bestow upon viable human fetuses the legal status of "human being" under Oklahoma law, cannot and shall not be used as the basis for bringing homicide charges against either a woman who chooses a lawful abortion or a physician who performs a lawful abortion.

Having determined that an unborn viable fetus is a "human being" under Section 691, we turn to the question of whether our ruling is applicable to Hughes or should be prospective only. . . .

Due process requires that a criminal statute give fair warning of the conduct which it prohibits. . . . Hence, the question of whether the judicial construction of a statute may be applied retroactively rests on the foreseeability of the court's action. Specifically, did Hughes have "fair warning" that her conduct would result in a criminal charge?

We believe that our construction of Section 691 as including viable fetuses was not foreseeable. . . . Because Hughes did not have fair warning that her conduct was criminal and subject to punishment under Section 691, our decision should not apply to her. . . .

In conclusion, we reject the born alive rule and hold that a viable human fetus is a "human being" against whom a homicide as defined in Section 691 may be committed. The fetus that suffered fatal injuries as a result of Hughes's drunk driving was viable. However, Hughes may not be convicted for having caused its death because she could not have foreseen this Court's decision to abolish the born alive rule and effectively render her actions homicidal. Accordingly, today's ruling will apply wholly prospectively to those homicides which occur after its date. . . .

## Notes and Questions

1. In *Keeler v. Superior Court of Amador County*, 470 P.2d 617 (Cal. 1970), the California Supreme Court adhered to the common law "born alive" rule described in *Hughes* and ruled that the defendant could not be prosecuted for the criminal homicide of a viable fetus. Keeler became upset when he encountered his estranged wife, who was pregnant by another man. He kneed her in the abdomen after declaring, "I'm going to stomp it out of you." The fetus, which had developed to the stage of viability, died *in utero* and was delivered stillborn. Keeler was charged with murder through an indictment alleging that "he did 'unlawfully kill a human being . . . with malice aforethought.'"

   In dismissing the murder indictment, the California Supreme Court noted the long history of the common law "born alive" rule and explained as follows: "[T]he courts cannot go so far as to create an offense by enlarging a statute, by inserting or deleting words, or by giving the terms used false or unusual meanings. . . . Whether to extend liability for murder in California is a determination solely within the province of the Legislature."

   Is *Hughes* or *Keeler* the more persuasive decision? Which decision adopts the preferable policy regarding the scope of criminal homicide laws? Which adopts the more defensible position regarding the role of a court *vis-à-vis* a state legislature? *See also Commonwealth v. Morris*, 142 S.W.3d 654 (Ky. 2004) (overruling precedent and abrogating common law "born alive" rule by ruling that an unborn viable fetus is a human being for purposes of law of criminal homicide); *Commonwealth v. Booth*, 766 A.2d 843 (Pa. 2001) (declining to modify common law "born alive" rule judicially, concluding that legislature is appropriate body to consider such action).

2. What reasons justified the "born alive" rule at common law? To what extent do those reasons have continuing vitality today? Should the question of when a "human being" exists for purposes of the criminal law be treated as a *normative* issue? Should its resolution depend on scientific or technological capabilities in identifying and being able to sustain life forms? On the need to achieve certainty and avoid arbitrary application of the law? On some combination of these considerations?

3. If Keeler had not succeeded in causing the viable fetus his estranged wife was carrying to die *in utero* and a live birth had occurred and the baby had died thereafter as a result of the prenatal injuries, could he have been successfully prosecuted for murder? If the answer to this question is yes, is the result in *Keeler* defensible? Does it "reward" offenders who succeed in inflicting such serious prenatal injuries that a stillbirth occurs? *See generally, State v. Cotton*, 5 P.3d 918 (Ariz. App. 2000) (criminal homicide statute applies to killing of a child who is born alive, although injuries were inflicted before birth); *Jones v. Commonwealth*, 830 S.W.2d 877 (Ky. 1992) (although Kentucky followed common law "born alive" rule at the time, defendant was properly convicted of manslaughter when injuries were inflicted prior to birth, and the baby was born alive and subsequently died as a result of those injuries); *State v. Aiwohi*, 123 P.3d 1210 (Haw. 2005) (discussing cases sustaining criminal homicide convictions where child dies from injuries inflicted by another person prior to birth but declining to uphold criminal homicide conviction where prenatal injuries were inflicted by drug use by child's mother); *State v. Deborah J.Z.*, 596 N.W.2d 490 (Wis. App.), *rev. den.*, 604 N.W.2d 570 (Wis. 1999) (reaching same result as in *State v. Aiwohi, supra*, involving mother whose alcohol use while pregnant resulted in child's death).

4. If, under the rule announced in *Hughes*, a viable fetus is a "human being" for criminal homicide purposes, what are the implications for other areas of the law? Would the fetus have inheritance or other property rights? Would a civil action for wrongful death be sustainable? Would a woman whose life or health would be seriously jeopardized by giving birth ever be allowed to abort a viable fetus?

5. If a viable fetus is a "human being" under *Hughes*, should a nonviable fetus or an embryo also qualify?

6. A number of other states have enacted statutes that make the killing of an unborn child murder, manslaughter, or the crime of feticide. Following *Keeler*, California amended its penal code to define murder as "the unlawful killing of a human being, or a fetus, with malice aforethought." Cal. Penal Code § 187(a). In *People v. Davis*, 872 P.2d 591 (Cal. 1994), the California Supreme Court ruled that a fetus need not be viable to support a murder conviction under the statute, although its ruling was prospective only. *See also* Ill. Stat. ch. 720 § 5/9-1.2 (making intentional homicide of an unborn child a crime punishable by life imprisonment and defining unborn child as "any individual of the human species from fertilization until birth." *Id.* at § 5/9-1.2(3)(b)); Ariz. Stat. § 13-1103(A)(5) (defining as manslaughter "[k]nowingly or recklessly causing the death of an unborn child at any stage of its development. . . ."); Ind. Stat. § 35-42-1-6 ("A person who knowingly or intentionally terminates a human pregnancy with an intention other than to produce a live birth or to remove a dead fetus

*(Continues)*

(*Continued*)

commits feticide, a Class C felony. This section does not apply to [a lawful] abortion. . . ."); N.Y. Penal Law § 125.00 ("Homicide means conduct which causes the death of a person or an unborn child with which a female has been pregnant for more than twenty-four weeks. . . .").

**7.** When has a child been "born alive" for purposes of the common law rule? In *People v. Chavez*, 176 P.2d 92 (Cal. App. 1947), the California Court of Appeals rejected the traditional rule requiring a complete separation of mother and child, including evidence of independent respiration and circulatory activity on the infant's part. It opted for a more flexible test:

> [A] viable child in the process of being born is a human being within the meaning of the homicide statutes, whether or not the process has been fully completed. It should at least be considered a human being where it is a living baby and where in the natural course of events a birth which is already started would naturally be successfully completed.

## Death

### *State v. Guess*, 715 A.2d 643 (Ct. 1998)

[Following a confrontation in a New Haven convenience store, the defendant, Barry Guess, shot Melvin McKoy in the head. The victim was taken to Yale New Haven Hospital. On admission, he was "in a coma with a heart rate of forty and no respiratory function. There was no evidence of brain stem function." He was placed on a ventilator and a respirator. He could not breathe on his own and his heart could not beat without life support systems. "The victim's parents authorized that the machines be disconnected and the victim was pronounced dead." The defendant was convicted of murder and sentenced to 50 years in prison.

[On appeal, the defendant argued that the victim was still alive—as evidenced by his heart and respiratory activity—when the life support systems were removed. He consequently urged that he was not responsible for causing the victim's death.]

Barnard, Presiding Justice. . . .

On appeal to the Appellate Court, the defendant argued that, because the legislature had not adopted the **Uniform Determination of Death Act** (*see* Footnote 4), and because the legislature did not define death in the Penal Code to include brain death, the court, in determining who or what caused the victim's death, must use a common-law definition of death, which does not include brain death, but rather depends solely upon the cessation of circulatory and respiratory functions of the body. (*See* Footnote 5.) On the basis of the evidence, the Appellate Court concluded that "because traditional principles of causation allow this defendant to be found guilty of murder despite the action of another of removing the victim from life support systems," it did not need to decide whether to adopt judicially in the criminal setting a definition of death that includes brain death. . . .

We begin with the defendant's claim that this court has accepted a common-law definition of death that is limited to the cessation of the respiratory and circulatory systems, and that the legislature, in drafting the Penal Code without providing a statutory definition of death, intended that the common-law definition then in existence apply. In other words, according to the defendant, by failing to define the term in the Penal Code, the legislature contemplated reliance on a common-law definition frozen in time.

Although death has typically been discussed in terms of cessation of the heart and respiratory systems, the defendant has pointed to no case, and we have found none, in which this court has *expressly* defined death in such terms. Perhaps the fact that death has not been legally defined merely reflects the fact that, until now, no reason existed for this traditional medical definition to engender legal controversy. Indeed, only recently have medical science and technology evolved to the point where a person's heartbeat and respiration may be sustained mechanically even in the face of an irreversible loss of all brain functions, and where machines that artificially maintain cardiorespiratory functions have come into widespread use.

"Traditionally, in criminal prosecutions for homicide, when the fact or time of death was at issue, the common law defined death as 'the cessation of life' and set a medical standard of the stoppage of the circulatory and respiratory systems." Annot., 42 A.L.R. 4th 742, 745 (1985). According to pre-1960 medical standards, the cessation of life was determined by the stoppage of the circulatory and respiratory systems. *Id*. The criteria of the stoppage of the circulatory and respiratory systems were cast into flux as the medical community gained a better appreciation of human physiology. "These traditional legal rules on death became troubling during the 1960's, following medical advances in three areas: (1) improved organ transplant technology, creating a need for 'fresh' organs, (2) the ability of external respiratory and circulatory machines to maintain these bodily functions artificially for longer and longer periods, and (3) the enhanced ability to detect and monitor brain activity." Annot., 42 A.L.R. 4th, supra, at 746. As a result, a new medical definition of "death" centering on brain activity was developed.

The criteria by which the medical community determines brain death, first established in 1968 by the Ad Hoc Committee of the Harvard Medical School to Examine the Definition of Brain Death (Harvard Committee), include: (1) a total lack of responsivity to externally applied stimuli; (2) no spontaneous muscular movements or respiration; and (3) no reflexes, as measured by fixed, dilated pupils and lack of ocular, pharyngeal and muscle-tendon reflexes. Additionally, the Harvard Committee emphasized that these tests could be confirmed by a flat or isoelectric electroencephalogram reading, that the tests should be conducted twenty-four hours apart, and that hypothermia or the use of central nervous system depressants should be excluded as causative factors. . . .

This evolution in medicine has spawned concomitant developments in the law. In 1970, Kansas became the first state to adopt by statute a brain-based definition of "death." Kan. Stat. Ann. § 77-202 (1970). Then, in 1980, the National Conference of Commissioners on Uniform State laws, in concert with the American Medical Association and the American Bar Association, drafted the Uniform Determination of Death Act, which created alternative standards for determining death: either the traditional irreversible cessation of circulatory and respiratory functions, or the irreversible cessation of all functions of the entire brain including the brain stem, "a determination to be made in accordance with accepted medical standards." Annot., 42 A.L.R. 4th supra, at 747. Black's Law Dictionary also responded to the advances in medical science in 1979 when it included for the first time a definition of "brain death" that provides in pertinent part: "Characteristics of brain death consist of: (1) unreceptivity and unresponsiveness to externally applied stimuli and internal needs; (2) no spontaneous movements or breathing; (3) no reflex activity; and (4) a flat electroencephalograph reading after 24 hour period of observation. . . ." Black's Law Dictionary (5th Ed. 1979). . . .

When the legislature adopted the homicide statutes, it was also content to defer to the prevailing medical judgment as to the criteria by which to determine death.

This legislative deference does not mean, however, that it intended to freeze the definition of death at that point in time. "'[T]he common law of today is not a frozen mold of ancient ideas, but such law is active and dynamic and thus changes with the times and growth of society to meet its needs.'" *Swafford v. State,* 421 N.E.2d at 600. We have searched unsuccessfully for evidence that the legislature intended to render immutable the criteria by which to determine death. In the absence of any such indication, we are loath to limit the criteria to a fixed point in the past. . . .

"Principles of law which serve one generation well may, by reason of changing conditions, disserve a later one. . . . The adaptability of the common law to the changing needs of passing time has been one of its most beneficent characteristics." (Internal quotation marks omitted.) *Jolly, Inc. v. Zoning Board of Appeals,* 237 Conn. 184, 196, 676 A.2d 831 (1996). . . . As it has become "clear in medical practice that the traditional 'vital signs'—breathing and heartbeat—are not independent indicia of life, but are, instead, part of an integration of functions in which the brain is dominant," our focus must shift from those traditional "vital signs" to recognize cessation of brain functions as criteria for death following this medical trend.

The question of when death has occurred carries significant legal ramifications including, but not limited to, issues of: inheritance; criminal and civil liability; termination of mechanical support; liability under insurance contracts; and exposure of physicians to medical malpractice claims. Although we are examining this issue in the context of a homicide, the defendant's position that we are wedded to a definition of death from the time the Penal Code was adopted juxtaposes uncomfortably with the medical community's capacity to sustain heartbeat and respiration through artificial means. The defendant acknowledges that brain death became the medically accepted standard for determining death some time ago. . . . This continuing evolution in the science of determining brain death would be lost were we to adopt the defendant's position that we are wedded to a definition of death that existed when the Penal Code was enacted but that has been abandoned long ago by the medical profession. In the absence of an overwhelming impediment, reasonable evolving medical standards must continue to play a dominant role. (*See* Footnote 11.)

We also reject the defendant's contention that only the legislature can determine whether brain death should be recognized in law. The legislature clearly has the authority to define by statute the standards by which a hospital may determine death. We are, however, confident in our parallel authority,

*(Continues)*

(*Continued*)

as a matter of common law, to determine what constitutes death and to embrace the advances made in medical science and technology during the last three decades. . . .

We conclude that our recognition of brain-based criteria for determining death is not unfaithful to any prior judicial determinations. "Death remains the single phenomenon identified at common law; the supplemental criteria are merely adapted to account for the 'changed conditions' that a dead body may be attached to a machine so as to exhibit demonstrably false indicia of life. It reflects an improved understanding that in the complete and irreversible absence of a functioning brain, the traditional loci of life—the heart and the lungs—function only as a result of stimuli originating from outside of the body and will never again function as part of an integrated organism." *People v. Eulo,* 63 N.Y.2d at 356, 472 N.E.2d 286. Because the trial court at the hearing in probable cause reasonably found that the defendant's act of shooting the victim caused extensive brain damage, leaving the victim with no evidence of brain function, the court properly found that the state had established probable cause to charge the defendant with the crime of murder.

The judgment of the Appellate Court is affirmed.

## Notes and Questions

1. Can an argument be made that a legislature, and not a court, is the appropriate institution to define "death" for purposes of criminal homicide? *See State v. Olson,* 435 N.W.2d 530 (Minn. 1989) (declining to resolve whether a cessation of brain functioning qualified as "death" as that term was used in criminal homicide statutes because it was not necessary to the case decision and because the legislature "should first be given an opportunity to consider the legal implications of brain death. The legislature, with its broad based representation, its committee hearings, and its floor debates, presents the kind of public forum this issue deserves.").

2. Did the court in *Guess* have to define death as the cessation of brain functioning to uphold the defendant's murder conviction? In *State v. Velarde,* 734 P.2d 449 (Utah 1986), after accepting "brain death" as the test for when death transpires, the court noted that "even if the support systems were removed prematurely, defendant would still be responsible for [the victim's] death since intervening medical error is not a defense to a defendant who has inflicted a mortal wound upon another." *See generally, Kusmider v. State,* 688 P.2d 957 (Alaska App. 1984) (presented in Chapter 3).

## Conclusion

All criminal homicides involve the unlawful killing of a human being by another person. The harm caused—a death—remains constant. No distinctions are drawn between young and old victims, the healthy or infirm, or those who have lived their lives well and those who have not. The law governing criminal homicide, accordingly, looks to other factors to distinguish offenders' culpability.

In this chapter we reviewed a variety of approaches adopted by the criminal law to reflect the relative seriousness of different kinds of killings. The law has evolved over time in its continuing efforts to make principled distinctions between homicides. At very early common law, all killings were treated alike. During medieval times, courts granted benefit of clergy to spare some criminal offenders the death penalty. After Parliament declared that killings committed with "malice prepensed" or malice aforethought were no longer clergyable, creative and sympathetic judges began to recognize that some unlawful homicides could be committed without malice. Thus arose the distinction between murder (an unlawful killing with malice) and manslaughter (an unlawful killing without malice). The distinction between these crimes gave rise to the corresponding need to define the concept of malice.

As we have seen, giving meaning to malice proved to be no easy task. The common law standard for negating malice involving intentional killings depended on "the heat of passion on adequate provocation." This test relied on the "reasonable man," the absence of reasonable "cooling time," the merits of different forms of "provocation," and other ambiguous concepts. Unintentional killings—for example, those committed during the perpetration or attempted perpetration of felonies and "depraved heart" homicides—also were defined as entailing malice and hence constituted murder. By statute, with Pennsylvania taking the lead, several jurisdictions later divided murder into first and second degree, with the death penalty available only for the former.

The MPC reconceptualized the law of criminal homicide in several significant respects. The MPC makes no use of the concepts of "malice" or "provocation" to distinguish murder and manslaughter. The MPC also dispenses with the felony-murder rule and degrees of murder. Murders are unlawful killings committed purposely, knowingly, or recklessly manifesting extreme indifference to human life. Manslaughter is criminally reckless homicide or a purposeful or knowing killing committed under the influence of extreme mental or emotional disturbance for which there is reasonable explanation or excuse. The "reasonable man" standard gives way under the MPC to an assessment of reasonableness "from the viewpoint of a person in the actor's situation under the circumstances as he believes them to be." The drafters of the MPC anticipated modern death penalty law by requiring the sentencing authority to examine and balance aggravating and mitigating factors associated with a crime and the offender before deciding whether capital punishment is appropriate.

The law of criminal homicide presents a host of difficult yet intriguing issues on matters involving culpability. These issues range from assessing offenders' blameworthiness, to resolving such fundamental questions as how to delineate the beginning and ending of human life. Murder and manslaughter also serve as excellent vehicles to study the operation of the basic principles of criminal law, such as *mens rea*, harm, and causation. Knowledge of the different rules that have been created to help make distinctions between homicides is important. However, understanding the underlying philosophical and principled dimensions of the law pertaining to murder and manslaughter is absolutely essential to understanding the formulation, application, and relative wisdom of those rules. This chapter should help illuminate the history and rationale supporting the law's judgments about the relative seriousness of different kinds of criminal homicide.

## Key Terms

| | |
|---|---|
| "born alive" rule | murder |
| deliberation | premeditation |
| depraved-heart murder | second-degree murder |
| felony-murder rule | Uniform Determination of |
| first-degree murder | Death Act |
| homicide | vicarious liability |
| involuntary manslaughter | voluntary manslaughter |
| malice | |

## Review Questions

1. What was the crucial distinguishing factor between *murder* and *manslaughter* in medieval England?
2. How does the MPC approach in distinguishing murder from manslaughter differ from the traditional common law approach?
3. What role does malice have in determining the level of criminal homicide?
4. Under the MPC, what elements are necessary to establish the commission of *murder*?
5. How does *murder* differ from *manslaughter* under the MPC?
6. How do *voluntary manslaughter* and *involuntary manslaughter* differ in the eyes of the law? Be specific by listing necessary elements of each.
7. Briefly define *premeditation* and *deliberation* as they pertain to murder, and give an example of each.
8. Why is specific intent critical in a prosecution of first-degree murder?
9. How does *vicarious liability* relate to the *felony-murder rule*?
10. What is the intent of the Uniform Determination of Death Act?

## Notes

1. Stephen JF. *A History of the Criminal Law of England*, vol. 3. New York: Burt Franklin, 1883:2.
2. Stephen, vol. 1, p. 463.
3. *Model Penal Code and Commentaries (Official Draft and Revised Comments)*, Part n, vol. 1, art. 210.2. Philadelphia: The American Law Institute, 1980:13–14.
4. LaFave WR. *Criminal Law*, 3rd ed. St. Paul, MN: West Publishing, 2000:655.
5. *Model Penal Code and Commentaries*, Part II, vol. 1, art. 210, p. 1.
6. *Model Penal Code and Commentaries*, Part II, vol. 1, art. 210.3, pp. 49–50.
7. *See* Keedy ER. History of the Pennsylvania statute creating degrees of murder. 97 *Univ. Penn. Law Rev.* 1949;97:759, 772–773.
8. LaFave, p. 396.
9. *Model Penal Code and Commentaries*, Part II, vol. 1, art. 210.2, p. 31.
10. *Model Penal Code and Commentaries*, Part II, vol. 1, art. 210.2, p. 31, n. 74.
11. *Model Penal Code and Commentaries*, Part II, vol. 1, art. 210.2, pp. 36–39.
12. *See* Sudduth T. Comment: The *Dillon* dilemma: finding proportionate felony-murder punishments. *Calif. Law Rev.* 1984;72:1299; Tomkovicz JJ. The endurance of the felony-murder rule: a study of the forces that shape our criminal law. 51 *Wash & Lee Law Rev.* 1994;51:1429.

## Footnotes for *State v. Guess*, 715 A.2d 643 (Ct. 1998)

4. The Uniform Determination of Death Act (1980) § 1, provides: "An individual who has sustained either (1) irreversible cessation of circulatory and respiratory functions, or (2) irreversible cessation of all functions of the entire brain, including the brain stem, is dead. A determination of death must be made in accordance with accepted medical standards."
5. The term "death" is defined as "[t]he cessation of life; the ceasing to exist; defined by physicians as a total stoppage of the circulation of the blood, and a cessation of the animal and vital functions consequent thereon, such as respiration, pulsation, etc." Black's Law Dictionary (4th Ed. 1968).
11. We note that several other jurisdictions, faced with the changes in medicine, chose judicially to adopt a brain death standard in homicide cases.

# Rape and Sexual Assault

## Chapter Objectives

- Learn the definition of forcible rape
- Know the implications of the marital rape exception
- Understand the concept of resistance and forcible compulsion
- Understand the purpose and application of *rape shield laws* and *rape trauma syndrome*
- Know the purpose behind statutory rape statutes

At the outset, it is important to recognize several empirical realities about rape and sexual assault more generally. First, the crimes pose an enormous problem in American society. In 2006 alone more than 272,000 individuals were the victims of rape or sexual assault.[1] The number was considerably higher in 1999 and 2000, averaging more than 322,000 annually.[2] Second, the victims of rape and sexual assault are overwhelmingly female. Between 1992 and 2000, "[f]emale victims accounted for 94% of all completed rapes, 91% of all attempted rapes, and 89% of all completed and attempted sexual assaults."[3] Finally, it is well established that rapes are perhaps the most underreported of all serious crimes. Roughly 60 percent of all rapes and sexual assaults are not reported to law enforcement.[4] Although experts offer numerous explanations for this underreporting, most agree that the trauma and stigma that too often attend the formal investigation and prosecution of sex crimes lead many victims to fear being "revictimized" by the justice system.

In approaching the cases and materials in this chapter, keep in mind the many social, physical, psychological, and emotional harms suffered by the victims of rape and sexual assault. At the same time, because these crimes are so serious, defendants accused of committing them face extremely serious consequences on conviction. Few, if any, other areas of the substantive criminal law approach sex crimes in terms of the highly personal and sensitive issues involved. The issues raised here are both difficult and challenging, and their careful analysis and discussion are important to our study.

## Forcible Rape

Perhaps more than any other type of crime, rape reflects the imprint of evolving social values, practices, and understandings. At early common law and for many centuries thereafter, rape was defined as "the carnal knowledge of a woman forcibly and against her will."[5] "Carnal knowledge" required penetration (however slight—emission or ejaculation was not necessary) by the offender's penis of the victim's vagina. Today, in most jurisdictions a rape victim does not have to be a woman; although comparatively infrequent, males can be victims of rape and other forms of sexual assault. In addition to being gender neutral, modern laws typically recognize that other body parts and objects, and not just a rapist's penis, can be used to commit a sexual assault. Similarly, sexual assaults are not necessarily restricted to offensive contact with the victim's sex organs.

Another significant change in the definition and prosecution of rape over time concerns the role of force. Rape has always required that the sexual act occur without the consent of the victim, a requirement satisfied when the victim is unconscious, asleep, or mentally impaired or, as we shall see, when the sexual act was committed by certain types of fraud. For many centuries, however, the victim was also required to "resist to her utmost," a requirement often entailing actual physical resistance to the sexual advance. A mere verbal objection from the victim did not suffice on the reasoning that a "woman jealous of her chastity, shuddering at the bare thought of dishonor" would be expected to resist her assailant strenuously. *See State v. Burgdorf,* 53 Mo. 65 (1873). In recent decades American jurisdictions have noticeably relaxed this requirement, in response to claims that physical resistance can serve to worsen the injuries suffered by victims. Moreover, although active resistance might plainly manifest a lack of consent, the absence of resistance does not, by the same token, as readily signal affirmative consent.[6]

The crime of rape, according to Professor Susan Estrich, can be conceived of as occurring in two forms.[7] First, is "real rape," in which the victim and the rapist are unknown to each other. Second, is what has come to be known as "date" or "acquaintance rape," which involves at least some degree of familiarity between the victim and offender. Under the first scenario the main challenge for police is to identify and arrest the assailant, often a task made more difficult because of the anonymity of the suspect and the trauma suffered by the victim. With acquaintance rape the identity of the perpetrator is known, but police and prosecutors face the challenge of sorting out the often-conflicting facts generated by the preexisting relationship. Usually, the defendant will admit that sexual intercourse occurred but will assert that it was consensual, taking the act outside the bounds of the criminal law.

Our focus here is on this second type of rape, for two reasons. First, the legal liability in "real rape" cases, as suggested, frequently turns on issues of identification, a challenge sometimes facilitated in recent times by DNA testing. In comparison, the factual and legal issues relating to "acquaintance rape," as we shall see, are quite complicated. The second reason is empirical. Contrary to the view of many, rapes committed by non-strangers are far more common than rapes committed by strangers.[8]

## Actus Reus

It was then about 1 A.M. Pat accompanied Rusk across the street into a totally dark house. She followed him up two flights of stairs. She neither saw nor heard anyone in the building. Once they ascended the stairs, Rusk unlocked the door to his one-room apartment, and turned on the light. According to Pat, he told her to sit down. She sat in a chair beside the bed. Rusk sat on the bed. After Rusk talked for a few minutes, he left the room for about one to five minutes. Pat remained seated in the chair. She made no noise and did not attempt to leave. She said that she did not notice a telephone in the room. When Rusk returned, he turned off the light and sat down on the bed. Pat asked if she could leave; she told him that she wanted to go home and "didn't want to come up." She said, "'Now, [that] I came up, can I go?'" Rusk, who was still in possession of her car keys, said he wanted her to stay.

Rusk then asked Pat to get on the bed with him. He pulled her by the arms to the bed and began to undress her, removing her blouse and bra. He unzipped her slacks and she took them off after he told her to do so. Pat removed the rest of her clothing, and then removed Rusk's pants because "he asked me to do it." After they were both undressed Rusk started kissing Pat as she was lying on her back. Pat explained what happened next:

> "I was still begging to him to please let, you know, let me leave. I said, 'you can get a lot of other girls down there, for what you want,' and he just kept saying, 'no'; and then I was really scared, because I can't describe, you know, what was said. It was more the look in his eyes; and I said, at that point—I didn't know what to say; and I said, 'If I do what you want, will you let me go without killing me?' Because I didn't know, at that point, what he was going to do; and I started to cry; and when I did, he put his hands on my throat, and started lightly to choke me; and I said, 'If I do what you want, will you let me go?' And he said, yes, and at that time, I proceeded to do what he wanted me to."

Pat testified that Rusk made her perform oral sex and then vaginal intercourse.

Immediately after the intercourse, Pat asked if she could leave. She testified that Rusk said, "'Yes,'" after which she got up and got dressed and Rusk returned her car keys. She said that Rusk then "walked me to my car, and asked if he could see me again; and I said, 'Yes'; and he asked me for my telephone number; and I said, 'No, I'll see you down Fells Point sometime,' just so I could leave." Pat testified that she "had no intention of meeting him again." She asked him for directions out of the neighborhood and left.

On her way home, Pat stopped at a gas station, went to the ladies room, and then drove "pretty much straight home and pulled up and parked the car." At first she was not going to say anything about the incident. She explained her initial reaction not to report the incident: "I didn't want to go through what I'm going through now [at the trial]." As she sat in her car reflecting on the incident, Pat said she began to "wonder what would happen if I hadn't of done what he wanted me to do. So I thought the right thing to do was to go report it, and I went from there to Hillendale to find a police car." She reported the incident to the police at about 3:15 A.M. Subsequently, Pat took the police to Rusk's apartment, which she located without any great difficulty. . . .

In argument before us on the merits of the case, the parties agreed that the issue was whether, in light of the principles of [*Hazel v. State*, 157 A.2d 922 (Md. 1960)] there was evidence before the jury legally sufficient to prove beyond a reasonable doubt that the intercourse was "[b]y force or threat of force against the will and without the consent" of the victim in violation of Art. 27, § 463(a)(1). . . .

The vaginal intercourse once being established, the remaining elements of rape in the second degree under § 463(a)(1) are, as in a prosecution for common law rape (1) force—actual or constructive, and (2) lack of consent. The terms in § 463(a)(1)—"force," "threat of force," "against the will" and "without the consent"—are not defined in the statute, but are to be afforded their "judicially determined meaning" as applied in cases involving common law rape. . . .

[*Hazel*] recognized that force and lack of consent are distinct elements of the crime of rape. [However, *Hazel* also] made it clear that lack of consent could be established through proof that the victim submitted as a result of fear of imminent death or serious bodily harm. In addition, if the actions and conduct of the defendant were reasonably calculated to induce this fear in the victim's mind, then the element of force is present. *Hazel* recognized, therefore, that the same kind of evidence may be used in establishing both force and non-consent, particularly when a threat rather than actual force is involved. . . .

*Hazel* did not expressly determine whether the victim's fear must be "reasonable." Its only reference to reasonableness related to whether "the acts and threats of the defendant were reasonably calculated to create in the mind of the victim . . . a real apprehension, due to fear, of imminent bodily harm. . . ."

Manifestly, the Court was there referring to the calculations of the accused, not to the fear of the victim. While *Hazel* made it clear that the victim's fear had to be genuine, it did not pass upon whether a real but unreasonable fear of imminent death or serious bodily harm would suffice. The vast majority of jurisdictions have required that the victim's fear be reasonably grounded in order to obviate the need for either proof of actual force on the part of the assailant or physical resistance on the part of the victim. We think that, generally, this is the correct standard. . . .

We think the reversal of Rusk's conviction by the Court of Special Appeals was in error. . . . In view of the evidence adduced at the trial, the reasonableness of Pat's apprehension of fear was plainly a question of fact for the jury to determine. . . . [I]t is readily apparent to us that the trier of fact could rationally find that the elements of force and non-consent had been established and that Rusk was guilty of the offense beyond a reasonable doubt. Of course, it was for the jury to observe the witnesses and their demeanor, and to judge their credibility and weigh their testimony. Quite obviously, the jury disbelieved Rusk and believed Pat's testimony. From her testimony, the jury could have reasonably concluded that the taking of her car keys was intended by Rusk to immobilize her alone, late at night, in a neighborhood with which she was not familiar; that after Pat had repeatedly refused to enter his apartment, Rusk commanded in firm tones that she do so; that Pat was badly frightened and feared that Rusk intended to rape her; that unable to think clearly and believing that she had no other choice in the circumstances, Pat entered Rusk's apartment; that once inside Pat asked permission to leave but Rusk told her to stay; that he then pulled Pat by the arms to the bed and undressed her; that Pat was afraid that Rusk would kill her unless she submitted; that she began to cry and Rusk then put his hands on her throat and began "'lightly to choke'" her; that Pat asked him if he would let her go without killing her if she complied with his demands; that Rusk gave an affirmative response, after which she finally submitted.

Just where persuasion ends and force begins in cases like the present is essentially a factual issue, to be resolved in light of the controlling legal precepts. That threats of force need not be made in any particular manner in order to put a person in fear of bodily harm is well established. *Hazel, supra*; *Dumer v. State*, 219 N.W.2d 592 (Wis. 1974). Indeed, conduct, rather than words, may convey the threat. *See People v. Benavidez*, 63 Cal. Rptr. 357 (1967). . . . That a victim did not scream out for help or attempt to escape, while bearing on the question of consent, is unnecessary where she is restrained by fear of violence. *See People v. Merritt*, 381 N.E.2d 407 (Ill. 1978). . . .

Considering all of the evidence in the case, with particular focus upon the actual force applied by Rusk to Pat's neck, we conclude that the jury could rationally find that the essential elements of second degree rape had been established and that Rusk was guilty of that offense beyond a reasonable doubt.

[Judgment reversed.]

Cole, Judge, dissenting. . . .

I find it incredible for the majority to conclude that on these facts, without more, a woman was forced to commit oral sex upon the defendant and then to engage in vaginal intercourse. In the absence of any verbal threat to do her grievous bodily harm or the display of any weapon and threat to use it, I find it difficult to understand how a victim could participate in these sexual activities and not be willing.

What was the nature and extent of her fear anyhow? She herself testified she was "fearful that maybe I had someone following me." She was afraid because she didn't know him and she was afraid he was going to "rape" her. But there are no acts or conduct on the part of the defendant to suggest that these fears were created by the defendant or that he made any objective, identifiable threats to her which would give rise to this woman's failure to flee, summon help, scream, or make physical resistance.

As the defendant well knew, this was not a child. This was a married woman with children, a woman familiar with the social setting in which these two actors met. It was an ordinary city street, not an isolated spot. He had not forced his way into her car; he had not taken advantage of a difference in years or any state of intoxication or mental or physical incapacity on her part. He did not grapple with her. She got out of the car, walked with him across the street and followed him up the stairs to his room. She certainly had to realize that they were not going upstairs to play Scrabble.

Once in the room she waited while he went to the bathroom where he stayed for five minutes. In his absence, the room was lighted but she did not seek a means of escape. She did not even "try the door" to determine if it was locked. She waited.

*(Continues)*

Upon his return, he turned off the lights and pulled her on the bed. There is no suggestion or inference to be drawn from her testimony that he yanked her on the bed or in any manner physically abused her by this conduct. As a matter of fact there is no suggestion by her that he bruised or hurt her in any manner, or that the "choking" was intended to be disabling.

He then proceeded to unbutton her blouse and her bra. He did not rip her clothes off or use any greater force than was necessary to unfasten her garments. He did not even complete this procedure but requested that she do it, which she did "because he asked me to." However, she not only removed her clothing but took his clothes off, too.

Then for a while they lay together on the bed kissing, though she says she did not return his kisses. However, without protest she then proceeded to perform oral sex and later submitted to vaginal intercourse. After these activities were completed, she asked to leave. They dressed and he walked her to her car and asked to see her again. She indicated that perhaps they might meet at Fells Point. He gave her directions home and returned to his apartment where the police found him later that morning.

The record does not disclose the basis for this young woman's misgivings about her experience with the defendant. The only substantive fear she had was that she would be late arriving home. The objective facts make it inherently improbable that the defendant's conduct generated any fear for her physical well-being.

In my judgment the State failed to prove the essential element of force beyond a reasonable doubt and, therefore, the judgment of conviction should be reversed.

Judges Smith and Digges have authorized me to state that they concur in the views expressed herein.

## *In re M.T.S.*, 609 A.2d 1266 (N.J. 1992)

Handler, J.

Under New Jersey law a person who commits an act of sexual penetration using physical force or coercion is guilty of second-degree sexual assault. The sexual assault statute does not define the words "physical force." The question posed by this appeal is whether the element of "physical force" is met simply by an act of nonconsensual penetration involving no more force than necessary to accomplish that result.

That issue is presented in the context of what is often referred to as "acquaintance rape." The record in the case discloses that the juvenile, a seventeen-year-old boy, engaged in consensual kissing and heavy petting with a fifteen-year-old girl and thereafter engaged in actual sexual penetration of the girl to which she had not consented. There was no evidence or suggestion that the juvenile used any unusual or extra force or threats to accomplish the act of penetration. . . .

On Monday, May 21, 1990, fifteen-year-old C.G. was living with her mother, her three siblings, and several other people, including M.T.S. and his girlfriend. A total of ten people resided in the three-bedroom town-home at the time of the incident. M.T.S., then age seventeen, was temporarily residing at the home with the permission of . . . C.G.'s mother; he slept downstairs on a couch. C.G. had her own room on the second floor. At approximately 11:30 P.M. on May 21, C.G. went upstairs to sleep after having watched television with her mother, M.T.S., and his girlfriend. When C.G. went to bed, she was wearing underpants, a bra, shorts, and a shirt. At trial, C.G. and M.T.S. offered very different accounts concerning the nature of their relationship and the events that occurred after C.G. had gone upstairs. The trial court did not credit fully either teenager's testimony.

C.G. stated that earlier in the day, M.T.S. had told her three or four times that he "was going to make a surprise visit up in [her] bedroom." She said that she had not taken M.T.S. seriously and considered his comments a joke because he frequently teased her. She testified that M.T.S. had attempted to kiss her on numerous other occasions and at least once had attempted to put his hands inside of her pants, but that she had rejected all of his previous advances.

C.G. testified that on May 22, at approximately 1:30 A.M., she awoke to use the bathroom. As she was getting out of bed, she said, she saw M.T.S., fully clothed, standing in her doorway. According to C.G., M.T.S. then said that "he was going to tease [her] a little bit." C.G. testified that she "didn't think anything of it"; she walked past him, used the bathroom, and then returned to bed, falling into a "heavy" sleep within fifteen minutes. The next event C.G. claimed to recall of that morning was waking up with M.T.S. on top of her, her underpants and shorts removed. She said "his penis was into [her] vagina." As soon as C.G. realized what had happened, she said, she immediately slapped M.T.S.

once in the face, then "told him to get off [her], and get out." She did not scream or cry out. She testified that M.T.S. complied in less than one minute after being struck; according to C.G., "he jumped right off of [her]." She said she did not know how long M.T.S. had been inside of her before she awoke.

C.G. said that after M.T.S. left the room, she "fell asleep crying" because "[she] couldn't believe that he did what he did to [her]." She explained that she did not immediately tell her mother or anyone else in the house of the events of that morning because she was "scared and in shock." According to C.G., M.T.S. engaged in intercourse with her "without [her] wanting it or telling him to come up [to her bedroom]." By her own account, C.G. was not otherwise harmed by M.T.S.

At about 7:00 A.M., C.G. went downstairs and told her mother about her encounter with M.T.S. earlier in the morning and said that they would have to "get [him] out of the house." While M.T.S. was out on an errand, C.G.'s mother gathered his clothes and put them outside in his car; when he returned, he was told that "[he] better not even get near the house." C.G. and her mother then filed a complaint with the police.

According to M.T.S., he and C.G. had been good friends for a long time, and their relationship "kept leading on to more and more." He had been living at C.G.'s home for about five days before the incident occurred; he testified that during the three days preceding the incident they had been "kissing and necking" and had discussed having sexual intercourse. The first time M.T.S. kissed C.G., he said, she "didn't want him to, but she did after that." He said C.G. repeatedly had encouraged him to "make a surprise visit up in her room."

M.T.S. testified that at exactly 1:15 A.M. on May 22, he entered C.G.'s bedroom as she was walking to the bathroom. He said C.G. soon returned from the bathroom, and the two began "kissing and all," eventually moving to the bed. Once they were in bed, he said, they undressed each other and continued to kiss and touch for above five minutes. M.T.S. and C.G. proceeded to engage in sexual intercourse. According to M.T.S., who was on top of C.G., he "stuck it in" and "did it [thrust] three times, and then the fourth time [he] stuck it in, that's when [she] pulled [him] off of her." M.T.S. said that as C.G. pushed him off, she said, "stop, get off," and he "hopped off right away."

According to M.T.S., after about one minute, he asked C.G. what was wrong; she replied with a back-hand to his face. He recalled asking C.G. what was wrong a second time, and her replying, "how can you take advantage of me or something like that." M.T.S. said that he proceeded to get dressed and told C.G. to calm down, but that she then told him to get away from her and began to cry. Before leaving the room, he told C.G., "I'm leaving . . . I'm going with my real girlfriend, don't talk to me . . . I didn't want nothing to do with you or anything, stay out of my life . . . don't tell anybody about this . . . it would just screw everything up." He then walked downstairs and went to sleep.

On May 23, 1990, M.T.S. was charged with conduct that if engaged in by an adult would constitute second-degree sexual assault of the victim, contrary to *N.J.S.A.* 2C:14-2c(1). . . .

Following a two-day trial on the sexual assault charge, M.T.S. was adjudicated delinquent. . . .

The New Jersey Code of Criminal Justice, *N.J.S.A.* 2C:14-2c(1), defines "sexual assault" as the commission "of sexual penetration" "with another person" with the use of "physical force or coercion." An unconstrained reading of the statutory language indicates that both the act of "sexual penetration" and the use of "physical force or coercion" are separate and distinct elements of the offense. *See Medical Soc. v. Department of Law & Pub. Safety,* 575 A.2d 1348 (N. J. 1990) (declaring that no part of a statute should be considering meaningless or superfluous). Neither the definitions section of *N.J.S.A.* 2C:14-1 to -8, nor the remainder of the Code of Criminal Justice provides assistance in interpreting the words "physical force." . . .

The parties offer two alternative understandings of the concept of "physical force" as it is used in the statute. The State would read "physical force" to entail any amount of sexual touching brought about involuntarily. A showing of sexual penetration coupled with a lack of consent would satisfy the elements of the statute. The Public Defender urges an interpretation of "physical force" to mean force "used to overcome lack of consent." That definition equates force with violence and leads to the conclusion that sexual assault requires the application of some amount of force in addition to the act of penetration. . . .

[After reviewing historic changes in the "force" requirement, the Court examined changes made to New Jersey sexual assault law by the State legislature.] The Legislature's concept of sexual assault and the role of force was significantly colored by its understanding of the law of assault and battery. As a general matter, criminal battery is defined as "the unlawful application of force to the person of another." . . . The application of force is criminal when it results in either (a) a physical injury or

(*Continues*)

(b) an offensive touching. . . . Any unauthorized touching of another [is] a battery." *Perna v. Pirozzi,* 457 A.2d 431 (N.J. 1983). Thus, by eliminating all references to the victim's state of mind and conduct, and by broadening the definition of penetration to cover not only sexual intercourse between a man and a woman but a range of acts that invade another's body or compel intimate contact, the Legislature emphasized the affinity between sexual assault and other forms of assault and battery. . . .

Because the statute eschews any reference to the victim's will or resistance, the standard defining the role of force in sexual penetration must prevent the possibility that the establishment of the crime will turn on the alleged victim's state of mind or responsive behavior. We conclude, therefore, that any act of sexual penetration engaged in by the defendant without the affirmative and freely-given permission of the victim to the specific act of penetration constitutes the offense of sexual assault. Therefore, physical force in excess of that inherent in the act of sexual penetration is not required for such penetration to be unlawful. The definition of "physical force" is satisfied under *N.J.S.A.* 2C:14-2c(1) if the defendant applies any amount of force against another person in the absence of what a reasonable person would believe to be affirmative and freely-given permission to the act of sexual penetration.

Under the reformed statute, permission to engage in sexual penetration must be affirmative and it must be given freely, but that permission may be inferred either from acts or statements reasonably viewed in light of the surrounding circumstances. Persons need not, of course, expressly announce their consent to engage in intercourse for there to be affirmative permission. Permission to engage in an act of sexual penetration can be and indeed often is indicated through physical actions rather than words. Permission is demonstrated when the evidence, in whatever form, is sufficient to demonstrate that a reasonable person would have believed that the alleged victim had affirmatively and freely given authorization to the act. . . .

Today the law of sexual assault is indispensable to the system of legal rules that assures each of us the right to decide who may touch our bodies, when, and under what circumstances. The decision to engage in sexual relations with another person is one of the most private and intimate decisions a person can make. Each person has the right not only to decide whether to engage in sexual contact with another, but also to control the circumstances and character of that contact. No one, neither a spouse, nor a friend, nor an acquaintance, nor a stranger, has the right or the privilege to force sexual contact.

We emphasize as well that what is now referred to as "acquaintance rape" is not a new phenomenon. Nor was it a "futuristic" concept in 1978 when the sexual assault law was enacted. Current concern over the prevalence of forced sexual intercourse between persons who know one another reflects both greater awareness of the extent of such behavior and a growing appreciation of its gravity. Notwithstanding the stereotype of rape as a violent attack by a stranger, the vast majority of sexual assaults are perpetrated by someone known to the victim. . . . Similarly, contrary to common myths, perpetrators generally do not use guns or knives and victims generally do not suffer external bruises or cuts. Although this more realistic and accurate view of rape only recently has achieved widespread public circulation, it was a central concern of the proponents of reform in the 1970s. . . .

In a case such as this one, in which the State does not allege violence or force extrinsic to the act of penetration, the factfinder must decide whether the defendant's act of penetration was undertaken in circumstances that led the defendant reasonably to believe that the alleged victim had freely given affirmative permission to the specific act of sexual penetration. Such permission can be indicated either through words or through actions that, when viewed in the light of all the surrounding circumstances, would demonstrate to a reasonable person affirmative and freely-given authorization for the specific act of sexual penetration.

In applying that standard to the facts in these cases, the focus of attention must be on the nature of the defendant's actions. . . . The role of the factfinder is [to decide] whether the defendant's belief that the alleged victim had freely given affirmative permission was reasonable. . . .

In short, in order to convict under the sexual assault statute in cases such as these, the State must prove beyond a reasonable doubt that there was sexual penetration and that it was accomplished without the affirmative and freely-given permission of the alleged victim. . . . If there is evidence to suggest that the defendant reasonably believed that such permission had been given, the State must demonstrate either that defendant did not actually believe that affirmative permission had been freely-given or that such a belief was unreasonable under all of the circumstances. Thus, the State bears the burden of proof throughout the case.

. . . Under the reformed statute, a person's failure to protest or resist cannot be considered or used as justification for bodily invasion.

We acknowledge that cases such as this are inherently fact sensitive and depend on the reasoned judgment and common sense of judges and juries. The trial court concluded that the victim had not expressed consent to the act of intercourse, either through her words or actions. We conclude that the record provides reasonable support for the trial court's disposition.

Accordingly, we reverse the judgment of the Appellate Division and reinstate the disposition of juvenile delinquency for the commission of second-degree sexual assault.

## Notes and Questions

1. Under Maryland law the prosecution in *Rusk* needed to prove beyond a reasonable doubt: (a) an act of intercourse; (b) "by force or threat of force"; that was (c) against the "will" or "consent" of the victim. Despite the fact that force and nonconsent are legally distinct elements, courts typically look to the same facts to evaluate nonconsent and force and define the elements interchangeably. However, analytically and conceptually, the prosecution must establish both actual/threatened force and lack of the victim's consent, and you should keep their differences in mind.

   Assume you serve as defense counsel for Rusk. Given that the act of intercourse was admitted, what facts would you emphasize to convince the fact-finder that the prosecution failed to establish your client's guilt beyond a reasonable doubt?

2. Under Maryland law "threat of force" is a legally sufficient basis to overcome the victim's "will." If fear suffices, and this fear must be "reasonable," how are lines to be drawn? Should a "look" in the defendant's eyes suffice? Should it matter that the victim experiencing such a "look" actually was in fear, perhaps as a result of a prior victimization, when another person might not be so affected? Who has the better argument in *Rusk*, the majority or the dissenting opinion? Note that the Maryland Court of Appeals divided 4–3 regarding the sufficiency of the evidence in this case. The lower court had reversed Rusk's rape conviction by vote of 8–5. *Rusk v. State*, 406 A.2d 624 (Md. App. 1979).

3. In *In re M.T.S.*, New Jersey adopted a far less demanding standard, one that still requires lack of consent yet nothing more than the physical force associated with the act of penetration. The rule adopted by the New Jersey court sends a clear message: Potential sexual assailants are put on notice that the "force" requirement is satisfied "if the defendant applies any amount of force" in the absence of "affirmative and freely given permission to the act of sexual penetration." Although perhaps laudable for its certainty, does the "affirmative" permission requirement give rise to practical problems? Does it assist the trier of fact that permission can also be inferred, in the court's words, "through physical actions rather than words"? What does the *M.T.S.* court mean by this statement and what possible line-drawing problems might arise, given the steamy context in which such "signals" might be sent (and received)? Why wasn't the "heavy petting" consented to in *M.T.S.* a sufficient signal to proceed to sexual intercourse?

4. The controversy relating to the "resistance requirement" springs from an age-old question in the substantive criminal law: Should the law only reflect social norms or instead seek to alter or promote the evolution of social norms? Sex crimes represent an especially difficult realm. As one court has observed, "[m]any men have been conditioned to believe that initial refusals [by women] are an essential part of the 'mating game' ritual which dictates that women must resist somewhat to make themselves more attractive to men. . . ." *Deborah S. v. Diorio*, 583 N.Y.S.2d 872, 877 (Misc. 1992).

   If there is any truth to this view, should the law take into account the influence of this historic pattern of socialization? Is the law so forgiving with respect to other crimes examined in this book? Alternatively, should the law play a prescriptive role and punish those who have succumbed to such socialization? What social costs attach to pursuing either route?

5. Sexual assault remains a serious problem, with untold personal consequences. However, does a decision like *M.T.S.* go too far? Does it require us to micro-manage sexuality—highly intimate behaviors that perhaps do not lend themselves to clear lines of impropriety? If society wishes to go down this road, which gender should shoulder the burden of communication? Professor Donald Dripps frames the issue as follows:[9]

   > The burden of asking permission can be placed on the man, or the burden of expressing refusal can be placed on the woman. Granting that gender prejudice is implicated by either choice, I think the second alternative is superior, because sexual encounters ought not to be lived or analyzed as sequences of particular touchings.

   Do you agree with Professor Dripps's conclusion? Is it consistent with *M.T.S.*?

6. Should the victim's behavior immediately after the sexual encounter be considered relevant to whether a rape may have occurred? In one case the court found it significant that the victim took the time to smoke a cigarette after intercourse and further told the defendant that it was "up to him"

*(Continues)*

(Continued)

> when he asked whether she was ready to leave the premises. *People v. Bain*, 283 N.E.2d 701, 703 (Ill. 1972). Should such behaviors have any bearing on the issue of whether the legal elements were satisfied at the time of the alleged sexual assault?
>
> 7. Does the physical resistance requirement have any value? To the extent that physical resistance serves as an outward manifestation of a lack of consent, some have argued that the resistance requirement is helpful in making the consent determination. At the same time a blanket legal requirement would create obvious and perhaps quite dangerous problems. Should it matter that we do not require resistance to be exhibited with respect to other crimes that also involve lack of consent (*e.g.*, battery and robbery)?

## Mens Rea

An important issue in the substantive law of rape involves the accused's mental state. Given that consent plays such a central role in rape prosecutions, must the state prove that the defendant engaged in the sexual act while aware that the victim did not consent? In most jurisdictions the answer is no—rape is viewed as a "general intent" crime, as opposed to one requiring that the defendant possess a specific intent to have sex without the consent of the victim. (See Chapter 3, which discusses general and specific intent.)

A similar issue arises in the context of the defense of "mistake." Can the defendant avoid liability because he was mistaken about the victim's willingness to have sex? Under the majority rule, a reasonable mistake as to the victim's consent can exculpate the defendant. Indeed, the New Jersey court in *M.T.S.*, discussed above, adopted such a view. According to the *M.T.S.* court,

> If there is evidence to suggest that the defendant reasonably believed that . . . permission had been given, the State must demonstrate either that the defendant did not actually believe that affirmative permission had been freely-given or that such belief was unreasonable under all of the circumstances. *In re M.T.S.*, 609 A.2d 1266, 1279 (N.J. 1992).

Under this approach, the fact-finder must assess whether "the words or conduct of the complainant under all the circumstances would justify a reasonable belief that [the victim] had consented." *State v. Smith*, 554 A.2d 713, 717 (Conn. 1989).

Jurisdictions not adopting the majority view pursue one of two routes. First, a few take an especially hard-line approach, allowing liability even when the defendant operated under an honest and reasonable mistake about the victim's consent. *See, e.g., Commonwealth v. Williams*, 439 A.2d 765 (Pa. 1982). On the other extreme, some courts have held that even an unreasonable belief concerning the victim's consent can serve to exculpate the defendant, as long as the belief is honestly held. *See, e.g., Reynolds v. State*, 664 P.2d 621 (Alaska App. 1983) (holding that the state must at least prove that the defendant acted "recklessly" regarding the victim's consent). Such was the view of the British House of Lords in the infamous case of *Director of Public Prosecutions v. Morgan*, 2 All E.R. 347 (H.L. 1975). In *Morgan* a married woman was aroused from her sleep by three men, who were drinking companions of her husband. She was "held by her limbs" while each of the three had sexual intercourse with her, all the while making "her opposition to what was being done very plain indeed." The three male defendants raised a mistake of consent defense, testifying that the husband had suggested that the encounter take place and that "they must not be surprised if his wife struggled a bit, since she was 'kinky' and this was the only way in which she could get 'turned on.'" The House of Lords agreed that in a rape prosecution, the defendants' honest (*i.e.*, actual) belief about whether the alleged victim had consented should be controlling and would constitute a defense if accepted by the jury.

## Special Note: The "Marital Rape" Exemption

One of the many noteworthy aspects of the evolving history of the law of rape involves what is known as the "**marital rape**" exemption. The exemption stems from the ancient common law that viewed the legal institution of marriage as barring the state from prosecuting a husband for rape, even if intercourse was accomplished against the wife's consent.

By the mid-1980s, however, many states reconsidered the rationale underlying the marital rape exemption. The next case, *State v. Smith*, was one of the first decisions to reevaluate legal and social tenets underpinning the marital rape exemption.

### State v. Smith, 426 A.2d 38 (N.J. 1981)

Pashman, J.

Since the enactment in New Jersey of the new Code of Criminal Justice, *N.J.S.A.* 2C:1-1 to 984, *L.* 1978, *c.* 95, no person can claim that a sexual assault committed after the effective date of the code, September 1, 1979, is exempt from prosecution because the accused and victim were husband and wife. The Criminal Code expressly excludes marriage to the victim as a defense against prosecution of sexual crimes. *N.J.S.A.* 2C:14-5(b). The criminal acts alleged in this case, however, occurred before the effective date of the Code. The issue before the Court is whether a defendant can be charged with and convicted of raping his wife under the former statute, *N.J.S.A.* 2A:138-1. We hold that, at least under the circumstances of this case, he can.

## I.

The State alleged that on October 1, 1975, defendant Albert Smith broke into the apartment of his estranged wife, Alfreda Smith, and repeatedly beat and raped her. On that date the accused and victim were legally married. They had been married for seven years but had lived separately for approximately one year. [The court then summarized the record, concluding that the accused and the victim were not legally separated, and neither had filed a complaint for divorce.] . . .

At the time of the alleged incident on October 1, 1975, defendant and his wife lived in different cities. The State accuses defendant of arriving at his wife's apartment at about 2:30 A.M., breaking through two doors to get inside, and once there threatening, choking and striking her. According to the State, over a period of a few hours he repeatedly beat her, forced her to have sexual intercourse and committed various other atrocities against her person. As a result of these alleged attacks, Alfreda Smith required medical care at a hospital.

After hearing testimony on December 11, 1975, and January 5, 1976, the Essex County Grand Jury returned an indictment charging defendant with four separate counts—atrocious assault and battery, private lewdness, impairing the morals of a minor, and a rape. Defendant moved to dismiss the rape charge on the ground that he was legally married to the victim at the time of the incident. The trial judge reluctantly granted the motion. He believed that the common law included a marital exemption from the crime of rape, which was implicitly incorporated into this State's statutory definition of rape from early Revolutionary times to the present. Although the trial judge expressed unequivocal disapproval of such an anachronistic rule of law, he considered it the prerogative of the Legislature to change it. . . .

We granted the State's petition for certification, 82 *N.J.* 292, 412 *A.2d* 798 (1980), to consider the reach of our former rape statute.

## II.

The rape statute under which defendant was charged provided in part:

> Any person who has carnal knowledge of a woman forcibly against her will . . . is guilty of a high misdemeanor and shall be punished by a fine of not more than $5,000, or by imprisonment for not more than 30 years, or both. . . . [*N.J.S.A.* 2A:138-1 (repealed)]

* * *

The first State Constitution of New Jersey provided for a limited incorporation of English common and statutory law in existence at that time. . . . The question in this case is whether a marital exemption from rape was, at the time of defendant's conduct, one such rule. To answer that question, we must first consider whether there actually existed a marital exemption rule under pre-Revolutionary common law.

## A.

Sir Matthew Hale, a seventeenth century English jurist, wrote a treatise on English law which is invariably cited as authority for the rule. Hale discussed the crime of rape and possible defenses, stating:

> But the husband cannot be guilty of a rape committed by himself upon his lawful wife, for by their mutual matrimonial consent and contract the wife hath given up herself in his kind unto her husband, which she cannot retract. [Hale, *History of the Pleas of the Crown* *629]

Hale cited no authority for this proposition and we have found none in earlier writers. Thus the marital exemption rule expressly adopted by many of our sister states has its source in a bare, extra-judicial declaration made some 300 years ago. Such a declaration cannot itself be considered a definitive and binding statement of the common law, although legal commentators have often restated the rule since the time of Hale without evaluating its merits.

We need not decide, however, the broad question of whether a marital exemption existed under English common law. The narrower question here is whether such a marital exemption, even if it existed, would have applied inflexibly for as long as a marriage continued to exist in the legal sense. We think not.

*(Continues)*

*(Continued)*

We believe that Hale's statements concerning the common law of spousal rape derived from the nature of marriage at a particular time in history. Hale stated the rule in terms of an implied matrimonial consent to intercourse which the wife could not retract. This reasoning may have been persuasive during Hale's time, when marriages were effectively permanent, ending only by death or an act of Parliament. Since the matrimonial vow itself was not retractable, Hale may have believed that neither was the implied consent to conjugal rights. Consequently, he stated the rule in absolute terms, as if it were applicable without exception to all marriage relationships. In the years since Hale's formulation of the rule, attitudes towards the permanency of marriage have changed and divorce has become far easier to obtain. The rule, formulated under vastly different conditions, need not prevail when those conditions have changed. . . .

To summarize our view of the marital exemption under English common law, we think that the existence of the rule is not as obvious as the lower courts here or courts in other jurisdictions have believed. . . . . The fact that many jurisdictions have mechanically applied the rule, without evaluating its merits under changed conditions, does not mean that such blind application was part of the "principles of the common law" adopted in this State. . . . Therefore, we decline to apply mechanically a rule whose existence is in some doubt and which may never have been intended to apply to the factual situation presented by this case.

## B.

Having determined that the common law did not include an absolute marital exemption from prosecution for rape under all conditions, we next consider what rule of exemption, if any, did exist as the law in this State at the time of these alleged criminal acts.

The inquiry must begin with enactment of the State's first rape statute in 1796. That statute provided in part:

> That any person who shall have carnal knowledge of a woman, forcibly and against her will, . . . shall, on conviction, be adjudged guilty of a high misdemeanor. . . . [L. 1796, An Act for the Punishment of Crimes, § 8]

This statutory language is almost identical to N.J.S.A. 2A:138-1, the applicable statute in this case. The definition of **forcible rape** seems to have come directly from Blackstone's eighteenth century definition. . . . Although discussing other statements by Hale, Blackstone never mentions a marital exemption. Thus, we do not know whether the 1796 Legislature intended to incorporate implicitly any form of a marital exemption within the statutory definition of rape. However, we need not answer that question for the purposes of this case. Our concern here is whether a marital exemption, even if it existed in this State, covered defendant's conduct. . . .

[W]e will assume for the purpose of our inquiry that some form of a marital exemption did exist in eighteenth century New Jersey and next consider whether that exemption was the rule in 1975, when defendant allegedly raped his wife.

A common law rule of marital exemption was probably based on three major justifications, which might have constituted the common law principles adopted in this State. The first of these, the notion that a woman was the property of her husband or father . . . was never valid in this country. Rape laws may originally have protected a woman's chastity and therefore her value to her father or husband. In this State, however, rape statutes have always aimed to protect the safety and personal liberty of women. Thus, the common law "principle" that a wife is her husband's chattel does not support the view that our rape statute included a marital exemption.

Second, the common law once included the concept that a husband and wife were one person, that after marriage a man and woman no longer retained separate legal existence. As a result of this concept, some have argued that a husband could not be convicted of, in effect, raping himself. This argument does not take into account how, in spite of marital unity, a husband could always be convicted of other crimes upon his wife, such as assault and battery. . . .

Furthermore, even if the argument had validity at one time, the "principle" of marital unity was discarded in this State long before the commission of defendant's alleged crime. *See N.J.S.A.* 37:2-1 to -30 (Married Woman's Acts, enacted in the nineteenth and early twentieth centuries, giving married women rights to sue and be sued, own property, and enter contracts separately from their husbands). . . .

The third and most prevalent justification for the exemption rule is the one utilized by Hale himself—that upon entering the marriage contract a wife consents to sexual intercourse with her husband. This irrevocable consent negates the third essential element of the crime of rape, lack of consent. . . . We cannot say with certainty whether such a rationale was justified even in the seventeenth century. . . .

More importantly, this implied consent rationale, besides being offensive to our valued ideals of personal liberty, is not sound where the marriage itself it not irrevocable. If a wife can exercise a legal right to separate from her husband and eventually terminate the marriage "contract," may she not also revoke a "term" of that contract, namely, consent to intercourse? Just as a husband has no right to imprison his wife because of her marriage vow to him . . . he has no right to force sexual relations upon her against her will. If her repeated refusals are a "breach" of the marriage "contract," his remedy is in a matrimonial court, not in violent or forceful self-help. . . .

Since the common law exemption supposedly operated by negating an essential element of the crime—lack of consent—it could not be applied where marital consent to sexual intercourse could be legally revoked. By 1975 our matrimonial laws recognized the right of a wife to withdraw consent prior to the dissolution of a marriage and even prior to a formal judicial order of separation.

This conclusion is consistent with those cases or statutes which have refused to apply the exemption rule after a judicial decree of separation has been issued. Such cases have reasoned that a legal process which originally created the marital state has stepped in to terminate it. . . . But the wife could not rely on her unilateral decision outside the legal process to mark the end of the marriage, for unless she could allege and prove proper grounds, a court would probably decline to terminate the marriage. . . . The legal setting in this State is different. Since the advent of the no-fault ground for divorce, *see N.J.S.A.* 2A:34-2(d), any spouse may make a unilateral decision to end the marriage. By separating from her husband and living apart for 18 months, a wife is entitled to a divorce without further proof of proper grounds. The corollary of this right is that a wife can refuse sexual intercourse with her husband during the period of separation. If a wife has a right to refuse intercourse, or deny consent, then a husband's forceful carnal knowledge of his wife clearly includes all three essential elements of the crime of rape. He cannot defend by asserting that there was no lack of consent because he was still legally married to the victim.

Our no-fault divorce law has existed since 1971, several years before the parties here separated and before the alleged acts of defendant. Defendant cannot claim that in 1975 his wife consented to have sexual intercourse with him although they were living separately. . . .

With none of the three major common law justifications viable under our laws at the time of defendant's conduct, it would be irrational to believe that a common law marital exemption for rape endured in this State while other laws had changed so dramatically. . . . Therefore, we hold that this State did not have a marital exemption rule for rape in 1975 that would have applied to this defendant and prevented his indictment and conviction on the charge of raping his wife.

## Notes and Questions

1. Today, the marital rape exemption has been largely, but not totally, abandoned.[10] Idaho law, for instance, provides that "[n]o person shall be convicted of rape for any act . . . with that person's spouse, except under [specified] circumstances. . . ." *See* Idaho Code § 18-6107 (1999). In South Carolina, subject to certain exceptions, "[a] person cannot be guilty of criminal sexual conduct . . . if the victim is the legal spouse unless the couple is living apart"; the law also requires that the victim's spouse report the violation within 30 days, in contrast to the ordinary rape reporting requirement of 1 year. *See* S.C. Code Ann. § 16-3-658 (1999).

   Is the marital exemption defensible, given the emotional turbulence that often accompanies marriages? Again, is this an area that is simply too complex and sensitive for the criminal law? More practically, some commentators suggest that the threat of prosecution only worsens already unstable marriages, making reconciliation even less likely. Others worry that the threat of criminal prosecution becomes a bargaining chip or object of threat used by embattled spouses. Are these arguments persuasive in the face of the obvious personal autonomy issues at stake for the victimized spouse?

2. What is the magnitude of harm associated with marital rape? Does it compare to rape involving unmarried acquaintances or rape committed by a stranger?[11]

3. The decision in *People v. Liberta,* 474 N.E.2d 567 (N.Y. 1984) questioned the continued use of the marital rape exemption on another basis. Liberta had been legally separated from his wife and was subject to a court order to keep away from her at the time that he allegedly raped her. Because he was living separately from his wife, Liberta could be prosecuted for rape, despite the fact that New York law recognized the marital rape exemption for men living with their wives. On appeal of his rape conviction, Liberta successfully argued that the law violated the Fourteenth Amendment's Equal Protection Clause (see Chapter 2) because it burdened "some, but not all males (all but those within the 'marital exemption')." Liberta did not have his conviction overturned as a result of this ruling. Rather, the New York Court of Appeals invalidated the marital rape exemption in its entirety.

## Use of Trickery and Psychological Coercion

It is well established that either threatened or actual physical force can suffice to overcome the will of the victim, hence invalidating any expression of "consent." Should the criminal law condemn, as well, behaviors calculated to obtain sex by fraud or deceit, or the threat of unpleasant but nonviolent consequences? Although comparatively infrequent, such behaviors have been addressed by appellate courts over the years, raising an array of intriguing legal questions.

### Trickery

---

### *Boro v. Superior Court*, 163 Cal. App.3d 1224, 210 Cal. Rptr. 122 (1985)

Newsom, Associate Justice.

By timely petition filed with this court, petitioner Daniel Boro seeks a writ of prohibition to restrain further prosecution of Count II of the information on file against him . . . charging him with a violation of Penal Code section 261, subdivision (4), rape: "an act of sexual intercourse accomplished with a person not the spouse of the perpetrator, under any of the following circumstances: . . . (4) Where a person is at the time unconscious of the nature of the act, and this is known to the accused." . . .

In relevant part the factual background may be summarized as follows. Ms. R., the rape victim, was employed as a clerk at the Holiday Inn in South San Francisco when, on March 30, 1984, at about 8:45 A.M., she received a telephone call from a person who identified himself as "Dr. Stevens" and said that he worked at Peninsula Hospital.

"Dr. Stevens" told Ms. R. that he had the results of her blood test and that she had contracted a dangerous, highly infectious and perhaps fatal disease; that she could be sued as a result; that the disease came from using public toilets; and that she would have to tell him the identity of all her friends who would then have to be contacted in the interest of controlling the spread of the disease.

"Dr. Stevens" further explained that there were only two ways to treat the disease. The first was a painful surgical procedure—graphically described—costing $9,000, and requiring her uninsured hospitalization for six weeks. A second alternative, "Dr. Stevens" explained, was to have sexual intercourse with an anonymous donor who had been injected with a serum which would cure the disease. The latter, nonsurgical procedure would only cost $4,500. When the victim replied that she lacked sufficient funds the "doctor" suggested that $1,000 would suffice as a down payment. The victim thereupon agreed to the non-surgical alternative and consented to intercourse with the mysterious donor, believing "it was the only choice I had."

After discussing her intentions with her work supervisor, the victim proceeded to the Hyatt Hotel in Burlingame as instructed, and contacted "Dr. Stevens" by telephone. The latter became furious when he learned Ms. R. had informed her employer of the plan, and threatened to terminate his treatment, finally instructing her to inform her employer she had decided not to go through with the treatment. Ms. R. did so, then went to her bank, withdrew $1,000 and, as instructed, checked into another hotel and called "Dr. Stevens" to give him her room number.

About a half hour later the defendant "donor" arrived at her room. When Ms. R. had undressed, the "donor," petitioner, after urging her to relax, had sexual intercourse with her.

At the time of penetration, it was Ms. R.'s belief that she would die unless she consented to sexual intercourse with the defendant: as she testified, "My life felt threatened, and for that reason and that reason alone did I do it."

Petitioner was apprehended when the police arrived at the hotel room, having been called by Ms. R.'s supervisor. Petitioner was identified as "Dr. Stevens" at a police voice lineup by another potential victim of the same scheme. . . .

Our research discloses sparse California authority on the subject. A victim need not be totally and physically unconscious in order that section 261, subdivision (4) apply. In *People v. Minkowski* (1962), 23 Cal.Rptr. 92, the defendant was a physician who "treated" several victims for menstrual cramps. Each victim testified that she was treated in a position with her back to the doctor, bent over a table, with feet apart, in a dressing gown. And in each case the "treatment" consisted of the defendant first inserting a metal instrument, then substituting an instrument which "felt different"—the victims not realizing that the second instrument was in fact the doctor's penis. The precise issue before us was never tendered in *People v. Minkowski* because the petitioner there *conceded* the sufficiency of evidence to support the element of consciousness.

The decision is useful to this analysis, however, because it exactly illustrates certain traditional rules in the area of our inquiry. Thus, as a leading authority has written, "if deception causes a

---

misunderstanding as to the fact itself (**fraud in the *factum***) there is no legally-recognized consent because what happened is not that for which consent was given; whereas consent induced by fraud is as effective as any other consent, so far as direct and immediate legal consequences are concerned, if the deception relates not to the thing done but merely to some collateral matter (**fraud in the inducement**)." (Perkins & Boyce, Criminal Law (3d ed. 1982) ch. 9, § 3, p. 1079.)

The victims in *Minkowski* consented, not to sexual intercourse, but to an act of an altogether different nature, penetration by medical instrument. The consent was to a pathological, and not a carnal, act, and the mistake was, therefore, in the *factum* and not merely in the inducement.

Another relatively common situation in the literature on this subject—discussed in detail by Perkins (*supra*, at p. 1080) is the fraudulent obtaining of intercourse by impersonating a spouse. As Professor Perkins observes, the courts are not in accord as to whether . . . the crime of rape is thereby committed. "[T]he disagreement is not in regard to the underlying principle but only as to its application. Some courts have taken the position that such a misdeed is fraud in the inducement on the theory that the woman consents to exactly what is done (sexual intercourse) and hence there is no rape; other courts, with better reason it would seem, hold such a misdeed to be rape on the theory that it involves fraud in the *factum* since the woman's consent is to an innocent act of marital intercourse while what is actually perpetrated upon her is an act of adultery." . . .

In California, of course, we have by statute adopted the majority view that such fraud is in the *factum,* not the inducement, and have thus held it to vitiate consent. It is otherwise, however, with respect to the conceptually much murkier statutory offense with which we here deal, and the language of which has remained essentially unchanged since its enactment.

The language itself could not be plainer. It defines rape to be "an act of sexual intercourse" with a non-spouse, accomplished where the victim is "at the time unconscious of the nature of the act. . . ." (§ 261, subd. (4).) Nor, as we have just seen, can we entertain the slightest doubt that the Legislature well understood how to draft a statute to encompass fraud in the *factum* (§ 261, subd. (5)) and how to specify certain fraud in the inducement as vitiating consent. Moreover, courts of this state have previously confronted the general rule that fraud in the inducement does not vitiate consent. . . . If the Legislature . . . had desired to correct the [gap in the law,] (*see* Footnote 5) it could certainly have done so.

To so conclude is not to vitiate the heartless cruelty of petitioner's scheme, but to say that it comprised crimes of a different order than a violation of section 261, subdivision (4). . . .

## Notes and Questions

1. The California court in *Boro* draws a critical distinction between "fraud in the *factum*" and "fraud in the inducement." The harms ultimately suffered by victims in both scenarios could well be similar. What justifies the legal distinction? Does the distinction, in effect, make an exception for gullible victims? What message does it send to potential victims and defendants? Do you agree with the reasoning and the result in *Boro*? Following *Boro*, the California legislature enacted the following statute: "Every person who induces another person to engage in sexual intercourse, sexual penetration, oral copulation, or sodomy when his or her consent is procured by false or fraudulent representation or pretense that is made with the intent to create fear, and which does induce fear, and that would cause a reasonable person in like circumstances to act contrary to the person's free will, and does cause the victim to so act, is punishable by" one to four years imprisonment. Cal. Penal Code § 266c (2001). Would Boro be guilty under this provision? Is such legislation a step in the right direction? What does it mean to "cause a reasonable person in like circumstances to act contrary to the person's free will"?

2. Perhaps because of the perceived unfairness in cases such as *Boro,* the fine legal distinctions made there with respect to fraud are showing signs of wear. In a notable Tennessee case, for instance, a 17-year-old boy sought out a physician in the hope of obtaining steroids. *See State v. Tizard*, 897 S.W.2d 732 (Tenn. Crim. App. 1994). On several office visits the doctor fondled the boy's penis, one time resulting in ejaculation. The boy testified that "he was embarrassed and thought what the defendant did . . . was not right" yet continued the visits without telling his parents because he "wanted steroids."

   The doctor was convicted of sexual battery and on appeal argued that the conviction was erroneous because the boy was not under any misunderstanding as to the sexual encounter, and indeed manifested

*(Continues)*

(*Continued*)

his willingness by his repeated visits to the doctor. The Tennessee court disagreed, adopting a broad definition of fraud. According to the *Tizard* court

> the definition of fraud . . . is not limited to any particular type of fraud. [F]raud comprises "anything calculated to deceive, including all acts, omissions, and concealments involving a breach of legal or equitable duty, trust, or confidence justly reposed, resulting in damage to another, or by which an undue . . . advantage is taken of another." *Id.* at 741 (citation omitted).

Although the touching was carried out, in the court's words, under the fraudulent "guise of medical examination," does the victim's testimony undercut the conclusion that he was tricked? In light of *Tizard*, are there advantages to the distinction used in *Boro*?

3. In *State v. Bolsinger*, 709 N.W.2d 560 (Iowa 2006), the defendant, a counselor in a juvenile correction facility for delinquent boys, touched the genitals of several of the boys "saying he was checking for bruises, scratches, hernias, and testicular cancer. The testimony of the boys revealed that Bolsinger had asked permission to touch them in this way . . . ." Alleging that Bolsinger undertook this activity for sexual gratification, the State prosecuted him for sexual abuse in the third degree, among other crimes. Sexual abuse in the third degree required proof that the defendant "performs a sex act . . . by force or against the will of the other person . . . ." Iowa Code § 709.4 (1). If the distinction is preserved between fraud in the *factum* and fraud in the inducement, should Bolsinger's conviction under this statute be upheld? The Iowa Supreme Court overturned the conviction, reasoning as follows:

> . . . [I]f the boys had consented to acts such as massaging their legs and instead Bolsinger had touched their genital area, this would clearly be fraud in fact; they would have consented to one act but subjected to a different one. That is not the case, however. We conclude that the consents given here were based on fraud in the inducement, not on fraud in fact, as the victims were touched in exactly the manner represented to them. The consents, therefore, were not vitiated.

4. Assume that a man slips into a darkened bedroom and a woman lying in the bed, believing the man to be her husband, consents to engage in intercourse. She thereafter learns that the man was not her husband and protests that she would not have consented had she known the man was an imposter. Assuming that the man fraudulently misrepresented himself as the woman's husband, is he guilty of rape? Would this situation present a case of fraud in the *factum* or fraud in the inducement? Perhaps surprisingly, several cases of this nature have been reported. Courts have taken different approaches in resolving them. Some conclude that because the woman knowingly consented to the act of sexual intercourse, the circumstances present a case of fraud in the inducement and hence there has been no rape. *See, e.g., Suliveres v. Commonwealth*, 865 N.E.2d 1086 (Mass. 2007); *People v. Hough*, 607 N.Y.S.2d 884 (Misc. 1994). Others reason that fraud in the *factum* has occurred because the woman thought she was consenting to "the fact" of having sexual intercourse with her husband or a regular sexual partner and not to having sexual intercourse with a stranger or an imposter. *See, e.g., United States v. Hughes*, 48 M.J. 214 (U.S. Court Appeals for the Armed Forces 1998); *Pinson v. State*, 518 So.2d 1220 (Miss. 1988).[12] Note that the Model Penal Code defines the crime of Gross Sexual Imposition to include when a man has sexual intercourse with a woman and "he knows that she is unaware that a sexual act is being committed upon her or that she submits because she mistakenly supposes that he is her husband." MPC § 213.1(2)(c).

## Psychological Coercion

### *Commonwealth v. Mlinarich*, 542 A.2d 1336 (Pa. 1988)

Nix, Chief Justice.

In the instant appeal we have agreed to consider the Commonwealth's contention that the threats made by an adult guardian to a fourteen-year-old girl to cause her to be recommitted to a juvenile detention facility supplies the "forcible compulsion" element of the crime of rape. For the reasons that follow, we are constrained to conclude that they do not and that the appellee's convictions of rape and attempted rape may not be permitted to stand.

The facts of the instant matter are no longer open to dispute. . . . [T]he complainant was living with her brother Gary and his wife and child. . . . When a diamond ring belonging to his wife disappeared, Gary asked the complainant if she had taken it, which she admitted. She asserted, however, that she had "lost" the ring, which prompted Gary to file criminal charges against her to teach her a lesson, apparently believing that the experience would lead to the recovery of the ring. As a result, the complainant was committed by court order to the custody of the Cambria County Detention Home.

Appellee, Joseph Mlinarich, . . . was sixty-three years old and suffered from emphysema and heart trouble. . . . Appellee and his wife had known the complainant's family for approximately six years, and the complainant had done housework for appellee's wife. After the complainant was committed to the detention home, appellee's wife suggested that the complainant live with her and appellee. The complainant's father considered this to be an acceptable arrangement, and, after a juvenile hearing, the complainant was released into the custody of appellee's wife pending further proceedings.

On May 28, 1981, the complainant's fourteenth birthday, she and appellee were watching television in the living room. Appellee told her to remove her outer garments and sit on his lap.

She complied, and appellee fondled her for approximately four minutes, during which time the victim "told him he shouldn't do that. . . ." Appellee engaged in similar conduct towards the complainant "[a] couple times a week," over her protestations, desisting only if she began to cry. Appellant's wife was always out of the house during these and subsequent episodes.

In mid and late June of 1981, the perverse character of appellee's unwanted attentions escalated. During one incident, which led to a charge of attempted rape, appellee asked the victim to disrobe and, when she did not remove her bra and under garments, he ordered her to undress completely. When she refused, appellee threatened to send her back to the detention home if she did not comply. The complainant obeyed, and appellee removed his clothing. When she insisted that she "did not want to do anything," appellee repeated his threat to "send [her] back." Appellee then proceeded with an unsuccessful attempt at penetration, during which the complainant experienced pain and "scream[ed], holler[ed]" and cried. A similar encounter on June 19, 1981, resulted in a second charge of attempted rape. Appellee, in yet another attempt to achieve penetration, finally succeeded on June 26, 1981.

Appellee also successfully engaged the complainant in oral intercourse on June 29 and July 1, 1981. The same threat was repeated on those occasions. Finally, on July 2, 1981, when appellee "asked [her] to do that again, and [she] wouldn't," appellee engaged in verbal abuse of the victim which convinced her to leave appellee's home and report his reprehensible conduct to her father.

Appellee was subsequently arrested and charged with rape as well as multiple counts of attempted rape, involuntary deviate sexual intercourse, corruption of a minor, indecent exposure, and endangering the welfare of a minor. . . .

## I.

Much of the confusion in this matter has resulted from the attempt to focus upon the words "**forcible compulsion**" out of the context in which it was used by the legislature. When viewed in proper context, the meaning of the phrase at issue becomes clear and the legislative scheme readily apparent. For the reasons that follow, we conclude that the term "forcible compulsion" includes both physical force as well as psychological duress. We are constrained to reject the contention that "forcible compulsion" was intended by the General Assembly, in this context, to be extended to embrace appeals to the intellect or the morals of the victim. . . .

## II.

In this setting we will now undertake to ascertain the legislative intent in its use of the term "forcible compulsion." Webster's Third New International Dictionary gives the following as a primary meaning of the noun "compulsion": "an act of compelling: a driving by force, power, pressure or necessity. . . ." The legislative use of the adjective "forcible" was obviously an effort to describe the particular type of compulsion required. . . . The term "forcible compulsion" does not describe either the intensity of the force nor does it tell us the source of the opposition or resistance that must be overcome. Fortunately, the impact of the force has been described as one sufficient to overcome the "resistance by a person of reasonable resolution." . . . Thus the only question remaining is the source of the opposition or resistance, *i.e.*, the will or the intellect. . . .

The critical distinction is where the compulsion overwhelms the will of the victim in contrast to a situation where the victim can make a deliberate choice to avoid the encounter even though the

*(Continues)*

*(Continued)*

alternative may be an undesirable one. Indeed, the victim in this instance apparently found the prospect of being returned to the detention home a repugnant one. Notwithstanding, she was left with a choice and therefore the submission was a result of a deliberate choice and was not an involuntary act. This is not in any way to deny the despicable nature of appellant's conduct or even to suggest that it was not criminal. We are merely constrained to recognize that it does not meet the test of "forcible compulsion" set forth in subsections (1) and (2) of sections 3121 and 3123. . . .

## Notes and Questions

1. What, precisely, is the distinction between "compulsion [that] overwhelms the will" and making a "deliberate choice to avoid the encounter even though the alternative may be an undesirable one"? Would succumbing to a pointed gun, for example, represent compulsion that overwhelms the will? A deliberate choice to avoid an undesirable alternative? Both? Under the facts in *Mlinarich,* can you make an argument in support of the view that the victim's choice was not "deliberate"?

2. Psychological duress in connection with sexual assault has also arisen in cases involving threatened arrest by actual or purported police officers. For instance, in *Commonwealth v. Caracciola,* 569 N.E.2d 774 (Mass. 1991), the defendant falsely told the female victim he was a police officer and instructed her to get off the street. The victim then went to a bus station to make a phone call, after which the defendant approached her again and told her to get into his car. After doing so, the victim began to cry, prompting the defendant to threaten that if she did not stop crying he would "lock [her] up for more things than [he] was planning on." After driving to a parking lot, the two had sexual intercourse.

   The Massachusetts Supreme Judicial Court concluded that the facts supported the conclusion that intercourse was achieved "by force and against [the victim's] will." The court concluded that the unsupported threat to arrest made by the defendant, who was visibly armed, justified the indictment for rape. One justice disagreed, stating

   > There was no suggestion in the victim's testimony that she submitted to intercourse because of a threat of bodily injury. The victim's testimony was clear. She submitted in order to avoid arrest, prosecution, and resulting imprisonment. . . . Of course, I do not contend that rape is limited solely to the use of physical force. . . . My contention is that rape consists of a use of physical force or a threat to do bodily injury—related concepts—in order to compel a person to submit to sexual intercourse. *Id.* at 781 (O'Connor, J. dissenting).

   Is the dissent's argument persuasive? Does limiting coercion to that involving physical force make sense? Is there any reason to distinguish the situation in *Caracciola* from the approach taken with the defense of duress more generally, which customarily requires threat of physical harm (see Chapter 5)?

## Prosecuting Rape

Rape prosecutions typically are highly emotional proceedings that often depend on difficult factual determinations concerning whether the sexual encounter was consensual. This section examines two of the many challenges attending rape prosecutions and how the justice system has attempted to reconcile the competing interests involved.

We first examine what have come to be known as "rape shield laws," provisions designed to limit the admission at trial of information regarding the complaining witness's sexual history and reputation for chasteness. In barring such information the justice system has had to weigh the competing demands of (1) the defendant's Sixth Amendment constitutional right to cross-examine his or her accuser and (2) the complainant's interest in avoiding the public airing of sensitive private information, which arguably has little or no relevance to the issue of guilt and carries the risk of unduly prejudicing the fact-finder on the issue of consent.

The second issue we explore concerns the admission of expert testimony on a psychological phenomenon known as "**rape trauma syndrome,**" which is said to account for post-rape behaviors among some victims that lay persons might consider inconsistent with the occurrence of rape. For instance, rape victims have been known to delay for hours, days, weeks, or longer in seeking help and reporting their trauma to law enforcement. Such delay, on its face, might seem inexplicable and incongruous with the state's allegation that a rape occurred. Increasingly, courts have been asked by the prosecution to admit expert testimony on rape trauma syndrome to explain how the consequences of rape can include such responses. Decisions reached by the courts, once again, reflect a competing balance of rights and interests, in this context involving society's interests in the reporting and prosecution of rape, victims' interests, the contributions of social science, accurate fact-finding, and the rights of the accused.

## Rape Shield Laws

Under the traditional common law, an alleged rape victim could be cross-examined about sexual behavior with the defendant or third parties as well as her general reputation for engaging in sexual activity. The law was premised on the simple idea that the alleged victim's past behaviors should influence the determination of whether she consented to the act of sexual intercourse being prosecuted as rape. Underlying this assumption was a distinction between a victim "who has already submitted herself to the lewd embraces of another, and the coy and modest female severely chaste and instinctively shuddering at the thought of impurity." *People v. Abbott*, 19 Wend. 192, 196 (N.Y. 1838).

In the 1970s, however, legislatures and courts began to rethink the admissibility of such evidence. They did so for two basic reasons. First, questions arose over the usefulness of information about complaining witnesses' past conduct and reputation: Why should the fact that a woman had engaged in consensual sex in the past influence the determination of whether she consented in the present case? And, even if relevant to some extent, does such information risk prejudicing the fact-finder and distracting attention from whether consent was given in the present case? Second, jurisdictions became concerned that rape prosecutions all too often "tried the victim," resulting in even greater reluctance of rape victims to report their victimization to law enforcement. Responding to these concerns, "**rape shield laws**" were put in place to limit the admission of evidence relating to complaining witnesses' prior sexual conduct and reputation. Today, such laws are in effect nationwide.[13]

The following case, *State v. Sheline*, provides a good example of a rape shield law in action, including the several exceptional circumstances that can permit admission of evidence concerning the alleged victim's prior sexual behavior.

---

### *State v. Sheline*, 955 S.W.2d 42 (Tenn. 1997)

Anderson, Chief Justice.

The question presented by this appeal is whether the trial court correctly applied the Tennessee "rape shield" rule when it excluded evidence of a rape victim's prior sexual conduct.

#### BACKGROUND

The defendant, Stephen Tracy Sheline, and the victim were 23-year-old university students who met one evening at a local bar where both had been drinking. During the evening, Sheline approached the victim, who knew he was a student. He told her he needed a ride, and she agreed to give him a ride home to his fraternity house. When Sheline said he no longer lived at the fraternity and was between apartments, the victim said she offered to let him sleep on the living room couch in her apartment, which she shared with three other women. Sheline testified that she simply offered to let him stay at her apartment.

The victim testified that when they arrived at her apartment, the defendant pushed her against a wall and attempted to lift her skirt. She resisted and told Sheline that he was to sleep on the living room couch. According to the victim, she said goodnight and went into her bedroom, and then went into the adjacent bathroom. When she emerged, Sheline was in her bedroom wearing only his pants. Although the victim resisted, he put her on the bed and pushed up her skirt. She fought him, "tried to wiggle out from underneath" him, and "told him to stop." Sheline, however, held the victim's hands down at her sides and vaginally penetrated her with his penis. When the victim was able to get away, she ran from her apartment to a neighbor's apartment for help. The neighbor testified that the victim was crying hysterically.

Sheline's version was different. He said that he met the victim at the bar, and that an acquaintance told him that the victim "thought he was cute and that she wanted to kiss him." He said he spent most of the evening with his friends, but the victim talked to him periodically and also hugged him. According to his version, when the bar closed, the victim asked if he wanted a ride home, and he agreed. As they drove towards his fraternity house, he told her he was between apartments and she offered to let him stay at her apartment. When they arrived at the victim's apartment, he said he sat down on the bed while the victim went into the bathroom. Sheline testified that when she emerged from the bathroom, they started kissing and touching, and they removed their shirts. He said he performed oral sex on the victim and then penetrated her with his penis. They stopped having intercourse when the victim said, "I don't think we can do this anymore." They continued to kiss and hug for several minutes, and then Sheline fell asleep.

As a result of a police investigation, the defendant was indicted for rape. His defense was that the sexual intercourse was consensual. [Outside the jury's presence], the defendant proffered the testimony of two witnesses under Tenn. R. Evid. 412.

*(Continues)*

(*Continued*)

The first witness, Eric Gray, testified during the offer of proof that he and the victim had known each other for a year, were good friends, and that he had engaged in sexual intercourse with her on two occasions. The first time, he said he and the victim were drinking heavily at a bar and then returned to Gray's apartment where they engaged in oral sex and sexual intercourse. The second time, he and the victim went to a basketball game, to a fraternity party, and then to the victim's apartment, where they engaged in sexual intercourse.

The second witness, Gary Jindrak, testified during the offer of proof that he and the victim talked in the bar on the same evening she met the defendant. He said they kissed and the victim put her arm around him. He testified that later she asked him to go home with her, but he declined. Jindrak testified that he never had sexual relations with the victim at any time.

The defendant argued to the trial court that he was entitled to present the testimony of Gray and Jindrak because the victim's "promiscuity" was relevant to his defense of consent. The trial court ruled that the proffered testimony of Gray and Jindrak was inadmissible. The trial then proceeded on the proof of the victim and the defendant and other supporting witnesses. After deliberating, the jury accredited the victim's testimony and found the defendant guilty of rape.

On appeal, the Court of Criminal Appeals, in a two to one decision, reversed the rape conviction on the grounds that the trial court erred in excluding the evidence of the first sexual episode described by Gray, as well as the events on the evening in question described by Jindrak. . . .

We granted the State's application for permission to appeal to review the applicability in this case of the rape shield provisions set forth in Tenn. R. Evid. 412.

## TENNESSEE RAPE SHIELD LAW

Rape shield laws were adopted in response to anachronistic and sexist views that a woman who had sexual relations in the past was more likely to have consented to sexual relations with a specific criminal defendant. Those attitudes resulted in two rape trials at the same time—the trial of the defendant and the trial of the rape victim based on her past sexual conduct. It has been said that the victim of a sexual assault is actually assaulted twice—once by the offender and once by the criminal justice system. The protections in rape shield laws recognized that intrusions into the irrelevant sexual history of a victim were not only prejudicial and embarrassing but also a practical barrier to many victims reporting sexual crimes. As the comments to Tennessee's rape shield law (Rule 412) observe:

> [T]he rule . . . takes into account that the public's interest in prosecuting and convicting people guilty of various sexual offenses is frustrated when sexual assault victims refuse to report the offenses or to testify about them at trial because of the possible admission of evidence of their sexual history. Moreover, the rule seeks to minimize the likelihood that evidence of the alleged victim's sexual history may cause the jury to be unfairly prejudiced against the victim.

At the same time, rape shield laws recognize those circumstances in which the admission of such evidence, despite its potentially embarrassing nature, must be admitted to preserve an accused's right to a fair trial.

As does nearly every jurisdiction, Tennessee's "rape shield" rule limits the admissibility of evidence about the prior sexual behavior of a victim of a sexual offense, and establishes procedures for determining when evidence is admissible. . . .

[When] the victim's sexual behavior was with persons other than the defendant, it may be admissible under the rule only

   (i) to rebut or explain scientific or medical evidence,
   (ii) to prove or explain the source of semen, injury, disease, or knowledge of sexual matters, or
   (iii) to prove consent *if the evidence is of a pattern of sexual behavior so distinctive and so closely resembling the defendant's version of the alleged encounter with the victim that it tends to prove that the victim consented to the act charged or behaved in such a manner as to lead the defendant reasonably to believe that the victim consented.* Tenn. R. Evid. 412(c)(4) (emphasis added).

As with other evidentiary rulings, the admissibility of the evidence rests in the discretion of the trial court.

When evidence is offered under the rule, the trial court must conduct a non-public hearing and consider the admissibility of the proposed evidence, and if otherwise admissible, then determine whether its probative value outweighs its unfair prejudice to the victim. Tenn. R. Evid. 412(d). As with other evidentiary rulings, the admissibility of the evidence rests in the discretion of the trial court. . . .

The State argues that the testimony of Gray and Jindrak did not establish a pattern of distinctive sexual behavior by the victim that so closely resembled the defendant's version of events as to be relevant on the issue of consent. The defendant, however, maintains that the Court of Criminal Appeals correctly interpreted Rule 412(c)(4)(iii), and that excluding the evidence denied his rights under the Tennessee and United States Constitution.

As we consider these arguments, we first turn to the language of Rule 412(c)(4), which requires a pattern of sexual behavior. Although there is no Tennessee case on the issue, it is clear that a "pattern" of sexual conduct requires more than one act of sexual conduct. . . . The plain language of the rule speaks of "specific instances" of sexual conduct with "persons" other than the defendant. Moreover, other jurisdictions have consistently observed that a "pattern" of sexual conduct denotes repetitive or multiple acts and not just an isolated occurrence. . . .

The rule also requires that the pattern consist of sexual behavior so distinctive and so closely resembling the defendant's version that it tends to prove that the victim consented to the act charged. The advisory comments to Rule 412 use the word "signature" cases to describe the distinctive behavior required. As the Illinois Supreme Court has said, the sexual conduct must be "so unusual, so outside the normal, that it had distinctive characteristics which make it the complainant's modus operandi." *People v. Sandoval*, 552 N.E.2d 726, 738 (Ill. 1990). . . . Moreover, to have probative value on the issue of consent, the pattern of distinctive sexual conduct must closely resemble the defendant's version of facts. . . .

Applying the rule and the foregoing principles to the facts of this case, we conclude that there was no *pattern* of prior sexual behavior on the victim's part. Gray testified that on one occasion in the past he had oral sex and sexual intercourse with the victim after the couple met in a bar where both had been drinking. On the second occasion, he and the victim had sexual intercourse after attending a basketball game and a fraternity party. On the other hand, Jindrak testified only that the victim kissed him and asked him to go home with her. For the Jindrak testimony to be considered part of a "pattern," a court would have to presume that the victim would have consented to engage in sexual intercourse with Jindrak. We are unwilling to make such a presumption. The two separate occasions of sexual intercourse described by Gray, therefore, do not establish the required pattern of sexual conduct.

Moreover, there was nothing so distinctive about the testimony of either Gray or Jindrak that so closely resembled the defendant's version of the events that took place between he and the victim. The described acts could hardly be characterized as signature cases. As Judge Peay observed in his dissent, to decide otherwise is to hold that any woman who has sexual intercourse with someone she meets in a bar is automatically engaged in "distinctive" sexual conduct. Other courts have rejected such a conclusion as well:

> We observe, however, that a victim's sexual history with others only goes to show a generalized attitude toward sex and says little if anything about her attitude toward a specific act of sex with the defendant. . . . We conclude that the evidence of [the victim's] sexual advances, over a period of years, toward some of the men she had met in a bar is not relevant, in itself, to establish that she consented to have sex with [the defendant]. *State v. Peite*, 839 P.2d 1223, 1229 (Id. 1992).

Accordingly, we hold that the evidence proffered by the defendant did not establish a pattern of distinctive sexual behavior as required for admissibility under Tenn. R. Evid. 412(c)(4)(iii), and it was properly excluded by the trial court. . . .

## Notes and Questions

1. Rape trials can entail a "swearing contest" between the defendant and the victim, presenting varied accounts of a physical encounter witnessed only by the parties involved. Given this reality, should the defendant be permitted to introduce information barred by rape shield laws? Indeed, why is the information barred? If relevance is the main concern, many people—including jurors—might consider the victim's prior behaviors most helpful in trying to make sense of the conflicting evidence before them. Is the information therefore barred because it is perhaps "too relevant," and hence prejudicial, to the average juror, or even the judge (in the event of a "bench trial") who, of course, also is a member of society?

*(Continues)*

2. In *Wood v. Alaska*, 957 F.2d 1544 (9th Cir. 1991), the defendant was convicted of rape. The trial court had barred information that the victim had (a) posed for Penthouse magazine, (b) acted in X-rated movies and other sexual performances, and (c) provided the defendant with Penthouse photos and discussed her past sexual experiences with him. In upholding the preclusion of the evidence, the *Wood* court drew a narrow distinction between the possible purposes of the evidence.

First, "[t]he fact that [the victim] was willing to pose for Penthouse or act in sexual movies and performances says virtually nothing about whether she would have sex with Wood. It only tends to show that she was willing to have sex, not that she was willing to have sex with this particular man at this particular time." *Id.* at 1550. But still, the court reasoned that the victim's alleged discussions with Wood were relevant because the "communications could be interpreted as sexually provocative acts." *Id.* at 1551. The court nevertheless barred the evidence, reasoning: "Showing sexually provocative photographs to a platonic friend may be . . . unusual, but in this case it might only indicate that [the victim] wanted to show off her modeling experience." *Id.* at 1553. Do you agree with the result?

3. Assume that Sheila suffers from crack cocaine addiction and in the past has bartered sexual favors in return for crack. Further assume that Jim, a drug dealer, has sex with Sheila. He knows of her previous barter practice yet does not provide her with any cocaine. Later, Sheila files a rape complaint. If the case were prosecuted in a jurisdiction with a rape shield law similar to that in *Sheline*, would Jim's attorney be permitted to cross-examine Sheila about her barter system? Why or why not? *See State v. Johnson*, 632 A.2d 152 (Md. 1993).

4. In a jurisdiction with a rape shield law in place, should the defendant be permitted to introduce evidence that the alleged victim had previously made "demonstrably false" accusations of rape? *See State v. Walton*, 715 N.E.2d 824 (Ind. 1999).

## "Rape Trauma Syndrome"

### *People v. Taylor, People v. Banks*, 552 N.E.2d 131 (N.Y. 1991)

Wachtler, Chief Judge.

In these two cases, we consider whether expert testimony that a complaining witness has exhibited behavior consistent with "rape trauma syndrome" is admissible at the criminal trial of the person accused of the rape. Both trial courts admitted the testimony and the Appellate Division affirmed in both cases. We now affirm in *People v. Taylor*, 536 N.Y.S.2d 825, and reverse in *People v. Banks*, 536 N.Y.S.2d 316. . . .

### I. *People v. Taylor*

On July 29, 1984, the complainant, a 19-year-old Long Island resident, reported to the town police that she had been raped and sodomized at gunpoint on a deserted beach near her home. The complainant testified that at about nine that evening she had received a phone call from a friend, telling her that he was in trouble and asking her to meet him at a nearby market in half an hour. Twenty minutes later, the same person called back and changed the meeting place. The complainant arrived at the agreed-upon place, shut off the car engine and waited. She saw a man approach her car and she unlocked the door to let him in. Only then did she realize that the person who had approached and entered the car was not the friend she had come to meet. According to the complainant, he pointed a gun at her, directed her to nearby Clarke's Beach, and once they were there, raped and sodomized her.

The complainant arrived home around 11:00 P.M., woke her mother and told her about the attack. Her mother then called the police. Sometime between 11:30 P.M. and midnight, the police arrived at the complainant's house. At that time, the complainant told the police she did not know who her attacker was. She was taken to the police station where she described the events leading up to the attack and again repeated that she did not know who her attacker was. At the conclusion of the interview, the complainant was asked to step into a private room to remove the clothes that she had been wearing at the time of the attack so that they could be examined for forensic evidence. While she was alone with her mother, the complainant told her that the defendant John Taylor, had been her attacker. The time was approximately 1:15 A.M. The complainant had known the defendant for years, and she later testified that she happened to see him the night before the attack at a local convenience store.

Her mother summoned one of the detectives and the complainant repeated that the defendant had been the person who attacked her. The complainant said that she was sure that it had been the defendant because she had had ample opportunity to see his face during the incident. The complainant subsequently identified the defendant as her attacker in two separate lineups. He was arrested on July 31, 1984, and was indicted by the Grand Jury on one count of rape in the first degree, two counts of sodomy in the first degree and one count of sexual abuse in the first degree.

The defendant's first trial ended without the jury being able to reach a verdict. At his second trial, the Judge permitted Eileen Treacy, an instructor at the City University of New York, Herbert Lehman College, with experience in counseling sexual assault victims, to testify about rape trauma syndrome. . . .

The prosecutor introduced this testimony for two separate purposes. First, Treacy's testimony on the specifics of rape trauma syndrome explained why the complainant might have been unwilling during the first few hours after the attack to name the defendant as her attacker where she had known the defendant prior to the incident. Second, Treacy's testimony that it was common for a rape victim to appear quiet and controlled following an attack, responded to evidence that the complainant had appeared calm after the attack and tended to rebut the inference that because she was not excited and upset after the attack, it had not been a rape. At the close of the second trial, the defendant was convicted of two counts of sodomy in the first degree and one count of attempted rape in the first degree and was sentenced to an indeterminate term of 7 to 21 years on the two sodomy convictions and 5 to 15 years on the attempted rape conviction.

## II. *People v. Banks*

On July 7, 1986, the defendant Ronnie Banks approached the 11-year-old complainant, who was playing with her friends in the City of Rochester. The complainant testified that the defendant told her to come to him and when she did not, he grabbed her by the arm and pulled her down the street. According to the complainant, the defendant took her into a neighborhood garage where he sexually assaulted her. The complainant returned to her grandmother's house, where she was living at the time. The next morning, she told her grandmother about the incident and the police were contacted. The defendant was arrested and charged with three counts involving forcible compulsion—rape in the first degree, sodomy in the first degree and sexual abuse in the first degree—and four counts that were based solely on the age of the victim—rape in the second degree, sodomy in the second degree, sexual abuse in the second degree and endangering the welfare of a child.

At trial, the complainant testified that the defendant had raped and sodomized her. In addition, she and her grandmother both testified about the complainant's behavior following the attack. Their testimony revealed that the complainant had been suffering from nightmares, had been waking up in the middle of the night in a cold sweat, had been afraid to return to school in the fall, had become generally more fearful and had been running and staying away from home. Following the introduction of this evidence, the prosecution sought to introduce expert testimony about the symptoms associated with rape trauma syndrome.

Clearly, the prosecution, in an effort to establish that forcible sexual contact had in fact occurred, wanted to introduce this evidence to show that the complainant was demonstrating behavior that was consistent with patterns of response exhibited by rape victims. The prosecutor does not appear to have introduced this evidence to counter the inference that the complainant consented to the incident, since the 11-year-old complainant is legally incapable of consent (Penal Law § 130.05[3][a]). Unlike *Taylor*, the evidence was not offered to explain behavior exhibited by the victim that the jury might not understand; instead, it was offered to show that the behavior that the complainant had exhibited after the incident was consistent with a set of symptoms commonly associated with women who had been forcibly attacked. The clear implication of such testimony would be that because the complainant exhibited these symptoms, it was more likely than not that she had been forcibly raped.

The Judge permitted David Gandell, an obstetrician-gynecologist on the faculty of the University of Rochester, Strong Memorial Hospital, with special training in treating victims of sexual assault, to testify as to the symptoms commonly associated with rape trauma syndrome. After Gandell had described rape trauma syndrome he testified hypothetically that the kind of symptoms demonstrated by the complainant were consistent with a diagnosis of rape trauma syndrome. At the close of the trial, the defendant was acquitted of all forcible counts and was convicted on the four statutory counts.

*(Continues)*

(Continued)

He was sentenced to indeterminate terms of 3½ to 7 years on the rape and sodomy convictions and to definite one-year terms on the convictions of sexual abuse in the second degree and endangering the welfare of a child.

## III. Rape Trauma Syndrome

In a 1974 study rape trauma syndrome was described as "the acute phase and long-term reorganization process that occurs as a result of forcible rape or attempted forcible rape. This syndrome of behavioral, somatic, and psychological reactions is an acute stress reaction to a life-threatening situation" (Burgess & Holmstrom, *Rape Trauma Syndrome*, 131 Am. J. Psychiatry 981, 982 [1974]). . . .

According to Burgess and Holmstrom, the rape victim will go through an acute phase immediately following the incident. The behavior exhibited by a rape victim after the attack can vary. While some women will express their fear, anger and anxiety openly, an equal number of women will appear controlled, calm, and subdued (Burgess & Holmstrom, *op. cit.*, at 982). Women in the acute phase will also experience a number of physical reactions. These reactions include the actual physical trauma that resulted from the attack, muscle tension that could manifest itself in tension headaches, fatigue, or disturbed sleep patterns, gastrointestinal irritability and genitourinary disturbance (*id.*). Emotional reactions in the acute phase generally take the form of fear, humiliation, embarrassment, fear of violence and death, and self-blame (*id.* at 983).

As part of the long-term reorganizational phase, the victim will often decide to make a change in her life, such as a change of residence. At this point, the woman will often turn to her family for support (*id.*). Other symptoms that are seen in this phase are the occurrence of nightmares and the development of phobias that relate to the circumstances of the rape (*id.*, at 984). For instance, women attacked in their beds will often develop a fear of being indoors, while women attacked on the street will develop a fear of being outdoors (*id.*).

While some researchers have criticized the methodology of the early studies of rape trauma syndrome, Burgess and Holmstrom's model has nonetheless generated considerable interest in the response and recovery of rape victims and has contributed to the emergence of a substantial body of scholarship in this area. . . . The question before us today, then, is whether the syndrome, which has been the subject of study and discussion for the past 16 years, can be introduced before a lay jury as relevant evidence in these two rape trials.

We realize that rape trauma syndrome encompasses a broad range of symptoms and varied patterns of recovery. Some women are better able to cope with the aftermath of sexual assault than other women. It is also apparent that there is no single typical profile of a rape victim and that different victims express themselves and come to terms with the experience of rape in different ways. We are satisfied, however, that the relevant scientific community has generally accepted that rape is a highly traumatic event that will in many women trigger the onset of certain identifiable symptoms. . . .

We are aware that rape trauma syndrome is a therapeutic and not a legal concept. Physicians and rape counselors who treat victims of sexual assault are not charged with the responsibility of ascertaining whether the victim is telling the truth when she says that a rape occurred. That is part of the truth-finding process implicated in a criminal trial. We do not believe, however, that the therapeutic origin of the syndrome renders it unreliable for trial purposes. Thus, although we acknowledge that evidence of rape trauma syndrome does not by itself prove that the complainant was raped, we believe that this should not preclude its admissibility into evidence at trial when relevance to a particular disputed issue has been demonstrated.

## IV. The Law

Having concluded that evidence of rape trauma syndrome is generally accepted within the relevant scientific community, we must now decide whether expert testimony in this area would aid a lay jury in reaching a verdict. "[E]xpert opinion is proper when it would help to clarify an issue calling for professional or technical knowledge, possessed by the expert and beyond the ken of the typical juror" (*DeLong v. County of Erie*, 60 N.Y.2d 296). . . .

Because cultural myths still affect common understanding of rape and rape victims and because experts have been studying the effects of rape upon its victims only since the 1970's, we believe that patterns of response among rape victims are not within the ordinary understanding of the lay juror. For that reason, we conclude that introduction of expert testimony describing rape trauma syndrome may under certain circumstances assist a lay jury in deciding issues in a rape trial. . . .

Among those States that have allowed such testimony to be admitted, the purpose for which the testimony was offered has proven crucial. A number of States have allowed testimony of rape trauma syndrome to be admitted where the defendant concedes that sexual intercourse occurred, but contends that it was consensual. . . .

Other States have permitted the admission of this testimony where it was offered to explain behavior exhibited by the complainant that might be viewed as inconsistent with a claim of rape. . . .

Having concluded that evidence of rape trauma syndrome can assist jurors in reaching a verdict by dispelling common misperceptions about rape, and having reviewed the different approaches taken by the other jurisdictions that have considered the question, we too agree that the reason why the testimony is offered will determine its helpfulness, its relevance and its potential for prejudice. In the two cases now before us, testimony regarding rape trauma syndrome was offered for entirely different purposes. We conclude that its admission at the trial of John Taylor was proper, but that its admission at the trial of Ronnie Banks was not.

As noted above, the complaining witness in *Taylor* had initially told the police that she could not identify her assailant. Approximately two hours after she first told her mother that she had been raped and sodomized, she told her mother that she knew the defendant had done it. The complainant had known the defendant for years and had seen him the night before the assault. We hold that under the circumstances present in this case, expert testimony explaining that a rape victim who knows her assailant is more fearful of disclosing his name to the police and is in fact less likely to report the rape at all was relevant to explain why the complainant may have been initially unwilling to report that the defendant had been the man who attacked her. Behavior of this type is not within the ordinary understanding of the jury and testimony explaining this behavior assists the jury in determining what effect to give to the complainant's initial failure to identify the defendant (*see, People v. Keindl, supra*). This evidence provides a possible explanation for the complainant's behavior that is consistent with her claim that she was raped. As such, it is relevant.

Rape trauma syndrome evidence was also introduced in *Taylor* in response to evidence that revealed the complainant had not seemed upset following the attack. We note again in this context that the reaction of a rape victim in the hours following her attack is not something within the common understanding of the average lay juror. Indeed, the defense would clearly want the jury to infer that because the victim was not upset following the attack, she must not have been raped. This inference runs contrary to the studies cited earlier, which suggest that half of all women who have been forcibly raped are controlled and subdued following the attack (Burgess & Holmstrom, *op. cit.*, at 982). Thus, we conclude that evidence of this type is relevant to dispel misconceptions that jurors might possess regarding the ordinary responses of rape victims in the first hours after their attack. We do not believe that evidence of rape trauma syndrome, when admitted for that express purpose, is unduly prejudicial.

The admission of expert testimony describing rape trauma syndrome in *Banks*, however, was clearly error. As we noted earlier, this evidence was not offered to explain behavior that might appear unusual to a lay juror not ordinarily familiar with the patterns of response exhibited by rape victims. We conclude that evidence of rape trauma syndrome is inadmissible when it inescapably bears solely on proving that a rape occurred, as was the case here.

Although we have accepted that rape produces identifiable symptoms in rape victims, we do not believe that evidence of the presence, or indeed of the absence, of those symptoms necessarily indicates that the incident did or did not occur. Because introduction of rape trauma syndrome evidence by an expert might create such an inference in the minds of lay jurors, we find that the defendant would be unacceptably prejudiced by the introduction of rape trauma syndrome evidence for that purpose alone. We emphasize again that the therapeutic nature of the syndrome does not preclude its admission into evidence under all circumstances. We believe, however, that its usefulness as a fact-finding device is limited and that where it is introduced to prove the crime took place, its helpfulness is outweighed by the possibility of undue prejudice. Therefore, the trial court erred in permitting the admission of expert testimony regarding rape trauma syndrome under the facts present in *Banks*.

## Notes and Questions

1. Testimony on "rape trauma syndrome" presents a dilemma to the justice system that differs from that of rape shield laws. Although shield laws signify a desire to limit the amount of information put before the fact-finder, expert testimony on rape trauma syndrome entails the opposite: a desire to provide fact-finders with *more* information about the alleged victim. Despite increasing acceptance of rape trauma syndrome by the courts, some observers have raised concerns over its scientific reliability.[14]

*(Continues)*

2. Most often, the prosecution introduces rape trauma syndrome expert testimony to help rebut a claim of consent by the defendant. Increasingly, however, defendants have sought to introduce rape trauma syndrome testimony to establish that the alleged victim's behavior was inconsistent with the occurrence of rape (*i.e.*, that his or her post-rape behavior is not symptomatic of rape trauma syndrome).

   In *Henson v. State*, 535 N.E.2d 1189 (Ind. 1989), the defendant sought to introduce expert testimony that the victim's post-rape conduct (including dancing and drinking in a bar) was inconsistent with a that of a person having suffered a forcible rape, as had been alleged. The *Henson* court held that such defensive use of rape trauma syndrome testimony was proper, given that the prosecution could use rape trauma syndrome testimony. Should this reciprocal use of rape trauma syndrome be permitted? For a discussion of this question see Davis.[15]

## Statutory Rape

"**Statutory rape**" involves sexual intercourse with an underage person, as opposed to another adult (as with common law rape). Statutory rape derives its name from the fact that legislatures—as opposed to the judiciary, which has developed the common law over the centuries—stepped in to enact statutes to create a new type of offense. In contrast to common law rape, with statutory rape "consent" to the sexual act is immaterial, because the law presumes that minors lack the legal capacity to consent to the alleged sex act.

The defining feature of statutory rape is chronological age of the victim. As a result, in most jurisdictions statutory rape in effect is a strict liability offense (*see* Chapters 3 and 10). However, statutory rape carries significantly greater stigma and punishment than other strict liability offenses (which typically involve public safety and health) which are punished as misdemeanors.

### *Garnett v. State*, 632 A.2d 797 (Md. 1993)

Murphy, Chief Judge.

Maryland's "statutory rape" law prohibiting sexual intercourse with an underage person is codified in Maryland Code (1957, 1992 Repl.Vol.) Art. 27, § 463, which reads in full:

"Second degree rape.
   (a) *What constitutes.*—A person is guilty of rape in the second degree if the person engages in vaginal intercourse with another person:
      (1) By force or threat of force against the will and without the consent of the other person; or
      (2) Who is mentally defective, mentally incapacitated, or physically helpless, and the person performing the act knows or should reasonably know the other person is mentally defective, mentally incapacitated, or physically helpless; or
      (3) Who is under 14 years of age and the person performing the act is at least four years older than the victim.
   (b) *Penalty.*—Any person violating the provisions of this section is guilty of a felony and upon conviction is subject to imprisonment for a period of not more than 20 years."

Subsection (a)(3) represents the current version of a statutory provision dating back to the first comprehensive codification of the criminal law by the Legislature in 1809. Now we consider whether under the present statute, the State must prove that a defendant knew the complaining witness was younger than 14 and, in a related question, whether it was error at trial to exclude evidence that he had been told, and believed, that she was 16 years old.

### I.

Raymond Lennard Garnett is a young retarded man. At the time of the incident in question he was 20 years old. He has an I.Q. of 52. His guidance counselor from the Montgomery County public school system, Cynthia Parker, described him as a mildly retarded person who read on the third-grade level, did arithmetic on the 5th-grade level, and interacted with others socially at school at the level of someone 11 or 12 years of age. . . .

In November or December 1990, a friend introduced Raymond to Erica Frazier, then aged 13; the two subsequently talked occasionally by telephone. On February 28, 1991, Raymond, apparently wishing to call for a ride home, approached the girl's house at about nine o'clock in the evening. Erica opened her bedroom window, through which Raymond entered; he testified that "she just told me to get a ladder and climb up her window." The two talked, and later engaged in sexual intercourse. Raymond left at about 4:30 A.M. the following morning. On November 19, 1991, Erica gave birth to a baby, of which Raymond is the biological father.

Raymond was tried before the Circuit Court for Montgomery County (Miller, J.) on one count of second degree rape under § 463(a)(3) proscribing sexual intercourse between a person under 14 and another at least four years older than the complainant. At trial, the defense twice proffered evidence to the effect that Erica herself and her friends had previously told Raymond that she was 16 years old, and that he acted with that belief. The trial court excluded such evidence as immaterial. . . .

The court found Raymond guilty. . . .

The precise legal issue here rests on Raymond's unsuccessful efforts to introduce into evidence testimony that Erica and her friends had told him she was 16 years old, the age of consent to sexual relations, and that he believed them. Thus the trial court did not permit him to raise a defense of reasonable mistake of Erica's age, by which defense Raymond would have asserted that he acted innocently without a criminal design. At common law, a crime occurred only upon the concurrence of an individual's act and his guilty state of mind. *Dawkins v. State,* 547 A.2d 1041 (Md. 1988). In this regard, it is well understood that generally there are two components of every crime, the *actus reus* or guilty act and the *mens rea* or the guilty mind or mental state accompanying a forbidden act. The requirement that an accused have acted with a culpable mental state is an axiom of criminal jurisprudence. . . .

To be sure, legislative bodies since the mid-19th century have created strict liability criminal offenses requiring no *mens rea*. Almost all such statutes responded to the demands of public health and welfare arising from the complexities of society after the Industrial Revolution. Typically misdemeanors involving only fines or other light penalties, these strict liability laws regulated food, milk, liquor, medicines and drugs, securities, motor vehicles and traffic, the labeling of goods for sale, and the like. . . . Statutory rape, carrying the stigma of felony as well as a potential sentence of 20 years in prison, contrasts markedly with the other strict liability regulatory offenses and their light penalties. . . .

We think it sufficiently clear . . . that Maryland's second degree rape statute defines a strict liability offense that does not require the State to prove *mens rea;* it makes no allowance for a mistake-of-age defense. The plain language of § 463, viewed in its entirety, and the legislative history of its creation lead to this conclusion. . . .

Section 463(a)(3) prohibiting sexual intercourse with underage persons makes no reference to the actor's knowledge, belief, or other state of mind. As we see it, this silence as to *mens rea* results from legislative design. . . .

This interpretation is consistent with the traditional view of statutory rape as a strict liability crime designed to protect young persons from the dangers of sexual exploitation by adults, loss of chastity, physical injury, and, in the case of girls, pregnancy. . . .

Maryland's second degree rape statute is by nature a creature of legislation. Any new provision introducing an element of *mens rea*, or permitting a defense of reasonable mistake of age, with respect to the offense of sexual intercourse with a person less than 14, should properly result from an act of the Legislature itself, rather than judicial fiat. Until then, defendants in extraordinary cases, like Raymond, will rely upon the tempering discretion of the trial court at sentencing.

Judgment affirmed. . . .

## Notes and Questions

1. Most jurisdictions refuse to recognize mistake-of-age as a defense in statutory rape cases. *See, e.g., State v. Yanez,* 716 A.2d 759 (R.I. 1998). Does this approach seem appropriate? Some jurisdictions allow a qualified mistake defense—for instance, if the defendant has a "reasonable belief" that the victim is of legal age. *See, e.g., People v. Hernandez,* 393 P.2d 673 (Cal. 1964). Do you agree that the harms posed by underage sexual activity warrant imposition of strict liability, especially given the very serious penalties associated with statutory rape?

2. Today, three main approaches are encompassed by the criminal law's regulation of underage sex. First, some states focus on the difference in age between the parties. These states in effect condone sex between minors but impose criminal penalties when an adult is involved or when the defendant is a certain number of years older than the victim. *See, e.g., State v. Jason B.,* 729 A.2d 760 (Conn. 1999) (upholding conviction of 16-year-old boy for engaging in consensual sexual activity with 14-year-old girl when age difference was "two years, three months and seven days," ruling that such separation qualified as "more than two years" under governing statute).[16]

*(Continues)*

(*Continued*)

Second, some states make the age of the victim the determining factor in whether the offense is classified as a misdemeanor or a felony. For instance, in Maryland, where the age of consent is 15, intercourse with a 14-year-old girl is a felony, whereas the same act with a 15-year-old girl is a misdemeanor (assuming that the defendant is 4 or more years older than his partner).

Third, several states tie criminal liability to the nature of the sexual act and the age of the victim. In Michigan, for instance, any sexual contact with a child less than 13 years of age is forbidden; 13- to 15-year-olds are presumed capable of consenting to activities like "petting" but lack legal capacity to consent to intercourse; at age 16 consent to intercourse can be recognized. For a more detailed discussion of the various approaches see Guerrina.[17]

3. The states also vary with respect to the minimum age of lawful consent. Although most states set the age of consent at 13 or 14, some fix the age level at 17 or 18 years.[18] With such laws, society uses chronological age as a proxy for immaturity, giving the justice system a bright-line rule that avoids the need for delving into the particular characteristics of the involved individuals.

4. Historically, statutory rape has been justified by two basic rationales: (1) that youths lack the capacity to provide meaningful consent to sex and (2) that youths should be protected against sexually exploitive behaviors by adults. However, given the widespread changes in sexuality in American society, are these rationales still persuasive? In Georgia 17-year-old Genarlow Wilson received a 10-year prison term in 2005 for having "consensual" oral sex with his 15-year-old girlfriend. At the time of Wilson's conviction, Georgia law (which subsequently was amended) required that "aggravated child molestation," which covered Wilson's conduct, be punished by a minimum of 10 years and a maximum of 30 years in prison. Amid widespread public sentiment that this punishment was unjust, the Georgia Supreme Court ruled (4–3) that the mandatory minimum 10-year prison sentence was disproportionate to the crime and violated the state and federal constitutional prohibitions against cruel and unusual punishments (*see Ewing v. California*, 538 U.S. 11 (2003), which we considered in Chapter 2). *Humphrey v. Wilson*, 652 S.E.2d 501 (Ga. 2007). Has the law relating to statutory rape been outpaced by changes in the larger society, especially with respect to sex occurring among older adolescents?

5. As discussed earlier, the law of sexual assault can provide an intriguing benchmark for sociological analysis. The U.S. Supreme Court's decision in *Michael M. v. Superior Court of Sonoma County*, 450 U.S. 464 (1981) (discussed in Chapter 2) amply illustrates this fact. There, the Court permitted the State of California to target young men alone for statutory rape prosecution, finding that boys and girls are not "similarly situated" and hence need not be treated the same by the criminal law for purposes of the Equal Protection Clause of the Fourteenth Amendment. The Court reasoned that

> [o]nly women may become pregnant, and they suffer disproportionately the profound physical, emotional and psychological consequences of sexual activity. . . . Because virtually all of the significant harmful and inescapably identifiable consequences of teenage pregnancy fall on the young female, a legislature acts well within its authority when it elects to punish only the participant who, by nature, suffers few of the consequences of his conduct. . . . Moreover, the risk of pregnancy itself constitutes a substantial deterrence to young females. *Id.* at 471–73.

Although today statutory rape laws increasingly are gender neutral, the Court's decision has continued to inspire discussion, perhaps most notably within the feminist community. Should basic, irrefutable biological differences between the sexes be reflected in the substantive criminal law? Are single-sex statutory rape laws beneficial or do they reflect outmoded notions that females must be "protected" and lack the capacity to consent to sex?[19]

Also, given the historically high rates of teenage pregnancy, is there reason to question the accuracy of the Court's assessment that "pregnancy itself constitutes a substantial deterrence to young females"? On the other hand, given the prevalence of teenage pregnancy, especially involving fathers beyond the age of maturity, is some possible deterrence better than none at all?

## Grading and Punishing Sex Crimes

### Grading in Sex Crimes

Over time, jurisdictions have identified different gradations of seriousness in sex crimes with corresponding differences in punishments. The New Jersey statute at issue in *M.T.S.* reflects one such approach. In New Jersey, defendants are prosecuted for "aggravated sexual assault" and "sexual assault." Aggravated sexual assault is punishable by 10 to 20 years in prison and involves sexual penetration under a variety of especially serious circumstances, such as when the victim is less than 13 years of age or the assault entails the use of force and the infliction of "severe personal injury" on the victim. *See* N.J. Stat. Ann. § 2C: 14-2(a) (West 1999). Sexual assault carries a prison term of 5 to 10 years and involves behavior that is less blameworthy, such as in *M.T.S.* where the defendant used "physical force or coercion, but the victim [did] not sustain severe personal injury." *See* N.J. Stat. Ann. § 2C: 14-2(c). New Jersey additionally makes illegal "criminal sexual contact" and "lewdness," which are also punished less severely. *See* N.J. Stat. Ann. §§ 2C: 14-3 to 4.

## Punishing Aggravated Rape

In recognition of its serious emotional and physical consequences, a rape conviction today carries heavy punishment, ranging from life to a very substantial term of years in prison. Perhaps surprisingly, some have suggested that such severe punishments are unwarranted, with feminists in particular at times asserting that singling rape out for special treatment can serve to perpetuate harmful stereotypes and myths of feminine weakness. They buttress such arguments by pointing to the lesser punishments imposed on those convicted of aggravated batteries, which also involve unconsented physical violations. What do you think?

In the past several decades this debate has played out in the context of the death penalty, society's "ultimate sanction." As we saw in Chapter 2, in *Coker v. Georgia*, 433 U.S. 584 (1977), the U.S. Supreme Court held that the imposition of a death sentence for the rape of an adult woman was "grossly out of proportion to the severity of the crime," and thus violated the Eighth Amendment to the U.S. Constitution as a "cruel and unusual punishment." In reaching its result the *Coker* Court stated as follows:

> Rape is without doubt deserving of serious punishment; but in terms of moral depravity and of the injury to the person and to the public, it does not compare with murder, which does involve the unjustified taking of a human life. . . . The murderer kills; the rapist, if no more than that, does not. Life is over for the victim of the murderer; for the rape victim, life is not over and normally not beyond repair. We have the abiding conviction that the death penalty, which is "unique in its severity and irrevocability," is an excessive penalty for the rapist who, as such, does not take human life. *Id.* at 598.

More recently, in *Kennedy v. Louisiana*, 128 S. Ct. 2641 (2008), the Supreme Court by a 5–4 majority reached the same conclusion with respect to the crime of raping a child. That being said, in *Kennedy* the justices were less willing to embrace this reasoning in its entirety. The majority opinion conceded as follows:

> It must be acknowledged that there are moral grounds to question a rule barring capital punishment for a crime against an individual that did not result in death. These facts illustrate the point. Here the victim's fright, the sense of betrayal, and the nature of her injuries caused more prolonged physical and mental suffering than, say, a sudden killing by an unseen assassin. The attack was not just on her but on her childhood. For this reason, we should be most reluctant to rely upon the language of the plurality in *Coker*, which posited that, for the victim of rape, "life may not be nearly so happy as it was" but it is not beyond repair. Rape has a permanent psychological, emotional, and sometimes physical impact on the child. We cannot dismiss the years of long anguish that must be endured by the victim of child rape. 128 S. Ct., at 2658.

If the harm inflicted on a rape victim is serious, as it certainly is, why should it matter whether it is of equivalent magnitude to the loss of life? Is it necessarily the case that the threshold level of harm caused by a crime must be death to justify punishment of death? If so, why did the Court in *Kennedy* stress that its judgment prohibiting the death penalty applied only to crimes against persons and did not necessarily reach crimes such as espionage, treason, and major drug offenses (see Chapter 2)?

## Conclusion

Rape and sexual assault rank among the most serious crimes, causing enormous emotional and physical harm that can last throughout victims' lifetimes. At the same time, the criminal prosecution of sex crimes raises a great many controversial legal issues, which at once pose significant challenges to the operation of the criminal justice system and reflect age-old yet evolving views of sexual behavior and human interaction. For these reasons, this area of the law has been, and doubtless will continue to be, among the most fluid within the substantive criminal law and also among the most interesting.

## Key Terms

forcible compulsion
forcible rape
fraud in the *factum*
fraud in the inducement
marital rape

psychological coercion
rape shield laws
rape trauma syndrome
statutory rape

## Review Questions

1. List the elements of forcible rape. In your response, address whether or not a man can be the victim of rape.
2. What does Susan Estrich mean by "real rape" and why does she distinguish it from acquaintance rape?
3. Rape typically is considered to be a "general intent" crime. In this context, what is the difference between rape and the "specific intent" requirement as in the case of murder?
4. Briefly describe *fraud in the factum* and *fraud in the inducement* and give an example of each.
5. Suppose a man enters the darkened bedroom of his identical twin brother and engages in sexual intercourse with the wife of his twin. She does not discover until the next day that she had engaged in intercourse with her brother-in-law and not her husband. She protests. Would this be an example of *trickery, psychological coercion,* or *forcible compulsion,* and why?
6. Briefly describe the purposes of rape shield laws. Do you believe such laws are achieving those purposes?
7. Rape trauma syndrome is not considered to be a legal concept; rather it is a therapeutic term. Describe what is meant by *rape trauma syndrome* and explain its importance in context of a rape trial.
8. How does *statutory rape* differ from "acquaintance rape" and "real rape"?

## Notes

1. *See* U.S. Department of Justice, Bureau of Justice Statistics. *Criminal Victimization.* Washington, DC: U.S. Dept. of Justice. 2006:3.
2. *Id.*
3. U.S. Department of Justice, Bureau of Justice Statistics. *Rape and Sexual Assault: Reporting to Police and Medical Attention, 1992–2000.* Washington, DC: U.S. Dept. of Justice. 2002:1.
4. *See id.,* p. 2; U.S. Department of Justice, 2006, p. 5.

5. Blackstone W. *Commentaries on the Laws of England* (London: T. Cadell and W. Davies 1809/1979) vol.4:210.

6. *See* Dressler J. *Understanding Criminal Law,* 4th ed. Newark, NJ: Lexis/Nexis. 2006:631–632.

7. *See* Estrich S. *Real Rape.* Cambridge, MA: Harvard University Press, 1987.

8. *See* U.S. Department of Justice. *Sourcebook of Criminal Justice Statistics,* Table 3.14.2005, available at http://www.albany.edu/sourcebook/pdf/t3142005.pdf (reporting that 75.8% of completed rapes involve non-strangers); Berliner D. Rethinking the reasonable belief defense to rape. *Yale Law J.* 1991;100;2687 (citing studies indicating ranges from 46 to 81 percent).

9. Dripps DA. Beyond rape: An essay on the difference between the presence and the absence of nonconsent. *Columbia Law Rev.* 1992; 92:1780, 1793, note 14.

10. *See* Vetterhoffer D. No means no: Weakening sexism in rape law by legitimizing post-penetration rape. *St. Louis Univ. Law J.* 2005; 49:1229, 1240–1241; Anderson MJ. Marital immunity, intimate relationships, and improper inferences: A new law on sexual offenses by intimates. *Hastings Law J.* 2003;54:1465, 1468–1473.

11. *See* Jackson L. Marital rape: A higher standard is in order. *William & Mary J. Women Law* 1994;1:183, 194 (footnotes omitted) ("Data demonstrates that rape in marriage is actually more emotionally traumatic than any other kind of rape and carries with it longer lasting emotional effects. Victims of marital rape also tend to suffer greater physical harm than victims of non-marital rape and are in fact often victims of the most brutal and life-threatening rapes.").

12. *See generally* LaFave WR. *Criminal Law,* 3rd ed. St. Paul, MN: West Group. 2000:767; Dressler, p. 636.

13. *See* Byrnes CT. Putting the focus where it belongs: Mens rea, consent, force, and the crime of rape. *Yale J. Law Femin.* 1998;10:277, 305, note 17.

14. *See* Stefan S. The protection racket: Rape trauma syndrome, psychiatric labeling, and law. *Northwest. Univ. Law Rev.* 1994;88:1271.

15. Davis KM. Note: Rape, resurrection, and the quest for truth: The law and science of rape trauma syndrome in constitutional balance with the rights of the accused. *Hastings Law J.* 1998;48:1511, 1526–1529.

16. Malone MJ. Age gap in teenage sex continues to stir debate. *New York Times,* § 14CN, p. 7, June 24, 2007.

17. Guerrina B. Comment: Mitigating the punishment for statutory rape. *Univ. Chicago Law Rev.* 1998;65:1251, 1255–1256.

18. *See* Phipps CA. Children, adults, sex and the criminal law: In search of reason. *Seton Hall Legislative J.* 1997;22:1.

19. For an overview of this debate, see Note on Feminist legal analysis and sexual autonomy: Using statutory rape laws as an illustration. *Harvard Law Rev.* 1999;112:1065.

## Footnote to *Boro v. Superior Court,* 163 Cal. App.3d 1224, 210 Cal. Rptr. 122 (1985)

5. It is not difficult to conceive of reasons why the Legislature may have consciously wished to leave the matter where it lies. Thus, as a matter of degree, where consent to intercourse is obtained by promises of travel, fame, celebrity and the like—ought the liar and seducer to be chargeable as a rapist? Where is the line to be drawn?

# Other Crimes Against the Person

## Chapter Objectives

- Distinguish between *assault* and *battery*
- Distinguish between *simple assault and battery* and *aggravated assault and battery*
- Distinguish between *stalking* and *harassment*
- Explain the differences between *false imprisonment* and *kidnapping*
- Define the crime of *custodial interference*

One purpose behind the criminal law is to protect individuals against bodily harm imposed by others. Without the criminal law and its enforcement, stronger, larger, more aggressive individuals could assert their will over other members of society through the use of physical force and intimidation. To discourage this behavior, all American jurisdictions have laws prohibiting the unjustified use or threat of physical force against another.

Under the common law assault and battery were two distinct crimes. In general terms a battery involved an offensive touching, whereas an assault did not involve physical contact. Although many modern statutes have merged assault and battery into a single crime, it is still important to understand the separate underlying common law concepts and elements behind each offense.

## Battery

The common definition of **battery** as an "offensive touching" sounds simple enough. Although this may be true to some extent, when we look at the specific elements that need to be proven, the issues become more complex than might be expected at first. To help us develop a more comprehensive understanding of battery, let us look at the specific elements of the crime.

### Actus Reus

The *actus reus* of battery is an unjustified offensive touching. This is somewhat misleading in that the actor does not actually have to physically touch the victim for a battery to occur. Consider the case where a person stabs somebody with a knife or shoots another person with a gun. Although the actor did not physically touch the victim, his or her actions caused an object to touch and physically injure the victim. A person also may be guilty of battery if he or she sets in motion the events that cause an injury to another person. So, if I sic my German Shepherd on a person or if I whip my horse so that it tramples another person (and I am acting with the statutorily required *mens rea* and the potential injuries in fact occur), I have committed battery.

### Mens Rea

The mental state required of the actor for a criminal battery varies between jurisdictions. All jurisdictions provide that a person who commits an unjustified offensive touch, either intentionally or knowingly, has committed battery. In addition, most jurisdictions provide that a person who performs an unjustified offensive touching either through recklessness or criminal negligence may be convicted of battery as well. For instance, the Model Penal Code requires a *mens rea* of purposely, knowingly, or recklessly for battery.

### Harm Required

For a battery to be committed, harm or injury must occur. Obviously, when a person physically injures another through his or her offensive touching, sufficient harm exists for battery. What if an offensive touching does not cause a physical injury but rather causes only psychological damage? Is this harm sufficient for battery? Consider the case of *Lynch v. Commonwealth*.

---

### *Lynch v. Commonwealth,* 109 S.E. 427 (Va. 1921)

Kelly, P.

The defendant, Frank Lynch . . . was found guilty of assault and battery, and sentenced to a term in jail and the payment of a fine. . . .

*(Continues)*

---

(*Continued*)

On the afternoon of March 30, 1920, Mrs. Mary Martin was at home alone with her two small children, the elder of them being only two years of age. She heard a noise or knocking at the back door, and, on going to see what it meant, was met at the door by Frank Lynch, who said to her, "Say, Mrs. Martin, I want to kiss a white woman; I want to see what it is like to kiss a white woman." She replied, "No, sir"; and he thereupon put his hand upon her shoulder and said, "I didn't mean to insult you." At this juncture, she told him "to get out," and he left.

The evidence was certified in narrative form. We have stated it from the commonwealth's standpoint as fully as it appears in the record. Was it sufficient to support the verdict? . . .

The surrounding circumstances, such as time and place, the relationship or sex of the parties, the subject-matter of the words used, may have a most important bearing in determining whether a particular act constitutes an assault and battery. The law upon the subject is intended primarily to protect the sacredness of the person, and secondarily to prevent breaches of the peace. . . .

To constitute battery there must be some touching of the person of another, but not every such touching will amount to the offense. Whether it does or not will depend, not upon the amount of force applied, but upon the intent of the actor.

"A battery is the unlawful touching of the person of another by the aggressor himself, or by some substance set in motion by him. . . . The intended injury may be to the feelings or mind as well as to the corporeal person. . . . The law cannot draw the line between different degrees of force, and therefore totally prohibits the first and lowest stage of it."

"Any touching by one of the person or clothes of another in rudeness or in anger is an assault and battery." *Engelhardt v. State,* 7 South. 154.

In *Goodrum v. State,* 60 Ga. 509, a case holding that it is an assault and battery for a man, without any innocent excuse, to put his arm around the neck of another's wife against her will, the court said:

> To touch a virtuous wife in the way of illicit love is a far greater outrage than to touch her in anger, and equally a breach of the peace. It is violence proceeding from lust, instead of violence proceeding from rage. It issues from the passion, which, unrestrained, culminates in rape, instead of from the passion which culminates in homicide.

Mr. Minor defines battery as "The actual infliction of corporeal hurt on another (*e.g.,* the least touching of another's person), willfully or in anger, whether by the party's own hand or by some means set in motion by him." Minor's Synopsis Criminal Law, 77.

The above definition by Mr. Minor does not expressly mention rudeness or insult, but by using the word "willfully" clearly implies that those elements, as well as anger, may become a test of the offense. The word "willfully" is variously defined in legal parlance, and may mean, among other things, "designedly," "intentionally," or "perversely." Its correct application in a particular case will generally depend upon the character of the act involved and the attending circumstances.

Having in view the foregoing definitions and explanations of the offense, we are unable to say that the verdict of the jury in the instant case was not warranted by the evidence. Upon the case as made by the commonwealth, the accused, without invitation or excuse, entered the home of the prosecutrix, grossly insulted her, and almost simultaneously placed his hand upon her shoulder. This was an unlawful touching of her person, and hence, in contemplation of law, a battery.

It is argued that the wrongful intention of the accused in touching the prosecutrix is disproved by his declaration, "I did not mean to insult you," followed almost immediately by his departure from the room. This is perhaps one way, but it is certainly not the only way, in which the jury might under the evidence have fairly weighed his conduct. The exact order of events must here be observed. He entered the house with an evil purpose, which he communicated to Mrs. Martin. She promptly rejected his insulting suggestion. Not stopping with that, he proceeded to place his hand on her shoulder and say, "I did not mean to insult you." Was this a cessation of his overtures and a genuine apology, or was it an impudent experimental attempt on his part to continue the conversation in the hope of overcoming her objection? He had not yet shown any purpose of leaving the house. That followed promptly, to be sure, but only after he had added one insult to another by placing his hand upon her, and after her second rebuff, which assumed the form of an order to get out of the house. If it be conceded that any man, after having thus insulted a woman, could lawfully touch her, even in connection with an apology (and such a concession does not seem to us at all well warranted), there is certainly enough in the case to have justified the jury in finding that what he actually did rendered him guilty of a rude insult and a willful violation of the sanctity of her person, amounting in law to an assault and battery. . . .

The judgment complained of is right and should be affirmed.

## Notes and Questions

1. What was the *actus reus*? The *mens rea*? The harm that resulted? Do you agree that Lynch committed battery?
2. The Model Penal Code and many modern statutes require a physical injury for a battery to have occurred. In these jurisdictions statutes covering sexual contact and conduct apply to incidents such as the one in *Lynch*.
3. In general, consent is not a defense to a charge of battery. Can a person consent to being the victim of a battery by participating in an activity where batteries often occur? Or, can a person be charged with committing battery when he or she harms someone during an athletic activity that is inherently dangerous? Consider the following situations. Assume a defensive player in football causes a serious injury to an opponent by tackling him by his face mask, an action against the rules of the game. Has he committed assault? What if a hockey player breaks another player's nose by punching him during a fight in the heat of battle? What if he hits the other player over the head with a hockey stick during the same fight and causes serious brain damage? What if the fight occurred before the start of the game?[1]

## Assault

There are two types of **assault**: *attempted battery assault* and *threatened battery assault*. **Attempted battery assault** occurs when a person attempts to commit a battery but fails to do so. It requires an actor to have the specific intent to actually carry out a battery and to take a substantial step toward doing so but in the end fail to commit it. Shooting at a person and missing or throwing a punch at a person who ducks out of the way exemplify attempted battery assaults.

The second type of assault, **threatened battery assault**, occurs when a person places another in fear of imminent serious bodily harm. Although it requires more than just words, it does not require an actual intent to injure. In addition, threatened battery assault can occur only if the victim is aware of the threatened harm and reasonably feels in fear of serious bodily injury. If Joe pulls a gun on Tom while Tom's back is turned and then changes his mind and puts the gun away before Tom turns around, Joe has not committed assault because Tom was not placed in fear. On the other hand, if Joe tells Tom not to turn around or he will shoot and then decides to put the gun away, Joe has committed assault as long as Tom reasonably believed Joe had a gun and was placed in fear.

### *State v. Simuel,* 94 Wash. App. 1021, 1999 WL 106909 (Wash. App. Div. 1)

Webster, J. . . .

Appellant Raymond Simuel was charged with second degree assault with a deadly weapon against Wayne Wendel, an employee at the 21 Club. Employees had asked Simuel to leave the 21 Club after he stepped behind the cash register counter. The State's witnesses testified that Simuel complied, but as he walked towards the front door, he reached into his coat pocket, pulled out a box-cutting utility knife and took a swipe at Wendel. Wendel jumped out of the way of the knife. Employees immediately tackled Simuel and subdued him until the police arrived. . . .

The jury found Simuel guilty of second degree assault. . . .

"In an alternative means case, where a single offense may be committed in more than one way, there must be jury unanimity as to guilt for the single crime charged." *State v. Kitchen,* 110 Wash.2d 403, 410, 756 P.2d 105 (1988) (emphasis in original). But jury unanimity is not required as to the means by which the crime was committed so long as substantial evidence supports each alternative means. A reviewing court must determine whether a rational trier of fact could have found each means of committing the crime proved beyond a reasonable doubt. Furthermore, "[i]f one of the alternative means upon which a charge is based fails and there is only a general verdict, the verdict cannot stand unless the reviewing court can determine that the verdict was founded upon one of the methods with regard to which substantial evidence was introduced." *State v. Bland,* 71 Wash.App. 345, 354, (1993).

The jury here was provided with all three alternative means under the common law definitions of assault. (*See* Footnote 1.) Simuel contends that there was insufficient evidence to support one of the

*(Continues)*

alternative means—actual battery with a knife. The State concedes that the actual battery instruction should not have been given because it was not supported by the evidence. However, the State argues that such error was essentially "harmless" under *State v. Bland,* 71 Wash.App. at 354. We agree with the State.

In *Bland,* the defendant argued that the State failed to prove one of the alternative means for two counts of second degree assault with a deadly weapon. We agreed that there was not substantial evidence to support the actual battery alternative means for one of the counts of second degree assault, but we nevertheless upheld that conviction because it was clear that the verdict was based on the defendant's use of the gun to threaten the victim. In other words, "[the defendant] did not touch, strike, or cut [the victim] with the gun. There is no contention that he did. Thus, the first alternative means of committing assault—touching, striking, or cutting— clearly could not apply to count 1." We were therefore able to determine that the verdict was based on the reasonable fear and apprehension means, for which there was substantial evidence. *Id.* We noted that in effect, the State made an election between the alternate means, even though the State was not required to do so.

Likewise, in the present case, no evidence was presented to suggest that Simuel touched, struck, or cut Wendel with the knife. In fact, it became apparent during closing argument that the State did not contend that Simuel committed assault by actual battery. Rather, the prosecutor explicitly stated that Simuel instead committed assault by causing fear and apprehension:

[Prosecuting Attorney]:

Because if you look to the way assault is defined, it's very, very specific. There's all sorts of ways to commit an assault. You can take a bat and crack it over somebody['s] head, and obviously, that is an assault. You can punch somebody in the nose, and that's an assault. And that's kind of the assaults that we're all used to, when people hit each other, cut each other.

Well, in the law, an assault is also defined very technically as creating fear in another individual. If I grabbed a bat and came up to Juror No. 14 and held it up to her head as though I were going to swing it and I caused fear on her part, then that, in fact, is an assault, and that's what this instruction says. Because it says an assault is also an act with unlawful force done with the intent to create in another apprehension and fear of bodily injury and which, in fact, creates in another a reasonable apprehension and imminent fear of bodily injury even though the actor did not actually intend to inflict bodily injury.

\* \* \*

If I came up to Juror No. 14 with a baseball bat in my hand as though I was going to swing at her head, even though I do not swing, if I create fear in her that I may swing and that I may cause her bodily injury, then I have committed an assault. And that is what the Defendant did to Wayne Wendel.

Because if you remember what the Defendant did, according to the witnesses in this case, is that he took this knife with the blade exposed and he swung it at him between his chin and his throat. It was approximately a foot away. And if you remember right, Wayne Wendel said that he had to jump back. He had to jump back about a foot, and this was about a foot away. The Defendant, by swinging this knife at that man's throat, in between his chin and his throat, was the direct testimony, committed an assault. He created in Wayne Wendel an apprehension, a fear, a concern, about being caused bodily injury.

\* \* \*

And again, we get back to the same facts all over again, and that is what do we have here? We have a razor-sharp knife. How was it used? We have someone swinging wildly at another person's face. Where on their face is that person swinging? Swinging between his chin and his throat. How far away is that person swinging? About a foot away. How was that contact avoided? By the victim in this case jumping back a foot.

\* \* \*

Now, if everything is as straightforward as I have put it, the Defendant assaulted Wayne Wendel by slashing at him between his chin and his throat a foot away. . . . 5VRP 504–506, 511–512.

As in Bland, the State clearly "elected" to pursue only the reasonable fear and apprehension means, rather than the actual battery means.

The prosecutor's closing argument focused exclusively on the fear and apprehension means, and the record is completely devoid of any evidence that Simuel touched, struck, or cut Wendel with the knife. Thus, although there was insufficient evidence to support actual battery, we are satisfied that the jury could not have relied on that alternate means of assault. We further conclude that given the testimony at trial, the remaining two alternative means (attempted battery and causing fear and apprehension) are well supported by substantial evidence. As such, there is no danger that the jury's verdict rested on an unsupported alternative means. Because we can determine from our record that the verdict was founded upon one of the methods supported by substantial evidence, reversal is not required.

### Notes and Questions

1. What did Simuel do that constituted assault? What kind of assault was this?
2. Does it make sense to you for assault and battery to be two separate statutory crimes? Is this merely a meaningless common law carryover or is there good reason to have them be two distinct offenses? The Model Penal Code and many states, including Washington (where Simuel's trial was held), have merged assault and battery into one crime. Section 211.1 of the Model Penal Code reads as follows:

   (1) *Simple Assault.* A person is guilty of assault if he:
      (a) attempts to cause or purposely, knowingly or recklessly causes bodily harm to another; or
      (b) negligently causes bodily injury to another with a deadly weapon; or
      (c) attempts by physical menace to put another in fear of imminent serious bodily injury.

   Paragraph (a) includes battery as well as attempted battery assault, whereas paragraph (c) covers threatened battery assault. Does this statutory construction make sense to you? In your opinion, what were the reasons for this construction?

## Aggravated Assault

Assaults and batteries are generally categorized as either simple or aggravated. Initially, assaults are presumed to be simple; however, there are several ways by which they can become aggravated. The two most common modes of **aggravated assault and battery** are (1) a battery committed or attempted to be committed with the intent to kill or inflict serious bodily harm, or (2) an assault or battery committed with a dangerous or deadly weapon/instrument.

The question that often arises in aggravated assault and battery prosecutions is what is a dangerous or deadly weapon. Obviously a gun or a knife qualifies, but what about more innocuous objects, such as a pin? A chair? Fists? Teeth? Consider the following case.

### *People v. Owusu*, 712 N.E.2d 1228, 609 N.Y.S.2d 863 (1999)

Wesley, J.

In this case we are called upon to decide whether an individual's teeth can constitute a "dangerous instrument" within the meaning of Penal Law § 10.00(13). While the use of an object to produce injury is an appropriate analytical vehicle to determine whether an object is dangerous, the statute's ordinary meaning, its legislative history and our jurisprudence persuade us that an individual's body part does not constitute an instrument under the statute. We therefore reverse the order of the Appellate Division.

Defendant Maxwell Owusu is charged in a 13-count indictment. . . . Four counts of the indictment contain the aggravating factor that defendant used or threatened to use a dangerous instrument.

These charges stem from an incident in which defendant forced his way into his estranged wife's apartment and became embroiled in a fight with another man. During the fight, defendant bit the victim's finger so severely that nerves were severed. . . .

In *People v. Carter* (53 N.Y.2d 113), we stated that "'any instrument, article or substance,' no matter how innocuous it may appear to be when used for its legitimate purpose, becomes a dangerous instrument when it is used in a manner which renders it readily capable of causing serious physical injury." The People point to *Carter* and the recodification of the Penal Law in 1967 as clear indications that the meaning of "dangerous instrument" should be subject to case-by-case

*(Continues)*

(*Continued*)

functional inquiries into the use of instruments, articles or substances (see, Penal Law § 10.00[13]). (*See* Footnote 2.) . . .

The starting point of statutory interpretation is, of course, plain meaning (*Council of the City of New York v. Giuliani,* 1999 WL 02634). In our view, the plain meaning of instrument in this context is a device or object which is capable of causing harm as defined by the statute. One's hands, teeth and other body parts are not, in common parlance, "instruments."

The Penal Law and our jurisprudence have long recognized that how an object is used determines if it is "dangerous." Neither the Legislature nor the courts, however, have classified a person's hands, teeth or other body part as a weapon or instrument. Contrary to the People's position, the definition of a dangerous weapon under former law parallels the current definition of a dangerous instrument. The prior statute included any "other instrument or thing likely to produce grievous bodily harm . . ." (former Penal Law § 242 [4]; see also, former Penal Law § 1897). The 1937 Law Revision Commission Report noted that killing "with bare fists cannot be said to be effecting death with a 'dangerous weapon,' and . . . , a fatal shooting . . . may be termed killing with a 'dangerous instrument' as a matter of law" (1937 Report of N.Y. Law Revision Commission, p. 728). The Commission noted that "[b]eyond this statement there is no absolute certainty," but the extreme positions, at least, were considered well defined (*id.*).

In working on the recodification of the Penal Law, the Law Revision Commission noted that the proposed "dangerous instrument" provision was meant to ". . . includ[e] assaults committed with knives, crowbars, etc., as well as those committed with firearms, blackjack, metal knuckles, etc. [*i.e.,* the enumerated devices]" (Commission Staff Comments on Changes in the New Penal Law Since the 1964 Study Bill, McKinney's Revised Penal Law Special Pamphlet, p. 272). There is no indication that the purpose was to expand the definition of dangerous instrument, as it was then understood, to include the human body itself, and indeed the specific reference to obviously dangerous objects such as "knives" and "crowbars" suggests that such an expansive reading was not at all intended. Thus, our analysis is not premised on placing an exclusion in the statute's definition; rather, it is founded on the well-documented legislative history that a body part was never considered a dangerous weapon or instrument. . . .

The proper statutory interpretation can only be reached upon careful objective historical and structural analysis. In light of this long history, reflected in the staff comments of both the 1937 and 1964 Law Revision Commissions, the best evidence suggests that the Legislature always intended that the "dangerous instrument" concept be limited to external objects.

Our jurisprudence also reflects that, although the Penal Law invoked criminal liability for innocuous objects capable of causing physical harm, a person's body was never considered to fall within the statute's scope. In *People v. Adamkiewicz* (298 N.Y. 176), this Court assessed whether a particular item was dangerous in terms of its use in a given circumstance. There, the Court was construing Penal Law § 1897 (carrying and use of dangerous weapon). The statute set forth two categories of "dangerous weapon[s]": one, an itemized list of objects, and the second enumerating various types of items as well as "any other dangerous or deadly instrument, or weapon." The Court held that the second category was broad enough to include an ice pick "because of its obviously inherent dangerous and lethal character."

The following year, in *People v. Vollmer* (299 N.Y.2d 347), this Court held that the defendant could not be found to have used a dangerous instrument under the first degree manslaughter statute when he beat a man to death with his bare hands, because "[w]hen the Legislature talks of a 'dangerous weapon,' it means something quite different from the bare fist of an ordinary man. . . ." The People would have us ignore *Vollmer* (and the historical development of the dangerous instrument concept) on the basis that the 1967 revision of the Penal Law greatly expanded the dangerous instrument concept. As indicated, however, that is simply not the case.

Nor can the argument be made that by the "ordinary man" language in *Vollmer* the Court meant to leave the door open to the possibility that the hands of a boxer or martial arts expert could constitute dangerous instruments (see, Dissent, at 6–7). This would create an interesting anomaly itself, insofar as the defendant in *Vollmer* beat the victim to death with his "ordinary" hands. An "extraordinary man" rule would create increased criminal liability for use of a dangerous instrument where a heavyweight champion merely threatens a blow, but not where an ordinary man beats another to death. *Vollmer* sensibly avoided such a strained interpretation by concluding fists are not dangerous instruments. This conclusion was reached not because the defendant's fists, as utilized under those circumstances, were not readily capable of causing death (they in fact did), but because they were simply his hands, nothing more.

In every case where this Court has been called upon to decide whether something constitutes a dangerous instrument, the focus has been on an object. In *People v. Galvin* (65 N.Y.2d 761), for example, the defendant grasped the victim's head and smashed it into a sidewalk, causing severe injuries. The Court held that the sidewalk constituted the dangerous instrument (see also, *People v. Curtis*, 89 N.Y.2d 1003 [belt wrapped around defendant's fist]; *People v. Vasquez*, 88 N.Y.2d 561 [wadded-up paper towel stuffed in victim's mouth]; *People v. Cwikla*, 46 N.Y.2d 434 [handkerchief used to gag an elderly victim, choking her to death]). Thus, our "use-oriented approach" has always been directed at understanding if an instrument is (or can be) "dangerous"; it has not been used as a guide to determine if the means by which the victim was injured was an "instrument." . . .

Increased criminal liability arises from the use or threatened use of a dangerous instrument because the actor has upped the ante by employing a device to assist in the criminal endeavor (see, *Commonwealth v. Davis*, 10 Mass App Ct 190, 196–197, [holding that teeth can not constitute "dangerous weapon"]). Mr. Owusu's teeth came with him. In our view, that alone should not expose him to criminal liability beyond that measured by the extent of his victim's injury, and the Supreme Court properly dismissed or reduced those counts of the indictment predicated upon defendant's use of a dangerous instrument (see generally, Annotation, Parts of Human Body as Dangerous Weapons, 8 ALA 4th 1269 [noting that "the main line of authority . . . is to the effect that in no circumstances can fists or teeth be found to constitute deadly or dangerous weapons . . ."]; *Commonwealth v. Davis, supra,* 10 Mass App Ct, at 194, [noting that "[t]he clear weight of authority is to the effect that bodily parts alone cannot constitute a dangerous weapon . . ."]). . . .

Our decision does not plow new ground—the interpretation and application of the statute is [sic] firmly rooted both in the efforts of the Legislature and in our jurisprudence. Indeed, the dissenter concedes that in 1949 this Court unanimously ruled that a man's hands—even though they killed another—were not dangerous instruments. Did the *Vollmer* Court overlook this purported long standing, "all encompassing meaning"? We think not. The Court then (and now) did not engage in judicial legislation, it simply tried to make sense of a statutory term whose meaning was the focus of the case. The Court looked to its precedents and the legislative history of the statute, as we have done. To now change the statute's sweep will unsettle the careful statutory scheme of increased criminal liability arising from the use, or threatened use, of a dangerous instrument, and will result in applications that are without common sense.

Accordingly, the order of the Appellate Division should be reversed and the order of the Supreme Court reinstated.

BELLACOSA, J. (dissenting): . . .

The only question in this purely statutory interpretation appeal, affecting four serious counts of the indictment against defendant, is whether defendant's teeth constituted a "dangerous instrument," as that term is defined in the pertinent statutes of the Penal Law.

The Majority rules not only that defendant's teeth did not constitute an "instrument" in this case, but also that, as a matter of law, under no circumstances may any part of any person's natural body ever qualify as a "dangerous instrument"—the full statutory phrase. By this holding, the Majority substantially subtracts from the statute, while materially augmenting the breadth of *People v. Vollmer* (299 NY 347), and disregarding the import of *People v. Carter* (53 N.Y.2d 113, 115). . . . The reversal analysis is, thus, substantially outweighed by an array of countervailing authorities supporting the affirmance that I urge.

In sum, the Majority's per se holding contradicts:

- the plain, unqualified words of the statute;
- the more cogent legislative history, if that source must be used;
- well-settled statutory interpretation precedents that circumscribe judicial activity reserved to the legislative realm;
- practical context and realistic application of the prosecutorial tool;
- and last, but not least, commonsense.

Yet, the Majority summarily dismisses these complementary affirmance points, while this lone dissenter maintains that each ground is quite cogent in supporting a more measured resolution of this case and its key issue.

*(Continues)*

(*Continued*)

## II.

I begin my assessment with the legislatively-decreed plain language definition of a "dangerous instrument": "'Dangerous instrument' means any instrument, article or substance, including a 'vehicle' as that term is defined in this section, which, under the circumstances in which it is used, attempted to be used or threatened to be used, is readily capable of causing death or other serious physical injury" (Penal Law § 10.00[13]). The Majority finds it necessary to search beneath the embracive and unconditional statutory language. Its eschewal of the orderly, threshold plain language rubric, in judicial statutory interpretive analysis, is seriously unsettling in and of itself.

Anything that can be used to cause death or serious injury fits within the meaning of the sweeping statutory words, the only controlling definition. Undeniably, too, the statutory formula contains no exclusion for parts of the human body or for anything else for that matter. Thus, carving a substantive and categorical exclusion out of the statute is beyond this Court's allocated role in the distribution of lawmaking authority; it seems to me that what is being done here is nothing less than the functional equivalent of judicial legislation.

## III.

In addition to the plain language direction as to how to discern the meaning of this statute, the legislative history of Penal Law § 10.00(13), when courts must turn to that analytical step, offers useful instruction. The history does not support the Majority's departure from the case-by-case application of the pertinent statutes as to whether particular human body parts used by particular defendants may be prosecuted at all. The Majority mistakenly stresses that the Penal Law Revision Commission, in working on the recodification of the Penal Law in the 1960s, noted that the proposed "dangerous instrument" provision was meant to "includ[e] assaults committed with knives, crowbars, etc. as well as those committed with firearms, blackjack, metal knuckles, etc." (Majority Opn, at p 5.)

Using that list, the Majority writes in an unrestricted, new and constrictive boundary of its own making, and reaches a conclusion that "the specific reference to obviously dangerous objects such as 'knives' and 'crowbars' suggests" that an expansive reading of the definition to include parts of the human body was not intended (Majority Opn, at p 4). This turns the analytical process inside out. The judicial interpretive inquiry that should be dispositive here is whether the Legislature used plain language to reflect its intent on the question of excluding all body parts from the words "any instrument." The Legislature did not do so, or even so indicate. That should bind the courts to the standard interpretive method and canons of construction.

In any event, the secondary source materials relied on by the Majority plainly reflect an illustrative list, not an exhaustive or exclusive one. Indeed, that limited list of "obviously dangerous" objects, up to now, has supported this Court's meticulous, common-law approach to determining such issues on a case-by-case, fact-intensive basis. Thus, other innately innocent or innocuous items such as paper towels, a sidewalk, rubber boots and even a handkerchief have consistently over many years been found to meet the definition of a "dangerous instrument," when used in a manner which rendered them capable; of causing serious physical injury. . . .

## IV.

I turn next to *People v. Vollmer* (299 N.Y. 347). The Majority relies on an excerpted sentence from the case: "When the Legislature talks of a 'dangerous weapon,' it means something quite different from the bare fist of an ordinary man." Yet, the holding of that case itself falls far short of the sweep of this case flying on the wing of that one. Instead of espousing an escalation to an absolute bar against parts of the human body ever being considered as dangerous instruments, *Vollmer* precisely confines the reach of its holding to "an ordinary" person's use. If the *Vollmer* Court had intended to exclude the fists of any person, it would not have included the telling adjective. Tacking on the "ordinary" qualifier plainly denotes that the Court considered it possible that the fist of an extraordinary person—such as a boxer, a martial arts expert or even a comparatively large or strong person—could constitute a "dangerous instrument," under the statute and in case-by-case interstitial developmental circumstances. Thus, neither the Legislature's words ("any instrument") nor this Court's words ("ordinary" person) support the crimp now squeezed onto the operation of the pertinent statute by this transmutive holding.

Additionally, *Vollmer* is particularly limited to bare fists. While it may take special skill or disproportionate size or strength to render a fist a dangerous instrument under the circumstances in which it is used, "ordinary" folks are capable of using natural teeth to inflict serious injury, even upon larger or stronger victims. Mike Tyson's fists, we may all agree, were not intended to be covered by *Vollmer*; Evander Holyfield might then legitimately wonder about the implied extrapolation from this case holding that Tyson's (or anyone else's) "choppers" could never be deemed or used as an "instrument" of dangerous propensities or properties.

Since *Vollmer* leaves open the possibility that the bare fists of an extraordinary person might constitute a "dangerous instrument," it sensibly follows—though rejected by the Majority as a matter of law—that teeth, naturally (as in the instant case) or artificially enhanced or replaced, could also be used in a manner that might constitute a "dangerous instrument," as a matter of fact and proper proof. Yet, the Majority's projection of *Vollmer* now takes hands and teeth, and every other human part, out of the realm of realistically possible prosecutions, as a matter of law.

## V.

Other courts have probed whether teeth can constitute a dangerous instrument under respective and different statutory formulations. Some refuse to draw an artificial line, viewing the categorical approach as an "exercise in empty formalism" (*United States v. Sturgis*, 48 F.3d 784, 788, [4th Cir]; *United States v. Moore*, 846 F.2d 1163 [8th Cir]). The Fourth Circuit aptly reasoned:

> The test of whether a particular object was used as a dangerous weapon is not so mechanical that it can be readily reduced to a question of law. Rather, it must be left to the jury to determine whether, under the circumstances of each case, the defendant used some instrumentality, object, or (in some instances) a part of his body to cause death or serious injury. This test clearly invites a functional inquiry into the use of the instrument rather than a metaphysical reflection on its nature. (*United States v. Sturgis, supra.*)

\* \* \*

In sum, the strain of the Majority's formalistic logic falls over backwards from the extreme conclusion that body parts can never be "dangerous instruments." If ever Justice Oliver Wendell Holmes' aphorism applied—"[t]he life of the law has not been logic, it has been experience"—this case proves his lesson (Holmes, The Common Law, 1 [1881]). . . .

## VII.

Having tried to present my own thesis and analysis, and to address the Majority's multi-pronged rationale for overriding the plain language of Penal Law § 10.00(13), I return to where all courts are obligated to start: the legislatively-decreed utterance of the definition of "dangerous instrument." It is "*any instrument, article or substance* . . . which, under the circumstances in which it is used, attempted to be used or threatened to be used, is readily capable of causing death or other serious physical injury" (emphasis added). These legislative words are supple and elastic, not constrictive and rigid. If natural teeth can never be "any instrument, article or substance," the Legislature should remove them from the sensible operation of the statute—and surely knows how to do that.

Judicial work is an art, not a science. The craft traces, delves and shapes words—the Legislature's and the Judiciary's. Unless or until the Legislature exercises its prerogative (*see* Footnote 2) of performing major surgery on a statute, "the Court should construe [the statute] so as to give effect to the plain meaning of the words used."

In sum, I have searched the statute itself in vain for words, or even import, of substantial and categorical limitation. Likewise, I have searched the Majority's reversal rationale. I come up equally empty as to why the Court today chooses to retreat from its well-settled statutory interpretation methodology and precedents to decide this kind of case in this way. There is a compelling necessity to express this dissenting view, especially in this statutory construction case because a blanket bar is being created. Moreover, the Legislature is entitled to have the Majority's declaration brought to its more pointed attention for amendatory consideration, if that Branch of lawmaking government so desires. The standard mechanism for accomplishing this is the casting of my dissenting vote, through this Opinion that the Majority misconstrues the Legislature's original intent and misaligns the array of authorities that support an affirmance of a correct and nimble Appellate Division Majority decision.

*(Continues)*

**Notes and Questions**

1. Does the majority opinion partake in "judicial legislation"?
2. Under the dissent's theory, could anything be considered a deadly or dangerous weapon depending on how it is used? Why or why not?

## Stalking and Harassment

In the late 1980s several high-profile murders, including the 1989 killing of actress Rebecca Schaeffer, occurred after the killers aggressively stalked their victims. In response to these tragedies, in 1990 California became the first state to criminalize **stalking** in the United States. Today every state has statutes that make stalking a criminal offense. In general, these statutes criminalize a clear pattern of behavior in which a person follows, harasses, or threatens another person, putting that person in fear for his or her safety.

The enforcement of anti-stalking statutes has not been without controversy. As discussed in *State v. Evans*, the prosecution of alleged stalking is often centered around circumstantial evidence of an offender's truly frightening behavior. As you read *Evans*, consider at what point Evans' conduct ceases to be legal and turns into stalking.

### *State v. Evans*, 671 N.W.2d 720 (Iowa, 2003)

Streit, Justice. . .

### I. Background and Facts

The evidence, as viewed most favorably to the state, reveals the following facts. Hubert Gene Evans has a foot fetish. He is also, it appears, a published photographer. (*See* Footnote 1.) While a nursing student at Scott County Community College in the late 1990s, Evans asked Rebecca Arnold if she would let him photograph her feet. Evans told Arnold he took pictures for a magazine in New York and had helped other women become "big models." Arnold declined Evans' offer.

In 1998, Evans called Arnold at her house, and left a message with Arnold's father. Evans called a second time. Arnold's father told him not to call her again.

In 2000, Arnold saw Evans in a drug store parking lot. As Arnold made her way from her car to the store, Evans asked her if he could take pictures of her feet. He told her he would like her "legs bare and apart." A nervous and repulsed Arnold again declined Evans' offer, and went into the store.

On August 7, 2001, Evans called Arnold's house and left his phone number and a message with Arnold's mother. In the message, Evans said he had a surprise for Arnold. Arnold did not return Evans' phone call. Worried Evans would discover where she lived, Arnold called the police the next day. At a later date, this incident resulted in a charge of harassment, of which a jury found Evans not guilty.

On August 31, 2001, Evans went to Arnold's house. Evans asked Arnold if she would be willing to play a dominant female role in a new pictorial he was producing about Dred Scott. Arnold told Evans she thought it was weird. Evans replied, "Well, I do weird things because I'm a weird guy." Evans offered Arnold $100 simply to read his script. Arnold repeatedly said she wasn't interested. After Evans left, Arnold called the police. This encounter, too, would later result in another harassment charge against Evans. On this count, a jury found him guilty.

After this incident, Arnold saw Evans' phone number on her caller identification screen four or five times.

On September 21, 2001, Arnold went to a car wash. While standing at a change machine, Evans approached her. Evans told her he noticed she had gotten a new car. As she fled the car wash, Arnold, attempting to mislead Evans, told him the car wasn't hers. Arnold left without getting her change or washing her car. That evening, Evans called Arnold at home. Arnold "told him [she] wasn't interested" and immediately hung up.

On September 25, 2001, Evans and a friend went to Arnold's house. A frightened Arnold asked her mother to answer the door and to tell Evans she wasn't home. Evans told Arnold's mother to let

Arnold know he had dropped by and wanted to take photographs of her. As Evans left, Arnold's mother noticed Evans was wearing "red strap high heels" and had painted his toenails red. Eventually, the State charged Evans with a third count of harassment for this second incident at Arnold's home, and a jury again found him guilty.

On October 12, 2001, Arnold saw a woman standing outside her front door. When Arnold opened the door, Evans, who apparently had been bent over in front of the woman, "pop[ped] up" and asked Arnold for a drink of water. Arnold cussed at Evans, told him to "get off my property," and slammed the door shut. Arnold called the police. She then went to her window in order to try to catch Evans' license plate. Evans smiled and waved at Arnold. This incident did not result in a harassment charge, but was used to prove Evans stalked Arnold.

Evans was arrested for all these incidents and tried by jury for first-offense stalking (Count I) and three counts of first-degree harassment (Counts II-IV). *See* Iowa Code §§ 708.11, 708.7 (2001). Whereas the stalking charge drew upon the entire course of the defendant's conduct which culminated with Evans' appearance at Arnold's house on October 12, the three harassment charges were focused upon three discrete incidents, as indicated above: 1) on August 7, 2001, when Evans left a message for Arnold indicating he had a "surprise" for her (Count II); 2) on August 31, 2001, when Evans asked Arnold to read his "script" (Count III); and 3) on September 25, 2001, when Evans went to Arnold's house in high heels and with painted toenails (Count IV). The harassment charges were elevated to the first-degree because Evans had previously been convicted of harassment four times. *See id.* § 708.7(2) (enhancing harassment charge to the first-degree where defendant has three or more convictions for harassment within the previous ten years).

A jury convicted Evans of stalking and two counts of first-degree harassment. The jury acquitted Evans of Count II. The judge sentenced Evans to a term not to exceed two years of prison for each conviction, to be served consecutively. He also ordered Evans to pay $3000 in fines.

On appeal, Evans claims there is insufficient evidence to support the convictions . . .

## 1. Harassment

Evans was convicted of two counts of first-degree **harassment** in violation of Iowa Code sections 708.7(2) and 708.7(1)(*b*). In order to address Evans' sufficiency challenge, we review Iowa Code section 708.7(1)(*b*), which states, in relevant part:

> A person commits harassment when the person, purposefully and without legitimate purpose, has personal contact with another person, with the intent to threaten, intimidate, or alarm that other person . . . .

Evans claims that, as a published photographer, his sole purpose in contacting Arnold was to take her picture. Moreover, the defendant maintains he never threatened Arnold; he points out we have yet to sustain a conviction for harassment in the absence of a showing of a threat on the part of the defendant, and therefore ought not to do so here. We are satisfied a reasonable jury could find the defendant guilty on the charges of harassment beyond a reasonable doubt. Threatening the victim is not a necessary element of our harassment statute. To conclude otherwise would ignore the disjunctive language of section 708.7(1)(*b*), which forbids a person from "personally and without legitimate purpose . . . [having] personal contact with another person, with the intent to threaten, intimidate, *or* alarm that other person." . . .

Evans correctly points out harassment is a specific intent crime. Intent is "seldom capable of direct proof," however, and "a trier of fact may infer intent from the normal consequences of one's actions."

Given the prior history between the two parties, we believe the evidence was sufficient to permit a reasonable jury to find the defendant acted with the intent to alarm or intimidate Arnold in the two incidents for which he was convicted of harassment. By August 31, when Evans showed up at Arnold's home uninvited and asked Arnold if she would be willing to play a dominant female role in a new pictorial he was producing about Dred Scott, Arnold had repeatedly rebuffed Evans' offers. By September 25, Evans had made several more unsuccessful attempts to persuade Arnold to pose for him, including the incident at the car wash where Arnold hurriedly ran away from Evans without washing her car or picking up her change. It should not have been unexpected on Evans' part that his conduct would alarm Arnold.

In the context of the history between these two parties, we think a reasonable jury could find, at the very least, Evans intended to alarm Arnold when he showed up at her front door asking to take

*(Continues)*

her picture on August 31. This behavior would easily cause Arnold to feel frightened, disturbed, or in danger; such is the natural consequence of Evans' acts, from which the requisite intent for harassment may be inferred. This inference becomes even stronger by the September 25 incident, in light of the additional unsuccessful attempts which occurred in the interim and the fact Evans appeared wearing red, strappy high heels with matching toenails. There is ample evidence here of a ratcheting up of bizarre and alarming behavior. For both incidents, a rational jury could conclude Evans intended to alarm Arnold.

## 2. Stalking

Evans also claims there was insufficient evidence presented at trial to sustain his conviction for stalking Arnold. Evans contends Arnold never explicitly told him to leave her alone before October 12; rather, he alleges, Arnold only told him she wasn't interested, which, according to Evans, means something entirely different. Moreover, Evans points out there is no evidence the police told Evans to leave Arnold alone nor any allegation Evans violated a protective order. And lastly, absent any threats on his part, Evans maintains there is no evidence a reasonable person would fear bodily injury or death because of his behavior. Again, we must reject Evans' claims.

Iowa's stalking statute states:

A person commits stalking when all of the following occur:

  a. The person purposefully engages in a course of conduct directed at a specific person that would cause a reasonable person to fear bodily injury to, or the death of, that specific person or a member of the specific person's immediate family.
  b. The person has knowledge or should have knowledge that the specific person will be placed in reasonable fear of bodily injury to, or the death of, that specific person or a member of that specific person's immediate family by the course of conduct.
  c. The person's course of conduct induces fear in the specific person of bodily injury to, or the death of, the specific person or a member of the specific person's immediate family.

Proof of all three elements, beyond a reasonable doubt, is necessary to sustain a conviction under Iowa's stalking statute.

We think a reasonable jury could find the prosecution met its burden on all three elements of Iowa Code section 708.11(2) in the present case. Although the defendant never threatened Arnold, making a threat is not an element of our stalking statute. *See State v. Limbrecht,* 600 N.W.2d 316, ("Instead of targeting a 'credible threat,' the statute criminalizes a 'course of conduct' that may or may not include threats."). Nor is it necessary the defendant receive official notice from the police his behavior is causing the victim fear. Given the persistent, repeated, and sexual nature of Evans' questioning, including his unexpected arrivals at Arnold's home, we think a rational jury could find Evans guilty of stalking Arnold.

We think the record clearly shows Evans purposefully engaged in a course of conduct directed at Arnold which would cause a reasonable person to fear bodily injury, and Evans knew Arnold would be placed in such fear. The record discloses a course of conduct in which Arnold repeatedly said she was not interested in Evans' disturbing offers. On two occasions, she fled from Evans. Importantly, Evans' behavior escalated over time, culminating in two uninvited visits to Arnold's home, even though Arnold did not tell Evans where she lived. On the first visit, Evans wore red strappy heels with painted toenails to match; on the second, he hid in front of his companion and jumped out when Arnold answered the door. Given Evans' repeated and ever-escalating attempts to photograph Arnold, her steadfast refusals, and the bizarre sexual nature of Evans' proposals, a rational jury could find a reasonable person would fear injury. Moreover, there is sufficient evidence Evans knew Arnold would have such fears. Nor is there any question Arnold, in fact, feared Evans would injure her. Arnold testified at trial she now takes a different route to work, parks her car in a secure lot, and is afraid to answer the door or even the telephone.

## Notes and Questions

  1. Stalking is largely an anticipatory offense. By criminalizing stalking, states enable police to arrest a person before acts of harassment and pestering turn into acts of violence.
  2. If you were prosecuting a case involving stalking, how would you prove that the defendant knew or should have known that his or her conduct was placing the victim in fear? Should this involve a

reasonable person standard of fear, or should it be case specific depending on the mental state of each victim?

3. The National Center for Victims of Crime maintains the Stalking Resource Center, which operates an excellent website that discusses many issues related to stalking (*see* http://www.ncvc.org/src/main.aspx?dbID=DB_ncvc696). The website contains links to the federal and state anti-stalking statutes in effect across the country. It also has links to current data and research as well as resources available to stalking victims.

## False Imprisonment and Kidnapping

**False imprisonment** is traditionally defined as the unlawful confinement of a person. It does not require that a person actually be confined in a specific place, but rather it involves the unlawful exercise or show of force used to compel a person to go where he or she does not want to go or stay where he or she does not want to stay.

**Kidnapping** can be thought of as aggravated false imprisonment. Under the common law, kidnapping was defined as the forcible abduction or stealing away of a man, woman, or child from his own country and sending him into another.[2] Thus, under the common law, an abduction was not a kidnapping unless the victim was taken to another country.

Although modern statutes no longer require that a person be removed from a country for an act to be kidnapping, just what is sufficient movement or asportation of a person to constitute kidnapping has not always been clear, as illustrated in *State v. Masino.*

### *State v. Masino,* 466 A.2d 955 (N.J. 1983)

Clifford, J. . . .

At 2:00 A.M. on September 23, 1979, L.F. and her friend Cindy went to a disco club, where a man later identified as the defendant asked L.F. to dance. She accepted, but soon changed her mind because she was too tired. Because defendant said he felt insulted, L.F. apologized and invited him to accompany Cindy and her for coffee. Defendant, in his car, followed L.F. and Cindy in the victim's car. On the way Cindy changed her mind about the coffee and asked L.F. to drive her home. After leaving Cindy, L.F. pulled out from Cindy's driveway and drove a short distance down the road to where defendant had parked. L.F. rolled down her window and told defendant she had decided to go home, whereupon defendant asked for directions to the Garden State Parkway. Moments later he entered the passenger door of L.F.'s car and began kissing her. She rebuffed him and insisted that she had to go home. Defendant got out but reappeared on the driver's side, announced "O.K. bitch, now you're going to get it," punched her several times in the face and tried to pry her from her grasp on the steering wheel. Weakened by the blows, L.F. finally let go and alighted from the car. She tried to flee across the road toward Cindy's house but defendant caught her, thrashed her, and warned he would kill her if she made any noise. He dragged her back across the street and down to the pond where he threatened to drown her as he repeatedly thrust her face under water. He then ripped off some of her clothes, sexually assaulted her, beat her again, stripped her completely, and fled with her clothes. L.F. crawled to the street and searched in vain for Cindy's house. After running naked down several streets she finally located her friend's house, from where the police were called.

A jury convicted defendant of, among other things, kidnapping. . . . The kidnapping statute, N.J.S.A. 2C:13-1, reads:

a. Holding for ransom, reward or as a hostage. A person is guilty of kidnapping if he unlawfully removes another from the place where he is found or if he unlawfully confines another with the purpose of holding that person for ransom or reward or as a shield or hostage.

b. Holding for other purposes. A person is guilty of kidnapping if he unlawfully removes another from his place of residence or business, or a substantial distance from the vicinity where he is found, or if he unlawfully confines another for a substantial period, with any of the following purposes:
   (1) To facilitate commission of any crime or flight thereafter;
   (2) To inflict bodily injury on or to terrorize the victim or another; or
   (3) To interfere with the performance of any governmental or political function. . . .

*(Continues)*

*(Continued)*

In its charge the trial court instructed the jury that to convict defendant of kidnapping it had to find that defendant had moved his victim a "substantial distance." The court defined "substantial" as "an ample or considerable amount, quantity, size and so forth . . ." but added that "substantial" referred to a "relative" distance that must have some "bearing on the evil at hand."

The Appellate Division reversed and vacated the kidnapping conviction. Noting that "[t]here was no evidence of the distance defendant moved [the victim] except that it was across the street and to a nearby pond," the court said that while "substantial distance" is normally a jury question, this case presented insufficient evidence from which a jury could make such a finding. Implicit in that holding is the idea that a kidnapping prosecution can fail as a matter of law solely because the distance of the abduction is too small. . . .

## II.

Blackstone defined kidnapping as the "forcible abduction or stealing away of a man, woman, or child from their own country and sending them into another." W. Blackstone, Commentaries 291. From approximately 1790 to 1898 New Jersey's kidnapping statute followed Blackstone's definition but broadened its reach by adding that removal could be from this state into another state or country. In 1898 the law was changed to provide that kidnapping occurred when the asportation of the victim was to any point within this state, or into another state or country. The immediate predecessor to our current kidnapping law, N.J.S.A. 2A:118-1, reads in part:

Any person who kidnaps or steals or forcibly takes away a man, woman or child, and sends or carries, or with intent to send or carry, such man, woman or child to any other point within this state, or into another state, territory or country . . . is guilty of a high misdemeanor, and shall be punished by imprisonment for life, or for such other term of not less than 30 years as the court deems proper.

In discussing the asportation requirement of common law kidnapping the drafters of the Model Penal Code recalled that the removal of the victim from the protection of his friends and his sovereign was kidnapping's prime danger. Asportation out of the state or country contemplated a displacement significant "not only because of the distance and difficulties of repatriation; but especially because the victim was removed beyond the reach of English law and effective aid of his associates." Model Penal Code § 212.1 Comment (Tent. Draft No. 11, 1960), at 13 [Draft]. According to the drafters, the original concept of kidnapping was expanded because, among other reasons, experience had demonstrated that "distance and isolation could be achieved within the realm, and that even distance was not essential to isolating a victim from the law and his friends." Thus the common law asportation requirement evolved so that the asportation began to signify more than distance; it signified isolation and vulnerability to continued harm. The forlorn state of the victim remained the paramount evil of kidnapping.

The elimination of the requirement that the victim be moved a fixed distance gave rise to such highly questionable results as convictions for kidnapping when the victim was forced to move from room to room as a robber ransacked his house. For instance, in *People v. Chessman,* 38 Cal.2d 166, 192, (1951), the court construed the phrase "kidnaps or carries away" to mean the act of forcibly moving the victim any distance whatever, no matter how short or for what purpose, declaring that "[i]t is the fact, not the distance of forcible removal which constitutes kidnapping in this State." In *People v. Wein,* 50 Cal.2d 383, (1958), the court found kidnapping by a defendant who, in the course of robbing and raping several women in their respective homes, forced them to move from room to room, sometimes as little as five feet.

This approach rendered the asportation requirement meaningless. Worse, it meant that kidnapping's harsh sentence, even when the movement constituting the "kidnapping" was simply incidental to the underlying crime of rape or robbery, was available to state public outcry or prosecutorial zeal. The potential for abusive prosecution became evident.

California abandoned the *Chessman/Wein* line of cases in *People v. Daniels,* 80 Cal.Rptr. 897, (1969), which involved a series of rapes and robberies. The defendants in that case would enter a victim's apartment, force her at knife point to another room, rob her and rape her. The court reversed the kidnapping convictions on the ground that the movement in the defendants' crimes was an integral part of the underlying rape or robbery. The court, however, did not attempt to define asportation in quantitative terms such as distance or duration; it held that the asportation must substantially increase the risk of harm over and above that necessarily present in the underlying rape or robbery, and must not be merely incidental to the rape or robbery.

New York's approach to the asportation requirement similarly fluctuated. In *People v. Florio,* 301 N.Y. 46 (1950), the court affirmed kidnapping convictions of defendants who had lured a woman to their car, driven from Manhattan to Queens, and then raped her. While conceding that "detention inevitably occurring during the immediate act of commission of such a crime as rape or robbery would not form a basis for a separate crime of kidnapping," the court held that the circumstances of the case warranted conviction.

Several years later, in *People v. Levy,* 256 N.Y.S.2d 793 (1965), the Court of Appeals strictly construed the asportation requirement in reversing kidnapping convictions of defendants who had accosted a husband and wife, forced them into a car, and driven 27 blocks in a 20 minute span while robbing them. The court noted that the breadth of kidnapping "could literally overrun several other crimes, notably robbery and rape, and in some circumstances, assault, since detention and sometimes confinement, against the will of the victim, frequently accompany these crimes. . . . It is a common experience in robbery, for example, that the victim be confined briefly at gunpoint or bound and detained, or moved into and left in another room or place." The court then overruled *Florio* and limited the "application of the kidnapping statute to 'kidnapping' in the conventional sense in which that term has now come to have acquired meaning."

More recently, in *People v. Miles,* 297 N.Y.S.2d 913, 922 (1969), the New York Court of Appeals elucidated its Levy holding:

> In short, the Levy . . . rule was designed to prevent gross distortion of lesser crimes into a much more serious crime by excess of prosecutorial zeal. It was not designed to merge "true" kidnappings into other crimes merely because the kidnappings were used to accomplish ultimate crimes of lesser or equal or greater gravity. Moreover, it is the rare kidnapping that is an end in itself; almost invariably there is another ultimate crime.

New Jersey was once quite expansive in its concept of kidnapping and ostensibly followed the Chessman/Wein rule. In *State v. Dunlap,* 61 N.J.Super. 582 (1961), the court cited Chessman in holding that kidnapping occurred when the defendants dragged the victim from her parked car into theirs and raped her while driving around the immediate vicinity of the abduction. The Court found no requirement that the victim be taken to any specific destination. The Law Division in *State v. Kress,* 105 N.J.Super. 514, 522 (1969), cited Chessman for the proposition that any movement satisfied the asportation requirement. In that case, however, the defendant bank robber used his victim as a shield, a distinguishing feature of that case and one specifically covered under section (a) of the current statute. In *State v. Ginardi,* 111 N.J.Super. 435 (App.Div. 1970), the defendant forced his way into the victims' car, and while one victim drove, he raped the other. This atrocity continued for 1½ hours over "substantial distances," a circumstance that obviated the necessity of the court deciding whether asportation for a slight distance incidental to the underlying crime would ever be enough to establish kidnapping.

This Court squarely addressed the elements of kidnapping in *State v. Hampton,* 61 N.J. 250 (1972). The victim in that case had just entered her apartment when the defendant crept up behind her, forced her to leave the apartment and to drive him around the town and its environs for about an hour. When she tried to escape, he shot her. The Court upheld his kidnapping conviction:

> In the present case we are satisfied that the conviction for kidnapping was justified. Defendant's act in forcing Mrs. Rayborn to leave her apartment at gun point and to drive him in her car about the countryside at night for about an hour, frequently threatening her with being shot, and in fact causing her to believe he intended to shoot her to the point where in her fright she did risk death to get out of the car, was not an integral part of the commission of the single crime of breaking and entering. The forcible detention could reasonably be considered a separate event deliberately undertaken and warranting separate prosecution. The asportation and detention were not incidental to the underlying crime, and they substantially increased the risk of harm to the victim beyond that normally inherent in breaking and entering. Under the circumstances, the jury was justified in finding guilt of the separate crime of kidnapping. 61 N.J. at 175–76. . . .

As noted above, a literal application of the Chessman/Wein rule allowed for abusive prosecution. The drafters of the Model Penal Code, after which our criminal code is modeled, wrote, "[t]he criminologically non-significant circumstance that the victim was detained or moved incident to [an underlying] crime determines whether the offender lives or dies." Accordingly, the drafters required that the victim be moved a "substantial" distance, and that he be moved from the "vicinity" rather

*(Continues)*

than the "place" where he is found so as "to preclude kidnapping convictions based on trivial changes of location having no bearing on the evil at hand." This philosophy likewise shaped the drafting of our kidnapping statute. It is clear, however, that neither the Model Penal Code nor N.J.S.A. 2C:13-1(b) contemplated "substantial distance" as a linear measurement. The isolation of the victim was viewed as the heart of the offense.

In attempting to ascertain what our legislature envisioned as a "substantial distance" we can begin by noting what is not meant: a movement is not "substantial" simply because it facilitates a crime. As applied to this case, N.J.S.A. 2C:13-1 requires that the victim be moved a substantial distance and that the movement facilitate a crime. To interpret movement in terms of facilitating a crime would be tautological.

More sensible is the interpretation that views a "substantial distance" as one that isolates the victim and exposes him or her to an increased risk of harm. The drafters of the Model Penal Code focused on such isolation:

> [I]f the offense is properly defined so as to be limited to substantial isolation of the victim from his normal environment, it reaches a form of terrifying and dangerous aggression not otherwise adequately punished. . . . A disposition to violence or theft in an actor who takes the trouble to set the scene so that he will have a relatively free hand to deal with his isolated victim is obviously more likely to lead to more dangerous consequences. Draft, *supra*, at 15.

Pennsylvania, which has also adopted the Model Penal Code's language in its kidnapping statute, analyzed the phrase "substantial distance" in *Commonwealth v. Hughes*, 399 A.2d 694 (Pa.Super. 1979). The defendant in that case forced the victim into his car and drove two miles to a woods where he raped her. Noting that the definition of "substantial" could not be "confined to a given linear distance nor a certain time period," the court held that "the legislature intended to exclude from kidnapping the incidental movement of a victim during commission of a crime which does not substantially increase the risk of harm to the victim."

We believe our legislature adopted N.J.S.A. 2C: 13-1 with the same intent. We arrive at this conclusion by weighing the potential for abusive prosecution against the terror of kidnapping and against the increased risk of harm to isolated victims. The above-mentioned comments from the drafters of the Model Penal Code buttress this conclusion. Any argument that our legislature intended to soften its treatment of kidnappers is foreclosed by reference to an early draft of 2C: 13-1, subsequently rejected, that discussed a downgrading provision: "We propose to maximize the kidnapper's incentive to return the victim alive by making first degree penalties apply only when the victim is not 'released alive in a safe place'. . . . Certainly those formulations which authorize extreme penalties unless the victim is 'liberated unharmed' are unsatisfactory. . . ." As it turned out, of course, the legislature did ultimately authorize first degree sentences of 15 to 30 years unless the victim was released unharmed. N.J.S.A. 2C: 13-1 (c). It is evident that the legislature intended harsh treatment for kidnappers; it is further evident that by maximizing the kidnapper's incentive to return the victim unharmed, the legislature realized that the risk of harm attendant upon isolation is the principal danger of the crime.

It would certainly be convenient to fix a linear distance for asportation. It would also be arbitrary and irrational, especially when juxtaposed with kidnappings from a home or business, removal from which has no requisite "substantial distance." See N.J.S.A. 2C:13-1(b).

We hold that one is transported a "substantial distance" if that asportation is criminally significant in the sense of being more than merely incidental to the underlying crime. That determination is made with reference not only to the distance traveled but also to the enhanced risk of harm resulting from the asportation and isolation of the victim. That enhanced risk must not be trivial.

Applying those principles to this case, we are satisfied that a jury could have properly determined that the defendant removed his victim a substantial distance. By dragging her from the roadside to the pond's edge behind a row of trees defendant isolated her and arguably obtained a "free hand to deal with his isolated victim." After assaulting and threatening to drown L.F., defendant stripped her, thereby impeding her ability to follow him from the area and call attention to her plight. She was left beaten, exposed to the elements, and hidden from passersby.

The trial court's charge to the jury, while not a paragon of clarity, allowed the jury fair consideration of these factors. It referred to the distance as being "relative" and instructed that the movement had to have some "bearing on the evil at hand." In determining that the charge under scrutiny did not fall so far short of the test we have made explicit today as to amount to reversible error, we emphasize that a more punctilious charge in the future should explain "substantial distance" in terms of sufficient criminal significance that is more than incidental to the underlying crime and that substantially increases the risk of harm to the victim. The jury should be instructed that if the victim is removed only a slight distance from the vicinity where he or she is found and such movement does not create the isolation and increased risk of harm that are at the heart of N.J.S.A. 2C:13-1(b), then it should not convict.

Furthermore, we underscore the need for strict adherence by prosecutors and trial courts to the elements of kidnapping examined in this opinion. Today's decision is not to be read as a crack in the door against overzealous or creative prosecution for kidnapping nor as encouragement for use of a kidnapping charge as some sort of "bonus" count in an indictment.

So much of the judgment of the Appellate Division as reversed the kidnapping conviction is reversed, and the judgment of conviction of kidnapping is hereby reinstated.

## Notes and Questions

1. Many modern statutes have eliminated the asportation requirement for kidnapping. Instead, they require that the kidnappers confine the victim for an improper purpose. For example, Section 5-11-102 of the Criminal Code of Arkansas provides as follows:

    (a) A person commits the offense of kidnapping if, without consent, he restrains another person so as to interfere substantially with his liberty with the purpose of:
    (1) Holding him for ransom or reward, or for any other act to be performed or not performed for his return or release; or
    (2) Using him as a shield or hostage; or
    (3) Facilitating the commission of any felony or flight thereafter; or
    (4) Inflicting physical injury upon him, or of engaging in sexual intercourse, deviate sexual activity, or sexual contact with him; or
    (5) Terrorizing him or another person; or
    (6) Interfering with the performance of any governmental or political function.

2. Under the Arkansas statute, what would be the result in *Masino*? Why?
3. California law holds that movement of a victim that is incidental to another crime, such as robbery or rape, and does not increase the risk of harm to the victim is not sufficient to give rise to kidnapping. Penal Law § 208. Therefore a defendant who broke into a house, grabbed the victim, and dragged her through the house while looking to see if anyone else was home and then proceeded to rob and rape the victim was not guilty of kidnapping because the asportation was incidental to the other crimes. *People v. Daniels*, 80 Cal. Rptr. 897 (1969).

## Custodial Interference

A crime related to kidnapping is custodial interference. **Custodial interference**, also referred to as child stealing or interference with custody, involves keeping a child away from his or her custodial parent. Although custodial interference statutes provide that this can be accomplished through threats, the statutes also provide that it is a crime to keep a child away from his or her custodial parent through enticements.

Section 212.4 of the Model Penal Code provides as follows:

A person commits an offense if he knowingly or recklessly takes or entices any child under the age of 18 from the custody of its parent, guardian, or other lawful custodian, when he has no privilege to do so.

Custodial interference becomes a difficult issue when it involves parents who are either separated or in the process of getting divorced. At times, children are used as pawns to help either or both parents obtain what they want from the divorce or inflict pain on their soon-to-be former spouse. Although this is a problem for divorce courts, it can become a problem for criminal courts as well.

## State v. Fitouri, 893 P.2d 556 (Or. App. 1995)

Haselton, Judge. . . .

In 1992, defendant, his wife, Jeanette Dieringer, and their four-year-old son lived together in Beaverton. On June 27, 1992, while Dieringer was at a weekend retreat, defendant, without informing Dieringer of his intentions, took the couple's son from Oregon to Frankfurt, Germany, and, ultimately, to Tripoli, Libya, where his family resides. When Dieringer came home on June 28, she found defendant,

*(Continues)*

the child, and most of their belongings were gone. After finding credit card statements showing the purchase of airline tickets, Dieringer began to suspect that defendant had taken their son to Libya. However, at least until October 1992, Dieringer's efforts to contact defendant and their son, or to otherwise positively establish their whereabouts, were unsuccessful.

In early August, Dieringer petitioned for, and was granted, temporary custody of the child. In mid-September, the state obtained two separate indictments against defendant, each of which charged defendant with a single count of custodial interference in the first degree. The custodial interference statute, ORS 163.257, provides:

"(1) A person commits the crime of custodial interference in the first degree if the person violates ORS 163.245 and:
   "(a) Causes the person taken, enticed or kept from the lawful custodian or in violation of a valid joint custody order to be removed from the state; or
   "(b) Exposes that person to a substantial risk of illness or physical injury."

ORS 163.245 provides:

"(1) A person commits the crime of custodial interference in the second degree if, knowing or having reason to know that the person has no legal right to do so, the person takes, entices or keeps another person from the other person's lawful custodian or in violation of a valid joint custody order with intent to hold the other person permanently or for a protracted period of time."

The first of the two indictments alleged that defendant had committed the criminal acts necessary for a conviction for custodial interference in the first degree "on and between June 26 and September 2, 1992." The second indictment charged that defendant had committed those acts "on or between September 3 and September 10, 1992."

On October 1, 1992, Dieringer was finally able to contact defendant in Libya by telephone. Dieringer tape-recorded that conversation and four subsequent telephone conversations she had with defendant. Those recordings show that Dieringer never informed defendant of the custody order or the indictments during those conversations, even when defendant expressed concerns about the consequences of his actions. Defendant said nothing during the course of those recorded conversations that would suggest that, as of the dates alleged in the indictments, he knew that Dieringer had been awarded custody by court order.

In December 1992, defendant returned to the United States with the child. He was immediately arrested and charged with custodial interference.

At trial, at the close of the state's evidence, defendant moved for judgment of acquittal on both indictments. He argued that a parent who is a lawful custodian cannot be liable for custodial interference under ORS 163.245, except for a knowing violation of a joint custody order. Relying on that construction, defendant argued that because the state's evidence was insufficient to show that he knew, as of the dates alleged in the two indictments, that a custody order had awarded Dieringer custody, no rational trier of fact could find that he had the state of mind necessary for a conviction for custodial interference.

The state responded that defendant's reading of the statute was too narrow and that, regardless of the custody order or defendant's knowledge of that order, he could be liable if he knew, or had reasons to know, that his conduct violated Dieringer's parental rights.

The trial court adopted defendant's construction of ORS 163.245. Consequently, it granted defendant's motion for a judgment of acquittal on the count charged in the first (June 26–September 2) indictment, on the ground that no custody order designating Dieringer as the child's lawful custodian was in effect during the greater part of the time period alleged. However, the court denied defendant's motion with respect to the second indictment, concluding that the state had presented sufficient evidence showing that defendant knew that an order granting custody to Dieringer was in effect during the time period (September 3–10) alleged in that indictment.

In submitting the remaining charge to the jury, the court adopted the state's proposed instructions, which employed the general terms of ORS 163.245 and which did not refer to the custody order, much less charge that, to convict, the jury must find that defendant knew or had reason to know of that order. Neither counsel meaningfully addressed the existence of a custody order, or defendant's knowledge thereof, in their closing arguments. In short, although the trial court based its rulings on defendant's narrower construction of ORS 163.245, that construction was never conveyed to the jury. The jury subsequently convicted defendant, and judgment was entered on that verdict.

Defendant assigns error to the trial court's denial of his motion for judgment of acquittal on the count charged in the second indictment. He contends that that ruling was erroneous because there was insufficient evidence to support a finding that, as of the dates alleged, he knew that a custody order had been issued granting Dieringer custody of their child, and nevertheless kept the child in Libya. Although the state disputes that contention, its primary response is a reiteration of its legal position at trial: ORS 163.245 applies to cases where one parent takes and keeps their child away from the other, even in the absence of a custody order. If the statute is so interpreted, the state argues, there was ample evidence to support the conviction.

We conclude that the state's construction of ORS 163.245 is correct and, thus, affirm. *State v. West*, 688 P.2d 406 (Or.App 1984). In *West*, we addressed the meaning of the pivotal statutory phrase, "knowing or having reason to know that the person has no legal right to do so." ORS 163.245(1). There, the parents shared custody pursuant to a joint custody order, and the defendant mother absconded with the child to another state. We affirmed the mother's conviction for custodial interference, holding that her status as one of the child's lawful custodians did not preclude a finding that she violated [the statute]. In particular, we concluded that, although the mother had a right to the physical custody of her child, she had no legal right to remove the child from the state, because in doing so, she necessarily infringed on the father's equal right to custody:

> Clearly, the primary focus of the statute is the protection of the rights and interests of the two victims of the offense: the child and the 'lawful custodian' from whom the child is 'taken, enticed or kept.' The focus is not on the legal status of the one who does the taking, enticing or keeping from.
>
> An award of 'joint custody' means that both parties share the legal responsibility for parental decision-making. . . . [T]he parents simultaneously and continuously share rights and responsibilities. . . .
>
> When defendant removed the child from the state and failed to disclose her whereabouts, she was infringing on rights and responsibilities of the father. The emotional and financial costs suffered by him in trying to locate his daughter are among the primary evils that the statute was intended to deter. 688 P.2d 406.

Defendant attempts to distinguish *West* on the ground that the parents had equal custodial rights pursuant to a joint custody order. That distinction, while accurate, is irrelevant to *West's* essential rationale. That rationale—that one parent has no legal right to infringe upon the rights of another with whom he or she lawfully shares custody—applies equally to parents who share custody without a joint custody order.

We thus conclude that the state's construction of ORS 163.257 is correct. Regardless of the existence of a joint custody order, a child's lawful custodian has no legal right to take or keep a child from another who has equal custodial rights to the child.

Consistent with that construction of ORS 163.245, the state was not obliged to prove that defendant knew, on and between September 3 and September 10, that he was keeping the child away from Dieringer in violation of a joint custody order. Rather, the state was required only to show that defendant knew that his actions infringed on Dieringer's equal custodial rights.

Viewing the evidence in the light most favorable to the state, we conclude that a rational trier of fact could have found that the prosecution met its burden of proof. In particular, the state presented Dieringer's testimony that: (1) defendant never informed her that he intended to take the child out of the country, much less for an indefinite period; (2) defendant removed bills, business documents, and pictures of defendant and the child from the family apartment before leaving; (3) defendant left a note that stated that he had taken the child to Wenatchee, Washington, for "a week or two"; (4) defendant did not contact Dieringer and inform her of his whereabouts; and (5) defendant returned to the United States from Libya with the child only after Dieringer assured him that he would not be punished for his actions. From that evidence, the jury could reasonably view defendant's actions as evincing subterfuge and flight, motivated by his knowledge that he had no legal right to take the child, because doing so infringed on Dieringer's equal custodial rights. The trial court correctly denied defendant's motion for a judgment of acquittal.

Affirmed.

## Notes and Questions

1. Does this decision seem fair? Why or why not?
2. How do the Oregon statutes differ from Section 212.4 of the Model Penal Code? If the Model Penal Code were the law in Oregon, would the result have been different? Why or why not?

## Conclusion

Although homicides receive a lot of attention in the media, other violent crimes are much more common, everyday occurrences in every community in the country. Whether they involve domestic violence, bar brawls, or child stealing, it is non-homicidal violent crimes that make up over 90 percent of all violent crimes committed. Understanding what the elements of these crimes are, as well as the rationale behind them, is critical for anyone who wants to understand the workings of the criminal law.

## Key Terms

aggravated assault and battery
assault
attempted battery assault
battery
custodial interference
false imprisonment
harassment
kidnapping
simple assault
stalking
threatened battery assault

## Review Questions

1. Briefly describe what is meant in the text when it states, ". . . the actor does not actually have to physically touch the victim for a battery to occur." Be sure to list at least one example that would constitute a *battery* when actual touching did not occur.

2. Because *battery* is defined as an unjustified offensive touching, do you believe a player in a major sporting event such as the National Football League or the National Hockey League could ever be charged with battery during game play? Why or why not?

3. Create a scenario that would be deemed *attempted battery assault* and another that would constitute *threatened battery assault* and explain the difference.

4. Because assaults are presumed to be *simple*, it is important to know what constitutes *aggravated assault*. Identify the two most common ways by which *simple assault* can become *aggravated*.

5. It can be argued that throughout the years many instances of assault and battery, and even murder, were precipitated by the *stalking* of the victim by the suspect. Why do you believe it took until 1990 for the first stalking law to "hit the books"?

6. Discuss the relevance of the victim of harassment and/or stalking to involve the police in building a "paper trail." Specifically, do you believe that by involving the police a victim increases the probability of a conviction?

7. How does *false imprisonment* differ from *kidnapping*?

8. Take a position regarding asportation of a kidnapping victim and support your position using any of the cases cited in the text.

9. List the elements required to establish a case of *custodial interference in the first degree* in Oregon.

## Notes

1. *See* White, DV. Note: Sports violence as criminal assault: Development of the doctrine by Canadian courts. *Duke Law J.* 1986;6:1030–1054.
2. *Blackstone Commentaries* 4:219.

## Footnote to *State v. Simuel,* 94 Wash. App. 1021, 1999 WL 106909 (Wash. App. Div. 1)

1. Because "assault" is not defined in the criminal code, Washington courts apply the common law definitions: (1) an attempt, with unlawful force, to inflict bodily injury upon another, (2) an unlawful touching with criminal intent, or (3) putting another in apprehension of harm whether or not the actor intends to inflict or is incapable of inflicting that harm.

## Footnotes to *People v. Owusu,* 712 N.E.2d 1228, 609 N.Y.S.2d 863 (1999)

2. Penal Law § 10.00(13) provides: "Dangerous instrument means any instrument, article or substance . . . which, under the circumstances in which it is used, attempted to be used or threatened to be used, is readily capable of causing death or serious physical injury."

2. With a delicious wordplay particularly apt for this case, the Great Bard provides a masterful conversation among a well known rebel band on how law can be made. In the Second Part of Henry VI, Act 4, Scene 7, ll.7–19, the following ensues:

> Dick [The Butcher]. Only that the laws of England may come out of your mouth

> * * *

> [Jack] Cade. I have thought upon it; it shall be so. Away! Burn all the records of the realm: my mouth shall be the parliament of England.
> Holl[and], (Aside.) Then we are like to have biting statutes, unless his [Jack Cade's] teeth be pulled out. [The Yale Shakespeare (emphasis added).]

## Footnote to *State v. Evans,* 671 N.W.2d 720 (Iowa, 2003)

1. Evans testified his photographs were last printed in *Leg Sex* in 2000.

# Property Crimes and Crimes Against Habitation

## CHAPTER 9

## Chapter Objectives

- Understand the historical development of larceny
- Understand the elements of *larceny*
- Explain the modifications made in modern property crime statutes
- Explain the common law elements of burglary and arson
- Explain the reforms contained in modern burglary and arson statutes
- Understand the elements of *embezzlement*
- Distinguish between *embezzlement* and *theft by false pretenses*
- Understand the offense of *receiving stolen property*
- Define the elements of *robbery* and *extortion*

## Larceny

To fully understand property crimes, one must first have an understanding of their historical development. Larceny was the first common law property crime. It dealt specifically with the taking of property from a person. The fact that the property had to be taken from the victim's possession, whether by stealth or force, limited the law's application. It did not include incidents where a bailee (such as a shipper) refused to return property that was entrusted to him or her by a person or when a person gave another some property as a result of fraudulent activity. Because the property was voluntarily given to the scoundrel, it was not taken from his or her possession and therefore beyond the scope of larceny.

It was largely up to legislatures, whether Parliament in England or state bodies in the United States, to get around the limitations of the common law. In the process new crimes were created to fill in the gaps left by larceny. Over time, laws that went beyond the limits of larceny were enacted. Statutes dealing with extortion, embezzlement, and fraud expanded the reach of the law to protect individuals from various forms of thievery. Today, the theft statutes in most states contain language that encompasses larceny as well as other crimes of thievery.

The seven elements of common law **larceny** are

1. A trespassory
2. taking

3. and asportation (carrying away)
4. of the personal property
5. of another
6. with the intent
7. to permanently deprive.

Although modern statutes eased the requirements of what acts can constitute larceny, most statutes still require the state to prove these items before a person can be convicted of larceny. The elements can be broken down into three distinct groups: the act required, the object of the act, and the state of mind of the actor. When broken down in this manner, the elements are easily understood. The *act required* consists of a trespassory taking and asportation. The *object of the taking* must be personal property of another. The *state of mind* of the actor must be that of having the intent to permanently deprive the owner of the property. Although in the abstract the elements of larceny might be easily comprehended, it is the details involved in each of the elements that require explanation for true understanding.

As you read the material in this chapter, it is important to keep in mind that the specific elements of crimes vary from jurisdiction to jurisdiction. Although most states follow the principles of the common law elements, some states eliminated some of the elements whereas others added additional elements.

### Act Required

As stated above, the act required for larceny is trespassory taking and asportation. Under the old common law a taking was trespassory only if property was taken from another's actual possession. If a person was not in actual possession of an object, there was no larceny. For example, if Jim left his horse at the trough while he went to school, Bob's taking the horse was not larceny because Jim was not in actual possession of the horse at the time of the taking.

Over the years courts expanded the meaning of possession to include situations where a person was not in actual possession of an object but still exercised control and dominion over it. This is known as **constructive possession**. In our horse example, Jim, although not in actual possession of his horse, still maintained control and dominion over it; therefore he was still in constructive possession of it. Accordingly, under modern laws Bob would have committed larceny.

The second and third elements, the taking and asportation, should be viewed together. Under the common law, taking required control and dominion over the property, whereas asportation referred to the moving or carrying away of the property. The distance the property was moved was irrelevant. As long as the thief took control and dominion of the property and moved it, the elements of larceny were satisfied.

Modern statutes still require a taking to have occurred before a person can be found guilty of larceny. With the advent of self-service shipping in the 20th century, courts have been faced with the question of when property has been wrongfully taken and thus the person liable for larceny as opposed to merely picking up an item as an innocent shopper. This question was addressed in *People v. Gasparik*.

---

## *People v. Gasparik*, 420 N.E.2d 40, 438 N.Y.S.2d 242 (1981)

Cooke, Chief Judge. . . .

Defendant was in a department store trying on a leather jacket. Two store detectives observed him tear off the price tag and remove a "sensormatic" device designed to set off an alarm if the jacket were carried through a detection machine. There was at least one such machine at the exit of each floor. Defendant placed the tag and the device in the pocket of another jacket on the merchandise rack. He took his own jacket, which he had been carrying with him, and placed it on a table. Leaving his own jacket, defendant put on the leather jacket and walked through the store, still on the same floor, bypassing several cash registers. When he headed for the exit from that floor, in the direction of the main floor, he was apprehended by security personnel. At trial, defendant denied removing the price tag and the sensormatic device from the jacket, and testified that he was looking for a cashier without a long line when he was stopped. The court, sitting without a jury, convicted defendant of petit larceny. Appellate Term affirmed. . . .

Larceny at common law was defined as a trespassory taking and carrying away of the property of another with intent to steal it. The early common-law courts apparently viewed larceny as defending society against breach of the peace, rather than protecting individual property rights, and therefore placed heavy emphasis upon the requirement of a trespassory taking. Thus, a person such as a bailee who had rightfully obtained possession of property from its owner could not be guilty of larceny. The result was that the crime of larceny was quite narrow in scope.

Gradually, the courts began to expand the reach of the offense, initially by subtle alterations in the common-law concept of possession. Thus, for instance, it became a general rule that goods entrusted to an employee were not deemed to be in his possession, but were only considered to be in his custody, so long as he remained on the employer's premises. And, in the case of *Chisser* (Raym. Sir. T. 275, 83 Eng.Rep. 142), it was held that a shop owner retained legal possession of merchandise being examined by a prospective customer until the actual sale was made. In these situations, the employee and the customer would not have been guilty of larceny if they had first obtained lawful possession of the property from the owner. By holding that they had not acquired possession, but merely custody, the court was able to sustain a larceny conviction.

As the reach of larceny expanded, the intent element of the crime became of increasing importance, while the requirement of a trespassory taking became less significant. As a result, the bar against convicting a person who had initially obtained lawful possession of property faded. In *King v. Pear* (1 Leach 212, 168 Eng.Rep. 208), for instance, a defendant who had lied about his address and ultimate destination when renting a horse was found guilty of larceny for later converting the horse. Because of the fraudulent misrepresentation, the court reasoned, the defendant had never obtained legal possession. Thus, "**larceny by trick**" was born.

Later cases went even further, often ignoring the fact that a defendant had initially obtained possession lawfully, and instead focused upon his later intent. The crime of larceny then encompassed, not only situations where the defendant initially obtained property by a trespassory taking, but many situations where an individual, possessing the requisite intent, exercised control over property inconsistent with the continued rights of the owner. During this evolutionary process, the purpose served by the crime of larceny obviously shifted from protecting society's peace to general protection of property rights. . . .

This evolution is particularly relevant to thefts occurring in modern self-service stores. In stores of that type, customers are impliedly invited to examine, try on, and carry about the merchandise on display. Thus in a sense, the owner has consented to the customer's possession of the goods for a limited purpose. That the owner has consented to that possession does not, however, preclude a conviction for larceny. If the customer exercises dominion and control wholly inconsistent with the continued rights of the owner, and the other elements of the crime are present, a larceny has occurred. Such conduct on the part of a customer satisfies the "taking" element of the crime.

It is this element that forms the core of the controversy in these cases. The defendants argue, in essence, that the crime is not established, as a matter of law, unless there is evidence that the customer departed the shop without paying for the merchandise.

Although this court has not addressed the issue, case law from other jurisdictions seems unanimous in holding that a shoplifter need not leave the store to be guilty of larceny. This is because a shopper may treat merchandise in a manner inconsistent with the owner's continued rights—and in a manner not in accord with that of prospective purchaser—without actually walking out of the store. Indeed, depending upon the circumstances of each case, a variety of conduct may be sufficient to allow the trier of fact to find a taking. It would be well-nigh impossible, and unwise, to attempt to delineate all the situations which would establish a taking. But it is possible to identify some of the factors used in determining whether the evidence is sufficient to be submitted to the fact finder.

In many cases, it will be particularly relevant that defendant concealed the goods under clothing or in a container. Such conduct is not generally expected in a self-service store and may in a proper case be deemed an exercise of dominion and control inconsistent with the store's continued rights. Other furtive or unusual behavior on the part of the defendant should also be weighed. Thus, if the defendant surveys the area while secreting the merchandise or abandons his or her own property in exchange for the concealed goods, this may evince larcenous rather than innocent behavior. Relevant too is the customer's proximity to or movement towards one of the store's exits. Certainly it is highly probative of guilt that the customer was in possession of secreted goods just a few short steps from the door or moving in that direction. Finally, possession of a known shoplifting device actually used to conceal merchandise, such as a specially designed outer garment or false bottomed carrying case, would be all but decisive.

Of course, in a particular case, anyone or any combination of these factors may take on special significance. And there may be other considerations, not now identified, which should be examined. So long as its bears upon the principal issue—whether the shopper exercised control wholly inconsistent with the owner's continued rights—any attending circumstance is relevant and may be taken into account.

## V.

Under these principles, there was ample evidence . . . to raise a factual question as to the defendant's guilt. . . .

[The] defendant removed the price tag and sensor device from a jacket, abandoned his own garment, put the jacket on and ultimately headed for the main floor of the store. Removal of the price tag and sensor device, and careful concealment of those items, is highly unusual and suspicious conduct for a shopper. Coupled with defendant's abandonment of his own coat and his attempt to leave the floor, those factors were sufficient to make out a prima facie case of a taking. . . .

## VII.

In sum, in view of the modern definition of the crime of larceny, and its purpose of protecting individual property rights, a taking of property in the self-service store context can be established by evidence that a customer exercised control over merchandise wholly inconsistent with the store's continued rights. Quite simply, a customer who crosses the line between the limited right he or she has to deal with merchandise and the store owner's rights may be subject to prosecution for larceny. Such a rule should foster the legitimate interests and continued operation of self-service shops, a convenience which most members of the society enjoy.

Accordingly, . . . the order of the Appellate Term should be affirmed.

## Notes and Questions

1. Under a strict reading of the common law, do you believe Mr. Gasparik was guilty of larceny? Why?
2. Section 223.2 of the Model Penal Code states, "A person is guilty of theft if he unlawfully takes or exercises control over movable property of another with purpose to deprive him thereof." Under the Model Penal Code definition of theft, is Gasparik guilty?

## Object of the Taking

The item stolen must be the personal property of another for larceny to occur. Under the common law, only tangible, personal property was included. Items such as services, stocks, checks, and other documents were excluded from common law larceny (except for the value of the paper, ink, etc.).

Modern statutes have broadened the scope of larceny to include intangible personal property. In addition, items such as gas, electricity, and cable television service are subject to theft statutes. One item that has caused the courts difficulty in recent years is the unauthorized use of computers.

---

### *Lund v. Commonwealth*, 232 S.E.2d 745 (Va. 1977)

I'Anson, Chief Justice.

Defendant, Charles Walter Lund, was charged in an indictment with the theft of keys, computer cards, computer printouts and using 'without authority computer operation time and services of Computer Center Personnel at Virginia Polytechnic Institute and State University (V.P.I. or University) . . . with intent to defraud, such property and services having a value of one hundred dollars or more.' Code §§ 18.1-100 and 18.1-118 were referred to in the indictment as the applicable statutes. Defendant pleaded not guilty and waived trial by jury. He was found guilty of grand larceny and sentenced to two years in the State penitentiary. The sentence was suspended, and defendant was placed on probation for five years.

Defendant was a graduate student in statistics and a candidate for a Ph.D. degree at V.P.I. The preparation of his dissertation on the subject assigned to him by his faculty advisor required the use of computer operation time and services of the computer center personnel at the University. His faculty advisor neglected to arrange for defendant's use of the computer, but defendant used it without obtaining the proper authorization.

The computer used by the defendant was leased on an annual basis by V.P.I. from the IBM Corporation. The rental was paid by V.P.I. which allocates the cost of the computer center to various departments within the University by charging it to the budget of that department. This is a bookkeeping entry, and no money actually changes hands. The departments are allocated 'computer credits (in dollars) back for their use (on) a proportional basis of their (budgetary) allotments.' Each department manager receives a monthly statement showing the allotments used and the running balance in each account of his department.

An account is established when a duly authorized administrator or 'department head' fills out a form allocating funds to a department of the University and an individual. When such form is received, the computer center assigns an account number to this allocation and provides a key to a locked post office box which is also numbered to the authorized individual and department. The account number and the post office box number are the access code which must be provided with each request before the computer will process a 'deck of cards' prepared by the user and delivered to computer center personnel. The computer print-outs are usually returned to the locked post office box. When the product is too large for the box, a 'check' is placed in the box, and it is used to receive the print-outs at the 'computer center main window.'

Defendant came under surveillance on October 12, 1974, because of complaints from various departments that unauthorized charges were being made to one or more of their accounts. When confronted by the University's investigator, defendant initially denied that he had used the computer service, but later admitted that he had. He gave to the investigator seven keys for boxes assigned to other persons. One of these keys was secreted in his sock. He told the investigating officer he had been given the keys by another student. A large number of computer cards and print-outs were taken from defendant's apartment.

The director of the computer center testified that the unauthorized sum spent out of the accounts associated with the seven post office box keys, amounted to $5,065. He estimated that on the basis of the computer cards and print-outs obtained from the defendant, as much as $26,384.16 in unauthorized computer time had been used by the defendant. He said, however, that the value of the cards and print-outs obtained from the defendant was 'whatever scrap paper is worth.'

Defendant testified that he used the computer without specific authority. He stated that he knew he was a large computer user, but, because he was doing work on his doctoral dissertation, he did not consider this use excessive or that 'he was doing anything wrong.'

Four faculty members testified in defendant's behalf. They all agreed that computer time 'probably would have been' or 'would have been' assigned to defendant if properly requested. Dr. Hinkleman, who replaced defendant's first advisor, testified that the computer time was essential for the defendant

---

to carry out his assignment. He assumed that a sufficient number of computer hours had been arranged by Lund's prior faculty advisor.

The head of the statistics department, at the time of the trial, agreed with the testimony of the faculty members that Lund would have been assigned computer time if properly requested. He also testified that the committee which recommended the awarding of degrees was aware of the charges pending against defendant when he was awarded his doctorate by the University.

The defendant contends that his conviction of grand larceny of the keys, computer cards, and computer print-outs cannot be upheld under the provisions of Code § 18.1-100 because (1) there was no evidence that the articles were stolen, or that they had a value of $100 or more, and (2) computer time and services are not the subject of larceny under the provisions of Code §§ 18.1-100 or 18.1-118.

Code § 18.1-100 (now § 18.2-95) provides as follows:

'Any person who:

(1) Commits larceny from the person of another of money or other thing of value of five dollars or more, or
(2) Commits simple larceny not from the person of another of goods and chattels of the value of one hundred dollars' or more, shall be deemed guilty of grand larceny. . . .'

Section 18.1-118 (now § 18.2-178) provides as follows:

'If any person obtain, by any false pretense or token, from any person, with intent to defraud, money or other property which may be the subject of larceny, he shall be deemed guilty of larceny thereof. . . .'

The Commonwealth concedes that the defendant could not be convicted of grand larceny of the keys and computer cards because there was no evidence that those articles were stolen and that they had a market value of $100 or more. The Commonwealth argues, however, that the evidence shows the defendant violated the provisions of § 18.1-118 when he obtained by false pretense or token, with intent to defraud, the computer print-outs which had a value of over $5,000.

Under the provisions of Code § 18.1-118, for one to be guilty of the crime of larceny by false pretense, he must make a false representation of an existing fact with knowledge of its falsity and, on that basis, obtain from another person money or other property which may be the subject of larceny, with the intent to defraud.

At common law, larceny is the taking and carrying away of the goods and chattels of another with intent to deprive the owner of the possession thereof permanently. Code § 18.1-100 defines grand larceny as a taking from the person of another money or other thing of value of five dollars or more, or the taking not from the person of another goods and chattels of the value of $100 or more. The phrase 'goods and chattels' cannot be interpreted to include computer time and services in light of the often repeated mandate that criminal statutes must be strictly construed.

At common law, labor or services could not be the subject of the crime of false pretense because neither time nor services may be taken and carried away. It has been generally held that, in the absence of a clearly expressed legislative intent, labor or services could not be the subject of the statutory crime of false pretense. Some jurisdictions have amended their criminal codes specifically to make it a crime to obtain labor or services by means of false pretense. We have no such provision in our statutes.

Furthermore, the unauthorized use of the computer is not the subject of larceny. Nowhere in Code §18.1-100 or § 18.1-118 do we find the word 'use.' The language of the statutes connotes more than just the unauthorized use of the property of another. It refers to a taking and carrying away of a certain concrete article of personal property. There it was held that the unauthorized use of machinery and spinning facilities of another to process wool did not constitute larceny under New York's false pretense statute.

We hold that labor and services and the unauthorized use of the University's computer cannot be construed to be subject of larceny under the provisions of Code §§ 18.1-100 and 18.1-118.

The Commonwealth argues that even though the computer print-outs had no market value, their value can be determined by the cost of the labor and services that produced them. We do not agree.

The cost of producing the print-outs is not the proper criterion of value for the purpose here. Where there is no market value of an article that has been stolen, the better rule is that its actual value should be proved. . . .

Here the evidence shows that the print-outs had no ascertainable monetary value to the University or the computer center. Indeed, the director of the computer center stated that the printouts had no

*(Continues)*

more value than scrap paper. Nor is there any evidence of their value to the defendant, and the value to him could only be based on pure speculation and surmise. Hence, the evidence was insufficient to convict the defendant of grand larceny under either Code § 18.1–100 or §18.1-118.

For the reasons stated, the judgment of the trial court is reversed, and the indictment is quashed.

### Notes and Questions

1. Why did the court find that using the computer was not larceny? Doesn't computer time have value? In *State v. McGraw*, 408 N.E.2d 552 (Ind. 1985), the Indiana Supreme Court decided a case of improper computer use at the workplace much like the *Lund* court. In dissent, however, Justice Pivarnik wrote, "Time and use are at the very core of the value of a computer system." *Id.* at 555. Do you agree? If so, would Lund be guilty of larceny? What factors might enter into your decision?

2. A number of cases involving the wrongful use of items, equipment, or personnel to produce a product or service are not larceny. Therefore an Air Force sergeant's use of men under his control to paint his house was not considered larceny in *Chappell v. United States*, 270 F.2d 274 (9th Cir. 1959).

   The Model Penal Code circumvented this by specifically including a section that deals with theft of services. Model Penal Code § 223.7 states as follows:

   > A person is guilty of theft if he purposely obtains services which he knows are available only for compensation, by deception or threat, or by false token, or other means to avoid payment for the service.

## State of Mind

To be convicted of larceny, a person must have the specific intent to permanently deprive the owner of his or her property. Merely intending to take an object is not sufficient; the actor must intend to permanently deprive the owner of it. Assume Tim goes to the university bookstore and puts his blue backpack on a shelf with 50 other backpacks. After shopping Tim picks up an identical backpack from the shelf and leaves the store. Has Tim committed larceny? No. Although he did intend to take the backpack, he did not have the specific intent to permanently deprive the owner of it.

A person who takes another's property with the intent to return it to the owner within a reasonable time has not committed larceny either. Although the taking was wrongful, because the intent to permanently deprive the owner was absent, larceny was not committed. This distinction is discussed in *State v. Hanson*.

### *State v. Hanson*, 446 A.2d 372 (Vt. 1982)

Barney, Chief Justice.

The hapless defendant and a companion came to rest in a Walden snowbank in the early morning hours of March 13, 1980, after a journey which began in Rutland the night before and took them between times to Burlington, Waterbury, Stowe and Montpelier in a succession of vehicles driven at speeds of up to 110 miles per hour. The defendant was subsequently charged with grand larceny of the car he was apprehended in, and with breaking and entering in the nighttime with intent to commit larceny. After trial by jury he was convicted of both crimes and now appeals that conviction on two grounds: error in the jury instruction and insufficient evidence to convict.

The facts of this case flavor the issues. The defendant and his friend, suspecting that the police were looking for them, fled Rutland in a pick-up truck they had taken without consent. They headed for Waterbury, where the defendant sought to pick up some medication. The truck was low on gas so they stopped at a service station, filled the tank, and left without paying. Still, even with gas, the truck was unsatisfactory. It drove too slowly and had no radio.

The pair stopped at a Chevrolet dealership in Waterbury to get another vehicle. The first car they took after admitting themselves to the garage turned out to have a faulty transmission, and after driving a short distance they returned it. The second car they attempted to take was in excellent condition, but somehow, after starting the engine, they managed to lock both doors with the keys inside and the motor running. Finally a third car was chosen, its keys were located, and license plates were taken from the first car and put in place. The two then headed for Montpelier, the defendant driving the car and his friend following in the truck, which they planned to abandon elsewhere.

This third car too was low on gas, so once again they stopped at a service station, filled up, and left without paying. This time the attendant called the police, who arrived to investigate.

Meanwhile the defendant and his friend had abandoned the truck on a back road near Montpelier, and, thinking they were in the clear, continued on. As it happened, however, the road circled back toward Montpelier and the unfortunate pair came out near the same service station they had just visited. Seeing the police, the defendant made a U-turn and the chase began, ending abruptly when defendant's car hit a patch of ice, catapulting it into the snowbank.

The first question raised on appeal is whether the trial court was obligated to instruct the jury that it could find the defendant guilty of the unlawful taking of tangible property as a lesser included offense of grand larceny.

Our law in this area has been clearly spelled out: in order for a defendant to be entitled to jury instruction on a lesser offense than that for which he is charged, the elements of the lesser offense must necessarily be included within the greater offense. What remains in any specific case is an examination of each offense to determine if the requisite identity of elements exists.

Our grand larceny statute provides as follows:

A person who steals from the actual or constructive possession of another, other than from his person, money, goods, [or] chattels . . . shall be imprisoned not more than ten years or fined not more than $500.00, or both, if the money or other property stolen exceeds $100.00 in value.

Stealing has been defined at common law, and identified in our case law, as:

[T]he taking and removal, by trespass, of personal property, which the trespasser knows to belong to another, with the felonious intent to deprive him of his ownership therein. *State v. Grant,* 373 A.2d 847, 849–50 (Vt. 1977); *State v. Levy,* 35 A.2d 853, 854 (Vt. 1944).

This larcenous intent has in turn been further defined in Vermont as an intent to take and keep property of another wrongfully, *State v. Reed,* 253 A.2d 227, 231 (Vt. 1969), so that the trespasser may appropriate it for his own purposes. Thus, larceny was not committed here unless the defendant intended to permanently separate the owner from his property, or at least deliberately act so as to make it unlikely that the owner and his property would be reunited. Wharton's Criminal Law § 363, at 333–34 (1980).

By contrast unlawful taking of tangible personal property, demands that:

A person who, without the consent of the owner, takes and carries away . . . any tangible personal property with the intent of depriving the owner temporarily of the lawful possession of his property shall be fined not more than $100.00. . . .

The language of this statute leaves little room for doubt that the criminal intent involved in this offense is a different intent than that which is implicated in the crime of grand larceny, or that the two intents are mutually exclusive. Still, the defendant argues that the lesser offense is included in the greater offense because anyone who intends to cause a permanent deprivation must also intend to cause a temporary deprivation, "some" time being a necessary component of "all" time. The suggestion appears to be that the intent to permanently deprive is but a succession of lesser intents to temporarily deprive. We do not agree. . . .

Larceny specifically requires an intent to steal at the very moment the property in question is taken into possession by the defendant. Since, as we have noted, stealing requires an intent to keep the property permanently, it is this intent which must be present when the property is taken or the crime of larceny has not been committed, and whatever other offense may have been committed with the same property but a different intent, it was not larceny, nor any offense included within it.

Next the defendant claims error in the court's denial of his motion for acquittal based on a failure to prove intent to steal. His argument is that because the State introduced no direct evidence establishing his intent to permanently deprive the owner of his car, a larceny conviction cannot stand. . . .

Even had the evidence in this case all been circumstantial, we have rejected the notion that it must have been so cogent as to exclude every reasonable hypothesis consistent with innocence, and have endorsed as our test, where sufficiency is in question, whether the evidence, viewed in the light most favorable to the State, is sufficient to convince a reasonable trier of fact that the defendant is guilty beyond a reasonable doubt.

Here, the evidence was more than circumstantial, even on the element of intent. The defendant's companion testified that neither he nor the defendant intended to return the car to its rightful owner when they took it. The question of what they did intend was properly a matter for the jury to decide, based on all the evidence before it.

Judgment affirmed.

*(Continues)*

## Notes and Questions

1. If Hanson took three cars and a truck, why was he convicted of stealing only one car?
2. Note that two kinds of intent are required for larceny: Both the general intent to commit the act of stealing and the specific intent to permanently deprive the owner of the property taken must be proven by the prosecution. Were both forms of intent present in *Hanson*?
3. What if Joe intends to steal Wendy's book. Joe takes the book intending to keep it forever. After reading the first 10 pages Joe decides it is not that good and decides to return the book to Wendy. To Joe's chagrin, after he returns the book to Wendy, she calls the police. The police arrest Joe for the larceny of the book. Joe's defense is that he returned the book and therefore did not have the intent to permanently deprive. Is Joe guilty? Yes. Once he took the book intending at the time to permanently deprive Wendy of it, the crime was completed. What he did with the book after is immaterial. Joe is guilty. On the other hand, if Joe took Wendy's book only intending to read it and return it once he completed it, he would not be guilty of larceny.

## Embezzlement

Embezzlement is a crime created by statute designed to fill in the gaps left under common law larceny. **Embezzlement** occurs when a person converts property entrusted to him or her to his or her own use in a manner that demonstrates his or her intent to deprive another person use of or access to the property. If a person is in lawful possession of property and subsequently acts in a way to seriously interfere with the property in manner not contemplated by the owner, the person has committed embezzlement. The elements and some of the nuances of embezzlement statutes are laid out in *State v. Frasher*.

### *State v. Frasher*, 265 S.E.2d 43 (W. Va. 1980)

Miller, Justice:

Eddie A. Frasher appeals his embezzlement conviction from the Circuit Court of Kanawha County. His primary assignments of error are that the State failed to prove that the true owner's property was embezzled and that he was an agent within the meaning of the embezzlement statute. . . .

The defendant Frasher operated a business known both as Frasher's Mail Service and Frasher License and Title Service. The chief function of the business was to expedite for automobile dealers the obtaining of automobile titles for the dealers' purchasers.

The transactions here involved an arrangement between Frasher's business and an automobile dealer, Parkersburg Datsun, Inc. (Datsun). The record establishes that the defendant's wife would obtain the title applications and the dealer's check covering the title taxes and license fees and deliver the documents to Frasher in Charleston, who would present them to the Department of Motor Vehicles, obtain the title certificates and license plates, and return them through his wife to the dealer.

Datsun would make its check payable either to the defendant Frasher, his wife or to the business itself for the total amount of title taxes and license fees due on the various title applications turned over to the Frashers. These checks would then be deposited by the Frashers and, in turn, new checks would be drawn by the defendant on his business account made payable to the Department of Motor Vehicles to cover the fees and tax costs.

The basis for the embezzlement charge was that there had been an alteration of the title applications by the defendant which lowered the stated value of the motor vehicles, thereby reducing the amount of the license taxes due. As a consequence, it was alleged that the defendant paid a lower amount of title tax and retained the difference. The indictment was substantially in accord with the statutory form and alleged that the defendant embezzled the sum of $1,546.75 of "bullion, money, bank notes, drafts, securities for money and other effects and property of and belonging to the said Parkersburg Datsun, Inc."

### I.

The threshold question raised by the defendant is whether the embezzlement indictment was insufficient in charging that he embezzled from Datsun as distinguished from Datsun's purchasers. The defendant urges that Datsun was a mere conduit for paying to the Department of Motor Vehicles (Department) the title taxes and license fees owed by the individual purchasers, and that the indictment was insufficient in that it had to charge embezzlement from the purchasers themselves. . . .

The record demonstrates that Datsun added the automobile title taxes and fees to the purchase price of the vehicle and charged the purchaser the total sum. Datsun did not earmark the tax money or segregate it in any manner, but commingled it with other funds in its bank account. Datsun's office manager wrote checks on the Datsun bank account to pay to Frasher the taxes and fees due the Department. The defendant had no agreement with the individual purchasers to pay the title taxes and fees, but solely with Datsun.

For all practical purposes, therefore, the checks presented to Frasher represented Datsun's money, or at least money held by Datsun as a fiduciary for the automobile purchaser. On these facts, to characterize Datsun as simply a "conduit" is meaningful only if the law requires that an embezzlement can occur only from the actual owner of the property, and not simply from the individual or entity that is the ostensible legal possessor of the property.

We considered our embezzlement statute in *State v. Moyer*, 52 S.E. 30 (W.Va. 1905). . . . :

> "(I)n order to constitute the crime of embezzlement, it is necessary to show, (1) the trust relation of the person charged, and that he falls within that class of persons named; (2) that the property or thing claimed to have been embezzled or converted is such property as is embraced in the statute; (3) that it is the property of another person; (4) that it came into the possession, or was placed in the care, of the accused, under and by virtue of his office, place of employment; (5) that his manner of dealing with or disposing of the property, constituted a fraudulent conversion and an appropriation of the same to his own use; and (6) that the conversion of the property to his own use was with the intent to deprive the owner thereof."

We also stated in *Moyer* that embezzlement is a purely statutory offense, since:

> "It was unknown to the common law, and the statute was enacted for the purpose of supplying what were regarded as defects in the common law of larceny, so as to reach and punish for the fraudulent conversion of money or property which could not be reached by the common law. . . . The distinction between embezzlement and larceny is that embezzlement is the wrongful conversion of property without trespass, or where the original taking and possession is lawful. In order to constitute the offense, it is necessary that the property embezzled should come lawfully into the hands of the party embezzling, and by virtue of the position of trust he occupies to the person whose property he takes. . . . " (52 S.E. at 32.)

*Moyer* demonstrates that the hallmark of embezzlement is the trust relationship and the subsequent conversion or appropriation of the entrusted property. As in larceny, the taking of the property need not be from the actual owner of the property, but may be from one who has lawful possession of it. In *State v. Workman*, 114 S.E. 276 (W.Va. 1922), a county commissioner was charged with embezzling county funds. He and another commissioner had voted to authorize the purchase of a farm and had an arrangement with the owner to receive a kickback from the proceeds. Workman argued that he never had the funds in his actual possession, since the sheriff honored the commission's warrant from funds over which the sheriff had exclusive control and which were on deposit in a local bank. We held actual possession not to be necessary: "That fund was under the care and management of the county court, consisting of its president and the two commissioners. . . . It makes no difference that the fund, when appropriated, was not in their actual possession. . . ." (114 S.E. at 277). . . .

In a reverse factual pattern, the federal courts have held that where money or property belonging to the United States and granted to third parties is embezzled, the money or property is still that of the United States if the Government retains substantial control of it or a property interest in it.

In *United States v. Maxwell*, 588 F.2d at 573, the court criticized as a "technical commercial law analysis" the approach requiring a direct link between the alleged embezzler and the true owner. It stated that "the relevant inquiry becomes whether the federal government maintains a property interest in the funds in (the third party's) account" (588 F.2d at 573). The court concluded that the money in question, federal funds for aid to college students, retained its federal character despite its having been deposited in the college's account because the Government had "a reversionary interest evidenced by the power to reallocate money already allocated to an institution if the institution does not anticipate using that money before the end of the period for which it is available" (588 F.2d at 573). This approach is obviously analogous to the special property interest analysis made by this Court in *State v. Heaton, supra*. . . .

On the basis of the foregoing law, we hold that the indictment in the present case validly charged the defendant with embezzling money from Datsun.

*(Continues)*

(Continued)

## II.

Defendant also contends that the State failed to prove that he, as distinguished from his wife, was an "agent" of Datsun within the meaning of the statutory phrase "any agent, clerk or servant," and failed to establish that he converted money, as opposed to checks, to his own use. . . .

Most courts have concluded that an "agent" under an embezzlement statute, which employs the terms "agent," "clerk," "servant," "officer," or like terminology, can be anyone entrusted with property by virtue of his position, and not simply an agent within the strict definition of the common law. *State v. Moreno*, 240 A.2d 871 (Conn. 1968). In *State v. Moreno, supra*, 240 A.2d at 875, the point was expressed in this fashion:

> "If at the time of a fraudulent conversion the accused was an agent for a particular purpose only and the property appropriated was entrusted to him by virtue of such agency, embezzlement is committed. It is, not the extent of the authority conferred, but the fact of the relationship which constitutes the essential element of the crime of embezzlement." (Citations omitted.)

Here, the State's evidence demonstrates that the title and licensing service was operated by the defendant and that his wife worked for him. The checks from Datsun covering the various title taxes and licensing fees were made payable either to the business, or to the defendant or his wife. The checks were, however, under the defendant's control, since they were placed in the business bank account. It was from this account that the misappropriation occurred, since the defendant, after lowering the amount of value of the vehicles on the title applications in order to pay reduced title taxes, drew checks on this account payable to the Department for the amounts then due. As *in Fraley*, the defendant here was entrusted with Datsun's checks for a specific purpose, to make a proper disbursement from his own account for the amount due on each title transfer. . . .

We find no error in the trial, and thus affirm the judgment of the Circuit Court of Kanawha County. Affirmed.

### Notes and Questions

1. The key difference between embezzlement and larceny is that with embezzlement there is no trespassory taking; the property has already been entrusted to the thief, whereas in larceny the property is taken directly from its owner.

## False Pretenses

The key element of common law larceny is that it requires a trespassory taking. Embezzlement statutes filled a gap when a person entrusted his or her property to a trustee and the trustee converted the property to his or her own use. Another area not covered by common law larceny involved instances where an owner of property voluntarily parted with title to it as a result of false statements or deception. The statutory crime of false pretenses was created to cover such situations.

For a person to be convicted of **false pretenses** the state must prove the following seven elements:

1. A representation
2. of a material fact
3. relied on by the victim
4. which the defendant knows to be false
5. and which he intends to
6. and does in fact cause the victim to
7. pass title of his property.

Difficulties arise in false pretense cases in a number of areas. These areas include puffing by salesmen, statements of opinion, and omissions.

## *Commonwealth v. Reske,* 684 N.E.2d 631 (Mass. App., Norfolk App. Ct. 1997)

Kass, Justice. . . .

Ronald Nellon, who was borderline retarded (*see* Footnote 2) and whose subnormal intellectual capacity was readily apparent, had come into an inheritance of $142,409. With money in his pocket, Nellon was able to indulge one of his dearest wants: to buy trucks. During the period of the sales, June 8, 1992, to July 17, 1992, Reske was the general manager of Quirk Chevrolet in Braintree (Quirk). He established the terms of sale in the six purchases that Nellon made.

In the first transaction, that which occurred on June 8, 1992, salesmen for the dealership had fixed a price of $17,566, yielding a profit of $1,145. That was very close to what David Quirk, the dealership's principal, had testified was the average or normal profit on a truck sale. A salesman for Quirk and Nellon got so far as to sign an agreement at that price. Reske reviewed the contract (one might think that a general manager would review a sales contract before it was signed) and raised the price $2,000 by increasing the "price of unit" $1,000 and decreasing the trade-in allowance $1,000. Now the profit was $3,145. Nellon contentedly signed the new agreement, although it was markedly to his disadvantage.

The second transaction occurred on July 1, 1992, and produced a profit of $4,943, four times the norm, by adding $2,610 to the sticker price (the manufacturer's suggested retail price) and "low balling" the allowance on the truck—purchased at another dealership on June 13, 1992, and with 175 miles on the odometer—that Nellon was trading in.

The third transaction, on July 8, 1992, was $2,700 over the sticker price and gave Nellon a trade-in allowance of $7,775 on the truck he bought from Quirk a week earlier and that had 109 miles on the odometer. The profit from this transaction was $7,288, six times the norm.

The fourth transaction occurred the very next day, i.e., July 9, 1992. Nellon received a trade-in allowance of $5,876 on the truck he had bought the day before for $14,625. The odometer on that vehicle read 26 miles. The profit margin on this transaction came to $4,313.

The fifth transaction again occurred on the day following the previous one. Nellon received a trade-in allowance of $5,530 on the truck for which the day before he had paid $13,818. The odometer reading was 20 miles. Profit on this sale came to $5,085.

The sixth transaction occurred on July 17, 1992. Reske allowed a trade-in of $4,925 on a truck that Nellon had purchased from another dealer for $13,470. At the time of trade, that vehicle had 125 miles on its odometer. The value of the truck traded in was placed on Quirk's books at $9,500. The profit on this transaction was $6,077.

A prosecution on the theory of larceny by false pretenses—that aspect of the larceny statute under which the prosecution proceeded—requires proof that: (1) a false statement of fact was made; (2) the defendant knew or believed the statement to be false when he made it; (3) the defendant intended that the person to whom he made the false statement would rely on it; and (4) the person to whom the false statement was made did rely on it and, consequently, parted with property.

The core of Reske's defense is that he made no false statements of fact, that prices are a matter of opinion, and—the defense comes to this—it is not a crime to gull a willing dupe. It is commonplace that prices, unless regulated, are subject to market fluctuation. That does not mean, however, that there is no such thing as a false statement about the value of an item that is for sale.

Here, the finder of fact may infer a false statement as to the value of the six trucks sold by Quirk to Nellon from the inordinate profit margin, from the manifestly unrealistic trade-in allowances, and from the inflation over sticker prices. There was evidence of market norms, that those norms had been exceeded by manipulation of trade-ins and base prices, that those norms were so exceeded that the dealership thought it proper to make restitution. A salesman under Reske's supervision quit his job because he thought what was being done to Nellon was so immoral. An experienced dealer, Reske knew that the prices he was charging Nellon had no relation to what he customarily charged. This could be inferred from his doctoring of the prices in the first contract, prepared in accordance with customary prices and profit margins by salesmen who reported to him. False statements of value could also be inferred from evidence given by an experienced car dealer that sales above sticker price were not normal, nor was it normal to take $7,000 off the price of a vehicle that had been driven only for 24 hours and 33 miles. . . .

Reske wanted those false values to be accepted by Nellon. He expected them to be accepted because Nellon's impaired cognitive capacities rendered him a gullible mark. The days may be over when caveat emptor was a byword of capitalism and a sucker had only himself to blame. Even at the apogee of economic Darwinism, however, there was an underlying assumption that the mark had normal mental capacity. . . .

The dissent advances the somewhat curious proposition that, unless a statute defines as a crime the selling or buying of property to or from a manifestly incompetent person at greatly inflated profit margins, which any person of normal understanding would reject as unacceptable, then there is no crime. The law is not so inelastic. Larceny by false pretenses is a crime of ancient lineage, originally enacted in 1757 by Parliament to supplement the common law. Our own statute descends from that source. It is a statute broad in scope and its application is not "limited to cases against which ordinary

*(Continues)*

skill and diligence cannot guard; . . . one of its principal objects is to protect the weak and credulous from the wiles and stratagems of the artful and cunning." *Commonwealth v. Drew*. The opinion continues, "but there must be some limit, and it would seem to be unreasonable to extend it to those who, having the means in their own hands [in that case a bank] to protect themselves. It may be difficult to draw a precise line of discrimination applicable to every possible contingency, and we think it safer to leave it to be fixed in each case as it may occur."

Precisely so. Here the mark's disability was the key necessary to Reske's being able to relieve him of his money through a false pretense of ordinary dealing in the automobile market. Nellon was quintessentially weak and credulous. Reske crossed the prohibitory line, and did so following a repetitive pattern.

As to the fourth element of the crime, the unfortunate Nellon did rely on Reske's inflated and deflated figures and paid whatever he was asked for, thus, parting with at least $23,651, which, in the aggregate, could not honestly have been charged. As to each of the six transactions, the amount of overcharge was substantially above $250.

Proof of the elements of larceny by false pretenses were, thus, put before the trial judge and he rightly denied the motion for a required finding of not guilty.

Judgments affirmed.

Gillerman, Justice (dissenting).

I have no doubt that there was larceny in the defendant's heart as he rewrote the first sales contract and wrote up the remaining five contracts for Nellon to sign. But it is fundamental to criminal law that "there can be no criminal liability for bad thoughts alone. . . ." LaFave and Scott, *Substantive Criminal Law* § 1.2(b) at 10 (1986). I dissent because I conclude that the defendant's conduct, however exploitive and unfair it may have been, does not fall within the crime of larceny by false pretenses.

1. There was no evidence of a false statement of material fact. The gist of the crime of larceny by false pretenses, as the majority points out, is proof that a false statement of fact was made by the defendant upon which the victim relied. The Commonwealth failed to introduce any direct evidence that the defendant made any false statements to Nellon. Neither Nellon nor the defendant testified, nor was there a witness who testified to what he heard the two men say to each other, and there was no other evidence of what the defendant said to Nellon, or of what Nellon said to the defendant.

Nevertheless, the judge found that there were misrepresentations of fact made by the defendant to Nellon. He found that the statements on the purchase contracts (which he referred to as "invoices . . . prepared by the defendant") "contain false statements of fact because they consistently undervalued the value of the trade-in and consistently overvalued the Maroney sticker prices [*i.e.*, the manufacturer's suggested retail price, commonly known as the 'sticker price']."

The purchase contracts prepared by the defendant and signed by Nellon are barren of any words of representation. The contracts merely recorded the price of the vehicle sold, the trade-in allowance, the miscellaneous costs of the transaction, and the mathematical calculation of the cash due on delivery of the vehicle purchased.

The judge's argument proves too much. He ruled, in substance, that the terms of the transactions were so unfair that the terms themselves were "false statements." But it makes much more sense to say that if the terms were that unfair, no competent adult would accept those terms regardless of what was said. I return to this point below; it is enough to say where, as here, a contract merely recites the financial terms of a transaction, it is unreasonable to conclude that those financial terms may be taken as a representation as to the fair value of the vehicles sold and bought. A car salesperson who offers and sells a vehicle to a competent adult in excess of the sticker price, or who offers and buys a used vehicle for less than blue book has not, without more, committed larceny; this is so because the offeror has made no false statement of fact. . . .

2. The exploitation of an elderly or disabled person is not, without more, a crime in Massachusetts. The cases dealing with larceny by false pretenses have focused on the false representations of the defendant, not the victim's perception of those events. In recent years, a number of States, recognizing the special vulnerability of elderly and disabled persons, have enacted statutes involving criminal penalties which are designed to discourage the exploitation of such persons.

The Florida statute is the most precise and illuminates the problem at hand. It is a crime in Florida for a person to obtain or use the funds of an elderly or disabled person if the accused "knows or reasonably should know that the elderly person or disabled adult lacks the capacity to consent." Fla Stat. § 825.103 (1996 Supp.). . . .

The point need not be labored. Until the Legislature decides to enact appropriate remedial legislation, such as that enacted in Florida, protecting a person in Nellon's circumstances from exploitation by unethical persons, Nellon's lack of capacity to understand and consent to the unfair transactions imposed upon him by the defendant has no bearing on the crime with which the defendant is charged, and the position of the majority, in my view, cannot be maintained. See *Commonwealth v. Drew,* 36 Mass. 179, (1837). In *Drew,* the defendant opened and thereafter maintained a checking account as part of his plan to obtain money from the bank at a time when there were insufficient funds in his account. The defendant, without making any statement to the bank, "merely drew and presented his checks and they were paid." *Id.* As planned, there were not sufficient funds in the account to cover the check. The verdict was set aside. The defendant could not be charged with having obtained money from the bank by false pretenses because he made no "representation of some fact or circumstance, calculated to mislead, which . . . [was] not true." . . . *Ibid.* "It is not the policy of the law to punish criminally mere private wrongs. . . . [I]t would be inexpedient and unwise to regard every private fraud as a legal crime."

The judgments should be reversed and the findings set aside.

## Notes and Questions

1. Do you agree with the majority or the dissent? Why?
2. What did Reske say that was false? Why do you believe the court ruled the way it did? Is this good or bad for the criminal justice system?
3. Normally, a misrepresentation for false pretenses requires an affirmative action. This can be either spoken or written words or some kind of affirmative conduct. Mere silence is not enough. However, if the actor created a false impression through words or action, the actor will be liable if he or she fails to correct it before the other person acts on it. In addition, if a person acts to prevent another from acquiring information that would affect his or her judgment, the former is liable.

## Receiving Stolen Property

It is not only a crime to acquire property by larceny, embezzlement, or false pretenses, but it is also illegal to knowingly receive or possess stolen property. Although statutes prohibiting the receipt of stolen property are primarily aimed at large-scale fencing organizations, it is often the average citizen who buys an item at a flea market or garage sale that is caught in the law's web.

The crime of **receiving stolen property** consists of a person who receives stolen property knowing it to be stolen and having the intent to permanently deprive the true owner of that property. Although some states require that actual knowledge that property is stolen be proven, others only require that an actual belief that it is stolen is necessary. In either instance the element of knowledge may be proven solely through circumstantial evidence, as illustrated in *State v. Morgan*.

## *State v. Morgan,* 861 S.W.2d 221 (Mo. App. 1993)

Crandall, Presiding Judge.

Defendant, Ricky L. Morgan, appeals from a judgment of conviction for receiving stolen property. He was sentenced as a persistent offender to imprisonment for a term of seven years. . . .

The evidence at trial, viewed in a light most favorable to the verdict, disclosed that in 1990, William Worthen operated a marina located at Horseshoe Lake in Illinois. He owned several boats including an Alumacraft john boat which he stored at the marina. Worthen had secured the boat by running a chain through a handle attached to the front of the boat.

On June 18, 1990, Worthen discovered several of his boats including the john boat had been stolen. The front handle of the john boat had been broken off and was still attached to the security chain. The original purchase price was approximately $600, and the boat was less than six months old when it was stolen. The boat had two serial-numbered identification tags attached to it.

On July 29, 1990, a police officer went to a home in Webster Groves to investigate a report of a stolen boat. The officer found defendant at that location with a john boat which fit the description of Mr. Worthen's stolen boat. The two serial-numbered tags had been pried off the boat. After the officer arrested defendant, defendant told him that he had purchased the john boat for $150 from an unknown man at an unknown gas station somewhere in St. Charles, Missouri.

*(Continues)*

(Continued)

Mr. Worthen identified the boat as his stolen boat. He inspected the boat at the Webster Groves police station. He recognized the oar locks on the boat and the special reinforcing plates that he had fabricated and installed on the boat himself. He also tested the broken handle against the handle stub on the recovered boat which was a perfect fit.

On direct appeal, defendant claims the trial court erred in denying his motion for judgment of acquittal because the state failed to make a submissible case.

To establish a prima facie case of receiving stolen property the prosecution must prove that defendant received stolen property with the purpose of depriving the owner of his or her interest and with the knowledge or belief that the property had been stolen. Here it is conceded that the property was stolen, the issue is whether there was sufficient evidence to prove that the defendant knew or believed that the property had been stolen.

A defendant's knowledge is seldom susceptible of proof by direct evidence. The mental state is generally established by inference from facts or circumstances in evidence.

In this case, defendant was found in possession of the stolen john boat within two months after it was stolen. Certainly the jury had a right to disbelieve defendant's evasive explanation for possession of the stolen boat. While recent, unexplained possession of stolen property does not give rise to an inference that the possessor is guilty of receiving stolen property, it is evidence that the jury is entitled to consider bearing on defendant's guilt together with other facts and circumstances in the case.

Defendant's statement was that he purchased the boat for $150. The evidence was that the boat had been worth $600 six months earlier. Whether the boat had depreciated to 25% of its original value, or whether the price defendant paid was inadequate, was a question of fact for the jury. If defendant bought the stolen boat at an inadequate price, that fact can be sufficient to support an inference of guilty knowledge.

Finally, the fact that the serial number identification tags had been pried off is sufficient to support an inference of guilty knowledge.

We conclude that the evidence, viewed in a light most favorable to the verdict, was sufficient to support defendant's conviction. . . .

## Notes and Questions

1. Do you believe the court infers too much knowledge on the part of Morgan? If Morgan had been an average citizen without a criminal past, would the court have decided the case differently?
2. In your opinion, should receiving or possessing stolen property be a crime? Looking back, have you ever possessed stolen property?
3. The court lists a number of factors that can be considered in determining whether a person possessing stolen property had knowledge that it was stolen. Can you think of any other factors that might tend to prove or disprove knowledge?

## Consolidated Theft Statutes

Realizing the specific language requirements under the common law property crime laws discussed above, many states have enacted what are known as "**consolidated theft statutes**." These statutes tend to combine several of the common law property crimes under one, overarching statute. The Iowa theft statute is a good example of this approach.

I.C.A. § 714.1 Theft defined

A person commits theft when the person does any of the following:

1. Takes possession or control of the property of another, or property in the possession of another, with the intent to deprive the other thereof.
2. Misappropriates property which the person has in trust, or property of another which the person has in the person's possession or control, whether such possession or control is lawful or unlawful, by using or disposing of it in a manner which is inconsistent with or a denial of the trust or of the owner's rights in such property, or conceals found property, or appropriates such property to the person's own use, when the owner of such property is known to the person. . . .
3. Obtains the labor or services of another, or a transfer of possession, control, or ownership of the property of another, or the beneficial use of property of another, by deception. . . .
4. Exercises control over stolen property, knowing such property to have been stolen, or having reasonable cause to believe that such property has been stolen, unless the person's purpose is to promptly restore it to the owner or to deliver it to an appropriate public officer. . . .
6. Makes, utters, draws, delivers, or gives any check, share draft, draft, or written order on any bank, credit

union, person, or corporation, and obtains property, the use of property, including rental property, or service in exchange for such instrument, if the person knows that such check, share draft, draft, or written order will not be paid when presented. . . .

This approach, similar to the approach adopted in the Model Penal Code, eases the need to meet the stringent requirements of the common law property crimes while maintaining their basic elements. The Iowa statute is only one example of consolidated theft statutes. Different states include different crimes in their statutes. In addition, some states still maintain the strict common law divisions and elements.

## Robbery

Robbery is a crime whose definition has changed very little under the common law and under modern statutes. It can generally be defined as larceny through the use or threatened use of immediate force. The five elements of **robbery** under the common law are

1. The taking and asportation
2. of another's property
3. from his or her person or in his or her presence
4. by force or threatened imminent force
5. with the intent to permanently deprive the owner of the property.

Notice that robbery does not require the actual use of force. A person who holds up a bank with a gun does not need to shoot anybody to be guilty of robbery. All that is required is the threatened use of imminent force. The threatened harm, however, must be immediate in nature. A threat of some future harm constitutes extortion, not robbery; extortion is discussed in the next section.

Just how much force is needed for an act to constitute a robbery? Consider the case of *State v. Smalls*.

---

### *State v. Smalls*, 708 A.2d 737 (N.J. Super. 1998)

Wallace, J.A.D.

Defendants James Smalls and Gregory Cousar were indicted for one count of robbery. A jury found both defendants guilty of second degree robbery, N.J.S.A. 2C:15-1. The trial judge sentenced Smalls to an extended term of fifteen years with a six year parole disqualification and sentenced Cousar to a ten year term with a five year parole disqualification. These appeals, calendared separately, are consolidated for the purposes of this opinion. . . .

On April 25, 1995, at approximately 2:30 P.M., the victim, Jariatu Sesay, exited a check cashing store in Jersey City. She was approached by Cousar, dressed in a grey suit and carrying a newspaper. Cousar informed Sesay that he was from Zimbabwe and had been evicted by his landlord. Sesay offered to give him the telephone number of the International Institute, an organization which assists foreigners who do not have a place to stay. Cousar rejected Sesay's offer and asked her to accompany him to Jones Street in Jersey City to see a priest. Sesay was a mental health volunteer. She suspected that Cousar might be mentally ill. Sesay was then approached by Smalls who chided her on her unwillingness to help another foreigner. Sesay told Smalls that she was willing to pay cab fare for Cousar to go to the Institute, but that he did not want her help.

While Sesay was talking to Smalls, Cousar began circling behind her. Sesay began to feel uncomfortable. She sensed that the two men knew each other. Sesay felt a bump above her jacket pocket. She believed it was Cousar, but she did not see him bump her. Immediately thereafter, Cousar told Smalls they had to go and the two men walked away quickly.

After the two left, Sesay checked her jacket and realized that her wallet was missing. She chased the men down Kennedy Boulevard, and after being told that they were in a grocery store, confronted them there. She grabbed the two men by their collar and demanded the return of her green card and school identification card. The two men returned to her nearly $60 "bit by bit" and "crumpled up." A short while later Detective Brian Gomm arrived and arrested both defendants.

With this background, we address defendants' challenges to their conviction for robbery. N.J.S.A. 2C:15-1 provides:

a. Robbery defined. A person is guilty of robbery if, in the course of committing a theft, he:
  (1) Inflicts bodily injury or uses force upon another; or
  (2) Threatens another with or purposely puts him in fear of immediate bodily injury; or
  (3) Commits or threatens immediately to commit any crime of the first or second degree.

An act shall be deemed to be included in the phrase "in the course of committing a theft" if it occurs in an attempt to commit theft or in immediate flight after the attempt or commission. [N.J.S.A. 2C:15-1 (a).]

*(Continues)*

*(Continued)*

A person is guilty of theft if he unlawfully takes movable property of another with purpose to deprive him thereof. N.J.S.A. 2C:20-3. The crime of theft becomes robbery, in part, when the defendant inflicts bodily injury or uses force upon another in the course of committing a theft. The critical issue is whether a bump under the circumstances here was sufficient evidence of force to raise this pick pocket offense to a second degree robbery.

Both the State and the defendants rely upon *State v. Sein,* 124 N.J. 209, 590 A.2d 665 (1991), in support of their respective contentions. In *Sein,* the Court examined whether "the sudden snatching of a purse from the grasp of its owner involves enough force to elevate the offense from theft from a person to robbery as defined by N.J.S.A. 2C:15-1a(1)." *Id.* at 210, 590 A.2d 665. The Court recognized that the question of the amount of force necessary to take property from a person "to warrant the more serious penalties associated with robbery has vexed those courts that have considered the question." *Id.* at 213, 590 A.2d 665. The Court reviewed the legislative history of the robbery statute and concluded that our Legislature intended to adopt the majority rule which has been set forth as follows:

> [A] simple snatching or sudden taking of property from the person of another does not of itself involve sufficient force to constitute robbery, though the act may be robbery where a struggle ensues, the victim is injured in the taking, or the property is so attached to the victim's person or clothing as to create resistance to the taking. *Id.* at 213–214, 590 A.2d 665.

The Court also addressed the Legislature's intention in adding the phrase "or uses force" in the 1981 amendment to N.J.S.A. 2C:15-1a(1). The Court concluded that the amendment was "intended to clarify that the type of force required to support a robbery conviction under the pre-Code statute still would be sufficient to elevate a theft to a robbery." *Sein, supra,* 124 N.J. at 216, 590 A.2d 665. The Court further explained:

> Although the Committee Statement refers to a "purse snatching" as an example of the conduct the amendment was intended to cover, it goes on to state that snatchings rising to the level of robbery include only those that involve "some degree of force to wrest the object" from the victim. To "wrest" is to "pull, force, or move by violent wringing or twisting movements." Webster's Third New International Dictionary 2640 (1971). The Legislature apparently determined that the violence associated with "wresting" is deserving of more severe punishment. It did not, however, intend to eliminate the requirement that robbery by use of force include force exerted "upon another." *Id.* at 216–17, 590 A.2d 665.

The Court also looked to the Commentary to the Code definition of "theft" in reaching its conclusion. The Court noted that in discussing theft under N.J.S.A. 2C:20-3, the Legislature made clear the following:

> The crime here defined may be committed in many ways, *i.e.,* by a stranger acting by stealth or snatching from the presence or even the grasp of the owner or by a person entrusted with the property as agent, bailee, trustee, fiduciary or otherwise. New Jersey Penal Code: Final Report of the New Jersey Criminal Law Revision Commission § 2C:20-3 commentary 2 at 222 (Oct. 1971).

The theft statute thus includes purse-snatchings from the grasp of an owner, while the robbery statute includes purse-snatchings that involve some degree of force to wrest the object from the victim. The only way to reconcile the two statutes is to hold that robbery requires more force than that necessary merely to snatch the object. *Sein, supra,* 124 N.J. at 217, 590 A.2d 665.

Here, there was no struggle, no shoving or pushing, and no wrestling in order to take the victim's wallet. Moreover, the slight bumping of the victim that occurred did not even alert the victim that something was awry. This was a pickpocketing by defendants. It was not until both defendants had walked away that the victim decided to straighten her jacket and then realized her wallet was missing.

In *Sein,* the victim was aware that her purse had been snatched from under her arm, but the Court concluded that robbery required more than the force necessary merely to snatch the object. Here, the slight bump of the victim did not even alert her that her wallet had been removed from her jacket. We perceive even less force in this pickpocketing than in the purse snatching in *Sein,* which was found not to be a robbery. A fortiori, we are convinced that the theft of the victim's wallet here does not constitute robbery under N.J.S.A. 2C:15-1a(1).

We also reject the State's argument that there was sufficient evidence to warrant submission of the robbery charge to the jury as a theft committed while putting the victim "in fear of immediate bodily injury." A thief commits a second degree robbery if he or she threatens another with bodily injury regardless of its seriousness. The victim here stated that the presence of the two defendants close to her, one in front and one behind, made her fearful. Further, she said she was relieved when defendant left, but she did not want to give anybody the impression that she was afraid of them.

A cautious person, however, may exhibit fear in many settings that are not criminal. The focus of the robbery is on the conduct of the accused, rather than on the characteristics of the victim. Here, the victim had conversations with first Cousar and, later, after Smalls arrived, with Smalls. At no time did either defendant threaten the victim. Although the conduct of Cousar in holding his newspaper and referring to rent receipts made the victim believe Cousar may have had a mental problem, there was no evidence that either defendant purposely put her "in fear of immediate bodily injury." N.J.S.A. 2C:15-1a(2). Cousar had requested help from the victim while Smalls later chided the victim for not helping Cousar. Eventually, the two men were, in front of and behind the victim. To be sure, no special words and/or conduct are required to make out a threat or to purposely put someone in fear of immediate bodily injury, but the totality of the circumstances presented must be considered. While there may be circumstances where conduct alone, without threats by one or more persons, may be sufficient to justify a conclusion that the persons purposely placed the victim in fear of immediate bodily injury, this is not such a case. We leave that to another day.

In sum, viewing the State's evidence in its entirety with the benefit of all legitimate favorable inferences, there was insufficient evidence before the jury to support the conclusion that defendants used force upon the victim or purposely put her in fear of immediate bodily injury. . . .

We, therefore, reverse the robbery conviction as to each defendant and remand for the entry of an amended judgment of conviction of theft, and for resentencing. . . .

Reversed and remanded.

### Notes and Questions

1. The court in *Smalls* follows the majority view regarding purse snatchings. Some states, however, have different rules regarding robbery and purse snatchings. For example, Kentucky takes the extreme position that the snatching of a purse in and of itself is enough force to constitute robbery. *Jones v. Commonwealth*, 112 Ky. 689, 66 S.W. 633 (1902). The rule in Massachusetts is if the force used is sufficient to make the victim aware that the purse is being taken, adequate force for a robbery conviction has been used. *Commonwealth v. Jones*, 362 Mass. 83, 283 N.E.2d 840 (1972). At the other extreme, Model Penal Code § 222.1 states that the force used must cause a serious bodily injury to constitute robbery. Which of the four levels of force do you believe is appropriate to require in a robbery statute?

2. Do not forget that robbery requires either the use of force or the threat of force. Accordingly, no force need actually be used. As long as force is threatened and the victim is placed in fear of immediate bodily harm, the force element of robbery is fulfilled.

3. Most robbery statutes distinguish between simple robbery and aggravated robbery. The factors that turn simple robbery into aggravated robbery are the use of a deadly or dangerous weapon or the infliction of serious bodily injury.

## Extortion

**Extortion**, often referred to as "blackmail," is the obtaining of an item through the threat of a future harm. The difference between extortion and robbery is that robbery involves a threat of *imminent* harm, whereas extortion involves a threat of *future* harm. Extortion statutes are often quite broad with regard to the nature of the threat. The threat made can be to physically harm the victim, obtain or damage the victim's property, or cause damage to the victim's reputation at some time in the future.

An area where the states differ in their treatment of extortion involves the nature of the property being sought by the extorter.

### *State v. Crone*, 545 N.W.2d 267 (Iowa 1996)

Ternus, Justice. . . .

In 1987 and 1988, Crone and Forman had a tumultuous relationship that resulted in Crone's conviction of domestic assault and criminal mischief. Although their relationship had been over for some time, they found themselves drinking at Maxie's, a restaurant and tavern in Iowa City, on the evening of September 26, 1989. Forman worked at Maxie's and had stopped to check her work schedule.

*(Continues)*

She stayed to socialize, however, because she was depressed over her current boyfriend's recent incarceration. Crone was there celebrating his receipt of funds from a lawsuit settlement.

Forman became quite intoxicated but remembers a stranger telling her that Crone wanted to apologize to her for his past actions. She eventually joined Crone in a booth. Forman recalls very little of what happened after that until she woke up in an unfamiliar mobile home the next morning. She was unclothed from the waist down and was lying beside Crone who was also naked. Forman hurriedly left the trailer and returned home.

Crone testified Forman, at her request, had accompanied him to the trailer where he was staying with a friend. Once there, they began to cuddle and kiss. Crone left to go to the bathroom; when he returned, Forman was naked from the waist down. It was then Crone who devised a plan to slander Forman. As Forman lay sleeping, Crone took pictures of her. One photograph showed Forman lying in a fetal position; her face was recognizable and she had no clothes on below her waist. The second picture was of Forman's naked buttocks, with Crone's erect penis nearby.

The next day, Crone constructed more than fifty 8½ by 11-inch fliers which displayed Crone's photographs of Forman and a message giving Forman's name, place of employment and phone number and inviting one and all to call her. He then proceeded to Maxie's with the pictures and fliers where he showed them to his friends and the head waitress. When Forman arrived at Maxie's to work that afternoon, Crone approached her and asked her to meet with him privately. She felt harassed and so agreed to meet him later, even though she did not plan to do so.

When Forman returned home that evening after work, she began to receive phone calls from Crone. Crone wanted Forman to meet him; this time Forman refused. She asked Crone to leave her and her family alone. Crone responded that he had some illicit pictures and that if Forman did not meet with him, he would talk to her boyfriend, her family and her friends, and would circulate the pictures. Eventually, Forman stopped answering the phone and some of Crone's calls were taped on Forman's answering machine.

In early October 1989, the fliers made by Crone appeared in a variety of public places: tacked on telephone poles, distributed throughout the University of Iowa student union and placed on the windshields of parked cars. Copies were also mailed to Forman's friends and family, as well as her mother's co-workers.

Crone fled the state and was later found in Idaho. In March 1994, he waived extradition and returned to Iowa. The State charged Crone with extortion in violation of Iowa Code section 711.4(3) (1989) and sexual abuse in the third degree, in violation of Iowa Code sections 709.1(1) and 709.4(1) (1989). Following a jury trial, Crone was convicted of extortion and assault with intent to commit sexual abuse. He was sentenced to consecutive terms of five years and two years respectively. Crone appeals only his extortion conviction and sentence.

On appeal, Crone claims the evidence is insufficient to prove two elements of extortion: (1) that Crone threatened Forman, and (2) that Crone acted for the purpose of obtaining something of value for himself or another. In response to the State's argument that error was not preserved on these issues, Crone claims his trial counsel was ineffective for failing to make these arguments in support of a motion for judgment of acquittal. In addition to the sufficiency-of-the-evidence claim, Crone also asserts his trial counsel was ineffective for not objecting to evidence of Crone's steroid use. . . .

Crone claims his trial counsel was ineffective because he did not raise the precise sufficiency-of-the-evidence claims now asserted on appeal. Crone also complains his trial attorney did not object to the admission of evidence of Crone's past use of steroids. To successfully prove his ineffective-assistance-of-counsel claim, Crone must prove his counsel failed to perform an essential duty and this failure resulted in prejudice. . . .

We first consider Crone's claim of ineffective assistance of counsel as it relates to the sufficiency of the evidence to prove extortion. Our review of the elements of this crime and the record made at trial convinces us a motion for judgment of acquittal based on the grounds urged on appeal would not have been successful. Therefore, trial counsel was not ineffective for failing to include these grounds in his motion for judgment of acquittal. Our analysis follows.

> A. Elements of crime. Iowa Code section 711.4 defines the crime of extortion:
>
> A person commits extortion if the person does any of the following with the purpose of obtaining for oneself or another anything of value, tangible or intangible, including labor or services: . . .
>
> 3. Threatens to expose any person to hatred, contempt, or ridicule. . . .

The State had the burden to prove that (1) Crone threatened to expose Forman to hatred, contempt or ridicule (2) for the purpose of obtaining for himself or another (3) anything of value, tangible or intangible. Crone contends the State failed to prove him guilty beyond a reasonable doubt because there was insufficient evidence of a threat or of an attempt to gain anything of value.

B. Existence of a threat. The State argued at trial that Crone threatened to circulate the illicit photographs of Forman if she did not meet with him. Crone claims on appeal that there is no evidence he actually threatened Forman. Forman testified, however, that when Crone called her the day after their encounter, "[h]e told me if I didn't meet him, he was going to talk to my boyfriend, my family, my friends, and start circulating these pictures." Forman said she interpreted Crone's phone calls to her as a threat. . . .

Threats need not be explicit; they may be made by innuendo or suggestion. It is only necessary that the threat be definite and understandable by a reasonable person of ordinary intelligence. We think there is substantial evidence to support a jury finding that Crone threatened Forman. A reasonable person of ordinary intelligence would have understood Crone's statements on the telephone to be a promise of a retaliatory act if Forman did not meet with him. Therefore, there is sufficient evidence of a threat by Crone.

C. Attempt to gain something of value. . . . We find substantial evidence that Crone's threats were intended to extort a personal meeting with Forman. If such a meeting had value to Crone, his threats are encompassed in the statute. The fact that Forman did not want to meet with Crone and that Crone intended to compel her to do so against her will is not a defense under the new statute. Thus, we now examine Crone's argument that a meeting with Forman had no value to him.

Crone claims he had nothing to gain from a meeting with Forman because by the time he called Forman at home, he had already embarrassed her by showing the fliers and photographs to his friends at Maxie's. We think the jury could have found otherwise.

Because the legislature did not define the word "value," we use its ordinary meaning. The dictionary defines value in several ways, two of which are potentially applicable here. One meaning is "the monetary worth of something." Webster's Third New Int'l Dictionary 2530 (1993). Another meaning is "relative worth, utility, or importance." *Id.* Although the word "value" has two different meanings, that fact does not necessarily mean the term is ambiguous as used in section 711.4. We examine the text and structure of the statute, giving effect to its object and purpose, to decide whether reasonable minds would disagree on the meaning of the word "value" as used in the statute.

The word "value" is modified in section 711.4: "anything of value, tangible or intangible." Iowa Code § 711.4 (1993). We think this language clearly points to the broader definition of "value,"—"relative worth, utility, or importance," not the narrow definition—"the monetary worth of something." Consequently, we conclude reasonable minds would not disagree on which meaning was intended by the legislature: The statute is unambiguous. . . .

Applying that definition here, we examine the record to determine whether there was substantial evidence a face-to-face meeting with Forman was important to Crone. We think there was. Crone was relentless in his efforts to exact such a meeting. He called Forman numerous times; he consistently asked her to meet with him. We reject Crone's argument that any meeting could not have had any value to him because he had already shown the photographs. It is not necessary that the jury be able to discern exactly why a meeting with Forman was important to Crone; it is enough that the evidence convinced the jury that for whatever reason a meeting with Forman was worth something to him. Such a conclusion is not unreasonable despite the fact that Crone had already shown the pictures and fliers to others. His wish for a personal meeting may have been as simple as a desire to witness the look on Forman's face when she saw the photographs.

We conclude there was sufficient evidence that Crone threatened Forman with the purpose of obtaining something of value for himself. Therefore, Crone's trial counsel was not ineffective for failing to challenge the jury's findings on these elements in his motion for judgment of acquittal. . . .

AFFIRMED.

## Notes and Questions

1. What threats did Crone make? What was he seeking to obtain? Do you agree with the court that these acts constitute extortion?
2. Extortion is a specific intent crime. The extorter must intend to obtain the item he or she is demanding. So, if the extorter can prove that he or she was only joking or had no intent to obtain the item requested, the person would not be guilty of extortion. This defense, however, is quite difficult to prove.

## Crimes Against Habitation

Few things in life are more sacred than a person's home. The saying "a man's home is his castle" dates back hundreds of years. This sentiment was expressed in two common law crimes: burglary and arson. Under the common law both of these crimes dealt specifically with the invasion of a person's home. Although modern statutes greatly expanded the acts that may constitute burglary or arson, the basic elements contained in the common law definitions are still present today.

## Burglary

Burglary was a crime created under the common law in an effort to prevent intruders from entering a person's house during the nighttime. The six elements of common law **burglary** are

1. Breaking
2. and entering
3. the dwelling of
4. another
5. at night
6. with the intent to commit a felony.

Under the common law the elements involved in the crime of burglary were interpreted very narrowly. Each element had to be satisfied for a crime to have been committed. As

such, it is important that we consider how each element is defined and interpreted.

### Breaking

Under the common law, for a breaking to have occurred an opening or breach had to be created by the intruder. A person entering through an open door or window would not be guilty of burglary because there was already an opening for the intruder to enter through. However, a constructive breaking was said to have occurred if the entry was accomplished through trickery or fraud. It is important to note that many modern statutes, including the Model Penal Code, have omitted the breaking requirement for burglary. Under these statutes, as long as there is an illegal entry as discussed below, a person may be guilty of burglary.

### Entry

Under both common law and statutory burglary, there must be an illegal entry by the intruder. This does not require that the person completely enter the building with his or her entire body. Rather, entry by any part of the body or even by an instrument that is intended to be used in the commission of a crime is all that is necessary. Modern statutes also extend the entry requirement to cover situations where a person was initially invited to enter the building but then refused to leave when asked.

---

### *State v. Miller*, 954 P.2d 925 (Wa. App. 1998)

Burchard, Judge.

Washington law provides that a person commits the crime of burglary when he enters or remains unlawfully in a building with intent to commit a crime therein. James C. Miller entered an open self-service car wash, broke into several coin boxes and took the money from them. He was properly found guilty of theft. However, he did not commit the crime of burglary and his convictions for that crime and the related crime of making or having burglar tools must be reversed and dismissed. Because the car wash was open to the public, Mr. Miller's entry and remaining were not unlawful.

Jim's Car Wash in Clarkston, Washington, was open for business 24 hours a day including the time of this incident. The car wash consisted of wash bays or stalls completely open at each end (without doors), a roof, side walls and concrete floor. Attached to a side wall in each bay was the washing apparatus and a coin box secured with a padlock.

During the early morning hours of October 18, 1994, Mr. Miller in the company of Timothy Burke drove his truck to Jim's Car Wash. After washing the truck, Mr. Miller used bolt cutters and other tools to remove the locks from coin boxes in three separate wash bays. He opened the coin boxes and took the contents. Police responding to an alarm interrupted Mr. Miller and caught him at the car wash with the bolt cutters, coins and pieces of the padlocks. . . .

Mr. Miller contends that his conduct does not constitute burglary. The State contends that Mr. Miller is guilty under the "remains unlawfully" alternative because his intent to commit a crime negated any license, invitation or privilege to remain or that Mr. Miller is guilty because he exceeded the scope of any license, invitation or privilege and committed burglary when he entered the coin boxes by cutting off the locks, reaching in and removing the contents.

MR. MILLER DID NOT ENTER OR REMAIN UNLAWFULLY IN THE CAR WASH.

RCW 9A.52.030 defines burglary in the second degree as follows:

(1) A person is guilty of burglary in the second degree if, with intent to commit a crime against a person or property therein, he enters or remains unlawfully in a building other than a vehicle or a dwelling.

RCW 9A.52.010(3) defines the circumstances under which a person enters or remains unlawfully:

A person "enters or remains unlawfully" in or upon premises when he is not then licensed, invited, or otherwise privileged to so enter or remain.

---

> A license or privilege to enter or remain in a building which is only partly open to the public is not a license or privilege to enter or remain in that part of a building which is not open to the public.

Relying on *State v. Collins*, 751 P.2d 837 (Wash. 1988), and *State v. Thomson*, 861 P.2d 492 (Wash.App. 1993), the State argues that Mr. Miller's entry into adjacent stalls with intent to commit a crime was an unlawful entry because it violated the implied limitation of purpose of any license, invitation or privilege granted by the owner. Alternately relying on *State v. Deitchler*, 876 P.2d 970 (Wash.App. 1994), the State argues that Mr. Miller remained unlawfully in the whole car wash when he formed the intent to commit a crime in violation of the implied limitation of purpose of any license, invitation or privilege granted by the owner.

The State's logic is that no owner would grant entry for the purpose of committing a crime. Any license, invitation or privilege is only granted for a legitimate purpose (like washing a vehicle for the required fee). Therefore, any entry or remaining for an illegitimate or criminal purpose violates the license, invitation or privilege and is unlawful. The State's argument is not supported by the statute or by case law and would lead to results far outside the legislative intent. For example, under this theory every shoplifting inside a building would be elevated from a misdemeanor to the class B felony of second-degree burglary. Most other indoor crimes might also be elevated to burglary.

In the present case, it is immaterial whether Mr. Miller formulated the intent to steal the contents of the coin boxes before he entered the car wash or after he was already present. Washington law does not provide that entry or remaining in a business open to the public is rendered unlawful by the defendant's intent to commit a crime. The authorities cited by the State do not support its position. Burglary requires proof of an unlawful entry or remaining and intent to commit a crime. At least in the context of an open car wash, proof of intent to commit a crime does not establish the other element—unlawful entry.

Washington courts have never held that violation of an implied limitation as to purpose is sufficient to establish unlawful entry or remaining. In *State v. McDaniels*, 692 P.2d 894 (Wash.App. 1984), the defendant was convicted of burglary for entering an open church with the intent to steal a coat. While Mr. McDaniels' purpose was a factor considered by the court, he could not have been convicted but for the fact that the public invitation to enter was revoked when he was earlier confronted by church members who forced him to leave. They told Mr. McDaniels they did not believe his explanation that he was waiting for a friend. The obvious understanding of the McDaniels court was that his first entry into the open church, before being confronted, was not unlawful regardless of his purpose.

This principle was further clarified by the Supreme Court in *Collins*. Mr. Collins was invited into the private residence of Charlotte and Ellah Dungey because he was lost and needed to use the telephone located in the front room. After making the call, he dragged the two women into a bedroom and assaulted them. The court held Mr. Collins had remained unlawfully because he exceeded implied limitations on the scope of his invitation. The court rejected the California rule that the defendant's criminal intent upon entry may render that entry unlawful under the burglary statute:

> We decline, however, to adopt the California rule preferring instead a case by case approach. While the formation of criminal intent per se will not always render the presence of the accused unlawful, that presence may be unlawful because of an implied limitation on, or revocation of, his privilege to be on the premises.

The court found that implied limitations were placed upon Mr. Collins' entry. He was only invited into the living room for the purpose of using the telephone and for as long as it took to use the telephone. "No reasonable person could construe this as a general invitation to all areas of the house for any purpose." *Id.* at 261, 751 P.2d 837.

In addition, a privilege or invitation can be revoked:

> A second theory, likewise to be applied on a case by case basis, supports the same result. Once Collins grabbed the two women and they resisted being dragged into the bedroom, any privilege Collins had up to that time was revoked. *Id.* at 261, 751 P.2d 837.

Nothing in *Collins* supports the argument that the harboring of criminal intent is in itself sufficient to violate an implied limitation or to establish revocation of any license, invitation or privilege.

We hold that, in some cases, depending on the actual facts of the case, a limitation on or revocation of the privilege to be on the premises may be inferred from the circumstances of the case. That neither renders part of the statute superfluous nor converts all indoor crimes into burglaries. . . .

DOES THE BREAKING OF A SMALL COIN BOX FOR THE PURPOSE OF STEALING THE CONTENTS CONSTITUTE BURGLARY?

*(Continues)*

*(Continued)*

This question was persuasively settled in *Deitchler*. RCW 9A.04.110(5) defines a building as follows:

"Building," in addition to its ordinary meaning, includes any dwelling, fenced area, vehicle, railway car, cargo container, or any other structure used for lodging of persons or for carrying on business therein, or for the use, sale or deposit of goods; each unit of a building consisting of two or more units separately secured or occupied is a separate building.

Judge Morgan concluded that a structure or space within a large building is not a separate building unless the large building has "two or more units separately secured or occupied" and the structure or space being considered is one of those "units." *Deitchler*, 75 Wash.App. at 137, 876 P.2d 970. Here the separate stalls and coin boxes were not "separate buildings" any more than the police evidence locker in *Deitchler* was a "separate building." . . .

Jim's Car Wash was occupied by a single tenant and was not a building consisting of two or more units separately secured or occupied. As a result the separate stalls and coin boxes were not "units" of the car wash nor "separate buildings" inside the car wash, and the charge of burglarizing the separate stalls or the coin boxes cannot be sustained.

Cases from other jurisdictions decided under a variety of statutes are not always consistent. However, burglary is ordinarily considered only in the context of a structure large enough to accommodate a human being. A structure too small for a human being to live in or do business in is not a "building" or "structure" for the purpose of burglary. The coin boxes at Jim's Car Wash fit that description. They were too small to constitute "buildings" under the burglary statutes. . . .

We conclude Mr. Miller was guilty of theft and sustain that conviction. However, he did not commit the crime of burglary and that charge must be reversed and dismissed. . . .

## Notes and Questions

1. As indicated by *Miller*, many state statutes have removed the breaking and entering requirement by providing that remaining unlawfully in a building originally entered legally can give rise to burglary. So, if a person enters a store during business hours, hides in a bathroom until the store closes, and then emerges to steal merchandise, the person has committed burglary.

2. Under the common law and statutes that still require breaking and entering, the breaking and entering do not need to occur at the same time. A person who breaks a window on Tuesday and enters a house through the broken window on Friday has satisfied the breaking and entering requirement.

3. Bob enters an office building during business hours to see his accountant. While walking down the hall he comes across an office with its door open and a Rolex watch sitting on the desk. Seeing no one in the office, Bob walks in, takes the Rolex off of the desk, and leaves the building. Has Bob committed burglary? Under the common law because there was no breaking, the building was not a dwelling, and the event did not occur at nighttime, the answer is no. Under most modern statutes, however, the answer is yes. As discussed below, most modern statutes have removed the type of building and time of day requirements from the elements of burglary. In addition, although Bob did not commit a breaking and he was legally in an office building that was open to the public, he did not have authority to enter the empty office. Under most modern statutes, entry into a separate, private area of a building without permission or an invitation can give rise to a burglary charge.

## Intent

Burglary is a specific intent crime. It requires the intent to perform the *actus reus* (breaking and entering a building) as well as the intent to commit a crime once inside. It is not necessary that the crime be carried out. As long as the intruder intended to commit a crime at the time he or she entered the building, the *mens rea* requirements of burglary are met.

## *Bruce v. Commonwealth*, 469 S.E.2d 64 (Va. App. 1996)

Elder, Judge.

Donnie Lee Bruce (appellant) appeals conviction for breaking and entering his estranged wife's residence armed with a deadly weapon, with the intent to commit assault, in violation of Code § 18.2-91. Appellant contends that the evidence was insufficient to prove the elements of the charge. Disagreeing with appellant, we affirm his conviction.

# I.

## FACTS

Appellant and Deborah Bruce (Deborah), although married, lived in separate residences during late 1993. Deborah lived with the couple's son, Donnie Bruce, Jr. (Donnie) and Donnie's girlfriend at Greenfield Trailer Park in Albemarle County, Virginia. . . .

On December 5, 1993, at approximately 2:00 P.M., Deborah, Donnie, and Donnie's girlfriend left their residence. Earlier that morning, Donnie told appellant that Deborah would not be home that afternoon. Upon departing, Donnie and Deborah left the front door and front screen door closed but unlocked. The front door lacked a knob but had a handle which allowed the door to be pulled shut or pushed open.

After Deborah, Donnie, and Donnie's girlfriend left their residence, a witness observed appellant drive his truck into the front yard of the residence and enter through the front door without knocking. Appellant testified, however, that he parked his truck in the lot of a nearby supermarket and never parked in front of the residence. Appellant stated that the front screen door was open and that the front door was open three to four inches when he arrived. Appellant testified that he gently pushed the front door open to gain access and entered the residence to look for Donnie.

While preparing to leave the residence, appellant answered a telephone call from a man with whom Deborah was having an affair. The conversation angered appellant, and he threw Deborah's telephone to the floor, breaking it. Appellant stated that he then exited through the residence's back door, leaving the door "standing open," and retrieved a .32 automatic gun from his truck, which was parked in the nearby supermarket parking lot. Appellant returned to the residence through the open back door. Appellant, who testified that he intended to shoot himself with the gun, went to Deborah's bedroom, lay on her bed, and drank liquor.

When Deborah, Donnie, and Donnie's girlfriend returned to their residence, appellant's truck was not parked in the front yard. Upon entering the residence, Donnie saw that someone was in the bathroom, with the door closed and the light on. When police arrived soon thereafter, they found appellant passed out on Deborah's bed and arrested him.

On May 24, 1994, a jury in the Circuit Court of Albemarle County convicted appellant of breaking and entering a residence, while armed with a deadly weapon, with the intent to commit assault. Appellant appealed to this Court.

# II.

## PROOF OF REQUISITE ELEMENTS

In order to convict appellant of the crime charged, the Commonwealth had to prove that appellant broke and entered into his wife's residence with the intent to assault her with a deadly weapon. Under the facts of this case, the Commonwealth satisfied this burden.

**Breaking**, as an element of the crime of burglary, may be either actual or constructive. . . . Actual breaking involves the application of some force, slight though it may be, whereby the entrance is effected. Merely pushing open a door, turning the key, lifting the latch, or resort to other slight physical force is sufficient to constitute this element of the crime. "Where entry is gained by threats, fraud or conspiracy, a constructive breaking is deemed to have occurred." *Jones v. Commonwealth* 349 S.E.2d 414, 416–17 (Va.App. 1986). "[A] breaking, either actual or constructive, to support a conviction of burglary, must have resulted in an entrance contrary to the will of the occupier of the house." *Johnson v. Commonwealth,* 275 S.E.2d at 595.

Appellant's initial entry into Deborah's residence constituted an actual breaking and entering. Sufficient credible evidence proved that appellant applied at least slight force to push open the front door and that he did so contrary to his wife's will. However, as the Commonwealth concedes on brief, appellant did not possess the intent to assault his wife with a deadly weapon at this time. The Commonwealth bears the burden of "proving beyond a reasonable doubt each and every constituent element of a crime before an accused may stand convicted of that particular offense." The Commonwealth therefore had to prove appellant intended to assault his wife when he re-entered the residence with his gun.

We hold that the Commonwealth presented sufficient credible evidence to prove the crime charged. On the issue of intent, the jury reasonably could have inferred that the phone call from Deborah's boyfriend angered appellant, resulting in his destruction of the telephone and the formation of an intent to commit an assault with a deadly weapon upon Deborah. Viewed in the light most favorable

*(Continues)*

to the Commonwealth, credible evidence proved that appellant exited the back door of the residence, leaving the door open, moved his truck to a nearby parking lot, and re-entered the residence carrying a gun with the intent to assault Deborah.

Well-established principles guide our analysis of whether appellant's exit and re-entry into the residence constituted an actual or constructive breaking. As we stated above, an "[a]ctual breaking involves the application of some force, slight though it may be, whereby the entrance is effected." "In the criminal law as to housebreaking and burglary, [breaking] means the tearing away or removal of any part of a house or of the locks, latches, or other fastenings intended to secure it, or otherwise exerting force to gain an entrance, with criminal intent. . . ." Black's Law Dictionary 189 (6th ed. 1990). Virginia, like most of our sister states, follows the view that "breaking out of a building after the commission of a crime therein is not burglary in the absence of a statute so declaring." Am.Jur.2d Burglary § 14, at 329 (1964) (footnote omitted). In this case, appellant exited the back door of the residence on his way to retrieve the gun from his truck. In doing so, the appellant did not break for the purpose of escaping or leaving. Rather, by opening the closed door, he broke in order to facilitate his re-entry. At the time he committed the breaking, he did so with the intention of reentering after retrieving his firearm. Although appellant used no force to effect his re-entry into the residence, he used the force necessary to constitute a breaking by opening the closed door on his way out. Even though no prior case involves facts similar to the instant case, the breaking and the entry need not be concomitant, so long as the intent to commit the substantive crime therein is concomitant with the breaking and entering.

Sound reasoning supports the conclusion that a breaking from within in order to facilitate an entry for the purpose of committing a crime is sufficient to prove the breaking element of burglary. The gravamen of the offense is breaking the close or the sanctity of the residence, which can be accomplished from within or without. A breaking occurs when an accomplice opens a locked door from within to enable his cohorts to enter to commit a theft or by leaving a door or window open from within to facilitate a later entry to commit a crime. Professor LaFave states, "if one gained admittance without a breaking but committed a breaking once inside, there could be no burglary unless there then was an entry through this breaking . . . [and] the entry may be separate in time from the breaking." Wayne R. LaFave, Handbook on Criminal Law § 96, at 711 (1972).

Accordingly, a breaking occurred when appellant opened the back door of the victim's residence, even though the breaking was accomplished from within. Thus, because the evidence was sufficient to prove an intent to commit assault at the time of the breaking and the entering, the Commonwealth proved the elements of the offense. Thus, we affirm appellant's conviction.

## Notes and Questions

1. Consider the following case. Carl's brother asks Carl to go with him to a friend's house and retrieve a lost coat. To Carl's surprise, when they arrive at the house, Carl's brother kicks the door down. The brothers then enter the house where they are arrested for burglary. Is Carl guilty of burglary? The appellate court, in reversing Carl's conviction, stated that because Carl entered the house with the intent to retrieve the lost coat and not to commit a crime, he did not have the necessary specific intent to commit a crime to be convicted of burglary.

2. Under the common law, a person had to have the intent to commit a felony (as well as the other required elements) when he or she entered a dwelling to be guilty of burglary. This requirement led to illogical results. For example, if a person did not form the intent to commit a felony until after he or she had already broken into the dwelling, the person did not commit burglary no matter what he or she did once inside. This problem has been made moot by the inclusion of wrongfully remaining language that is now part of most modern burglary statutes.

3. If a person broke into a dwelling intending to commit a misdemeanor rather than a felony, he was not guilty of burglary. Most modern statutes have relaxed these requirements and only require that the person intend to commit a crime to give rise to liability.

## Graded Burglary

Most modern statutory schemes have a graded system of burglary laws. Typically, burglary of a residence is more serious than nonresidential burglary, and burglaries where a weapon is involved or where people are injured are more serious than noninjury or weapon burglaries. These more serious burglaries are called aggravated burglary in some states, whereas other states simply break burglary down into differing degrees.

# Arson

**Arson** was defined under the common law as the malicious burning of a dwelling of another. This meant actually setting the building on fire. Setting off an explosion that damaged or even destroyed a building but did not light the building on fire was not arson under the common law. Modern statutes, as typified by Model Penal Code § 220.1, include explosions in their definition of arson. Furthermore, no actual burning of the structure is required. Any damage, such as smoke or structural damage, will do.

In most states the *mens rea* for arson is general intent. Arson is not a specific intent crime; the general intent to start the fire is all that is needed. Accordingly, it is not necessary for the person to have the specific intent to burn the building.

---

## *People v. Lockwood,* 608 N.E.2d 132 (Ill. 1st App. Dist. 1993)

Justice McMorrow delivered the opinion of the court: . . .

The evidence adduced at trial is as follows. Gloria Snow testified that she lived in the first floor apartment of a building on North Harding Avenue in Chicago with her two daughters, Robresta and Aurelius; and that her granddaughter was visiting on the night of January 12, 1990. Shortly after 10 P.M., defendant, who had been dating Robresta, came to the apartment and told her that he wanted to retrieve some clothes he had left there and to speak with Robresta. Gloria informed him that Robresta was not at home, but allowed him entrance and walked back to Robresta's room with him. After gathering his clothes, defendant told Gloria that he was going to wait there for Robresta. Gloria responded that he could not stay and noted that he smelled of beer. He assured her that he was "fine," but she repeated that she wanted him to leave. As defendant was leaving, Robresta arrived. They went into Robresta's room to talk, but within a short time they began fighting. When Gloria went to the room, she saw defendant "hitting [Robresta], shoving her around." After a struggle, Gloria came between him and Robresta and finally pressured defendant to leave. She then locked both the outer door to the building and the apartment door.

When the doorbell rang, Gloria looked out her living room windows from which the street and outer door to the vestibule of the building are visible, and saw defendant at the front door "laying" on the bell. Gloria knocked on the window and told him to stop ringing the bell and to leave. Defendant responded that he was going to stay there all night if necessary. While continuing to ring the bell, he said that if she did not allow Robresta to come out and talk to him, "I'm gonna take all of you out. I will get you out," and then he left. Approximately 30 minutes later, defendant returned and began ringing the bell again. He told Gloria to send Robresta out, but Gloria refused. Defendant then went to his car and took out a brown bag. He came to the front of the house and began shaking liquid on the sidewalk in front of the door and under the front windows. He also shook some liquid into the vestibule through the mail slot. He then gathered some leaves and paper and put them in a pile near the corner of the door, removed a lighter from the glove box of his car, walked back to the house and set the papers and leaves on fire. Gloria saw flames from the pile of debris and smelled strong gasoline fumes. She hollered to Robresta to call the police. Defendant went to his car and started it, but a police car drove up, preventing him from leaving. One of the officers ran to the house and stomped out the fire. . . .

Defendant first contends that his conviction for aggravated arson was improper because the State failed to prove an essential element of the offense. Specifically, with reliance on *People v. Oliff* (1935), 361 Ill.237, 197 N.E. 777, defendant asserts that the State was required to prove that the building was burned. He argues that the evidence established only that a fire was started on the ground outside of the front door and that it was stomped out by a police officer before any burning damage occurred.

In *Oliff* the defendant argued that the charred and scorched condition of the ceiling, walls and floor of a building was not sufficient to constitute a burning necessary to prove arson. The supreme court held that for there to be a burning there must be a wasting of the fibers of the wood, no matter how small in extent. The *Oliff* court further stated that the charring of a wall or floor is sufficient to constitute a burning. . . .

Although we believe that the evidence, reviewed below, was sufficient to satisfy the definition of a burning under *Oliff,* defendant's argument that the "essential element of burning" was not proved, misapprehends the law. The statutes in effect when *Oliff* and *Lueder* were decided defined arson as the burning of a structure. However, under the current statute, arson occurs when, by fire or explosive, a person knowingly damages the property of another without the person's consent. Aggravated arson occurs when in the course of committing arson, a person knowingly damages, partially or totally, any building or structure of another without his consent, and he knows or reasonably should know that one or more persons are present in the building. Neither provision uses any form of the word "burn."

*(Continues)*

At trial, the State presented testimony that the building was damaged by the fire. Gloria Snow testified that defendant gathered papers and leaves in the corner of the front entrance, shook liquid on them and set them afire. She further stated that defendant also poured gasoline into the mail slot and that there was a fire in the vestibule, onto which her daughter, Robresta, threw a pot of water, and that the floor was blackened. Kimberly Buckhalter and her mother, Katie Buckhalter, both testified that the front door was burned, the tile in the vestibule was blackened and a portion of the throw rug was burned. Officer Kostecki testified that when he arrived at the scene he saw "the front door . . . on fire." He extinguished the fire not only by stomping on it with his feet but also by striking at it with a rag he found nearby. Officer Cruz testified that when he drove up, the front door of the building appeared to be on fire.

Defendant also argues that the testimony that there was a fire in the hallway was contradicted by Officer Kostecki who testified that he did not see any fire inside. We note, however, that it was not until after Kostecki extinguished the fire outside that he entered the building. He testified that he was met at the doorway by a young woman carrying a pot. The vestibule smelled strongly of gasoline and the rug was saturated. Although he did not know whether the liquid on the floor was water, gasoline or both, his testimony is supportive of the residents' testimony that Robresta threw a pot of water onto the hallway floor to extinguish a fire there.

Further, the State introduced photographs of the premises taken after cleaning efforts by the residents. It appears from the photographs in the record that the wood on the bottom of the door and on the frame around the door was charred and scorched, rather than merely "smudged and discolored" by smoke as defendant argues. The inside of the door opposite the damaged outside section also appears scorched and blackened, and the vestibule tile at that location was also blackened. Viewing the evidence in the light most favorable to the prosecution, as we are bound to do, we find that the evidence was sufficient to sustain the jury's verdict.

## Notes and Questions

1. As with burglary, many of the common law requirements for arson have either been eliminated or retained for grading purposes. For instance, the Washington statutes on arson read as follows:

   9A.48.020. Arson in the first degree

   (1) A person is guilty of arson in the first degree if he knowingly and maliciously:
      (a) Causes a fire or explosion which is manifestly dangerous to any human life, including firemen; or
      (b) Causes a fire or explosion which damages a dwelling; or
      (c) Causes a fire or explosion in any building in which there shall be at the time a human being who is not a participant in the crime; or
      (d) Causes a fire or explosion on property valued at ten thousand dollars or more with intent to collect insurance proceeds.
   (2) Arson in the first degree is a class A felony.

   9A.48.030. Arson in the second degree

   (1) A person is guilty of arson in the second degree if he knowingly and maliciously causes a fire or explosion which damages a building, or any structure or erection appurtenant to or joining any building, or any wharf, dock, machine, engine, automobile, or other motor vehicle, watercraft, aircraft, bridge, or trestle, or hay, grain, crop, or timber, whether cut or standing or any range land, or pasture land, or any fence, or any lumber, shingle, or other timber products, or any property.
   (2) Arson in the second degree is a class B felony.

## Conclusion

Most crimes committed in the United States fall into the category of property offenses. Despite the seemingly common sense nature of the components of property crimes, the statutory elements that define property offenses have evolved significantly over time. As electronic and computer technology continue to develop, it is likely that the manner in which the criminal law defines and handles items involving identification theft, computer trespass, and other types of acts involving takings and trespass that have yet to be considered will continue to adjust. The materials in this chapter provide a foundation to consider and evaluate property offenses in the present and in the future.

## Key Terms

arson

breaking

burglary

consolidated theft statutes

constructive possession

embezzlement

extortion

false pretenses

larceny

larceny by trick

receiving stolen property

robbery

## Review Questions

1. List the seven elements of larceny at common law.
2. The elements of larceny are broken down into three groups. Name them.
3. Could a charge of larceny be levied against someone who takes property that does not belong to him or her without permission of the owner even if the owner is not present?
4. Discuss the requirements for a charge of larceny from a store. Be specific as to the intent of the law, and be sure you address whether or not a person must leave the store with the merchandise to constitute larceny.
5. At common law, is it possible for time and services to meet the requirements of "the object of taking"?
6. Make a distinction between larceny and embezzlement.
7. List the seven elements necessary to establish a claim of *theft by false pretenses*.
8. When charging a person with the crime of *receiving stolen property*, must the state be able to prove knowledge on the part of the accused or will lesser proof suffice? Explain.
9. List the five elements of *robbery* at common law.
10. List the six elements of *burglary* at common law.
11. Because burglary is a "specific intent" crime, could one be charged with burglary if discovered before the commission of theft or another "intended" crime? Why or why not?

## Footnote to *Commonwealth v. Reske,* 684 N.E.2d 631 (Mass. App., Norfolk App. Ct. 1997)

2. He had an IQ of 79 and could not comprehend numbers over 100.

# 10 White Collar Crimes

## Chapter Objectives

- Learn what types of crimes are considered to be *white collar crimes*
- Understand the distinction between *white collar crimes* and *street crimes*
- Learn what constitutes *insider trading*
- Understand why environmental crimes are considered *white collar* crimes
- Learn how workplace conditions can lead to charges under white collar crime guidelines
- Understand how product liability becomes a white collar crime issue

In this chapter we examine a diverse category of crimes that share a basic characteristic: The crimes emanate from business organizations (or those acting in their employ) and often involve an underlying financial motive. Although the harms addressed here (*e.g.*, the dumping of toxic pollutants) have been around for centuries, only in the last several decades has the criminal law intensively singled them out for punishment. In this sense **white collar crimes** can be distinguished from the vast majority of other offenses addressed in this book, such as murder, sexual assault, and burglary, which have ancient origins in Anglo-American common law.

The increasing receptiveness of government to criminalize business-related wrongdoing springs from two central realities. First, staggering economic and social costs are associated with white collar criminal activity. Each year Americans suffer untold financial losses from unscrupulous behaviors, ranging from securities fraud violations to unscrupulous telemarketing scams—losses that experts anticipate will escalate radically with the increasing use of computers and the Internet. Perhaps less direct, but surely no less significant, are the antitrust violations, massive health care frauds, and widespread financial theft such as that in the savings and loan industry. They cause losses to society at large that run into the billions of dollars. The harm of white collar crime is not limited to the economic realm. Whereas the dumping of toxic wastes or the violation of workplace safety laws might have an ultimate economic motive, the human toll of such misdeeds is enormous in terms of sickness, injury, and even death. In short, the sheer volume of the societal ill effects bred by white collar crime warrants legal intervention.

The second reason for the contemporary awakening to white collar crime stems from its relatively recent intellectual vintage. It was not until the landmark work of Professor Edwin Sutherland in the late 1930s and early 1940s in American corporations that legal thinkers came to conceptualize the harms caused by society's elite as worthy of criminal recognition. Sutherland offered a fresh perspective on criminal wrongdoing, which for centuries had focused almost exclusively on **"street" crimes**, not typically carried out by more wealthy members of society. Sutherland instead focused on the criminal acts of organizations and those of "high social status in the course of [their] occupation."[1] Sutherland saw close parallels between the socially disdainful acts of corporations and those of street criminals, and he explained the relative inattention paid by the law to the former on the basis of the disproportionate political power wielded by "white collar" criminals. In the decades that followed, and most especially after the Watergate scandal of the early 1970s, public outrage over "crime in the suites" (as opposed to the "streets") has supported a marked increase in the policing and prosecuting of what has come to be known as white collar crime.

We first examine the unique theoretical and practical challenges associated with white collar criminal liability that arise from the fact that a corporation is a fictitious entity, really existing only on paper. As a result, the prosecution of organizations raises intriguing questions with respect to the application of traditional penal justifications—retribution, deterrence, and incapacitation—with their traditional orientation toward individual actors. Later, we address several especially important aspects of white collar criminal activity, in particular financial frauds, environmental crimes, and "corporate homicide."

## Problems of Punishment

### Theoretical Challenges

Although there is no disputing that the acts of corporations can impose social harms, the fictitious nature of the corporation poses a vexing problem: How is blame to be apportioned and under what rationale is punishment to be imposed? The first case we consider, *New York Central & Hudson River Railroad*, provides one of the first judicial discussions in American law of this issue.

## New York Central & Hudson River Railroad v. United States, 212 U.S. 481, 29 S. Ct. 304, 53 L. Ed. 613 (1909)

Mr. Justice Day delivered the opinion of the Court.

[The railroad, its traffic manager, and its assistant traffic manager were indicted for financial improprieties under the federal Elkins Act. The railroad and the assistant manager were convicted and required to pay sizeable fines.]

\* \* \*

The principal attack in this court is upon the constitutional validity of certain features of the Elkins act: That act, among other things, provides:

(1) That anything done or omitted to be done by a corporation common carrier subject to the act to regulate commerce, and the acts amendatory thereof, which, if done or omitted to be done by any director or officer thereof . . . acting for or employed by such corporation, would constitute a misdemeanor . . . shall also be held to be a misdemeanor committed by such corporation, and upon conviction thereof it shall be subject to like penalties as are prescribed in said acts. . . .

\* \* \*

In construing and enforcing the provisions of this section, the act, omission or failure of any officer, agent or other person acting for or employed by any [corporation], acting within the scope of his employment, shall in every case be also deemed to be the act, omission or failure of such [corporation] as well as of that person.

It is contended that these provisions of the law are unconstitutional because Congress has no authority to impute to a corporation the commission of criminal offenses, or to subject a corporation to a criminal prosecution by reason of the things charged. The argument is that to thus punish the corporation is in reality to punish the innocent stockholders, and to deprive them of their property without opportunity to be heard, consequently without due process of law. And it is further contended that these provisions of the statute deprive the corporation of the presumption of innocence, a presumption which is part of due process in criminal prosecutions. . . . As no action of the board of directors could legally authorize a crime, and as indeed the stockholders could not do so, the arguments come to this: that owing to the nature and character of its organization and the extent of its power and authority, a corporation cannot commit a crime of the nature charged in this case.

Some of the earlier writers on common law held the law to be that a corporation could not commit a crime. . . . In Blackstone's Commentaries . . . we find it stated: "A corporation cannot commit treason, or felony, or other crime in its corporate capacity, though its members may in their distinct individual capacities." The modern authority, universally, so far as we know, is the other way. In considering the subject, Bishop's New Criminal Law, § 417, devotes a chapter to the capacity of corporations to commit crime, and states the law to be: "Since a corporation acts by its officers and agents their purposes, motives, and intent are just as much those of the corporation as are the things done. If, for example, the invisible, intangible essence of air, which we term a corporation, can level mountains, fill up valleys, lay down iron tracks, and run railroad cars on them, it can intend to do it, and can act therein as well viciously as virtuously." Without citing the state cases holding the same view, we may note *Telegram Newspaper Company v. Commonwealth,* 172 Massachusetts, 294, in which it was held that a corporation was subject to punishment for criminal contempt, and the court, speaking by Mr. Chief Justice Field, said: "We think that a corporation may be liable criminally for certain offenses of which a specific intent may be a necessary element. There is no more difficulty in imputing to a corporation a specific intent in criminal proceedings than in civil. A corporation cannot be arrested and imprisoned in either civil or criminal proceedings, but its property may be taken either as compensation for a private wrong or as punishment for a public wrong." . . .

It is true that there are some crimes, which in their nature cannot be committed by corporations. But there is a large class of offenses . . . wherein the crime consists in purposely doing the things prohibited by statute. In that class of crimes we see no good reason why corporations may not be held responsible for and charged with the knowledge and purposes of their agents, acting within the

*(Continues)*

*(Continued)*

authority conferred upon them. . . . If it were not so, many offenses might go unpunished and acts be committed in violation of law, where, as in the present case, the statute requires all persons, corporate or private, to refrain from certain practices forbidden in the interest of public policy. . . .

[The Elkins Act] does not embrace things impossible to be done by a corporation; its objects are to prevent favoritism, and to secure equal rights to all in interstate transportation, and one legal rate, to be published and posted and accessible to all alike. . . .

We see no valid objection in law, and every reason in public policy, why the corporation which profits by the transaction, and can only act through its agents and officers, shall be held punishable. . . . While the law should have regard to the rights of all, and to those of corporations no less than to those of individuals, it cannot shut its eyes to the fact that the great majority of business transactions in modern times are conducted through these bodies, and particularly that interstate commerce is almost entirely in their hands, and to give them immunity from all punishment because of the old and exploded doctrine that a corporation cannot commit a crime would virtually take away the only means of effectually controlling the subject-matter and correcting the abuses aimed at. . . .

We find no error in the proceedings of the Circuit Court, and its judgment is *Affirmed*.

## Notes and Questions

1. On what public policy basis does the Supreme Court justify imposing liability on the corporate entity?
2. *Hudson* underscores the "bottom line" effect of corporate punishment. Is it just to hold individual corporate shareholders financially responsible for the misdeeds of the corporation?
3. To the extent the stigma associated with the criminal sanction has a deterrent effect, in what ways does the effect of stigmatization on the corporation differ from that experienced by the average "street" criminal? Can the criminal sanction of a corporation be viewed as a "cost of doing business"? Is there reason to believe that corporate structures and incentives "muffle" the deterrent?

## Practical Challenges

Beyond the obvious basic theoretical challenges presented, the punishment of corporate misbehavior poses a litany of more practical issues. As discussed below, these issues differ in form depending on whether society seeks to single out the corporate entity or the individual corporate actor for punishment.

## Liability

### Corporate Liability

The main challenge in imposing liability on the corporate entity itself arises from fairness concerns relating to the use of vicarious liability (discussed in Chapter 11). Of course, a corporation itself does not "act" in any technical sense, nor does it possess a mental state. Rather, it is the employees (or agents) of the corporation that conduct the affairs of the corporate enterprise. As a result, the criminal law is forced to improvise some theoretical basis to connect the acts and mental state of the agent to the principal (the corporate entity).

For **corporate liability** to arise, several requirements must be met. First and foremost the employee committing the illegal act must be acting within the "scope" of his or her employment (*i.e.*, performing authorized duties). However, even if the illegal acts are committed within the agent's scope of employment, not all agents are considered worthy of representing the corporation for liability purposes.

The common law and the Model Penal Code (MPC) highlight two different approaches on this issue. Under the common law corporations can be liable for the acts of their agents regardless of the agent's position in the corporate hierarchy. In a classic application of the common law rule, the criminal liability of a steamship company for polluting the waters was upheld notwithstanding that the employee dumping the refuse into the water was a lowly kitchen worker. *See Dollar Steamship Co. v. United States*, 101 F.2d 638 (9th Cir. 1939). The MPC adopts a much stricter approach. Under the MPC the illegal act must be "authorized, requested, commanded, performed or recklessly tolerated by the board of directors or by a high managerial agent acting in behalf of the corporation within the scope of his office or employment." MPC § 2.07(1)(c). The MPC's approach thus requires the direct involvement of corporate higher-ups, in effect allowing the corporation to escape liability as long as supervisors exhibit some diligence in the policing of corporate wrongdoing.

A second requirement for corporate liability is that the agent's behavior must benefit the corporation in some manner. The corporation need not actually *receive* the benefit nor must the benefit be enjoyed *entirely* by the corporation (indeed, most employee acts may be said to have an element of self-interest). At the same time, however, the illegal acts committed cannot be expressly contrary to corporate interests. The U.S. Court of Appeals for the Fifth Circuit stated the following: "the corporation does not acquire that knowledge or possess the requisite 'state of mind essential for responsibility,' through the activities of unfaithful servants whose conduct was undertaken to advance the interest of parties other than their corporate employer." *Standard Oil Co. v. United States*, 307 F.2d 120, 129 (5th Cir. 1962).

Finally, special problems can arise when the substantive crime the corporation is alleged to have committed involves some form of mental culpability. In contrast to strict liability offenses (which are relatively more common in the white collar area), culpability-based crimes require that the law impute a wrongful intent to a legal fiction, based on the actions of the individual human agents that do its bidding. The courts use two basic approaches to impute liability. First, a theory based on "collective knowledge" permits liability if employees knew or had reason to know that a crime was being committed. This approach is based on the idea that corporations should not be permitted to escape liability because they tend to compartmentalize knowledge, "subdividing the elements of specific duties and operations into smaller components." *United States v. Bank of New England, N.A.*, 821 F.2d 844, 856 (1st Cir. 1987). The corporation is thus presumed to have acquired the "collective knowledge" of its workers and is held responsible for their failure to heed the law.

Second, the "willful blindness" doctrine permits liability when the corporation deliberately disregards wrongdoing. In *Bank of New England*, for instance, the bank was held liable for violating federal currency reporting laws when a teller carried out several large transactions with an individual customer in a single day, in flagrant disregard of Department of Treasury reporting requirements.

---

*Francis T. Cullen et al., "Corporate Crime Under Attack."*

## Corporate Criminals or Criminal Corporations?

Students of corporate social control have asked increasingly whether individual executives or corporate entities should be the object of criminal sanctions. Is it best to speak about "corporate criminals" or "criminal corporations"? The most common answer is that the preferred statutory scheme provides both individual and entity liability; the appropriateness of each is to be determined case by case by the prosecutor. Accordingly, a wide range of methods, apart from the traditional practice of fining businesses, have been proposed or tried to penalize corporate entities more effectively: mandatory community service, managerial intervention, government contract proscription, equity fines, and especially formal publicity.

Some critics, however, have argued against the application of criminal sanctions to corporate entities, primarily on two grounds. First, they challenge the deterrent effect of the sanction essentially because "corporations don't commit crimes, people do"; second, they question the retributive function because corporate criminal sanctions may actually punish innocent shareholders (by reducing the value of their shares) and consumers (by increasing the costs of goods and services). Although a detailed analysis of each objection is beyond the scope of this chapter, a few comments are in order. . . . We suggest that, in many instances, sanctioning the corporation is the most prudent and equitable policy, and thus that prosecutors' options should not be confined to imposing individual criminal liability.

The critics' first objection—that people, not corporations, commit crimes—ignores the reality that the labyrinthian structure of many modern corporations often makes it extremely difficult to pinpoint individual responsibility for specific decisions. Even in cases where employees who carried out criminal activities can be identified, controversial questions remain. John S. Martin, a former U.S. Attorney who actively prosecuted corporate and white-collar crime cases, comments that when individual offenders can be identified they "often turn out to be lower-level corporate employees who never made a lot of money, who never benefited personally from the transaction, and who acted with either the real or mistaken belief that if they did not commit the acts in question their jobs might be in jeopardy." Further, says Martin, "they may have believed that their superior was aware and approved of the crime, but could not honestly testify to a specific conversation or other act of the superior that would support an indictment of the superior." Thus a thorough investigation may well lead a prosecutor to conclude that indictments against individuals simply cannot be justified, even though the corporation benefited from a clear violation of a criminal statute. Such a result would disserve the deterrent function.

The existence of corporate criminal liability also provides an incentive for top officers to supervise middle- and lower-level management more closely. **Individual liability**, in the absence of corporate liability, encourages just the opposite: top executives may take the attitude of "don't tell me, I don't want to know." In the words of Peter Jones, former chief legal counsel at Levi Strauss, "a fundamental law of organizational physics is that bad news does not flow upstream." Only when directives come from the upper echelon of the corporation "will busy executives feel enough pressure to prevent activities that seriously threaten public health and safety." For a similar reason, proponents of the conservative "Chicago School" of law and economic thought advocate corporate rather than individual sanctioning: a firm's control mechanisms will be more efficient than the state's in deterring misconduct by its agents and will bring about adequate compliance with legal standards as long as the costs of punishment outweigh the potential benefits.

The second objection—that the cost of corporate criminal fines is actually borne by innocent shareholders and consumers—also seems unfounded. With regard to shareholders, whether individual or institutional, incidents of corporate criminal behavior may give the owners the right to redress the diminution of their interest by filing a derivative suit against individual officers and/or members of the board of directors. Although the cost and the uncertainty of winning such a suit may be high, shareholders must regard this cost as one of the risks incurred when they invest in securities. Just as shareholders may occasionally be enriched unjustly through undetected misbehavior by their company, it is only fair to expect them to bear a part of the burden on those occasions when illegality is discovered and duly sanctioned.

Finally, it is simplistic, if not untenable, to argue that corporate criminal fines will simply be passed on to the consuming public through higher prices. . . . If we assume that competition exists in the offending corporation's industry, the firm cannot simply decide to raise its prices to absorb the fine or the costs related to the litigation. If it does so, it risks becoming less competitive and suffering such concomitant problems as decreased profits, difficulty in securing debt and equity financing, curtailed expansion, and the loss of investors to more law-abiding corporations.

## Notes and Questions

1. Do you find the arguments made by Cullen et al. convincing?

2. Recently, the federal government announced the first criminal conviction in relation to a massive food poisoning. The defendant, Odwalla, Inc., was forced to pay a $1.5 million fine for selling bacteria-tainted apple juice in violation of federal food safety laws, which resulted in the death of an infant and injuries to numerous others. Odwalla also was placed on court-supervised probation for 5 years, requiring it to provide a detailed plan demonstrating its safety precaution.[2] Odwalla had already paid, as a result of private civil lawsuits, more than $12 million to those injured and suffered substantial lost profits. Presumably, Odwalla "got the message" even before the federal government imposed its harsh punishment. Would it have been best to allow the private sector (*i.e.*, those directly injured) alone to seek monetary redress from Odwalla? In a broader sense, will the government's criminal prosecution perhaps discourage others from entering the food production business, to the ultimate detriment of all consumers?

3. Since 1991 the U.S. Federal Sentencing Guidelines have permitted the levy of punishments on actual corporations for violations of federal law. The stated purpose of the Guidelines is to provide "just punishment, adequate deterrence, and incentives for organizations to maintain internal mechanisms for preventing, detecting, and reporting criminal conduct." U.S. *Sentencing Guidelines Manual* Ch. 8, Intro. Comment (1998). The Guidelines, to varying degrees, emphasize a "carrot and stick" approach. The "carrot" comes with the incentives provided to corporations to implement internal monitoring and compliance programs, which can lessen the degree of punishment imposed in the event an illegality occurs. The "stick" is evidenced by the punishments prescribed upon conviction, ranging from fines to community service to actual imprisonment of corporate representatives.

*Source:* Reprinted from *Corporate Crime Under Attack: The Ford Pinto Case and Beyond,* Francis T. Cullen et al., with permission. Copyright © 1987 Matthew Bender & Company, Inc., a member of the LexisNexis Group. All rights reserved.

The following case illustrates the flexible approach required when the law seeks to impose criminal sanctions on corporations.

---

## *Allegheny Bottling Co. v. United States,* 695 F. Supp. 856 (E.D. Va. 1988)

### SENTENCING OPINION

Doumar, District Judge.

On May 19, 1988, Allegheny Bottling Company (Allegheny Pepsi), Morton M. Lapides, and James J. Harford were found guilty of price-fixing, after a seven-week trial, in violation of section 1 of the Sherman Act, 15 U.S.C. § 1. Odis R. Allen pled guilty on March 22, 1988. This opinion is limited to setting forth the reasons for the sentence and the terms of probation imposed upon Allegheny Pepsi.

### I. BACKGROUND

This case arises from a conspiracy between Mid-Atlantic Coca-Cola Bottling Company (Mid-Atlantic Coke) and defendant Allegheny Bottling Company, formerly Allegheny Pepsi-Cola Bottling Company. The conspiracy began with Allegheny Pepsi, initially through its chairman of the board, with its president and its executive vice president. . . .

Allegheny Pepsi was found guilty by a jury of a price-fixing conspiracy which occurred in the Norfolk, Richmond and Baltimore areas. Morton M. Lapides, the chairman of the board of Allegheny Pepsi, was found guilty by a jury. James Sheridan, the president of Allegheny Pepsi, had previously pled guilty, and Armand Gravely, a Richmond area manager, previously had been found guilty by a jury. . . . Allegheny Pepsi was so permeated with the conspiracy that the lower ranking employees of the company were completely aware of the well known fact that there was a price-fixing agreement between the Pepsi and Coke distributors.

Mid-Atlantic Coca-Cola Bottling Company (Mid-Atlantic Coke) pled guilty to price-fixing in two jurisdictions and has been fined $1,000,000 in each of the jurisdictions under a plea bargain agreement. Defendant James J. Harford was the president of Mid-Atlantic Coke. Odis R. Allen, a vice president of Mid-Atlantic Coke and one time in charge of the areas encompassing both Richmond, Virginia, and Norfolk, Virginia, pled guilty before the trial began. . . .

The pre-conspiracy period was characterized by intense competition between Coke and Pepsi, and by frequent "deep discounting," *i.e.,* discounting below the prices offered by the companies in periodic letters mailed by each of the companies to its customers and known as promotional letters. The conspiracy began in 1982 and involved an agreement to adhere to the prices established in promotional letters published by Mid-Atlantic Coke and Allegheny Pepsi. . . . In this way, the companies ended a period of intense competition, stabilized the market, and maintained higher prices for their cola products.

It is difficult, if not impossible, to identify with particularity the specific persons who paid higher prices because of the conspiracy. . . . It does appear from the evidence, however, that the profit gained by the defendant companies through price-fixing over the previously established competitive prices is far in excess of the $1,000,000 fine provided by the **Sherman Act** for punishment of such conspiracies. Through the agreement, the companies received what appears to be an increase in revenue of between ten and twelve million dollars as shown in the pre-sentence reports.

* * *

The Lord Chancellor of England said some two hundred years ago, "Did you ever expect a corporation to have a conscience, when it has no soul to be damned, and no body to be kicked?" Two hundred years have passed since the Lord Chancellor espoused this view, and the whole era of what is and is not permitted or what is or is not prohibited, has changed both in design and in application. Certainly, this Court does not expect a corporation to have a conscience, but it does expect it to be ethical and abide by the law. This Court will deal with this company no less severely than it will deal with any individual who similarly disregards the law.

## II. SENTENCE IMPOSED

For the reasons stated, Allegheny Bottling Company is sentenced to three (3) years imprisonment and a fine of One Million Dollars ($1,000,000.00). Execution of the sentence of imprisonment is suspended and all but $950,000.00 of said fine is suspended, and the defendant is placed on probation for a period of three (3) years.

As special conditions of the probation, in addition to the normal terms of probation, the defendant, Allegheny Bottling Company shall provide:

(a) An officer or employee of Allegheny of comparable salary and stature to Jerry Pollino, Allegheny Pepsi-Cola Bottling company's (Allegheny Pepsi's) former Vice President of Sales, to perform forty (40) hours of community service per week in the Baltimore, Maryland, area for a two (2) year period without compensation to the defendant.

(b) An officer or employee of Allegheny of comparable salary and stature to Armand Gravely, Allegheny Pepsi's former Richmond Division Vice President and Manager, to perform forty (40) hours of community service per week for a one (1) year period without compensation to the defendant in the Richmond, Virginia, area.

(c) An officer or employee of Allegheny of comparable salary and stature to Stanley Fabian, Allegheny Pepsi's former Vice President and Norfolk Division Manager, to perform forty (40) hours of community service per week in the Norfolk, Virginia, area for a one (1) year period without compensation to the defendant.

(d) An officer or employee of Allegheny of comparable salary and stature to James Sheridan, Allegheny Pepsi's former President, to perform forty (40) hours of community service per week for a two (2) year period without compensation to the defendant in the areas of Maryland and Virginia formerly served by the Richmond, Norfolk and Baltimore Divisions of Allegheny Pepsi.

It is the further condition of such probation that the company shall not dispose of any of its franchises, capital assets or plants or facilities in the Norfolk, Richmond or Baltimore areas, without specific permission of this Court through the probation officer.

*(Continues)*

That portion of the fine that is not suspended ($950,000.00) shall be paid within 10 days from the date hereof. All of the community service shall be performed under the direction of the probation office and subject to the approval of the Court. In no event is Allegheny Bottling Company to receive any form of compensation for the community service performed.

## III. DISCUSSION

This case was brought pursuant to the Sherman Act, which provides in part:

> Every contract, combination in the form of trust or otherwise, or conspiracy, in restraint of trade . . . is hereby declared to be illegal. Every person who shall make any contract or engage in any combination or conspiracy hereby declared to be illegal shall be deemed guilty of a felony, and, on conviction thereof, shall be punished by a fine not exceeding one million dollars if a corporation, or if any other person, one hundred thousand dollars, *or by imprisonment not exceeding three years*, or by both said punishments in the discretion of the court. 15 U.S.C. § 1 (emphasis added).

A crucial issue in this case is whether the imprisonment term in this statute applies to corporations. Before the Court addresses that issue, however, another issue will be disposed of—the lingering idea that a corporation cannot be imprisoned. This Court today specifically holds that a corporation can be "imprisoned" under the Sherman Act, contrary to the traditional view. . . .

The term "imprisonment" is defined by Webster to include "constraint of a person either by force or by such other coercion as restrains him within limits against his will." *Webster's Third New International Dictionary* at 1137. It is similarly defined as "forcible restraint of a person against his will." *The Random House College Dictionary* at 669. The key to corporate imprisonment is this: imprisonment simply means *restraint*. . . .

Corporate imprisonment requires only that the Court restrain or immobilize the corporation. . . . When this sentence was contemplated, the United States Marshal for the Eastern District of Virginia, Roger Ray, was contacted. When asked if he could imprison Allegheny Pepsi, he stated that he could. He stated that he restrained corporations regularly for bankruptcy court. He stated that he could close the physical plant itself and guard it. He further stated that he could allow employees to come and go and limit certain actions or sales if that is what the Court imposes. . . .

Cases in the past have *assumed* that corporations cannot be imprisoned, without any cited authority for that proposition. . . . This Court, however, has been unable to find any case which actually held that corporate imprisonment is illegal, unconstitutional or impossible. Considerable confusion regarding the ability of courts to order a corporation imprisoned has been caused by courts mistakenly thinking that imprisonment necessarily involves incarceration in jail. *See, e.g., Melrose Distillers, Inc. v. United States*, 359 U.S. 271, 274 (1959). But since imprisonment of a corporation does not necessarily involve incarceration, there is no reason to continue the assumption, which has lingered in the legal system unexamined and without support, that a corporation cannot be imprisoned. Since the Marshal can restrain the corporation's liberty and has done so in bankruptcy cases, there is no reason that he cannot do so in this case as he himself has so stated prior to the imposition of this sentence.

Corporate imprisonment not only promotes the purposes of the Sherman Act, but also promotes the purposes of sentencing. The purposes of sentencing, according to the United States Sentencing Commission, include incapacitating the offender, deterring crime, rehabilitating the offender, and providing just punishment. *United States Sentencing Comm. Guideline Manual* 1.1 (Oct. 1987). The corporate imprisonment imposed today is specifically tailored to meet each of these purposes.

Incapacitation or restraint of the liberty of Allegheny is certainly accomplished by corporate imprisonment. No more price-fixing by Allegheny Pepsi will occur when the United States Marshal prevents the corporation from acting.

Deterrence of price-fixing will also be achieved by corporate imprisonment of Allegheny Pepsi. Other corporations will be deterred by seeing that violation of the Sherman Act could lead to consequences sufficiently severe as to dictate that the potential cost of price-fixing is seldom worth the potential gain.

The deterrence achieved by corporate imprisonment may in fact be the only effective deterrent to price-fixing, especially regarding corporations as large as the ones in this case. A corporation the size of Allegheny Pepsi will, through price-fixing, make many times the maximum fine by the time the price-fixing is detected. Knowing this, the fine will be ineffective as a deterrent even if the probability of detection is considered great. The fine becomes simply a cost of doing business, which is small in comparison with the potential profits. Corporate imprisonment, in contrast, prevents the cost-benefit analysis to economically justify price-fixing. The corporate decision-makers would know that, if caught, the corporation would lose more than they could gain.

## IV. CONCLUSION

For the reasons discussed herein, the Court has ORDERED the sentence as shown in the judgment and commitment order.

### Notes and Questions

1. Is it possible to "imprison" a corporation? Are you convinced by the court's analysis in *Allegheny Bottling*?
2. The *Allegheny Bottling* court, as part of its sentence, required four corporate officers to perform community service. Why do you believe the court imposed this requirement? Is it a fair or effective sanction?
3. The price fixing was so endemic in *Allegheny Bottling* that even low-level employees were aware of the wrongdoing. Should this fact be relevant? Professor Pamela Bucy has argued that in trying to gauge corporate intent for purposes of imposing criminal liability, the law should broaden "the concept of intent beyond the context of individual actors to focus on corporate intent. Such a focus should begin by acknowledging that each organization has an identifiable character or 'ethos.'"[3] This "corporate ethos," she argues, should be a basis for liability only if it encourages the criminal conduct at issue. Is this a viable alternative to the approaches reflected in the common law and the MPC, discussed above?

## Individual Liability

Although the criminal law frequently targets the fictional corporate entity for liability, it is important to recognize that punishment is also imposed on individual corporate actors. The following case, *United States v. Park*, illustrates the practical difficulties involved in imposing individual liability in response to corporate misbehaviors.

### *United States v. Park*, 421 U.S. 658, 95 S. Ct. 1903, 44 L. Ed. 2d 489 (1975)

Mr. Chief Justice Burger delivered the opinion of the Court.

* * *

Acme Markets, Inc., is a national retail food chain with approximately 36,000 employees, 874 retail outlets, 12 general warehouses, and four special warehouses. Its headquarters, including the office of the president, respondent Park, who is chief executive officer of the corporation, are located in Philadelphia, Pa, In a five-count information filed in the United States District Court for the District of Maryland, the Government charged Acme and respondent with violations of the Federal Food, Drug and Cosmetic Act. Each count of the information alleged that the defendants had received food that had been shipped in interstate commerce and that, while the food was being held for sale in Acme's Baltimore warehouse following shipment in interstate commerce, they caused it to be held in a building accessible to rodents and to be exposed to contamination by rodents. These acts were alleged to have resulted in the food's being adulterated within the meaning of 21 U.S.C. §§ 342(a)(3) and (4) in violation of 21 U.S.C. § 331(k).

Acme pleaded guilty to each count of the information. Respondent pleaded not guilty. The evidence at trial demonstrated that in April 1970 the Food and Drug Administration (FDA) advised respondent by letter of insanitary conditions in Acme's Philadelphia warehouse. In 1971 the FDA found that similar conditions existed in the firm's Baltimore warehouse. An FDA consumer safety

*(Continues)*

*(Continued)*

officer testified concerning evidence of rodent infestation and other insanitary conditions discovered during a 12-day inspection of the Baltimore warehouse in November and December 1971. He also related that a second inspection of the warehouse had been conducted in March 1972. On that occasion the inspectors found that there had been improvement in the sanitary conditions, but that "there was still evidence of rodent activity in the building and in the warehouses and we found some rodent-contaminated lots of food items."

The Government also presented testimony by the Chief of Compliance of the FDA's Baltimore office, who informed respondent by letter of the conditions at the Baltimore warehouse after the first inspection. There was testimony by Acme's Baltimore division vice president, who had responded to the letter on behalf of Acme and respondent and who described the steps taken to remedy the insanitary conditions discovered by both inspections. The Government's final witness, Acme's vice president for legal affairs and assistant secretary, identified respondent as the president and chief executive officer of the company and read a bylaw prescribing the duties of the chief executive officer. He testified that respondent functioned by delegating "normal operating duties," including sanitation, but that he retained "certain things, which are the big, broad, principles of the operation of the company," and had "the responsibility of seeing that they all work together." . . .

Respondent was the only defense witness. He testified that, although all of Acme's employees were in a sense under his general direction, the company had an "organizational structure for responsibilities for certain functions" according to which different phases of its operation were "assigned to individuals who, in turn, have staff and departments under them." He identified those individuals responsible for sanitation, and related that upon receipt of the January 1972 FDA letter, he had conferred with the vice president for legal affairs, who informed him that the Baltimore division vice president "was investigating the situation immediately and would be taking corrective action and would be preparing a summary of the corrective action to reply to the letter." Respondent stated that he did not "believe there was anything [he] could have done more constructively than what [he] found was being done." . . .

On cross-examination, respondent conceded that providing sanitary conditions for food offered for sale to the public was something that he was "responsible for in the entire operation of the company," and he stated that it was one of many phases of the company that he assigned to "dependable subordinates." Respondent was asked about and, over the objections of his counsel, admitted receiving, the April 1970 letter addressed to him from the FDA regarding insanitary conditions at Acme's Philadelphia warehouse. He acknowledged that, with the exception of the division vice president, the same individuals had responsibility for sanitation in both Baltimore and Philadelphia. Finally, in response to questions concerning the Philadelphia and Baltimore incidents, respondent admitted that the Baltimore problem indicated the system for handling sanitation "wasn't working perfectly" and that as Acme's chief executive officer he was responsible for "any result which occurs in our company."

\* \* \*

The jury found respondent guilty on all counts of the information, and he was subsequently sentenced to pay a fine of $50 on each count.

\* \* \*

[The Court reviewed its prior decision in *United States v. Dotterweich*, 320 U.S. 277 (1943), which also interpreted the Food and Drug Act. *Dotterweich* held that the public safety purposes of the Act warranted the imposition of individual criminal liability on the president of a corporation alleged to have violated the Act. This was because "the circumstances of modern industrialism . . . are largely beyond self-protection." *Id.* at 280. "In the interest of the larger good [the Act] puts the burden of acting at hazard upon a person otherwise innocent but standing in responsible relation to a public danger." *Id.* at 281.]

The rationale of the interpretation given the Act in *Dotterweich,* as holding criminally accountable the persons whose failure to exercise the authority and supervisory responsibility reposed in them by the business organization resulted in the violation complained of, has been confirmed in our subsequent cases. Thus, the Court has reaffirmed the proposition that "the public interest in the purity of its food is so great as to warrant the imposition of the highest standard of care on distributors." *Smith v. California,* 361 U.S. 147, 152 (1959). In order to make "distributors of food the strictest censors of their merchandise," *ibid.,* the Act punishes "neglect where the law requires care, or inaction where it imposes a duty."

[*Morissette v. United States, supra,* 342 U.S. 255 (1952)]. "The accused, if he does not will the violation, usually is in a position to prevent it with no more care than society might reasonably expect and no more exertion than it might reasonably exact from one who assumed his responsibilities." *Id.* at 256. . . .

Thus *Dotterweich* and the cases which have followed reveal that in providing sanctions which reach and touch the individuals who execute the corporate mission . . . the Act imposes not only a positive duty to seek out and remedy violations when they occur but also, and primarily, a duty to implement measures that will insure that violations will not occur. The requirements of foresight and vigilance imposed on responsible corporate agents are beyond question demanding, and perhaps onerous, but they are no more stringent than the public has a right to expect of those who voluntarily assume positions of authority in business enterprises whose services and products affect the health and well-being of the public that supports them. . . .

The Act does not, as we observed in *Dotterweich,* make criminal liability turn on "awareness of some wrongdoing" or "conscious fraud." The duty imposed by Congress on responsible corporate agents is, we emphasize, one that requires the highest standard of foresight and vigilance, but the Act, in its criminal aspect, does not require that which is objectively impossible. The theory upon which responsible corporate agents are held criminally accountable for "causing" violations of the Act permits a claim that a defendant was "powerless" to prevent or correct the violation. . . . If such a claim is made, the defendant has the burden of coming forward with evidence, but this does not alter the Government's ultimate burden of proving beyond a reasonable doubt the defendant's guilt, including his power, in light of the duty imposed by the Act, to prevent or correct the prohibited condition. Congress has seen fit to enforce the accountability of responsible corporate agents dealing with products which may affect the health of consumers by penal sanctions cast in rigorous terms, and the obligation of the courts is to give them effect so long as they do not violate the Constitution.

[The Court then addressed whether the jury was properly instructed on the proof required.]

Reading the entire charge satisfies us that the jury's attention was adequately focused on the issue of respondent's authority with respect to the conditions that formed the basis of the alleged violations. Viewed as a whole, the charge did not permit the jury to find guilt solely on the basis of respondent's position in the corporation; rather, it fairly advised the jury that to find guilt it must find respondent "had a responsible relation to the situation," and "by virtue of his position . . . had . . . authority and responsibility" to deal with the situation. The situation referred to could only be "food . . . held in unsanitary conditions in a warehouse with the result that it consisted, in part, of filth or . . . may have been contaminated with filth." . . .

The record in this case reveals that the jury could not have failed to be aware that the main issue for determination was not respondent's position in the corporate hierarchy, but rather his accountability, because of the responsibility and authority of his position, for the conditions which gave rise to the charges against him. . . .

Respondent testified in his defense that he had employed a system in which he relied upon his subordinates, and that he was ultimately responsible for this system. He testified further that he had found these subordinates to be "dependable" and had "great confidence" in them. By this and other testimony respondent evidently sought to persuade the jury that, as the president of a large corporation, he had no choice but to delegate duties to those in whom he reposed confidence, that he had no reason to suspect his subordinates were failing to insure compliance with the Act, and that, once violations were unearthed, acting through those subordinates he did everything possible to correct them. . . .

[The purpose of the evidence was to demonstrate that respondent was on notice] that he could not rely on his system of delegation to subordinates to prevent or correct insanitary conditions at Acme's warehouses, and that he must have been aware of the deficiencies of this system before the Baltimore violations were discovered. The evidence was therefore relevant since it served to rebut respondent's defense that he had justifiably relied upon subordinates to handle sanitation matters.

[The Court affirmed respondent's conviction and sentence.]

## Notes and Questions

1. Clearly, Mr. Park did not "intend" to violate the federal Food and Drug Act. Indeed, although "responsible" for the company's operation, his position in the corporate hierarchy was such that he likely had no specific knowledge of the sanitation problem, which happened many miles away from his office. What justifies the imposition of liability under such circumstances?

(*Continues*)

2. Note that a mere $250 fine was at issue. What would account for Mr. Park appealing his convictions all the way to the U.S. Supreme Court, which surely cost several times the fine amount?

3. Does the outcome in Park lead to efficient corporate management? To the extent that delegation to subordinates constitutes effective management practice, will the outcome in *Park* discourage corporate leaders from delegating oversight responsibilities? What alternative, if any, would you suggest?

4. In many instances corporations have in place insurance policies that shield corporate officers from being required to pay fines on their own. Should such a practice be prohibited? Why or why not?

## Fraud

Each year those engaged in fraudulent business enterprises swindle Americans out of countless millions of dollars. **Fraud** ranges from telemarketing scams that cheat the unwary of their funds, to fraudulent billing in the health care industry, to deceptive practices in the sale and offering of stocks. Although individual victims of fraud have historically been permitted to sue for civil damages, increasingly fraud is becoming the subject of criminal actions and the focus of intense concern within the state and federal governments. Recently, for instance, the U.S. Congress conducted hearings into the deceptive practices in the magazine subscription and sweepstakes industry, which annually defrauds elderly victims in particular of millions of dollars.[4] The following case, *United States v. Guadagna,* provides an example of the use of federal criminal law in the effort to combat telemarketing fraud.

### *United States v. Guadagna*, 183 F.3d 122 (2d Cir. 1999)

Feinberg, Circuit Judge:

\* \* \*

The criminal case against Mullen grew out of the investigation into the activities of Rocco F. Guadagna, one of Mullen's co-defendants. Guadagna owned five corporations, collectively called "RFG Group" by the government, all of which engaged in sweepstakes telemarketing. Between January 1991 and March 1992, Mullen worked for RFG Group in a variety of management and sales positions.

Through its various corporations, RFG Group marketed an assortment of products—water and air filters, vitamins, fire retardant spray, cosmetics, cleaning supplies and promotional items, such as pens and key chains—using one basic scheme: sales people cold-called potential purchasers and told them that they had been selected for a special promotion by the product's "sponsor." . . . The potential customer was told that he or she had been chosen to participate in an expensive promotion that entitled the customer to a valuable prize, to be determined at random. Before any product was even mentioned, the sales person described the various prizes, which included some combination of a new car, a Hawaiian vacation, art, a big screen TV, a substantial cash award and jewelry. The salesperson then explained that because the promotion was so expensive, the sponsor asked that the customer listen to a sales pitch for the sponsor's product.

Although the sales person never explicitly said that a purchase *had* to be made for the customer to be entered in the contest, the sales pitch was intentionally worded to imply that such a connection existed. . . . The pitch book used by the sales people contained many pages of rebuttal script, used to overcome customer objections to purchasing any product to win a prize. Only one paragraph in the pitch script, which covers 32 pages in the joint appendix, acknowledged that there was no requirement that a person purchase anything to have access to the "valuable prizes," and this portion of the script was apparently not read to the customer except in response to a customer who "elected" not to purchase a product but wanted the bonus.

Such extensive effort might not normally be necessary to sell cosmetics or cleaning products. . . . However, RFG Group was at a competitive disadvantage, as its products were sold at an enormous markup from their wholesale cost. For example, RFG sold a six-month supply of vitamins for between $259 and $399, even though the vitamins cost RFG less than $12; RFG sold the Pure-N-Simple Water Filter for between $499 and $799, although it cost them only $56; and RFG sold a one-year supply of Valentina skin products for between $599 and $899, after purchasing that quantity for $81. Because the products were so expensive, RFG Group sales people had to lead customers to believe that the prizes for which they were eligible were worth at least the cost of the products purchased. The sales people suggested that the prizes were so valuable that the products were actually a bargain.

In fact, although the prize list did contain prizes that were worth more than the prices charged for the products, the valuable prizes were never randomly awarded to customers. Although customers were often led to believe that they *had already won* the most valuable of the listed prizes, in fact the only prizes randomly awarded were those worth little more than the products. The vacation prize was actually a certificate for lodging at a motel in Hawaii, worth about $45; the jewelry was a diamond pendant worth about $90 or a diamond and sapphire tennis bracelet worth about $40–50; and the art was a $47 lithograph or a ceramic dolphin statue. Cars, cash and big screen televisions were never randomly awarded, despite sales pitch promises to the contrary. . . .

Thomas Mullen worked for RFG Group for about 14 months—from mid-January 1991 through mid-March 1992. At this time Mullen was in his mid-to-late 20s. However, he was not new to the industry and, by industry standards, he was not young. Prior to working at RFG Group, Mullen had worked for Vita Systems, another telemarketing company in Buffalo. Apparently, he was good at his job, because at some point in late 1990 he was recruited by RFG Group to work as a shift manager. . . .

[In June 1995, Mullen and 28 co-defendants were charged in a multi-count indictment. Mullen himself was charged with several counts of wire fraud. The jury ultimately found Mullen guilty of two counts, both of which were addressed by the Court of Appeals. The discussion here focused on sales made by Mullen to Phyllis Birckbichler, involving one such conviction.]

In order to prove a defendant guilty of wire fraud, the government must establish the existence of a scheme to defraud, that money or property were the object of the scheme, and that defendant used interstate wires in furtherance of that scheme. *United States v. Zagari*, 111 F.3d 307, 327 (2d Cir. 1997). Essential to a scheme to defraud is fraudulent intent. *United States v. D'Amato*, 39 F.3d 1249, 1257 (2d Cir. 1994). It is not sufficient that defendant realizes that the scheme is fraudulent and that it has the capacity to cause harm to its victims. *United States v. Gabriel*, 125 F.3d 89, 97 (2d Cir. 1997). Instead, the proof must demonstrate that the defendant had a "conscious knowing intent to defraud . . . [and] that the defendant contemplated or intended some harm to the property rights of the victim." *United States v. Leonard*, 61 F.3d 1181, 1187 (5th Cir. 1995). . . .

However, direct proof of defendant's fraudulent intent is not necessary. Intent may be proven through circumstantial evidence, including by showing that defendant made misrepresentations to the victim(s) with knowledge that the statements were false. *United States v. Smith*, 133 F.3d 737, 743 (10th Cir. 1997). "Misrepresentations amounting only to a deceit are insufficient," *United States v. Starr*, 816 F.2d 94, 98 (2d Cir. 1987), as "the deceit must be coupled with a contemplated harm to the victim." *Id.* However, "[w]hen the 'necessary result' of the . . . scheme is to injure others, fraudulent intent may be inferred from the scheme itself." *D'Amato*, 39 F.3d at 1257. Therefore, a jury may bring to its analysis of intent on individual counts all the circumstantial evidence it has received on the scheme and the purpose of the scheme in which the defendant allegedly participated. . . .

Count six charged Mullen with wire fraud based on a phone call made to Phyllis Birckbichler in March 1992. Ms. Birckbichler testified at Mullen's trial on behalf of the government. During direct examination, she testified that when she received the call from Mullen, she had already made three purchases for $1259, $660, and $1829. However, she had received sales calls since the $1829 purchase, made in July 1991, and declined to buy additional products. . . .

Ms. Birckbichler's testimony about the sales call was fairly detailed. She explained that she recalled the conversation clearly even five years later because she received it on her cellular phone while in the bathroom at her son's apartment. She testified that Mullen told her that she had "hit it big" in "jackpot nine." She recalled that when she expressed skepticism about winning the car, based on her prior experiences with [RFG], he indicated that he could "do the job" for her. He also told her that he was the boss of the person who had last called her, from whom she did not purchase. She testified that he implied so strongly that she had won the car that she "couldn't infer that he meant anything other than that [she] had won the car." Finally, she testified that he coached her on what to say to the verification department when she received the call to verify her order, since "it would mean his job" if she told them that she "was promised any certain prize."

On cross-examination, Mullen's attorney asked Ms. Birckbichler if there was "a point in time when . . . you kind of realized that maybe I'm not going to win one of the big prizes?" Ms. Birckbichler responded

*(Continues)*

"[y]es." She also said that she was familiar with the Publisher's Clearing House sweepstakes and understood that "everybody that enters into that thing doesn't get one of those big checks." . . .

* * *

The evidence showed that Mullen had been at the RFG Group for almost the full term of his 14-month employment when he called Ms. Birckbichler. The jurors heard the pitches from the pitch book and the "13 'Guaranteed to Close'm' Close$," authored by Mullen. They heard testimony that there was no contest or jackpot, that there were no sponsors, and that the calls were not prize notifications at all, but just attempts to sell extremely overpriced products. They heard that customers' photos were requested for "promotional advertising purposes," but, when received, were tacked to "mooch boards" to be laughed at by RFG employees. They heard that Mullen had laughed at the customers and their gullibility. They heard that Mullen had assisted another telemarketer, Don Ivey, in some of his more outrageous misrepresentations: In one case, Ivey pretended to be calling from a helicopter and Mullen made the whirring sound of the blades. Another time, Ivey claimed that he was calling from a submarine while Mullen provided sonar "blips" in the background. They heard Ms. Birckbichler's testimony that Mullen had implied so strongly that she had won a car that she could not help believing that she had won a car. Finally, they heard her testify that Mullen coached her on how to answer the questions from the verification department, including the warning that "it would mean his job" if she told the verifier that she was promised any particular prize. Based on this substantial body of evidence, the jury was entitled to conclude that when Mullen called Ms. Birckbichler and told her that she had "hit it big" in a nonexistent sweepstakes, he knew exactly what he was doing—misleading someone by suggesting that she had won a car and that she was somehow required to purchase overpriced air filters to claim her prize. . . .

[Judgment affirmed.]

## Notes and Questions

1. Do you agree that Mullen deserved to be held criminally liable? Some might argue that the criminal law should play no role in policing such misdeeds, asserting that the victims, after all, have willingly entered the marketplace—where *caveat emptor* (Latin for "let the buyer beware") otherwise prevails. Do you agree that fraud victims should be left to fend for themselves with civil lawsuits? Does the outcome in *Guadagna* in a sense discourage potential victims from being vigilant when they enter the marketplace?

2. What standard should be used to assess gullibility? Should the law be designed to protect the most gullible? Indeed, did not Ms. Birckbichler acknowledge some level of awareness, in relation to the Publisher's Clearinghouse Sweepstakes, that not every entry is a "winner"? In what way was she defrauded? How different was Mullen's behavior from "puffery," a staple of the sales industry, which is permitted?

3. It is estimated that fraudulent telemarketing costs consumers in the United States and Canada alone $40 billion annually. The industry's success stems in large part from its anonymity, which permits unscrupulous telemarketers, situated in "boiler rooms," to take liberties with the truth otherwise not likely possible. Recent data also suggest that telemarketing fraud disproportionately victimizes the elderly, a population perhaps more trusting and hence vulnerable to the slick scams and schemes common to the trade.[5]

## Securities Fraud

Fraud in the securities industry causes enormous harm to individual investors and creates an atmosphere of distrust that hinders investor confidence in the fairness and integrity of the financial markets. To combat these threats, Congress enacted a complex expanse of laws and regulations, originating in the concerted effort to clean up the industry in the wake of the Great Depression of the 1930s. The laws embodied in the **Securities Act of 1934**, in particular, are designed to guard against securities fraud involving (1) material omissions and misstatements in the sale or purchase of securities and (2) "insider trading" (*i.e.*, the improper use of sensitive, non-public information in the sale or purchase of securities). Although the U.S. Securities and Exchange Commission (SEC) has jurisdiction over civil and regulatory proceedings, the U.S. Department of Justice has sole jurisdiction over criminal proceedings.

Here, we examine **insider trading**, a topic that was thrust into the public limelight in the 1980s when government investigations revealed large-scale insider schemes in several of the nation's major brokerage houses. You might recall the names of Ivan Boesky and Michael Milliken, two of the more high-profile targets of the ensuing prosecutions.

Insider trading is prosecuted under Section 10(b) of the 1934 Securities Exchange Act, and SEC Rule 10b-5. Rule 10b-5, in particular, provides as follows:

It shall be unlawful for any person, directly or indirectly, by the use of any means or instrumentality of interstate commerce, or of the mails or of any facility of any national securities exchange.

(a) To employ any device, scheme, or artifice to defraud, [or] . . .

(c) To engage in any act, practice, or course of business which operates or would operate as a fraud or deceit upon any person, in connection with the purchase or sale of any security.

Under the federal securities laws, potential liability is not limited to buyers or sellers of securities; rather, it broadly extends to any deception "in connection with the purchase or sale of any security." *See United States v. O'Hagan*, 521 U.S. 642, 651 (1998). A violation arises when a "corporate insider" uses private information to carry out a securities transaction. Use of such non-public information is criminal because it violates what is called a "fiduciary relationship," a trust-based relationship between the corporation and its shareholders. *Id.* at 652. Use of insider information, in short, provides an unfair advantage relative to persons who lack the information, whether the persons are current shareholders who might otherwise sell their stock or others who might buy the stock if they had the information. As a result, if an insider learns of material information, he or she must either (1) disclose the information to all concerned or (2) refrain from trading on the security at issue. *Id.*

As the following case illustrates, the range of persons possibly targeted by the federal securities laws is quite broad.

---

## United States v. Carpenter, 791 F.2d 1024 (2d Cir. 1987)

Pierce, Circuit Judge:

\* \* \*

### BACKGROUND

Defendants Kenneth P. Felis ("Felis"), and R. Foster Winans ("Winans") appeal from judgments of conviction for federal securities fraud in violation of section 10(b) of the 1934 Act and Rule 10b-5 . . . in connection with certain securities trades conducted on the basis of material, nonpublic information regarding the subject securities contained in certain articles to be published in the *Wall Street Journal*. . . .

Since March 1981, Winans was a *Wall Street Journal* reporter and one of the writers of the "Heard on the Street" column (the "Heard" column), a widely read and influential column in the *Journal*. Carpenter worked as a news clerk at the *Journal* from December 1981 through May 1983. Felis, who was a stockbroker at the brokerage house of Kidder Peabody, had been brought to that firm by another Kidder Peabody stockbroker, Peter Brant ("Brant"), Felis' longtime friend who later became the government's key witness in this case.

Since February 2, 1981, it was the practice of Dow Jones, the parent company of the *Wall Street Journal*, to distribute to all new employees "The Insider Story," a forty-page manual with seven pages devoted to the company's conflicts of interest policy. . . .

Notwithstanding company policy, Winans participated in a scheme with Brant and later Felis and Carpenter in which Winans agreed to provide the two stockbrokers (Brant and Felis) with securities-related information that was scheduled to appear in "Heard" columns; based on this advance information the two brokers would buy or sell the subject securities. Carpenter, who was involved in a private, personal, nonbusiness relationship with Winans, served primarily as a messenger between the conspirators. . . .

During 1983 and early 1984, defendants made prepublication trades on the basis of their advance knowledge of approximately twenty-seven *Wall Street Journal* "Heard" columns, although not all of those columns were written by Winans. Generally, Winans would inform Brant of the subject of an article the day before its scheduled publication. Winans usually made his calls to Brant from a pay phone, and often used a fictitious name. The net profits from the scheme approached $690,000. . . .

*(Continues)*

# I.

Although the facts render the securities fraud issue herein one of first impression, we do not write on a clean slate in assessing whether this case falls within the purview of the "misappropriation" theory of section 10(b) and Rule 10b-5 thereunder. . . .

It is clear that defendant Winans, as an employee of the *Wall Street Journal,* breached a duty of confidentiality to his employer by misappropriating from the *Journal* confidential prepublication information, regarding the timing and content of certain newspaper columns, about which he learned in the course of his employment. We are presented with the question of whether that unlawful conduct may serve as the predicate for the securities fraud charges herein. . . .

[Appellants] argue, the misappropriation theory may be applied only where the information is misappropriated by corporate insiders or so-called quasi-insiders. . . . Thus, appellants would have us hold that it was not enough that Winans breached a duty of confidentiality to his employer, the *Wall Street Journal,* in misappropriating and trading on material nonpublic information; he would have to have breached a duty to the corporations or shareholders thereof whose stock they purchased or sold on the basis of that information.

\* \* \*

We do not say that merely using information not available or accessible to others gives rise to a violation of Rule 10b-5. That theory of 10b-5 liability has been rejected. . . . There are disparities in knowledge and the availability thereof at many levels of market functioning that the law does not presume to address. . . . Whatever may be the legal significance of merely using one's privileged or unique position to obtain material, nonpublic information, here we address specifically whether an employee's use of such information in breach of a duty of confidentiality to an employer serves as an adequate predicate for a securities violation. Obviously, one may gain a competitive advantage in the marketplace through conduct constituting skill, foresight, industry and the like. Certainly this is as true in securities law as in antitrust, patent, trademark, copyright and other fields. But one may not gain such advantage by conduct constituting secreting, stealing, purloining or otherwise misappropriating material nonpublic information in breach of an employer-imposed fiduciary duty of confidentiality. Such conduct constitutes chicanery, not competition; foul play, not fair play. Indeed, underlying section 10(b) and the major securities laws generally is the fundamental promotion of "'the highest ethical standards . . .' in every facet of the securities industry." *SEC v. Capital Gains Research Bureau,* 375 U.S. 180, 186–187 (1963) (quoting *Silver v. New York Stock Exchange,* 373 U.S. 341, 366 (1963)). We think the broad language and important objectives of section 10(b) and Rule 10b-5 render appellants' conduct herein unlawful. . . .

\* \* \*

Winans "misappropriated—stole, to put it bluntly—valuable nonpublic information entrusted to him in the utmost confidence." . . . The information misappropriated here was the *Journal's* own confidential schedule of forthcoming publications. It was the advance knowledge of the timing and content of these publications, upon which appellants, acting secretively, reasonably expected to and did realize profits in securities transactions. Since section 10(b) has been found to proscribe fraudulent trading by insiders or outsiders, such conduct constituted fraud and deceit, as it would had Winans stolen material nonpublic information from traditional corporate insiders or quasi-insiders. . . . Felis' liability as a tippee derives from Winans' liability given the district court's finding of the requisite scienter on Felis' part. . . .

Nor is there any doubt that this "fraud and deceit" was perpetrated "upon a[ny] person" under section 10(b) and Rule 10b-5. It is sufficient that the fraud was committed upon Winans' employer. . . . Appellants Winans, and Felis and Carpenter by their complicity, perpetrated their fraud "upon" the *Wall Street Journal,* sullying its reputation and thereby defrauding it "as surely as if they took [its] money." . . .

As to the "in connection with" standard, the use of the misappropriated information for the financial benefit of the defendants and to the financial detriment of those investors with whom appellants traded supports the conclusion that appellants' fraud was "in connection with" the purchase or sale of securities under section 10(b) and Rule 10b-5. We can deduce reasonably that those who purchased or sold securities without the misappropriated information would not have purchased or sold, at least at the transaction prices, had they had the benefit of that information. . . . Certainly the protection of investors is the major purpose of section 10(b) and Rule 10b-5. *Id.* at 235. Further, investors are endangered equally by fraud by non-inside misappropriators as by fraud by insiders. . . .

[Defendants' convictions for security fraud are affirmed.]

## Notes and Questions

1. It appears that the *Wall Street Journal*, Winans' employer, would not have violated the law if it traded on the information at issue. Why would this be so?
2. Was the "insider" information actually related to the securities at issue? Isn't it true that the information, really only data concerning the strengths and weaknesses of companies, was available to the public at large? Why should use of such information be criminal? Would the typical, non-insider investor be liable for analyzing the same available information?
3. In *Carpenter,* advance knowledge of the publication date of the "Heard on the Street" column seems to constitute insider information. Why is this so?
4. In *United States v. O'Hagan,* cited above, the U.S. Supreme Court approved of the "misappropriation" theory of securities fraud, used in *Carpenter.* Under this theory liability arises when a person "misappropriates confidential information for securities trading purposes, in breach of a duty owed to the source of the information." *United States v. O'Hagan,* 521 U.S. 642, 652 (1998). Liability thus turns on the "deception of those who entrusted him with access to the confidential information." *Id.* The misappropriation theory, in short, covers "outsiders" who breach a duty to the entity owning the information (in *Carpenter,* the *Wall Street Journal*)—not a duty owed to current or potential shareholders.

    The other theory of securities fraud is the "classical" theory. It targets use of non-public information by (1) "true insiders"—such as the officers and directors of the corporation—and (2) "temporary" insiders—such as lawyers, consultants, and accountants who gain access to information as a result of their work. Under the classical theory, such insiders in effect become part of the corporation, which of course owes a direct fiduciary duty to shareholders. The classical theory also covers "tippees," persons who buy or sell securities on the basis of nonpublic information gained from insiders (the "tippers").

## Environmental Crimes

One would think that fouling the air, water, and soil on which we all depend has always been the source of criminal concern. However, it was not until the 1970s, in the wake of the despoliation at Love Canal in upstate New York, that criminal enforcement of laws designed to protect the environment became significant in number. In the succeeding decades environmental debacles at such now-infamous sites as Bhopal, India (release of toxic gas); Rocky Flats, Colorado (nuclear waste); and Prince William Sound, Alaska (massive oil spill) have convinced most Americans of the devastating costs associated with environmental harm.

Although most environmental enforcement efforts are civil and regulatory in nature, the criminal law surely plays a key role. Indeed, over the past several years the federal government in particular has dramatically increased resources dedicated to the criminal investigation and prosecution of environmental crimes.[6] This is especially so with respect to intentional harms inflicted upon the environment. *See, e.g., United States v. Weitzenhoff,* 35 F.3d 1275 (9th Cir. 1994) (affirming defendants' felony convictions, and prison terms, for "knowingly" violating the federal Clean Water Act). More common, however, are reckless or negligent acts that result in harm to the environment, also the subject of criminal enforcement. The following case, *United States v. Hanousek,* underscores the seriousness with which the government approaches even such nonintentional acts.

## *United States v. Hanousek,* 176 F.3d 1116 (6th Cir. 1999)

DAVID R. THOMPSON, Circuit Judge:

Edward Hanousek, Jr., appeals his conviction and sentence for negligently discharging a harmful quantity of oil into a navigable water of the United States, in violation of the Clean Water Act, 33 U.S.C. §§ 1319(c)(1)(A) & 1321(b)(3). . . .

*(Continues)*

## FACTS

Hanousek was employed by the Pacific & Arctic Railway and Navigation Company (Pacific & Arctic) as roadmaster of the White Pass & Yukon Railroad, which runs between Skagway, Alaska, and Whitehorse, Yukon Territory, Canada. As roadmaster, Hanousek was responsible under his contract "for every detail of the safe and efficient maintenance and construction of track, structures and marine facilities of the entire railroad . . . and [was to] assume similar duties with special projects."

One of the special projects under Hanousek's supervision was a rock-quarrying project at a site alongside the railroad referred to as "6-mile," located on an embankment 200 feet above the Skagway River. . . . The project involved blasting rock outcroppings alongside the railroad, working the fractured rock toward railroad cars, and loading the rock onto railroad cars with a backhoe. Pacific & Arctic hired Hunz & Hunz, a contracting company, to provide the equipment and labor for the project.

At 6-mile, a high-pressure petroleum products pipeline owned by Pacific & Arctic's sister company, Pacific & Arctic Pipeline, Inc., runs parallel to the railroad at or above ground level, within a few feet of the tracks. To protect the pipeline during the project, a work platform of sand and gravel was constructed on which the backhoe operated to load rocks over the pipeline and into railroad cars. . . .

On the evening of October 1, 1994, Shane Thoe, a Hunz & Hunz backhoe operator, used the backhoe on the work platform to load a train with rocks. After the train departed, Thoe noticed that some fallen rocks had caught the plow of the train as it departed and were located just off the tracks in the vicinity of the unprotected pipeline. . . . Thoe moved the backhoe off the work platform and drove it down alongside the tracks between 50 to 100 yards from the work platform. While using the backhoe bucket to sweep the rocks from the tracks, Thoe struck the pipeline causing a rupture. The pipeline was carrying heating oil, and an estimated 1,000 to 5,000 gallons of oil were discharged over the course of many days into the adjacent Skagway River, a navigable water of the United States.

Following an investigation, Hanousek was charged with one count of negligently discharging a harmful quantity of oil into a navigable water of the United States, in violation of the Clean Water Act, 33 U.S.C. §§ 1319(c)(1)(A) & 1321(b)(3). Hanousek was also charged with one count of conspiring to provide false information to United States Coast Guard officials who investigated the accident, in violation of 18 U.S.C. §§ 371,1001.

The criminal provisions of the CWA constitute public welfare legislation. . . . Public welfare legislation is designed to protect the public from potentially harmful or injurious items, *see Staples v. United States,* 511 U.S. 600, 607, (1994), and may render criminal "a type of conduct that a reasonable person should know is subject to stringent public regulation and may seriously threaten the community's health or safety," *see Liparota v. United States,* 471 U.S. 419, 433, (1985).

It is well established that a public welfare statute may subject a person to criminal liability for his or her ordinary negligence without violating due process. *See United States v. Balint,* 258 U.S. 250, 252–53 (1922) ("[W]here one deals with others and his mere negligence may be dangerous to them, as in selling diseased food or poison, the policy of the law may, in order to stimulate proper care, require the punishment of the negligent person though he be ignorant of the noxious character of what he sells."); *see also Morissette v. United States,* 342 U.S. 246, 256, (1952) ("The accused, if he does not will the violation, usually is in a position to prevent it with no more care than society might reasonably expect and no more exertion than it might reasonably exact from one who assumed his responsibilities."); *United States v. Dotterweich,* 320 U.S. 277, 281, (1943) ("In the interest of the larger good it puts the burden of acting at hazard upon a person otherwise innocent but standing in responsible relation to a public danger."); *Staples,* 511 U.S. at 607 n. 3, (reiterating that public welfare statutes may dispense with a "mental element"). . . .

[We] conclude that section 1319(c)(1)(A) does not violate due process by permitting criminal penalties for ordinary negligent conduct. . . .

The government presented evidence at trial that Hanousek was responsible for the rock-quarrying project at 6-mile; that the project involved the use of heavy equipment and machinery along the 1,000-foot work site; that Hanousek directed the daily activities of Hunz & Hunz employees and equipment; and that it was customary to protect the pipeline with railroad ties and fill when using heavy equipment in the vicinity of the pipeline. The government also presented evidence that when work initially began in April, 1994, Hunz & Hunz covered an approximately 300-foot section of the pipeline with railroad ties, sand, and ballast material to protect the pipeline; that after Hanousek took over responsibility for the project in May, 1994, no further sections of the pipeline along the work site were protected; and that the section of the pipeline where the rupture occurred was not protected with railroad ties, sand or ballast. Finally, the government presented evidence that although the rock quarrying work had been completed in the location of the rupture, rocks would sometimes fall off the loaded railroad cars as they proceeded through the completed sections of the work site; that no policy prohibited the use of backhoes off the work platform for other activities; that a backhoe operator ruptured the unprotected pipeline while using a backhoe to remove a rock from the railroad tracks; and that a harmful quantity of oil was discharged into the Skagway River.

The totality of this evidence is sufficient to support Hanousek's conviction for negligently discharging a harmful quantity of oil into a navigable water of the United States, in violation of 33 U.S.C. §§ 1319(c)(1)(A) & 1321 (b)(3).

## F. Sentencing

Based on an offense level [under the U.S. Federal Sentencing Guidelines] of 12 and a criminal history category of 1, the district court sentenced Hanousek to 6 months in prison, 6 months in a halfway house, and 6 months of supervised release. . . .
Affirmed.

## Notes and Questions

1. Hanousek was subject to imprisonment based on a law that required mere "negligent" conduct on his part, an unusually low standard of culpability for the criminal law. Moreover, as "roadmaster," he—not his employer, the Pacific & Arctic Railway Navigation Company—was personally liable. Is this fair? Should the "public welfare" nature of the statute at issue resolve the question? Was Hanousek morally innocent?

2. *Hanousek* illustrates the uniqueness of criminal environmental enforcement in yet another respect. Environmental crimes, unlike run-of-the-mill "street crimes," often have as their byproduct some socially advantageous outcome. For instance, the rock-quarrying project undertaken by Hanousek in itself had social and economic benefit, unlike a common street robbery or burglary. If society wishes persons such as Hanousek to continue to provide the social good associated with their work, does it make sense to subject them to individual criminal penalties? Would it make more sense to limit liability to the corporation? Is *Hanousek* an instance of over-criminalization? Does society risk over-deterrence?

    Also, even short of altogether dissuading individuals or entities from pursuing environmentally risky pursuits, does a precedent like *Hanousek* encourage possible unnecessary preventive measures? Who, in the end, would pay for such additional measures? Is this approach, in the language of economics, "efficient"?

3. If you believe that criminal punishments are inappropriate for environmental offenders, what alternative sanctions, if any, would you advocate?

Although states play an important role in the criminal enforcement of environmental laws, by far the federal government undertakes the larger share of environmental prosecutions. To do so, federal prosecutors have at their disposal a veritable arsenal of laws, which are of unparalleled complexity and scope. One of the primary laws is the **Resource Conservation and Recovery Act (RCRA)**, initially enacted by Congress in 1976. RCRA is designed to regulate and monitor hazardous waste from the "cradle to the grave"—including its initial generation and subsequent treatment, storage, transportation, and disposal. The following case provides some insight into this complex area of the environmental laws.

## *United States v. Johnson & Towers, Inc.*, 741 F.2d 662 (3d Cir. 1984)

Sloviter, Circuit Judge.

Before us is the government's appeal from the dismissal of three counts of an indictment charging unlawful disposal of hazardous wastes under the Resource Conservation and Recovery Act. In a question of first impression regarding the statutory definition of "person," the district court concluded that the Act's criminal penalty provision imposing fines and imprisonment could not apply to the individual defendants. We will reverse.

*(Continues)*

(*Continued*)

# I.

The criminal prosecution in this case arose from the disposal of chemicals at a plant owned by Johnson & Towers in Mount Laurel, New Jersey. In its operations the company, which repairs and overhauls large motor vehicles, uses degreasers and other industrial chemicals that contain chemicals such as methylene chloride and trichlorethylene, classified as "hazardous wastes" under the Resource Conservation and Recovery Act (RCRA), 42 U.S.C. §§ 6901–6987 (1982) and "pollutants" under the Clean Water Act, 33 U.S.C. §§ 1251–1376 (1982). During the period relevant here, the waste chemicals from cleaning operations were drained into a holding tank and, when the tank was full, pumped into a trench. The trench flowed from the plant property into Parker's Creek, a tributary of the Delaware River. Under RCRA, generators of such wastes must obtain a permit for disposal from the Environmental Protection Agency (E.P.A.). The E.P.A. had neither issued nor received an application for a permit for Johnson & Towers' operations.

The indictment named as defendants Johnson & Towers and two of its employees, Jack Hopkins, a foreman, and Peter Angel, the service manager in the trucking department. According to the indictment, over a three-day period federal agents saw workers pump waste from the tank into the trench, and on the third day observed toxic chemicals flowing into the creek. . . .

Johnson & Towers pled guilty to the RCRA counts. Hopkins and Angel pled not guilty, and then moved to dismiss counts 2, 3, and 4. The court concluded that the RCRA criminal provision applies only to "owners and operators," *i.e.*, those obligated under the statute to obtain a permit. Since neither Hopkins nor Angel was an "owner" or "operator," the district court granted the motion as to the RCRA charges. . . .

We hold that section 6928(d)(2)(A) covers employees as well as owners and operators of the facility who knowingly treat, store, or dispose of any hazardous waste, but that the employees can be subject to criminal prosecution only if they knew or should have known that there had been no compliance with the permit requirement of section 6925.

# II.

The single issue in this appeal is whether the individual defendants are subject to prosecution under RCRA's criminal provision, which applies to:

> *[a]ny person* who—
>
> . . . .
>
> > (2) knowingly treats, stores, or disposes of any hazardous waste identified or listed under this subchapter either (A) without having obtained a permit under section 6925 of this title . . . or
> >
> > (B) in knowing violation of any material condition or requirement of such permit. 42 U.S.C. § 6928(d) (emphasis added).

The permit provision in section 6925, referred to in section 6928(d), requires "each person owning or operating a facility for the treatment, storage, or disposal of hazardous waste identified or listed under this subchapter to have a permit" from the E.P.A. . . .

We conclude that in RCRA, no less than in the Food and Drugs Act, Congress endeavored to control hazards that, "in the circumstances of modern industrialism, are largely beyond self-protection." *United States v. Dotterweich*, 320 U.S. at 280. It would undercut the purposes of the legislation to limit the class of potential defendants to owners and operators when others also bear responsibility for handling regulated materials. The phrase "without having obtained a permit *under section 6925*" (emphasis added) merely references the section under which the permit is required and exempts from prosecution under section 6928(d)(2)(A) anyone who has obtained a permit; we conclude that it has no other limiting effect. Therefore we reject the district court's construction limiting the substantive criminal provision by confining "any person" in section 6928(d)(2)(A) to owners and operators of facilities that store, treat or dispose of hazardous waste, as an unduly narrow view of both the statutory language and the congressional intent.

# III.

## A.

Since we must remand this case to the district court because the individual defendants are indeed covered by section 6928(d)(2)(A), it is incumbent on us to reach the question of the requisite proof as to individual defendants under that section. . . .

We focus again on the statutory language:

[a]ny person who—

    (2) *knowingly* treats, stores, or disposes of any hazardous waste identified or listed under this subchapter either—

        (A) without having obtained a permit under section 6925 of this title . . . or

        (B) *in knowing violation* of any material condition or requirement of such permit. 42 U.S.C. § 6928(d) (1982) (emphasis added).

If the word "knowingly" in section 6928(d)(2) referred exclusively to the acts of treating, storing or disposing, as the government contends, it would be an almost meaningless addition since it is not likely that one would treat, store or dispose of waste without knowledge of that action. At a minimum, the word "knowingly," which introduces subsection (A), must also encompass knowledge that the waste material is hazardous. . . .

Whether "knowingly" also modifies subsection (A) presents a somewhat different question. . . .

[The court thereafter concluded that the "knowingly" requirement applied to (A).]

[I]n light of our interpretation of section 6928(d)(2)(A), it is evident that the district court will be required to instruct the jury, *inter alia,* that in order to convict each defendant the jury must find that each knew that Johnson & Towers was required to have a permit, and knew that Johnson & Towers did not have a permit. Depending on the evidence, the district court may also instruct the jury that such knowledge may be inferred.

In summary, we conclude that the individual defendants are "persons" within section 6928(d)(2)(A), that all the elements of that offense must be shown to have been knowing, but that such knowledge, including that of the permit requirement, may be inferred by the jury as to those individuals who hold the requisite responsible positions with the corporate defendant. For the foregoing reasons, we will reverse the district court's order dismissing portions of counts 2, 3 and 4 of the indictment, and we will remand for further proceedings consistent with this opinion.

## Notes and Questions

1. The defendants in *Johnson & Towers* risked possible felony punishment. Does this reality make the court's reliance by analogy on the Food and Drug Act, under which defendants face punishment for misdemeanors, less persuasive?

2. Of course, under most circumstances "ignorance of the law is no excuse." But at the same time the defendants in *Johnson & Towers* were prosecuted pursuant to RCRA, an expansive statutory scheme one court has derided as "mind-numbing" in its complexity. *American Mining Congress v. EPA,* 824 F.2d 1177, 1189 (D.C. Cir. 1987). Should this complexity make the criminal prosecution of RCRA violations less appealing on grounds of fairness and public policy?[7]

3. The courts have repeatedly rejected defense efforts to avoid liability by arguing that the specific substance at issue was not "hazardous waste," as defined by RCRA. For criminal liability to arise, a defendant need only be aware (1) of its actions and (2) the "general hazardous character" of the waste. *United States v. Goldsmith,* 978 F.2d 643, 646 (11th Cir. 1992). The specific wastes identified as "hazardous" by the Environmental Protection Agency number in the several hundreds and are subject to constant reevaluation and modification. Does this uncertainty influence your view on whether criminal intervention is appropriate?

## Death and Injury

It is now widely recognized that despite their many positive contributions to society, corporations can play a direct role in consumer and worker safety-related injury and deaths. In one relatively recent example, owners of a chicken-processing plant in rural Hamlet, North Carolina, were concerned about theft among their low-skilled processors of "chicken nuggets." To prevent such theft, the owners locked the several exit doors that were legally required. When a fire engulfed the plant in

September 1991, 25 workers were killed by smoke inhalation and another 56 injured, with bloody footprints left on the locked doors by those desperately trying to batter the doors down with their feet. According to congressional hearings on the Hamlet tragedy, the chicken plant owners locked the doors after suffering, at most, $245 in stolen chicken nuggets. (For a good overview of the Hamlet tragedy and its aftermath *see* Aulette and Michalowski.[8])

*Cornellier v. Black,* discussed next, provides yet another tragic instance of workplace death.

## *Cornellier v. Black,* 425 N.W.2d 21 (Wis. Ct. App. 1988)

* * *

Eich, Judge.

The facts are not in dispute. Cornellier is an officer and director of Pyro Science Development Corporation, a fireworks manufacturer. On March 23, 1983, an employee at Pyro's Milton plant plugged a fan into an electrical outlet, generating sparks which caused a fire and explosion that killed another plant employee, Dennis Whitt, and injured several other people. After an investigation, Rock County authorities issued a criminal complaint charging Cornellier, who was alleged to be the day-to-day director of operations at the plant, with homicide by reckless conduct. The complaint, which will be discussed in detail later in this opinion, generally based the charge on Cornellier's knowledge of substantial fire and explosion hazards at the plant, many of them in violation of [U.S. Occupational Safety and Health Act (OSHA)] requirements, and his failure to take any steps to correct them.

* * *

Homicide by reckless conduct is defined by sec. 940.06(2), Stats., as "an act which creates a situation of unreasonable risk and high probability of death . . . and which demonstrates a conscious disregard for the safety of another and a willingness to take known chances of perpetrating an injury." The statute also provides that the definition is to be understood as embracing "all of the elements of what was heretofore known as gross negligence in the criminal law of Wisconsin."

In addition to describing how the fire and explosion occurred, and how it caused Whitt's death, the complaint alleges the following facts: (1) Cornellier controlled Pyro Corporation and ran the day-to-day operations at the Milton plant; (2) he was aware that Pyro was engaged in manufacturing fireworks without a permit in a structure that did not meet state and local safety requirements, and that, as a result, the manufacturing operations were "illegal"; (3) less than three weeks before the fatal fire and explosion, Cornellier was convicted of six violations of safety ordinances in connection with the manufacture of fireworks at a nearby Milton plant—including storing more than 500 pounds of explosives in a building and manufacturing fireworks in an open frame structure; (4) sometime before the explosion, Cornellier was advised by a business associate of safety concerns at the plant—concerns which had been described by another associate as "a disaster, bags of chemicals left open, everything was a mess, it was very apparent . . . that the risks were very high"; and (5) following the explosion and fire, the United States Department of labor reported nine separate violations of federal safety standards at the plant, including lack of adequate precautions against ignition of flammable vapors, lack of a safe means of escape from the building, mishandling of explosive materials in a manner hazardous to life, failure to implement safety practices to protect employees from explosion and fire, and generally hazardous and unsafe equipment and wiring throughout the building.

Cornellier argues first that the complaint does not identify him as committing any "act," and thus cannot state probable cause because not only is that word used in the statute, but cases involving criminal gross negligence all deal with an affirmative "act." We are not persuaded.

As a general rule, "[t]he requirement of an overt act . . . is not inherently necessary for criminal liability. . . . Omissions are as capable of producing consequences as overt acts." *State v. Williquette,* 385 N.W.2d 145, 151 (Wis. 1986). Indeed, the *Williquette* court "expressly reject[ed] the . . . claim that an act of commission, rather than omission, is a necessary element of a crime. The essence of criminal conduct is the requirement of a wrongful 'act.' This element, however, is satisfied by overt acts, as well as omissions to act where there is a legal duty to act" *Id.* at 150 (citation omitted).

As indicated, sec. 940.06, Stats., is intended to embrace all of the elements of gross negligence. . . . It is just as much an "act" to deliberately or recklessly refrain from performing a known legal duty as it is to negligently perform that duty. We conclude, therefore, that the statute, impliedly, if not directly, acknowledges that the crime of reckless homicide may be committed by omission, as well as commission.

* * *

We believe the allegations permit the reasonable inference that Cornellier was aware of the multitude of extremely dangerous conditions at the Milton plant prior to, and on the date of, the fire and explosion, and that he did nothing to correct those conditions or safeguard his employees from the known dangers. Similarly, one may infer from the complaint that Cornellier's failure to provide safe storage of explosive materials and a safe electrical system was a substantial factor in causing the explosion—that is, that it was "a factor actually operating and which had substantial effect in producing the death as a natural result" Wis J I—Criminal 1160. The complaint was adequate to establish probable cause.

Writ denied.

## Notes and Questions

1. Pyro Science was engaged in a legitimate commercial enterprise. Also, isn't it true that the death resulted from an "accident"? Clearly, civil and regulatory sanctions can and should play a role in combating such dangerous conditions, but what benefits come from pursuing criminal prosecution?

2. Another example of criminal enforcement in the safety realm arose recently when the state attorney in Miami-Dade County, Florida filed murder and manslaughter charges in relation to the 1996 Valujet airplane crash that killed 110 passengers. In particular, the prosecutor charged the aviation maintenance company that had allegedly improperly stored and labeled the oxygen canisters thought responsible for the crash. The charges are believed to be the first ever brought in an airline accident.[9] The prosecution raises an important yet subtle question: Is the use of the criminal law counterproductive under the circumstances? Will the threat of possible criminal prosecution make it less likely that in future accidents involved parties will cooperate, at the expense of survivors? What are the benefits of such a prosecution? Do they outweigh the possible negative effects?

## Harmful Products

One of the best-known examples of alleged "corporate death" from **harmful products** arose in 1978 in connection with the Ford Motor Company's Pinto automobile, when three people who were traveling in a 1973 Pinto were struck from behind by a van. Under normal circumstances the accident would have been no more than a "fender bender." In this instance, however, although the van incurred only minor damage, the Pinto was almost immediately engulfed in flames, killing all three passengers. As it turns out, the fire was caused by the placement of the gas tank at the rear end of the Pinto, which greatly increased the risk of explosion in the event of a rear-end collision. Furthermore, it appeared that Ford was aware of the risk, as a result of an investigative exposé in a national magazine and warnings from the federal government relative to the Pinto. Finally, internal Ford memoranda highlighted what appeared as a callous financial motive on the part of the company, indicating that Ford (1) was sensitive to the risks posed but situated the tank in the rear of the Pinto to save money and (2) decided not to initiate a nationwide recall of Pintos, reasoning that monetary costs from civil law suits for any eventual deaths would be less than that associated with a recall.

With pressure building to hold Ford accountable, a county prosecutor in Indiana secured three indictments against the company for "reckless homicide." Ford put on a vigorous defense, as would be expected. After a 10-week trial and 25 ballots, however, the jury acquitted Ford of the criminal charges. (The Pinto saga is retold in detail in Cullen et al.[10])

At the time of the prosecution, Ford was the fourth-largest corporate entity in the world, providing an enormous number of jobs and significantly contributing to the overall benefit of the nation's economy. At the same time, however, the facts presented at trial indicated that Ford's business decisions and practices embodied something less than ideal "corporate citizenship." Perhaps even more troubling, Ford seemed immune to traditional forms of social control, such as federal agency intervention and the threat of expensive civil suits brought by injured Pinto passengers. Do you agree that criminal prosecution was warranted? If you were the county prosecutor, would you have sought indictments? Does the criminal law have any place in the corporate boardroom?

## Conclusion

The subject of this chapter, white collar crime, betrays any concise description. To varying degrees the offenses discussed share the common characteristic of originating in the organizational realm and typically involve an economic motive. This motive can come in direct form, such as when a fraudulent telemarketer steals millions from unwary victims, or in indirect form, such as when a company engages in "midnight dumping" of toxic waste to avoid legal (and more costly) disposal of the waste. Either way, the social and economic tolls of white collar crime are staggering, prompting government to target such activity for criminal enforcement, along with the traditional "index crimes" discussed elsewhere in this book. As a result of this emphasis, white collar crime continues to evolve as a fertile area of the substantive criminal law and as a scholarly subdiscipline, posing an intriguing array of practical and theoretical issues not encountered elsewhere in the field of criminal law.

## Key Terms

corporate liability

fraud

harmful products

individual liability

insider trading

Resource Conservation and
 Recovery Act (RCRA)

Securities Act of 1934

Sherman Act

"street" crimes

white collar crimes

## Review Questions

1. White collar crimes are sometimes referred to as "elite crimes." Explain this term, and compare and contrast *white collar crimes* with *"street" crimes.*

2. What requirements must be met for corporate liability to arise?

3. What is the intent of the Sherman Act?

4. Is it possible for corporate officers to be charged individually with white collar crime? Explain your answer.

5. List the elements that constitute *insider trading.*

6. Give an example of an environmental crime and explain how it meets the criteria to be considered a *white collar crime.*

7. Explain how workplace conditions can lead to a charge under white collar crime criteria.

8. Product liability falls under the white collar crime classification. Even though Ford Motor Company was not held criminally liable in the Pinto case, other major corporations have been held liable in similar cases. Find a case in which a major corporation has been held criminally liable in a product liability case and discuss it.

## Notes

1. Sutherland EH. White collar criminality. *Am. Soc. Rev.* 1949;5:1, 9.

2. *See* Belluck P. Juice poisoning case brings guilty plea and a huge fine. *New York Times,* July 24, 1998, at A12.

3. *See* Bucy P. Organizational sentencing guidelines: The cart before the horse. *Washington Univ. Law Q.* 1993;71:329.

4. *See* Eaton S. You are probably not a winner: House subcommittee hears horror stories of sweepstakes promoters who prey on the elderly. *Cleveland Plain Dealer,* August 5, 1999, at A1.

5. *See* Reznek S. Fraudulent telemarketing: crime and punishment. *Mich. Bar J.* 1998:1210.

6. *See* Watson T. Today's EPA: You pollute, we prosecute. *USA Today,* May 21, 1998, at A5.

7. *See* Brickey KF. Environmental law at the crossroads: The intersection of environmental and criminal law theory. *Tulane Law Rev.* 1997;71:487.

8. Aulette JR, Michalowski R. Fire in Hamlet: A case study of state-corporate crime. In: Tunnell, KD, ed. *Political Crime in Contemporary America: A Critical Approach.* New York: Garland Publishing; 1993:171–206.

9. *See* Wald ML. Murder charges filed by Florida in Valujet crash. *New York Times,* July 14, 1999, at A1.

10. Cullen FT et al. *Corporate Crime Under Attack: The Ford Pinto Case and Beyond.* Cincinnati, OH: Anderson Publishing; 1987:145–308.

# Liability for the Conduct of Another

## CHAPTER 11

## Chapter Objectives

- Distinguish between *principals* and *accessories after the fact* in criminal complaints
- Learn the importance of *accomplice liability* and *accessory liability* in criminal matters
- Learn the common law rules for accomplice liability
- Understand the rules for accomplice liability under modern statutes
- Understand the concept of vicarious liability

Since its origin the substantive criminal law has targeted for liability individuals other than those personally or directly engaged in criminal activity. Indeed, liability has extended to persons involved before, during, and even after the principal crime in question.

At common law four basic categories of criminal actors existed:

1. **Principal in the first degree**: One who is the main criminal perpetrator.
2. **Principal in the second degree**: One who is present (actively or constructively) at the time of the offense and aids, counsels, or otherwise encourages the main criminal actor. Traditionally, this category has been synonymous with "aiding and abetting."
3. **Accessory before the fact**: One who is not present at the time of the offense yet prior thereto aids, counsels, or otherwise encourages the main criminal actor. One who "incites" the criminal action of another would fall in this category.
4. **Accessory after the fact**: One who is aware of the commission of the crime by the main criminal actor and purposefully provides help to the actor in the name of avoiding apprehension and conviction. This category of actors might be called "criminal protectors."

Today, the distinctions drawn among the four categories described above are less important than they were at common law. In particular, the first three categories are typically lumped together for legal purposes, yet category 4 remains distinct. The rationale behind this shift is that accessories after the fact engage in behaviors that differ in kind from principals and accessories before the fact. Their wrongdoing is mainly geared toward preventing the detection and prosecution of the main actor(s) committing the crime, as opposed to being actively engaged in the commission of the crime itself.

As a result, most often today both types of principals and accessories before the fact are deemed principals, resulting in identical punishment. *See, e.g.,* 18 U.S.C. § 2(a) ("Whoever commits an offense against the United States or aids, abets, counsels, commands, induces, or procures its commission, is punishable as a principal"). Meanwhile, accessories after the fact are prosecuted separately (and usually less harshly) because their behaviors are less blameworthy in the eyes of the law insofar as they seek to stymie police, not commit the actual crime. In short, this chapter examines two basic theories of criminal liability: **accomplice liability** (often referred to as "aiding and abetting") and accessory liability.

This chapter also explores a related basis for criminal liability: **vicarious liability**. With vicarious liability the wrongdoing of an "agent" results in the criminal liability of the "principal," based on a theory known as ***respondant superior*** (literally, "let the master answer"). Liability of this sort is purely relationship based, relying on the theory that a principal (a "master") should have legal responsibility for the wrongful act of his or her agent. Vicarious liability typically arises in the context of business-related wrongdoing, such as when a tavern employee provides liquor to persons after legal "closing time."

## Accomplice Liability

### *Actus Reus*

> ### *State v. Vaillancourt,* 453 A.2d 1327 (N.H. 1982)
>
> #### PER CURIAM.
>
> The only issue presented in this case is whether the Trial Court (*Wyman, J.*) erred in ruling that the indictment against the defendant was sufficient on its face. We hold that the court's ruling was erroneous, and we reverse.
>
> *(Continues)*

(*Continued*)

The factual backdrop of the case involves an attempted burglary on the morning of December 8, 1980, at the O'Connor residence in Manchester. On that day, a neighbor observed two young men, allegedly the defendant, David W. Vaillancourt, and one Richard Burhoe, standing together on the O'Connors' front porch. The men were ringing the doorbell and conversing with one another. Because they remained on the porch for approximately ten minutes, the neighbor became suspicious and began to watch them more closely. She saw them walk around to the side of the house where Burhoe allegedly attempted to break into a basement window. The defendant allegedly stood by and watched his companion, talking to him intermittently while the companion tried to pry open the window. The neighbor notified the police, who apprehended the defendant and Burhoe as they were fleeing the scene.

Shortly thereafter, a grand jury indicted the defendant for accomplice liability under RSA 626:8 III. The indictment alleged, in pertinent part, as follows:

> [T]hat David W. Vaillancourt . . . with the purpose of promoting and facilitating the commission of the offense of attempted burglary, did purposely aid Richard Burhoe . . . *by accompanying him to the location of said crime and watching* as the said Richard Burhoe [attempted to commit the crime of burglary]. . . . (Emphasis added.)

Prior to trial, the defendant filed a motion to dismiss, claiming that the indictment failed to allege criminal conduct on his part. The trial court denied the motion, and a jury subsequently found the defendant guilty as charged. The defendant now contests the sufficiency of his indictment.

The defendant bases his argument on the axiomatic principle that an indictment must allege some criminal activity. *See State v. Thresher*, 442 A.2d 578, 581 (N.H. 1982); *State v. Taylor*, 431 A.2d 775, 778 (N.H. 1981). He specifically contends that his indictment was insufficient because, even if the facts alleged in it were true, they would not have satisfied the elements necessary for accomplice liability or for any other crime. We agree.

The crime of accomplice liability under RSA 626:8 III(a) requires the actor to have solicited, aided, agreed to aid, or attempted to aid the principal in planning or committing the offense. The crime thus necessitates some active participation by the accomplice. We have held that knowledge and mere presence at the scene of a crime could not support a conviction for accomplice liability because they did not constitute sufficient affirmative acts to satisfy the *actus reus* requirement of the accomplice liability statute. *See State v. Goodwin*, 395 A.2d 1234, 1236 (N.H. 1978); *State v. Shippee*, 349 A.2d 587, 588 (N.H. 1975).

In the instant case, other than the requisite *mens rea*, the State alleged only that the defendant aided Burhoe "by accompanying him to the location of the crime and watching. . . ." Consistent with our rulings with respect to "mere presence," we hold that accompaniment and observation are not sufficient acts to constitute "aid" under RSA 626:8 III(a). We conclude that the trial court erred in upholding the defendant's indictment.

Reversed.

BOIS, Justice, with whom BROCK, Justice, joins, dissenting:

I cannot accept the majority's conclusion that accompaniment and observation are insufficient acts to constitute "aid" under the accomplice liability statute, RSA 626:8 III(a). Although I agree that "mere presence" would be an insufficient factual allegation, *see State v. Goodwin*, 395 A.2d 1234, 1236 (N.H. 1978), the indictment in this case alleged more than "mere presence." As the majority concedes, the indictment alleged the requisite *mens rea*. It also alleged accompaniment, which connotes presence *and* some *further connection* between the accomplice and the principal. While not a customary form of assistance, "accompaniment with the purpose of aiding" implies the furnishing of moral support and encouragement in the performance of a crime, thereby "aiding" a principal in the commission of an offense. *Cf. id.* at 1236–37 (jury could reasonably have concluded that defendant's presence facilitated and encouraged principal to commit rape). I would therefore hold that the indictment in this case sufficiently alleged criminal conduct on the part of the defendant.

## Notes and Questions

1. As *Vaillancourt* establishes, "mere presence" at the scene of a crime does not make one an accomplice. Yet another illustration of this principle is found in an Indiana case in which the defendant, his wife and two children, and a passenger named Rootes picked up a hitchhiker while traveling in the family car. Rootes thereafter robbed the hitchhiker at knife point, with the defendant (who was driving) doing nothing to intervene. The Indiana Supreme Court reversed the defendant's conviction as an

accomplice to robbery on the rationale that there was no evidence demonstrating that "the appellant aided and abetted in the alleged crime. While he was driving his car, nothing was said nor did he act in any manner to indicate his approval or countenance of the robbery." *State v. Pace,* 224 N.E.2d 312 (Ind. 1967).

2. How much help must the defendant give the perpetrator for there to be liability? Not much, it seems. One late 19th century judge put it as follows:

> The assistance given . . . need not contribute to the criminal result in the sense that but for it the result would not have ensued. It is quite sufficient if it facilitated a result that would have transpired without it. It is quite enough if the aid merely renders it easier for the principal actor to accomplish the end intended by him and the aider and abettor, though in all human possibility the end would have been attained without it. *State ex rel. Attorney General v. Talley,* 15 So. 722 (Ala. 1894).

However, even if a "but-for" causal relation need not be shown, it is still necessary that the behavior in question have had some effect. For instance, in *State v. Ulvinen,* 313 N.W.2d 425 (Minn. 1981), the court addressed the role of verbal encouragement in accomplice liability. The *Ulvinen* court rejected liability on the basis of a mother's unsubtle remark that it "would be best for the kids" or that "it will be best" if her son killed his wife, which he did. According to the court, there was "no evidence that her remark had any influence on her son's decision to kill his wife." The mother's words amounted to mere "passive approval."

3. Most often, accomplice liability is premised on the active involvement of the defendant in one form or another. At times, however, a failure to act—when the defendant has a duty to act—can lead to criminal liability. (See Chapter 3 for a discussion of *actus reus* and acts or omissions more generally.) Child abuse prosecutions provide the most common example of this scenario. In *State v. Walden,* 293 S.E.2d 780 (N.C. 1982), for instance, a mother (defendant) was present while her boyfriend savagely beat her 1-year-old son. On appeal, the North Carolina Supreme Court upheld the mother's conviction for aiding and abetting felonious assault. The court stated the following:

> It remains the law that one may not be found to be an aider and abettor, and thus guilty as a principal, solely because he is present when a crime is committed. . . . But we hold that the failure of a parent who is present to take all steps reasonably possible to protect the parent's child from an attack by another person constitutes an act of omission by the parent showing the parent's consent and contribution to the crime being committed.

Be aware, however, that not all jurisdictions impose liability under such circumstances. *See, e.g., Commonwealth v. Raposo,* 595 N.E.2d 773 (Mass. 1992) (rejecting liability based on the failure of a parent to prevent the sexual abuse of her daughter).

For an example of an omission leading to liability in the non–child abuse context, see *Powell v. United States,* 2 F.2d 47 (4th Cir. 1924), where the court held that a train conductor during the Prohibition Era aided and abetted the transportation of illegal liquor because he failed to report the plainly visible, poorly hidden liquor.

## Mens Rea

## State v. Gladstone, 474 P.2d 274 (Wash. 1970)

HALE, Justice.

A jury found defendant Bruce Gladstone guilty of aiding and abetting one Robert Kent in the unlawful sale of marijuana. Deferring imposition of sentence, the court placed defendant on probation. He appeals the order deferring sentencing contending that the evidence as a matter of law was insufficient to sustain a verdict of guilty. His point, we think, is well taken.

One who aids or abets another in the commission of a crime is guilty as a principal under RCW 9.01.030, which says:

> Every person concerned in the commission of a felony, gross misdemeanor or misdemeanor, whether he directly commits the act constituting the offense, or aids or abets in its commission, and whether present or absent; and every person who directly or indirectly counsels, encourages, hires, commands, induces or otherwise procures another to commit a felony, gross misdemeanor or misdemeanor, is a principal,

*(Continues)*

(*Continued*)

and shall be proceeded against and punished as such. The fact that the person aided, abetted, counseled, encouraged, hired, commanded, induced or procured, could not or did not entertain a criminal intent, shall not be a defense to any person aiding, abetting, counseling, encouraging, hiring, commanding, inducing or procuring him.

Gladstone's guilt as an aider and abettor in this case rests solely on evidence of a conversation between him and one Douglas MacArthur Thompson concerning the possible purchase of marijuana from one Robert Kent. There is no other evidence to connect the accused with Kent who ultimately sold some marijuana to Thompson. . . .

Even if it were accorded all favorable inferences, there appears at this point a gap in the evidence which we feel as a matter of law is fatal to the prosecution's cause. Neither on direct examination nor under cross-examination did Thompson testify that he knew of any prior conduct, arrangements or communications between Gladstone and Kent from which it could be even remotely inferred that the defendant had any understanding, agreement, purpose, intention or design to participate or engage in or aid or abet any sale of marijuana by Kent. Other than to obtain a simple map from Gladstone and to say that Gladstone told him Kent might have some marijuana available, Thompson did not even establish that Kent and the defendant were acquainted with each other. Testimony of the brief conversation and Gladstone's very crude drawing consisting of 8 penciled lines indicating where Kent lived constitute the whole proof of the aiding and abetting presented. . . .

That vital element—a nexus between the accused and the party whom he is charged with aiding and abetting in the commission of a crime—is missing. The record contains no evidence whatever that Gladstone had any communication by word, gesture or sign, before or after he drew the map, from which it could be inferred that he counseled, encouraged, hired, commanded, induced or procured Kent to sell marijuana to Douglas Thompson as charged, or took any steps to further the commission of the crime charged. He was not charged with aiding and abetting Thompson in the purchase of marijuana, but with Kent's sale of it. . . .

This court has recognized the necessity of proof of a nexus between aider and abettor and other principals to sustain a conviction. In *State v. Hinkley*, 315 P.2d 889 (Wash. 1958), amplifying the term *abet*, we said:

Although the word "aid" does not imply guilty knowledge or felonious intent, the word "abet" includes *knowledge* of the wrongful purpose of the perpetrator, as well as counsel and encouragement in the crime.

and approved the instruction that:

To abet another in the commission of a crime implies a consciousness of guilt in instigating, encouraging, promoting or aiding in the commission of such criminal offense.

It would be a dangerous precedent indeed to hold that mere communications to the effect that another might or probably would commit a criminal offense amount to an aiding and abetting of the offense should it ultimately be committed.

There being no evidence whatever that the defendant ever communicated to Kent the idea that he would in any way aid him in the sale of any marijuana, or said anything to Kent to encourage or induce him or direct him to do so, or counseled Kent in the sale of marijuana, or did anything more than describe Kent to another person as an individual who might sell some marijuana, or would derive any benefit, consideration or reward from such a sale, there was no proof of an aiding and abetting, and the conviction should, therefore, be reversed as a matter of law. Remanded with directions to dismiss.

## Notes and Questions

1. Do you agree with the reasoning and result reached in *Gladstone*? Is the court stretching the nexus requirement too far?
2. It is often difficult to draw the critical distinction between "knowing" and "purposeful" assistance. In New York this difficulty is lessened due to the availability of a specific statute prohibiting criminal "facilitation," a distinct yet less serious crime that punishes those who knowingly assist law breakers. Under this statute a person is guilty of **facilitation** when "believing it *probable* that he is rendering aid to a person who intends to commit a crime, he engages in conduct which provides such person with the means or opportunity for the commission thereof and which in fact aids such person to commit a felony." New York Penal Law § 115.00 (McKinney's 1998) (emphasis added).

## Sharing a Common Purpose

One of the central controversies with respect to the *mens rea* of accomplice liability turns on the distinction between purposeful versus knowing assistance. In other words, should it be legally sufficient that the defendant merely was aware that he or she assisted the crime in question, or should proof be required that he or she behaved with the purpose of actually aiding the crime? For instance, should accomplice liability arise if the defendant provides a weapon to another, with the knowledge that it might be used in a crime? In most jurisdictions there would be no liability under such a scenario because a defendant must (1) intend to perform the helpful act and (2) intend that the act promote commission of the actual crime. *See State v. Harrison,* 425 A.2d 111, 113 (Conn. 1979).

The rule was succinctly stated by the famous judge, Learned Hand, who said that an accomplice must "in some sort associate himself with the venture, . . . participate in it as something that he wishes to bring about, . . . seek by his action to make it succeed." *United States v. Peoni,* 100 F.2d 401, 420 (2d Cir. 1938). In sum, the individual must have a "purposive attitude towards" the criminal outcome. *Id.*

The decision in *United States v. Giovannetti,* 919 F.2d 1223 (7th Cir. 1990), illustrates the practical, deterrence-oriented approach often taken in analyzing this question. There, Janis rented a house to Merino (an undercover agent) with the knowledge that Merino would use the property for illegal gambling in which Giovannetti was also involved. Janis was successfully prosecuted for aiding and abetting the gambling, a conviction upheld on appeal. In upholding the conviction the court first distinguished Janis's case from a situation in which no aiding and abetting liability would arise:

> A stationer who sells an address book to a woman he knows to be a prostitute is not an aider and abettor. . . . He can hardly be said to be seeking by his action to make her venture succeed, since the transaction has very little to do with that success and his "livelihood will not be affected appreciably by whether her venture succeeds or fails. And . . . punishing him would not reduce the amount of prostitution—the prostitute would simply . . . shop for address books among stationers who did not know her trade.
>
> [However,] [i]f a gambling enterprise to succeed needs to enlist a landlord who knows the purpose of the rental, punishing him will make life significantly more difficult for the enterprise. This prospect makes it easier to defend

the imposition of aider and abettor liability than it was in our hypothetical example of the prostitute's purchase of an address book. *Id.* at 1227.

Assuming that purpose to aid can be established, does the government need to prove that the defendant shared with the main perpetrator the entire scope of the crime ultimately committed? For instance, presume that A encourages B to threaten a pedestrian with his fist and steal his wallet. In the course of carrying out the robbery, however, B whips out a pistol and shoots the victim, causing serious but nonfatal harm. Should A be liable for the robbery and aggravated battery? Under the common law liability would exist for all "natural and probable consequences" of the proposed crime. The California Supreme Court summarized this view as follows:

> [an aider and abettor's] knowledge that an act which is criminal was intended, and his action taken with the intent that the act be encouraged or facilitated, are sufficient to impose liability on him for any reasonably foreseeable offense committed as a consequence by the perpetrator. It is the intent to encourage and bring about conduct that is criminal, not the specific intent that is an element of the target offense, which . . . must be found by the jury. *People v. Croy,* 710 P.2d 392, 398 n.5 (Cal. 1985).

*See also* Kan. Stat. Ann. § 21-3205(2) (1998) (stating that an accomplice to the main offense "is also liable for any other crime committed in pursuance of the intended crime if reasonably foreseeable by such person as a probable consequence of committing or attempting to commit the crime intended").

It should be noted, however, that not all jurisdictions endorse such a broad view of liability. Under federal law, for example, courts typically require that the accomplice know that the principal would use a weapon under the hypothetical robbery offered above. *See, e.g., United States v. Dinkane,* 17 F.3d 1192 (9th Cir. 1994). The Model Penal Code (MPC) reflects a similarly narrow view. *See* MPC § 2.06(3)(a) (requiring that an accomplice have "the purpose of promoting or facilitating the commission of the offense" carried out by the principal).

## Defense of Withdrawal or Abandonment

What happens if the defendant has second thoughts about the criminal enterprise? Typically, one can make out a defense to accomplice liability on the basis of **abandonment** or withdrawal, but only under limited circumstances. The MPC, for instance, provides that a defense is available if the alleged accomplice "terminates his complicity prior to the commission of the offense and (i) wholly deprives it of effectiveness in the commission of the offense; or (ii) gives timely warning to the law enforcement authorities or otherwise makes proper effort to prevent commission of the offense." *See* MPC § 2.06(6)(c).

An example of behavior deemed insufficient to warrant a defense is found in *People v. Cooper,* 332 N.E.2d 453 (Ill. App. 1975), where a defendant and two others broke into a woman's house to carry out a robbery. The defendant helped tie up the victim, but then had second thoughts and left the house without taking any valuables. During the ensuing course of events inside the house, however, the victim was killed, resulting in murder prosecutions of the defendant and his two cohorts. The court rejected the defendant's abandonment defense because he had done nothing to free the victim.

## Distinguishing Accomplice Liability from Solicitation and Conspiracy

Because, by definition, an accomplice associates with another in an underlying criminal act, the law of accomplice liability is frequently confused with the law of solicitation and conspiracy (discussed at length in Chapter 12). It is important, however, to recognize that the crimes differ in important respects.

With conspiracy, an agreement to commit a crime must be proven by the government; with accomplice liability, aid itself—not an express agreement—suffices. Furthermore, with accomplice liability the target crime must of course be committed; conspiracy liability can arise even if the target crime is not ultimately committed.

Solicitation is more closely related because it targets for liability those who instigate others to commit criminal acts. Once again, however, accomplice liability requires that the perpetrator actually commit the target crime. With solicitation, the crime solicited does not actually need to be committed; it is only necessary "that the actor with intent that another person commit a crime, have enticed, advised, incited, ordered or otherwise encouraged that person to commit a crime."[2]

## Accessory Liability

As noted at the outset, the criminal law has always targeted those who render aid to felons after the commission of a crime. For accessory liability to exist, three requirements must be satisfied. First, the underlying felony must in fact have been completed at the time aid is provided (a mistaken belief of completion does not suffice). Second, the accessory must actually know, not merely suspect, that the felony was committed. Third, the aid must be given principally, but not solely, to hinder the apprehension, conviction, or punishment of the felon. In this latter regard, "it is not sufficient that the felon was actually aided if the assistance was given out of charity or for some similar motive."[3]

As a result of these requirements, only certain acts qualify as a basis for accessory liability. For instance, liability usually does not arise from a mere failure to report felonious activity. On the other hand, harboring a known felon, aiding a felon's escape, concealing or destroying evidence, or providing false information to divert the attention of police all can qualify. In general, the sort of behavior leading to accessory liability concerns that which obstructs justice. In the words of MPC § 242.3, accessorial conduct is that done "to hinder the apprehension, prosecution, conviction or punishment of another for crime."

### *United States v. Graves,* 143 F.3d 1185 (9th Cir. 1998)

Reinhardt, Circuit Judge:

This case requires us to construe the federal statute governing the offense of accessory after the fact; in doing so, we contrast its provisions with those contained in conspiracy and aiding and abetting statutes. *See* 18 U.S.C. §§ 2(a), 3, & 371. We must decide in particular whether the crime of being an accessory after the fact to a felon in possession of a firearm requires that the defendant have knowledge not only that the person who committed the primary offense ("the offender") possessed a gun (which may ordinarily be perfectly legal), but also that he was a felon (which renders the possession illegal under federal law). 18 U.S.C. § 922(g)(1). . . .

#### Background

The facts are generally undisputed. On the night of September 9, 1995, Lyndon Lloyd Graves accompanied his friend Shawn Prince to a party at an apartment on Treasure Island Naval Base in San Francisco, California. Upon their arrival, Prince entered the apartment and immediately began yelling at his girlfriend, trying to convince her to go home with him. During the argument, Prince brandished a gun. Although Graves attempted to persuade his friend to leave the party, he was unsuccessful in his efforts, and went outside to wait in his truck.

Soon thereafter, naval security officers arrived, handcuffed Prince, and placed him in the backseat of a police cruiser. The security officers searched Prince, but did not find the gun that he had used. While the officers were busy taking witness statements, Graves got out of his truck and, aided by a remarkable display of law enforcement negligence, opened the door of the cruiser and helped Prince escape. According to a neighbor who observed the events, Prince and Graves ran down the street into a nearby cul-de-sac, where Graves threw a rag-like object into a trash can. Based on the neighbor's report, naval security officers located the trash can and retrieved a Smith and Wesson .357 magnum revolver.

At some point the same night, Prince and Graves returned to the apartment and were arrested. Both were taken to the naval security station house and each received citations for carrying a loaded weapon in a public place, drawing and exhibiting a weapon in a rude or threatening manner, and possession of a deadly weapon. Because Prince used an alias, the security officers did not discover that he had a prior felony conviction. Both men were immediately released from custody. Once Prince's status as a felon came to light, however, he was charged with being a felon in possession of a firearm. Several weeks after Prince was indicted, and after he had failed to surrender pursuant to notice of the issuance of a federal arrest warrant, an FBI agent observed Graves driving a car with Prince in the passenger seat. The agent activated his lights and siren, and after a slow, mile-long chase, Graves pulled over. Graves and Prince were both taken into custody.

In a superseding indictment, Graves was charged with being an accessory after the fact to Prince in connection with the felon in possession offense, in violation of 18 U.S.C. § 3, and with aiding and abetting Prince's escape, in violation of 18 U.S.C. §§ 2 & 751. Graves and Prince were tried jointly.

* * *

After several hours of deliberation, the jury delivered a note posing the question: "[M]ust we find that Graves knew that Prince had a prior felony conviction . . . ?" The district court responded: "No." Graves was convicted on both counts, and sentenced to a 27-month term of imprisonment, to be followed by a two-year term of supervised release. He appeals only the conviction on the accessory after the fact count. . . .

## I.

As with any question of statutory construction, we look first to the statute itself. It provides in relevant part:

> Whoever, *knowing that an offense against the United States has been committed,* receives, relieves, comforts or assists the offender in order to hinder or prevent his apprehension, trial or punishment, is an accessory after the fact. 18 U.S.C. § 3 (emphasis added).

Because it is clear that the statute contains a knowledge requirement, the question is what, precisely, the defendant must know in order to "know[ ] that an offense against the United States has been committed." Obviously, the defendant must know at least some facts regarding the offender and the primary offense. Only when he has knowledge of the offense can he be guilty of comforting or assisting the offender in order to hinder his apprehension or trial.

In this case, there is no dispute that the government had to prove that Graves knew that Prince possessed a gun. The government contends, however, that Graves was required to know nothing more. Its basic position is that knowledge of Prince's felon status was unnecessary. . . .

More forcefully, the government maintains that the law governing accessories after the fact should be the same as the law governing aiders, abettors, and conspirators; and, because aiders, abettors, and conspirators do not need to be aware of the *principals* status as a felon, accessories do not need to be aware that the *offender* had previously been convicted of a felony. *See United States v. Canon,* 993 F.2d 1439, 1442 (9th Cir. 1993) (holding that an aider and abettor to the crime of felon in possession of a firearm does not need to be aware of the principal's status as a felon).

The government asserts that the only difference between the offenses of conspiracy or aiding and abetting, and the offense of accessory after the fact is the timing of the commission of the "facilitating" offense in relation to the underlying or primary offense. In other words, the government says, the only difference is that one occurs before or simultaneously with the principal offense and the other occurs afterwards. The government is plainly in error: its argument fails to recognize (1) the clear differences in language between the statute governing aiding and abetting and conspiracy, and (2) the fundamental distinctions in the nature of the two types of crimes.

## A.

Although the government urges us not to impose a knowledge requirement on accessories that differs from the knowledge requirement for aiders and abettors or conspirators, we cannot ignore the obvious

(Continues)

differences in the statutes both with respect to language and the very nature of the offenses. As to the first, the accessory after the fact statute contains an express knowledge requirement that is independent of and separate from the knowledge requirement for the primary offense. By contrast, the statutes governing conspiracy and aiding and abetting impose no knowledge requirement apart from the *mens rea* requirement for the underlying offense. *Compare* 18 U.S.C. § 3 [accessory after the fact] ("Whoever, *knowing that an offense against the United States has been committed,* receives, relieves, comforts, or assists the offender in order to hinder or prevent his apprehension, trial or punishment, is an accessory after the fact") (emphasis added), *with* 18 U.S.C. § 2(a) [aider and abettor] ("Whoever commits an offense against the United States or aids, abets, counsels, commands, induces or procures its commission, is punishable as a principal."), *and* 18 U.S.C. § 371 [conspiracy] (making it an offense for "two or more persons [to] conspire either to commit any offense against the United States, or to defraud the United States"). . . .

Indeed, as we have previously recognized, a defendant who is accused of being an accessory after the fact must be shown to have had actual knowledge of each element of the underlying offense. *See United States v. Burnette,* 698 F.2d 1038, 1052 (9th Cir. 1983). In *Burnette,* we upheld the conviction of an accessory after the fact on the ground that "the jury was clearly instructed that [the defendant] must have had actual knowledge of each element of [the armed bank robbery statute], including knowledge of the use of a gun, in order to be guilty as an accessory after the fact to armed bank robbery."

## B.

Equally important, there is a critical difference between the nature of the crime of being an accessory after the fact and the nature of the crime of being either an aider and abettor or a conspirator. One who acts as an accessory after the fact does not participate in the commission of the primary offense. Instead, an accessory is one who provides assistance to the offender by helping him to avoid apprehension or prosecution after he has already committed an offense. Accordingly, an accessory does not incur liability as a *principal. See United States v. Innie,* 7 F.3d 840, 851 (9th Cir. 1993) (underscoring the fact that "an accessory after the fact is not liable as a principal"). For this reason, the offense of accessory after the fact is considered a far less serious offense than the primary offense, and the accessory, who has not played a role in the actual commission of an offense, is considered far less culpable than one who has participated in the commission of the offense. In fact, the accessory after the fact statute specifically limits the punishment for that offense to *no more than half* of the punishment prescribed for the primary offense. 18 U.S.C. § 3; *see also* U.S.S.G. § 2X3.1 (establishing the offense level for the crime of accessory after the fact).

The contrary is true with respect to both the conspiracy and aiding and abetting offenses, which derive their elements solely from the underlying offense, impose liability on the conspirator and aider and abettor as though they were principals, and provide for the *same* punishment for conspirators and aiders and abettors as for principals. . . .

A natural reading of the language—"knowing that an offense against the United States has been committed"—inevitably leads to the conclusion that, at the very least, an accessory after the fact must be aware of conduct on the part of the offender that satisfies the essential elements of a particular offense against the United States.

## II.

Having decided that the accessory after the fact statute requires the government to establish that the defendant was aware of the facts that constitute the essential elements of the offender's crime, we turn to the question whether the government met its burden with respect to Graves. . . .

The evidence at trial did not, of course, support a finding that Graves knew Prince was a felon in possession. The only evidence that suggested that Graves knew Prince's possession was even "unlawful" was the fact that Graves had disposed of the gun in the trash can and that he tried to help Prince escape. Because Graves had observed Prince brandishing the gun at the party, an obviously unlawful act, and knew that Prince had been arrested for *that* conduct, the only reasonable inference is that Graves acted in order to help Prince avoid prosecution for the offense he witnessed. As noted earlier, the judge specifically told the jury that they did not need to find that Graves knew of Prince's prior felony and that they could convict him as an accessory after the fact as long as they found he knew Prince possessed the gun "unlawfully." Most important of all, however, there is absolutely *no* evidence that Graves knew Prince was a felon, and the government did not try to establish such knowledge on his part. . . .

Accordingly, Graves's conviction must be reversed for insufficiency of the evidence.

### Notes and Questions

1. For accessory liability to arise, the *Graves* court states that "the accessory must know that the offender had engaged in the conduct that constitutes the federal offense—though not necessarily that such conduct constitutes a federal offense." *United States v. Graves*, 143 F.3d 1185, 1186 (9th Cir. 1998). Do you agree with the court's interpretation of the federal accessory statute? Do you agree with the outcome?

## Hindering Prosecution

When considering accessory liability, it is important to recognize that today many states have complemented the traditional common law of accessory liability by implementing statutory provisions that significantly broaden the reach of the law. For its part, the MPC contains an omnibus provision that prohibits "**Hindering Apprehension or Prosecution**," which encompasses many specific behaviors reflected in the statutory law of jurisdictions. *See* MPC § 242.3. This section provides as follows:

> A person commits an offense if, with the purpose to hinder the apprehension, prosecution, conviction or punishment of another for crime, he:
>
> (1) harbors or conceals the other; or
> (2) provides or aids in providing a weapon, transportation, disguise or other means of avoiding apprehension or effecting escape; or
> (3) conceals or destroys evidence of the crime, or tampers with a witness, informant, document or other source of information, regardless of its admissibility in evidence; or
> (4) warns the other of impending discovery or apprehension, except that this paragraph does not apply to a warning given in connection with an effort to bring another into compliance with the law; or
> (5) volunteers false information to a police officer.

This offense is a felony in the third degree if the conduct which the actor knows has been charged or is liable to be charged against the person aided would constitute a felony of the first or second degree. Otherwise it is a misdemeanor. *Id.*

Also, although not commonly in effect today, it is worth mentioning that the common law made it an offense to "misprision a felony."[4] Today, federal criminal law contains such a provision, which states as follows:

> Whoever, having knowledge of the actual commission of a felony cognizable by a court of the United States, conceals and does not as soon as possible make known the same to some judge or other person in civil or military authority under the United States, shall be fined under this title or imprisoned not more than three years, or both. 18 U.S.C. § 4 (1999).

As should be evident, misprision of felony differs in a basic way from accessory law in that misprision involves only concealment of criminal activity, while accessory liability requires the commission of an affirmative act by the alleged accessory to aid evasion. *See Butler v. United States*, 481 A.2d 431, 444–45 I (D.C. Ct. App. 1984) (interpreting District of Columbia law).

## Failed or Deferred Prosecutions

Today, unlike at common law, in most jurisdictions the government can prosecute both accomplices and accessories, even if the main perpetrator is acquitted or for some reason evades prosecution. A practical reason exists for this seeming incongruity. As noted by the Arizona Supreme Court:

> whether or not the principal is convicted or acquitted in a separate trial can have no bearing on the trial of the aider and abettor, if the evidence shows the latter guilty. Society is no less injured by the illegal acts of the aider and abettor even though the principal himself escapes conviction. In order to convict an aider and abettor, justice demands no more than [that] . . . the evidence convincingly show that a crime was committed by the principal. *State v. Spillman*, 468 P.2d 376, 378 (Az. 1970).[5]

Aside from the distinct harm associated with the behaviors targeted by accessory and accomplice liability, the courts are sensitive to the reality that juries sometimes reach inconsistent results, at times stemming from motives of lenience or compromise. As the U.S. Supreme Court stated several years ago in upholding the liability of an accomplice, "different juries may reach different results. . . . While symmetry of results may be intellectually satisfying, it is not required." *Standefer v. United States*, 447 U.S. 10, 25 (1980).

## Vicarious Liability

As noted at the outset of the chapter, vicarious liability permits the wrongful acts of one person to result in the liability of another, purely by virtue of the supervisory-like relationship the latter has over the former. Thus unlike other areas of the criminal law, vicarious liability does not require any specific act or omission by the defendant. Rather, liability arises solely on the basis of the supervisory responsibility that the defendant has relative to the individual committing the act or omission.

Another important distinction is that vicarious liability quite often involves application of "strict" liability. These concepts are related but differ in important ways. With strict liability the defendant must merely commit an illegal act (or omission) and no *mens rea* is required for liability to arise, a very rare occurrence in the criminal law given the threatened loss of liberty and stigma it carries. (Strict liability is discussed at greater length in Chapter 3.) With vicarious liability the defendant does not need to even commit an act for liability to arise, because the *actus reus* of another party substitutes for that of the defendant.

The following case, *State v. Beaudry*, illustrates the application of vicarious liability, and the important role it plays in the control of corporate misbehavior (which itself is discussed at greater length in Chapter 10).

---

## *State v. Beaudry*, 365 N.W.2d 593 (Wis. 1985)

Abrahamson, Justice.

The jury found the defendant, Janet Beaudry, the agent designated by the corporation pursuant to sec. 125.04(6)(a), Stats. 1981–82, of the alcoholic beverage laws, guilty of the misdemeanor of unlawfully remaining open for business after 1:00 A.M. in violation of sec. 125.68(4)(c), which prohibits premises having a "Class B" license from remaining open between the hours of 1:00 A.M. and 8:00 A.M. The tavern manager, not the defendant personally, had kept the tavern open, giving drinks to friends, contrary to his instructions of employment. On the basis of the judgment the defendant was ordered to pay a fine in the amount of $200.00.

The defendant raises two issues on appeal: (1) whether the statutes impose vicarious criminal liability on the designated agent of a corporate licensee for the conduct of a corporate employee who violates sec. 125.68(4)(c), Stats. 1981–82, and (2) whether there is sufficient evidence to support the verdict in the case. Before we discuss these two issues, we shall set forth the facts.

### I.

The facts are undisputed. The defendant, Janet Beaudry, and her husband, Wallace Beaudry, are the sole shareholders of Sohn Manufacturing Company, a corporation which has a license to sell alcoholic beverages at the Village Green Tavern in the village of Elkhart Lake, Sheboygan county. Janet Beaudry is the designated agent for the corporate licensee pursuant to sec. 125.04(6)(a), Stats. 1981–82.

Janet Beaudry's conviction grew out of events occurring during the early morning hours of February 9, 1983. At approximately 3:45 A.M., a deputy sheriff for the Sheboygan County Sheriff's Department drove past the Village Green Tavern. He stopped to investigate after noticing more lights than usual inside the building and also seeing two individuals seated inside. As he approached the tavern, he heard music, saw an individual standing behind the bar, and saw glasses on the bar. Upon finding the tavern door locked, the deputy sheriff knocked and was admitted by Mark Witkowski, the tavern manager. The tavern manager and two men were the only persons inside the bar. All three were drinking. The deputy sheriff reported the incident to the Sheboygan county district attorney's office for a formal complaint.

At about noon on February 9, the tavern manager reported to Wallace Beaudry about the deputy's stop earlier that morning. After further investigation Wallace Beaudry discharged the tavern manager on February 11.

On March 2, 1983, the Sheboygan County Sheriff's Department served the defendant with a summons and a complaint charging her with the crime of keeping the tavern open after hours contrary to sec. 125.68(4)(c), Stats., and sec. 125.11(1), Stats. The tavern manager was not arrested or charged with an offense arising out of this incident.

The case was tried before a jury on May 20, 1983. At trial Janet Beaudry testified that she was not present at the tavern the morning of February 9. Wallace Beaudry testified that Janet Beaudry had delegated to him, as president of Sohn Manufacturing, the responsibilities of business administration associated with the Village Green Tavern; that he had hired Mark Witkowski as manager; that he had informed Witkowski that it was his duty to abide by the liquor laws; and that he never authorized Witkowski to remain open after 1:00 A.M., to throw a private party for his friends, or to give away liquor to friends.

Witkowski testified that he had served drinks after hours to two men. During cross-examination Witkowski confirmed that Wallace Beaudry had never authorized him to stay open after hours; that he had been instructed to close the tavern promptly at the legal closing time; that he knew it was illegal to serve liquor after 1:00 A.M. to anyone, including friends; that his two friends drank at the bar before 1:00 A.M. and had paid for those drinks; that he was having a good time with his friends before closing hours and wanted to continue partying and conversing with them after 1 A.M.; that after closing hours he was simply using the tavern to have a private party for two friends; that he did not charge his friends for any of the liquor they drank after 1:00 A.M.; and that by staying open he was trying to benefit not Wallace Beaudry but himself.

At the close of evidence, the jury was instructed that the law required the premises to be closed for all purposes between 1:00 A.M. and 8:00 A.M. and that if the jury found that there were patrons or customers on the premises after 1:00 A.M., it must find the premises open contrary to statute.

The jury was also instructed regarding Janet Beaudry's liability for the conduct of the tavern manager: As designated agent of the corporation, the defendant had full authority over the business and would be liable for the tavern manager's violation of the closing hour statute if he was acting within the scope of his employment. The instructions, which are pattern instructions Wis.J.I.Cr. 440 (1966), describe what activities are within the scope of employment and what are outside the scope of employment. . . .

The state's prosecution of the defendant under the criminal laws rests on a theory of vicarious liability, that is respondeat superior. Under this theory of liability, the master (here the designated agent) is liable for the illegal conduct of the servant (here the tavern manager). (*See* Footnote 4.) The defendant asserts, contrary to the position of the state, that the statutes do not impose vicarious criminal liability on her as designated agent of the corporation for the tavern manager's illegal conduct.

While the focus in this case is on the defendant's vicarious criminal liability, it is helpful to an understanding of vicarious liability to compare it with the doctrine of strict liability. Strict liability allows for criminal liability absent the element of *mens rea* found in the definition of most crimes. *See* LaFave & Scott, *Criminal Law*, § 31 (1972). . . . Thus under strict liability the accused has engaged in the act or omission; the requirement of mental fault, *mens rea*, is eliminated. . . .

Vicarious liability, in contrast to strict liability, dispenses with the requirement of the *actus reus* and imputes the criminal act of one person to another. LaFave & Scott, *Criminal Law*, section 32 (1972).

Inasmuch as the natural person licensee is subject to vicarious criminal liability for the conduct of her or his employee who illegally sells alcoholic beverages, it logically follows that a corporation licensee should be similarly liable for the illegal conduct of its employee. But in this case the defendant is not the corporation licensee; the defendant is the designated agent of the corporation. The question for the present case, therefore, is whether a designated corporate agent is subject to vicarious criminal liability for the illegal conduct, *i.e.*, remaining open after closing hours, of the tavern manager who is an employee of the corporation. . . .

We agree with the court of appeals that the legislature intended to impose such liability on the designated agent. The court of appeals correctly noted that unless vicarious criminal liability was imposed on the designated agent, a natural person licensee could avoid criminal liability by simply incorporating the business. . . . The legislature could not have intended to allow a natural person licensee to avoid criminal liability by incorporating the business. . . .

While legislative purpose and public policy support our holding that a designated agent is vicariously liable for the conduct of the corporate employee who violates the closing hours law, the defendant's final argument is that due process requires blameworthy conduct on the part of the defendant as a prerequisite to criminal liability. Although the imposition of criminal liability for faultless conduct does not comport with the generally accepted premise of Anglo-American criminal justice that criminal liability is based on personal fault, this court and the United States Supreme Court have upheld statutes imposing criminal liability for some types of offenses without proof that the conduct was knowing or wilful or negligent.

We turn now to the question of whether the evidence supports the verdict that the tavern manager was acting within the scope of his employment. As we stated previously, the jury was instructed that the defendant is liable only for the acts of the tavern manager that were within the scope of his employment. Thus the defendant is not liable for all the acts of the tavern manager, only for those acts within the scope of employment.

The application of the standard of scope of employment limits liability to illegal conduct which occurred while the offending employee was engaged in some job-related activity and thus limits the accused's vicarious liability to conduct with which the accused has a factual connection and with which the accused has some responsible relation to the public danger envisaged by the legislature.

We agree with the conclusion reached by the court of appeals. The credibility of the bar manager's testimony was a matter for the jury. *Braatz v. Continental Casualty Co.*, . . . 76 N.W.2d 303 (Wis. 1956).

*(Continues)*

*(Continued)*

The bar manager's testimony which supports the defendant's position that the manager was acting outside the scope of employment was based on a statement the bar manager gave defendant's counsel the night before trial. The jury may not have believed this testimony which was favorable to the defendant. Considering that the conduct occurred on the employer's premises and began immediately after "closing time"; that the employee had access to the tavern after hours only by virtue of his role as an employee of the corporate licensee, which role vested him with the means to keep the tavern open; and that the defendant may anticipate that employees may be tempted to engage in such conduct; the jury could conclude that the tavern manager's conduct was sufficiently similar to the conduct authorized as to be within the scope of employment. The jury could view the tavern manager's conduct as more similar to that of an employee to whom the operation of the business had been entrusted and for whose conduct the defendant should be held criminally liable than to that of an interloper for whose conduct the defendant should not be held liable.

For the reasons set forth, we affirm the decision of the court of appeals affirming the conviction. Decision of the court of appeals is affirmed.

### Notes and Questions

1. Does the result in *Beaudry* seem fair? Does the majority's decision seem justified under the facts and relevant law? What explains the outcome reached by the majority?

2. For many years vicarious liability has commonly been used to control misbehavior by businesses, which of course lack the mental apparatus and physical capacity to act themselves. Of late, however, vicarious criminal liability is being used in a distinctly different context—to induce parents to exercise greater control over their children.

   Not surprisingly, this approach has been the subject of judicial scrutiny. In *State v. Ackers*, 400 A.2d 38 (N.H. 1979), for instance, the New Hampshire Supreme Court addressed the constitutionality of a state law that made parents criminally liable for the violation by their children of laws pertaining to the use of off-road recreational vehicles. While recognizing that vicarious liability applied in the business context, the *Ackers* court rejected its use "merely because of [one's] status as a parent." *Id.* at 40. In concluding that the law violated constitutional substantive due process, the court stated the following:

   > Parenthood lies at the foundation of our civilization. The continuance of the human race is entirely dependent upon it. . . . Considering the nature of parenthood, we are convinced that the status of parenthood cannot be made a crime. This, however, is the effect of [the statute]. Even if a parent has been as careful as anyone could be, even if the parent has forbidden the conduct, and even if the parent is justifiably unaware of the activities of the child, criminal liability is still imposed. . . . There is no other basis for criminal responsibility other than the fact that a person is the parent of one who violates the law. *Id.*

   Even accepting the court's view of the essential biological and social role of parenthood, can you identify possible deterrent benefits to such a statute? Should it matter that the punishment for such a violation is relatively minor, such as a monetary fine? Do the potential public benefits of the use of vicarious liability under such circumstances warrant the intrusion on parental rights?

## Conclusion

The offenses examined in this chapter can best be thought of as theories of criminal liability rather than substantive crimes. Through accomplice and accessory liability, as well as vicarious liability, society criminalizes conduct that although not necessarily directly harmful in itself, nevertheless contributes to a social harm. Although in relative terms the offenses are prosecuted relatively infrequently, they pose interesting practical and theoretical challenges and once again highlight the potentially broad reach of the substantive criminal law.

## Key Terms

abandonment
accessory after the fact
accessory before the fact
accomplice liability
facilitation
Hindering Apprehension
   or Prosecution
principal in the first degree
principal in the second degree
*respondant superior*
vicarious liability

## Review Questions

1. List the four basic categories of criminal actors at common law. What common distinctions of criminals are used today?

2. Make a distinction between *principals* and *accessories after the fact* in the eyes of the courts.

3. What are the elements of *accomplice liability* under New Hampshire law?

4. What three requirements must be satisfied for *accessory liability* to exist?

5. What is *vicarious liability*? Differentiate between *strict liability* and *vicarious liability*.

6. Can prosecutors charge accomplices and accessories, even if the main perpetrator in a case is acquitted, or for some reason evades prosecution? Explain your answer.

## Notes

1. Weiss B. What were they thinking? The mental states of the aider and abettor and the causer under federal law. *Fordham Law Rev.* 2002; 70:1341.

2. LaFave WR, Scott AW Jr. *Criminal Law,* 2nd ed., § 6.1. St. Paul, MN: West Publishing,1986:486.

3. LaFave and Scott, § 6.9, p. 597.

4. *See* LaFave and Scott, § 6.9(b), pp. 600–601.

5. *See also* LaFave and Scott, § 6.9(a), p. 599, stating that "most jurisdictions now only require that the fact of the completed felony be proved in the trial of the accessory after the fact."

## Footnote to *State v. Beaudry,* 365 N.W.2d 593 (Wis. 1985)

4. Vicarious liability should be contrasted with liability for one's own acts as a party to a crime: that is, for directly committing the crime, for aiding and abetting the commission of a crime, or for being a party to a conspiracy to commit the crime. Sec. 939.05, Stats. 1981–82.

   It is apparently undisputed that the tavern manager violated the closing hour statute and could have been prosecuted as a party to the crime.

# CHAPTER

## 12 Inchoate Crimes

## Chapter Objectives

- Understand what *inchoate crimes* are
- Learn what constitutes an *attempt*
- Understand the defenses of *impossibility* and *abandonment*
- Know the elements that constitute *conspiracy* and *solicitation*
- Learn the purpose of the *RICO Act*
- Understand defenses to a charge of *attempt*

**Inchoate crimes** (pronounced "in-co-hate") are unique in the substantive criminal law. As discussed at length in Chapter 3 and elsewhere, the criminal law typically concerns itself with the harms associated with illegal acts that are consummated. With inchoate crimes—attempt, conspiracy, and solicitation—the law criminalizes behaviors that have not yet necessarily culminated in a tangible social harm. In the words of the drafters of the Model Penal Code (MPC), inchoate crimes "deal with conduct that is designed to culminate in the commission of a substantive offense, but has failed in the discrete case to do so or has not yet achieved its culmination because there is something that the actor or another still must do."[1]

Although inchoate crimes have diverse substantive requirements and differing theoretical justifications, they all reflect a basic perceived value in what might be called "preemptive justice." They target for punishment behaviors that if allowed to continue unabated would result in illegal behavior and greater social harm. Although no tangible, social harm has yet occurred, society feels justified in stepping in to ensure that the target of the inchoate offense (*e.g.*, homicide or robbery) does not occur. As you consider the inchoate crimes of attempt, conspiracy, and solicitation, asks yourself whether such preemptive intervention is necessary and appropriate.

## Attempt

At the outset it is useful to conceive of attempts as coming in two basic forms. First, **actual attempts** occur when the defendant fully puts into effect a planned criminal enterprise but is unsuccessful (*e.g.*, the defendant unsuccessfully poisons an archenemy). Second, **interrupted attempts** occur when the defendant plans and carries out some part of the crime but for some reason fails to carry out the remaining steps (*e.g.*, the defendant experiences a change of heart at the last minute or a police officer comes into the vicinity and the planned crime is aborted).

The law of attempt is not justified on traditional penal theories of deterrence or retribution. Rather, its justification is eminently practical: It provides law enforcement a basis to intervene and in effect short-circuit the larger criminal aims of a suspect.

### *Actus Reus* of Attempt

Attempt law poses unique line-drawing problems, which can significantly affect whether liability arises. Most important among these is the distinction between "mere preparation" and actual attempt. The former does not lead to liability, whereas the latter most certainly does. Because the criminal law is wary of imposing liability for mere bad thoughts, it requires an act of some kind as an external manifestation of the defendant's mental determination to commit the object crime. However, what kinds of acts and how close the defendant must come to completing the crime before attempt liability arises continue to vex the courts.

Indeed, writing over a century ago, the venerable Oliver Wendell Holmes,[2] before he became a member of the U.S. Supreme Court, captured this difficulty with characteristic insight:

> [L]ighting a match with intent to set fire to a haystack has been held to amount to a criminal attempt to burn it, although the defendant blew out the match on seeing that he was watched. . . . [But] [i]f a man starts from Boston to Cambridge for the purpose of committing a murder when he gets there, but is stopped by the draw[bridge] and goes home, he is no more punishable than if he had sat in his chair and resolved to shoot somebody, but on second thought had given up the notion.
>
> Eminent judges have puzzled where to draw the line, or even to state the principle on which it should be drawn, between the two sets of cases. But the principle is believed to be similar to that on which all other lines are drawn by the law. Public policy, that is to say, legislative considerations, are at the bottom of the matter; the considerations being, in this case, the nearness of the danger, the greatness of the harm and the degree of apprehension felt.

Today, jurisdictions usually take either of two basic approaches to distinguishing preparation from attempt. First, there is the *substantial step* approach, endorsed by the MPC. Section 5.01(c) of the MPC requires "an act or omission constituting a substantial step in a course of conduct planned to culminate in the [defendant's] commission of the crime." The MPC requires that such a step be "strongly corroborative of the actor's criminal purpose" and identifies several specific behaviors, including "lying in wait" for another and "searching for or following the contemplated victim of the crime," which if strongly corroborative of the actor's criminal purpose, "shall not be held insufficient as a matter of law." *See* MPC § 5.01(2).

Second, numerous jurisdictions use the *proximity* approach. As its name implies, under this approach the defendant must come close—physically and temporally—to consummating the crime. It can best be understood as focusing analysis on what remains to be done to complete the crime, as opposed to the "substantial step" approach, which focuses on the steps a defendant has already taken. The classic decision using the proximity approach is *People v. Rizzo,* 158 N.E.2d 888 (N.Y. 1927). In *Rizzo,* four men targeted Rao, a payroll delivery person, for robbery. Police arrested the four men while they were driving around New York City in unsuccessful pursuit of Rao. The four were subsequently convicted of attempted robbery.

The New York Court of Appeals, using the proximity approach, reversed the convictions on the following reasoning:

> Rao was not found; the defendants were still looking for him; no attempt to rob him could be made, at least until he came into sight. . . . Apparently no money had been drawn from the bank for the payroll by anybody at the time of the arrest. In a word, these defendants had planned to commit a crime and were looking around the city for an opportunity to commit it, but the opportunity fortunately never came. [One] would not be guilty of an attempt . . . to commit murder if he armed himself and started out to find the person whom he had planned to kill but could not find him. So here these defendants were not guilty of an attempt to commit robbery in the first degree when they had not found or reached the presence of the person they intended to rob. *Rizzo,* 158 N.E. at 889.

*See also Commonwealth v. Peaslee,* 59 N.E. 55 (Mass. 1901) (reversing conviction for attempted arson despite the fact that defendant had placed combustibles in targeted building and all that remained was ignition—the critical "last act" necessary).

The following case, *United States v. Harper,* which uses the substantial step test contained in the MPC, highlights the difficult distinctions that must be drawn in the attempt analysis.

## United States v. Harper, 33 F.3d 1143 (9th Cir. 1994)

Canby, Circuit Judge.

### BACKGROUND

Police officers in Buena Park, California, found appellants and one other codefendant, Carlos Munoz, sitting in a rented car in the parking lot of the Home Savings Bank shortly after 10:00 P.M. on the evening of September 21, 1992. The officers searched the defendants, the vehicle and the surrounding area. They found two loaded handguns—a .44 caliber Charter Arms Bulldog and a .357 magnum Smith and Wesson—under a bush located five or six feet from the car, where a witness had earlier seen one of the car's occupants bending over. In the car, the police discovered a roll of duct tape in a plastic bag, a stun gun, and a pair of latex surgical gloves. They found another pair of latex surgical gloves in the pocket of Munoz's sweat pants. They also found six rounds of .357 magnum ammunition in the pocket of his shorts, which he wore under his sweat pants. Some of this ammunition came from either the same box or the same lot as the ammunition in the loaded .357 magnum handgun. The defendants had a total of approximately $182 in cash among them and Sharrieff was carrying an automated teller machine (ATM) card which bore the name of Kimberly Ellis.

Harper had used the ATM card belonging to Kimberly Ellis shortly before 9:00 P.M. that evening in an ATM at the Buena Park branch of the Bank of America, which was located adjacent to the Home Savings parking lot in which the defendants were parked. The ATM's camera photographed Harper. Harper had requested a twenty dollar withdrawal from the ATM, but had not removed the cash from the cash drawer. This omission had created what is known as a "bill trap." When a bill trap occurs, the ATM shuts itself down and the ATM supply company that monitors the ATM contacts its ATM service technicians to come and repair the ATM. These facts were known to Harper, who had previously worked for both Bank of America and one of its ATM service companies.

On the basis of this evidence, Harper, Sharrieff and Munoz were indicted for conspiracy to rob a federally insured bank, attempted bank robbery, and carrying a firearm during and in relation to a crime of violence. The prosecution's theory was that Harper had intentionally caused the bill trap to summon the ATM service technicians who would have to open the ATM vault to clear the trap. At that time, the

*(Continues)*

theory went, the defendants planned to rob the technicians of the money in the ATM. The three defendants were convicted of all charges. . . .

To obtain a conviction for attempted bank robbery the prosecution must prove (1) culpable intent and (2) conduct constituting a substantial step toward the commission of the crime. *United States v. Still,* 850 F.2d 607, 608 (9th Cir. 1988), . . . *United States v. Buffington,* 815 F.2d 1292, 1301 (9th Cir. 1987). Here, there was sufficient evidence to permit a jury to find that the defendants intended to rob the Bank of America. We conclude, however, that under the law of this circuit there was insufficient evidence that the defendants took a substantial step toward commission of the robbery.

It is admittedly difficult to draw the line between mere preparation to commit an offense, which does not constitute an attempt, and the taking of a substantial step toward commission of the crime, which does. Various theories have been propounded for determining when the activities of one who intends to commit a crime ripen into an attempt, *see* W. LaFave & A. Scott, *Substantive Criminal Law,* § 6.2 at 31–39 (1986), and they yield varying results in a case like this. We must draw our guidance from our own precedent, however, and we conclude that the activities of the defendants, viewed in the light most favorable to the prosecution, fail to qualify as an attempt.

Our primary authorities are *Buffington* and *Still.* In *Buffington,* the defendants had driven past the supposed target bank twice. One of the three male defendants then entered a store near the bank and observed the bank. The two other defendants, one dressed as a woman, exited their vehicle in the bank parking lot, stood near the vehicle and focused their attention on the bank. They were armed. We held that the evidence was insufficient both as to the defendants' intent to rob the bank, and as to the existence of conduct constituting a substantial step towards the commission of the crime. With regard to the latter element, we observed:

> Not only did appellants not take a single step toward the bank, they displayed no weapons and no indication that they were about to make an entry. Standing alone, their conduct did not constitute that requisite "appreciable fragment" of a bank robbery, nor a step toward commission of the crime of such substantiality that, unless frustrated, the crime would have occurred. *Id.,* 815 F.2d at 1303.

The same may be said of the defendants in this case. True, Harper had left money in the ATM, causing a bill trap that would eventually bring service personnel to the ATM. That act, however, is equivocal in itself. The robbery was in the future and, like the defendants in *Buffington,* the defendants never made a move toward the victims or the Bank to accomplish the criminal portion of their intended mission. They had not taken a step of "such substantiality that, unless frustrated, the crime would have occurred." *Id.* Their situation is therefore distinguishable from that of the defendant in *United States v. Moore,* 921 F.2d 207 (9th Cir. 1990), upon which the government relies. In *Moore,* the defendant was apprehended "walking toward the bank, wearing a ski mask, and carrying gloves, pillowcases and a concealed, loaded gun." *Id.* at 209. These actions were a true commitment toward the robbery, which would be in progress the moment the would-be robber entered the bank thus attired and equipped. That stage of the crime had not been reached by Harper and Sharrieff; their actual embarkation on the robbery lay as much as 90 minutes away from the time when Harper left money in the ATM, and that time had not expired when they were apprehended.

*Still* provides further support for our conclusion. There, we relied upon *Buffington* to reverse a similar conviction for attempted bank robbery. The defendant in that case was seen sitting in a van approximately 200 feet from the supposed target bank. In the van he had a fake bomb, a red pouch with note demanding money attached to it, a police scanner, and a notebook containing drafts of the note. He also had been seen putting on a blonde wig while sitting in the van. The defendant's intent was clear; he told police that he had been about to rob the bank and "[t]hat's what [he] was putting the wig on for." *Still,* 850 F.2d at 608. We concluded, however, that the evidence was insufficient to establish that the defendant had taken a substantial step toward commission of the offense. "Our facts do not establish either actual movement toward the bank or actions that are analytically similar to such movement." *Id.* at 610. Defendant Still, like the defendants here, was sitting in his vehicle when the police approached. As in *Still,* we conclude that the crime was too inchoate to constitute an attempt.

When criminal intent is clear, identifying the point at which the defendants' activities ripen into an attempt is not an analytically satisfying enterprise. There is, however, a substantial difference between causing a bill trap, which will result in the appearance of potential victims, and moving toward such victims with gun and mask, as in *Moore.* Making an appointment with a potential victim is not of itself such a commitment to an intended crime as to constitute an attempt, even though it may make a later attempt possible. Little more happened here; this case is more like *Buffington* and *Still* than it is like *Moore*. Accordingly, we reverse the appellants' convictions for attempted bank robbery.

## Notes and Questions

1. Do you agree with the *Harper* court's conclusion? Was the conduct of the defendants "too inchoate"?
2. Of the two main approaches, the MPC's substantial step approach is viewed as being more advantageous to law enforcement. Why is this so?
3. The requirement of some form of perpetrated social harm is a mainstay of the criminal law. Where is the harm with attempts? Is the harm of a nonphysical nature, jeopardizing society as a whole? Does the harm posed, if any, differ with respect to "actual" versus "interrupted" attempts (as distinguished at the outset)? For example, insofar as a failed ("actual") attempt can result in the actor being discouraged from trying again, can the same be said of "interrupted" attempts? Would your response be influenced by your view of the relative importance of retribution, deterrence, and incapacitation as rationales of the criminal law?

Another important issue arising with attempt law concerns the required specificity of the alleged criminal undertaking by the accused. In this regard, consider the facts of *People v. Smith*, 593 N.E.2d 533 (Ill. 1992). In *Smith,* the defendant bought a one-way train ticket from Chicago to Waukegan and, upon his arrival, hailed a taxi. Defendant requested that he be taken to Genesee Street so that he could find a particular jewelry store. At one point the driver pointed to a jewelry store, which the defendant stated was "not the one" because it was "Mexican." The driver then inquired whether they should ask a police officer for directions to the jewelry store, but the defendant declined. Shortly thereafter the defendant pulled out a gun, robbed the driver, and stole the cab, only to soon abandon it. When observed by police, the defendant began running and dropped his gun as well as some money and a blue pillowcase. Police eventually found the defendant hiding in a trailer; the defendant was convicted of attempted robbery of the jewelry store.

On appeal, however, the conviction was reversed on the reasoning that the defendant failed to take a substantial step toward commission of the robbery:

> [T]here is no evidence as to the location of the jewelry store, even though defendant did drive along the street where he expected to find it. Nor is there any evidence on how close defendant came to the jewelry store, or the name of the jewelry store. In essence, the State is charging defendant with attempted armed robbery of an unnamed, undetermined jewelry store. Although the police maintained that defendant had told them that he "he knew what the building looked like," there is no evidence the defendant ever identified the building. Presumably he did not. Moreover, besides the "Mexican" jewelry store previously mentioned, there is no evidence in the record of a jewelry store on Genesee Street. Thus, we believe that it would be improper to conclude that defendant came "within a dangerous proximity to success." *Smith*, 593 N.E. 2d at 537 (citation omitted).

The following case discusses the level of precision regarding the specific steps a person takes toward completion of a crime required in charging a person with an attempt.

## *United States v. Resendiz-Ponce*, 549 U.S. 102, 127 S.Ct. 782, 166 L.Ed.2d 591 (2007)

Justice Stevens delivered the opinion of the Court.

A jury convicted respondent Juan Resendiz-Ponce, a Mexican citizen, of illegally attempting to reenter the United States. Because the indictment failed to allege a specific overt act that he committed in seeking reentry, the Court of Appeals set aside his conviction and remanded for dismissal of the indictment. . . .

### I.

Respondent was deported twice, once in 1988 and again in 2002, before his attempted reentry on June 1, 2003. On that day, respondent walked up to a port of entry and displayed a photo identification of his cousin to the border agent. Respondent told the agent that he was a legal resident and that he was

*(Continues)*

traveling to Calexico, California. Because he did not resemble his cousin, respondent was questioned, taken into custody, and ultimately charged with a violation of 8 U. S. C. § 1326(a). The indictment alleged:

> On or about June 1, 2003, JUAN RESENDIZ-PONCE, an alien, knowingly and intentionally attempted to enter the United States of America at or near San Luis in the District of Arizona, after having been previously denied admission, excluded, deported, and removed from the United States at or near Nogales, Arizona, on or about October 15, 2002, and not having obtained the express consent of the Secretary of the Department of Homeland Security to reapply for admission. . . .

Respondent moved to dismiss the indictment, contending that it "fail[ed] to allege an essential element, an overt act, or to state the essential facts of such overt act." The District Court denied the motion and, after the jury found him guilty, sentenced respondent to a 63-month term of imprisonment.

The Ninth Circuit reversed, reasoning . . . respondent's indictment was fatally flawed because it nowhere alleged "any specific overt act that is a substantial step" toward the completion of the unlawful reentry. The panel majority explained:

> The defendant has a right to be apprised of what overt act the government will try to prove at trial, and he has a right to have a grand jury consider whether to charge that specific overt act. Physical crossing into a government inspection area is but one of a number of other acts that the government might have alleged as a substantial step toward entry into the United States. The indictment might have alleged the tendering a bogus identification card; it might have alleged successful clearance of the inspection area; or it might have alleged lying to an inspection officer with the purpose of being admitted. . . . A grand jury never passed on a specific overt act, and Resendiz was never given notice of what specific overt act would be proved at trial. . . .

## II.

At common law, the attempt to commit a crime was itself a crime if the perpetrator not only intended to commit the completed offense, but also performed "'some open deed tending to the execution of his intent.'" W. LaFave, Substantive Criminal Law § 11.2(a), p. 205 (2d ed. 2003) (quoting E. Coke, Third Institute 5 (6th ed. 1680)). More recently, the requisite "open deed" has been described as an "overt act" that constitutes a "substantial step" toward completing the offense. As was true at common law, the mere intent to violate a federal criminal statute is not punishable as an attempt unless it is also accompanied by significant conduct.

The Government does not disagree with respondent's submission that he cannot be guilty of attempted reentry in violation of 8 U. S. C. § 1326(a) unless he committed an overt act qualifying as a substantial step toward completion of his goal. Nor does it dispute that "[a]n indictment must set forth each element of the crime that it charges." It instead contends that the indictment at bar implicitly alleged that respondent engaged in the necessary overt act "simply by alleging that he 'attempted to enter the United States.'" We agree.

Not only does the word "attempt" as used in common parlance connote action rather than mere intent, but more importantly, as used in the law for centuries, it encompasses both the overt act and intent elements. Consequently, an indictment alleging attempted illegal reentry under § 1326(a) need not specifically allege a particular overt act or any other "component par[t]" of the offense. See *Hamling* v. *United States*, 418 U. S. 87, 119 (1974). Just as it was enough for the indictment in *Hamling* to allege that the defendant mailed "obscene" material in violation of 18 U. S. C. § 1461, it was enough for the indictment in this case to point to the relevant criminal statute and allege that "[o]n or about June 1, 2003," respondent "attempted to enter the United States of America at or near San Luis in the District of Arizona."

In *Hamling*, we identified two constitutional requirements for an indictment: "first, [that it] contains the elements of the offense charged and fairly informs a defendant of the charge against which he must defend, and, second, [that it] enables him to plead an acquittal or conviction in bar of future prosecutions for the same offense." 418 U. S., at 117. In this case, the use of the word "attempt," coupled with the specification of the time and place of respondent's attempted illegal reentry, satisfied both. Indeed, the time-and-place information provided respondent with more adequate notice than would an indictment describing particular overt acts. After all, a given defendant may have approached the border or lied to a border-patrol agent in the course of countless attempts on innumerable occasions. For the same reason, the time-and-date specification in respondent's indictment provided ample protection against the risk of multiple prosecutions for the same crime.

Respondent nonetheless maintains that the indictment would have been sufficient only if it had alleged any of three overt acts performed during his attempted reentry: that he walked into an inspection area; that he presented a misleading identification card; or that he lied to the inspector. Individually and cumulatively, those acts tend to prove the charged attempt—but none was essential to the finding of guilt in this case. All three acts were rather part of a single course of conduct culminating in the charged "attempt." As Justice Holmes explained in *Swift & Co. v. United States*, 196 U. S. 375, 396 (1905), "[t]he unity of the plan embraces all the parts." . . .

[W]e reverse the judgment of the Court of Appeals and remand the case for further proceedings consistent with this opinion.

## Notes and Questions

**1.** What was the specific act Resendiz-Ponce was accused of doing? Do you agree that alleging the attempt itself is sufficient to give him notice as to the specific unlawful acts that made up his criminal conduct?

## Mens Rea of Attempt

Attempt is a specific intent crime. As a result, for there to be liability it is not sufficient that the government merely prove that some steps were taken toward consummation of the target crime. Rather, the government must also prove that the defendant intended to (a) perform the act alleged and (b) accomplish the desired criminal result. Invariably, given the anticipatory nature of attempts, the requisite mental state parallels that required of the target crime. As one noted authority states,[3]

> The crime of attempt does not exist in the abstract, but rather only exists in relation to other offenses; a defendant must be charged with an attempt to commit a specifically designated crime, and it is to that crime one must look in identifying the kind of intent required. For example, if the charge is attempted theft and theft is defined as requiring an intent to permanently deprive the owner of his property, then that same intent must be established to prove the attempt. It is not enough that the defendant intended to do some unspecified criminal act.

Thus one cannot be convicted of "attempted murder" in the absence of evidence showing an actual attempt to end the life of another. *See, e.g., Commonwealth v. Thacker,* 114 S.E. 504 (Va. 1922) (reckless shooting into the tent of another at night does not serve as basis for attempted murder prosecution). This requirement, as the *Thacker* court recognized, creates something of a paradox:[4]

> Thus, to commit murder, one need not intend to take life, but to be guilty of an attempt to murder, he must so intend. It is not sufficient that his act, had it proved fatal, would have been murder. We have seen that the unintended taking of life may be murder, yet there can be no attempt to murder without the specific intent to commit it. . . . For example, if one from a housetop recklessly throws down a billet of wood upon the sidewalk . . . and it falls upon a person passing by and kills him, this would be common law murder, but if instead of killing, it inflicts only a slight injury, the party could not be convicted of an assault with intent to commit murder since, in fact, the murder was not intended.

In short, an attempt to prosecute for a specific intent crime cannot be successful in the absence of the requisite specific intent. But, what about crimes involving reckless or negligent mental states? Despite the seeming illogic, courts have on occasion in fact upheld attempt liability under such circumstances. *See, e.g., People v. Thomas,* 729 P.2d 972 (Colo. 1986) (upholding attempted reckless manslaughter conviction because the defendant intentionally engaged in reckless conduct).

## *People v. Terrell*, 459 N.E.2d 1337 (Ill. 1984)

Thomas J. Moran, Justice.

\* \* \*

The evidence revealed that on August 7, 1980, at approximately 6:15 A.M., an anonymous telephone call was received by the Kankakee City police. The caller stated that two men, armed with guns, were hiding behind a service station. This report was dispatched and was responded to by Officer Whitehead in one patrol car, and Officers Pepin and Rokus, who were patrolling the area, in another car.

A diagram, entered into evidence, shows that the service station is located on the southwest corner of Erzinger and Maple streets. The first building south of the station, facing Maple Street, is a construction company. Further south is a tool company. To the rear of the buildings is a large grassy lot which extends to an alley running parallel to the buildings.

*(Continues)*

(*Continued*)

Officer Whitehead arrived at the scene within minutes of the radio dispatch and only seconds before Officers Rokus and Pepin. Whitehead pulled into the alley and onto the empty lot behind the station, where he immediately observed a man, crouched in the weeds, 20 to 30 feet from the station. As the officer got out of his car, the defendant, who he saw carrying a gun, jumped up from the weeds, ran towards the fence, climbed to the other side and proceeded south down Maple Street. Officer Whitehead testified that the defendant disposed of the gun sometime before he scaled the fence, although he could not remember seeing it being dropped.

Twelve to fifteen minutes after he was initially observed, Officer Pepin discovered the defendant hiding in the weeds behind the tool company, approximately 280 feet from the service station. The defendant had removed his shirt and was lying on it. A black nylon stocking with a knot in the end of it was found in his pocket. Although the defendant claimed that he was going to the gas station to buy cigarettes, the officer found no money on defendant's person. Officer Whitehead, later, positively identified the defendant as the man he had observed with the gun.

The defendant maintains that the State's evidence is insufficient to establish the two essential elements of the offense of attempt. Section 8-4(a) of the Criminal Code of 1961 (Ill.Rev.Stat.1979, ch. 38, par. 8-4(a)) provides:

> A person commits an attempt when, with *intent* to commit a specific offense, he does any act which constitutes a *substantial step* toward the commission of that offense. (Emphasis added.)

Section 18-2(a) of the Criminal Code of 1961 (Ill.Rev.Stat.1979, ch. 38, par. 18-2(a)) provides:

> A person commits armed robbery when he or she [takes property from the person or presence of another by the use of force or by threatening imminent use of force] while he or she carries on or about his or her person, or is otherwise armed with a dangerous weapon.

We find that the facts and circumstances of this case are sufficient to prove that the defendant possessed the requisite intent to commit a specific armed robbery and that he took a substantial step toward the commission of that armed robbery. For the reasons to follow, therefore, we affirm the judgment of the appellate court.

It is well established that, to obtain a conviction for attempt, the State must prove that the defendant intended to commit a specific offense. . . . The intent to commit a criminal offense need not be expressed, but may be inferred from the conduct of the defendant and the surrounding circumstances (*People v. Mulcahey* 381 N.E.2d 254 (Ill. 1978)). . . .

While the defendant, in the instant case, does not deny the presence of "some" criminal intent, he maintains that the evidence fails to "imply a design to commit an armed robbery at the station." He suggests a list of alternative targets and offenses which includes the crime of burglary as opposed to armed robbery. In addition, he finds it significant that the State failed to establish that the gas station was open when he was initially discovered.

We find this argument unpersuasive. It is unreasonable to expect a trier of fact to infer intent to commit burglary, rather than armed robbery when confronted with a suspect who was seen carrying a loaded revolver and in possession of a ladies' stocking but no burglary tools. In addition, the trial court could reasonably infer that the service station was the object of defendant's plan. The defendant was observed in close proximity to the station, by Officer Whitehead, as he arrived on the scene. This observation was in conformity with the initial tip from the telephone caller who specifically indicated that the suspects were hiding behind the service station. As for the victim necessary for an armed robbery, the trier of fact may reasonably have inferred that the defendant was awaiting the attendant's arrival before taking the final step in his plan. Furthermore, although armed robbery requires a victim, "[i]t [is] not . . . a defense to a charge of attempt that because of a misapprehension of the circumstances it would have been impossible for the accused to commit the offense attempted." (Ill.Rev.Stat.1979, ch. 38, par. 8-4(b).) It is not a defense to the charge of attempted armed robbery, therefore, to take a substantial step toward the commission of the armed robbery only to find the victim, the attendant, not yet present.

The evidence presented to the trier of fact, in this case, revealed a defendant who had concealed himself in the weeds in close proximity to a service station, which was about to open, while in possession of a stocking mask and a fully loaded revolver. In further support of an inference of criminal intent, defendant's efforts to elude the police as well as his weak excuse for his presence at the scene, could also be properly considered. (*People v. Harris* 288 N.E.2d 385 Ill. 1972). Faced with these facts, we find it incredulous that defendant had any intent other than the armed robbery of the service station.

We turn next to what has been described as "one of the most troublesome problems" in the area of inchoate offenses: "when preparation to commit an offense ceases and perpetration of the offense [attempt] begins." (Ill.Ann.Stat., ch. 38, par. 8-4(a), Committee Comments, at 512 (Smith-Hurd 1972).) Answering this question requires an analysis of the second statutory element of the offense of attempt—"any act which constitutes a substantial step toward the commission of that offense." (Ill.Rev.Stat.1979, ch. 38, par. 8-4(a).) Although it is not necessary that a defendant complete the last proximate act in order to be convicted of attempt, our cases have held that mere preparation is not a substantial step. . . . It would be an impossible task to compile a definitive list of acts which, if performed, constitute a substantial step toward the commission of every crime. Such a determination can only be accomplished by evaluating the facts and circumstances of the particular case. . . .

When defining "attempt" it becomes problematic deciding when to allow the police to intervene in an unfolding course of criminal conduct. While caution must be exercised to avoid punishment for inconclusive acts, prevention of an intended crime is necessary. . . . It should not be necessary to subject victims to face to face confrontation with a lethal weapon in order to make a positive finding of the essential element of a substantial step. . . .

The record reflects no error which requires reversal. We find the evidence presented in this case is sufficient to establish, beyond a reasonable doubt, defendant's intent to commit armed robbery of an individual within the service station and that he took a substantial step toward the commission of the armed robbery. The judgment of the appellate court is affirmed.

*Judgment affirmed.*

## Notes and Questions

1. *Terrell* highlights the usefulness of attempt to law enforcement. Even Terrell himself admitted he was up to no good early that morning. But what precisely was he seeking to do? Given the facts, does the court's decision strike you as legally defensible? Was there, in Holmes' words, the required "dangerous proximity"? At the same time, what alternative did the police have under the circumstances?

## Defense of "Impossibility"

One of the most intellectually challenging and confusing aspects of attempt law involves what is known as the "impossibility" defense. The law recognizes two types of impossibility: **legal impossibility**, which can constitute a complete defense to a charge of attempt, and **factual impossibility**, which never constitutes a defense. The distinction between the two has bedeviled criminal law scholars and practitioners for decades. One of the better judicial efforts to capture the slippery distinction was provided over 20 years ago by the U.S. Court of Appeals for the Fifth Circuit:

Legal impossibility occurs when the actions which the defendant performs or sets in motion, even if fully carried out as he desires, would not constitute a crime. Factual impossibility occurs when the objective of the defendant is proscribed by the criminal law but a circumstance unknown to the actor prevents him from bringing about that objective. *United States v. Oviedo*, 525 F.2d 881, 883 (5th Cir. 1976).

The following case illustrates these complexities and how the MPC has addressed the issue.

### *People v. Dlugash*, 41 N.Y.2d 725, 363 N.E.2d 1155, 395 N.Y.S.2d 419 (1977)

Jasen, Judge.

On December 22, 1973, Michael Geller, 25 years old, was found shot to death in the bedroom of his Brooklyn apartment. The body, which had literally been riddled by bullets, was found lying face up on the floor. An autopsy revealed that the victim had been shot in the face and head no less than seven times. Powder burns on the face indicated that the shots had been fired from within one foot of the victim. Four small caliber bullets were recovered from the victim's skull. The victim had also been critically wounded in the chest. One heavy caliber bullet passed through the left lung, penetrated the heart chamber, pierced the left ventricle of the heart upon entrance and again upon exit, and lodged in the victim's torso. A second bullet entered the left lung and passed through to the chest, but without reaching the heart area. Although the second bullet was damaged beyond identification, the bullet tracks indicated that these wounds were also inflicted by a bullet of heavy caliber. A tenth bullet, of unknown caliber, passed through the thumb of the victim's left hand. The autopsy report listed the

*(Continues)*

cause of death as "(m)ultiple bullet wounds of head and chest with brain injury and massive bilateral hemothorax with penetration of (the) heart." Subsequent ballistics examination established that the four bullets recovered from the victim's head were .25 caliber bullets and that the heart-piercing bullet was of .38 caliber.

Detective Joseph Carrasquillo of the New York City Police Department was assigned to investigate the homicide. On December 27, 1973, five days after the discovery of the body, Detective Carrasquillo and a fellow officer went to the defendant's residence in an effort to locate him. The officers arrived at approximately 6:00 P.M. The defendant answered the door and, when informed that the officers were investigating the death of Michael Geller, a friend of his, defendant invited the officers into the house. Detective Carrasquillo informed defendant that the officers desired any information defendant might have regarding the death of Geller and, since defendant was regarded as a suspect, administered the standard preinterrogation warnings. The defendant told the officers that he and another friend, Joe Bush, had just returned from a four- or five-day trip "upstate someplace" and learned of Geller's death only upon his return. Since Bush was also a suspect in the case and defendant admitted knowing Bush, defendant agreed to accompany the officers to the station house for the purposes of identifying photographs of Bush and of lending assistance to the investigation. Once again, Carrasquillo administered the standard preinterrogation statement of rights. The defendant then proceeded to relate his version of the events which culminated in the death of Geller. Defendant stated that, on the night of December 21, 1973, he, Bush and Geller had been out drinking. Bush had been staying at Geller's apartment and, during the course of the evening, Geller several times demanded that Bush pay $100 towards the rent on the apartment. According to defendant, Bush rejected these demands, telling Geller that "you better shut up or you're going to get a bullet." All three returned to Geller's apartment at approximately midnight, took seats in the bedroom, and continued to drink until sometime between 3:00 and 3:30 in the morning. When Geller again pressed his demand for rent money, Bush drew his .38 caliber pistol, aimed it at Geller and fired three times. Geller fell to the floor. After the passage of a few minutes, perhaps two, perhaps as much as five, defendant walked over to the fallen Geller, drew his .25 caliber pistol, and fired approximately five shots in the victim's head and face. Defendant contended that, by the time he fired the shots, "it looked like Mike Geller was already dead." After the shots were fired, defendant and Bush walked to the apartment of a female acquaintance. Bush removed his shirt, wrapped the two guns and a knife in it, and left the apartment, telling Dlugash that he intended to dispose of the weapons. Bush returned 10 or 15 minutes later and stated that he had thrown the weapons down a sewer two or three blocks away. . . .

At approximately 9:00 P.M., the defendant repeated the substance of his statement to an Assistant District Attorney. Defendant added that at the time he shot at Geller, Geller was not moving and his eyes were closed. While he did not check for a pulse, defendant stated that Geller had not been doing anything to him at the time he shot because "Mike was dead."

Defendant was indicted by the Grand Jury of Kings County on a single count of murder in that, acting in concert with another person actually present, he intentionally caused the death of Michael Geller. At the trial, there were four principal prosecution witnesses: Detective Carrasquillo, the Assistant District Attorney who took the second admission, and two physicians from the office of the New York City Chief Medical Examiner. For proof of defendant's culpability, the prosecution relied upon defendant's own admissions as related by the detective and the prosecutor. From the physicians, the prosecution sought to establish that Geller was still alive at the time defendant shot at him. Both physicians testified that each of the two chest wounds, for which defendant alleged Bush to be responsible, would have caused death without prompt medical attention. Moreover, the victim would have remained alive until such time as his chest cavity became fully filled with blood. Depending on the circumstances, it might take 5 to 10 minutes for the chest cavity to fill. Neither prosecution witness could state, with medical certainty, that the victim was still alive when, perhaps five minutes after the initial chest wounds were inflicted, the defendant fired at the victim's head.

The defense produced but a single witness, the former Chief Medical Examiner of New York City. This expert stated that, in his view, Geller might have died of the chest wounds "very rapidly" since, in addition to the bleeding, a large bullet going through a lung and the heart would have other adverse medical effects. "Those wounds can be almost immediately or rapidly fatal or they may be delayed in there, in the time it would take for death to occur. But I would say that wounds like that which are described here as having gone through the lungs and the heart would be fatal wounds and in most cases they're rapidly fatal."

The jury found the defendant guilty of murder. The defendant then moved to set the verdict aside. He submitted an affidavit in which he contended that he "was absolutely, unequivocally and positively certain that Michael Geller was dead before (he) shot him." Further, the defendant averred that he was in fear for his life when he shot Geller. "This fear stemmed from the fact that Joseph Bush, the admitted killer of Geller, was holding a gun on me and telling me, in no uncertain terms, that if I didn't shoot the dead body I, too, would be killed." This motion was denied.

On appeal, the Appellate Division reversed the judgment of conviction on the law and dismissed the indictment. The court ruled that "the People failed to prove beyond a reasonable doubt that Geller had been alive at the time he was shot by defendant; defendant's conviction of murder thus cannot stand." Further, the court held that the judgment could not be modified to reflect a conviction for attempted murder because "the uncontradicted evidence is that the defendant, at the time that he fired the five shots into the body of the decedent, believed him to be dead, and . . . there is not a scintilla of evidence to contradict his assertion in that regard."

Preliminarily, we state our agreement with the Appellate Division that the evidence did not establish, beyond a reasonable doubt, that Geller was alive at the time defendant fired into his body. To sustain a homicide conviction, it must be established, beyond a reasonable doubt, that the defendant caused the death of another person. The People were required to establish that the shots fired by defendant Dlugash were a sufficiently direct cause of Geller's death. While the defendant admitted firing five shots at the victim approximately two to five minutes after Bush had fired three times, all three medical expert witnesses testified that they could not, with any degree of medical certainty, state whether the victim had been alive at the time the latter shots were fired by the defendant. Thus, the People failed to prove beyond a reasonable doubt that the victim had been alive at the time he was shot by the defendant. Whatever else it may be, it is not murder to shoot a dead body. Man dies but once.

* * *

The concept that there could be criminal liability for an attempt, even if ultimately unsuccessful, to commit a crime is comparatively recent. The modern concept of attempt has been said to date from *Rex v. Scofield* (Cald 397), decided in 1784. In that case, Lord Mansfield stated that "(t)he intent may make an act, innocent in itself, criminal; nor is the completion of an act, criminal in itself, necessary to constitute criminality. Is it no offence to set fire to a train of gunpowder with intent to burn a house, because by accident, or the interposition of another, the mischief is prevented?" (Cald, at p. 400). The Revised Penal Law now provides that a person is guilty of an attempt to commit a crime when, with intent to commit a crime, he engages in conduct which tends to effect the commission of such crime. The revised statute clarified confusion in the former provision which, on its face, seemed to state that an attempt was not punishable as an attempt unless it was unsuccessful.

The most intriguing attempt cases are those where the attempt to commit a crime was unsuccessful due to mistakes of fact or law on the part of the would-be criminal. A general rule developed in most American jurisdictions that legal impossibility is a good defense but factual impossibility is not. Thus, for example, it was held that defendants who shot at a stuffed deer did not attempt to take a deer out of season, even though they believed the dummy to be a live animal. The court stated that there was no criminal attempt because it was no crime to "take" a stuffed deer, and it is no crime to attempt to do that which is legal. (*State v. Guffey*, 262 S.W.2d 152 (Mo.App.); see, also, *State v. Taylor*, 133 S.W.2d 336 (no liability for attempt to bribe a juror where person bribed was not, in fact, a juror).) These cases are illustrative of legal impossibility. A further example is Francis Wharton's classic hypothetical involving Lady Eldon and her French lace. Lady Eldon, traveling in Europe, purchased a quantity of French lace at a high price, intending to smuggle it into England without payment of the duty. When discovered in a customs search, the lace turned out to be of English origin, of little value and not subject to duty. The traditional view is that Lady Eldon is not liable for an attempt to smuggle.

On the other hand, factual impossibility was no defense. For example, a man was held liable for attempted murder when he shot into the room in which his target usually slept and, fortuitously, the target was sleeping elsewhere in the house that night. Although one bullet struck the target's customary pillow, attainment of the criminal objective was factually impossible. *State v. Moretti*, 244 A.2d 499, presents a similar instance of factual impossibility. The defendant agreed to perform an abortion, then a criminal act, upon a female undercover police investigator who was not, in fact, pregnant. The court sustained the conviction, ruling that "when the consequences sought by a defendant are forbidden by the law as criminal, it is no defense that the defendant could not succeed in reaching his goal because of circumstances unknown to him." On the same view, it was held that men who had sexual intercourse

*(Continues)*

with a woman, with the belief that she was alive and did not consent to the intercourse, could be charged for attempted rape when the woman had, in fact, died from an unrelated ailment prior to the acts of intercourse.

The New York cases can be parsed out along similar lines. One of the leading cases on legal impossibility is *People v. Jaffe,* 78 N.E. 169, in which we held that there was no liability for the attempted receipt of stolen property when the property received by the defendant in the belief that it was stolen was, in fact under the control of the true owner. Similarly, in *People v. Teal,* 89 N.E. 1086, a conviction for attempted subornation of perjury was overturned on the theory that the testimony attempted to be suborned was irrelevant to the merits of the case. Since it was not subornation of perjury to solicit false, but irrelevant, testimony, "the person through whose procuration the testimony is given cannot be guilty of subornation of perjury and, by the same rule, an unsuccessful attempt to that which is not a crime when effectuated, cannot be held to be an attempt to commit the crime specified."(89 N.E. at p. 1088.) Factual impossibility, however, was no defense. Thus, a man could be held for attempted grand larceny when he picked an empty pocket.

As can be seen from even this abbreviated discussion, the distinction between "factual" and "legal" impossibility was a nice one indeed and the courts tended to place a greater value on legal form than on any substantive danger the defendant's actions posed for society. The approach of the draftsmen of the Model Penal Code was to eliminate the defense of impossibility in virtually all situations. Under the code provision, to constitute an attempt, it is still necessary that the result intended or desired by the actor constitute a crime. However, the code suggested a fundamental change to shift the locus of analysis to the actor's mental frame of reference and away from undue dependence upon external considerations. The basic premise of the code provision is that what was in the actor's own mind should be the standard for determining his dangerousness to society and, hence, his liability for attempted criminal conduct.

In the belief that neither of the two branches of the traditional impossibility arguments detracts from the offender's moral culpability, the Legislature substantially carried the code's treatment of impossibility into the 1967 revision of the Penal Law. Thus, a person is guilty of an attempt when, with intent to commit a crime, he engages in conduct, which tends to effect the commission of such crime. It is no defense that, under the attendant circumstances, the crime was factually or legally impossible of commission, "if such crime could have been committed had the attendant circumstances been as such person believed them to be." (Penal Law, s 110.10.) Thus, if defendant believed the victim to be alive at the time of the shooting, it is no defense to the charge of attempted murder that the victim may have been dead.

Turning to the facts of the case before us, we believe that there is sufficient evidence in the record from which the jury could conclude that the defendant believed Geller to be alive at the time defendant fired shots into Geller's head. Defendant admitted firing five shots at a most vital part of the victim's anatomy from virtually point blank range. Although defendant contended that the victim had already been grievously wounded by another, from the defendant's admitted actions, the jury could conclude that the defendant's purpose and intention was to administer the coup de grace. The jury never learned of defendant's subsequent allegation that Bush had a gun on him and directed defendant to fire at Geller on the pain of his own life. Defendant did not testify and this statement of duress was made only in a postverdict affidavit, which obviously was never placed before the jury. In his admissions that were related to the jury, defendant never made such a claim. Nor did he offer any explanation for his conduct, except for an offhand aside made casually to Detective Carrasquillo. Any remaining doubt as to the question of duress is dispelled by defendant's earlier statement that he and Joe Bush had peacefully spent a few days together on vacation in the country. Moreover, defendant admitted to freely assisting Bush in disposing of the weapons after the murder and, once the weapons were out of the picture, defendant made no effort at all to flee from Bush. Indeed, not only did defendant not come forward with his story immediately, but when the police arrived at his house, he related a false version designed to conceal his and Bush's complicity in the murder. All of these facts indicate a consciousness of guilt which defendant would not have had if he had truly believed that Geller was dead when he shot him.

Defendant argues that the jury was bound to accept, at face value, the indications in his admissions that he believed Geller dead. Certainly, it is true that the defendant was entitled to have the entirety of the admissions, both the inculpatory and the exculpatory portions, placed in evidence before the trier of facts. However, the jury was not required to automatically credit the exculpatory portions of the admissions. The general rule is, of course, that the credibility of witnesses is a question of fact and the jury may choose to believe some, but not all, of a witness' testimony. The general rule applies with equal force to proof of admissions. Thus, it has been stated that "where that part of the declaration which discharges the party making it is in itself highly improbable or is discredited by other evidence the (jury) may believe one part of the admission and reject the other." In *People v. Miller*, 286 N.Y.S. 702, 706, relied upon by defendant, Justice Lewis (later Chief Judge) concluded that the damaging aspects of an admission should not be accepted and the exculpatory portion rejected "unless the latter is disputed by other evidence in the case, or is so improbable as to be unworthy of belief." In this case, there is ample other evidence to contradict the defendant's assertion that he believed Geller dead. There were five bullet wounds inflicted with stunning accuracy in a vital part of the victim's anatomy. The medical testimony indicated that Geller may have been alive at the time defendant fired at him. The defendant voluntarily left the jurisdiction immediately after the crime with his coperpetrator. Defendant did not report the crime to the police when left on his own by Bush. Instead, he attempted to conceal his and Bush's involvement with the homicide. In addition, the other portions of defendant's admissions make his contended belief that Geller was dead extremely improbable. Defendant, without a word of instruction from Bush, voluntarily got up from his seat after the passage of just a few minutes and fired five times point blank into the victim's face, snuffing out any remaining chance of life that Geller possessed. Certainly, this alone indicates a callous indifference to the taking of a human life. His admissions are barren of any claim of duress and reflect, instead, an unstinting co-operation in efforts to dispose of vital incriminating evidence. Indeed, defendant maintained a false version of the occurrence until such time as the police informed him that they had evidence that he lately possessed a gun of the same caliber as one of the weapons involved in the shooting. From all of this, the jury was certainly warranted in concluding that the defendant acted in the belief that Geller was yet alive when shot by defendant.

The jury convicted the defendant of murder. Necessarily, they found that defendant intended to kill a live human being. Subsumed within this finding is the conclusion that defendant acted in the belief that Geller was alive. Thus, there is no need for additional fact findings by a jury. Although it was not established beyond a reasonable doubt that Geller was, in fact, alive, such is no defense to attempted murder since a murder would have been committed "had the attendant circumstances been as (defendant) believed them to be." (Penal Law, s 110.10.) The jury necessarily found that defendant believed Geller to be alive when defendant shot at him.

The Appellate Division erred in not modifying the judgment to reflect a conviction for the lesser included offense of attempted murder. An attempt to commit a murder is a lesser included offense of murder and the Appellate Division has the authority, where the trial evidence is not legally sufficient to establish the offense of which the defendant was convicted, to modify the judgment to one of conviction for a lesser included offense which is legally established by the evidence. Thus, the Appellate Division, by dismissing the indictment, failed to take the appropriate corrective action. Further, questions of law were erroneously determined in favor of the appellant at the Appellate Division. While we affirm the order of the Appellate Division to the extent that the order reflects that the judgment of conviction for murder cannot stand, a modification of the order and a remittal for further proceedings is necessary.

## Notes and Questions

1. Would you say the aforementioned facts establish legal or factual impossibility? What specific facts would you advance in support of your argument?

As noted, distinguishing legal from factual impossibility has perplexed students of the law for generations. One reason for this is that in a sense there exist two forms of legal impossibility: "pure" and "hybrid."[5] Pure legal impossibility indisputably constitutes a defense, because it is based on the legality principle (*see* Chapter 2). Suppose, for example, that Jim strikes a match in bold disregard of the warning to "close cover before striking," thinking it will land him in jail. Jim, of course, could not be criminally prosecuted. Why? Simply because such an act is not criminalized.

Hybrid impossibility exists if the actor's goal is illegal "but commission of the offense is impossible due to a factual mistake . . . regarding the legal status of some attendant circumstances . . . ."[6] Illustrative of hybrid legal impossibility

is the *Guffey* decision, cited above, in which the defendant shot a stuffed deer thinking he was poaching a live deer out of season. Guffey's act, if it involved a live deer, would have been a crime. However, his factual mistake precluded liability. Similarly, in *Jaffe*, also cited above, there was no attempt liability when the defendant received property thinking it was stolen when it was not.

In the end, whether legal impossibility is available as a defense largely turns on how the facts are characterized by the court. As Professor Dressler notes, "[u]ltimately, any case of hybrid legal impossibility may reasonably be characterized as factual impossibility."[7] Compared with legal impossibility, *factual impossibility* remains clear-cut. One who tries to pick a pocket that turns out to be empty is a thief—the defendant's factual mistake in no way serves as a defense.

## Defense of Abandonment

As noted at the outset, attempts come in two basic varieties: failed and interrupted. With the defense of impossibility we addressed the former. In this section we address the latter and consider whether attempt liability can perhaps be avoided by a defendant who has already taken one or more steps toward achieving the ultimate criminal objective. The MPC, for instance, makes available the defense of **abandonment** and provides as follows:

> When the actor's conduct would otherwise constitute an attempt . . . it is an affirmative defense that he abandoned his effort to commit the crime or otherwise prevented its commission, under circumstances manifesting a complete and voluntary renunciation of his criminal purpose. MPC § 5.01(4).

The extent and nature of the acts already taken by the defendant are thus critical to whether attempt liability can be avoided on the basis of abandonment. Plainly, one accused of a "failed" attempt cannot raise an abandonment defense;

such an actor has manifested a clear desire to consummate the object crime, only to have some undesired element prevent the desired result. *See, e.g., State v. Smith,* 409 N.E.2d 1199 (Ind. App. 1980) (upholding attempted murder conviction despite the fact that defendant rushed the victim to the hospital after stabbing him). As the Maryland Court of Appeals has noted, "a voluntary abandonment of an attempt which has proceeded beyond mere preparation . . . does not expiate the guilt of, or forbid the punishment for, the crime already committed." *State v. Wiley,* 207 A.2d 478 1480 (Md. Ct. App. 1965).

The defense of abandonment is therefore available only to those accused of "interrupted" attempts. But here, too, abandonment might not be available. This is because the actor's conduct can simply pass the point of no return in terms of triggering liability. When circumstances indicate that the defendant has gone beyond the realm of "mere preparation" into that of attempt yet has done nothing to cancel the logical inference that the crime will come to pass if not interrupted, there can be no abandonment.

The following case, *State v. Mahoney,* illustrates the sometimes difficult distinctions that must be drawn in evaluating the defense of abandonment.

---

### *State v. Mahoney,* 264 Mont. 89, 870 P.2d 65 (Mont. 1994)

Nelson, Justice.

On or about May 29, 1989, at approximately 7:00 P.M., Chris Mahoney went to a Town Pump in Billings where Beth Brandt was working the night shift. Having previously purchased a soft drink and having left the store, he returned at about 8:30 P.M. and engaged Ms. Brandt in conversation. Mahoney was in the store about 9:00 P.M., closing time, when Ms. Brandt asked him to leave so she could close the store. Mahoney walked out of the store followed by Ms. Brandt who then hung up a "closed" sign near the door. She walked back inside through the door, and, as she turned to lock the door from the inside, the defendant pushed it open, pushing Ms. Brandt away from the door in the process. He demanded she lock the door, and then prevented her attempts to leave through the door.

Ms. Brandt attempted to scream, but Mahoney covered her mouth with his hand and forced her to the floor near the cash register. As she struggled, the defendant kept telling her to lock the door. He forced her into a squatting position facing away from him on the floor. Mahoney then produced a knife from his coat and began stabbing her. After stabbing her repeatedly and in that process severely lacerating her neck, exposing her carotid artery, he restrained her, forcibly, partially disrobed her and, ". . . still dressed, began to rub his penis up and down against the back of the victim's hands as she held them in front of her genitals."

When he saw a large amount of blood from her wounds, he stopped his actions, went around the checkout counter and called the Billings police, reporting that he had cut a clerk and that an ambulance was needed. According to the affidavits in support of the information and amended information subsequently filed, when the police arrived at the scene, Mahoney was cooperative, Miranda warnings were read, and he provided a factual account of the incident.

\* \* \*

Mahoney [was charged] with attempted deliberate homicide and attempted sexual intercourse without consent. . . .

[Several months later] a proceeding was held in which Mahoney, represented by counsel, withdrew his plea of not guilty and entered a plea of guilty to the offenses charged. Mahoney was interrogated by the District Court Judge about his understanding of the consequences of his guilty plea, and he was questioned about his understanding of the "Acknowledgement of Waiver of Rights by Plea of Guilty," which he had read, discussed with his attorney and signed. The District Court concluded that Mahoney's change of plea was knowingly and voluntarily made and accepted his plea of guilty. There was no plea agreement.

After a presentence report was filed, Mahoney appeared with counsel and was sentenced to 40 years in the State Prison for the crime of attempted deliberate homicide, 18 years for the crime of attempted sexual intercourse without consent and an additional eight years for the use of a weapon. The sentences were ordered to be served consecutively, Mahoney was designated a dangerous offender and conditions were imposed in the event of his parole.

Mahoney subsequently filed a motion to withdraw his guilty plea on July 10, 1992. The District Court denied his motion on June 29, 1993, and Mahoney's notice of appeal was filed on August 26, 1993.

\* \* \*

## Voluntary Abandonment of the Crimes Charged

Mahoney argues that he could not be found guilty of the crimes charged because he voluntarily abandoned his criminal efforts. The Montana "attempt" statute in effect when Mahoney pled guilty provides that, "[a] person commits the offense of attempt when, with the purpose to commit a specific offense, he does *any* act towards the commission of such offense." (Emphasis added.) Section 45-4-103(1), MCA (1987). *State v. Ribera* (1979), 183 Mont. 1, 11. Further:

> This Court has stated that an overt act "must reach far enough towards the accomplishment of the desired result to amount to the commencement of the consummation." In addition, the Court stated that "there must be at least some appreciable fragment of the crime committed, and it must be in such progress that it will be consummated unless interrupted by circumstances independent of the will of the attempter." *Ribera*, 597 P.2d at 1170.

In this case, Mahoney accosted Ms. Brandt, prevented her escape from the store and then stabbed his struggling victim twelve times, causing serious injuries, including damage to her lungs, liver and kidneys. It can hardly be argued that this is not "at least some appreciable fragment of the crime [of deliberate homicide] committed [and that his actions reached] far enough towards the accomplishment of the desired result to amount to the commencement of the consummation." *Ribera*, 597 P.2d at 1170.

Moreover, Mahoney forcibly pulled down Ms. Brandt's pants, lifted her shirt up, cut her bra straps and attempted to have sexual intercourse with her. Notwithstanding that she was seriously wounded from the stabbing, Ms. Brandt resisted the defendant's attack by holding her hands in front of her genitals to prevent him from having sexual intercourse. Mahoney himself stated in the proceeding to change his plea to guilty that his intent was to have sexual intercourse with the victim. These actions by Mahoney likewise unequivocally established that at least some fragment of the crime of sexual intercourse without consent was committed and that such actions reached far enough towards the accomplishment of the desired result to amount to the commencement of the consummation. *Ribera*, 597 P.2d at 1170.

There is no doubt from the conduct of the defendant in this case that, with the purpose to commit the crimes of deliberate homicide and sexual intercourse without consent, he did ". . . act[s] toward the commission of such offense[s]." Section 45-4-103(1), MCA (1987).

However, Subsection (4) of § 45-4-103, MCA (1987), provides a defense to the offense of attempt. It is on this defense which Mahoney relies for his claim that he could not be found guilty of the crimes to which he pled guilty. Section 45-4-103(4), MCA (1987), provides that:

> A person shall not be liable under this section if, under circumstances manifesting a voluntary and complete renunciation of his criminal purpose, he avoided the commission of the offense attempted by abandoning his criminal effort.

Mahoney offers his telephone call to the police as evidence that he voluntarily and completely renounced his criminal purpose by abandoning his efforts to commit deliberate homicide and sexual intercourse without consent.

The amended information discloses that after preventing her escape from the store and after stabbing the victim numerous times, Mahoney forcibly tried to disrobe the struggling Ms. Brandt, and "then began to rub his penis up and down against the back of the victim's hands as she held them in front of her genitals. When he saw a large amount of blood from her wounds he stopped, went around the checkout counter and called Billings police."

We are assisted in answering the question of whether Mahoney voluntarily and completely renounced his criminal purpose and abandoned his criminal effort by the decision of a sister state which has extensively explored this issue.

*(Continues)*

(*Continued*)

The State of Michigan recognizes the affirmative defense of voluntary abandonment but qualifies the definition of voluntary abandonment, stating:

> Abandonment is not "voluntary" when the defendant fails to complete the attempted crime because of *unanticipated difficulties, unexpected resistance,* or circumstances which increase the probability of detention or apprehension. (Emphasis added.) *People v. McNeal* (1986), 393 N.W.2d 907, 912, citing *People v. Kimball* (1981), 311 N.W.2d 343, 349.

The defendant in *Kimball* was charged with and convicted of attempted unarmed robbery. He argued that he voluntarily abandoned his criminal enterprise before the crime was consummated; therefore, he could not be found guilty of attempt.

The *Kimball* court concluded that voluntary abandonment was a defense to a prosecution for criminal attempt. However, the court stated emphatically, that abandonment was not "voluntary" when the criminal endeavor was not completed because of unanticipated difficulties or unexpected resistance. Although Michigan's attempt statute differs from ours, we agree with the approach set forth in *Kimball.*

In arriving at its conclusion that voluntary abandonment is a defense to a criminal attempt, the *Kimball* court extensively reviewed authoritative commentary on criminal attempt law, citing Perkins, Criminal Law (2d ed), ch. 6, § 3, p. 590, among others. Perkins states that "although a criminal plan has proceeded far enough to support a conviction of criminal attempt, it would be sound to recognize the possibility of a *locus penitentiae* so long as *no substantial harm has been done* and *no act of actual danger* committed." *Kimball,* 311 N.W.2d at 347. (Underlined emphasis added.) Perkins prefaces the above cited statement in his treatise on criminal law, by commenting that there are limitations to the use of abandonment as a defense to the crime of attempt. "Attempted murder cannot be purged after the victim has been wounded, no matter what may cause the plan to be abandoned." "Perkins" at 590.

We agree with Perkins' logic, particularly in this case, where the defendant prevented the victim's escape, stabbed her twelve times causing grievous injury and then tried to forcibly rape her. At this point, substantial harm had been done and acts of actual danger had, indeed, been committed. Moreover, Mahoney's actions represent at least "some appreciable fragment of the crime [to be] committed." *Ribera,* 597 P.2d at 1170.

Nor did Mahoney voluntarily and completely renounce his criminal purpose by abandoning his attempt to commit the crimes of deliberate homicide and sexual intercourse without consent, as required by § 45-4-103, MCA (1987). Even after the victim had been repeatedly stabbed, she was able to thwart the defendant's attack by holding her hands over her genitals, trying to prevent him from engaging in sexual intercourse with her. As stated in *Kimball* and *McNeal,* an abandonment is not voluntary if the defendant fails to complete the crime because of "unanticipated difficulties [or] unexpected resistance." *Kimball,* 311 N.W.2d at 349; *McNeal,* 393 N.W.2d at 912.

Mahoney did not abandon his criminal conduct until he met with unanticipated difficulties and unexpected resistance. He only called the police after observing that the victim was bleeding profusely from the wounds which he inflicted and because she struggled and successfully protected herself from being raped. Mahoney's conduct is not a manifestation of voluntary and complete renunciation of criminal purpose and an abandonment of criminal effort. That he did not actually consummate the crimes of sexual intercourse without consent and deliberate homicide was due not to any voluntary renunciation of criminal purpose on his part but to good fortune that the victim was not killed by the stabbing and the simple fact that circumstances occasioned by his brutal attack and the victim's continued resistance made further criminal effort impracticable. Mahoney's self-serving and conclusory arguments to the contrary exalt form over substance.

Under the test enunciated above, *i.e.,* that there is no voluntary and complete renunciation of criminal purpose and abandonment of criminal effort, where substantial harm has been done and acts of actual danger have been committed, or where the defendant fails to complete the attempted crime because of unanticipated difficulties, unexpected resistance or circumstances which increase the probability of detention or apprehension; under § 45-4-103(1) and (4), MCA (1987), and our prior case law; and under the facts of this case, we conclude that Mahoney committed the offenses of attempted deliberate homicide and attempted sexual intercourse without consent and that he has failed to establish his abandonment defense. Accordingly, Mahoney's argument that he could not have been found guilty of the crimes of attempted deliberate homicide and attempted sexual intercourse without consent to which he pled guilty is without merit.

## Notes and Questions

1. The Commentary to the MPC offers two rationales in support of the abandonment defense. First, abandonment "tends to negative dangerousness" in that it signals a "lack of firm purpose" for criminal behavior (*i.e.*, the actor renounced his criminal intent while still wielding the power to commit the crime). Second, allowing abandonment to serve as a defense provides "actors with a motive for desisting from their criminal designs, thereby diminishing the risk that the substantive crime will be committed." *See* MPC, § 5.01(8).

   The latter rationale, especially, is distinctly pragmatic in nature and arguably quite sensible. But, at the same time, would society be better served by adopting an unforgiving, bright-line rule of no return? Are defendants accused of other crimes permitted to simply wipe the slate clean of their criminal activity? Is the crime of attempt sufficiently different in nature to warrant the exception? If so, why?

2. The particular motivation behind the actor's alleged renunciation of the criminal scheme can be critically important. More specifically, although abandonment must always be "voluntary," there are degrees and types of voluntariness. The MPC, for instance, bars the defense in the event the defendant (a) abandons the effort only because of some newly developed circumstance that makes apprehension more likely or otherwise makes the object crime more difficult to carry out or (b) abandons the object crime in the present in the interest of carrying it out at some future time on the same or another victim. *See* MPC § 5.01(4).

   Put another way, abandonment is available when the defendant abandons the object crime by virtue of his or her own "internal" decision but not the intervention of "external" forces. For example, in *Smith v. State,* 636 N.E.2d 124 (Ind. 1994), Smith abducted two juvenile males and drove them to a wooded area, where he handcuffed one to a tree and commanded the other to take his clothes off. The latter boy refused on two occasions and Smith then threatened him with a gun, only to relent when he heard tree branches snapping nearby. Smith was convicted of attempted criminal deviate sexual conduct. On appeal, the Indiana Supreme Court concluded that the facts failed to show abandonment because Smith did not have a "genuine change of heart." "To be considered voluntary, the decision to abandon must originate with the accused and not be the product of extrinsic factors that increase the probability of detection or make more difficult the accomplishment of the criminal purpose." *Id* at 127. For another example of the distinction, see *United States v. McDowell,* 705 F.2d 426 (11th Cir. 1983), where the court upheld the defendant's conviction for attempted purchase of cocaine from an undercover officer because he "spurned the deal not to renounce his criminal intent but because he doubted either the genuineness or quality of the cocaine." *Id* at 428.

   What possible jurisprudential and practical bases are there for this "internal" versus "external" distinction? Also, does the requirement make the fact-finder's job of distinguishing true voluntariness any easier?

# Conspiracy

The crime of **conspiracy** has its origins in the ancient fear of group-based behavior. The California Supreme Court articulated the "group danger" rationale of conspiracy as follows:

> The division of labor inherent in group association is seen to encourage the selection of more elaborate and ambitious goals and to increase the likelihood that the scheme will be successful. Moreover, the moral support of the group is seen as strengthening the perseverance of each member of the conspiracy, thereby acting to discourage any reevaluation of the decision to commit the offense which a single offender might undertake. And even if a single conspirator reconsiders and contemplates stopping the wheels which have been set in motion to attain the objective of the conspiracy, a return to the status quo will be much more difficult since it will entail persuasion of the other conspirators. *People v. Zamora,* 557 P.2d 75, 86-87 (Cal. 1976).

Like the law of attempt, conspiracy is preemptive in nature. However, with conspiracy, the fear over group-based action is such that the substantive criminal law allows police to intercede at an even earlier juncture. For conspiracy liability to exist, the government must prove (1) an agreement between or among at least two persons (the *actus reus*) and (2) an intent (the *mens rea*) to carry out either (a) an unlawful act *or* (b) a lawful act by unlawful means (the latter being a rare occurrence). In an increasing number of jurisdictions there must also be an "overt act" in furtherance of the conspiracy, which, as discussed below, can be quite easy for the government to establish.

## *Actus Reus*

### Agreement

The "gist" of any conspiracy is an **agreement**—a "meeting of the minds"—between at least two people with criminal designs. *State v. Carbone,* 91 A.2d 571 (N.J. 1952). The agreement itself can be rather vague, and because of the usual secret nature of conspiracies, guilt can be based on both circumstantial and direct evidence. According to the U.S. Supreme Court, "[t]he agreement need not be shown to have been explicit. It can instead be inferred from the facts and circumstances of the case." *Iannelli v. United States,*

420 U.S. 770, 777 n.10 (1975). However, one's mere presence during conspiratorial conversations or "[m]ere knowledge, approval of or acquiescence in the object or purpose of the conspiracy, without an intention and agreement to cooperate in the crime" does not result in conspiracy liability. *Cleaver v. United States,* 238 F.2d 766, 771 (10th Cir. 1956). Also, importantly, the criminal objective contemplated by the agreement does not need to actually be carried out for conspiracy liability to arise. *See, e.g., State v. St. Christopher,* 232 N.W.2d 798 (Minn. 1975).

## Unilateral or Bilateral Agreements

Yet another issue relating to the agreement requirement turns on the status of those involved in the alleged agreement. Given that conspiracy liability requires a "meeting of the minds," must the alleged conspirators have the capacity to agree? Consider the following scenario. A and B meet at a hotel room and concoct a plan to burglarize Jim's house the next morning. Later that night B acquires masks and guns so the burglary can take place. Unbeknownst to B, A is a police officer who is only pretending to go along with the scheme. Can B be prosecuted for conspiracy? Under the traditional rule, no, because A did not in fact agree—he was only faking agreement. This is called the bilateral approach. However, MPC § 5.03(1) adopts the unilateral approach, which would allow prosecution of B. The unilateral approach is followed in most jurisdictions today, on the rationale that B's culpability remains despite the factual mistake, as he has shown his clear desire to commit a crime. *See* MPC, Commentary to § 5.03, p. 400.

## "Overt Act"

As discussed, not all jurisdictions require proof of an overt act, something done (or not done—an omission if a duty exists) to achieve the contemplated goal of the conspiracy. In those jurisdictions without such a requirement, the conspiratorial agreement itself can establish guilt. Among those states that do require an overt act, the requirement primarily serves an evidentiary function. "The function of the overt act in a conspiracy prosecution is simply to manifest 'that the conspiracy is at work' . . . and is neither a project resting solely in the minds of the conspirators nor a fully completed operation no longer in existence." *Yates v. United States,* 354 U.S. 298, 1334 (1957) (citation omitted). Secondarily, the overt act requirement affords what one court called a *locus poenitentiae,* a boundary line of sorts that permits conspirators to cease their criminal design before they risk prosecution for conspiracy. *People v. Olson,* 42 Cal. Rptr. 760, 767 (Cal. 1965).

Three points are especially worthy of mention with respect to the overt act requirement. First, the courts typically are very lenient in their interpretation of the requirement, deeming virtually any act in furtherance sufficient. Indeed, courts usually require far less in the way of proof to establish an overt act than the "substantial step" requirement of attempt law, discussed earlier. One commentator characterized the difference as follows:[8]

> To obtain an attempt conviction, the prosecutor must prove that the actor performed an act beyond mere preparation. . . . To obtain a conspiracy conviction, however, the prosecutor need only prove that the conspirators agreed to undertake a criminal scheme or, at most, that they took an overt step in pursuance of the conspiracy. Even an insignificant act may suffice.

In short, as one court has noted, the overt act requirement is satisfied upon proof of mere "preparatory conduct which furthers the ability of the conspirators to carry out the agreement." *State v. Dent,* 869 P.2d 392, 398 (Wash. 1994).

Second, it is important to recognize that the act of one or more members of the conspiracy suffices to create liability for the entire group of conspirators. *See Blumenthal v. United States,* 332 U.S. 539 (1947). For instance, if A and B have a phone conversation about how to carry out a bank heist planned for the next day, then this overt act is imputed to co-conspirator C.

Third, the behavior alleged as an overt act need not be criminal or unlawful in itself to trigger liability. *See Braverman v. United States,* 317 U.S. 49 (1942). Thus, to continue with the previous example, if A legally purchases guns to be used in the bank heist, then the overt act requirement is satisfied.

The following case, *United States v. Brown,* provides an excellent example of the law's use of circumstantial evidence to satisfy the *actus reus* requirements of conspiracy.

## United States v. Brown, 776 F.2d 397 (2d Cir. 1985)

Friendly, Circuit Judge:

[The defendant and his co-defendant, Valentine, were charged with conspiracy to distribute heroin and with distribution of heroin. The prosecution stemmed from a purchase by undercover New York City police officer William Grimball of a "joint" of "D" (roughly $40 worth of heroin). Defendant was convicted of conspiracy but acquitted of the distribution count.]

Officer Grimball was the Government's principal witness. He testified that early in the evening of October 9, 1984, he approached Gregory Valentine on the corner of 115th Street and Eighth Avenue and asked him for a joint of "D". Valentine asked Grimball whom he knew around the street. Grimball asked if Valentine knew Scott. He did not. Brown "came up" and Valentine said, "He wants a joint, but I don't know him." Brown looked at Grimball and said, "He looks okay to me." Valentine then said, "Okay. But I am going to leave it somewhere and you [meaning Officer Grimball] can pick it up."

Brown interjected, "You don't have to do that. Just go and get it for him. He looks all right to me." After looking again at Grimball, Brown said, "He looks all right to me" and "I will wait right here."

Valentine then said, "Okay. Come on with me around to the hotel." Grimball followed him to 300 West 116th Street, where Valentine instructed him, "Sit on the black car and give me a few minutes to go up and get it." Valentine requested and received $40, which had been prerecorded, and then said, "You are going to take care of me for doing this for you, throw some dollars my way?," to which Grimball responded, "Yeah."

Valentine then entered the hotel and shortly returned. The two went back to 115th Street and Eighth Avenue, where Valentine placed a cigarette box on the hood of a blue car. Grimball picked up the cigarette box and found a glassine envelope containing white powder, stipulated to be heroin. Grimball placed $5 of prerecorded buy money in the cigarette box, which he replaced on the hood. Valentine picked up the box and removed the $5. Grimball returned to his car and made a radio transmission to the backup field team that "the buy had went down" and informed them of the locations of the persons involved. Brown and Valentine were arrested. Valentine was found to possess two glassine envelopes of heroin and the $5 of prerecorded money. Brown was in possession of $31 of his own money; no drugs or contraband were found on him. The $40 of marked buy money was not recovered, and no arrests were made at the hotel.

[The court then upheld the admission of expert testimony provided by Grimball that heroin dealers used "steerers," persons paid to facilitate heroin sales and to detect police.]

## Sufficiency of the Evidence

In considering the sufficiency of the evidence, we begin with some preliminary observations. One is that, in testing sufficiency, "the relevant question is whether, after viewing the evidence in the light most favorable to the prosecution, *any* rational trier of fact could have found the essential elements of the crime beyond a reasonable doubt" *Jackson v. Virginia*, 443 U.S. 307, 319 (1979) (emphasis in original). . . . *Jackson's* emphasis on "*any,*" while surely not going so far as to excise "rational," must be taken as an admonition to appellate judges not to reverse convictions because they would not have found the elements of the crime to have been proved beyond a reasonable doubt when other rational beings might do so.

The second observation is that since the jury convicted on the conspiracy count alone, the evidence must permit a reasonable juror to be convinced beyond a reasonable doubt not simply that Brown had aided and abetted the drug sale but that he had agreed to do so. . . . On the other hand, the jury's failure to agree on the aiding and abetting charge does not operate against the Government; even an acquittal on that count would not have done so. . . .

A review of the evidence against Brown convinces us that it was sufficient, even without Grimball's characterization of Brown as a steerer, although barely so.

Although Brown's mere presence at the scene of the crime and his knowledge that a crime was being committed would not have been sufficient to establish Brown's knowing participation in the conspiracy . . . the proof went considerably beyond that. Brown was not simply standing around while the exchanges between Officer Grimball and Valentine occurred. He came on the scene shortly after these began and Valentine immediately explained the situation to him. Brown then conferred his seal of approval on Grimball, a most unlikely event unless there was an established relationship between Brown and Valentine. Finally, Brown took upon himself the serious responsibility of telling Valentine to desist from his plan to reduce the risks by not handing the heroin directly to Grimball. A rational mind could take this as bespeaking the existence of an agreement whereby Brown was to have the authority to command, or at least to persuade. Brown's remark, "Just go and get it for him," permits inferences that Brown knew where the heroin was to be gotten, that he knew that Valentine knew this, and that Brown and Valentine had engaged in such a transaction before.

The mere fact that these inferences were not ineluctable does not mean that they were insufficient to convince a reasonable juror beyond a reasonable doubt. . . . When we add to the inferences that can be reasonably drawn from the facts to which Grimball testified the portion of his expert testimony about the use of steerers in street sales of narcotics, which was clearly unobjectionable once Grimball's qualifications were established, we conclude that the Government offered sufficient evidence, apart from Grimball's opinion that Brown was a steerer, for a reasonable juror to be satisfied beyond a reasonable doubt not only that Brown had acted as a steerer but that he had agreed to do so. (*See* Footnote 10.)

Affirmed.

## Mens Rea

Conspiracy can be said to require two different forms of intent. First, the defendant must intend to enter into an agreement (discussed above). Second, under the MPC and in the majority of jurisdictions, the defendant must intend to carry out the object of the conspiracy. *See United States v. United States Gypsum Co.*, 438 U.S. 422, 443 n.20 (1978). This second requirement begs an important question—whether, in the words of MPC, the defendant must have the "purpose to promote or facilitate" the crime or must merely know that it will be committed. This is the question taken up in the next case, *People v. Lauria*.

### *People v. Lauria,* 59 Cal. Rptr. 628 (Cal. 1967)

Fleming, Associate Justice.

In an investigation of call-girl activity the police focused their attention on three prostitutes actively plying their trade on call, each of whom was using Lauria's telephone answering service, presumably for business purposes.

On January 8, 1965, Stella Weeks, a policewoman, signed up for telephone service with Lauria's answering service. Mrs. Weeks, in the course of her conversation with Lauria's office manager, hinted broadly that she was a prostitute concerned with the secrecy of her activities and their concealment from the police. She was assured that the operation of the service was discreet and "about as safe as you can get." It was arranged that Mrs. Weeks need not leave her address with the answering service, but could pick up her calls and pay her bills in person.

On February 11, Mrs. Weeks talked to Lauria on the telephone and told him her business was modelling and she had been referred to the answering service by Terry, one of the three prostitutes under investigation. She complained that because of the operation of the service she had lost two valuable customers, referred to as tricks. Lauria defended his service and said that her friends had probably lied to her about having left calls for her. But he did not respond to Mrs. Weeks' hints that she needed customers in order to make money, other than to invite her to his house for a personal visit in order to get better acquainted. In the course of his talk he said "his business was taking messages."

On February 15, Mrs. Weeks talked on the telephone to Lauria's office manager and again complained of two lost calls, which she described as a $50 and a $100 trick. On investigation the office manager could find nothing wrong, but she said she would alert the switchboard operators about slip-ups on calls.

On April 1 Lauria and the three prostitutes were arrested. Lauria complained to the police that this attention was undeserved, stating that Hollywood Call Board had 60 to 70 prostitutes on its board while his own service had only 9 or 10, that he kept separate records for known or suspected prostitutes for the convenience of himself and the police. When asked if his records were available to police who might come to the office to investigate call girls, Lauria replied that they were whenever the police had a specific name. However, his service didn't "arbitrarily tell the police about prostitutes on our board. As long as they pay their bills we tolerate them." In a subsequent voluntary appearance before the Grand Jury Lauria testified he had always cooperated with the police. But he admitted he knew some of his customers were prostitutes, and he knew Terry was a prostitute because he had personally used her services, and he knew she was paying for 500 calls a month.

Lauria and the three prostitutes were indicated [sic] for conspiracy to commit prostitution, and nine overt acts were specified. Subsequently the trial court set aside the indictment as having been brought without reasonable or probable cause. (Pen.Code, § 995.) The People have appealed, claiming that a sufficient showing of an unlawful agreement to further prostitution was made.

To establish agreement, the People need show no more than a tacit, mutual understanding between coconspirators to accomplish an unlawful act. . . . Here the People attempted to establish a conspiracy by showing that Lauria, well aware that his codefendants were prostitutes who received business calls from customers through his telephone answering service, continued to furnish them with such service. This approach attempts to equate knowledge of another's criminal activity with conspiracy to further such criminal activity, and poses the question of the criminal responsibility of a furnisher of goods or services who knows his product is being used to assist the operation of an illegal business. Under what circumstances does a supplier become a part of a conspiracy to further an illegal enterprise by furnishing goods or services which he knows are to be used by the buyer for criminal purposes?

The two leading cases on this point face in opposite directions. In *United States v. Falcone*, 311 U.S. 205, . . . the sellers of large quantities of sugar, yeast, and cans were absolved from participation in a moonshining conspiracy among distillers who bought from them, while in *Direct Sales Co. v. United States*, 319 U.S. 703, . . . a wholesaler of drugs was convicted of conspiracy to violate the federal narcotic laws by selling drugs in quantity to a codefendant physician who was supplying them to addicts. The distinction between these two cases appears primarily based on the proposition that distributors of such dangerous products as drugs are required to exercise greater discrimination in the conduct of their business than are distributors of innocuous substances like sugar and yeast.

In the earlier case, *Falcone*, the sellers' knowledge of the illegal use of the goods was insufficient by itself to make the sellers participants in a conspiracy with the distillers who bought from them. Such knowledge fell short of proof of a conspiracy, and evidence on the volume of sales was too vague to support a jury finding that respondents knew of the conspiracy from the size of the sales alone.

In the later case of *Direct Sales*, the conviction of a drug wholesaler for conspiracy to violate federal narcotic laws was affirmed on a showing that it had actively promoted the sale of morphine sulphate in quantity and had sold codefendant physician, who practiced in a small town in South Carolina, more than 300 times his normal requirements of the drug, even though it had been repeatedly warned of the dangers of unrestricted sales of the drug. . . .

While *Falcone* and *Direct Sales* may not be entirely consistent with each other in their full implications, they do provide us with a framework for the criminal liability of a supplier of lawful goods or services put to unlawful use. Both the element of *knowledge* of the illegal use of the goods or services and the element of *intent* to further that use must be present in order to make the supplier a participant in a criminal conspiracy.

Proof of *knowledge* is ordinarily a question of fact and requires no extended discussion in the present case. The knowledge of the supplier was sufficiently established when Lauria admitted he knew some of his customers were prostitutes and admitted he knew that Terry, an active subscriber to his service, was a prostitute. . . . On this record we think the prosecution is entitled to claim positive knowledge by Lauria of the use of his service to facilitate the business of prostitution.

The more perplexing issue in the case is the sufficiency of proof of *intent* to further the criminal enterprise. The element of intent may be proved either by direct evidence, or by evidence of circumstances from which an intent to further a criminal enterprise by supplying lawful goods or services may be inferred. . . . When the intent to further and promote the criminal enterprise comes from the lips of the supplier himself, ambiguities of inference from circumstance need not trouble us. But in cases where direct proof of complicity is lacking, intent to further the conspiracy must be derived from the sale itself and its surrounding circumstances in order to establish the supplier's express or tacit agreement to join the conspiracy.

In the case at bench the prosecution argues that since Lauria knew his customers were using his service for illegal purposes but nevertheless continued to furnish it to them, he must have intended to assist them in carrying out their illegal activities. . . . Essentially, the People argue that knowledge alone of the continuing use of his telephone facilities for criminal purposes provided a sufficient basis from which his intent to participate in those criminal activities could be inferred. . . .

1. Intent may be inferred from knowledge, when the purveyor of legal goods for illegal use has acquired a stake in the venture. . . .

    In the present case, no proof was offered of inflated charges for the telephone answering services furnished the codefendants.

2. Intent may be inferred from knowledge, when no legitimate use for the goods or services exists. . . .

    However, there is nothing in the furnishing of telephone answering service which would necessarily imply assistance in the performance of illegal activities. Nor is any inference to be derived from the use of an answering service by women, either in any particular volume of calls, or outside normal working hours. . . .

3. Intent may be inferred from knowledge, when the volume of business with the buyer is grossly disproportionate to any legitimate demand, or when sales for illegal use amount to a high proportion of the seller's total business. In such cases an intent to participate in the illegal enterprise may be inferred from the quantity of the business done. . . .

No evidence of any unusual volume of business with prostitutes was presented by the prosecution against Lauria.

Inflated charges, the sale of goods with no legitimate use, sales in inflated amounts, each may provide a fact of sufficient moment from which the intent of the seller to participate in the criminal enterprise may be inferred. In such instances participation by the supplier of legal goods to the illegal enterprise may be inferred because in one way or another the supplier has acquired a special interest in the operation of the illegal enterprise. His intent to participate in the crime of which he has knowledge may be inferred from the existence of his special interest. . . .

*(Continues)*

When we review Lauria's activities in the light of this analysis, we find no proof that Lauria took any direct action to further, encourage, or direct the call-girl activities of his codefendants and we find an absence of circumstances from which his special interest in their activities could be inferred. Neither excessive charges for standardized services, nor the furnishing of services without a legitimate use, nor an unusual quantity of business with call girls, are present. The offense which he is charged with furthering is a misdemeanor, a category of crime which has never been made a required subject of positive disclosure to public authority. Under these circumstances, although proof of Lauria's knowledge of the criminal activities of his patrons was sufficient to charge him with that fact, there was insufficient evidence that he intended to further their criminal activities, and hence insufficient proof of his participation in a criminal conspiracy with his codefendants to further prostitution. Since the conspiracy centered around the activities of Lauria's telephone answering service, the charges against his codefendants likewise fail for want of proof.

## Notes and Questions

1. Are you persuaded by the distinctions drawn by the majority? Should Lauria have been held accountable regardless of the lack of his "direct action"?
2. Imagine that *Fantasy Soldier Magazine* publishes an article entitled "How To Be a Mad Bomber," which chronicles in detail how lethal bombs can be made, with the full knowledge that in fact some readers might use the information to build bombs and destroy buildings. Would there be liability under MPC § 5.03? Is this the proper outcome and why?

## Scope and Duration of Conspiracy

One of the most challenging aspects of conspiracy law concerns identifying the extent of liability among co-conspirators—a task complicated by the reality that conspiracies often entail multiple objectives and numerous participants and do not always occur in uninterrupted, seamless fashion. In this section we first examine the important issue of the scope of conspiracy liability. Later, we address the equally important question of how and when a conspiracy can be terminated.

### Scope

#### Criminal Objectives

It is not uncommon for co-conspirators to hatch a plan that involves several different crimes. For many years U.S. courts were split over the question whether to charge such conspirators with several or a single act of conspiracy. In the landmark decision of *Braverman v. United States*, 317 U.S. 49 (1942), the U.S. Supreme Court settled the question. In *Braverman*, the defendants were liquor bootleggers, each charged with seven conspiracy counts to violate seven different Internal Revenue Service criminal laws. The *Braverman* Court reversed the multiple convictions, adopting an approach that focused primarily on the nature of the agreement reached, not its objectives. According to the Court,

> Whether the object of a single agreement is to commit one or many crimes, it is in either case that agreement which constitutes the conspiracy. . . . The one agreement cannot be taken to be several agreements and hence several conspiracies because it envisages the violation of several statutes rather than one. *Id.* at 53.

MPC § 5.03(3) reflects a similar view, stating that "[i]f a person conspires to commit a number of crimes, he is guilty of only one conspiracy so long as such multiple crimes are the object of the same agreement or continuous conspiratorial relationship."

This basic rule—that a single agreement with several criminal objectives constitutes a single conspiracy—should be distinguished, however, from facts indicating the existence of several distinct conspiracies, which can pose multi-count conspiracy liability. For example, assume A, B, and C agree to rob two banks on Tuesday, a third on Wednesday, and separately agree to kill D on Wednesday. Under such facts, two separate conspiracy prosecutions, for robbery and murder, respectively, could be pursued. *See, e.g., United States v. Aguilera*, 179 F.3d 604 (8th Cir. 1999) (finding two distinct drug conspiracies, despite the fact that they overlapped in time and geographic locales).

Finally, under what is known as the *Pinkerton* rule, a conspirator is held liable both for (1) the crime that is the object of the conspiracy and (2) any other foreseeable crimes committed by his or her co-conspirators in furtherance of the conspiracy. *See Pinkerton v. United States*, 328 U.S. 640 (1946). In *Pinkerton*, for instance, two brothers (Daniel and Walter) conspired in a bootleg liquor business, which continued well after Daniel was imprisoned for unrelated offenses. The U.S. Supreme Court affirmed Daniel's liability as to the foreseeably related crimes carried out by Walter as a result of the brothers' agreement. Thus *Pinkerton* casts a very wide net of liability.

This said, however, it is important to keep in mind two points. First, *Pinkerton* creates liability only as to those crimes foreseeable *after* one becomes a member of the conspiracy, not criminal acts perpetrated by the conspiracy before the defendant's agreement to participate. *See, e.g., United States v. O'Campo*, 973 F.2d 1015 (1st Cir. 1992) (limiting conspirator's liability for illegal drug sales). Also, be aware that not all jurisdictions follow the *Pinkerton* rule. *See, e.g., State ex rel. Woods v. Cohen,* 844 P.2d 1147 (Az. 1992). As a result, in such jurisdictions a conspirator cannot be held responsible for separate criminal acts of co-conspirators unless accomplice liability is otherwise established (see Chapter 11).

### Parties Liable

Conspiracies, which by definition involve two or more participants, can involve extended and often shadowy social relationships with persons only being vaguely familiar to one another. As the social relationships become more attenuated, so too can the measure of cooperation and agreement among the parties, which in turn raises question over whether a conspiracy prosecution is justified. MPC § 5.03(2) deals with this potential problem by broadly requiring that if a defendant "knows that a person with whom he conspires to commit a crime has conspired with another person or persons to commit the same crime, he is guilty of conspiring with such other person or persons, whether or not he knows their identity, to commit such crime."

For their part, the courts have conceived of conspiracies as either "wheel" or "chain" conspiracies. Wheel conspiracies involve a single person or group of persons ("the hub"), providing connection with numerous individuals ("the spokes"). *See, e.g., Kotteakos v. United States,* 328 U.S. 750 (1946) (involving several fraudulent loans, benefiting different individuals, facilitated by a single actor). Chain conspiracies involve interaction extending down a line of different persons, each with a distinct function in the conspiracy (the "links"), who are largely unknown to one another. Most often these conspiracies involve the distribution of commodities such as illegal drugs or other contraband. *See, e.g., United States v. Bruno,* 105 F.2d 921 (2d Cir. 1939) (a drug smuggling and sale enterprise involving 88 persons, stretching from New York to Texas).

The critical issue, with respect to the evaluation of either type of conspiracy, is whether the alleged conspirators have a community of interest in achieving their larger criminal objective. Of the two types, chain conspiracies most often permit this legal inference, because of the greater interdependence of the conspirators in achieving the target crime (*i.e.,* the failure of one "link" can doom the entire conspiratorial objective). As the court stated with respect to the drug sale conspiracy alleged in *Bruno,*

> the smugglers knew that the middlemen must sell to retailers, and the retailers knew that the middlemen must sell to retailers, and the retailers knew that the middlemen must buy [from] importers of one sort or another. Thus the conspirators at one end of the chain knew that the unlawful business would not, and could not, stop with their buyers; and those at the other end knew that it had not begun with their sellers. *United States v. Bruno,* 105 F.2d 921, 922 (2d Cir. 1939).

Wheel conspiracies, on the other hand, are less likely to result in liability because the prosecution will have more difficulty showing community of interest, due to the lessened likelihood that they were aware of one another and shared a common criminal purpose. For instance, in *Kotteakos,*

cited above, which involved fraudulent loans secured under the federal National Housing Act, the conspiracies were deemed independent because each loan was an end in itself and each applicant cared only about his or her loan. There was no overarching, comprehensive plan shared by the separate loan applicants and hence no conspiracy.

### "Plurality" Requirement

Conspiracy is testament to the old adage that it takes (at least) "two to tango." However, the basic rationale of conspiracy, that collective action is more threatening, seems diminished in the event one's alleged co-conspirators escape liability. Not surprisingly, the criminal law takes account of the illogic and perceived unfairness of such an outcome. Typically, the acquittal of those with whom a defendant is alleged to have conspired compels that the defendant also avoid liability. *See United States v. Sheikh,* 654 F.2d 1057 (5th Cir. 1981). However, courts in recent times have carved out an exception to this general rule when the acquittal was in a separate trial. *See, e.g., State v. Soles,* 459 S.E.2d 4 (N.C. App. 1995).

Less uniformity exists when the government refuses, for whatever reason, to prosecute a particular conspirator. Some courts view the decision as equivalent to an acquittal and preclude the prosecution of the remaining conspirator(s); others do not regard the decision as a bar to prosecution.[9]

## Duration

Under the traditional common law approach, once an agreement is reached the legal taint of conspiracy itself arises and cannot be expunged by abandonment or other subsequent acts. As a result, under the common law, a conspirator who withdraws from the enterprise remains liable for the initial conspiratorial agreement but not for the later acts of the conspiracy coming after withdrawal. *See United States v. Read,* 658 F.2d 1225 (7th Cir. 1981). Withdrawal, under the common law, is thus only a partial, not complete, defense to conspiracy. Also, for withdrawal to be effective, "notice" of its occurrence must be communicated to all of one's co-conspirators.[10]

According to one court,

> [a defendant] must abandon the illegal enterprise in a manner reasonably calculated to reach coconspirators. Mere cessation of participation in the conspiracy is not enough. However, a defendant need not take action to stop, obstruct or interfere with the conspiracy in order to withdraw from it. *United States v. Lowell,* 649 F.2d 950, 957 (3d Cir. 1981).

The MPC, however, permits a suspect to avoid liability for even the initial conspiracy on the basis of *abandonment* or *renunciation.* As to abandonment, MPC § 5.03(7)(b) provides that "abandonment is presumed if neither the defendant nor anyone with whom he conspired does any overt act in pursuance of the conspiracy during the applicable statute of limitation." Importantly, the conspiracy is terminated *only* as to the person who abandons the conspiracy, and *only* when (1) "he advises those with whom he conspired of his abandonment" or (2) "he informs the law enforcement authorities of the existence of the conspiracy and his participation therein." MPC § 5.03(7)(c).

The **defense of renunciation** applies only to the individual member of the conspiracy and requires behavior of a more affirmative nature. The MPC states that for renunciation the actor must "thwart[ ] the success of the conspiracy, under circumstances manifesting a complete and voluntary renunciation of his criminal purpose." MPC § 5.03(6). The added requirement that the conspiracy be "thwarted" is thought justified on the rationale that the objective of the conspiracy will often be pursued by others notwithstanding that the conspiracy has been renounced by one actor. *Id.*, Commentary at 458.

## How Conspiracy Differs from Complicity

On its face, conspiracy shares much with accomplice liability (discussed in Chapter 11). What distinguishes conspiracy, however, is that an agreement must exist among the conspirators. With accomplice liability, an agreement is often present, but it is not necessary for liability to arise. Another distinction is that accomplice liability requires that the accused actually participate in the criminal objective. Conspiracy has no such requirement of direct individual participation; with conspiracy a mere agreement (and sometimes a *de minimis* "overt act" by anyone of the conspirators) suffices.

## Racketeer Influenced and Corrupt Organizations Act

Closely related to the ancient common law crime of conspiracy is the federal **Racketeer Influenced and Corrupt Organizations (RICO) Act**, enacted by the U.S. Congress in 1970 to combat organized crime. RICO originated out of concern over the suspected infiltration of legitimate businesses by criminal syndicates and the difficulty federal prosecutors were having in securing convictions against the shadowy enterprises by means of traditional conspiracy law. Although initially aimed at organizations like the Mafia, today RICO enjoys significant use in criminal prosecutions of a broad variety of other targets, such as otherwise legitimate businesses. Indeed, RICO has become a valued tool in federal prosecutors' battle against "white collar" crime (discussed at length in Chapter 10).

RICO is an extremely complex series of statutory provisions that entails both criminal and massive civil penalties as well as the forfeiture of money and property. RICO targets "enterprises" engaged in a "pattern" of "racketeering activity." *See* 18 U.S.C. § 1962. As already mentioned, both legal and illegal entities are "enterprises" within RICO's criminal scope. An "enterprise" can be either an individual or "a group of persons associated together for a common purpose of engaging in a course of conduct" *United States v. Turkette*, 452 U.S. 576, 583 (1981). The enterprise requirement is proven "by evidence of an ongoing organization, formal or informal, and by evidence that the various associates function as a continuing unit." *Id.* "Racketeering activities" include a very broad range of substantive crimes, including arson, murder, kidnapping, illegal drug dealing, and extortion. *See* 18 U.S.C. § 1961(1). A "pattern" is established if the prosecution can prove at least two such acts within a 10-year span. *Id.* § 1961(5).

Importantly, however, in addition to allowing prosecution of substantive RICO provisions (*see* U.S.C. § 1962(a)–(c)), RICO contains a separate conspiracy provision. *See* 18 U.S.C. § 1962(d). As a result, those involved in an enterprise can be convicted of conspiring to violate a substantive RICO provision, thereby providing the government with a much more powerful tool to combat loosely concerted criminal activity. As noted by one court,

> In the context of organized crime, [common law conspiracy] inhibited mass prosecutions because a single agreement or "common objective" cannot be inferred from the commission of highly diverse crimes by apparently unrelated individuals. RICO helps to eliminate this problem by creating a substantive offense which ties together these diverse parties and crimes. . . . Under [RICO], it is irrelevant that each defendant participated in the enterprise's affairs through different, even unrelated crimes, so long as we may reasonably infer that each crime was intended to further the conspiracy's affairs.
>
> Through RICO, Congress defined a new separate crime to help snare those who make careers of crime. Participation in the affairs of an enterprise through the commission of two or more ["racketeering activities"] is now an offense separate and distinct . . . So too is conspiracy to commit this new offense a crime. . . . The necessity which mothered this statutory invention was caused by the inability of the traditional criminal law to punish and deter organized crime. *United States v. Elliott*, 571 F.2d 880, 911 (5th Cir. 1978).

## Solicitation

The final inchoate crime we examine is **solicitation**. The law of solicitation permits the state to intervene in possible criminal behavior at a point even earlier than with attempts and conspiracies. Solicitation liability can arise upon the mere asking or encouraging of another to commit a crime, entailing neither "overt acts" nor "substantial steps." According to the MPC § 5.02(1),

> A person is guilty of solicitation to commit a crime if with the purpose of promoting or facilitating its commission he commands, encourages or requests another to engage in specific conduct which would constitute such crime or an attempt to commit such crime or which would establish his **complicity** in its commission or attempted commission.

In short, the actor must (1) advise or encourage another to commit a crime and (2) do so with the intent that the person solicited commits the acts constituting the particular crime.[11]

### Actus Reus

Liability for solicitation arises at the moment the actor, with the requisite intent, asks or encourages another to commit a crime, even if the person solicited refuses to carry out the criminal overture. But the alleged utterance must be something more than mere idle talk or rumination. For instance, no solicitation occurs if A mutters absent-mindedly in a crowded elevator that she "sure wished somebody would kill my husband."

Given that the request or encouragement is the core requirement of liability, what happens when the solicitation itself is not communicated? Under the MPC, solicitation liability arises even if "the actor fails to communicate with the person he solicits to commit a crime if his conduct was designed to effect such communication." MPC § 5.02(2). The traditional common law, however, requires that the solicitation be communicated, as the following case suggests.

---

## People v. Ruppenthal, 771 N.E.2d 1002 (Ill. App. 2002)

Presiding Justice GALLAGHER delivered the opinion of the court:

Following a bench trial, the defendant, Stephen Ruppenthal, was convicted of two counts of indecent solicitation of a child and was sentenced to two years of probation. Ruppenthal was arrested at O'Hare International Airport in Chicago after communicating on the Internet with a person whom he believed to be a 14-year-old girl named "Stacy," but who was actually a Cook County sheriff's detective. On appeal, defendant first argues that because he did not in fact solicit a minor but instead communicated with an adult, he should not have been charged with indecent solicitation of a child under section 11-6 of the Criminal Code of 1961. . . .

The following facts were presented at a stipulated bench trial. On April 26, 2000, Cook County sheriff's detective Michael Anton logged onto a Internet site titled "sex" and assumed the identity of a 14-year-old girl named "Stacy." Defendant, who was 53 years old and lived in Tomales, California, communicated privately with "Stacy" for about two hours using his home computer. The State entered into evidence a transcript of their on-line conversation, which is included in the record.

We relate only the pertinent portions. Defendant told "Stacy" that he was 46 years old. "Stacy" replied that she would "be 15 real soon." Defendant said his flight would stop in Chicago the following day and suggested they meet at O'Hare and "find a private place and be together." Defendant said he wanted to rub her chest and vagina and that they "would find a place in the airport without many people and sit together" and they could be under a blanket so no one would see what they did. Defendant told "Stacy" his flight number and asked her to meet him at his gate. Defendant said that if she was ready tomorrow, they could "find a place," perhaps a restroom, where he could put his penis in her vagina; he later stated, "I can't promise we can do it. It will depend if we can find a place." "Stacy" told defendant her last name was Hugh and she would wear her school uniform.

When defendant arrived at O'Hare, a female sheriff's detective dressed as "Stacy" met him at his gate. The female detective approached defendant and said his name. Defendant replied, "Yes, I thought it was you, but I wasn't sure. It's good that you wore your uniform. Then I knew it was you." After further conversation between the female detective and defendant, Detective Anton asked defendant who he was meeting. Defendant said he was meeting Stacy Hugh. Defendant was placed under arrest. After his arrest, defendant told Anton and a prosecutor that he thought "Stacy" was "about 15" years old. The remainder of defendant's statement was consistent with the evidence presented at trial. The trial court convicted defendant of two counts of indecent solicitation of a child with the intent to commit aggravated criminal sexual abuse, pursuant to section 11-6(c)(3) of the Code. Those counts were premised upon defendant's stated intent to touch the vagina of a minor child. The trial court acquitted defendant on a third count of indecent solicitation, stating that although the parties stipulated that defendant had suggested sexual intercourse, defendant's intent on that issue "seem[ed] to be unresolved."

On appeal, defendant first challenges the constitutionality of the statute under which he was convicted, arguing that the law violates his "first amendment right to freedom of thought and belief." He asserts that his conviction was "a function of one element only—his belief that he was speaking to a child" and that the only act that accompanied his belief was the communication with Detective Anton, an adult. Defendant claims that because he did not commit an illegal act, he was punished only for his "bad state of mind."

Section 11-6 of the Code provides:

A person of the age of 17 years and upwards commits the offense of indecent solicitation of a child if the person, with the intent that the offense of aggravated criminal sexual assault, criminal sexual assault, predatory criminal sexual assault of a child, or aggravated criminal sexual abuse be committed, knowingly solicits a child *or one whom he or she believes to be a child* to perform an act of sexual penetration or sexual conduct as defined in Section 12-12 of this Code. (Emphasis added.) 720 ILCS 5/11-6(a) (West 2000).

The statute defines "solicit" as:

to command, authorize, urge, incite, request, or advise another to perform an act by any means including, but not limited to, in person, over the phone, in writing, by computer, or by advertisement of any kind. 720 ILCS 5/11-6(b) (West 2000).

*(Continues)*

(*Continued*)

Child" is defined as "a person under 17 years of age. 720 ILCS 5/11-6(b) (West 2000).

The criminal act defined by the statute is knowingly soliciting a child or one believed to be a child to perform an act of sexual penetration or sexual conduct, with the intent that the conduct be committed. 720 ILCS 5/11-6(a) (West 2000). The offense of solicitation is complete when the principal offense is commanded, encouraged or requested with the intent that it be committed. *People v. Edwards*, 611 N.E.2d 1196, 1202 (1993) (discussing solicitation to commit murder). "Whether or not the actual crime took place is meaningless under the applicable statute. Defendant's offense was complete when the words at issue were spoken. . . ." *Edwards*, 611 N.E.2d at 1202.

Defendant argues that although he intended to commit a sexual act with a child, he committed no crime by speaking words of solicitation to an adult. We disagree. Defendant is being punished for his intent to engage in sexual activity with someone he admittedly believed to be under the age of 17 and his solicitation of that activity. The fact that defendant's words were transmitted to an adult does not negate defendant's belief that he was speaking to a minor, which is the culpable act defined by the statute. The specific intent required to prove the elements of the offense of solicitation can be inferred from the surrounding circumstances and acts of the defendant. Defendant's trip to Illinois for the admitted purpose of meeting a girl he knew to be "about 15" exhibited his intent to engage in the sexual activity discussed on the Internet. . . .

For all of the foregoing reasons, we affirm defendant's conviction and sentence.

## Mens Rea

As we see in *Ruppenthal*, for solicitation liability to arise, the defendant must not just encourage another to commit a crime; one also must do so with the intent that the crime actually be committed. *People v. Quentin,* which follows, addresses the important question of just how specific the intent must be.

## *People v. Quentin*, 296 N.Y.S.2d 443 (N.Y. 1968)

Bernard Thompson, Judge.

[Defendants faced several charges, including solicitation, based on a brochure they had circulated. The cover of the brochure depicted a couple engaged in sexual intercourse, while its interior contained extensive writing of a non-sexual nature. After pleading guilty, the defendants moved to withdraw their pleas. The portion of the decision excerpted below focuses upon the court's treatment of the solicitation charge, and the non-sexual content of the brochure in question.]

. . . The cover picture is unrelated to the balance of the pamphlet. It is obviously used only to attract attention and in so doing appeals to the prurient interest.

The rest of the brochure is worthy of note. The inside front cover concisely describes the philosophy of the defendants. In part it reads, "America is carnivorous. She eats the world for desert[sic]. Behind slick pictures of pretty-suburban-middle-churchgoing-family lie hamburgers seasoned with napalm, race crimes too brutal to recall, cultures plundered, and triviality elevated into a way of life. . . . The rich are rich because they are thieves and the poor because they are victims, and the future will condemn those who accept the present as reality. Break down the family, church, nation, city, economy. . . . Subversiveness saves us . . . our professors are spies; let us close the schools and flow into the streets. . . . Grow hair long and become too freaky to fit into the machine culture. What's needed is a generation of people who are freaky, crazy, irrational, sexy, angry, irreligious, childish and mad: people who burn draft cards, burn high school and college degrees: People who say: 'To hell with your goals.'; people who lure the youth with music, pot and acid: people who re-define reality, who re-define the normal . . . The white youth of America have more in common with Indians plundered, than they do with their own parents. Burn their house down, and you will be free."

This is followed by a paragraph entitled "How to make a fire bomb," and a recipe for Tryptamine, a psychedelic agent. The recipe ends with the statements, "This last (Tetrahydrofurane) is a very powerful reducing agent; wear safety glasses, add very cautiously, and perform this step with ventilation, away from flames ($H_2$ is evolved). The yield is about 40 grams of DMT, in tetrahydrofurane solution. This cannot be drunk or injected, but may be smoked by sprinkling on mint or cannabis leaves and letting the solution evaporate. It's evaporated when it starts smelling like DMT instead of tetrahydrofurane."

* * *

Count three (not so labeled) alleges a violation of Section 100.00 of the Penal Law which is criminal solicitation in the third degree. It charges that the defendants violated the section in that they attempted to cause persons to whom a brochure was distributed to possess a chemical compound known as DET and DMT which violates § 229 (429) of the Mental Hygiene Law. The brochure on one of its pages gives a formula for making a fire bomb. Below that is also a formula for making both DET and DMT. On the page with the formula is no other solicitation, request or advocacy concerning the drugs. The formula taken alone appears to be such as would be found in a chemistry book or encyclopedia. It is clear that Section 100.00 was intended to cover a situation where a particular person importunes another specified individual to do a specific act which constitutes a crime. The purpose was to hold the solicitor criminally responsible even if the one solicited fails to commit the act. It does not appear that Section 100.00 was designed to cover a situation where the defendant makes a general solicitation (however reprehensible) to a large indefinable group to commit a crime.

The defendants' motion is granted and all counts of the information are dismissed with leave to the District Attorney to file a new information.

## Notes and Questions

1. Do you agree with the *Quentin* court's analysis? At what point does a "general solicitation" become one that is subject to criminal liability? How specific must the criminal design be? Consider the bizarre facts of *People v. Vandelinder*, 481 N.W.2d 787, 788 (Mich Ct. App. 1992):

   The [solicitation] charges arose out of defendant's offer to an undercover police officer to pay $1,000 for the kidnapping, rape, and possible murder of his estranged wife. Defendant's alternative aims were either to reconcile with his wife or to get rid of her. He told the supposed kidnapper to videotape several people raping the victim in order to have some leverage against her if she agreed to his demands to reconcile on his terms. In the event she did not agree, the kidnapper was to kill the victim.

   Applying the common law, the Michigan court held that the aforementioned facts, involving what amounted to a form of conditional intent, sufficed for solicitation liability.

2. Compared with the rationales behind the law of attempt and conspiracy, the rationale behind solicitation continues to inspire significant controversy. Indeed, as suggested by *Quentin*, solicitation raises concern over whether the First Amendment free speech rights of defendants are being unduly "chilled." Beyond this constitutional concern, the core social harm associated with solicitation is thought by many to pale relative to attempt and conspiracy, calling into question whether solicitation itself should be criminalized. The Commentary to § 5.02 to the MPC summarizes as follows:

   It has been argued, on the one hand, that the conduct of the solicitor is not dangerous since between it and the commission of the crime that is his object is the resisting will of an independent moral agent. By the same token it is urged that the solicitor, manifesting his reluctance to commit the crime himself, is not a menace of significance. Against this is the view that a solicitation is, if anything, more dangerous than a direct attempt, since it may give rise to the cooperation among criminals that is a special hazard. Solicitation, may, indeed, be thought of as an attempt to conspire. Moreover, the solicitor, working his will through one or more agents, manifests an approach to crime more intelligent and masterful than the efforts of his hireling.

   Do you agree that solicitation is less serious than other inchoate crimes? In what way is a solicitation an "attempt to conspire," making it in effect a "double-inchoate" crime? Does this alone warrant that solicitations not be subject to criminal prosecution?

## Defense of Renunciation

Under the traditional common law, a solicitation once made cannot be renounced. The crime is complete at the point of solicitation, regardless of whether the solicitor has second thoughts. *People v. Gordon*, 120 Cal. Rptr. 840 (Cal. 1975), provides an example of this harsh rule. In *Gordon*, the defendant solicited a police officer to plant illegal drugs on another person. On the following day, however, the defendant informed the officer that she did not want to carry out the scheme, out of concern for her political reputation. Despite several overtures from the officer, the defendant refused to arrange a meeting between the third party and the officer so that the planting of drugs could be effectuated. The defendant's conviction for solicitation was upheld on appeal, with the court disregarding the defendant's subsequent equivocal behavior.

Be aware, however, that several jurisdictions, as well as the MPC, do permit individuals to escape liability if certain circumstances are satisfied. For instance, according to the MPC, "It is an affirmative defense that the actor, after soliciting another person to commit a crime, persuaded him not to do so or otherwise prevented the commission of the crime, under circumstances manifesting a complete and voluntary renunciation of his criminal purpose." MPC § 5.02(3).

## Defense of Impossibility

In light of the rationale supporting solicitation, should it matter that the illegal act solicited was legally impossible? The North Carolina Court of Appeals addressed this issue in *State v. Keen*, 214 S.E.2d 242 (N.C. App. 1975). The defendant in *Keen* solicited the help of an undercover police officer in the murder of his wife. The court upheld the defendant's conviction for solicitation, reasoning that the "crime of solicitation . . . is complete with the solicitation even though there could never have been an acquiescence in the scheme by the one solicited." *Id.* at 244. Does this result seem fair? What harm could come of the misbegotten request?

## Distinguishing Solicitation from Attempt and Conspiracy

Solicitation differs from attempt in important ways. As a basic rule, solicitation is complete when the request, encouragement, command, or incitement occurs. Attempt, on the other hand, typically requires the carrying out of some act (*e.g.*, a "substantial step") toward completion of the target crime. In short, a mere request, without more, that another commit a crime can amount to a solicitation. Such behavior would not typically lead to attempt liability. *See, e.g., State v. Grundy*, 886 P.2d 208 (Wash. 1994).

Likewise, as a general rule, solicitation requires less proof than conspiracy. First, conspiracy requires a "meeting of minds," whereas solicitation liability arises merely upon the making of the request (the person solicited need not agree to participate). Second, conspiracy, unlike solicitation, often requires an "overt act" (similar to the law of attempt). A solicitation, once again, is complete upon the request or incitement, with nothing more required. In this sense solicitation can be thought of as an "attempt to conspire." *See* MPC § 5.02, Commentary, p. 381. Of course, if the individual solicited agrees to the criminal overture, conspiracy liability might then arise, depending on the given facts and the law of the jurisdiction.

## Grading of Inchoate Crimes and the Role of "Merger"

### Grading

Jurisdictions vary in their approaches to the grading of inchoate crimes. The MPC, for instance, punishes attempts, solicitations, and conspiracies to the full extent of the object crime to which they are related. *See* MPC § 5.05. The only exception involves capital crimes or first-degree felonies, in which case the inchoate offense is punished as a second-degree felony. *Id.* The Commentary to the Code provides the following rationale for this approach:

> The theory of this grading system may be stated simply. To the extent that sentencing depends upon the antisocial disposition of the actor and the demonstrated need for a

corrective sanction, there is likely to be little difference in the gravity of the required measures depending on the consummation or the failure of the plan. It is only when and insofar as the severity of sentence is designed for general deterrent purposes that a distinction on this ground is likely to have reasonable force. It is doubtful, however, that the threat of punishment for the inchoate crime can add significantly to the net deterrent efficacy of the sanction threatened for the substantive offense that is the actor's object, which he, by hypothesis, ignores. Hence, there is a basis for economizing in use of the heaviest and most afflictive sanctions by removing them from the inchoate crimes. The sentencing provisions for second degree felonies, . . . should certainly suffice to meet whatever danger is presented by the actor. MPC, Commentary to sec. 5.05, 489–490.

What penal theory does the MPC's approach reflect? What approach would a proponent of strict retributive theory suggest? A utilitarian?

## Merger

Because of their intrinsic preparatory nature, the prosecution of inchoate crimes gets complicated when the object crime is actually consummated. Can the government prosecute a defendant both for the object crime committed and for an inchoate offense? The answer depends on both the nature of the object crime and which of the inchoate offense is supported by the facts. With respect to attempts, if one is convicted of a consummated crime, one cannot also be convicted of an attempt to commit that crime. The attempt "merges" into the completed crime. Of course, because attempt is typically a "lesser included offense," one can be convicted of an attempt, not the completed crime, if the facts fail to satisfy the elements of the completed crime.

As a practical matter, solicitation charges usually are brought only when the crime solicited was neither attempted nor completed. Otherwise, if the solicitation is acted upon—either as a failed or successful attempt—**merger** occurs. The solicitor can be prosecuted as an accomplice to the crime or, depending on the facts, as a conspirator along with the party agreeing to the solicitation.

The law handles conspiracy quite differently, however. Because conspiracy is seen as a separate and distinct crime in most jurisdictions and under the MPC, the completed crime and the conspiracy are subject to independent prosecutions and convictions (*i.e.*, there is no merger). Also, as a result, even if the conspirator is acquitted of the target crime, he or she can still be convicted of conspiracy.

While on the topic of merger and conspiracy, the so-called Wharton's rule warrants brief mention. Wharton's rule is named after Francis Wharton, a well-known criminal law scholar. Although rarely invoked, the rule bars conspiracy prosecutions for certain crimes that by definition cannot be committed by a person acting alone—for example, adultery, bigamy, dueling, or incest. Wharton's rule therefore represents "an exception to the general principle that a conspiracy and the substantive offense is its immediate end do not merge upon proof of the latter." *See Iannelli v. United States*, 420 U.S. 770,1781 (1975).

## Conclusion

The inchoate crimes discussed in this chapter share a unique characteristic: They impose liability for future criminal behaviors that have not yet come to pass. With attempt, the law sanctions those who actually try to commit criminal acts, whereas the law of conspiracy and solicitation targets those who either agree to carry out crimes or seek others to do their criminal bidding. Taken together, the inchoate crimes constitute an important and challenging area of the substantive criminal law, one distinguishable from other areas because the tangible social harm usually of concern to the criminal law has not yet come to pass.

## Key Terms

| | |
|---|---|
| abandonment | inchoate crimes |
| actual attempts | interrupted attempts |
| actual impossibility | legal impossibility |
| agreement | merger |
| complicity | Racketeer Influenced and |
| conspiracy | Corrupt Organizations |
| defense of renunciation | (RICO) Act |
| factual impossibility | solicitation |

## Review Questions

1. What differentiates *inchoate crimes* from other crimes?
2. Why does the text say that *inchoate crimes* are "preemptive justice"?
3. Attempts can come in two forms. Name and define both.
4. Because *attempt* is a specific intent crime, what steps must the government take to prove that the defendant committed an *attempt*?
5. What is the difference between *legal impossibility* and *actual impossibility*, and which *could* constitute a complete defense to the charge of attempt?
6. Draw a distinction between *conspiracy* and *complicity*.
7. Define *solicitation*.
8. With regards to the crime of *solicitation*, can a person renounce the solicitation, thus avoiding a criminal charge? Use both the MPC position and *People v. Gordon* when preparing your response.
9. What is the RICO Act, and what purpose does it serve?

## Notes

1. *See Model Penal Code and Commentaries*, article 5 introduction. Official Draft 1962 and Revised Comments. Philadelphia: American Law Institute, 1985, p. 293.
2. Holmes OW Jr. *The Common Law*. Boston: Little, Brown, 1881:67–69.
3. LaFave WR, Scott AW Jr. *Criminal Law*, 2nd ed., § 602. St. Paul, MN: West Publishing; 1986:501.
4. LaFave and Scott, p. 506.
5. *See* Dressler J. *Understanding Criminal Law*, § 27.07[0]. New York: Matthew Bender & Co.; 1995.
6. *See* Dressler, § 27.07[D], p. 373.
7. Dressler, § 27.07[D], p. 374.
8. Robbins IP. Double inchoate crimes. *Harv. J. Legislat.* 1989;26:1, 27–29.
9. LaFave and Scott, p. 561.
10. LaFave and Scott, p. 559.
11. *See* Boyce RN. *Criminal Law*, 3rd ed. Mineola, NY: Foundation Press; 1982:582–588.

## Footnote to *United States v. Brown,* 776 F.2d 397 (2d Cir. 1985)

10. We do not read *United States v. Tyler*, 758 F.2d 66 (2 Cir. 1985), as being to the contrary. The court read the evidence as showing "no more than that Tyler helped a willing buyer find a willing seller." *Id.* at 70. Since there was no basis for inferring a prior contact between Tyler, the introducer, and Bennett, the seller, Tyler could properly be convicted only as an aider or abettor, not as a conspirator. Here a jury could reasonably infer prior arrangements or an established working relationship between Brown and Valentine. Chief Judge Motley's opinion in *United States v. Jones*, 605 F.Supp. 513, also relied upon by Brown, is readily distinguishable. There was insufficient evidence for a jury to conclude beyond a reasonable doubt that Jones, the counterpart of Brown in that case, even knew that a narcotics transaction was going on. By contrast, Brown's conversation with Valentine in the presence of Grimball establishes that he knew precisely what the transaction was. Finally, the facts in *United States v. Cepeda*, 768 F.2d 1515 (2 Cir. 1985) (Oakes, J.), relied upon by the dissent, are not analogous to the situation here. In *Cepeda*, there was no evidence that there had been a sale, or that anyone other than the defendant was involved in the transaction. Thus, two elements of the charged conspiracy were called into question—the agreement with other persons and the intent to distribute. Here, there is no question that a sale occurred, and it is clear that Brown and at least one other person, Valentine, participated in the transaction. Thus, the only further inference required for a conviction—that Brown's participation was pursuant to an agreement with Valentine, with someone else, or with both—is much stronger. . . .

# Glossary

## A

**abandonment** A defense used by one who changes his or her mind before the completion of the crime. The courts have generally held that the abandonment defense may only be used in cases where it may be shown that the defendant has demonstrated a complete renunciation of his or her efforts before passing beyond the point of no return

**accessory after the fact** One who is aware of the commission of the crime by the main criminal actor and purposefully provides help to the actor in the name of avoiding apprehension and conviction. This category of actors might be called "criminal protectors"

**accessory before the fact** One who is not present at the time of the offense yet prior thereto aids, counsels, or otherwise encourages the main criminal actor. One who "incites" the criminal action of another would fall in this category

**accomplice liability** A situation wherein the actor solicited, aided, agreed to aid, or attempted to aid the principal in planning or committing the offense

**actual attempts** Cases in which a person puts into action a planned event, but it is unsuccessful; as in unsuccessfully poisoning another

*actus reus* The physical component of a crime; the act that accompanies the intent

**affirmative defense** Defenses in which the defendant admits to the essential elements of the crime charged but offers an excuse or justification that negates criminal responsibility

**aggravated assault and battery** Assaults and batteries are generally categorized as either *simple* or *aggravated*. Initially, assaults are presumed to be simple; however, there are several ways by which they can become aggravated. The two most common modes of aggravated assault and battery are (1) a battery committed or attempted to be committed with the intent to kill or inflict serious bodily harm, or (2) an assault or battery committed with a dangerous or deadly weapon/instrument

**aggravated rape** Enhanced "grading" of rape charges may be given in some jurisdictions based on aggravating circumstances, such as the youthful age of the victim or the severity of injuries suffered by the victim during the commission of the rape

**agreement** A "meeting of the minds" between at least two people with criminal designs, as in a conspiracy

**arson** Arson was defined under the common law as the malicious burning of a dwelling of another. This meant actually setting the building on fire. No actual burning of the structure is required. Any damage, such as smoke or structural damage, will suffice. In most states the *mens rea* for arson is general intent.

Arson is not a specific intent crime; the general intent to start the fire is all that is needed

**assault** An attempt by a person to strike the person of another. In many jurisdictions an assault is the event that leads to a battery, but in some jurisdictions there is no distinction between the assault and the battery

**attempted battery assault** Occurs when a person attempts to commit a battery but fails to do so. It requires an actor to have the specific intent to actually carry out a battery and to take a substantial step toward doing so but in the end fail to commit it. Shooting at a person and missing or throwing a punch at a person who ducks out of the way exemplify attempted battery assaults

## B

**battered woman syndrome** Refers to a "cycle of violence" that induces a state of "learned helplessness" and keeps a battered woman in the relationship; may end with a violent reaction, causing death to abuser

**battery** The unlawful touching of the person of another. Most jurisdictions require that the touching be knowing and intentional to constitute battery

**bench trial** A procedure by which a judge, rather than a jury, determines the guilt or innocence of a defendant

**"born alive" rule** As used in Chapter 6, the "born alive" rule is a legal principle used to apply homicide laws to cases that involve a child that has been born alive. This principle was necessitated by the lack of medical knowledge as to the viability of a child *in utero*

**breaking** Breaking, as an element of the crime of burglary, may be either actual or constructive. Actual breaking involves the application of some force, slight though it may be, whereby the entrance is effected. Merely pushing open a door, turning the key, lifting the latch, or other slight physical force is sufficient to constitute this element of the crime. A constructive breaking is deemed to have occurred where entry is gained by threats, fraud, or conspiracy

**burden of proof** Legal principle that dictates on whom (plaintiff or defendant) the requirement to establish one's case rests

**burglary** Burglary was a crime created under the common law in an effort to prevent intruders from entering a person's house during the nighttime. The elements of common law burglary are breaking and entering the dwelling of another at night with the intent to commit a felony. Under most statutes today, the elements of "night" and "felony" have been changed. Today, a burglary charge can be made if it can be shown that the perpetrator intended to commit any crime within the dwelling. Some statutes offer varying degrees of burglary depending on whether the dwelling was occupied, a business, or another distinction

# C

**castle doctrine** A person has a reasonable expectation of safety in his or her home and may defend him- or herself without retreat

**causation** A element required to demonstrate that a person's conduct "causes" the harm proscribed by law

**choice of evils** A justification defense (*see* necessity)

**commensurate desserts** Legal principle that the severity of punishment should be commensurate with the seriousness of the wrong; also known as *just desserts*

**complaint** A statement by the plaintiff setting forth the cause of action

**complicity** Conspiratorial conduct involving two or more persons who plan and attempt to commit a crime. Complicity is the effort by one person to assist another in the commission of the crime

**concurrence principle** A coexistence of two or more conditions for the elements of a crime or harm to be met

**consolidated theft statutes** These statutes combine several of the common law property crimes under one overarching statute, thus removing the requirement that specific language from various theft statutes be included in the charge

**conspiracy** An agreement between or among at least two persons and an intent to carry out either an unlawful act *or* a lawful act by unlawful means. In an increasing number of jurisdictions, there must also be an "overt act" in furtherance of the conspiracy

**constructive possession** Exists when a person does not have *actual possession* of an item but he or she still maintains dominion and control over the item

**corporate liability** Liability assigned to a corporate entity resulting from a criminal or civil case

**court of appeals** An intermediate-level court in which the outcome of cases from lower-level courts are appealed

**crime** A legally prohibited action that is deemed injurious to the public welfare

**criminal** Having to do with the nature of crime; an individual guilty of a felony or misdemeanor

**criminal negligence** Actions by an offender that demonstrate reckless disregard for another, and such action leads to injury or other harm

**cruel and unusual punishment** Sometimes seen as a vague and unspecified concept of the Eighth Amendment, a guarantee by the constitution that punishment shall not be excessive

**culpable omissions** The failure to act in a situation that results in harm to another

**custodial interference** Also referred to as child stealing or interference with custody, it involves keeping a child away from his or her custodial parent. Although custodial interference statutes provide that this can be accomplished through threats, they also provide that it is a crime to keep a child away from his or her custodial parent through enticements. The Model Penal Code provides that a person commits the offense of custodial interference if he or she takes or entices any child under the age of 18 from the custody of his or her parent, guardian, or other lawful custodian when that person has no privilege to do so

# D

**deadly force** A degree of force used to apprehend suspects that may result in death or bodily harm

**defense of renunciation** A defense made by a person who, after a solicitation, has a change of heart and communicates desire. Under common law a solicitation once made cannot be renounced because the solicitation has already occurred. Some legal jurisdictions, however, allow a defense of renunciation under specific circumstances

**deliberation** The process of carefully weighing matters one has pondered. In Chapter 6, deliberation includes considering the wisdom of going ahead with a proposed killing, the manner in which the killing will be accomplished, and the consequences that may be visited upon the killer if and when apprehended

**dependent intervening cause** Refers to a condition that must be present for the incident to occur, and will not absolve a defendant of criminal liability

**depraved-heart murder** A murder that results from the reckless actions of another. Conviction of depraved-heart murder usually requires proof that the defendant acted recklessly under circumstances manifesting extreme indifference to the value of human life

**district courts** County, state, or federal level trial courts having general and original jurisdiction

**duress** Conduct that otherwise would be criminal may be excused if committed under duress. At common law, several elements had to be satisfied to establish the defense of duress; a reasonable belief or fear; of a present, imminent, and impending threat of; death or serious injury; to him- or herself or another (sometimes limited to a close relative); from which there was no reasonable escape or alternative course of action; and the danger or threat was not brought about by the defendant's own conduct

# E

**embezzlement** Occurs when a person converts property entrusted to him or her to his or her own use in a manner that demonstrates his or her intent to deprive another person use of or access to the property

**entrapment** A legal defense in which a person claims to have committed an offense that he or she would not otherwise have committed but did so due to inducement or persuasion by law enforcement officials

**equal protection of law** Fourteenth Amendment clause guaranteeing all citizens, regardless of race, color, gender, class, origin, or religion, equal protection under the law

**ex post facto laws** Laws that retroactively make an act illegal or increase the penalty for an existing crime, held to be unconstitutional

**excuse defenses** A situation in which a defendant admits committing the accused act, and even that the act was wrong and the ensuing harm regrettable, but claims that his or her conduct should be excused because of extenuating circumstances such as insanity or duress

**extortion** Often referred to as "blackmail," extortion is the obtaining of an item through the threat of a future harm. The difference between extortion and robbery is that robbery involves a threat of *imminent* harm, whereas extortion involves a threat of *future* harm

# F

**facilitation** Occurs when a person who believes it *probable* that he or she is rendering aid to a person who intends to commit a crime, and he or she engages in conduct that provides such person with the means or opportunity for the commission thereof and which in fact aids such person to commit a felony

**factual impossibility** Factual impossibility occurs when the objective of the defendant is proscribed by the criminal law but a circumstance unknown to the actor prevents him or her from bringing about that objective

**fair notice** An ordinance or statute that clearly defines what conduct is prohibited or required

**false imprisonment** Traditionally defined as the unlawful confinement of a person. It does not require that a person actually be confined in a specific place, but rather it involves the unlawful exercise or show of force used to compel a person to go where he or she does not want to go or stay where he or she does not want to stay

**false pretenses** Involves instances where an owner of property voluntarily parts with title to property as a result of false statements or deception

**Federal Insanity Defense Reform Act** Created in 1984, and governing all trials in federal courts that involve the insanity defense, the Federal Insanity Defense Reform Act makes it more difficult for a defendant to be found insane than the tests under the Model Penal Code and M'Naghten. The defense must prove that the defendant was unable to appreciate the nature and quality or the wrongfulness of his or her acts at the time of the commission of the acts and that such diminished capacity was the result of a severe mental disease or defect

**felony** A serious crime punishable by more than 1 year in state or federal prison

**felony-murder rule** A law that holds as equally culpable all persons involved in any capacity in the commission of certain felonies if the victim of such crime dies as a result of its commission

**first-degree murder** First-degree murder is defined by most state statutes as a killing of a human being that was done willingly, deliberately, with premeditation and malice aforethought

**forcible compulsion** Includes both physical force and psychological duress to induce a person into action

**forcible rape** The carnal knowledge of a woman forcibly and against her will

**fraud** An act of deception with the intent to deprive another of something of value

**fraud in the *factum*** Deception that leads a person to misunderstand the facts in a given situation resulting in a decision and/or actions that would not have occurred in absence of the deception

**fraud in the inducement** Fraud in the inducement occurs when a person is induced to complete an act by a person who does not fully disclose material information about the act which would likely lead the induced party to decide otherwise

## G

**general deterrence** The idea that future crime can be prevented by inflicting punishment on one lawbreaker to deter others

**general intent crimes** Method of legal proof used in court that establishes that actions can indicate a criminal purpose (*e.g.*, trespassing or breaking and entering)

**Good Samaritan laws** Legal obligation in some jurisdictions to summon or render aid to others in distress

**guilty but mentally ill (GBMI)** Guilty but mentally ill (GBMI) is an alternative to not guilty by reason of insanity offered in some jurisdictions. Under a GBMI verdict the defendant is deemed to have been mentally ill at the time of the crime but is still criminally responsible for his or her conduct

## H

**harassment** Occurs when the person, purposefully and without legitimate purpose, has personal contact with another person, with the intent to threaten, intimidate, or alarm that other person

**harmful products** Products that have the potential to be harmful to humans either through their use or consumption

**Hindering Apprehension or Prosecution** A person hinders the apprehension, prosecution, conviction, or punishment of another for crime when he or she harbors or conceals the other; provides or aids in providing a weapon, transportation, disguise or other means of avoiding apprehension or effecting escape; conceals or destroys evidence of the crime, or tampers with a witness, informant, document or other source of information, regardless of its admissibility in evidence; volunteers false information to a police officer; or warns the other of impending discovery or apprehension. Exceptions are given in the event of a warning given in connection with an effort to bring another into compliance with the law

**homicide** The killing of a human being by another

## I

**imminent harm** An apparent, immediate threat to life or safety that must be dealt with instantly

**imperfect self-defense** Imperfect self-defense is not a complete defense, but it operates to negate malice, an element the State must prove to establish murder, and is used in cases that do not warrant either a murder conviction or an acquittal. The successful invocation of this doctrine does not completely exonerate the defendant but mitigates murder to voluntary manslaughter

**inchoate crimes** Crimes such as attempt, conspiracy, and solicitation that have not yet culminated in a criminal harm

**independent intervening cause** Breaks the chain of causation and absolve the defendant of criminal liability

**indictment** A formal accusation by a grand jury to initiate a criminal case

**individual deterrence** The idea that punishment inflicted on a specific person will keep that person from committing future crimes. This principle is also known as *specific deterrence*

**individual liability** Liability assigned to individual officers within a corporate structure resulting from a criminal or civil case

**infancy** This principle addresses consideration given by a court as to whether or not a person will be held responsible for his or her actions based on age

**information** A formal criminal charge brought by the authority of a public prosecutor rather than through the indictment of a grand jury

**insanity defense** A claim by a defendant that he or she lacks the mental capacity to make a distinction between right and wrong due to a biological or psychological cause

**insider trading** Trading (meaning the purchase, sale, or other exchange) of a security when information on which the trade is based on information not available to the general public. The intent of insider trading laws is to ensure that the fiduciary relationship between the corporation and its stockholders is not broken

**interrupted attempts** Cases in which a person plans and carries out a portion of a crime but does not complete the crime due to being interrupted by change of heart, discovery by another, or some other event

**involuntary intoxication** Involuntary intoxication is a defense to a crime. To establish involuntary intoxication as a defense, the defendant generally must show that he or she unknowingly or involuntarily ingested an intoxicating substance that prevented him or her from forming the *mens rea* required for conviction or understanding the difference between right and wrong

**involuntary manslaughter** Homicides that include reckless and criminally negligent behavior

## J

**justification defenses** Instances where the defendant admits committing the accused act but claims that the actions were appropriate under the circumstances

## K

**kidnapping** Kidnapping can be thought of as aggravated false imprisonment. Under the common law, kidnapping was defined as the forcible abduction or stealing away of a man, woman, or child from his or her own country and sending him or her into another. Thus under the common law an abduction was not a kidnapping unless the victim was taken to another country. Many modern statutes no longer require that a person be removed from a country for an act to be kidnapping

**knowingly** Conscious; intentional; deliberate

## L

**larceny** At common law, larceny is the taking of property from a person. Currently, statutes incorporate seven elements required before larceny may be proven: trespassory, taking, carrying away (asporting), personal property, of another, with the intent, to permanently deprive

**larceny by trick** The expansion of larceny to include situations in which the property owner was fraudulently deceived of his or her property

**legal impossibility** Occurs when the actions that the defendant performs or sets in motion, even if fully carried out as he or she desires, would not constitute a crime

**legality principle, the** A constitutional limitation on state or federal power to enforce criminal laws

## M

**malice** Malice describes a harmful intent and is a term that distinguishes murder from manslaughter

**marital rape** Marital rape is an exemption in rape law that stems from the ancient common law that viewed the legal institution of marriage as barring the state from prosecuting a husband for rape, even if intercourse was accomplished against the wife's consent. This exemption is generally no longer allowed in the United States

**mens rea** The mental factor necessary to prove criminality, including purpose, knowledge, recklessness, and negligence

**merger** Merger occurs with respect to attempts when a crime is consummated. The attempt "merges" into the completed crime

**misdemeanor** A minor crime punishable by usually less than 1 year in city/county jail and/or a fine

**mistake of fact** A mistake of fact involves a misunderstanding about the true nature of events (other than the law) that, in this context, results in the commission of an act that ordinarily would be a crime

**mistake of law** A mistake of law is presented when a person does not know the proper meaning of a law or otherwise misinterprets it

**M'Naghten rule** A rule that stems from a mid-1880s English case and is used in the United States as a standard to ensure that the defendant is presumed to be sane until it can be shown that he or she suffered from a defect in reason, a disease of the mind, or that he or she was not capable of knowing the nature of the crime for which he or she is charged

**Model Penal Code** Assembled in 1962 by the American Law Institute, the Model Penal Code was an attempt to codify American criminal law for the purpose of uniformity

**Model Penal Code test** A person is not responsible for criminal conduct, if at the time of such conduct, it results from mental disease or a defect, a lack of substantial capacity either to appreciate the criminality of his or her conduct, or a lack of ability to conform his or her conduct to the requirements of the law

**murder** An unlawful killing committed with malice

## N

**necessity** A situation in which a person commits what under normal circumstances would be a crime to prevent a greater harm from occurring

## O

**objective standard** A standard that is based on the perspectives of the average person on the street

**objective test** The objective test of entrapment focuses on the conduct of the police and essentially considers whether the police's actions would likely have induced a normal, law-abiding citizen to commit the crime

**offensive conduct** Behavior that has a tendency to provoke *others* to acts of violence or to in turn disturb the peace

**offensive speech** Speech that is not protected by the First Amendment to the Constitution

**overbreadth doctrine** Doctrine that allows invalidation of laws that punish people for engaging in constitutionally protected behaviors, such as expression of speech and religious rituals

## P

**police courts** A municipal court that tries those accused of violating city ordinances

**premeditation** Premeditation is the process of thinking about a proposed crime before engaging in the homicidal conduct. Premeditation is done in such a cool mental state, under such circumstances, and for such a period of time as to permit a "careful weighing" of the proposed decision

**principal in the first degree** The main criminal perpetrator in a crime

**principal in the second degree** One who is present (actively or constructively) at the time of the offense and aids, counsels, or otherwise encourages the main criminal actor. Traditionally, this category has been synonymous with "aiding and abetting"

**product test** First pioneered by the Supreme Court of New Hampshire in *State v. Pike*, 49 N.H. 399, 402 (1869), a test designed to facilitate full and complete expert testimony and to permit the jury to consider all relevant information, rather than restrict its inquiry to data relating to a sole symptom or manifestation of mental illness

**proportionality** The legal principle that a person may use only that amount of force that is reasonably necessary to repel his or her attacker

**proportionality of punishment** Legal principle that punishment imposed should fit the crime and not be excessive (*see* commensurate desserts)

**provocation** An action designed to result in a particular response, such as assault

**proximate cause** An act from which an injury or event results as a direct, uninterrupted consequence, one that without the injury would not have occurred

**psychological coercion** Mental or emotional inducements by a person that are not accompanied by physical force but that lead the victim to action he or she would not otherwise have taken

**punishment** A penalty imposed for committing an offence against legal rules

**purposel** Intentionally; deliberately; with a particular purpose specified

# R

**Racketeer Influenced and Corrupt Organizations (RICO) Act** The RICO Act was enacted by the U.S. Congress in 1970 to combat organized crime. RICO originated out of concern over the suspected infiltration of legitimate businesses by criminal syndicates and the difficulty federal prosecutors were having in securing convictions against the shadowy enterprises by means of traditional conspiracy law. Although initially aimed at organizations like the Mafia, today RICO has become a valued tool in federal prosecutors' battle against "white collar" crime

**rape shield laws** In rape cases, provisions designed to limit the admission at trial of information regarding the complaining witness's sexual history and reputation for chasteness

**rape trauma syndrome** Psychological phenomenon that is said to account for post-rape behaviors among some victims that laypersons might consider inconsistent with the occurrence of rape

**reasonable doubt** Doubt as would cause a reasonable and prudent person to pause and hesitate before taking the represented facts as true and relying and acting thereon

**receiving stolen property** Consists of a person who receives stolen property knowing it to be stolen and having the intent to permanently deprive the true owner of that property

**recklessness** Conscious disregard for the consequences of one's actions

**Resource Conservation and Recovery Act (RCRA)** Enacted by Congress in 1976, RCRA is designed to regulate and monitor hazardous waste from the "cradle to the grave"—including its initial generation and subsequent treatment, storage, transportation, and disposal

**respondant superior** Literally, "let the master answer." This term is used in liability cases, and liability of this sort is purely relationship based, relying on the theory that a principal (a "master") should have legal responsibility for the wrongful act of his or her agent. In business, this doctrine establishes that the owner of a business is responsible for the actions of his or her employees

**retributive justification** To punish in kind for wrongs committed; "an eye for an eye"

**right of association** As protected by the First Amendment, the right of association with others without prejudice is guaranteed

**right of privacy** The individual's right to decide in what way they will share or withhold beliefs or behaviors

**robbery** Larceny through the use or threatened use of immediate force. The elements of robbery under the common law are the taking and asportation of another's property from his or her person or in his or her presence by force or threatened imminent force with the intent to permanently deprive the owner of the property

# S

**second-degree murder** Charges of second-degree murder are reserved for cases that do not rise to the level of first-degree murder because they lack one or more of the requisite elements of *willingly*, *deliberately*, *premeditation*, or *malice aforethought*

**Securities Act of 1934** This is the federal law that created the Securities and Exchange Commission, which was established to regulate the exchange and issuance of securities in a manner free from fraud, misrepresentation, and manipulation

**self-defense** The justifiable use of force against another person to repel an unprovoked attack

**Sherman Act** Also known as the "Sherman Antitrust Act," an 1890 act of Congress that prohibited "trusts" from being established, thus creating monopolies. Monopolies, it is argued, inhibit free trade of commerce

**simple assault** A person is guilty of assault if he or she attempts to cause or purposely, knowingly, or recklessly causes bodily harm to another; or negligently causes bodily injury to another with a deadly weapon; or attempts by physical menace to put another in fear of imminent serious bodily injury

**solicitation** An inchoate crime that can arise upon the mere asking or encouraging of another to commit a crime, entailing neither "overt acts" nor "substantial steps"

**specific intent crimes** A special mental element required above and beyond any mental state required with respect to the *actus reus* of the crime (*e.g.*, breaking and entering with the intent to steal jewelry)

**stalking** Generally defined as conduct directed at a specific person that involves repeated (usually two or more occasions) visual or physical proximity, nonconsensual communication, or verbal, written, or implied threats, or a combination thereof, that would cause a reasonable person to fear for his or her safety

**stare decisis** Legal doctrine that a court will refer to precedent cases when determining the outcome of a present case

**status** In legal terms, status refers to one's condition as opposed to one's action; it is not illegal to be an alcoholic, but it is illegal to drive while intoxicated

**statutory rape** In contrast to common law rape, statutory rape makes sexual intercourse with persons under a given age illegal, thus removing "consent" to the sexual act as a material fact. Statutory rape laws presume that minors lack the legal capacity to consent to the alleged sex act

**"street" crimes** Assault-like, street-level offenses against a person, usually in urban areas

**subjective criterion** A principle that dictates whether a given act seemed reasonable to the actor at the time

**subjective test** The subjective test of entrapment requires a defendant establish that (a) government conduct induced the commission of the crime, which (b) the defendant was not predisposed to commit. If a defendant was predisposed to commit the crime, the entrapment defense will not be available. The subjective test, designed to allow the police to capture unwary criminals, shifts much of the emphasis away from the conduct of the police and toward the character and criminal history of the defendant

**substantive due process** A legal principle that deems that not only is one's due process guaranteed, but their *substantial* rights are also protected in a legal process

**superior courts** Higher level trial courts in which felony cases are resolved

**supreme court** The highest appellate court in a jurisdiction

# T

**threatened battery assault** Occurs when a person places another in fear of imminent serious bodily harm

**torts** Torts are cases involving personal injury and damage to or loss of property handled as civil action

**transferred intent** If an illegal act, although unintentional, results from the intent to commit a crime, that act is also considered in the case

**trickery** In legal context, trickery refers to actions that are intended to deceive

# U

**U.S. district court** Federal trial courts that only have jurisdiction over cases involving federal crimes or crimes alleged to have been committed against the U.S. government

**Uniform Determination of Death Act** The Uniform Determination of Death Act was created in 1980, and it provides that death must be made in accordance with accepted medical standards in cases where an individual has sustained either irreversible cessation of circulatory and respiratory functions or irreversible cessation of all functions of the entire brain, including the brainstem

**utilitarian justification** A principle that supposes benefit from punishment, such as deterrence, incapacitation, and rehabilitation

# V

**vagueness** A law that is not clearly defined

**vicarious liability** Permits the wrongful acts of one person to result in the liability of another, purely by virtue of the supervisory-like relationship the latter has over the former. Vicarious liability does not require any specific act or omission by the defendant. Rather, liability arises solely on the basis of the supervisory responsibility that the defendant has relative to the individual committing the act or omission

**voluntary intoxication** In cases requiring specific intent, some jurisdictions permit the jury to consider a defendant's voluntary intoxication. Some states do not permit the jury to consider evidence of voluntary intoxication under any circumstances, whereas others allow its consideration in cases involving first-degree murder. There is no constitutional obligation for them to do so

**voluntary manslaughter** Intentional homicide that does not rise to the level of murder

# W

**white collar crimes** A term coined by Professor Edwin Sutherland in the late 1940s signaling a distinction between crimes by corporate elites and "street" crimes. Subsequent research on crime distinctions has refined this term to include crimes by white collar workers as well as a crime committed by someone not considered to be a corporate elite but who is also other than a street criminal, such as embezzlement by a jewelry store employee

**writ of certiorari** An order by which a superior court can call up for review the record of a proceeding in an inferior court

# Y

**year-and-a-day rule** A limiting rule that assumes and assigns criminal responsibility to a defendant if a situation occurs within "a year and a day" of the commission of the crime that lead to the onset of the condition

# INDEX